S0-AQJ-402

Set Theoretical

38	$\in$	belongs to, is an element of
39	$\emptyset$	the empty set
40	$\subseteq$	is a subset of
40	$\subsetneqq$	is a proper subset of
40	$\supseteq$	is a superset of, contains
41	$\mathcal{P}(A)$	the power set (the set of all subsets) of A
43	$\cup$	union
43	$\cap$	intersection
45	$\setminus$	set difference
45	A^c	the complement of A
46	$[a, b]$	$\{x \in R \mid a \leq x \leq b\}$, where $a, b \in R$
46	$(a, b]$	$\{x \in R \mid a < x \leq b\}$, where $a, b \in R$
46	$(-\infty, b]$	$\{x \in R \mid x \leq b\}$, where $b \in R$
46	(a, ∞)	$\{x \in R \mid x > a\}$, where $a \in R$
47	$\oplus$	symmetric difference
48	$\times$	Cartesian (or direct) product
88	$|A|$	the cardinality of A (if A is finite, this is just the number of elements in A)

Equivalence Relations

57	$A/\sim$	quotient set of A (the set of equivalence classes of an equivalence relation on a set A)
57, 126	$\bar{a}$	equivalence class of a; also used to denote the congruence class of an integer $a \pmod{n}$

Partially Ordered Sets

63	$\preceq$	a partial order
66	$\vee$	least upper bound
66	$\wedge$	greatest lower bound

Functions

72	$f : A \to B$	f is a function from A to B
73	1–1	one-to-one
77	ι	the identity function
77	ι_A	the identity function on the set A
80	f^{-1}	inverse of the function f
82	$\circ$	composition (of functions)

Discrete
Mathematics
with Graph Theory

Discrete Mathematics
with Graph Theory
Second Edition

Edgar G. Goodaire
Memorial University of Newfoundland

Michael M. Parmenter
Memorial University of Newfoundland

Prentice Hall

PRENTICE HALL
Upper Saddle River, NJ 07458

Library of Congress Cataloging-in-Publication Data

Goodaire, Edgar G.
  Discrete mathematics with graph theory / Edgar G. Goodaire, Michael M. Parmenter.-
2nd ed.
      p. cm.
  Includes bibliographical references and index.
      ISBN 0-13-092000-2
      1. Mathematics. 2. Computer science—Mathematics. 3. Graph theory. I. Parmenter,
  Michael M. II. Title.
  QA39.3.G66 2002
  511—dc21   2001037448

Acquisitions Editor: George Lobell
Production Editor/Assistant Managing Editor: Bayani Mendoza de Leon
Vice-President/Director of Production and Manufacturing: David W. Riccardi
Executive Managing Editor: Kathleen Schiaparelli
Senior Managing Editor: Linda Mihatov Behrens
Manufacturing Buyer: Alan Fischer
Manufacturing Manager: Trudy Pisciotti
Marketing Manager: Angela Battle
Assistant Editor of Media: Vince Jansen
Managing Editor, Audio/Video Assets: Grace Hazeldine
Creative Director: Carole Anson
Paul Belfanti: Director of Creative Services
Interior/Cover Designer: John Christiana
Art Director: Maureen Eide
Editorial Assistant: Melanie Van Benthuysen
Cover Image: Wassily Kandinsky, "Entwurf zu Kreise im Kreis" 1923, Philadelphia Museum of Art/
            Corbis/Artists Right Society, NY

© 2002, 1998 by Prentice-Hall, Inc.
Upper Saddle River, NJ 07458

Printed in the United States of America

10   9   8   7   6   5   4   3

ISBN 0-13-092000-2

Pearson Education Ltd., *London*
Pearson Education Australia Pty. Limited, *Sydney*
Pearson Education Singapore, Pte. Ltd.
Pearson Education North Asia Ltd., *Hong Kong*
Pearson Education Canada, Ltd., *Toronto*
Pearson EducaciUn de Mexico, S.A. de C.V.
Pearson Education – Japan, *Tokyo*
Pearson Education Malaysia, Pte. Ltd.

To Linda   *E. G. G.*

To Brenda   *M. M. P.*

Only those who live with an author can appreciate the work that goes into writing. We are sincerely grateful for the loving encouragement and patience of our wives over a period of years while we have worked on two editions of this book.

Contents

Preface

To the Student from the Authors

Few people ever read a preface, and those who do often just glance at the first few lines. So we begin by answering the question most frequently asked by the readers of our manuscript: "What does [BB] mean?" Like most undergraduate texts in mathematics these days, answers to some of our exercises appear at the back of the book. Those which do are marked [BB] for "**B**ack of **B**ook." In this book, complete solutions, not simply answers, are given to almost all the exercises marked [BB]. So, in a sense, there is a free Student Solutions Manual at the end of this text.

We are active mathematicians who have always enjoyed solving problems. It is our hope that our enthusiasm for mathematics and, in particular, for discrete mathematics is transmitted to our readers through our writing.

The word "discrete" means separate or distinct. Mathematicians view it as the opposite of "continuous." Whereas, in calculus, it is continuous functions of a real variable that are important, such functions are of relatively little interest in discrete mathematics. Instead of the real numbers, it is the natural numbers $1, 2, 3, \ldots$ that play a fundamental role, and it is functions with domain the natural numbers that are often studied. Perhaps the best way to summarize the subject matter of this book is to say that discrete mathematics is the study of problems associated with the natural numbers.

You should never read a mathematics book or notes taken in a mathematics course the way you read a novel, in an easy chair by the fire. You should read at a desk, with paper and pencil at hand, verifying statements which are less than clear and inserting question marks in margins so that you are ready to ask questions at the next available opportunity.

Definitions and terminology are terribly important in mathematics, much more so than many students think. In our experience, the number one reason why students have difficulty with "proofs" in mathematics is their failure to understand what the question is asking. This book contains a glossary of definitions (often including examples) at the end as well as a summary of notation inside the front and back covers. We urge you to consult these areas of the book regularly.

As an aid to interaction between author and student, we occasionally ask you to "pause a moment" and think about a specific point we have just raised. Our Pauses are little questions intended to be solved on the spot, right where they occur, like this.

Pause 1 Where will you find [BB] in this book and what does it mean?

The answers to Pauses are given at the end of every section just before the exercises. So when a Pause appears, it is easy to cheat by turning over the page and looking at the answer, but that, of course, is not the way to learn mathematics!

 We believe that writing skills are terribly important, so, in this edition, we have highlighted some exercises where we expect answers to be written in complete sentences and good English.

Discrete mathematics is quite different from other areas in mathematics which you may have already studied, such as algebra, geometry, or calculus. It is much less structured; there are far fewer standard techniques to learn. It is, however, a rich subject full of ideas at least some of which we hope will intrigue you to the extent that you will want to learn more about them. Related sources of material for further reading are given in numerous footnotes throughout the text.

To the Student from a Student

I am a student at Memorial University of Newfoundland and have taken a course based on a preliminary version of this book. I spent one summer working for the authors, helping them to try to improve the book. As part of my work, they asked me to write an introduction for the student. They felt a fellow student would be the ideal person to prepare (warn?) other students before they got too deeply engrossed in the book.

There are many things which can be said about this textbook. The authors have a unique sense of humor, which often, subtly or overtly, plays a part in their presentation of material. It is an effective tool in keeping the information interesting and, in the more subtle cases, in keeping you alert. They try to make discrete mathematics as much fun as possible, at the same time maximizing the information presented.

While the authors do push a lot of new ideas at you, they also try hard to minimize potential difficulties. This is not an easy task considering that there are many levels of students who will use this book, so the material and exercises must be challenging enough to engage all of them. To balance this, numerous examples in each section are given as a guide to the exercises. Also, the exercises at the end of every section are laid out with easier ones at the beginning and the harder ones near the end.

Concerning the exercises, the authors' primary objective is not to stump you or to test more than you should know. The purpose of the exercises is to help clarify the material and to make sure you understand what has been covered. The authors intend that you stop and think before you start writing.

Inevitably, not everything in this book is exciting. Some material may not even seem particularly useful. As a textbook used for discrete mathematics and graph theory, there are many topics which must be covered. Generally, less exciting material is in the first few chapters and more interesting topics are introduced later. For example, the chapter on sets and relations may not captivate your attention, but it is essential for the understanding of almost all later topics. The chapter on principles of counting is both interesting and useful, and it is fundamental to a subsequent chapter on permutations and combinations.

This textbook is written to engage your mind and to offer a fun way to learn some mathematics. The authors do hope that you will not view this as a painful experience, but as an opportunity to begin to think seriously about various areas of modern mathematics. The best way to approach this book is with pencil, paper, and an open mind.

To the Instructor

Since the first printing of this book, we have received a number of queries about the existence of a solutions manual. Let us begin then with the assurance that a complete solutions manual does exist and is available from the publisher, for the benefit of instructors.

The material in this text has been taught and tested for many years in two one-semester courses, one in discrete mathematics at the sophomore level (with no graph theory) and the other in applied graph theory (at the junior level). We believe this book is more elementary and written with a far more leisurely style than comparable books on the market. For example, since students can enter our courses without calculus or without linear algebra, this book does not presume that students have backgrounds in either subject. The few places where some knowledge of a first calculus or linear algebra course would be useful are clearly marked. With one exception, this book requires virtually no background. The exception is Section 10.3, on the adjacency matrix of a graph, where we assume a little linear algebra. If desired, this section can easily be omitted without consequences.

The material for our first course can be found in Chapters 1 through 7, although we find it impossible to cover all the topics presented here in the thirty-three 50-minute lectures available to us. There are various ways to shorten the course. One possibility is to omit Chapter 4 (The Integers), although it is one of our favorites, especially if students will subsequently take a number theory course. Another solution is to omit all but the material on mathematical induction in Chapter 5, as well as certain other individual topics, such as partial orders (Section 2.5) and derangements (Section 7.4).

Graph theory is the subject of Chapters 9 through 15, and again we find that there is more material here than can be successfully treated in thirty-three lectures. Usually, we include only a selection of the various applications and algorithms presented in this part of the text. We do not always discuss the puzzles in Section 9.1, scheduling problems (Section 11.5), applications of the Max Flow–Min Cut Theorem, or matchings (Sections 15.3 and 15.4). Chapter 13 (Depth-First Search and Applications) can also be omitted without difficulty. In fact, most of the last half of this book is self-contained and can be treated to whatever extent the instructor may desire.

Chapter 8, which introduces the concepts of algorithm and complexity, seems to work best as the introduction to the graph theory course.

Wherever possible, we have tried to keep the material in various chapters independent of material in earlier chapters. There are, of course, obvious situations where this is simply not possible. It is necessary to understand equivalence relations (Section 2.4), for example, before reading about congruence in Section 4.4, and one must study Hamiltonian graphs (Section 10.2) before learning about facilities design in Section 14.3. For the most part, however, the graph theory material can be read independently of earlier chapters. Some knowledge of such basic notions as function (Chapter 3) and equivalence relation is needed in several places and, of course, many proofs in graph theory require mathematical induction (Section 5.1).

On the other hand, we have deliberately included in most exercise sets some problems which relate to material in earlier sections, as well as some which are based solely on the material in the given section. This opens a wide variety of possibilities to instructors as to the kind of syllabus they wish to follow and to the level of exercise that is most appropriate to their students. We hope students of our book will appreciate the complete solutions, not simply answers, provided for many of the exercises at the back. By popular demand, we have increased the number of [BB]s in this second edition by over 60%.

One of the main goals of this book is to introduce students in a rigorous, yet friendly, way to the "mysteries" of theorem proving. Sections 1.1 and 1.2 are intended as background preparation for this often difficult journey. Because many instructors wish to include more formal topics in logic, this edition includes sections on truth tables, the algebra of propositions, and logical arguments (Sections 1.3, 1.4, 1.5, respectively).

Supplements

There is a full Instructor's Solutions Manual (0130920126) free to faculty, available only through Prentice Hall's sales reps and home offices. In addition, there is a student website of activities available by November 1, 2001 at the following address: www.prenhall.com/goodaire. This site is free to all users/purchasers of this text.

New in the Second Edition

The most common (negative) criticism of our first edition was the short treatment of logic and the absence of truth tables. This problem has been remedied with Chapter 1 (previously Chapter 0) completely rewritten and expanded significantly to include new sections on truth tables, the algebra of propositions, and logical arguments. The text now includes more than enough material for instructors who wish to include a substantial unit on formal logic, while continuing to permit a shorter treatment dealing exclusively with the major points and jargon of proofs in mathematics.

The second most common complaint—and every student's favorite—was the shortage of answers in the back of the book. In fact, the first edition contained over 500 solved problems. For the second edition, however, this number has been increased to over 800.

Other features of the second edition include the following:

- A new section (12.5) on acyclic graphs and an algorithm of Bellman
- A shortest path algorithm due to Bellman and Ford which permits negative weight arcs, in the section (11.2) on digraphs
- Algorithms rewritten in a less casual way so as to more closely resemble computer code
- Review exercises at the end of every chapter
- Nonmathematical exercises, often requiring some research on behalf of the reader, asking that answers be written in good clear English, in order to encourage the development of sound writing and expository skills
- A new numbering scheme which will make searching much easier for our readers

Acknowledgments

This book represents the culmination of many years of work and reflects the comments and suggestions of numerous individuals. We acknowledge with gratitude the assistance and patience of our acquisitions editor, George Lobell, and his assistant, Melanie Van Benthuysen; our production editor, Bayani Mendoza de Leon; and all the staff at Prentice Hall who have helped with this project.

We thank sincerely the literally hundreds of students at Memorial University of Newfoundland and elsewhere who have used and helped us to improve this text. Matthew Case spent an entire summer carefully scrutinizing our work. Professors from far and wide have made helpful suggestions. For the current edition, we are especially indebted to Peter Booth, Clayton Halfyard, David Pike, and Nabil Shalaby at Memorial University of Newfoundland. David, in particular, gave us a lot of help for which we are most grateful.

Without exception, each of the reviewers Prentice Hall employed on this project gave us helpful and extensive criticism with just enough praise to keep us working. We thank, in particular, the reviewers of our first edition:

Amitabha Ghosh (Rochester Institute of Technology)

Akihiro Kanamori (Boston University)

Nicholas Krier (Colorado State University)
Suraj C. Kothari (Iowa State University)
Joseph Kung (University of North Texas)
Nachimuthu Manickam (Depauw University)

those of the second edition:

David M. Arnold (Baylor University)
Kiran R. Bhutani (The Catholic University of America)
Krzysztof Galicki (University of New Mexico)
Heather Gavias (Grand Valley State University)
Gabor J. Szekely (Bowling Green State University)

and those who prefer to remain anonymous.

While this book was eventually typeset by Prentice Hall, it was prepared on the first author's computer using the MathTime and MathTime Plus fonts of Y&Y, Inc. We wish to thank Mr. Louis Vosloo of Y&Y Support for his unusual patience and help.

Most users of this book have had, and will continue to have, queries, concerns, and comments to express. We are always delighted to engage in correspondence with our readers and encourage students and course instructors alike to send us e-mail anytime. Our addresses are edgar@math.mun.ca and michael1@math.mun.ca.

Answers to Pauses

1. [BB] is found throughout the exercises in this book. It means that the answer to the exercise which it labels can be found in the Back of the Book.

EXERCISES

There are no exercises in this Preface, but there are over two thousand exercises in the rest of the book!

E. G. Goodaire
M. M. Parmenter

edgar@math.mun.ca
michael1@math.mun.ca

Suggested Lecture Schedule

Discrete
Mathematics
with Graph Theory

1

Yes, There Are Proofs!

An Introduction to Logic

"How many dots are there on a pair of dice?" The question once popped out of the box in a game of trivia in which one of us was a player. A long pause and much consternation followed the question. After the correct answer was finally given, the author (a bit smugly) pointed out that the answer "of course" was 6×7, twice the sum of the integers from 1 to 6. "This is because," he declared, "the sum of the integers from 1 to n is $\frac{1}{2}n(n+1)$, so twice this sum is $n(n+1)$ and, in this case, $n = 6$."

"What?" asked one of the players.

At this point, the game was delayed for a considerable period while the author found pencil and paper and made a picture like that in Fig 1.1. If we imagine the dots on one die to be solid and on the other hollow, then the sum of the dots on two dice is just the number of dots in this picture. There are seven rows of six dots each—42 dots in all. What is more, a similar picture could be drawn for seven-sided dice showing that

$$2(1 + 2 + 3 + 4 + 5 + 6 + 7) = 7 \times 8 = 56$$

and generally,

$$(*) \qquad 2(1 + 2 + 3 + \cdots + n) = n \times (n + 1).$$

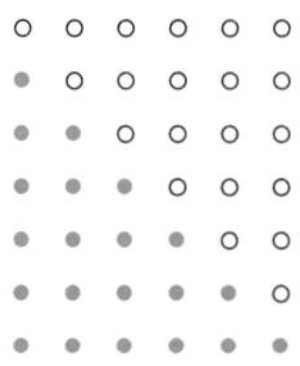

Figure 1.1

Sadly, that last paragraph is fictitious. Everybody was interested in, and most experimented with, the general observation in equation (*), but nobody (except an author) cared why. Everybody wanted to resume the game!

Pause 1 What is the sum $1 + 2 + 3 + \cdots + 100$ of the integers from 1 to 100?

"Are there proofs?" This is one of the first questions students ask when they enter a course in analysis or algebra. Young children continually ask "why" but, for whatever reason, as they grow older, most people only want the facts. We take the view that intellectual curiosity is a hallmark of advanced learning and that the ability to reason logically is an increasingly sought-after commodity in the world today. Since sound logical arguments are the essence of mathematics, the subject provides a marvellous training ground for the mind. The expectation that a math course will sharpen the powers of reason and provide instruction in clear thinking surely accounts for the prominence of mathematics in so many university programs today. So yes, proofs—reasons or convincing arguments probably sound less intimidating—will form an integral part of the discussions in this book.

In a scientific context, the term "statement" means an ordinary English statement of fact (subject, verb, and predicate in that order) which can be assigned a "truth value," that is, which can be classified as being either true or false. We occasionally say "mathematical" statement to emphasize that a statement must have this characteristic property of being true or false. The following are all mathematical statements:

"There are 168 primes less than 1000."

"Seventeen is an even number."

"$\sqrt{3}^{\sqrt{3}}$ is a rational number."

"Zero is not negative."

Each statement is certainly either true or false. (It is not necessary to know which!) On the other hand, the following are not mathematical statements:

"What are irrational numbers?"

"Suppose every positive integer is the sum of three squares."

The first is a question and the second a conditional; it would not make sense to classify either as true or false.

1.1 COMPOUND STATEMENTS

"And" and "Or"

A *compound statement* is a statement formed from two other statements in one of several ways, for example, by linking them with "and" or "or." Consider

$$\text{"}9 = 3^2 \text{ and } 3.14 < \pi.\text{"}$$

This is a compound statement formed from the simpler statements "$9 = 3^2$" and "$3.14 < \pi$." How does the truth of an "and" compound statement depend upon the truth of its parts? The rule is

> "p and q" is true if both p and q are true; it is false
> if either p is false or q is false.

Thus, "$-2^2 = -4$ and $5 < 100$" is true while "$2^2 + 3^2 = 4^2$ and $3.14 < \pi$" is false.

In the context of mathematics, just as in everyday English, one can build a compound statement by inserting the word "or" between two other statements. In everyday English, "or" can be a bit problematic because sometimes it is used in an inclusive sense, sometimes in an exclusive sense, and sometimes ambiguously, leaving the listener unsure about just what was intended. We illustrate with three sentences.

"To get into that college, you have to have a high school diploma or be over 25." (Both options are allowed.)

"That man is wanted dead or alive." (Here both options are quite impossible.)

"I am positive that either blue or white is in that team's logo." (Might there be both?)

Since mathematics does not tolerate ambiguities, we must decide precisely what "or" should mean. The decision is to make "or" inclusive: "or" always includes the possibility of both.

> "p or q" is true if p is true or q is true or both p and q are true; it is false only when both p and q are false.

Thus,

"$7 + 5 = 12$ or 571 is the 125th prime" and "25 is less than or equal to 25"

are both true sentences, while

"5 is an even number or $\sqrt{8} > 3$"

is false.

Implication

Many mathematical statements are *implications*; that is, statements of the form "p implies q," where p and q are statements called, respectively, the *hypothesis* and *conclusion*. The symbol $\rightarrow$ is read "implies," so

Statement 1: "2 is an even number $\rightarrow$ 4 is an even number"

is read "2 is an even number implies 4 is an even number."

In Statement 1, "2 is an even number" is the hypothesis and "4 is an even number" is the conclusion.

Implications often appear without the explicit use of the word "implies." To some ears, Statement 1 might sound better as

"If 2 is an even number, then 4 is an even number"

or

"2 is an even number only if 4 is an even number."

Whatever wording is used, common sense tells us that this implication is true.

Under what conditions will an implication be false? Suppose your parents tell you

Statement 2: "If it is sunny tomorrow, you may go swimming."

If it is sunny, but you are not allowed to go swimming, then clearly your parents have said something which is false. However, if it rains tomorrow and you are not allowed to go swimming, it would be unreasonable to accuse them of breaking their word. This example illustrates an important principle.

> The implication "$p \to q$" is false only when the hypothesis p is true and the conclusion q is false. In all other situations, it is true.

In particular, Statement 1 is true since both the hypothesis, "2 is an even number," and the conclusion, "4 is an even number," are true. Note, however, that an implication is true whenever the hypothesis is false (no matter whether the conclusion is true or false). For example, if it were to rain tomorrow, the implication contained in Statement 2 is true because the hypothesis is false. For the same reason, each of the following implications is true.

"If -1 is a positive number, then $2 + 2 = 5$."

"If -1 is a positive number, then $2 + 2 = 4$."

Pause 2   Think about the implication, "If $4^2 = 16$, then $-1^2 = 1$." Is it true or false?

The Converse of an Implication

The *converse* of the implication $p \to q$ is the implication $q \to p$. For example, the converse of Statement 1 is

"If 4 is an even number, then 2 is an even number."

Pause 3   Write down the converse of the implication given in Pause 2. Is this true or false?

Double Implication

Another compound statement which we will use is the double implication, $p \leftrightarrow q$, read "p *if and only if* q." As the notation suggests, the statement "$p \leftrightarrow q$" is simply a convenient way to express

$$\text{"}p \to q \text{ and } p \leftarrow q\text{."}$$

(We would be more likely to write "$q \to p$" than "$p \leftarrow q$.")

Putting together earlier observations, we conclude that

> The double implication "$p \leftrightarrow q$" is true if p and q have the same truth values; it is false if p and q have different truth values.

For example, the statement

"2 is an even number $\leftrightarrow$ 4 is an even number"

is true since both "2 is an even number" and "4 is an even number" are true. However,

"2 is an even number if and only if 5 is an even number"

is false because one side is true while the other is false.

Pause 4 Determine whether each of the following double implications is true or false.

(a) "$4^2 = 16 \leftrightarrow -1^2 = -1$."
(b) "$4^2 = 16$ if and only if $(-1)^2 = -1$."
(c) "$4^2 = 15$ if and only if $-1^2 = -1$."
(d) "$4^2 = 15 \leftrightarrow (-1)^2 = -1$."

Negation

The *negation* of the statement p is the statement which asserts that p is not true. We denote the negation of p by "$\neg p$" and say "not p." The negation of "x equals 4" is the statement "x does not equal 4." In mathematical writing, a slash (/) through a symbol is used to express the negation of that symbol. So, for instance, $\neq$ means "not equal." Thus, the negation of "$x = 4$" is "$x \neq 4$." In succeeding chapters, we shall meet other symbols like $\in$, $\subseteq$, and $|$, each of which is negated with a slash, $\notin$, $\not\subseteq$, $\nmid$.

Some rules for forming negations are a bit complicated because it is not enough just to say "not p": We must also understand what is being said! To begin, we suggest that the negation of p be expressed as

"It is not the case that p."

Then we should think for a minute or so about precisely what this means. For example, the negation of "25 is a perfect square" is the statement "It is not the case that 25 is a perfect square," which surely means "25 is not a perfect square." To obtain the negation of

"$n < 10$ or n is odd,"

we begin

"It is not the case that $n < 10$ or n is odd."

A little reflection suggests that this is the same as the more revealing "$n \geq 10$ and n is even."

The negation of an "or" statement is always an "and" statement and the negation of an "and" is always an "or." The precise rules for expressing the negation of compound statements formed with "and" and "or" are due to Augustus De Morgan (whom we shall meet again in Chapter 2).

> The negation of "p and q" is the assertion "$\neg p$ or $\neg q$."
> The negation of "p or q" is the assertion "$\neg p$ and $\neg q$."

For example, the negation of "$a^2 + b^2 = c^2$ and $a > 0$" is "Either $a^2 + b^2 \neq c^2$ or $a \leq 0$." The negation of "$x + y = 6$ or $2x + 3y < 7$" is "$x + y \neq 6$ and $2x + 3y \geq 7$."

The Contrapositive

The *contrapositive* of the implication "$p \to q$" is the implication "$(\neg q) \to (\neg p)$." For example, the contrapositive of

"If 2 is an even number, then 4 is an even number"

is

"If 4 is an odd number, then 2 is an odd number."

Pause 5   Write down the contrapositive of the implications in Pauses 2 and 3. In each case, state whether the contrapositive is true or false. How do these truth values compare, respectively, with those of the implications in these Pauses?

Quantifiers

The expressions "there exists" and "for all," which quantify statements, figure prominently in mathematics. The universal quantifier *for all* (and equivalent expressions such as *for every*, *for any*, and *for each*) says, for example, that a statement is true *for all* integers or *for all* polynomials or *for all* elements of a certain type. The following statements illustrate its use. (Notice how it can be disguised; in particular, note that "for any" and "all" are synonymous with "for all.")

"$x^2 + x + 1 > 0$ for all real numbers x."

"All polynomials are continuous functions."

"For any positive integer n, $2(1 + 2 + 3 + \cdots + n) = n \times (n + 1)$."

"$(AB)C = A(BC)$ for all square matrices A, B, and C."

Pause 6   Rewrite "All positive real numbers have real square roots," making explicit use of a universal quantifier.

The existential quantifier *there exists* stipulates the existence of a single element for which a statement is true. Here are some assertions in which it is employed.

"There exists a smallest positive integer."

"Two sets may have no element in common."

"Some polynomials have no real zeros."

Again, we draw attention to the fact that the ideas discussed in this chapter can arise in subtle ways. Did you notice the implicit use of the existential quantifier in the second of the preceding statements?

"There exists a set A and a set B such that A and B have no element in common."

Pause 7   Rewrite "Some polynomials have no real zeros" making use of the existential quantifier.

Here are some statements which employ both types of quantifiers.

"There exists a matrix 0 with the property that $A + 0 = 0 + A$ for all matrices A."

"For any real number x, there exists an integer n such that $n \leq x < n + 1$."

"Every positive integer is the product of primes."

"Every nonempty set of positive integers has a smallest element."

Sometimes it requires real thought to express in a useful way the negation of a statement which involves one or more quantifiers. It's usually helpful to begin with "It is not the case." Consider, for instance, the assertion

"For every real number x, x has a real square root."

A first stab at its negation gives

"It is not the case that every real number x has a real square root."

Now we must think about what this really means. Surely,

"There exists a real number which does not have a real square root."

It is helpful to observe that the negation of a statement involving one type of quantifier is a statement which uses the other. Specifically,

> The negation of "For all something, p" is the statement "There exists something such that $\neg p$."
>
> The negation of "There exists something such that p" is the statement "For all something, $\neg p$."

For example, the negation of

"There exist a and b for which $ab \neq ba$"

is the statement

"For all a and b, $ab = ba$."

1.1.1 REMARK ▶ The symbols $\forall$ and $\exists$ are commonly used for the quantifiers "for all" and "there exists", respectively. For example, we might encounter the statement

$$\forall x, \exists n \text{ such that } n > x$$

in a book in real analysis. We won't use this notation in this book, but it is so common that you should know about it.

What may I assume?

In our experience, when asked to prove something, students often wonder just what they are allowed to assume. For the rest of this book, the answer is any fact, including the result of any exercise, stated **earlier** in the book. This chapter is somewhat special because we are talking "about" mathematics and endeavoring to use only "familiar" ideas to motivate our discussion. In addition to basic college algebra, here is a list of mathematical definitions and facts which we assume and which the student is free also to assume in this chapter's exercises.

- The product of nonzero real numbers is nonzero.
- The square of a nonzero real number is a positive real number.
- An even integer is one which is of the form $2k$ for some integer k; an odd integer is one which is of the form $2k + 1$ for some integer k.
- The product of two even integers is even; the product of two odd integers is odd; the product of an odd integer and an even integer is even.
- A real number is rational if it is a common fraction, that is, the quotient $\frac{m}{n}$ of integers m and n with $n \neq 0$.

- A real number is irrational if it is not rational. For example, π and $\sqrt[3]{5}$ are irrational numbers.
- An irrational number has a decimal expansion which neither repeats nor terminates.
- A prime is a positive integer $p > 1$ which is divisible evenly only by ± 1 and $\pm p$, for example, 2, 3 and 5.

Answers to Pauses

1. By equation (*), twice the sum of the integers from 1 to 100 is 100×101. So the sum itself is $50 \times 101 = 5050$.
2. This is false. The hypothesis is true but the conclusion is false: $-1^2 = -1$, not 1.
3. The converse is "If $-1^2 = 1$, then $4^2 = 16$." This is true, because the hypothesis, "$-1^2 = 1$", is false.
4. **(a)** This is true because both statements are true.
   (b) This is false because the two statements have different truth values.
   (c) This is false because the two statements have different truth values.
   (d) This is true because both statements are false.
5. The contrapositive of the implication in Pause 2 is "If $-1^2 \neq 1$, then $4^2 \neq 16$." This is false because the hypothesis is true, but the conclusion is false. The contrapositive of the implication in Pause 3 is "If $4^2 \neq 16$, then $-1^2 \neq 1$." This is true because the hypothesis is false.
   These answers are the same as in Pauses 2 and 3. This is always the case, as we shall see in Section 1.2.
6. "For all real numbers $x > 0$, x has a real square root."
7. "There exists a polynomial with no real zeros."

EXERCISES

The symbol [BB] means that an answer can be found in the Back of the Book.

You are urged to read the final paragraph of this section—What may I assume?—before attempting these exercises.

1. Classify each of the following statements as true or false and explain your answers.

   (a) [BB] "$4 = 2 + 2$ and $7 < \sqrt{50}$."
   (b) "$4 \neq 2 + 2$ and $7 < \sqrt{50}$."
   (c) [BB] "$4 = 2 + 2 \rightarrow 7 < \sqrt{50}$."
   (d) "$4 = 2 + 2 \leftrightarrow 7 < \sqrt{50}$."
   (e) [BB] "$4 \neq 2 + 2 \rightarrow 7 < \sqrt{50}$."
   (f) "$4 \neq 2 + 2 \leftrightarrow 7 < \sqrt{50}$."
   (g) [BB] "$4 = 2 + 2 \rightarrow 7 > \sqrt{50}$."
   (h) "The area of a circle of radius r is $2\pi r$ or its circumference is πr^2."
   (i) "$2 + 3 = 5 \rightarrow 5 + 6 = 10$."

2. Classify each of the following statements as true or false and explain your answers.

   (a) [BB] If a and b are integers with $a - b \geq 0$ and $b - a \geq 0$, then $a = b$.
   (b) If a and b are integers with $a - b > 0$ and $b - a > 0$, then $a = b$.

3. Write down the negation of each of the following statements in clear and concise English. Do not use the expression "It is not the case that" in your answers.

   (a) [BB] Either $a^2 > 0$ or a is not a real number.
   (b) x is a real number and $x^2 + 1 = 0$.
   (c) [BB] $x = \pm 1$.
   (d) Every integer is divisible by a prime.
   (e) [BB] For every real number x, there is an integer n such that $n > x$.
   (f) There exist a, b, and c such that $(ab)c \neq a(bc)$.
   (g) [BB] There exists a planar graph which cannot be colored with at most four colors.

(h) For every $x > 0$, $x^2 + y^2 > 0$ for all y.

(i) [BB] For all integers a and b, there exist integers q and r such that $b = qa + r$.

(j) There exists an infinite set whose proper subsets are all finite.

4. Write down the converse and the contrapositive of each of the following implications.

(a) [BB] If $\frac{a}{b}$ and $\frac{b}{c}$ are integers, then $\frac{a}{c}$ is an integer.

(b) $x^2 = 1 \rightarrow x = \pm 1$.

(c) [BB] Every Eulerian graph is connected.

(d) $ab = 0 \rightarrow a = 0$ or $b = 0$.

(e) [BB] A square is a four-sided figure.

(f) If $\triangle BAC$ is a right triangle, then $a^2 = b^2 + c^2$.

5. Rewrite each of the following statements using the quantifiers "for all" and "there exists" as appropriate.

(a) [BB] Not all continuous functions are differentiable.

(b) For real x, 2^x is never negative.

(c) [BB] There is no largest real number.

(d) There are infinitely many primes.

(e) [BB] Every positive integer is the product of primes.

(f) All positive real numbers have real square roots.

6. Is it possible for both an implication and its converse to be false? Explain your answer.

1.2 PROOFS IN MATHEMATICS

Many mathematical theorems are statements that a certain implication is true. A simple result about real numbers says that if x is between 0 and 1, then $x^2 < 1$. In other words, for any choice of a real number between 0 and 1, it is a fact that the square of the number will be less than 1. We are asserting that the implication

Statement 3: "$0 < x < 1 \rightarrow x^2 < 1$"

is true. In Section 1.1, the hypothesis and conclusion of an implication could be any two statements, even statements completely unrelated to each other. In the statement of a theorem or a step of a mathematical proof, however, the hypothesis and conclusion will be statements about the same class of objects and the statement (or step) is the assertion that an implication is always true. The only way for an implication to be false is for the hypothesis to be true and the conclusion false. So the statement of a mathematical theorem or a step in a proof only requires proving that whenever the hypothesis is true, the conclusion must also be true.

When the implication "$\mathcal{A} \rightarrow \mathcal{B}$" is the statement of a theorem, or one step in a proof, $\mathcal{A}$ is said to be a *sufficient* condition for $\mathcal{B}$ and $\mathcal{B}$ is said to be a *necessary* condition for $\mathcal{A}$. For instance, the implication in Statement 3 can be restated "$0 < x < 1$ is sufficient for $x^2 < 1$" or "$x^2 < 1$ is necessary for $0 < x < 1$."

Pause 8 Rewrite the statement "A matrix with determinant 1 is invertible" so that it becomes apparent that this is an implication. What is the hypothesis? What is the conclusion? This sentence asks which is a necessary condition for what? What is a sufficient condition for what?

To prove that Statement 3 is true, it is **not** enough to take a single example, $x = \frac{1}{2}$ for instance, and claim that the implication is true because $\left(\frac{1}{2}\right)^2 < 1$. It is **not** better to take ten, or even ten thousand, such examples and verify the implication in each special case. Instead, a **general argument** must be given which works for all x between 0 and 1. Here is such an argument.

Assume that the hypothesis is true; that is, x is a real number with $0 < x < 1$. Since $x > 0$, it must be that $x \cdot x < 1 \cdot x$ because multiplying both sides of an inequality such as $x < 1$ by a positive number such as x preserves the inequality. Hence $x^2 < x$ and, since $x < 1$, $x^2 < 1$ as desired.

Now let us consider the converse of the implication in Statement 3:

Statement 4: "$x^2 < 1 \to 0 < x < 1$."

This is false. For example, when $x = -\frac{1}{2}$, the left-hand side is true, since $(-\frac{1}{2})^2 = \frac{1}{4} < 1$, while the right-hand side is false. So the implication fails when $x = -\frac{1}{2}$. It follows that this implication cannot be used as part of a mathematical proof. The number $x = -\frac{1}{2}$ is called a *counterexample* to Statement 4; that is, a specific example which proves that an implication is false.

There is a very important point to note here. To show that a theorem, or a step in a proof, is false, it is enough to find a single case where the implication does not hold. However, as we saw with Statement 3, to show that a theorem is true, we must give a proof which covers all possible cases.

Pause 9 State the contrapositive of "$0 < x < 1 \to x^2 < 1$." Is this true or false?

Theorems in mathematics don't have to be about numbers. For example, a very famous theorem proved in 1976 asserts that if $\mathcal{G}$ is a planar graph, then $\mathcal{G}$ can be colored with at most four colors. (The definitions and details are in Chapter 14.) This is an implication of the form "$\mathcal{A} \to \mathcal{B}$," where the hypothesis $\mathcal{A}$ is the statement that $\mathcal{G}$ is a planar graph and the conclusion $\mathcal{B}$ is the statement that $\mathcal{G}$ can be colored with at most four colors. The Four-Color Theorem states that this implication is true.

Pause 10 (For students who have studied linear algebra.) The statement given in Pause 8 is a theorem in linear algebra; that is, the implication is true. State the converse of this theorem. Is this also true?

A common expression in scientific writing is the phrase *if and only if* denoting both an implication and its converse. For example,

Statement 5: "$x^2 + y^2 = 0 \leftrightarrow (x = 0$ and $y = 0)$."

As we saw in Section 1.1, the statement "$\mathcal{A} \leftrightarrow \mathcal{B}$" is a convenient way to express the compound statement

"$\mathcal{A} \to \mathcal{B}$ and $\mathcal{B} \to \mathcal{A}$."

The sentence "$\mathcal{A}$ is a *necessary and sufficient* condition for $\mathcal{B}$" is another way of saying "$\mathcal{A} \leftrightarrow \mathcal{B}$." The sentence "$\mathcal{A}$ and $\mathcal{B}$ are *necessary and sufficient conditions* for $\mathcal{C}$" is another way of saying

"$(\mathcal{A}$ and $\mathcal{B}) \leftrightarrow \mathcal{C}$."

For example, "a triangle has three equal angles" is a necessary and sufficient condition for "a triangle has three equal sides." We would be more likely to hear "In order for a triangle to have three equal angles, it is necessary and sufficient that it have three equal sides."

To prove that "$\mathcal{A} \leftrightarrow \mathcal{B}$" is true, we must prove separately that "$\mathcal{A} \to \mathcal{B}$" and "$\mathcal{B} \to \mathcal{A}$" are both true, using the ideas discussed earlier. In Statement 5, the implication "$(x = 0$ and $y = 0) \to x^2 + y^2 = 0$" is easy.

Pause 11 Prove that "$x^2 + y^2 = 0 \to (x = 0$ and $y = 0)$."

As another example, consider

Statement 6: "$0 < x < 1 \leftrightarrow x^2 < 1$."

Is this true? Well, we saw earlier that "$0 < x < 1 \to x^2 < 1$" is true, but we also noted that its converse, "$x^2 < 1 \to 0 < x < 1$," is false. It follows that Statement 6 is false.

Pause 12 Determine whether "$-1 < x < 1 \leftrightarrow x^2 < 1$" is true or false.

Sometimes, a theorem in mathematics asserts that three or more statements are *equivalent*, meaning that all possible implications between pairs of statements are true. Thus

"The following are equivalent:

1. $\mathcal{A}$
2. $\mathcal{B}$
3. $\mathcal{C}$"

means that each of the double implications $\mathcal{A} \leftrightarrow \mathcal{B}$, $\mathcal{B} \leftrightarrow \mathcal{C}$, $\mathcal{A} \leftrightarrow \mathcal{C}$ is true. Instead of proving the truth of the six implications here, it is more efficient just to establish the truth, say, of the sequence

$$\mathcal{A} \to \mathcal{B} \to \mathcal{C} \to \mathcal{A}$$

of three implications. It should be clear that if these implications are all true, then any implication involving two of $\mathcal{A}$, $\mathcal{B}$, $\mathcal{C}$ is also true; for example, the truth of $\mathcal{B} \to \mathcal{A}$ would follow from the truth of $\mathcal{B} \to \mathcal{C}$ and $\mathcal{C} \to \mathcal{A}$.

Alternatively, to establish that $\mathcal{A}$, $\mathcal{B}$, and $\mathcal{C}$ are equivalent, we could establish the truth of the sequence

$$\mathcal{B} \to \mathcal{A} \to \mathcal{C} \to \mathcal{B}$$

of implications. Which of the two sequences a person chooses is a matter of preference but is usually determined by what appears to be the easiest way to argue. Here is an example.

PROBLEM 1. Let x be a real number. Show that the following are equivalent.

(1) $x = \pm 1$.
(2) $x^2 = 1$.
(3) If a is any real number, then $ax = \pm a$.

Solution. To show that these statements are equivalent, it is sufficient to establish the truth of the sequence

$$(2) \to (1) \to (3) \to (2).$$

$(2) \to (1)$: The notation means "assume (2) and prove (1)." Since $x^2 = 1$, $0 = x^2 - 1 = (x + 1)(x - 1)$. Since the product of real numbers is zero if and only if one of the numbers is zero, either $x + 1 = 0$ or $x - 1 = 0$; hence $x = -1$ or $x = +1$, as required.

(1) → (3): The notation means "assume (1) and prove (3)." Thus either $x = +1$ or $x = -1$. Let a be a real number. If $x = +1$, then $a \cdot x = a \cdot 1 = a$ while if $x = -1$, then $ax = -a$. In every case, $ax = \pm a$ as required.

(3) → (2): We assume (3) and prove (2). We are given that $ax = \pm a$ for any real number a. With $a = 1$, we obtain $x = \pm 1$ and squaring gives $x^2 = 1$, as desired.   ∎

The symbol ∎ marks the end of a proof. Some authors write "Q.E.D." from the Latin *quod erat demonstrandum* (which was to be proved) for the same purpose.

In the index (see *equivalent*), you are directed to other places in this book where we establish the equivalence of a series of statements.

Direct Proof

Most theorems in mathematics are stated as implications: "$\mathcal{A} \to \mathcal{B}$." Sometimes, it is possible to prove such a statement *directly*; that is, by establishing the validity of a sequence of implications:

$$\mathcal{A} \to \mathcal{A}_1 \to \mathcal{A}_2 \to \cdots \to \mathcal{B}.$$

PROBLEM 2.   Prove that for all real numbers x, $x^2 - 4x + 17 \neq 0$.

Solution.   We observe that $x^2 - 4x + 17 = (x - 2)^2 + 13$ is the sum of 13 and a number, $(x - 2)^2$, which is never negative. So $x^2 - 4x + 17 \geq 13$ for any x; in particular, $x^2 - 4x + 17 \neq 0$.   ∎

PROBLEM 3.   Suppose that x and y are real numbers such that $2x + y = 1$ and $x - y = -4$. Prove that $x = -1$ and $y = 3$.

Solution. $(2x + y = 1 \text{ and } x - y = -4) \to (2x + y) + (x - y) = 1 - 4$

$$\to 3x = -3 \to x = -1.$$

Also,

$$(x = -1 \text{ and } x - y = -4) \to (-1 - y = -4) \to (y = -1 + 4 = 3). \quad ∎$$

Many of the proofs in this book are direct. In the index (under *direct*), we guide you to several of these.

Proof by Cases

Sometimes a direct argument is made simpler by breaking it into a number of cases, one of which must hold and each of which leads to the desired conclusion.

PROBLEM 4.   Let n be an integer. Prove that $9n^2 + 3n - 2$ is even.

Solution.   **Case 1:** n is even.

The product of an even integer and any integer is even. Since n is even, $9n^2$ and $3n$ are even too. Thus $9n^2 + 3n - 2$ is even because it is the sum of three even integers.

Case 2: n is odd.

The product of odd integers is odd. In this case, since n is odd, $9n^2$ and $3n$ are also odd. The sum of two odd integers is even. Thus $9n^2 + 3n$ is even. The sum of even integers is even, so $9n^2 + 3n - 2 = (9n^2 + 3n) + (-2)$ is even. ∎

In the index, we guide you to other places in this book where we give proofs by cases. (See *cases*.)

Prove the Contrapositive

A very important principle of logic, foreshadowed by Pause 5, is summarized in the next theorem.

1.2.1 THEOREM ▶ "$A \to B$" is true if and only if its contrapositive "$\neg B \to \neg A$" is true.

Proof "$A \to B$" is false if and only if A is true and B is false; that is, if and only if $\neg A$ is false and $\neg B$ is true; that is, if and only if "$\neg B \to \neg A$" is false. Thus the two statements "$A \to B$" and "$\neg B \to \neg A$" are false together (and hence true together); that is, they have the same truth values. The result is proved. ∎

PROBLEM 5. If the average of four different integers is 10, prove that one of the integers is greater than 11.

Solution. Let A and B be the statements

A: "The average of four integers, all different, is 10."

B: "One of the four integers is greater than 11."

We are asked to prove the truth of "$A \to B$." Instead, we prove the truth of the contrapositive, "$\neg B \to \neg A$", from which the result follows by Theorem 1.2.1.

Call the given integers a, b, c, d. If B is false, then each of these numbers is at most 11 and, since they are all different, the biggest value for $a + b + c + d$ is $11 + 10 + 9 + 8 = 38$. So the biggest possible average would be $\frac{38}{4}$, which is less than 10, so A is false. ∎

Proof by Contradiction

Sometimes a direct proof of a statement A seems hopeless: We simply do not know how to begin. In this case, we can sometimes make progress by assuming that the negation of A is true. If this assumption leads to a statement which is obviously false (an "absurdity") or to a statement which contradicts something else, then we will have shown that $\neg A$ is false. So, A must be true.

PROBLEM 6. Show that there is no largest integer.

Solution. Let A be the statement "There is no largest integer." If A is false, then there is a largest integer N. This is absurd, however, because $N + 1$ is an integer larger than N. Thus $\neg A$ is false, so A is true. ∎

Remember that *rational number* just means common fraction, the quotient $\frac{m}{n}$ of integers m and n with $n \neq 0$. A number which is not rational is called *irrational*.

PROBLEM 7.   Suppose that a is a nonzero rational number and that b is an irrational number. Prove that ab is irrational.

Solution.   We are asked to prove the truth of the implication "$\mathcal{A} \to \mathcal{B}$," where $\mathcal{A}$ and $\mathcal{B}$ are the statements

$\mathcal{A}$: "a is a nonzero rational and b is irrational."
$\mathcal{B}$: "ab is irrational."

Suppose that this implication is false and remember that this occurs only when $\mathcal{A}$ is true and $\mathcal{B}$ is false; that is, a is rational, b is irrational, and ab is rational. So we know that $ab = \frac{m}{n}$ for integers m and n, $n \neq 0$ and $a = \frac{k}{\ell}$ for integers k and ℓ, $\ell \neq 0$ and $k \neq 0$ (because $a \neq 0$). Thus,

$$b = \frac{m}{na} = \frac{m\ell}{nk}$$

with $nk \neq 0$, so b is rational. This contradicts the fact that b is irrational. So "$\mathcal{A} \to \mathcal{B}$" is true.   ∎

Here is a well-known but nonetheless beautiful example of a proof by contradiction.

PROBLEM 8.   Prove that $\sqrt{2}$ is an irrational number.

Solution.   If the statement is false, then there exist integers m and n such that $\sqrt{2} = \frac{m}{n}$. If both m and n are even, we can cancel 2's in numerator and denominator until at least one of them is odd. Thus, *without loss of generality*, we may assume that not both m and n are even.

Square both sides of $\sqrt{2} = \frac{m}{n}$. We get $m^2 = 2n^2$, so m^2 is even. Since the square of an odd integer is odd, m must be even. Since not both m and n are even, it follows that n must be odd. Write $m = 2k$ and $n = 2\ell + 1$. Then from $m^2 = 2n^2$ we get

$$4k^2 = 2(4\ell^2 + 4\ell + 1) = 8\ell^2 + 8\ell + 2$$

and, dividing by 2,

$$2k^2 = 4\ell^2 + 4\ell + 1.$$

This is an absurdity because the left side is even and the right is odd.   ∎

---

8. "A matrix A has determinant $1 \to A$ is invertible."
   Hypothesis: A matrix A has determinant 1.
   Conclusion: A is invertible.

The invertibility of a matrix is a necessary condition for its determinant being equal to 1; determinant 1 is a sufficient condition for the invertibility of a matrix.

9. The contrapositive is "$x^2 \geq 1 \to (x \leq 0$ or $x \geq 1)$." This is true, and here is a proof.

   Assume $x^2 \geq 1$. If $x \leq 0$, we have the desired result, so assume $x > 0$. In that case, if $x < 1$, we have seen that $x^2 < 1$, which is not true, so we must have $x \geq 1$, again as desired.

10. The converse is "An invertible matrix has determinant 1." This is false: for example, the matrix $A = \begin{bmatrix} 0 & 1 \\ 1 & 0 \end{bmatrix}$ is invertible (the inverse of A is A itself), but $\det A = -1$.

11. Assume that $x^2 + y^2 = 0$. Since the square of a real number cannot be negative and the square of a **nonzero** real number is positive, if either $x^2 \neq 0$ or $y^2 \neq 0$, the sum $x^2 + y^2$ would be positive, which is not true. This means $x^2 = 0$ and $y^2 = 0$, so $x = 0$ and $y = 0$, as desired.

12. This statement is true. To prove it, we must show that two implications are true.

    ($\to$) First assume that $-1 < x < 1$. If $0 < x < 1$, then we saw in the text that $x^2 < 1$ while, if $x = 0$, clearly $x^2 = 0 < 1$. If $-1 < x < 0$, then $0 < -x < 1$ (multiplying an inequality by a negative number reverses it) so, by the argument in the text $(-x)^2 < 1$, that is, $x^2 < 1$. In all cases, we have $x^2 < 1$. Thus $-1 < x < 1 \to x^2 < 1$ is true.

    ($\leftarrow$) Next we prove that $x^2 < 1 \to -1 < x < 1$ is true. Assume $x^2 < 1$. If $x \geq 1$, then we would also have $x^2 = x \cdot x \geq x \cdot 1 = x \geq 1$, so $x^2 \geq 1$, which is not true. If we had $x \leq -1$, then $-x \geq 1$ and so $x^2 = (-x)^2 \geq 1$, which, again, is not true. We conclude that $-1 < x < 1$, as desired.

EXERCISES

The symbol [BB] means that an answer can be found in the Back of the Book.

You are urged to read the final paragraph of Section 1.1—What may I assume?—before attempting these exercises.

1. What is the hypothesis and what is the conclusion in each of the following implications?

   (a) [BB] The sum of two positive numbers is positive.

   (b) The square of the length of the hypotenuse of a right-angled triangle is the sum of the squares of the lengths of the other two sides.

   (c) All primes are even.

2. [BB] Determine whether or not the following implication is true.

   "x is an even integer $\to x + 2$ is an even integer."

3. State the converse of the implication in Exercise 2 and determine whether or not it is true.

4. Answer Exercise 2 with $\to$ replaced by $\leftrightarrow$. [*Hint*: Exercises 2 and 3.]

5. Let n be an integer greater than 1 and consider the statement "$\mathcal{A}$: $2^n - 1$ prime is necessary for n to be prime."

   (a) Write $\mathcal{A}$ as an implication.

   (b) Write $\mathcal{A}$ in the form "p is sufficient for q."

   (c) Write the converse of $\mathcal{A}$ as an implication.

   (d) Determine whether the converse of $\mathcal{A}$ is true or false.

6. [BB] A theorem in calculus states that every differentiable function is continuous. State the converse of this theorem.

   (For students who have taken calculus) Is the converse true or false? Explain.

7. Let n be an integer, $n \geq 3$. A certain mathematical theorem asserts that n statements $\mathcal{A}_1, \mathcal{A}_2, \ldots, \mathcal{A}_n$ are equivalent.

   (a) A student proves this by showing that $\mathcal{A}_1 \leftrightarrow \mathcal{A}_2$, $\mathcal{A}_2 \leftrightarrow \mathcal{A}_3, \ldots, \mathcal{A}_{n-1} \leftrightarrow \mathcal{A}_n$ are all true. How many implication proofs did the student write down?

(b) Another student proves the truth of $\mathcal{A}_1 \rightarrow \mathcal{A}_2$, $\mathcal{A}_2 \rightarrow \mathcal{A}_3, \ldots, \mathcal{A}_{n-1} \rightarrow \mathcal{A}_n$, and $\mathcal{A}_n \rightarrow \mathcal{A}_1$. How many implication proofs did this student write down?

(c) A third student wishes to find a proof which is different from that in 7(b) but uses the same number of implication proofs as in 7(b). Outline a possible proof for this student.

The next three exercises illustrate that the position of a quantifier is very important.

8. [BB] Consider the assertions

$\mathcal{A}$: "For every real number x, there exists an integer n such that $n \le x < n + 1$."

$\mathcal{B}$: "There exists an integer n such that $n \le x < n + 1$ for every real number x."

One of these assertions is true. The other is false. Which is which? Explain.

9. Answer Exercise 8 with $\mathcal{A}$ and $\mathcal{B}$ as follows.

$\mathcal{A}$: "There exists a real number y such that $y > x$ for every real number x."

$\mathcal{B}$: "For every real number x, there exists a real number y such that $y > x$."

10. Answer true or false and supply a direct proof or a counterexample to each of the following assertions.

(a) There exists an integer n such that nq is an integer for every rational number q.

(b) For every rational number q, there exists an integer n such that nq is an integer.

11. (a) [BB] Let a be an integer. Show that either a or $a+1$ is even.

(b) [BB] Show that $n^2 + n$ is even for any integer n.

12. Provide a direct proof that $n^2 - n + 5$ is odd, for all integers n.

13. [BB] Prove that $2x^2 - 4x + 3 > 0$ for any real number x.

14. Let a and b be integers. By examining the four cases

   i. a, b both even,
   ii. a, b both odd,
   iii. a even, b odd,
   iv. a odd, b even,

find a necessary and sufficient condition for $a^2 - b^2$ to be odd.

15. [BB] Let n be an integer. Prove that n^2 is even if and only if n is even.

16. Let x be a real number. Find a necessary and sufficient condition for $x + \dfrac{1}{x} \ge 2$. Prove your answer.

17. [BB] Prove that if n is an odd integer then there is an integer m such that $n = 4m + 1$ or $n = 4m + 3$. [*Hint:* Consider a proof by cases.]

18. Prove that if n is an odd integer, there is an integer m such that $n = 8m + 1$ or $n = 8m + 3$ or $n = 8m + 5$ or $n = 8m + 7$. (You may use the result of Exercise 17.)

19. Prove that there exists no smallest positive real number. [*Hint:* Find a proof by contradiction.]

20. [BB] (For students who have studied linear algebra) Suppose 0 is an eigenvalue of a matrix A. Prove that A is not invertible. [*Hint:* There is a short proof by contradiction.]

21. Let $n = ab$ be the product of positive integers a and b. Prove that either $a \le \sqrt{n}$ or $b \le \sqrt{n}$.

22. [BB] Suppose a and b are integers such that $a + b + ab = 0$. Prove that $a = b = 0$ or $a = b = -2$. Give a direct proof.

23. Suppose that a is a rational number and that b is an irrational number. Prove that $a + b$ is irrational.

24. [BB] Prove that the equations

$$
\begin{aligned}
2x + 3y - z &= 5 \\
x - 2y + 3z &= 7 \\
x + 5y - 4z &= 0
\end{aligned}
$$

have no solution. (Give a proof by contradiction.)

25. Find a proof or exhibit a counterexample to each of the following statements.

(a) [BB] $2x^2 + 3y^2 > 0$ for all real numbers x and y.

(b) a an even integer $\rightarrow \frac{1}{2}a$ an even integer.

(c) [BB] For each real number x, there exists a real number y such that $xy = 1$.

(d) If a and b are real numbers with $a + b$ rational, then a and b are rational.

(e) [BB] a and b real numbers with ab rational $\rightarrow a$ and b rational.

26. Suppose ABC and $A'B'C'$ are triangles with *pairwise equal* angles; that is, $\angle A = \angle A'$, $\angle B = \angle B'$, and $\angle C = \angle C'$. Then it is a well-known result in Euclidean geometry that the triangles have pairwise proportional sides (the triangles are *similar*). Does the same property hold for polygons with more than three sides? Give a proof or provide a counterexample.

27. (a) [BB] Suppose m and n are integers such that $n^2 + 1 = 2m$. Prove that m is the sum of squares of two integers.

(b) [BB] Given that $(4373)^2 + 1 = 2(9,561,565)$, write 9,561,565 as the sum of two squares.

28. Observe that for any real number x,

$$4x^4 + 1 = (2x^2 + 2x + 1)(2x^2 - 2x + 1).$$

(a) Use this identity to express $2^{4n+2} + 1$ (n a positive integer) and $2^{18} + 1$ as the product of two integers each greater than 1.

(b) Express $2^{36} - 1$ as the product of four integers each of which is larger than 1.

29. Prove that one of the digits $1, 2, \ldots, 9$ occurs infinitely often in the decimal expansion of π.

30. Prove that there exist irrational numbers a and b such that a^b is rational.

1.3 TRUTH TABLES

In previous sections, we presented a rather informal introduction to some of the basic concepts of mathematical logic. There are times, however, when a more formal approach can be useful. We begin to look at such an approach now.

Let p and q be statements. For us, remember that *statement* means a statement of fact which is either true or false. The compound statements "p or q" and "p and q," which were introduced in Section 1.1, will henceforth be written "$p \vee q$" and "$p \wedge q$," respectively.

$$p \vee q: \quad p \text{ or } q$$
$$p \wedge q: \quad p \text{ and } q.$$

The way in which the truth values of these compound statements depend upon those of p and q can be neatly summarized by tables called *truth tables*. Truth tables for $p \vee q$ and $p \wedge q$ are shown in Fig 1.2.

p	q	$p \vee q$
T	T	T
T	F	T
F	T	T
F	F	F

p	q	$p \wedge q$
T	T	T
T	F	F
F	T	F
F	F	F

Figure 1.2   Truth tables for $p \vee q$ (p or q) and $p \wedge q$ (p and q).

In each case, the first two columns show all possible truth values for p and q—each is either true (T) or false (F)—and the third column shows the corresponding truth value for the compound statement.

The truth table for the implication $p \to q$, introduced in Section 1.1, is shown on the left in Fig 1.3. On the right, we show the particularly simple truth table for "$\neg p$," the negation of p.

p	q	$p \to q$
T	T	T
T	F	F
F	T	T
F	F	T

p	$\neg p$
T	F
F	T

Figure 1.3   Truth tables for $p \to q$ (p implies q) and $\neg p$ (not p).

Truth tables for more complicated compound statements can be constructed using the truth tables we have seen so far. For example, the statement "$p \leftrightarrow q$," defined in Section 1.1 as "$(p \rightarrow q)$ and $(q \rightarrow p)$," is "$(p \rightarrow q) \wedge (q \rightarrow p)$." The truth values for $p \rightarrow q$ and $q \rightarrow p$ are shown in Fig 1.4. Focusing on columns 3 and 4 and remembering that $r \wedge s$ is true if and only if both r and s are true—see the truth table for "$\wedge$" shown in Fig 1.2—we obtain the truth table for $(p \rightarrow q) \wedge (q \rightarrow p)$, that is, for $p \leftrightarrow q$.

p	q	$p \rightarrow q$	$q \rightarrow p$	$(p \rightarrow q) \wedge (q \rightarrow p)$
T	T	T	T	T
T	F	F	T	F
F	T	T	F	F
F	F	T	T	T

Figure 1.4   The truth table for $p \leftrightarrow q$, p if and only if q.

It is the first two columns and the last column which are the most important, of course, so in future applications, we remember $p \leftrightarrow q$ with the simple truth table shown in Fig 1.5.

p	q	$p \leftrightarrow q$
T	T	T
T	F	F
F	T	F
F	F	T

Figure 1.5   The truth table for $p \leftrightarrow q$.

Here is another demonstration of how to analyze complex compound statements with truth tables.

EXAMPLE 9   Suppose we want the truth table for $p \rightarrow \neg(q \vee p)$.

p	q	$q \vee p$	$\neg (q \vee p)$	$p \rightarrow \neg(q \vee p)$
T	T	T	F	F
T	F	T	F	F
F	T	T	F	T
F	F	F	T	T

Although the answer is presented as a single truth table, the procedure is to construct appropriate columns one by one until the answer is reached. Here, columns 1 and 2 are used to form column 3—$q \vee p$. Then column 4 follows from column 3 and, finally, columns 1 and 4 are used to construct column 5, using the truth table for an implication in Fig 1.3.   ▲

When three statements p, q, and r are involved, eight rows are required in a truth table since it is necessary to consider the two possible truth values for r for each of the four possible truth values of p and q.

PROBLEM 10.   Construct a truth table for $(p \lor q) \leftrightarrow [((\neg p) \land r) \to (q \land r)]$.

Solution.

p	q	r	$\neg p$	$(\neg p) \land r$	$q \land r$	$((\neg p) \land r) \to (q \land r)$	$p \lor q$
T	T	T	F	F	T	T	T
T	T	F	F	F	F	T	T
T	F	T	F	F	F	T	T
T	F	F	F	F	F	T	T
F	T	T	T	T	T	T	T
F	T	F	T	F	F	T	T
F	F	T	T	T	F	F	F
F	F	F	T	F	F	T	F

$(p \lor q) \leftrightarrow [((\neg p) \land r) \to (q \land r)]$
T
T
T
T
T
T
T
F

Of course, it is only necessary to construct an entire truth table if a complete analysis of a certain compound statement is desired. We do not need to construct all 32 rows of a truth table to do the next problem.

PROBLEM 11.   Find the truth value of

$$[p \to ((q \land (\neg r)) \lor s)] \land [(\neg t) \leftrightarrow (s \land r)],$$

where p, q, r, and s are all true while t is false.

Solution.   We evaluate the expression step by step, showing just the relevant row of the truth table.

p	q	r	s	t	$\neg r$	$q \land (\neg r)$	$(q \land (\neg r)) \lor s$
T	T	T	T	F	F	F	T

$p \to [(q \land (\neg r)) \lor s]$	$\neg t$	$s \land r$	$(\neg t) \leftrightarrow (s \land r)$
T	T	T	T

$[p \to [((q \land (\neg r)) \lor s)] \land [(\neg t) \leftrightarrow (s \land r)]$
T

The truth value is true.

A notion that will be important in later sections is that of *logical equivalence*. Formally, statements A and B are logically equivalent if they have identical truth tables.

PROBLEM 12. Show that "$A\colon p \to (\neg q)$" and "$B\colon \neg(p \wedge q)$" are logically equivalent.

Solution. We simply observe that the final columns of the two truth tables are identical.

p	q	$\neg q$	$p \to (\neg q)$
T	T	F	F
T	F	T	T
F	T	F	T
F	F	T	T

p	q	$p \wedge q$	$\neg(p \wedge q)$
T	T	T	F
T	F	F	T
F	T	F	T
F	F	F	T

EXAMPLE 13 In Section 1.1, we defined the *contrapositive* of the statement "$p \to q$" as the statement "$(\neg q) \to (\neg p)$." In Theorem 1.2.1, we proved that these implications are logically equivalent without actually introducing the terminology. Here is how to establish the same result using truth tables.

p	q	$p \to q$
T	T	T
T	F	F
F	T	T
F	F	T

p	q	$\neg q$	$\neg p$	$(\neg q) \to (\neg p)$
T	T	F	F	T
T	F	T	F	F
F	T	F	T	T
F	F	T	T	T

▲

1.3.1 DEFINITIONS ▶ A compound statement that is always true, regardless of the truth values assigned to its variables, is a *tautology*. A compound statement that is always false is a *contradiction*.

Pause 13 Show that "$(p \wedge q) \to (p \vee q)$" is a tautology, while "$((\neg p) \wedge q) \wedge (p \vee (\neg q))$" is a contradiction.

The truth table for $p \leftrightarrow q$ appears in Fig 1.5. From this, we see immediately that two statements A and B are logically equivalent precisely when the statement $A \leftrightarrow B$ is a tautology.

Answers to Pauses 13.

p	q	$p \wedge q$	$p \vee q$	$(p \wedge q) \to (p \vee q)$
T	T	T	T	T
T	F	F	T	T
F	T	F	T	T
F	F	F	F	T

The final column shows that $(p \wedge q) \to (p \vee q)$ is true for all values of p and q, so this statement is a tautology.

p	q	$\neg p$	$(\neg p) \wedge q$	$\neg q$	$p \vee (\neg q)$	$((\neg p) \wedge q) \wedge (p \vee (\neg q))$
T	T	F	F	F	T	F
T	F	F	F	T	T	F
F	T	T	T	F	F	F
F	F	T	F	T	T	F

The final column shows that $((\neg p) \wedge q) \wedge (p \vee (\neg q))$ is false for all values of p and q, so this statement is a contradiction.

EXERCISES

The symbol [BB] means that an answer can be found in the Back of the Book.

1. Construct a truth table for each of the following compound statements.

   (a) [BB] $p \wedge ((\neg q) \vee p)$

   (b) $(p \wedge q) \vee ((\neg p) \to q)$

   (c) $\neg (p \wedge (q \vee p)) \leftrightarrow p$

   (d) [BB] $(\neg (p \vee (\neg q))) \wedge ((\neg p) \vee r)$

   (e) $(p \to (q \to r)) \to ((p \wedge q) \vee r)$

2. (a) If $p \to q$ is false, determine the truth value of $(p \wedge (\neg q)) \vee ((\neg p) \to q)$.

   (b) Is it possible to answer 2(a) if $p \to q$ is true instead of false? Why or why not?

3. [BB] Determine the truth value for
$$[p \to (q \wedge (\neg r))] \vee [r \leftrightarrow ((\neg s) \vee q)]$$
   when p, q, r and s are all true.

4. Repeat Exercise 3 in the case where p, q, r, and s are all false.

5. (a) Show that $q \to (p \to q)$ is a tautology.

   (b) Show that $[p \wedge q] \wedge [(\neg p) \vee (\neg q)]$ is a contradiction.

6. (a) [BB] Show that $[(p \to q) \wedge (q \to r)] \to (p \to r)$ is a tautology.

   (b) [BB] Explain in plain English why the answer to 6(a) makes sense.

7. Show that the statement
$$[p \vee ((\neg r) \to (\neg s))] \vee [(s \to ((\neg t) \vee p))$$
$$\vee ((\neg q) \to r)]$$
   is neither a tautology nor a contradiction.

8. Given that the compound statement A is a contradiction, establish each of the following.

   (a) [BB] If B is any statement, $A \to B$ is a tautology.

   (b) If B is a tautology, $B \to A$ is a contradiction.

9. (a) Show that the statement $p \to (q \to r)$ is not logically equivalent to the statement $(p \to q) \to r$.

   (b) What can you conclude from 9(a) about the compound statement $[p \to (q \to r)] \leftrightarrow [(p \to q) \to r]$?

10. If p and q are statements, then the compound statement $p \veebar q$ (often called the *exclusive or*) is defined to be true if and only if exactly one of p, q is true; that is, either p is true or q is true, but not both p and q are true.

   (a) [BB] Construct a truth table for $p \veebar q$.

   (b) Construct a truth table for $(p \veebar ((\neg p) \wedge q)) \vee q$.

   (c) [BB] Show that $(p \veebar q) \to (p \vee q)$ is a tautology.

   (d) Show that $p \veebar q$ is logically equivalent to $\neg (p \leftrightarrow q)$.

1.4 THE ALGEBRA OF PROPOSITIONS

In Section 1.3, we discussed the notion of logical equivalence and noted that two statements A and B are logically equivalent precisely when the statement $A \leftrightarrow B$ is a tautology.

More informally, when statements A and B are logically equivalent, we often think of statement B as just a rewording of statement A. Clearly then, it is of interest to be able to determine in an efficient manner when two statements are logically equivalent and when they are not. Truth tables will do this job for us, but, as you may already have noticed, they can become cumbersome rather easily. Another approach is first to gather together some of the fundamental examples of logically equivalent statements and then to analyze more complicated situations by showing how they reduce to these basic examples. Henceforth, we shall write $A \iff B$ to denote the fact that A and B are logically equivalent.

The word "proposition" is a synonym for "(mathematical) statement." Just as there are rules for addition and multiplication of real numbers—commutativity and associativity, for instance—there are properties of $\land$ and $\lor$ which are helpful in recognizing that a given compound statement is logically equivalent to another, often more simple, one.

Some Basic Logical Equivalences

1. **Idempotence:**

   (i) $(p \lor p) \iff p$
   (ii) $(p \land p) \iff p$

2. **Commutativity:**

   (i) $(p \lor q) \iff (q \lor p)$
   (ii) $(p \land q) \iff (q \land p)$

3. **Associativity:**

   (i) $((p \lor q) \lor r) \iff (p \lor (q \lor r))$
   (ii) $((p \land q) \land r) \iff (p \land (q \land r))$

4. **Distributivity:**

   (i) $(p \lor (q \land r)) \iff ((p \lor q) \land (p \lor r))$
   (ii) $(p \land (q \lor r)) \iff ((p \land q) \lor (p \land r))$

5. **Double Negation:** $\lnot (\lnot p) \iff p$
6. **De Morgan's Laws:**

   (i) $\lnot (p \lor q) \iff ((\lnot p) \land (\lnot q))$
   (ii) $\lnot (p \land q) \iff ((\lnot p) \lor (\lnot q))$

Property 6 was discussed in a less formal manner in Section 1.1.

It is clear that any two tautologies are logically equivalent and that any two contradictions are logically equivalent. Letting **1** denote a tautology and **0** a contradiction, we can add the following properties to our list.

7. **(i)** $(p \lor \mathbf{1}) \iff \mathbf{1}$
   (ii) $(p \land \mathbf{1}) \iff p$

8. **(i)** $(p \lor \mathbf{0}) \iff p$
   (ii) $(p \land \mathbf{0}) \iff \mathbf{0}$

9. **(i)** $(p \lor (\lnot p)) \iff \mathbf{1}$
   (ii) $(p \land (\lnot p)) \iff \mathbf{0}$

10. (i) $\neg 1 \iff 0$
    (ii) $\neg 0 \iff 1$

We can also add three more properties to our list.

11. $(p \to q) \iff [(\neg q) \to (\neg p)]$
12. $(p \leftrightarrow q) \iff [(p \to q) \land (q \to p)]$
13. $(p \to q) \iff [(\neg p) \lor q]$

Property 11 simply restates the fact, proven in Theorem 1.2.1, that a statement and its contrapositive are logically equivalent. The definition of "↔" gives Property 12 immediately. Property 13 shows that any implication is logically equivalent to a statement which does not use the symbol →.

Pause 14   Given the statement $p \to q$, show that its *converse*, $q \to p$, and its *inverse*, $[(\neg p) \to (\neg q)]$, are logically equivalent.

Pause 15   Show that $[\neg(p \leftrightarrow q)] \iff [(p \land (\neg q)) \lor (q \land (\neg p))]$.

PROBLEM 14.   Show that $(\neg p) \to (p \to q)$ is a tautology.

Solution.   Using Property 13, we have

$$[(\neg p) \to (p \to q)] \iff [(\neg p) \to ((\neg p) \lor q)]$$
$$\iff [(\neg(\neg p)) \lor ((\neg p) \lor q)]$$
$$\iff p \lor [(\neg p) \lor q]$$
$$\iff [p \lor (\neg p)] \lor q \iff 1 \lor q \iff 1. \quad \blacksquare$$

In the exercises, we ask you to verify all the properties of logical equivalence which we have stated. Some are absurdly simple. For example, to see that $(p \lor p) \iff p$, we need only observe that p and $p \lor p$ have the same truth tables:

p	$p \lor p$
T	T
F	F

Others require more work. To verify the second distributive property, for example, we would construct two truth tables.

p	q	r	$q \lor r$	$p \land (q \lor r)$
T	T	T	T	T
T	T	F	T	T
T	F	T	T	T
T	F	F	F	F
F	T	T	T	F
F	T	F	T	F
F	F	T	T	F
F	F	F	F	F

p	q	r	$p \wedge q$	$p \wedge r$	$(p \wedge q) \vee (p \wedge r)$
T	T	T	T	T	T
T	T	F	T	F	T
T	F	T	F	T	T
T	F	F	F	F	F
F	T	T	F	F	F
F	T	F	F	F	F
F	F	T	F	F	F
F	F	F	F	F	F

PROBLEM 15.   Express the statement $(\neg (p \vee q)) \vee ((\neg p) \wedge q)$ in simplest possible form.

Solution.   De Morgan's first law says that $\neg (p \vee q) \iff (\neg p) \wedge (\neg q)$. Thus

(*) $[(\neg (p \vee q)) \vee ((\neg p) \wedge q)] \iff [((\neg p) \wedge (\neg q)) \vee ((\neg p) \wedge q)]$,

since adjoining the same term to statements which have the same truth tables will produce statements with the same truth values. Using distributivity, the right side of (*) is

$$[((\neg p) \wedge (\neg q)) \vee ((\neg p) \wedge q)] \iff [(\neg p) \wedge ((\neg q) \vee q)]$$
$$\iff [(\neg p) \wedge \mathbf{1}] \iff \neg p,$$

so the given statement is logically equivalent simply to $\neg p$.   ∎

Pause 16   In the solution to Problem 15, we said "adjoining the same term to statements which have the same truth tables will produce statements with the same truth values." Make this claim into a formal assertion and verify it.

In Problems 14 and 15 we used, in a sneaky way, a very important principle of logic, which we now state as a theorem. The fact that we didn't really think about this at the time tells us that the theorem is easily understandable and quite painless to apply in practice.

1.4.1 THEOREM ▶   Suppose $\mathcal{A}$ and $\mathcal{B}$ are logically equivalent statements involving variables $p_1, p_2, \ldots, p_n$. Suppose that $\mathcal{C}_1, \mathcal{C}_2, \ldots, \mathcal{C}_n$ are statements. If, in $\mathcal{A}$ and $\mathcal{B}$, we replace p_1 by $\mathcal{C}_1$, p_2 by $\mathcal{C}_2$ and so on until we replace p_n by $\mathcal{C}_n$, then the resulting statements will still be logically equivalent.

Pause 17   Explain how Theorem 1.4.1 was used in Problem 15.

PROBLEM 16.   Show that

$$[(p \vee q) \vee ((q \vee (\neg r)) \wedge (p \vee r))] \iff \neg [(\neg p) \wedge (\neg q)].$$

Solution.   The left hand side is logically equivalent to

$$[(p \vee q) \vee (q \vee (\neg r))] \wedge [(p \vee q) \vee (p \vee r)]$$

by distributivity. Associativity and idempotence, however, say that

$$[(p \vee q) \vee (q \vee (\neg r))] \iff [p \vee (q \vee (q \vee (\neg r)))]$$
$$\iff [p \vee ((q \vee q) \vee (\neg r))]$$
$$\iff [p \vee (q \vee (\neg r))] \iff [(p \vee q) \vee (\neg r)],$$

while we also have

$$[(p \vee q) \vee (p \vee r)] \iff [(q \vee p) \vee (p \vee r)]$$
$$\iff [q \vee (p \vee (p \vee r))]$$
$$\iff [q \vee ((p \vee p) \vee r)]$$
$$\iff [q \vee (p \vee r)]$$
$$\iff [(q \vee p) \vee r] \iff [(p \vee q) \vee r].$$

Hence our expression is logically equivalent to

$$[((p \vee q) \vee (\neg r)) \wedge ((p \vee q) \vee r)] \iff [(p \vee q) \vee ((\neg r) \wedge r)]$$
$$\iff [(p \vee q) \vee \mathbf{0}] \iff p \vee q.$$

Finally, with the help of double negation and one of the laws of De Morgan, we obtain

$$p \vee q \iff \neg (\neg (p \vee q)) \iff \neg[(\neg p) \wedge (\neg q)]. \qquad \blacksquare$$

The next problem illustrates clearly why employing the basic logical equivalences discussed in this section is often more efficient than working simply with truth tables.

PROBLEM 17. Show that $[s \rightarrow (((\neg p) \wedge q) \wedge r)] \iff \neg[(p \vee (\neg (q \wedge r))) \wedge s]$.
Solution.

$$[s \rightarrow (((\neg p) \wedge q) \wedge r)] \iff [(\neg s) \vee (((\neg p) \wedge q) \wedge r)]$$
$$\iff [(\neg s) \vee ((\neg p) \wedge (q \wedge r))]$$
$$\iff [(\neg s) \vee (\neg (p \vee (\neg (q \wedge r))))]$$
$$\iff \neg[(s \wedge (p \vee (\neg (q \wedge r))))]$$
$$\iff \neg[((p \vee (\neg (q \wedge r))) \wedge s)]. \qquad \blacksquare$$

A primary application of the work in this section is reducing statements to logically equivalent simpler forms. There are times, however, when a different type of logically equivalent statement is required.

1.4.2 DEFINITION ▶ Let $n \geq 1$ be an integer. A compound statement based on variables $x_1, x_2, \ldots, x_n$ is said to be in *disjunctive normal form* if it looks like

$$(a_{11} \wedge a_{12} \wedge \cdots \wedge a_{1n}) \vee (a_{21} \wedge a_{22} \wedge \cdots \wedge a_{2n}) \vee \cdots \vee (a_{m1} \wedge a_{m2} \wedge \cdots \wedge a_{mn})$$

where, for each i and j, $1 \leq i \leq m$, $1 \leq j \leq n$, either $a_{ij} = x_j$ or $a_{ij} = \neg x_j$ and all *minterms* $a_{i1} \wedge a_{i2} \wedge \cdots \wedge a_{in}$ are distinct.

EXAMPLE 18 The statement $(x_1 \wedge x_2 \wedge x_3) \vee (x_1 \wedge (\neg x_2) \wedge (\neg x_3))$ is in disjunctive normal form (on the variables x_1, x_2, x_3). ▲

EXAMPLE 19 The statement $((p \wedge q) \vee r) \wedge ((p \wedge q) \vee (\neg q))$ is not in disjunctive normal form. One reason is that the minterms, $(p \wedge q) \vee r$ and $(p \wedge q) \vee (\neg q)$, involve the symbol $\vee$. This statement is logically equivalent to $(p \wedge q) \vee (r \wedge (\neg q))$, which is still not in disjunctive normal form because the minterms, $p \wedge q$ and $r \wedge (\neg q)$, don't contain all the variables. Continuing, however, our statement is logically equivalent to

$$(p \wedge q \wedge r) \vee (p \wedge q \wedge (\neg r)) \ \vee \ (p \wedge r \wedge (\neg q)) \vee ((\neg p) \wedge r \wedge (\neg q)),$$

which is in disjunctive normal form (on the variables p, q, r). ▲

As shown in Example 19, when writing a statement in disjunctive normal form, it is very useful to note that

$$(1) \qquad\qquad x \iff [(x \wedge y) \vee (x \wedge (\neg y))]$$

for any statements x and y. This follows from

$$x \iff (x \wedge \mathbf{1}) \iff [x \wedge (y \vee (\neg y))] \iff [(x \wedge y) \vee (x \wedge (\neg y))].$$

PROBLEM 20. Express $p \to (q \wedge r)$ in disjunctive normal form.

Solution. **Method 1:** We construct a truth table.

p	q	r	$q \wedge r$	$p \to (q \wedge r)$
T	T	T	T	T
T	T	F	F	F
T	F	T	F	F
F	T	T	T	T
T	F	F	F	F
F	T	F	F	T
F	F	T	F	T
F	F	F	F	T

Now focus attention on the rows for which the statement is true—each of these will contribute a minterm to our answer. For example, in row 1, p, q, and r are all T, so $p \wedge q \wedge r$ agrees with the T in the last column. In row 4, p is F, while q and r are both T. This gives the minterm $(\neg p) \wedge q \wedge r$. In this way, we obtain

$$(p \wedge q \wedge r) \vee ((\neg p) \wedge q \wedge r) \vee ((\neg p) \wedge q \wedge (\neg r))$$
$$\vee ((\neg p) \wedge (\neg q) \wedge r) \vee ((\neg p) \wedge (\neg q) \wedge (\neg r)).$$

Method 2: We have

$$[p \rightarrow (q \wedge r)]$$
$$\Longleftrightarrow [(\neg p) \vee (q \wedge r)]$$
$$\Longleftrightarrow [((\neg p) \wedge q) \vee ((\neg p) \wedge (\neg q)) \vee (q \wedge r)]$$
$$\Longleftrightarrow [((\neg p) \wedge q \wedge r) \vee ((\neg p) \wedge q \wedge (\neg r)) \vee ((\neg p) \wedge (\neg q) \wedge r)$$
$$\vee ((\neg p) \wedge (\neg q) \wedge (\neg r)) \vee (p \wedge q \wedge r) \vee ((\neg p) \wedge q \wedge r)]$$
$$\Longleftrightarrow [((\neg p) \wedge q \wedge r) \vee ((\neg p) \wedge q \wedge (\neg r)) \vee ((\neg p) \wedge (\neg q) \wedge r)$$
$$\vee ((\neg p) \wedge (\neg q) \wedge (\neg r)) \vee (p \wedge q \wedge r)],$$

omitting the second occurrence of $(\neg p) \wedge q \wedge r$ at the last step. We leave it to you to decide for yourself which method you prefer. ∎

Disjunctive normal form is useful in applications of logic to computer science, particularly in the construction of designs for logic circuits.

Answers to Pauses

14. This is just a restatement of Property 11, writing p instead of q and q instead of p.

15. This could be done with truth tables. Alternatively, we note that

$$\neg (p \leftrightarrow q) \Longleftrightarrow [\neg((p \rightarrow q) \wedge (q \rightarrow p))]$$
$$\Longleftrightarrow \neg[((\neg p) \vee q) \wedge ((\neg q) \vee p)]$$
$$\Longleftrightarrow \neg [((\neg p) \vee q) \vee (\neg((\neg q) \vee p))]$$
$$\Longleftrightarrow [(p \wedge (\neg q)) \vee (q \wedge (\neg p))].$$

16. The claim is that if $\mathcal{A} \Longleftrightarrow \mathcal{B}$, then $(\mathcal{A} \vee \mathcal{C}) \Longleftrightarrow (\mathcal{B} \vee \mathcal{C})$ for any other statement $\mathcal{C}$. Since $\mathcal{A} \Longleftrightarrow \mathcal{B}$, $\mathcal{A}$ and $\mathcal{B}$ have the same truth tables. Establishing $(\mathcal{A} \vee \mathcal{C}) \Longleftrightarrow (\mathcal{B} \vee \mathcal{C})$ requires four rows of a truth table.

$\mathcal{A}$	$\mathcal{B}$	$\mathcal{C}$	$\mathcal{A} \vee \mathcal{C}$	$\mathcal{B} \vee \mathcal{C}$
T	T	T	T	T
T	T	F	T	T
F	F	T	T	T
F	F	F	F	F

The last two columns establish our claim. A similar argument shows that if $\mathcal{A} \Longleftrightarrow \mathcal{B}$, then $(\mathcal{A} \wedge \mathcal{B}) \Longleftrightarrow (\mathcal{B} \wedge \mathcal{C})$ for any statement $\mathcal{C}$.

17. In applying the distributive property, we are using $\neg p$, $\neg q$ and q in place of p, q and r. Also, when applying Property 7, we use $\neg p$ instead of p.

EXERCISES

The symbol [BB] means that an answer can be found in the Back of the Book.

1. Verify each of the 13 properties of logical equivalence which appear in this section [BB; 1,3,5,7,9, 11,13].

2. Simplify each of the following statements.

   (a) [BB] $(p \land q) \lor (\neg((\neg p) \lor q))$
   (b) $(p \lor r) \to [(q \lor (\neg r)) \to ((\neg p) \to r)]$
   (c) $[(p \to q) \lor (q \to r)] \land (r \to s)$

3. Using truth tables, verify the following *absorption* properties.

   (a) [BB] $(p \lor (p \land q)) \iff p$
   (b) $(p \land (p \lor q)) \iff p$

4. Using the properties in the text together with the absorption properties given in Exercise 3, establish each of the following logical equivalences.

   (a) [BB] $[(p \lor q) \land (\neg p)] \iff [(\neg p) \land q]$
   (b) $[p \to (q \to r)] \iff [(p \land (\neg r)) \to (\neg q)]$
   (c) $[\neg(p \leftrightarrow q)] \iff [p \leftrightarrow (\neg q)]$
   (d) [BB] $\neg[(p \leftrightarrow q) \lor (p \land (\neg q))] \iff [(p \leftrightarrow (\neg q)) \land ((\neg p) \lor q)]$
   (e) $[(p \land (\neg q)) \land ((p \land (\neg q)) \lor (q \land (\neg r)))] \iff [p \land (\neg q)]$

5. Prove that the statements $(p \land (\neg q)) \to q$ and $(p \land (\neg q)) \to \neg p$ are logically equivalent. What simpler statement is logically equivalent to both of them?

6. In Exercise 10 of Section 1.3 we defined the *exclusive or* "$p \veebar q$" to be true whenever either p or q is true, but not both. For each of the properties discussed in this section (including those of absorption given in Exercise 3) determine whether or not the property holds with $\veebar$ replacing $\lor$ wherever it occurs [BB; 1,3,7,9,13].

7. Which of the following are in disjunctive normal form (on the appropriate set of variables)?

   (a) $(p \lor q) \land ((\neg p) \lor (\neg q))$
   (b) [BB] $(p \land q) \lor ((\neg p) \land (\neg q))$
   (c) [BB] $p \lor ((\neg p) \land q)$
   (d) $(p \land q) \lor ((\neg p) \land (\neg q) \land r)$
   (e) $(p \land q \land r) \lor ((\neg p) \land (\neg q) \land (\neg r))$

8. Express each of the following statements in disjunctive normal form.

   (a) [BB] $p \land q$
   (b) [BB] $(p \land q) \lor (\neg((\neg p) \lor q))$
   (c) $p \to q$
   (d) $(p \to q) \land (q \land r)$

9. Find out what you can about Augustus De Morgan and write a paragraph or two about him, in good English, of course!

1.5 LOGICAL ARGUMENTS

Proving a theorem in mathematics involves drawing a conclusion from some given information. The steps required in the proof generally consist of showing that if certain statements are true, then the truth of other statements must follow. Taken in its entirety, the proof of a theorem demonstrates that if an initial collection of statements—called *premises* or *hypotheses*—are all true, then the conclusion of the theorem is also true.

Different methods of proof were discussed informally in Section 1.2. Now we relate these ideas to some of the more formal concepts introduced in Sections 1.3 and 1.4. First, we define what is meant by a *valid argument*.

1.5.1 DEFINITIONS ▶   An *argument* is a finite collection of statements $A_1, A_2, \ldots, A_n$ called *premises* (or *hypotheses*) followed by a statement B called the *conclusion*. Such an argument is *valid* if, whenever $A_1, A_2, \ldots, A_n$ are all true, then B is also true.

It is often convenient to write an elementary argument in column form, like this.

$$\mathcal{A}_1$$
$$\mathcal{A}_2$$
$$\vdots$$
$$\frac{\mathcal{A}_n}{\mathcal{B}}$$

PROBLEM 21. Show that the argument

$$p \to \neg q$$
$$r \to q$$
$$\frac{r}{\neg p}$$

is valid.

Solution. We construct a truth table.

p	q	r	$\neg q$	$p \to \neg q$	$r \to q$	$\neg p$	
T	T	T	F	F	T	F	
T	T	F	F	F	T	F	
T	F	T	T	T	F	F	
F	T	T	F	T	T	T	★
T	F	F	T	T	T	F	
F	T	F	F	T	T	T	
F	F	T	T	T	F	T	
F	F	F	T	T	T	T	

Observe that row 4—marked with the star (★)—is the only row where the premises $p \to \neg q$, $r \to q$, r are all marked T. In this row, the conclusion $\neg p$ is also T. Thus the argument is valid. ∎

In Problem 21, we were a bit fortunate because there was only one row where all the premises were marked T. In general, in order to assert that an argument is valid when there are several rows with all premises marked T, it is necessary to check that the conclusion is also T in every such row.

Arguments can be shown to be valid without the construction of a truth table. For example, here is an alternative way to solve Problem 21.

Assume that all premises are true. In particular, this means that r is true. Since $r \to q$ is also true, q must also be true. Thus $\neg q$ is false and, because $p \to (\neg q)$ is true, p is false. Thus $\neg p$ is true as desired.

PROBLEM 22. Determine whether or not the following argument is valid.

If I like biology, then I will study it.
Either I study biology or I fail the course.
$$\overline{\text{If I fail the course, then I do not like biology.}}$$

Solution. Let p be "I like biology," q be "I study biology," and r be "I fail the course." In symbols, the argument we are to check becomes

$$p \to q$$
$$q \vee r$$
$$\overline{r \to (\neg p).}$$

This can be analyzed by a truth table.

p	q	r	$p \to q$	$q \vee r$	$\neg p$	$r \to (\neg p)$	
T	T	T	T	T	F	F	⋆
T	T	F	T	T	F	T	⋆
T	F	T	F	T	F	F	
F	T	T	T	T	T	T	⋆
T	F	F	F	F	F	T	
F	T	F	T	T	T	T	⋆
F	F	T	T	T	T	T	⋆
F	F	F	T	F	T	T	

The rows marked ⋆ are those in which the premises are true. In row 1, the premises are true, but the conclusion is F. The argument is not valid. ∎

The theorem which follows relates the idea of a valid argument to the notions introduced in Sections 1.3 and 1.4.

1.5.2 THEOREM ▶ An argument with premises $\mathcal{A}_1, \mathcal{A}_2, \ldots, \mathcal{A}_n$ and conclusion $\mathcal{B}$ is valid precisely when the compound statement $\mathcal{A}_1 \wedge \mathcal{A}_2 \wedge \cdots \wedge \mathcal{A}_n \to \mathcal{B}$ is a tautology.

Surely, this is not hard to understand. In order for the implication $\mathcal{A}_1 \wedge \mathcal{A}_2 \wedge \cdots \wedge \mathcal{A}_n \to \mathcal{B}$ to be a tautology, it must be the case that whenever $\mathcal{A}_1 \wedge \mathcal{A}_2 \wedge \cdots \wedge \mathcal{A}_n$ is true, then $\mathcal{B}$ is also true. But $\mathcal{A}_1 \wedge \mathcal{A}_2 \wedge \cdots \wedge \mathcal{A}_n$ is true precisely when each of $\mathcal{A}_1, \mathcal{A}_2, \ldots, \mathcal{A}_n$ is true, so the result follows from our definition of a valid argument.

In the same spirit as Theorem 1.4.1, we have the following important *substitution theorem*.

1.5.3 THEOREM ▶ **[Substitution]** Assume that an argument with premises $\mathcal{A}_1, \mathcal{A}_2, \ldots, \mathcal{A}_n$ and conclusion $\mathcal{B}$ is valid, and that all these statements involve variables $p_1, p_2, \ldots, p_m$. If $p_1, p_2, \ldots, p_m$ are replaced by statements $\mathcal{C}_1, \mathcal{C}_2, \ldots, \mathcal{C}_m$, the resulting argument is still valid.

Rules of Inference

Because of Theorem 1.5.3, some very simple valid arguments which regularly arise in practice are given special names. Here is a list of some of the most common *rules of inference*.

1. **Modus ponens:**
$$p$$
$$\frac{p \rightarrow q}{q}$$

2. **Modus tollens:**
$$p \rightarrow q$$
$$\frac{\neg q}{\neg p}$$

3. **Disjunctive syllogism:**
$$p \vee q$$
$$\frac{\neg p}{q}$$

4. **Chain rule:**
$$p \rightarrow q$$
$$\frac{q \rightarrow r}{p \rightarrow r}$$

5. **Resolution:**
$$p \vee r$$
$$\frac{q \vee (\neg r)}{p \vee q}$$

Pause 18 Verify modus tollens.

We illustrate how the rules of inference can be applied.

PROBLEM 23. Show that the following argument is valid.

$$(p \vee q) \rightarrow (s \wedge t)$$
$$\frac{[\neg ((\neg s) \vee (\neg t))] \rightarrow [(\neg r) \vee q]}{(p \vee q) \rightarrow (r \rightarrow q)}$$

Solution. One of the laws of De Morgan and the principle of double negation—see Section 1.4—tell us that

$$[\neg ((\neg s) \vee (\neg t))] \iff [(\neg (\neg s)) \wedge (\neg(\neg t))] \iff (s \wedge t).$$

Property 13 of logical equivalence as given in Section 1.4 says that $(\neg r) \vee q \iff (r \rightarrow q)$. Thus the given argument can be rewritten as

$$(p \vee q) \rightarrow (s \wedge t)$$
$$\frac{(s \wedge t) \rightarrow (r \rightarrow q)}{(p \vee q) \rightarrow (r \rightarrow q).}$$

The chain rule now tells us that our argument is valid. ∎

Pause 19 If a truth table were used to answer Problem 23, how many rows would be required?

Sometimes, rules of inference need to be combined.

PROBLEM 24.   Determine the validity of the following argument.

> If I study, then I will pass.
> If I do not go to a movie, then I will study.
> I failed.
> _____
> Therefore, I went to a movie.

Solution.   Let p, q and r be the statements

$$p: \text{``I study.''}$$

$$q: \text{``I pass.''}$$

$$r: \text{``I go to a movie.''}$$

The given argument is

$$p \to q$$
$$(\neg r) \to p$$
$$\underline{\neg q \qquad}$$
$$r$$

The first two premises imply the truth of $(\neg r) \to q$ by the chain rule. Since $(\neg r) \to q$ and $\neg q$ imply $\neg(\neg r)$ by modus tollens, the validity of the argument follows by the principle of double negation: $\neg(\neg r) \iff r$.   ∎

Answers to Pauses

18. While this can be shown with a truth table, we prefer an argument by words. Since $\neg q$ is true, q is false. Since $p \to q$ is true, p must also be false. Hence $\neg p$ is true and we are done.

19. There are five variables, each of which could be T or F, so we would need $2^5 = 32$ rows.

EXERCISES

The symbol [BB] means that an answer can be found in the Back of the Book.

  You are encouraged to use the result of any exercise in this set to assist with the solution of any other.

1. Determine whether or not each of the following arguments is valid.

(a) [BB] $p \to (q \to r)$
$$\underline{q \qquad\qquad}$$
$$p \to r$$

(b) [BB] $p \to q$
$$\underline{q \vee r \qquad}$$
$$r \to (\neg q)$$

(c) $p \to q$
$$\underline{r \to q}$$
$$r \to p$$

(d) $p \to q$
$$\underline{(q \vee (\neg r)) \to (p \wedge s)}$$
$$s \to (r \vee q)$$

2. Verify that each of the five rules of inference given in this section is a valid argument.

3. Verify that each of the following arguments is valid.

(a) [BB] $p \to r$
$$\underline{q \to r \qquad}$$
$$(p \vee q) \to r$$

(b)
$$p \to r$$
$$q \to s$$
$$\overline{(p \land q) \to (r \land s)}$$

(c)
$$p \lor q$$
$$(\neg p) \lor r$$
$$(\neg r) \lor s$$
$$\overline{q \lor s}$$

(d)
$$p \lor ((\neg q) \land r)$$
$$\neg (p \land s)$$
$$\overline{\neg (s \land (q \lor (\neg r)))}$$

4. Test the validity of each of the following arguments.

(a) [BB]
$$p \to q$$
$$(\neg r) \lor (\neg q)$$
$$r$$
$$\overline{\neg p}$$

(b) [BB]
$$p \lor (\neg q)$$
$$(t \lor s) \to (p \lor r)$$
$$(\neg r) \lor (t \lor s)$$
$$p \leftrightarrow (t \lor s)$$
$$\overline{(p \lor r) \to (q \lor r)}$$

(c)
$$p \lor (\neg q)$$
$$(t \lor s) \to (p \lor r)$$
$$(\neg r) \lor (t \lor s)$$
$$p \leftrightarrow (t \lor s)$$
$$\overline{(q \lor r) \to (p \lor r)}$$

(d)
$$[(p \land q) \lor r] \to (q \land r \land s)$$
$$[(\neg p) \land (\neg q)] \to (r \lor p)$$
$$[p \lor (\neg q) \lor r] \to (q \land s)$$
$$\overline{(p \land q) \leftrightarrow [(q \land r) \lor s]}$$

5. Determine the validity of each of the following arguments. If the argument is one of those listed in the text, name it.

(a) [BB] If I stay up late at night, then I will be tired in the morning.
I stayed up late last night.
_____
I am tired this morning.

(b) [BB] If I stay up late at night, then I will be tired in the morning.
I am tired this morning.
_____
I stayed up late last night.

(c) If I stay up late at night, then I will be tired in the morning.
I am not tired this morning.
_____
I did not stay up late last night.

(d) If I stay up late at night, then I will be tired in the morning.
I did not stay up late last night.
_____
I am not tired this morning.

(e) [BB] Either I wear a red tie or I wear blue socks.
I am wearing pink socks.
_____
I am wearing a red tie.

(f) Either I wear a red tie or I wear blue socks.
I am wearing blue socks.
_____
I am not wearing a red tie.

(g) [BB] If I work hard, then I earn lots of money.
If I earn lots of money, then I pay high taxes.
_____
If I pay high taxes, then I have worked hard.

(h) If I work hard, then I earn lots of money.
If I earn lots of money, then I pay high taxes.
_____
If I work hard, then I pay high taxes.

(i) If I work hard, then I earn lots of money.
If I earn lots of money, then I pay high taxes.
_____
If I do not work hard, then I do not pay high taxes.

(j) If I like mathematics, then I will study.
I will not study.
Either I like mathematics or I like football.
_____
I like football.

(k) Either I study or I like football.
If I like football, then I like mathematics.
_____
If I don't study, then I like mathematics.

(l) [BB] If I like mathematics, then I will study.
Either I don't study or I pass mathematics.
If I don't graduate, then I didn't pass mathematics.
_____
If I graduate, then I studied.

(m) If I like mathematics, then I will study.
Either I don't study or I pass mathematics.
If I don't graduate, then I didn't pass mathematics.
_____
If I like mathematics, then I will graduate.

6. [BB] Given the premises $p \to (\neg r)$ and $r \lor q$, either write down a valid conclusion which involves p and q only and is not a tautology or show that no such conclusion is possible.

7. [BB] Repeat Exercise 6 with the premises $(\neg p) \to r$ and $r \lor q$.

8. (a) [BB] Explain why two premises p and q can always be replaced by the single premise $p \wedge q$ and vice versa.

(b) Using 8(a), verify that the following argument is valid.

$$
\begin{array}{c}
p \wedge q \\
p \rightarrow r \\
s \rightarrow \neg q \\
\hline
(\neg s) \wedge r
\end{array}
$$

9. Let n be an integer greater than 1. Show that the following argument is valid.

$$
\begin{array}{c}
p_1 \rightarrow (q_1 \rightarrow r_1) \\
p_2 \rightarrow (q_2 \rightarrow r_2) \\
\vdots \\
p_n \rightarrow (q_n \rightarrow r_n) \\
q_1 \wedge q_2 \wedge \cdots \wedge q_n \\
\hline
(p_1 \rightarrow r_1) \wedge (p_2 \rightarrow r_2) \wedge \cdots \wedge (p_n \rightarrow r_n)
\end{array}
$$

[*Hint*: Exercise 8(a).]

10. [BB] What language is being used when we say "modus ponens" or "modus tollens"? Translate these expressions into English and explain.

REVIEW EXERCISES FOR CHAPTER 1

1. State, with a reason, whether each of the following statements is true or false.

(a) If a and b are integers with $a - b > 0$ and $b - a > 0$, then $a \neq b$.

(b) If a and b are integers with $a - b \geq 0$ and $b - a \geq 0$, then $a \neq b$.

2. Write down the negation of each of the following statements in clear and concise English.

(a) Either x is not a real number or $x > 5$.

(b) There exists a real number x such that $n > x$ for every integer n.

(c) If x, y, z are positive integers, then $x^3 + y^3 \neq z^3$.

(d) If a graph has n vertices and $n + 1$ edges, then its chromatic number is at most 3.

3. Write down the converse and contrapositive of each of the following implications.

(a) If a and b are integers, then ab is an integer.

(b) If x is an even integer, then x^2 is an even integer.

(c) Every planar graph can be colored with at most four colors.

4. Rewrite each of the following statements using the quantifiers "for all" and "there exists" as appropriate.

(a) Not all countable sets are finite.

(b) 1 is the smallest positive integer.

5. (a) Determine whether or not the following implication is true.

"x is a positive odd integer
$\rightarrow x + 2$ is a positive odd integer."

(b) Repeat (a) with $\rightarrow$ replaced by $\leftrightarrow$.

6. Let n be an integer. Prove that n^3 is odd if and only if n is odd.

7. With a proof by contradiction, show that there exists no largest negative rational number.

8. Construct a truth table for the compound statement $[p \wedge (q \rightarrow (\neg r))] \rightarrow [(\neg q) \vee r]$.

9. Determine the truth value of $[p \vee (q \rightarrow ((\neg r) \wedge s))] \leftrightarrow (r \wedge t)$, where p, q, r, s and t are all true.

10. Two compound statements $\mathcal{A}$ and $\mathcal{B}$ have the property that $\mathcal{A} \rightarrow \mathcal{B}$ is logically equivalent to $\mathcal{B} \rightarrow \mathcal{A}$. What can you conclude about $\mathcal{A}$ and $\mathcal{B}$?

11. (a) Suppose $\mathcal{A}$, $\mathcal{B}$, and $\mathcal{C}$ are compound statements such that $\mathcal{A} \iff \mathcal{B}$ and $\mathcal{B} \iff \mathcal{C}$. Explain why $\mathcal{A} \iff \mathcal{C}$.

(b) Give a proof of Property 11 which uses the result of Property 13.

12. Establish the logical equivalence $[(p \rightarrow q) \rightarrow r] \iff [(p \vee r) \wedge (\neg(q \wedge (\neg r)))]$.

13. Express $((p \vee q) \wedge r) \vee ((p \vee q) \wedge (\neg p))$ in disjunctive normal form.

14. Determine whether or not each of the following arguments is valid.

(a) $\neg((\neg p) \wedge q)$
$$\neg(p \wedge r)$$
$$\underline{r \vee s}$$
$$q \rightarrow s$$

(b) $p \vee (\neg q)$
$$(t \vee s) \rightarrow (p \vee r)$$
$$(\neg r) \vee (t \vee s)$$
$$\underline{p \leftrightarrow (t \vee s)}$$
$$(p \wedge r) \rightarrow (q \wedge r)$$

15. Discuss the validity of the argument

$$p \wedge q$$
$$\underline{(\neg p) \wedge r}$$
$$\text{Purple toads live on Mars.}$$

16. Determine the validity of each of the following arguments. If the argument is one of those listed in the text, name it.

(a) Either I wear a red tie or I wear blue socks.
Either I wear a green hat or I do not wear blue socks.

Either I wear a red tie or I wear a green hat.

(b) If I like mathematics, then I will study.
Either I don't study or I pass mathematics.
If I don't pass mathematics, then I don't graduate.

If I graduate, then I like mathematics.

2

Sets and Relations

2.1 SETS

Any branch of science, like a foreign language, has its own terminology. "Isomorphism," "cyclotomic," and "coset" are words one is unlikely to hear except in a mathematical context. On the other hand, quite a number of common English words—field, complex, function—have precise mathematical meanings quite different from their usual ones. Students of French or Spanish know that memory work is a fundamental part of their studies; it is perfectly obvious to them that without a firm grasp of the meaning of words, their ability to learn grammar and to communicate will be severely hindered. It is, however, not always understood by science students that they must memorize the terminology of their discipline with the same diligence that students of Russian memorize vocabulary. Without constant review of the meanings of words, one's understanding of a paragraph of text or the words of a teacher is very limited. We advise readers of this book to maintain and constantly review a mathematical vocabulary list.

What would it be like to delve into a dictionary if you didn't already know the meanings of some of the words in it? Most people, at one time or another, have gone to a dictionary in search of a word only to discover that the definition uses another unfamiliar word. Some reflection indicates that a dictionary can be of no use unless there are some words which are so basic that we can understand them without definitions. Mathematics is the same way. There are a few basic terms which we accept without definitions.

Most of mathematics is based upon the single undefined concept of *set*, which we think of as just a collection of things called *elements* or *members*. Primitive humans discovered the set of *natural numbers* with which they learned to count. The set of natural numbers, which is denoted with a capital boldface N or, in handwriting, with this symbol, $\mathbb{N}$, consists of the numbers $1, 2, 3, \ldots$ (the three dots meaning "and so on").[1] The elements of N are, of course, just the positive integers. The full set of *integers*, denoted Z or $\mathbb{Z}$, consists of the natural numbers, their negatives, and 0. We might describe this set by $\ldots, -3, -2, -1, 0, 1, 2, 3, \ldots$. Our convention, which is not universal, is that 0 is an integer, but **not** a natural number.

There are various ways to describe sets. Sometimes it is possible to list the elements of a set within braces.

- {egg1, egg2} is a set containing two elements, egg1 and egg2.
- $\{x\}$ is a set containing one element, x.
- N = $\{1, 2, 3, \ldots\}$ is the set of natural numbers.
- Z = $\{\ldots, -3, -2, -1, 0, 1, 2, 3, \ldots\}$ is the set of integers.

On other occasions, it is convenient to describe a set with so-called *set builder* notation. This has the format

$$\{x \mid x \text{ has certain properties}\},$$

which is read "the set of x such that x has certain properties." We read "such that" at the vertical line, "|."

More generally, we see

$$\{\text{some expression} \mid \text{the expression has certain properties}\}.$$

Thus, the set of odd natural numbers could be described as

$$\{n \mid n \text{ is an odd integer}, n > 0\}$$

or as

$$\{2k - 1 \mid k = 1, 2, 3, \ldots\}$$

or as

$$\{2k - 1 \mid k \in \text{N}\}.$$

The expression "$k \in$ N" is read "k belongs to N," the symbol $\in$ denoting set membership. Thus, "$m \in$ Z" simply records the fact that m is an integer. Recall that a slash (/) written over any mathematical symbol negates the meaning of that symbol. So, in the same way that $\pi \neq 3.14$, we have $0 \notin$ N.

The set of common fractions—numbers like $\frac{3}{4}$, $\frac{-2}{17}$, and 5 $(= \frac{5}{1})$, which are ratios of integers with nonzero denominators—is more properly called the set of *rational numbers* and is denoted Q or $\mathbb{Q}$. Formally,

$$\text{Q} = \{\tfrac{m}{n} \mid m, n \in \text{Z}, n \neq 0\}.$$

The set of all real numbers is denoted R or $\mathbb{R}$. To define the real numbers properly requires considerable mathematical maturity. For our purposes, we think

---

[1] Since the manufacture of boldface symbols such as N is a luxury not afforded users of chalk or pencil, it has long been traditional to use $\mathbb{N}$ on blackboards or in handwritten work as the symbol for the natural numbers and to call $\mathbb{N}$ a *blackboard bold* symbol.

of real numbers as numbers which have decimal expansions of the form $a.a_1a_2\ldots$ where a is an integer and $a_1, a_2, \ldots$ are integers between 0 and 9 inclusive. In addition to the rational numbers, whose decimal expansions terminate or repeat, the real numbers include numbers like $\sqrt{2}$, $\sqrt[3]{17}$, e, π, $\ln 5$, and $\cos\frac{\pi}{6}$ whose decimal expansions neither terminate nor repeat. Such numbers are called *irrational*. An irrational number is a number which cannot be written in the form $\frac{m}{n}$ with m and n both integers. Incidentally, it can be very difficult to decide whether or not a given real number is irrational. For example, it is unknown whether or not such numbers as $e + \pi$ or $\frac{e}{\pi}$ are irrational.

The *complex numbers*, denoted C or $\mathbb{C}$, have the form $a + bi$ where a and b are real numbers and $i^2 = -1$; that is,

$$C = \{a + bi \mid a, b \in R, i^2 = -1\}.$$

Sometimes people are surprised to discover that a set can be an element of another set. For example, $\{\{a, b\}, c\}$ is a set with two elements, one of which is $\{a, b\}$ and the other c.

Pause 1 Let S denote the set $\{\{a\}, b, c\}$. True or false?

(a) $a \notin S$.
(b) $\{a\} \in S$.

Equality of Sets

2.1.1 DEFINITION ▶ Sets A and B are *equal*, and we write $A = B$, if and only if A and B contain the same elements or neither set contains any element.

EXAMPLES 1
- $\{1, 2, 1\} = \{1, 2\} = \{2, 1\}$;
- $\{\frac{1}{2}, \frac{2}{4}, \frac{-3}{-6}, \frac{\pi}{2\pi}\} = \{\frac{1}{2}\}$;
- $\{t \mid t = r - s, \ r, s \in \{0, 1, 2\}\} = \{-2, -1, 0, 1, 2\}$.   ▲

The Empty Set

One set which arises in a variety of different guises is the set which contains no elements. Consider, for example, the set SMALL of people less than 1 millimeter in height, the set LARGE of people taller than the Eiffel Tower, the set

$$\mathfrak{PECULIAR} = \{n \in N \mid 5n = 2\},$$

and the set

$$S = \{n \in N \mid n^2 + 1 = 0\}.$$

These sets are all equal since none of them contains any elements. The unique set which contains no elements is called the *empty set*. Set theorists originally used 0 (zero) to denote this set, but now it is customary to use a 0 with a slash through it, Ø, to avoid confusion between zero and a capital "Oh."

Pause 2 True or false? $\{\emptyset\} = \emptyset$.

Subsets

2.1.2 DEFINITION ▶ A set A is a *subset* of a set B, and we write $A \subseteq B$, if and only if every element of A is an element of B. If $A \subseteq B$ but $A \neq B$, then A is called a *proper subset* of B and we write $A \subsetneq B$.

When $A \subseteq B$, it is common to say "A is contained in B" as well as "A is a subset of B." The notation $A \subset B$, which is common, unfortunately means $A \subsetneq B$ to some people and $A \subseteq B$ to others. For this reason, we avoid it, while reiterating that it is present in a lot of mathematical writing. Upon encountering it, the reader should make an effort to discover what the intended meaning is.

We occasionally see "$B \supseteq A$," read "B is a *superset* of A." This is an alternative way to express "$A \subseteq B$," A is a subset of B, just as "$y \geq x$" is an alternative way to express "$x \leq y$." We generally prefer the subset notation.

EXAMPLES 2
- $\{a, b\} \subseteq \{a, b, c\}$
- $\{a, b\} \subsetneq \{a, b, c\}$
- $\{a, b\} \subseteq \{a, b, \{a, b\}\}$
- $\{a, b\} \in \{a, b, \{a, b\}\}$
- $\mathbb{N} \subsetneq \mathbb{Z} \subsetneq \mathbb{Q} \subsetneq \mathbb{R} \subsetneq \mathbb{C}$ ▲

Note the distinction between $A \subsetneq B$ and $A \nsubseteq B$, the latter expressing the negation of $A \subseteq B$; for example,

$$\{a, b\} \subsetneq \{a, b, c\} \nsubseteq \{a, b, x\}.$$

2.1.3 PROPOSITION ▶ For any set A, $A \subseteq A$ and $\emptyset \subseteq A$.

Proof If $a \in A$, then $a \in A$, so $A \subseteq A$. The proof that $\emptyset \subseteq A$ is a classic model of proof by contradiction. If $\emptyset \subseteq A$ is false, then there must exist some $x \in \emptyset$ such that $x \notin A$. This is an absurdity since there is no $x \in \emptyset$. ∎

Pause 3 True or false?

(a) $\{\emptyset\} \in \{\{\emptyset\}\}$

(b) $\emptyset \subseteq \{\{\emptyset\}\}$

(c) $\{\emptyset\} \subseteq \{\{\emptyset\}\}$

(As Shakespeare once wrote, "Much ado about nothing.")

The following proposition is an immediate consequence of the definitions of "subset" and "equal sets," and it illustrates the way we prove two sets are equal in practice.

2.1.4 PROPOSITION ▶ If A and B are sets, then $A = B$ if and only if $A \subseteq B$ and $B \subseteq A$.

There are two assertions being made here.

($\rightarrow$) If $A = B$, then A is a subset of B and B is a subset of A.

($\leftarrow$) If A is a subset of B and B is a subset of A, then $A = B$.

Remember that another way to state Proposition 2.1.4 is to say that, for two sets to be equal, it is **necessary and sufficient** that each be a subset of the other.

Note the distinction between **membership**, $a \in b$, and **subset**, $a \subseteq b$. By the former statement, we understand that a is an element of the set b; by the latter, that a is a set each of whose elements is also in the set b.[2]

EXAMPLES 3 Each of the following assertions is true.

- $\{a\} \in \{x, y, \{a\}\}$
- $\{a\} \subsetneq \{x, y, a\}$
- $\{a\} \nsubseteq \{x, y, \{a\}\}$
- $\{a, b\} \subseteq \{a, b\}$
- $\emptyset \in \{x, y, \emptyset\}$
- $\emptyset \subseteq \{x, y, \emptyset\}$
- $\{\emptyset\} \notin \{x, y, \emptyset\}$

▲

The Power Set

An important example of a set, **all** of whose elements are themselves sets, is the *power set* of a set.

2.1.5 DEFINITION ▶ The *power set* of a set A, denoted $\mathcal{P}(A)$, is the set of all subsets of A:

$$\mathcal{P}(A) = \{B \mid B \subseteq A\}.$$

EXAMPLES 4
- If $A = \{a\}$, then $\mathcal{P}(A) = \{\emptyset, \{a\}\}$
- If $A = \{a, b\}$, then $\mathcal{P}(A) = \{\emptyset, \{a\}, \{b\}, \{a, b\}\}$
- $\mathcal{P}(\{a, b, c\}) = \{\emptyset, \{a\}, \{b\}, \{c\}, \{a, b\}, \{a, c\}, \{b, c\}, \{a, b, c\}\}$

▲

Answers to Pauses

1. Both statements are true. The set S contains the **set** $\{a\}$ as one of its elements, but not the element a.
2. This statement is false: $\{\emptyset\}$ is not the empty set for it contains one element, namely, the set $\emptyset$.
3. (a) True: $\{\{\emptyset\}\}$ is a set which contains the single element $\{\emptyset\}$.
   (b) True: The empty set is a subset of any set.
   (c) False: There is just one element in the set $\{\emptyset\}$ (namely, $\emptyset$) and this is not an **element** of the set $\{\{\emptyset\}\}$, whose only element is $\{\emptyset\}$.

EXERCISES

The symbol [BB] means that an answer can be found in the Back of the Book.

1. List the (distinct) elements in each of the following sets:

(a) [BB] $\{x \in \mathsf{R} \mid x^2 = 5\}$

(b) $\{x \in \mathsf{Z} \mid xy = 15 \text{ for some } y \in \mathsf{Z}\}$

(c) [BB] $\{x \in \mathsf{Q} \mid x(x^2 - 2)(2x + 3) = 0\}$

(d) $\{x + y \mid x \in \{-1, 0, 1\}, y \in \{0, 1, 2\}\}$

(e) $\{a \in \mathsf{N} \mid a < -4 \text{ and } a > 4\}$

---

[2]Note the use of lowercase letters for sets, which is not common but certainly permissible.

2. List five elements in each of the following sets:

   (a) [BB] $\{a + bi \mid a, b \in \mathsf{Z}, i^2 = -1\}$
   (b) $\{a + b\sqrt{2} \mid a \in \mathsf{N}, -b \in \{2, 5, 7\}\}$
   (c) $\{\frac{x}{y} \mid x, y \in \mathsf{R}, x^2 + y^2 = 25\}$
   (d) $\{n \in \mathsf{N} \mid n^2 + n \text{ is a multiple of } 3\}$

3. Let $A = \{1, 2, 3, 4\}$.

   (a) [BB] List all the subsets B of A such that $\{1, 2\} \subseteq B$.
   (b) List all the subsets B of A such that $B \subseteq \{1, 2\}$.
   (c) List all the subsets B of A such that $\{1, 2\} \nsubseteq B$.
   (d) List all the subsets B of A such that $B \nsubseteq \{1, 2\}$.
   (e) List all the subsets B of A such that $\{1, 2\} \subsetneq B$.
   (f) List all the subsets B of A such that $B \subsetneq \{1, 2\}$.

4. [BB] Let $A = \{\{a, b\}\}$. Are the following statements true or false? Explain your answer.

   (a) $a \in A$.
   (b) $A \in A$.
   (c) $\{a, b\} \in A$.
   (d) There are two elements in A.

5. Determine which of the following are true and which are false. Justify your answers.

   (a) [BB] $3 \in \{1, 3, 5\}$
   (b) $\{3\} \in \{1, 3, 5\}$
   (c) $\{3\} \subsetneq \{1, 3, 5\}$
   (d) [BB] $\{3, 5\} \nsubseteq \{1, 3, 5\}$
   (e) $\{1, 3, 5\} \subsetneq \{1, 3, 5\}$
   (f) $1 \in \{a + 2b \mid a, b \text{ even integers}\}$
   (g) $0 \in \{a + b\sqrt{2} \mid a, b \in \mathsf{Q}, b \neq 0\}$

6. Find the power sets of each of the following sets:

   (a) [BB] $\emptyset$
   (b) $\{\emptyset\}$
   (c) $\{\emptyset, \{\emptyset\}\}$

7. Determine whether each of the following statements is true or false. Justify your answers.

   (a) [BB] $\emptyset \subseteq \emptyset$
   (b) $\emptyset \subseteq \{\emptyset\}$
   (c) $\emptyset \in \emptyset$
   (d) $\emptyset \in \{\emptyset\}$
   (e) [BB] $\{1, 2\} \nsubseteq \{1, 2, 3, \{1, 2, 3\}\}$
   (f) $\{1, 2\} \in \{1, 2, 3, \{1, 2, 3\}\}$
   (g) $\{1, 2\} \subsetneq \{1, 2, \{\{1, 2\}\}\}$
   (h) [BB] $\{1, 2\} \in \{1, 2, \{\{1, 2\}\}\}$
   (i) $\{\{1, 2\}\} \subseteq \{1, 2, \{1, 2\}\}$

8. [BB] Let A be a set and suppose $x \in A$. Is $x \subseteq A$ also possible? Explain.

9. (a) List all the subsets of the set $\{a, b, c, d\}$ which contain:

   i. four elements
   ii. [BB] three elements
   iii. two elements
   iv. one element
   v. no elements

   (b) How many subsets of $\{a, b, c, d\}$ are there altogether?

10. (a) How many elements are in the power set of the power set of the empty set?

    (b) Suppose A is a set containing one element. How many elements are in $\mathcal{P}(\mathcal{P}(A))$?

11. (a) [BB] If A contains two elements, how many elements are there in the power set of A?

    (b) [BB] If A contains three elements, how many elements are there in the power set of A?

    (c) [BB] If a set A contains $n \geq 0$ elements, guess how many elements are in the power set of A.

12. Suppose A, B, and C are sets. For each of the following statements either prove it is true or give a counterexample to show that it is false.

    (a) [BB] $A \in B, B \in C \rightarrow A \in C$.
    (b) $A \subseteq B, B \subseteq C \rightarrow A \subseteq C$.
    (c) $A \nsubseteq B, B \nsubseteq C \rightarrow A \nsubseteq C$.
    (d) [BB] $A \in B, B \subseteq C \rightarrow A \in C$.
    (e) $A \in B, B \subseteq C \rightarrow A \subseteq C$.
    (f) $A \subseteq B, B \in C \rightarrow A \in C$.
    (g) $A \subseteq B, B \in C \rightarrow A \subseteq C$.

13. Suppose A and B are sets.

    (a) Answer true or false and explain: $A \nsubseteq B \rightarrow B \subsetneq A$.
    (b) Is the converse of the implication in (a) true or false? Explain.

14. Suppose A, B, and C are sets. Prove or give a counterexample which disproves each of the following assertions.

    (a) [BB] $C \in \mathcal{P}(A) \leftrightarrow C \subseteq A$.
    (b) $A \subseteq B \leftrightarrow \mathcal{P}(A) \subseteq \mathcal{P}(B)$.
    (c) $A = \emptyset \leftrightarrow \mathcal{P}(A) = \emptyset$.

2.2 OPERATIONS ON SETS

In this section, we discuss ways in which two or more sets can be combined in order to form a new set.

Union and Intersection

2.2.1 DEFINITIONS ▶ The *union* of sets A and B, written $A \cup B$, is the set of elements in A or in B (or in both). The *intersection* of A and B, written $A \cap B$, is the set of elements which belong to both A and B.

EXAMPLES 5
- If $A = \{a, b, c\}$ and $B = \{a, x, y, b\}$, then

$$A \cup B = \{a, b, c, x, y\}, \quad A \cap B = \{a, b\},$$

$$A \cup \{\emptyset\} = \{a, b, c, \emptyset\} \quad \text{and} \quad B \cap \{\emptyset\} = \emptyset.$$

- For any set A, $A \cup \emptyset = A$ and $A \cap \emptyset = \emptyset$. ▲

As with addition and multiplication of real numbers, the union and intersection of sets are *associative* operations. To say that set union is associative is to say that

$$(A_1 \cup A_2) \cup A_3 = A_1 \cup (A_2 \cup A_3)$$

for any three sets A_1, A_2, A_3. It follows that the expression

$$A_1 \cup A_2 \cup A_3$$

is unambiguous. The two different interpretations (corresponding to different insertions of parentheses) agree. The union of n sets $A_1, A_2, \dots, A_n$ is written

(1) $$A_1 \cup A_2 \cup A_3 \cup \cdots \cup A_n \quad \text{or} \quad \bigcup_{i=1}^{n} A_i$$

and represents the set of elements which belong to one or more of the sets A_i. The intersection of $A_1, A_2, \dots, A_n$ is written

(2) $$A_1 \cap A_2 \cap A_3 \cap \cdots \cap A_n \quad \text{or} \quad \bigcap_{i=1}^{n} A_i$$

and denotes the set of elements which belong to all of the sets.

Do not assume from the expression $A_1 \cup A_2 \cup A_3 \cup \cdots \cup A_n$ that n is actually greater than 3 since the first part of this expression—$A_1 \cup A_2 \cup A_3$—is present only to make the general pattern clear; a union of sets is being formed. The last term—A_n—indicates that the last set in the union is A_n. If $n = 2$, then $A_1 \cup A_2 \cup A_3 \cup \cdots \cup A_n$ means $A_1 \cup A_2$. Similarly, if $n = 1$, the expression $A_1 \cap A_2 \cap \cdots \cap A_n$ simply means A_1.

While parentheses are not required in expressions like (1) or (2), they are mandatory when both union and intersection are involved. For example, $A \cap (B \cup C)$ and $(A \cap B) \cup C$ are, in general, different sets. This is probably most easily seen by the use of the *Venn diagram* shown in Fig 2.1.

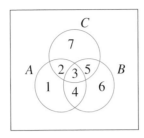

Figure 2.1   A Venn diagram.

The diagram indicates that A consists of the points in the regions labeled 1, 2, 3, and 4; B consists of those points in regions 3, 4, 5, and 6 and C of those in 2, 3, 5, and 7. The set $B \cup C$ consists of points in the regions labeled 3, 4, 5, 6, 2, and 7. Notice that $A \cap (B \cup C)$ consists of the points in regions 2, 3, and 4. The region $A \cap B$ consists of the points in regions 3 and 4; thus, $(A \cap B) \cup C$ is the set of points in the regions labeled 3, 4, 2, 5, and 7. The diagram enables us to see that, in general, $A \cap (B \cup C) \neq (A \cap B) \cup C$ and it shows how we could construct a specific counterexample. Namely, we could let A, B, and C be the sets

$$A = \{1, 2, 3, 4\}, \ B = \{3, 4, 5, 6\}, \ C = \{2, 3, 5, 7\}$$

as suggested by the diagram and then calculate

$$A \cap (B \cup C) = \{2, 3, 4\} \neq \{2, 3, 4, 5, 7\} = (A \cap B) \cup C.$$

There is a way to rewrite $A \cap (B \cup C)$. In Fig 2.1, we see that $A \cap B$ consists of the points in the regions labeled 3 and 4 and that $A \cap C$ consists of the points in 2 and 3. Thus, the points of $(A \cap B) \cup (A \cap C)$ are those of 2, 3, and 4. These are just the points of $A \cap (B \cup C)$ (as observed previously), so the Venn diagram makes it easy to believe that, in general,

(3) $A \cap (B \cup C) = (A \cap B) \cup (A \cap C).$

While pictures can be helpful in making certain statements seem plausible, they should not be relied upon because they can also mislead. For this reason, and because there are situations in which Venn diagrams are difficult or impossible to create, it is important to be able to establish relationships among sets without resorting to a picture.

PROBLEM 6.   Let A, B and C be sets. Verify equation (3) without the aid of a Venn diagram.

Solution.   As observed in Proposition 2.1.4, showing that two sets are equal is equivalent to showing that each is a subset of the other. Here this just amounts to expressing the meaning of $\cup$ and $\cap$ in words.

To show $A \cap (B \cup C) \subseteq (A \cap B) \cup (A \cap C)$, let $x \in A \cap (B \cup C)$. Then x is in A and also in $B \cup C$. Since $x \in B \cup C$, either $x \in B$ or $x \in C$. This suggests cases.

Case 1: $x \in B$.
In this case, x is in A as well as in B, so it's in $A \cap B$.

Case 2: $x \in C$.

Here x is in A as well as in C, so it's in $A \cap C$.

We have shown that either $x \in A \cap B$ or $x \in A \cap C$. By definition of union, $x \in (A \cap B) \cup (A \cap C)$, completing this half of our proof.

Conversely, we must show $(A \cap B) \cup (A \cap C) \subseteq A \cap (B \cup C)$. For this, let $x \in (A \cap B) \cup (A \cap C)$. Then either $x \in A \cap B$ or $x \in A \cap C$. Thus, x is in both A and B or in both A and C. In either case, $x \in A$. Also, x is in either B or C; thus, $x \in B \cup C$. So x is both in A and in $B \cup C$; that is, $x \in A \cap (B \cup C)$. This completes the proof. ∎

PROBLEM 7. For sets A and B, prove that $A \cap B = A$ if and only if $A \subseteq B$.

Solution. Remember that there are two implications to establish and that we use the symbolism ($\rightarrow$) and ($\leftarrow$) to mark the start of the proof of each implication.

($\rightarrow$) Here we assume $A \cap B = A$ and must prove $A \subseteq B$. For this, suppose $x \in A$. Then, $x \in A \cap B$ (because we are assuming $A = A \cap B$). Therefore, x is in A and in B, in particular, x is in B. This proves $A \subseteq B$.

($\leftarrow$) Now we assume $A \subseteq B$ and prove $A \cap B = A$. To prove the equality of $A \cap B$ and A, we must prove that each set is a subset of the other. By definition of intersection, $A \cap B$ is a subset of A, so $A \cap B \subseteq A$. On the other hand, suppose $x \in A$. Since $A \subseteq B$, x is in B too; thus, x is in both A and B. Therefore, $A \subseteq A \cap B$. Therefore, $A = A \cap B$. ∎

Pause 4 For sets A and B, prove that $A \cup B = B$ if and only if $A \subseteq B$.

Set Difference

2.2.2 DEFINITIONS ▶ The *set difference* of sets A and B, written $A \setminus B$, is the set of those elements of A which are not in B. The *complement* of a set A is the set $A^c = U \setminus A$, where U is some universal set made clear by the context.

EXAMPLES 8
- $\{a, b, c\} \setminus \{a, b\} = \{c\}$
- $\{a, b, c\} \setminus \{a, x\} = \{b, c\}$
- $\{a, b, \emptyset\} \setminus \emptyset = \{a, b, \emptyset\}$
- $\{a, b, \emptyset\} \setminus \{\emptyset\} = \{a, b\}$
- If A is the set {Monday, Tuesday, Wednesday, Thursday, Friday}, the context suggests that the universal set is the days of the week, so $A^c = $ {Saturday, Sunday}. ▲

Notice that $A \setminus B = A \cap B^c$ and also that $(A^c)^c = A$. For example, if $A = \{x \in \mathsf{Z} \mid x^2 > 0\}$, then $A^c = \{0\}$ (it being understood that $U = \mathsf{Z}$) and so

$$(A^c)^c = \{0\}^c = \{x \in \mathsf{Z} \mid x > 0\} \cup \{x \in \mathsf{Z} \mid x < 0\} = A.$$

You may have previously encountered standard notation to describe various types of intervals of real numbers.

2.2.3 INTERVAL NOTATION ▶

If a and b are real numbers with $a < b$, then

$$[a, b] = \{x \in \mathsf{R} \mid a \leq x \leq b\} \qquad \textbf{closed}$$

$$(a, b) = \{x \in \mathsf{R} \mid a < x < b\} \qquad \textbf{open}$$

$$(a, b] = \{x \in \mathsf{R} \mid a < x \leq b\} \qquad \textbf{half open}$$

$$[a, b) = \{x \in \mathsf{R} \mid a \leq x < b\} \qquad \textbf{half open}.$$

As indicated, a closed interval is one which includes both end points, an open interval includes neither, and a half open interval includes just one end point. A square bracket indicates that the adjacent end point is in the interval. To describe infinite intervals, we use the symbol ∞ (which is just a symbol) and make obvious adjustments to our notation. For example,

$$(-\infty, b] = \{x \in \mathsf{R} \mid x \leq b\},$$

$$(a, \infty) = \{x \in \mathsf{R} \mid x > a\}.$$

The first interval here is half open; the second is open.

Pause 5

If $A = [-4, 4]$ and $B = [0, 5]$, then $A \setminus B = [-4, 0)$. What is $B \setminus A$? What is A^c?

2.2.4 THE LAWS OF DE MORGAN ▶

The following two laws, of wide applicability, are attributed to Augustus De Morgan (1806–1871), who, together with George Boole (1815–1864), helped to make England a leading center of logic in the nineteenth century.[3]

$$(A \cup B)^c = A^c \cap B^c; \qquad (A \cap B)^c = A^c \cup B^c.$$

Readers should be struck by the obvious connection between these laws and the rules for negating *and* and *or* compound sentences described in Section 1.1. We illustrate by showing the equivalence of the first law of De Morgan and the rule for negating "$\mathcal{A}$ or $\mathcal{B}$."

PROBLEM 9. Prove that $(A \cup B)^c = A^c \cap B^c$ for any sets A, B, and C.

Solution. Let $\mathcal{A}$ be the statement "$x \in A$" and $\mathcal{B}$ be the statement "$x \in B$." Then

$$
\begin{aligned}
x \in (A \cup B)^c &\leftrightarrow \neg(x \in A \cup B) \\
&\leftrightarrow \neg(\mathcal{A} \text{ or } \mathcal{B}) & \text{definition of union} \\
&\leftrightarrow \neg\mathcal{A} \text{ and } \neg\mathcal{B} & \text{rule for negating "or"} \\
&\leftrightarrow x \in A^c \text{ and } x \in B^c & \\
&\leftrightarrow x \in A^c \cap B^c & \text{definition of intersection.}
\end{aligned}
$$

---

[3] As pointed out by Rudolf and Gerda Fritsch (*Der Vierfarbensatz*, B. I. Wissenschaftsverlag, Mannheim, 1994 and English translation, *The Four-Color Theorem*, by J. Peschke, Springer-Verlag, 1998), it was in a letter from De Morgan to Sir William Rowan Hamilton that the question giving birth to the famous "Four-Color Theorem" was first posed. See Section 14.2 for a detailed account of this theorem whose proof was found only quite recently, after over 100 years of effort!

The sets $(A \cup B)^c$ and $A^c \cap B^c$ contain the same elements, so they are the same. ∎

Symmetric Difference

2.2.5 DEFINITION ▶

The *symmetric difference* of two sets A and B is the set $A \oplus B$ of elements which are in A or in B, but not in both.

The notation $A \triangle B$ is used by some authors to denote symmetric difference although, as suggested, this is not our preference.

Notice that the symmetric difference of sets can be expressed in terms of previously defined operations. For example,

$$A \oplus B = (A \cup B) \setminus (A \cap B)$$

and

$$A \oplus B = (A \setminus B) \cup (B \setminus A).$$

EXAMPLES 10

- $\{a, b, c\} \oplus \{x, y, a\} = \{b, c, x, y\}$
- $\{a, b, c\} \oplus \emptyset = \{a, b, c\}$
- $\{a, b, c\} \oplus \{\emptyset\} = \{a, b, c, \emptyset\}$ ▲

PROBLEM 11. Use a Venn diagram to illustrate the plausibility of the fact that $\oplus$ is an associative operation; that is, use a Venn diagram to illustrate that for any three sets A, B, and C,

(4) $$(A \oplus B) \oplus C = A \oplus (B \oplus C).$$

Solution. With reference to Fig 2.1 again, $A \oplus B$ consists of the points in the regions labeled 1, 2, 5, and 6 while C consists of the points in the regions 2, 3, 5, and 7. Thus, $(A \oplus B) \oplus C$ is the set of points in the regions 1, 3, 6, and 7. On the other hand, $B \oplus C$ consists of the regions 2, 7, 4, and 6 and A, of regions 1, 2, 3, 4. Thus, $A \oplus (B \oplus C)$ also consists of the points in regions 1, 3, 6, and 7. ∎

As a consequence of (4), the expression $A \oplus B \oplus C$, which conceivably could be interpreted in two ways, is in fact unambiguous. Notice that $A \oplus B \oplus C$ is the set of points in an odd number of the sets A, B, C: Regions 1, 6, and 7 contain the points of just one of the sets while region 3 consists of points in all three. More generally, the symmetric difference $A_1 \oplus A_2 \oplus A_3 \oplus \cdots \oplus A_n$ of n sets $A_1, A_2, A_3, \ldots, A_n$ is well defined and, as it turns out, is the set of those elements which are members of an odd number of the sets A_i. (See Exercise 21 of Section 5.1.)

The Cartesian Product of Sets

There is yet another way in which two sets can be combined to obtain another.

2.2.6   DEFINITIONS ▶   If A and B are sets, the *Cartesian product* (or the *direct product*) of A and B is the set

$$A \times B = \{(a, b) \mid a \in A, b \in B\}.$$

(We say "A cross B" for "$A \times B$.") The Cartesian product $A \times A$ is denoted A^2. More generally,

$$A^n = \underbrace{A \times A \times \cdots \times A}_{n \text{ times}} = \{(a_1, a_2, \ldots, a_n) \mid a_i \in A \text{ for } i = 1, 2, \ldots, n\}.$$

The elements of $A \times B$ are called *ordered pairs* because their order is important: $(a, b) \neq (b, a)$ (unless $a = b$). The elements a and b are the *coordinates* of the ordered pair (a, b); the first coordinate is a and the second is b. The elements of A^n are called *n-tuples*.

Elements of $A \times B$ are equal if and only if they have the same first coordinates and the same second coordinates:

$$(a_1, b_1) = (a_2, b_2) \text{ if and only if } a_1 = b_1 \text{ and } a_2 = b_2.$$

EXAMPLE 12   Let $A = \{a, b\}$ and $B = \{x, y, z\}$. Then

$$A \times B = \{(a, x), (a, y), (a, z), (b, x), (b, y), (b, z)\}$$

and

$$B \times A = \{(x, a), (x, b), (y, a), (y, b), (z, a), (z, b)\}.$$

This example illustrates that, in general, the sets $A \times B$ and $B \times A$ are different.   ▲

EXAMPLE 13   The Cartesian plane, in which calculus students sketch curves, is a picture of $\mathsf{R} \times \mathsf{R} = \mathsf{R}^2 = \{(x, y) \mid x, y \in \mathsf{R}\}$. The adjective "Cartesian" is derived from Descartes,[4] as Cartesius was Descartes's name in Latin.   ▲

PROBLEM 14.   Let A, B, and C be sets. Prove that $A \times (B \cup C) \subseteq (A \times B) \cup (A \times C)$.

Solution.   We must prove that any element in $A \times (B \cup C)$ is in $(A \times B) \cup (A \times C)$. Since the elements in $A \times (B \cup C)$ are ordered pairs, we begin by letting $(x, y) \in A \times (B \cup C)$ (this is more helpful than starting with "$x \in A \times B$") and ask ourselves what this means. It means that x, the first coordinate, is in A and y, the second coordinate, is in $B \cup C$. Therefore, y is in either B or C. If y is in B, then, since x is in A, $(x, y) \in A \times B$. If y is in C, then, since x is in A, $(x, y) \in A \times C$. Thus, (x, y) is either in $A \times B$ or in $A \times C$; thus, (x, y) is in $(A \times B) \cup (A \times C)$, which is what we wanted to show.   ∎

---

[4]René Descartes (1596–1650), together with Pierre de Fermat, the inventor of analytic geometry, introduced the method of plotting points and graphing functions in R^2 with which we are so familiar today.

Pause 6 Let A, B, and C be three sets. Prove that $(A \times B) \cup (A \times C) \subseteq A \times (B \cup C)$. What can you conclude about the sets $A \times (B \cup C)$ and $(A \times B) \cup (A \times C)$? Why?

Pause 7 Let A and B be nonempty sets. Prove that $A \times B = B \times A$ if and only if $A = B$. Is this true if $A = \emptyset$?

Answers to Pauses

4. ($\rightarrow$) Suppose the first statement, $A \cup B = B$, is true. We show $A \subseteq B$. So let $x \in A$. Then x is certainly in $A \cup B$, by the definition of $\cup$. But $A \cup B = B$, so $x \in B$. Thus, $A \subseteq B$.

   ($\leftarrow$) Conversely, suppose the second statement, $A \subseteq B$, is true. We have to show $A \cup B = B$. To prove the sets $A \cup B$ and B are equal, we have to show each is a subset of the other. First, let $x \in A \cup B$. Then x is either in A or in B. If the latter, $x \in B$, and if the former, $x \in B$ because A is a subset of B. In either case, $x \in B$. Thus, $A \cup B \subseteq B$. Second, assume $x \in B$. Then x is in $A \cup B$ by definition of $\cup$. So $B \subseteq A \cup B$ and we have equality, as required.

5. $B \setminus A = (4, 5]$; $A^c = (-\infty, -4) \cup (4, \infty)$.

6. An element of $(A \times B) \cup (A \times C)$ is either in $A \times B$ or in $A \times C$; in either case, it's an ordered pair. So we begin by letting $(x, y) \in (A \times B) \cup (A \times C)$ and noting that either $(x, y) \in A \times B$ or $(x, y) \in A \times C$. In the first case, x is in A and y is in B; in the second case, x is in A and y is in C. In either case, x is in A and y is either in B or in C; so $x \in A$ and $y \in B \cup C$. Therefore, $(x, y) \in A \times (B \cup C)$, establishing the required subset relation. The reverse subset relation was established in Problem 14. We conclude that the two sets in question are equal; that is, $A \times (B \cup C) = (A \times B) \cup (A \times C)$.

7. ($\rightarrow$) Suppose that the statement $A \times B = B \times A$ is true. We prove $A = B$. So suppose $x \in A$. Since $B \neq \emptyset$, we can find some $y \in B$. Thus, $(x, y) \in A \times B$. Since $A \times B = B \times A$, $(x, y) \in B \times A$. So $x \in B$, giving us $A \subseteq B$. Similarly, we show that $B \subseteq A$ and conclude $A = B$.

   ($\leftarrow$) On the other hand, if $A = B$ is a true statement, then $A \times B = A \times A = B \times A$.

   Finally, if $A = \emptyset$ and B is any nonempty set, then $A \times B = \emptyset = B \times A$, but $A \neq B$. So $A \times B = B \times A$ does not mean $A = B$ in the case $A = \emptyset$.

EXERCISES

The symbol [BB] means that an answer can be found in the Back of the Book.

1. Let $A = \{x \in \mathbb{N} \mid x < 7\}$, $B = \{x \in \mathbb{Z} \mid |x - 2| < 4\}$ and $C = \{x \in \mathbb{R} \mid x^3 - 4x = 0\}$.

   (a) [BB] List the elements in each of these sets.
   (b) Find $A \cup C$, $B \cap C$, $B \setminus C$, $A \oplus B$, $C \times (B \cap C)$, $(A \setminus B) \setminus C$, $A \setminus (B \setminus C)$, and $(B \cup \emptyset) \cap \{\emptyset\}$.
   (c) List the elements in $S = \{(a, b) \in A \times B \mid a = b+2\}$ and in $T = \{(a, c) \in A \times C \mid a \leq c\}$.

2. Let $S = \{2, 5, \sqrt{2}, 25, \pi, \frac{5}{2}\}$ and $T = \{4, 25, \sqrt{2}, 6, \frac{3}{2}\}$.

   (a) [BB] Find $S \cap T$, $S \cup T$, and $T \times (S \cap T)$.
   (b) [BB] Find $\mathbb{Z} \cup S$, $\mathbb{Z} \cap S$, $\mathbb{Z} \cup T$, and $\mathbb{Z} \cap T$.

   (c) List the elements in each of the sets $\mathbb{Z} \cap (S \cup T)$ and $(\mathbb{Z} \cap S) \cup (\mathbb{Z} \cap T)$. What do you notice?
   (d) List the elements of $\mathbb{Z} \cup (S \cap T)$ and list the elements of $(\mathbb{Z} \cup S) \cap (\mathbb{Z} \cup T)$. What do you observe?

3. Let $A = \{(-1, 2), (4, 5), (0, 0), (6, -5), (5, 1), (4, 3)\}$. List the elements in each of the following sets.

   (a) [BB] $\{a + b \mid (a, b) \in A\}$
   (b) $\{a \mid a > 0$ and $(a, b) \in A$ for some $b\}$
   (c) $\{b \mid b = k^2$ for some $k \in \mathbb{Z}$ and $(a, b) \in A$ for some $a\}$

4. List the elements in the sets $A = \{(a, b) \in N \times N \mid a \le b, b \le 3\}$ and $B = \{\frac{a}{b} \mid a, b \in \{-1, 1, 2\}\}$.

5. For $A = \{a, b, c, \{a, b\}\}$, find

   (a) [BB] $A \setminus \{a, b\}$

   (b) $\{\emptyset\} \setminus \mathcal{P}(A)$

   (c) $A \setminus \emptyset$

   (d) $\emptyset \setminus A$

   (e) [BB] $\{a, b, c\} \setminus A$

   (f) $(\{a, b, c\} \cup \{A\}) \setminus A$

6. Find A^c (with respect to $U = R$) in each of the following cases.

   (a) [BB] $A = (1, \infty) \cup (-\infty, -2]$

   (b) $A = (-3, \infty) \cap (-\infty, 4]$

   (c) $A = \{x \in R \mid x^2 \le -1\}$

7. Let $n > 3$ and $A = \{1, 2, 3, \ldots, n\}$.

   (a) [BB] How many subsets of A contain $\{1, 2\}$?

   (b) How many subsets B of A have the property that $B \cap \{1, 2\} = \emptyset$?

   (c) How many subsets B of A have the property that $B \cup \{1, 2\} = A$?

   Explain your answers.

8. [BB] Let a and b be real numbers with $a < b$. Find $(a, b)^c$, $[a, b)^c$, $(a, \infty)^c$, and $(-\infty, b]^c$.

9. The universal set for this problem is the set of students attending Miskatonic University. Let

   • M denote the set of math majors
   • CS denote the set of computer science majors
   • T denote the set of students who had a test on Friday
   • P denote those students who ate pizza last Thursday.

Using only the set theoretical notation we have introduced in this chapter, rewrite each of the following assertions.

   (a) [BB] Computer science majors had a test on Friday.

   (b) [BB] No math major ate pizza last Thursday.

   (c) One or more math majors did not eat pizza last Thursday.

   (d) Those computer science majors who did not have a test on Friday ate pizza on Thursday.

   (e) Math or computer science majors who ate pizza on Thursday did not have a test on Friday.

10. Using only the set theoretical notation introduced in this chapter, express the negation and converse of statements (a), (c), and (d) in the previous question.

11. Let P denote the set of primes and E the set of even integers. As always, Z and N denote the integers and natural numbers, respectively. Find equivalent formulations of each of the following statements using the notation of set theory that has been introduced in this section.

   (a) [BB] There exists an even prime.

   (b) 0 is an integer but not a natural number.

   (c) Every prime is both a natural number and an integer.

   (d) Every prime except 2 is odd.

12. For $n \in Z$, let $A_n = \{a \in Z \mid a \le n\}$. Find each of the following sets.

   (a) [BB] $A_3 \cup A_{-3}$

   (b) $A_3 \cap A_{-3}$

   (c) $A_3 \cap (A_{-3})^c$

   (d) $\bigcap_{i=0}^{4} A_i$

13. [BB] In Fig 2.1, the region labeled 7 represents the set $C \setminus (A \cup B)$. What set is represented by the region labeled 2? By that labeled 3? By that labeled 4?

14. **(a)** [BB] Suppose A and B are sets such that $A \cap B = A$. What can you conclude? Why?

   (b) Repeat (a) assuming $A \cup B = A$.

15. [BB] Let $n \ge 1$ be a natural number. How many elements are in the set $\{(a, b) \in N \times N \mid a \le b \le n\}$? Explain.

16. Suppose A is a subset of $N \times N$ with the properties

   • $(1, 1) \in A$ and
   • $(a, b) \in A \to (a + 1, b)$ and $(a + 1, b + 1)$ are both in A.

   Do you think that $\{(m, n) \in N \times N \mid m \ge n\}$ is a subset of A? Explain. [*Hint*: A picture of A in the xy-plane might help.]

17. Let A, B, and C be subsets of some universal set U.

   (a) Prove that $A \cap B \subseteq C$ and $A^c \cap B \subseteq C \to B \subseteq C$.

   (b) [BB] Given that $A \cap B = A \cap C$ and $A^c \cap B = A^c \cap C$, does it follow that $B = C$? Justify your answer.

18. Let A, B, and C be sets.

   (a) Find a counterexample to the statement $A \cup (B \cap C) = (A \cup B) \cap C$.

   (b) Without using Venn diagrams, prove that $A \cup (B \cap C) = (A \cup B) \cap (A \cup C)$.

19. Use the first law of De Morgan to prove the second: $(A \cap B)^c = A^c \cup B^c$.

20. [BB] Use the laws of De Morgan and any other set theoretic identities discussed in the text to prove that $(A \setminus B) \setminus C = A \setminus (B \cup C)$ for any sets A, B, and C.

21. Let A, B, C, and D be subsets of a universal set U. Use set theoretic identities discussed in the text to simplify the expression $[(A \cup B)^c \cap (A^c \cup C)^c]^c \setminus D^c$.

22. Let A, B, and C be subsets of some universal set U. Use set theoretic identities discussed in the text to prove that $A \setminus (B \setminus C) = (A \setminus B) \cup (A \setminus C^c)$.

23. Suppose A, B, and C are subsets of some universal set U.

    (a) [BB] Generalize the laws of De Morgan by finding equivalent ways to describe the sets $(A \cup B \cup C)^c$ and $(A \cap B \cap C)^c$.

    (b) Find a way to describe the set $(A \cap (B \setminus C))^c \cap A$ without using the symbol ^{c} for set complement.

24. Let A and B be sets.

    (a) [BB] Find a necessary and sufficient condition for $A \oplus B = A$.

    (b) Find a necessary and sufficient condition for $A \cap B = A \cup B$.

    Explain your answers (with Venn diagrams if you wish).

25. Which of the following conditions imply that $B = C$? In each case, either prove or give a counterexample.

    (a) [BB] $A \cup B = A \cup C$

    (b) $A \cap B = A \cap C$

    (c) $A \oplus B = A \oplus C$

    (d) $A \times B = A \times C$

26. True or false? In each case, provide a proof or a counter-example.

    (a) $A \subseteq C, B \subseteq D \to A \times B \subseteq C \times D$.

    (b) $A \times B \subseteq C \times D \to A \subseteq C$ and $B \subseteq D$.

    (c) $A \subseteq C$ and $B \subseteq D$ if and only if $A \times B \subseteq C \times D$.

    (d) [BB] $A \cup B \subseteq A \cap B \to A = B$.

27. Show that $(A \cap B) \times C = (A \times C) \cap (B \times C)$ for any sets A, B, and C.

28. Let A, B, and C be arbitrary sets. For each of the following, either prove the given statement is true or exhibit a counterexample to prove it is false.

    (a) $A \setminus (B \cup C) = (A \setminus B) \cup (A \setminus C)$

    (b) $(A \setminus B) \times C = (A \times C) \setminus (B \times C)$

    (c) [BB] $(A \oplus B) \times C = (A \times C) \oplus (B \times C)$

    (d) $(A \cup B) \times (C \cup D) = (A \times C) \cup (B \times D)$

    (e) $(A \setminus B) \times (C \setminus D) = (A \times C) \setminus (B \times D)$

29. Find out what you can about George Boole and write a paragraph or two about him (in good English, of course).

2.3 BINARY RELATIONS

If A and B are sets, remember that the Cartesian product of A and B is the set $A \times B = \{(a, b) \mid a \in A, b \in B\}$. There are occasions when we are interested in a certain subset of $A \times B$. For example, if A is the set of former major league baseball players and $B = \mathsf{N} \cup \{0\}$ is the set of nonnegative integers, then we might naturally be interested in

$$\mathcal{R} = \{(a, b) \mid a \in A, b \in B, \text{player } a \text{ had } b \text{ career home runs}\}.$$

For example, (Hank Aaron, 755) and (Mickey Mantle, 536) are elements of $\mathcal{R}$.

2.3.1 DEFINITIONS ▶ Let A and B denote sets. A *binary relation from A to B* is a subset of $A \times B$. A *binary relation on A* is a subset of $A \times A$.

The empty set and the entire Cartesian product $A \times B$ are always binary relations from A to B, although these are generally not as interesting as certain nonempty proper subsets of $A \times B$.

EXAMPLES 15
- If A is the set of students who were registered at the University of Toronto during the Fall 2001 semester and B is the set {History, Mathematics, English, Biology}, then $\mathcal{R} = \{(a, b) \mid a \in A \text{ is enrolled in a course in subject } b\}$ is a binary relation from A to B.

- Let A be the set of surnames of people listed in the Seattle telephone directory. Then $\mathcal{R} = \{(a, n) \mid a$ appears on page $n\}$ is a binary relation from A to the set N of natural numbers.
- $\{(a, b) \mid a, b \in$ N, $\frac{a}{b}$ is an integer$\}$ and $\{(a, b) \mid a, b \in$ N, $a - b = 2\}$ are binary relations on N.
- $\{(x, y) \mid y = x^2\}$ is a binary relation on R whose graph the reader may recognize. ▲

Our primary intent in this section is to identify special properties of binary relations on a set, so, henceforth, all binary relations will be subsets of $A \times A$ for some set A.

2.3.2 DEFINITION ▶ A binary relation $\mathcal{R}$ on a set A is *reflexive* if and only if $(a, a) \in \mathcal{R}$ for all $a \in A$.

EXAMPLES 16
- $\{(x, y) \in$ R$^2 \mid x \leq y\}$ is a reflexive relation on R since $x \leq x$ for any $x \in$ R.
- $\{(a, b) \in$ N$^2 \mid \frac{a}{b} \in$ N$\}$ is a reflexive relation on N since $\frac{a}{a}$ is an integer, namely 1, for any $a \in$ N.
- $\mathcal{R} = \{(x, y) \in$ R$^2 \mid x^2 + y^2 > 0\}$ is not a reflexive relation on R since $(0, 0) \notin \mathcal{R}$. (This example reminds us that a reflexive relation must contain **all** pairs of the form (a, a): **Most** is not enough.) ▲

PROBLEM 17. Suppose $\mathcal{R} = \{(a, b) \in$ Z $\times$ Z $\mid a^2 = b^2\}$. Criticize and then correct the following "proof" that $\mathcal{R}$ is reflexive:

$$(a, a) \in \mathcal{R} \text{ if } a^2 = a^2.$$

Solution. The statement "$(a, a) \in \mathcal{R}$ if $a^2 = a^2$" is the implication "$a^2 = a^2 \to (a, a) \in \mathcal{R}$," which has almost nothing to do with what is required. To prove that $\mathcal{R}$ is reflexive, we must establish an implication of the form "something $\to \mathcal{R}$ is reflexive." Here is a good argument, in this case.

For any integer a, we have $a^2 = a^2$ and, hence, $(a, a) \in \mathcal{R}$. Therefore, $\mathcal{R}$ is reflexive. ∎

2.3.3 DEFINITION ▶ A binary relation $\mathcal{R}$ on a set A is *symmetric* if and only if

if $a, b \in A$ and $(a, b) \in \mathcal{R}$, then $(b, a) \in \mathcal{R}$.

EXAMPLES 18
- $\mathcal{R} = \{(x, y) \in$ R$^2 \mid x^2 + y^2 = 1\}$ is a symmetric relation on R since if $x^2 + y^2 = 1$, then $y^2 + x^2 = 1$ too: If $(x, y) \in \mathcal{R}$, then $(y, x) \in \mathcal{R}$.
- $\{(x, y) \in$ Z $\times$ Z $\mid x - y$ is even$\}$ is a symmetric relation on Z since if $x - y$ is even, so is $y - x$.
- $\mathcal{R} = \{(x, y) \in$ R$^2 \mid x^2 \geq y\}$ is not a symmetric relation on R. For example, $(2, 1) \in \mathcal{R}$ because $2^2 \geq 1$, but $(1, 2) \notin \mathcal{R}$ because $1^2 \ngeq 2$. ▲

Suppose $\mathcal{R}$ is a binary relation on $A =$ R^2. In this case, the elements of $\mathcal{R}$, being ordered pairs of elements of A, are ordered pairs of elements each of which is an ordered pair of real numbers. Consider, for example,

$$\mathcal{R} = \{((x, y), (u, v)) \in \mathsf{R}^2 \times \mathsf{R}^2 \mid x^2 + y^2 = u^2 + v^2\}.$$

This is a symmetric relation on R^2 since if $((x, y), (u, v)) \in \mathcal{R}$, then $x^2 + y^2 = u^2 + v^2$, so $u^2 + v^2 = x^2 + y^2$, so $((u, v), (x, y)) \in \mathcal{R}$.

Pause 8   Is this relation reflexive?

A binary relation on the real numbers (or on any subset of R) is symmetric if, when its points are plotted as usual in the Cartesian plane, the figure is symmetric about the line with equation $y = x$. The set $\{(x, y) \in \mathsf{R}^2 \mid x^2 + y^2 = 1\}$ is a symmetric relation because its points are those on the graph of the unit circle centered at the origin and this circle is certainly symmetric about the line with equation $y = x$.

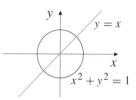

If a set A has n elements and n is reasonably small, a binary relation on A can be conveniently described by labeling with the elements of A the rows and the columns of an $n \times n$ grid and then inserting some symbol in row a and column b to indicate that (a, b) is in the relation.

EXAMPLE 19   The picture in Fig 2.2 describes the relation

$$\mathcal{R} = \{(1, 1), (1, 2), (1, 4), (2, 1), (2, 2), (3, 2), (3, 3), (4, 4)\}$$

on the set $A = \{1, 2, 3, 4\}$. This relation is reflexive (all points on the *main diagonal*—top left corner to lower right—are present), but not symmetric (the $\times$'s are not symmetrically located with respect to the main diagonal). For example, there is a $\times$ in row 1, column 4, but not in row 4, column 1. ▲

	1	2	3	4
1	×	×		×
2	×	×		
3		×	×	
4				×

Figure 2.2

2.3.4 DEFINITION ▶   A binary relation $\mathcal{R}$ on a set A is *antisymmetric* if and only if

if $a, b \in A$ and both $(a, b) \in \mathcal{R}$ and (b, a) are in $\mathcal{R}$, then $a = b$.

EXAMPLE 20 $\mathcal{R} = \{(x, y) \in \mathsf{R}^2 \mid x \leq y\}$ is an antisymmetric relation on R since $x \leq y$ and $y \leq x$ implies $x = y$; thus, $(x, y) \in \mathcal{R}$ and $(y, x) \in \mathcal{R}$ implies $x = y$. ▲

EXAMPLE 21   If S is a set and $A = \mathcal{P}(S)$ is the power set of S, then $\{(X, Y) \mid X, Y \in \mathcal{P}(S), X \subseteq Y\}$ is antisymmetric since $X \subseteq Y$ and $Y \subseteq X$ implies $X = Y$. ▲

EXAMPLE 22 $\mathcal{R} = \{(1, 2), (2, 3), (3, 3), (2, 1)\}$ is not antisymmetric on $A = \{1, 2, 3\}$ because $(1, 2) \in \mathcal{R}$ and $(2, 1) \in \mathcal{R}$ but $1 \neq 2$. ▲

Note that "antisymmetric" is not the same as "not symmetric." The relation in Example 22 is not symmetric but neither is it antisymmetric.

Pause 9   Why is this relation not symmetric?

Pause 10   Is the relation $\mathcal{R} = \{((x, y), (u, v)) \in \mathsf{R}^2 \times \mathsf{R}^2 \mid x^2 + y^2 = u^2 + v^2\}$ antisymmetric?

2.3.5 DEFINITION ▶   A binary relation $\mathcal{R}$ on a set A is *transitive* if and only if

if $a, b, c \in A$, and both (a, b) and (b, c) are in $\mathcal{R}$, then $(a, c) \in \mathcal{R}$.

EXAMPLE 23 $\mathcal{R} = \{(x, y) \in \mathsf{R}^2 \mid x \leq y\}$ is a transitive relation on R since if $x \leq y$ and $y \leq z$, then $x \leq z$: if (x, y) and (y, z) are in $\mathcal{R}$, then $(x, z) \in \mathcal{R}$. ▲

EXAMPLE 24 $\{(a, b) \in \mathsf{Z} \times \mathsf{Z} \mid \frac{a}{b}$ is an integer$\}$ is a transitive relation on Z since if $\frac{a}{b}$ and $\frac{b}{c}$ are integers, then so is $\frac{a}{c}$ because $\frac{a}{c} = \frac{a}{b} \cdot \frac{b}{c}$. ▲

EXAMPLES 25
- $\mathcal{R} = \{(x, y), (x, z), (y, u), (x, u)\}$ is a transitive binary relation on the set $\{x, y, z, u\}$ because there is only one pair of the form $(a, b), (b, c)$ belonging to $\mathcal{R}$ (namely (x, y) and (y, u)) and, for this pair, it is true that $(a, c) = (x, u) \in \mathcal{R}$.
- $\mathcal{R} = \{(a, b), (b, a), (a, a)\}$ is not transitive on $\{a, b\}$ since it contains the pairs (b, a) and (a, b), but not the pair (b, b).
- $\{(a, b) \mid a$ and b are people and a is an ancestor of $b\}$ is a transitive relation since if a is an ancestor of b and b is an ancestor of c, then a is an ancestor of c.
- $\mathcal{R} = \{(x, y) \in \mathsf{R}^2 \mid x^2 \geq y\}$ is not transitive on R because $(3, 4) \in \mathcal{R}$ $(3^2 \geq 4)$ and $(4, 10) \in \mathcal{R}$ $(4^2 \geq 10)$ but $(3, 10) \notin \mathcal{R}$ $(3^2 \ngeq 10)$. ▲

Answers to Pauses

8. The answer is yes. For any $(x, y) \in \mathsf{R}^2$, we have $x^2 + y^2 = x^2 + y^2$; in other words, $((x, y), (x, y)) \in \mathcal{R}$ for any $(x, y) \in \mathsf{R}^2$.
9. $(2, 3) \in \mathcal{R}$ but $(3, 2) \notin \mathcal{R}$.
10. No. For example, $((1, 2), (2, 1)) \in \mathcal{R}$ because $1^2 + 2^2 = 2^2 + 1^2$ and, similarly, $((2, 1), (1, 2)) \in \mathcal{R}$; however, $(1, 2) \neq (2, 1)$.

EXERCISES

The symbol [BB] means that an answer can be found in the Back of the Book.

1. [BB] Let B denote the set of books in a college library and S denote the set of students attending that college. Interpret the Cartesian product $S \times B$. Give a sensible example of a binary relation from S to B.

2. Let A denote the set of names of streets in St. John's, Newfoundland and B denote the names of all the residents of St. John's. Interpret the Cartesian product $A \times B$. Give a sensible example of a binary relation from A to B.

3. Determine which of the properties reflexive, symmetric, transitive apply to the following relations on the set of people.

(a) [BB] is a father of

(b) is a friend of

(c) [BB] is a descendant of

(d) have the same parents

(e) is an uncle of

4. With a table like that in Fig 2.2, illustrate a relation on the set $\{a, b, c, d\}$ which is

(a) [BB] reflexive and symmetric

(b) not symmetric and not antisymmetric

(c) not symmetric but antisymmetric

(d) transitive

Include at least six elements in each relation.

5. Let $A = \{1, 2, 3\}$. List the ordered pairs in a relation on A which is

(a) [BB] not reflexive, not symmetric, and not transitive

(b) reflexive, but neither symmetric nor transitive

(c) symmetric, but neither reflexive nor transitive

(d) transitive, but neither reflexive nor symmetric

(e) reflexive and symmetric, but not transitive

(f) reflexive and transitive, but not symmetric

(g) [BB] symmetric and transitive, but not reflexive

(h) reflexive, symmetric, and transitive

6. Is it possible for a binary relation to be both symmetric and antisymmetric? If the answer is no, why not? If it is yes, find all such binary relations.

7. [BB] What is wrong with the following argument which purports to prove that a binary relation which is symmetric and transitive must necessarily be reflexive as well?

> Suppose $\mathcal{R}$ is a symmetric and transitive relation on a set A and let $a \in A$. Then for any b with $(a, b) \in \mathcal{R}$, we have also $(b, a) \in \mathcal{R}$ by symmetry. Since now we have both (a, b) and (b, a) in $\mathcal{R}$, we have $(a, a) \in \mathcal{R}$ as well, by transitivity. Thus, $(a, a) \in \mathcal{R}$, so $\mathcal{R}$ is reflexive.

8. Determine whether or not each of the binary relations $\mathcal{R}$ defined on the given sets A are reflexive, symmetric, antisymmetric, or transitive. If a relation has a certain property, prove this is so; otherwise, provide a counterexample to show that it does not.

(a) [BB] A is the set of all English words; $(a, b) \in \mathcal{R}$ if and only if a and b have at least one letter in common.

(b) A is the set of all people. $(a, b) \in \mathcal{R}$ if and only if neither a nor b is currently enrolled at the Miskatonic University or else both are enrolled at MU and are taking at least one course together.

9. Answer Exercise 8 for each of the following relations:

(a) $A = \{1, 2\}$; $\mathcal{R} = \{(1, 2)\}$.

(b) [BB] $A = \{1, 2, 3, 4\}$; $\mathcal{R} = \{(1, 1), (1, 2), (2, 1), (3, 4)\}$.

(c) [BB] $A = \mathsf{Z}$; $(a, b) \in \mathcal{R}$ if and only if $ab \geq 0$.

(d) $A = \mathsf{R}$; $(a, b) \in \mathcal{R}$ if and only if $a^2 = b^2$.

(e) $A = \mathsf{R}$; $(a, b) \in \mathcal{R}$ if and only if $a - b \leq 3$.

(f) $A = \mathsf{Z} \times \mathsf{Z}$; $((a, b), (c, d)) \in \mathcal{R}$ if and only if $a - c = b - d$.

(g) $A = \mathsf{N}$; $\mathcal{R}$ is $\neq$.

(h) $A = \mathsf{Z}$; $\mathcal{R} = \{(x, y) \mid x + y = 10\}$.

(i) [BB] $A = \mathsf{R}^2$; $\mathcal{R} = \{((x, y), (u, v)) \mid x + y \leq u + v\}$.

(j) $A = \mathsf{N}$; $(a, b) \in \mathcal{R}$ if and only if $\frac{a}{b}$ is an integer.

(k) $A = \mathsf{Z}$; $(a, b) \in \mathcal{R}$ if and only if $\frac{a}{b}$ is an integer.

10. Let S be a set which contains at least two elements a and b. Let A be the power set of S. Determine which of the properties—reflexivity, symmetry, antisymmetry, transitivity—each of the following binary relations $\mathcal{R}$ on A possesses. Give a proof or counterexample as appropriate.

(a) [BB] $(X, Y) \in \mathcal{R}$ if and only if $X \subseteq Y$.

(b) $(X, Y) \in \mathcal{R}$ if and only if $X \subsetneq Y$.

(c) $(X, Y) \in \mathcal{R}$ if and only if $X \cap Y = \emptyset$.

11. Let A be the set of books for sale in a certain university bookstore and assume that among these are books with the following properties.

Book	Price	Length
U	\$10	100 pages
W	\$25	125 pages
X	\$20	150 pages
Y	\$10	200 pages
Z	\$ 5	100 pages

(a) [BB] Suppose $(a, b) \in \mathcal{R}$ if and only if the price of book a is greater than or equal to the price of book b **and** the length of a is greater than or equal to the length of b. Is $\mathcal{R}$ reflexive? Symmetric? Antisymmetric? Transitive?

(b) Suppose $(a, b) \in \mathcal{R}$ if and only if the price of a is greater than or equal to the price of b **or** the length of a is greater than or equal to the length of b. Is $\mathcal{R}$ reflexive? Symmetric? Antisymmetric? Transitive?

2.4 EQUIVALENCE RELATIONS

It is useful to think of a binary relation on a set A as establishing relationships between elements of A, the assertion "$(a, b) \in \mathcal{R}$" relating the elements a and b. Such relationships occur everywhere. Two people may be of the same sex, have the same color eyes, live on the same street. These three particular relationships are reflexive, symmetric, and transitive and hence *equivalence relations*.

2.4.1 DEFINITION ▶ An *equivalence relation* on a set A is a binary relation $\mathcal{R}$ on A which is reflexive, symmetric, and transitive.

Suppose A is the set of all people in the world and

$$\mathcal{R} = \{(a, b) \in A \times A \mid a \text{ and } b \text{ have the same parents}\}.$$

This relation is reflexive (every person has the same set of parents as himself/herself), symmetric (if a and b have the same parents, then so do b and a), and transitive (if a and b have the same parents, and b and c have the same parents, then a and c have the same parents) and so $\mathcal{R}$ is an equivalence relation. It may be because of examples like this that it is common to say "a is related to b," rather than "$(a, b) \in \mathcal{R}$," even for an abstract binary relation $\mathcal{R}$.

If $\mathcal{R}$ is a binary relation on a set A and $a, b \in A$, some authors use the notation $a\mathcal{R}b$ to indicate that $(a, b) \in \mathcal{R}$. In this section, we will usually write $a \sim b$ and, in the case of an equivalence relation, say "a is equivalent to b." Thus, to prove that $\mathcal{R}$ is an equivalence relation, we must prove that $\mathcal{R}$ is

reflexive: $a \sim a$ for all $a \in A$,
symmetric: if $a \in A$ and $b \in A$ and $a \sim b$, then $b \sim a$, and
transitive: if $a, b, c \in A$ and both $a \sim b$ and $b \sim c$, then $a \sim c$.

EXAMPLE 26 Let A be the set of students currently registered at the University of Southern California. For $a, b \in A$, call a and b equivalent if their student numbers have the same first two digits. Certainly $a \sim a$ for every student a because any number has the same first two digits as itself. If $a \sim b$, the student numbers of a and b have the same first two digits, so the student numbers of b and a have the same first two digits; therefore, $b \sim a$. Finally, if $a \sim b$ and $b \sim c$, then the student numbers of a and b have the same first two digits, and the student numbers of b and c have the same first two digits, so the student numbers of a and c have the same first two digits. Since $\mathcal{R}$ is reflexive, symmetric, and transitive, $\mathcal{R}$ is an equivalence relation on A. ▲

EXAMPLE 27 Let A be the set of all residents of the continental United States. Call a and b equivalent if a and b are residents of the same state (or district). The student should mentally confirm that $\sim$ defines an equivalence relation. ▲

EXAMPLE 28 (Congruence mod 3)[5] Define $\sim$ on the set Z of integers by $a \sim b$ if $a - b$ is divisible (evenly) by 3.[6] For any $a \in \mathsf{Z}$, $a - a = 0$ is divisible by 3 and so $a \sim a$.

---

[5]This is an example of an important equivalence relation called *congruence* to which we later devote an entire section, Section 4.4.

[6]Within the context of integers, *divisible* always means *divisible evenly*, that is, with remainder 0.

If $a, b \in \mathbf{Z}$ and $a \sim b$, then $a - b$ is divisible by 3, so $b - a$ (the negative of $a - b$) is also divisible by 3. Hence, $b \sim a$. Finally, if $a, b, c \in \mathbf{Z}$ with $a \sim b$ and $b \sim c$, then both $a - b$ and $b - c$ are divisible by 3, so $a - c$, being the sum of $a - b$ and $b - c$, is also divisible by 3. Thus, $\sim$ is an equivalence relation.   ▲

EXAMPLE 29   The relation $\leq$ on the real numbers—$a \sim b$ if and only if $a \leq b$—is not an equivalence relation on R. While it is reflexive and transitive, it is not symmetric: $4 \leq 5$ but $5 \nleq 4$.   ▲

Pause 11   Let A be the set of all people. For $a, b \in A$, define $a \sim b$ if either (i) both a and b are residents of the same state of the United States or (ii) neither a nor b is a resident of any state of the United States. Does $\sim$ define an equivalence relation?

Surely the three most fundamental properties of equality are

reflexivity: $a = a$ for all a;
symmetry: if $a = b$, then $b = a$; and
transitivity: if $a = b$ and $b = c$, then $a = c$.

Thus, equality is an equivalence relation on any set. For this reason, we think of equivalence as a weakening of equality. We have in mind a certain characteristic or property of elements and wish only to consider as different, elements which differ with respect to this characteristic. Little children may think of their brothers and sisters as the same and other children as "different." A statistician trying to estimate the percentages of people in the world with different eye colors is only interested in eye color; for her, two people are only "different" if they have different colored eyes. All drop-off points in a given neighborhood of town may be the "same" to the driver of a newspaper truck. An equivalence relation changes our view of the universe (the underlying set A); instead of viewing it as individual elements, attention is directed to certain groups or subsets. The equivalence relation "same parents" groups people into families; "same color eyes" groups people by eye color; "same neighborhood" groups newspaper drop-off points by neighborhood.

The groups into which an equivalence relation divides the underlying set are called *equivalence classes*. The equivalence class of an element is the collection of all things related to it.

2.4.2 DEFINITION ▶   If $\sim$ denotes an equivalence relation on a set A, the *equivalence class* of an element $a \in A$ is the set $\overline{a} = \{x \in A \mid x \sim a\}$. The set of all equivalence classes is called the *quotient set* of A *mod* $\sim$ and denoted $A/\sim$.

Since an equivalence relation is symmetric, it does not matter whether we write $x \sim a$ or $a \sim x$ in the definition of $\overline{a}$. The set of things related to a is the same as the set of things to which a is related.

For the equivalence relation in Example 26, the students who are related to a particular student x are those whose student numbers have the same first two digits as x's student number. For this equivalence relation, a typical equivalence class (a typical member of the quotient set) is the set of all students whose student numbers begin with the same first two digits. The set of all students has been grouped into smaller sets—the class of 87, for instance (all students whose numbers begin 87), the class of 90 (all students whose numbers begin 90), and so forth.

In Example 27, if x is a resident of the continental United States, then the people to whom x is related are those people who reside in the same state or district as x. The residents of Colorado, for example, form one equivalence class, as do the residents of Rhode Island, the residents of Florida, and so on. The quotient set is the set of (forty-nine) American states on the North American continent, together with the District of Columbia.

What are the equivalence classes for the equivalence relation which is congruence mod 3? What is $\bar{0}$, the equivalence class of 0, for instance? If $a \sim 0$, then $a - 0$ is divisible by 3; in other words, a is divisible by 3. Thus, $\bar{0}$ is the set of all integers which are divisible by 3. We shall denote this set 3Z. So $\bar{0} = 3Z$. What is $\bar{1}$? If $a \sim 1$, then $a - 1 = 3k$ for some integer k, so $a = 3k + 1$. Thus, $\bar{1} = \{3k+1 \mid k \in Z\}$, a set we denote 3Z+1. Similarly, $\bar{2} = \{3k+2 \mid k \in Z\} = 3Z+2$ and we have found all the equivalence classes for congruence mod 3.

Pause 12 Why?

In general, for given natural numbers n and r, $nZ + r$ is the set of integers of the form $na + r$ for some $a \in Z$:

(5)
$$nZ + r = \{na + r \mid a \in Z\}.$$

Also, we write nZ instead of $nZ + 0$:

(6)
$$nZ = \{na \mid a \in Z\}.$$

The even integers, for instance, can be denoted 2Z.

Pause 13 What are the equivalence classes for the equivalence relation described in Pause 11? How many elements does the quotient set contain?

PROBLEM 30. For (x, y) and (u, v) in R^2, define $(x, y) \sim (u, v)$ if $x^2 + y^2 = u^2 + v^2$. Prove that $\sim$ defines an equivalence relation on R^2 and interpret the equivalence classes geometrically.

Solution. If $(x, y) \in R^2$, then $x^2 + y^2 = x^2 + y^2$, so $(x, y) \sim (x, y)$: The relation is reflexive. If $(x, y) \sim (u, v)$, then $x^2 + y^2 = u^2 + v^2$, so $u^2 + v^2 = x^2 + y^2$ and $(u, v) \sim (x, y)$: The relation is symmetric. Finally, if $(x, y) \sim (u, v)$ and $(u, v) \sim (w, z)$, then $x^2 + y^2 = u^2 + v^2$ and $u^2 + v^2 = w^2 + z^2$. Thus, $x^2 + y^2 = u^2 + v^2 = w^2 + z^2$. Since $x^2 + y^2 = w^2 + z^2$, $(x, y) \sim (w, z)$, so the relation is transitive.

The equivalence class of (a, b) is

$$\overline{(a, b)} = \{(x, y) \mid (x, y) \sim (a, b)\} = \{(x, y) \mid x^2 + y^2 = a^2 + b^2\}.$$

For example, $\overline{(1, 0)} = \{(x, y) \mid x^2 + y^2 = 1^2 + 0^2 = 1\}$, which we recognize as the graph of a circle in the Cartesian plane with center $(0, 0)$ and radius 1. For general (a, b), letting $c = a^2 + b^2$, the equivalence class $\overline{(a, b)}$ is the set of points (x, y) satisfying $x^2 + y^2 = c$. So this equivalence class is the circle with center

$(0, 0)$ and radius $\sqrt{c}$. With one exception, the equivalence classes are circles with center $(0, 0)$. ∎

Pause 14   What is the exception?

2.4.3 PROPOSITION ▶   Let $\sim$ denote an equivalence relation on a set A. Let $a \in A$. Then for any $x \in A$, $x \sim a$ if and only if $\overline{x} = \overline{a}$.

Proof   ($\leftarrow$) Suppose $\overline{x} = \overline{a}$. We know $x \in \overline{x}$ because $x \sim x$, so $x \in \overline{a}$; thus, $x \sim a$. It is the implication $\rightarrow$ which is the substance of this proposition.

($\rightarrow$) Suppose that $x \sim a$. We must prove that the two sets $\overline{x}$ and $\overline{a}$ are equal. As always, we do this by proving that each set is a subset of the other. First suppose $y \in \overline{x}$. Then $y \sim x$ and $x \sim a$, so $y \sim a$ by transitivity. Therefore, $y \in \overline{a}$, so $\overline{x} \subseteq \overline{a}$. On the other hand, suppose $y \in \overline{a}$. Then $y \sim a$. Since we also have $a \sim x$, we have both $y \sim a$ and $a \sim x$; therefore, by transitivity, $y \sim x$. Thus, $y \in \overline{x}$ and $\overline{a} \subseteq \overline{x}$. Therefore, $\overline{a} = \overline{x}$. ∎

In each of the examples of equivalence relations which we have discussed in this section, different equivalence classes never overlapped. If a person is a resident of one state, he or she is not a resident of another. A student number cannot begin with 79 and also with 84. An integer which is a multiple of 3 is not of the form $3k+1$. These examples are suggestive of a result which is true in general.

2.4.4 PROPOSITION ▶   Suppose $\sim$ denotes an equivalence relation on a set A and $a, b \in A$. Then $\overline{a} \neq \overline{b}$ if and only if $\overline{a} \cap \overline{b} = \emptyset$.

Proof   ($\rightarrow$) Suppose $\overline{a} \neq \overline{b}$. We must prove $\overline{a} \cap \overline{b} = \emptyset$ and offer a proof by contradiction.

Suppose $\overline{a} \cap \overline{b} \neq \emptyset$. Then there is an element $x \in \overline{a} \cap \overline{b}$. Since $x \in \overline{a}$, $\overline{x} = \overline{a}$ by Proposition 2.4.3. Similarly, since $x \in \overline{b}$, we also have $\overline{x} = \overline{b}$. Thus, $\overline{a} = \overline{b}$, which is a contradiction.

($\leftarrow$) On the other hand, if $\overline{a} \cap \overline{b} = \emptyset$, then $a \in \overline{a}$ but $a \notin \overline{b}$, so $\overline{a} \neq \overline{b}$. ∎

If $\sim$ denotes an equivalence relation on A, reflexivity says that every element a in A belongs to some equivalence class, namely, to $\overline{a}$. In conjunction with Proposition 2.4.4, this observation says that the equivalence classes of any equivalence relation divide A into disjoint (that is, nonoverlapping) subsets which cover the entire set, just like the pieces of a jigsaw puzzle. We say that the equivalence classes "partition" A or "form a partition of" A. (The word *partition* is used as both verb and noun.)

2.4.5 DEFINITION ▶   A *partition* of a set A is a collection of disjoint nonempty subsets of A whose union is A. These disjoint sets are called *cells* (or *blocks*). The cells are said to *partition* A.

EXAMPLES 31   • Canada is partitioned into ten provinces and three territories.[7]

---

[7]Nunavut, created from the eastern half of the former Northwest Territories, joined Canada on April 1, 1999.

- Students are partitioned into groups according to the first two digits of their student numbers.
- The human race is partitioned into groups by eye color.
- A deck of playing cards is partitioned into four suits.
- If $A = \{a, b, c, d, e, f, x\}$, then $\{\{a, b\}, \{c, d, e\}, \{f\}, \{x\}\}$ is a partition of A. So is $\{\{a, x\}, \{b, d, e, f\}, \{c\}\}$. ▲

We have seen that the equivalence classes of an equivalence relation on a set A are disjoint sets whose union is A; each element $a \in A$ is in precisely one equivalence class, namely, $\bar{a}$. Thus, we have the following basic theorem about equivalence relations.

2.4.6 THEOREM ▶ The equivalence classes associated with an equivalence relation on a set A form a partition of A.

Not only does an equivalence relation determine a partition, but, conversely, any partition of a set A determines an equivalence relation, specifically, that equivalence relation whose equivalence classes are the cells of the partition. The partition of the integers into "evens" and "odds" corresponds to the equivalence relation which says two integers are equivalent if and only if they are both even or both odd. The partition

	a	g	b	d	e	f	c
a	×	×					
g	×	×					
b			×	×	×	×	
d			×	×	×	×	
e			×	×	×	×	
f			×	×	×	×	
c							×

$$\{\{a, g\}, \{b, d, e, f\}, \{c\}\}$$

of the set $\{a, b, c, d, e, f, g\}$ corresponds to the equivalence relation whose equivalence classes are $\{a, g\}$, $\{b, d, e, f\}$ and $\{c\}$, that is, to the equivalence relation described in the figure, where a cross in row x and column y is used to indicate $x \sim y$.

Pause 15 The suits "heart," "diamond," "club," "spade" partition a standard deck of playing cards. Describe the corresponding equivalence relation on a deck of cards.

The correspondence between equivalence relations and partitions provides a simple way to exhibit equivalence relations on small sets. For example, the equivalence relation defined in Fig 2.3 can also be described by listing its equivalence classes: $\{a, b\}$ and $\{c\}$.

	a	b	c
a	×	×	
b	×	×	
c			×

Figure 2.3   An equivalence relation with two equivalence classes, $\{a, b\}$ and $\{c\}$.

Answers to Pauses **11.** It sure does. First, every person is either a resident of the same state in the United States as himself/herself or not a resident of any U.S. state, so $a \sim a$

for all $a \in A$: $\sim$ is reflexive. Second, if $a, b \in A$ and $a \sim b$, then either a and b are residents of the same U.S. state (in which case, so are b and a) or else neither a nor b is a resident of any state in the United States (in which case, neither is b or a). Thus, $b \sim a$: $\sim$ is symmetric. Finally, suppose $a \sim b$ and $b \sim c$. Then either a and b are residents of the same U.S. state or neither is a resident of any U.S. state, and the same holds true for b and c. It follows that either all three of a, b, and c live in the same U.S. state, or none is a resident of a U.S. state. Thus, $a \sim c$: $\sim$ is transitive as well.

12. The equivalence class of 3 is $\{3k + 3 \mid k \in \mathsf{Z}\}$, but this is just the set $3\mathsf{Z}$ of multiples of 3. Thus $\overline{3} = \overline{0}$. The equivalence class of 4 is $\{3k + 4 \mid k \in \mathsf{Z}\}$, but $3k + 4 = 3(k + 1) + 1$, so this set is just $3\mathsf{Z} + 1$, the equivalence class of 1: $\overline{4} = \overline{1}$. In general, the equivalence class of an integer r is $3\mathsf{Z}$ if r is a multiple of 3, $3\mathsf{Z} + 1$ if r is of the form $3a + 1$, and $3\mathsf{Z} + 2$ if r is of the form $3a + 2$. Since every integer r is either a multiple of 3, or $3a + 1$ or $3a + 2$ for some a, the only equivalence classes are $3\mathsf{Z}$, $3\mathsf{Z} + 1$ and $3\mathsf{Z} + 2$.

13. The equivalence class of a consists of those people equivalent to a in the sense of $\sim$. If a does not live in any state of the United States, the equivalence class of a consists of all those people who also live outside any U.S. state. If a does live in a U.S. state, the equivalence class of a consists of those people who live in the same state. The quotient set has 51 elements, consisting of the residents of the 50 American states and the set of people who do not live in any U.S. state.

14. The equivalence class of $(0, 0)$ is the set $\{(0, 0)\}$ whose only element is the single point $(0, 0)$.

15. Two cards are equivalent if and only if they have the same suit.

EXERCISES

The symbol [BB] means that an answer can be found in the Back of the Book.

1. Let A be the set of all citizens of New York City. For $a, b \in A$, define $a \sim b$ if and only if

   (a) neither a nor b have a cell phone, or
   (b) both a and b have cell phones in the same exchange (that is, the first three digits of each phone number are the same).

   Show that $\sim$ defines an equivalence relation on A and find the corresponding equivalence classes.

2. Explain why each of the following binary relations on $S = \{1, 2, 3\}$ is not an equivalence relation on S.

   (a) [BB] $\mathcal{R} = \{(1, 1), (1, 2), (3, 2), (3, 3), (2, 3), (2, 1)\}$
   (b) $\mathcal{R} = \{(1, 1), (2, 2), (3, 3), (2, 1), (1, 2), (2, 3), (3, 1), (1, 3)\}$

   (c)

	1	2	3
1	×	×	×
2	×	×	
3			×

3. [BB] The sets $\{1\}, \{2\}, \{3\}, \{4\}, \{5\}$ are the equivalence classes for a well-known equivalence relation on the set $S = \{1, 2, 3, 4, 5\}$. What is the usual name for this equivalence relation?

4. [BB] For $a, b \in \mathsf{R} \setminus \{0\}$, define $a \sim b$ if and only if $\frac{a}{b} \in \mathsf{Q}$.

   (a) Prove that $\sim$ is an equivalence relation.
   (b) Find the equivalence class of 1.
   (c) Show that $\overline{\sqrt{3}} = \overline{\sqrt{12}}$.

5. For natural numbers a and b, define $a \sim b$ if and only if $a^2 + b$ is even. Prove that $\sim$ defines an equivalence relation on N and find the quotient set determined by $\sim$.

6. [BB] For $a, b \in \mathsf{R}$, define $a \sim b$ if and only if $a - b \in \mathsf{Z}$.

   (a) Prove that $\sim$ defines an equivalence relation on Z.
   (b) What is the equivalence class of 5? What is the equivalence class of $5\frac{1}{2}$?
   (c) What is the quotient set determined by this equivalence relation?

7. [BB] For integers a, b, define $a \sim b$ if and only if $2a + 3b = 5n$ for some integer n. Show that $\sim$ defines an equivalence relation on Z.

8. Define $\sim$ on Z by $a \sim b$ if and only if $3a + b$ is a multiple of 4.

   (a) Prove that $\sim$ defines an equivalence relation.
   (b) Find the equivalence class of 0.
   (c) Find the equivalence class of 2.
   (d) Make a guess about the quotient set.

9. For integers a and b, define $a \sim b$ if $3a + 4b = 7n$ for some integer n.

   (a) Prove that $\sim$ defines an equivalence relation.
   (b) Find the equivalence class of 0.

10. [BB] For $a, b \in \mathsf{Z} \setminus \{0\}$, define $a \sim b$ if and only if $ab > 0$.

    (a) Prove that $\sim$ defines an equivalence relation on Z.
    (b) What is the equivalence class of 5? What's the equivalence class of -5?
    (c) What is the partition of $\mathsf{Z} \setminus \{0\}$ determined by this equivalence relation?

11. For $a, b \in \mathsf{Z}$, define $a \sim b$ if and only if $a^2 - b^2$ is divisible by 3.

    (a) [BB] Prove that $\sim$ defines an equivalence relation on Z.
    (b) What is $\overline{0}$? What is $\overline{1}$?
    (c) What is the partition of Z determined by this equivalence relation?

12. Determine, with reasons, whether or not each of the following defines an equivalence relation on the set A.

    (a) [BB] A is the set of all triangles in the plane; $a \sim b$ if and only if a and b are congruent.
    (b) A is the set of all circles in the plane; $a \sim b$ if and only if a and b have the same center.
    (c) A is the set of all straight lines in the plane; $a \sim b$ if and only if a is parallel to b.
    (d) A is the set of all lines in the plane; $a \sim b$ if and only if a is perpendicular to b.

13. List the pairs in the equivalence relation associated with each of the following partitions of $A = \{1, 2, 3, 4, 5\}$.

    (a) [BB] $\{\{1, 2\}, \{3, 4, 5\}\}$
    (b) $\{\{1\}, \{2\}, \{3, 4\}, \{5\}\}$
    (c) $\{\{1, 2, 3, 4, 5\}\}$

14. (a) List all the equivalence relations on the set $\{a\}$. How many are there altogether?
    (b) Repeat (a) for the set $\{a, b\}$.
    (c) [BB] Repeat (a) for the set $\{a, b, c\}$.
    (d) Repeat (a) for the set $\{a, b, c, d\}$.

    (Remark: The number of partitions of a set of n elements grows rather rapidly. There are 52 partitions of a set of five elements, 203 partitions of a set of six elements, and 877 partitions of a set of seven elements.)

15. Define $\sim$ on R^2 by $(x, y) \sim (u, v)$ if and only if $x - y = u - v$.

    (a) [BB] Criticize and then correct the following "proof" that $\sim$ is reflexive.

    "If $(x, y) \sim (x, y)$, then $x - y = x - y$,

    which is true."

    (b) What is wrong with the following interpretation of symmetry in this situation?

    "If $(x, y) \in \mathcal{R}$, then $(y, x) \in \mathcal{R}$."

    Write a correct statement of the symmetric property (as it applies to the relation $\sim$ in this exercise).

    (c) Criticize and then correct the following "proof" that $\sim$ is symmetric.

    "$(x, y) \sim (u, v)$ if $x - y = u - v$.

    Then $u - v = x - y$. So $(u, v) \sim (x, y)$."

    (d) Criticize and correct the following "proof" of transitivity.

    "$(x, y) \sim (u, v)$ and $(u, v) \sim (w, z)$.

    Then $u - v = w - z$, so if $x - y = u - v$,

    then $x - y = w - z$. So $(x, y) \sim (w, z)$."

    (e) Why does $\sim$ define an equivalence relation on R^2?
    (f) Determine the equivalence classes of $(0, 0)$ and $(2, 3)$ and describe these geometrically.

16. [BB] For (x, y) and $(u, v) \in \mathsf{R}^2$ define $(x, y) \sim (u, v)$ if and only if $x^2 - y^2 = u^2 - v^2$. Prove that $\sim$ defines an equivalence relation on R^2. Describe geometrically the equivalence class of $(0, 0)$. Describe geometrically the equivalence class of $(1, 0)$.

17. Determine which of the following define equivalence relations in R^2. For those which do, give a geometrical interpretation of the elements of the quotient set.

(a) $(a, b) \sim (c, d)$ if and only if $a + 2b = c + 2d$.

(b) $(a, b) \sim (c, d)$ if and only if $ab = cd$.

(c) $(a, b) \sim (c, d)$ if and only if $a^2 + b = c + d^2$.

(d) $(a, b) \sim (c, d)$ if and only if $a = c$.

(e) $(a, b) \sim (c, d)$ if and only if $ab = c^2$.

18. Let $\sim$ denote an equivalence relation on a set A. Assume $a, b, c, d \in A$ are such that $a \in \bar{b}$, $c \in \bar{d}$, and $d \in \bar{b}$. Prove that $\bar{a} = \bar{c}$.

19. [BB] Let $A = \{1, 2, 3, 4, 5, 6, 7, 8, 9\}$. For $a, b \in A$, define $a \sim b$ if and only if ab is a perfect square (that is, the square of an integer).

(a) What are the ordered pairs in this relation?

(b) For each $a \in A$, find $\bar{a} = \{x \in A \mid x \sim a\}$.

(c) Explain why $\sim$ defines an equivalence relation on A.

20. [BB] In Exercise 19, let A be the set of all natural numbers and $\sim$ be as given. Show that $\sim$ defines an equivalence relation on A.

21. Repeat Exercise 19 for $A = \{1, 2, 3, 4, 5, 6, 7\}$ and the relation on A defined by $a \sim b$ if and only if $\frac{a}{b}$ is a power of 2, that is, $\frac{a}{b} = 2^t$ for some integer t, positive, negative or zero.

22. In Exercise 21, let A be the set of all natural numbers and $\sim$ be as given. Show that $\sim$ defines an equivalence relation on A.

23. Let $\mathcal{R}$ be an equivalence relation on a set S and let $\{S_1, S_2, \ldots, S_t\}$ be a collection of subsets of S with the property that $(a, b) \in \mathcal{R}$ if and only if a and b are elements of the same set S_i, for some i. Suppose that for each i, $S_i \not\subseteq \bigcup_{j \neq i} S_j$. Prove that $\{S_1, S_2, \ldots, S_t\}$ is a partition of S.

2.5 PARTIAL ORDERS

In the previous section, we defined an equivalence relation as a binary relation which possesses the three fundamental properties of equality—reflexivity, symmetry, transitivity. We mentioned that we view equivalence as a weak form of equality and employed a symbol, $\sim$, suggesting "equals." In an analogous manner, in this section, we focus on three fundamental properties of the order relation $\leq$ on the real numbers—reflexivity, antisymmetry, transitivity—and define a binary relation called *partial order*, which can be viewed as a weak form of $\leq$. We shall use the symbol $\preceq$ for a partial order to remind us of its connection with $\leq$ and, for the same reason, say that "a is less than or equal to b" whenever $a \preceq b$.

2.5.1 DEFINITIONS ▶ A *partial order* on a set A is a reflexive, antisymmetric, transitive relation on A. A *partially ordered set*, *poset* for short, is a pair $(A, \preceq)$, where $\preceq$ is a partial order on a set A.

Writing $a \preceq b$ to mean that (a, b) is in the relation, a partial order on A is a binary relation which is

reflexive:	$a \preceq a$ for all $a \in A$,
antisymmetric:	If $a, b \in A$, $a \preceq b$ and $b \preceq a$, then $a = b$, and
transitive:	If $a, b, c \in A$, $a \preceq b$ and $b \preceq c$, then $a \preceq c$.

It is convenient to use the notation $a \prec b$ (and to say "a is less than b") to signify $a \preceq b$, $a \neq b$, just as we use $a < b$ to mean $a \leq b$, $a \neq b$. Similarly, the meanings of $a \succeq b$ and $a \succ b$ should be apparent.

There is little purpose in making a definition unless there is at hand a variety of examples which fit the definition. Here then are a few examples of partial orders.

EXAMPLES 32
- The binary relation $\leq$ on the real numbers (or on any subset of the real numbers) is a partial order because $a \leq a$ for all $a \in \mathsf{R}$ (reflexivity), $a \leq b$ and $b \leq a$ implies $a = b$ (antisymmetry) and $a \leq b$, $b \leq c$ implies $a \leq c$ (transitivity).
- For any set S, the binary relation $\subseteq$ on the power set $\mathcal{P}(S)$ of S is a partial order because $X \subseteq X$ for any $X \in \mathcal{P}(S)$ (reflexivity), $X \subseteq Y$, $Y \subseteq X$ for $X, Y \in \mathcal{P}(S)$ implies $X = Y$ (antisymmetry) and $X \subseteq Y$, $Y \subseteq Z$ for $X, Y, Z \in \mathcal{P}(S)$ implies $X \subseteq Z$ (transitivity). ▲

EXAMPLE 33 (Lexicographic Ordering) Suppose we have some alphabet of symbols (perhaps the English alphabet) which is partially ordered by some relation $\preceq$. By "word," we mean any string of letters from this alphabet. For words $a = a_1 a_2 \cdots a_n$ and $b = b_1 b_2 \cdots b_m$, define $a \preceq b$ if

- a and b are identical, or
- $a_i \preceq b_i$ in the alphabet at the first position i where the words differ, or
- $n < m$ and $a_i = b_i$ for $i = 1, \ldots, n$. (This is the situation where word a, which is shorter than b, forms the initial sequence of letters in b.)

This ordering of words is called *lexicographic* because when the basic alphabet is the English alphabet, it is precisely how words are ordered in a dictionary; car $\preceq$ cat $\preceq$ catalog. ▲

The adjective "partial", as in "partial order" draws our attention to the fact that the definition does not require that every pair of elements be *comparable*, in the following sense.

2.5.2 DEFINITION ▶ If $(A, \preceq)$ is a partially ordered set, elements a and b of A are said to be *comparable* if and only if either $a \preceq b$ or $b \preceq a$.

If X and Y are subsets of a set S, it need not be the case that $X \subseteq Y$ or $Y \subseteq X$; for example, $\{a\}$ and $\{b, c\}$ are not comparable.

2.5.3 DEFINITION ▶ If $\preceq$ is a partial order on a set A and every two elements of A are comparable, then $\preceq$ is called a *total order* and the pair $(A, \preceq)$ is called a *totally ordered set*.

The real numbers are totally ordered by $\leq$ because for every pair a, b of real numbers either $a \leq b$ or $b \leq a$. On the other hand, the set of sets, $\{\{a\}, \{b\}, \{c\}, \{a, c\}\}$ is not totally ordered by $\subseteq$ since neither $\{a\} \subseteq \{b\}$ nor $\{b\} \subseteq \{a\}$.

Pause 16 Is lexicographic order on a set of words (in the usual sense) a total order?

Partial orders are often pictured by means of a diagram named after Helmut Hasse (1898–1979), for many years professor of mathematics at Göttingen.[8] In the *Hasse diagram* of a partially ordered set $(A, \preceq)$,

- there is a dot (or vertex) associated with each element of A;
- if $a \preceq b$, then the dot for b is positioned higher than the dot for a; and
- if $a \prec b$ and there is no intermediate c with $a \prec c \prec b$, then a line is drawn from a to b. (In this case, we say that the element b *covers* a.)

The effect of the last property here is to remove redundant lines. Two Hasse diagrams are shown in Fig 2.4. The reader should appreciate that these would be

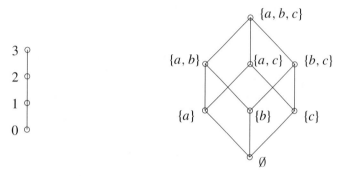

Figure 2.4   The Hasse diagrams for $(\{0, 1, 2, 3\}, \leq)$ and $(\mathcal{P}(\{a, b, c\}), \subseteq)$.

unnecessarily complicated were we to draw all lines from a to b whenever $a \preceq b$ instead of just those lines where b covers a. No knowledge of the partial order is lost by this convention: After all, if $a \preceq b$ and $b \preceq c$, then (by transitivity), $a \preceq c$, so if there is a line from a to b and a line from b to c, then we can correctly infer that $a \preceq c$, from the diagram. For example, in the diagram on the left, we can infer that $1 \preceq 3$ since $1 \preceq 2$ and $2 \preceq 3$. In the diagram on the right, we similarly infer that $\{b\} \preceq \{a, b, c\}$.

Pause 17 Suppose that in some Hasse diagram, a vertex c is "above" another vertex a, but there is no line from a to c. Is it the case that $a \preceq c$? Explain.

2.5.4 DEFINITIONS ▶ An element a of a poset $(A, \preceq)$ is *maximum* if and only if $b \preceq a$ for every $b \in A$ and *minimum* if and only if $a \preceq b$ for every $b \in A$.

In the poset $(\mathcal{P}(\{a, b, c\}), \subseteq)$, $\varnothing$ is a minimum element and the set $\{a, b, c\}$ a maximum element. In the poset $\{ \{a\}, \{b\}, \{c\}, \{a, c\} \}$ (with respect to $\subseteq$), there is neither a maximum nor a minimum because, for each of the elements $\{a\}$, $\{b\}$, $\{c\}$, and $\{a, c\}$, there is another of these with which it is not comparable.

---

[8]There is a fascinating account by S. L. Segal of the ambiguous position in which Hasse found himself during the Nazi period. The article, entitled "Helmut Hasse in 1934," appears in *Historia Mathematica* **7** (1980), 45–56.

If a poset has a maximum element, then this element is unique; similarly, a poset can have at most one minimum. (See Exercise 11.)

One must be careful to distinguish between maximum and **maximal** elements and between minimum and **minimal** elements.

2.5.5 DEFINITIONS ▶ An element a of a poset A is *maximal* if and only if,

$$\text{if } b \in A \text{ and } a \preceq b, \text{ then } b = a$$

and *minimal* if and only if,

$$\text{if } b \in A \text{ and } b \preceq a, \text{ then } b = a.$$

Thus, a **maximum** element is "bigger" (in the sense of $\preceq$) than every other element in the set while a **maximal** element is one which is not less than any other. Considering again the poset $\{\{a\}, \{b\}, \{c\}, \{a, c\}\}$, while there is neither a maximum nor a minimum, each of $\{a\}$, $\{b\}$, and $\{c\}$ is minimal while both $\{b\}$ and $\{a, c\}$ are maximal.

Pause 18 What, if any, are the maximum, minimum, maximal, and minimal elements in the poset whose Hasse diagram appears in Fig 2.5?

Figure 2.5

2.5.6 DEFINITIONS ▶ Let $(A, \preceq)$ be a poset. An element g is a *greatest lower bound* (abbreviated *glb*) of elements $a, b \in A$ if and only if

1. $g \preceq a$, $g \preceq b$, and
2. if $c \preceq a$ and $c \preceq b$ for some $c \in A$, then $c \preceq g$.

Elements a and b can have at most one glb (see Exercise 14). When this element exists, it is denoted $a \wedge b$, pronounced "a meet b."

An element ℓ is a *least upper bound* (abbreviated *lub*) of a and b if

1. $a \preceq \ell$, $b \preceq \ell$, and
2. if $a \preceq c$, $b \preceq c$ for some $c \in A$, then $\ell \preceq c$.

As with greatest lower bounds, a least upper bound is unique if it exists. The lub of a and b is denoted $a \vee b$, "a join b," if there is such an element.

EXAMPLES 34
- In the poset $(\mathsf{R}, \leq)$, the glb of two real numbers is the smaller of the two and the lub the larger.
- In the poset $(\mathcal{P}(S), \subseteq)$ of subsets of a set S, $A \wedge B = A \cap B$ and $A \vee B = A \cup B$. (See Exercise 12.) Remembering that $\vee$ means $\cup$ and $\wedge$ means $\cap$ in a poset of sets provides a good way to avoid confusing the symbols $\vee$ and $\wedge$ in a general poset. ▲

Pause 19 With reference to Fig 2.5, find $a \vee b$, $a \wedge b$, $b \vee d$ and $b \wedge d$, if these exist.

2.5.7 DEFINITION ▶ A poset $(A, \preceq)$ in which every two elements have a greatest lower bound in A and a least upper bound in A is called a *lattice*.

EXAMPLE 35 The posets described in Examples 34 are both lattices. ▲

Answers to Pauses

16. Sure it is; otherwise, it would be awfully hard to use a dictionary.

17. We can conclude $a \preceq c$ **only** if there is a sequence of intermediate vertices between a and c with lines between each adjacent pair. Look at the Hasse diagram $(\mathcal{P}(\{a, b, c\}), \subseteq)$ in Fig 2.4. Here we have $\{b, c\}$ above $\{a\}$, but $\{a\} \not\preceq \{b, c\}$ because these elements are incomparable. On the other hand, $\{a, b, c\}$ is above $\{a\}$ and we can infer that $\{a\} \preceq \{a, b, c\}$ because, for the intermediate vertex $\{a, c\}$, we have upward directed lines from $\{a\}$ to $\{a, c\}$ and from $\{a, c\}$ to $\{a, b, c\}$.

18. There is no maximum but a and b are maximal; d is both minimal and a minimum.

19. $a \vee b$ does not exist; $a \wedge b = c$; $b \vee d = b$; $b \wedge d = d$.

EXERCISES

The symbol [BB] means that an answer can be found in the Back of the Book.

1. Determine whether or not each of the following relations is a partial order and state whether or not each partial order is a total order.

  (a) [BB] For $a, b \in \mathsf{R}$, $a \preceq b$ means $a \geq b$.

  (b) [BB] For $a, b \in \mathsf{R}$, $a \preceq b$ means $a < b$.

  (c) $(\mathsf{R}, \preceq)$, where $a \preceq b$ means $a^2 \leq b^2$.

  (d) $(\mathsf{N} \times \mathsf{N}, \preceq)$, where $(a, b) \preceq (c, d)$ if and only if $a \leq c$.

  (e) $(\mathsf{N} \times \mathsf{N}, \preceq)$, where $(a, b) \preceq (c, d)$ if and only if $a \leq c$ and $b \geq d$.

  (f) $(\mathcal{W}, \preceq)$ where $\mathcal{W}$ is the set of all strings of letters from the alphabet ("words" real or imaginary) and $w_1 \preceq w_2$ if and only if w_1 has length not exceeding the length of w_2. (Length means number of letters.)

2. (a) [BB] List the elements of the set $\{11, 1010, 100, 1, 101, 111, 110, 1001, 10, 1000\}$ in lexicographic order, given $0 \preceq 1$.

  (b) Repeat part (a) assuming $1 \preceq 0$.

3. List all pairs (x, y) with $x \prec y$ in the partial orders described by each of the following Hasse diagrams.

(a) [BB]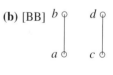

(b) [BB]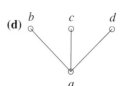

(c)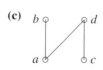

(d)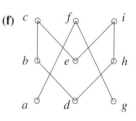

(e)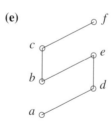

(f) (Hasse diagram with vertices c, f, i on top; b, e, h in middle; a, d, g on bottom)

4. [BB; (a), (b)] List all minimal, maximal, minimum, and maximum elements for each of the partial orders described in Exercise 3.

5. Draw the Hasse diagrams for each of the following partial orders.

   (a) $(\{1, 2, 3, 4, 5, 6\}, \leq)$

   (b) $(\{\{a\}, \{a, b\}, \{a, b, c\}, \{a, b, c, d\}, \{a, c\}, \{c, d\}\}, \subseteq)$

6. List all minimal, minimum, maximal, and maximum elements for each of the partial orders in Exercise 5.

7. [BB] In the poset $(\mathcal{P}(S), \subseteq)$ of subsets of a set S, under what conditions does one set B cover another set A?

8. Learn what you can about Helmut Hasse and write a short biographical note about this person, in good clear English of course!

9. (a) [BB] Prove that any finite (nonempty) poset must contain maximal and minimal elements.

   (b) Is the result of (a) true in general for posets of arbitrary size? Explain.

10. (a) Let $A = \mathbb{Z}^2$ and, for a $= (a_1, a_2)$ and b $= (b_1, b_2)$ in A, define a $\preceq$ b if and only if $a_1 \leq b_1$ and $a_1 + a_2 \leq b_1 + b_2$. Prove that $\preceq$ is a partial order on A. Is this partial order a total order? Justify your answer with a proof or a counterexample.

    (b) Generalize the result of part (a) by defining a partial order on the set $\mathbb{Z}^n$ of n-tuples of integers. (No proof is required.)

11. (a) [BB] Prove that a poset has at most one maximum element.

    (b) Prove that a poset has at most one minimum element.

12. Let S be a nonempty set and let A and B be elements of the power set of S. In the partially ordered set $(\mathcal{P}(S), \subseteq)$,

    (a) [BB] prove that $A \wedge B = A \cap B$;

    (b) prove that $A \vee B = A \cup B$.

13. Let a and b be two elements of a poset $(A, \preceq)$ with $a \preceq b$.

    (a) [BB] Show that $a \vee b$ exists, find this element, and explain your answer.

    (b) Show that $a \wedge b$ exists, find this element, and explain your answer.

14. (a) [BB] Prove that a glb of two elements in a poset $(A, \preceq)$ is unique whenever it exists.

    (b) Prove that a lub of two elements in a poset $(A, \preceq)$ is unique whenever it exists.

15. (a) If a and b are two elements of a partially ordered set $(A, \preceq)$, the concepts

$$\max(a, b) = \begin{cases} a & \text{if } b \preceq a \\ b & \text{if } a \preceq b \end{cases}$$

and

$$\min(a, b) = \begin{cases} a & \text{if } a \preceq b \\ b & \text{if } b \preceq a \end{cases}$$

do not make sense unless the poset is totally ordered. Explain.

    (b) Show that any totally ordered set is a lattice.

16. (a) [BB] Give an example of a partially ordered set which has a maximum and a minimum element but is **not** totally ordered.

    (b) Give an example of a totally ordered set which has no maximum or minimum elements.

17. Prove that in a totally ordered set, any maximal element is a maximum.

18. Suppose $(A, \preceq)$ is a poset containing a minimum element a.

    (a) [BB] Prove that a is minimal.

    (b) Prove that a is the only minimal element.

REVIEW EXERCISES FOR CHAPTER 2

1. If $A = \{x \in \mathbb{N} \mid x < 7\}$, $B = \{x \in \mathbb{Z} \mid |x - 5| < 3\}$ and $C = \{2, 3\}$, find $(A \oplus B) \setminus C$.

2. Let $A = \{x \in \mathbb{Z} \mid -1 \leq x \leq 2\}$, $B = \{2x - 3 \mid x \in A\}$ and $C = \{x \in \mathbb{R} \mid x = \frac{a}{b}, a \in A, b \in B\}$.

   (a) List the elements of A, B, and C.

   (b) List the elements of $(A \cap B) \times B$.

   (c) List the elements of $B \setminus C$.

   (d) List the elements of $A \oplus C$.

3. Let A, B, and C be sets. Are the following statements true or false? In each case, provide a proof or exhibit a counterexample.

   (a) $A \cap B = A$ if and only if $A \subseteq B$.

   (b) $(A \cap B) \cup C = A \cap (B \cup C)$.

4. Let $A = \{1\}$. Find $\mathcal{P}(\mathcal{P}(A))$.

5. Let A, B, C, and D be sets.

    (a) Give an example showing that the statement "$A \oplus (B \setminus C) = (A \oplus B) \setminus C$" is false in general.

    (b) Prove that the statement "$A \subseteq C, B \subseteq D \rightarrow A \times B \subseteq C \times D$" is true.

6. Give an example showing that the statement "$(A \times B) \subseteq (C \times D) \rightarrow A \subseteq C$ and $B \subseteq D$" is **false** is general.

7. Let A be a set.

    (a) What is meant by the term *binary relation on A*?

    (b) Suppose A has ten elements. How many binary relations are there on A?

8. For $a, b \in \mathsf{N}$, define $a \sim b$ if $a \leq 2b$. Determine whether or not $\sim$ is reflexive, symmetric, antisymmetric, transitive, an equivalence relation or a partial order on N.

9. Define a relation $\mathcal{R}$ on Z by $a\mathcal{R}b$ if $4a + b$ is a multiple of 5. Show that $\mathcal{R}$ defines an equivalence relation on Z.

10. Define a relation $\mathcal{R}$ on Z by $a\mathcal{R}b$ if $2a + 5b$ is a multiple of 7.

    (a) Prove that $\mathcal{R}$ defines an equivalence relation.

    (b) Is $\mathcal{R}$ a partial order? Explain your answer briefly.

11. Let $\sim$ denote an equivalence relation on a set A. Prove that $x \sim a \leftrightarrow \overline{x} = \overline{a}$ for any $x, a \in A$.

12. Let $\sim$ denote an equivalence relation on a set A. Assume $a, b, c, d \in A$ are such that $a \in \overline{b}, d \notin \overline{c}$ and $d \in \overline{b}$. Prove that $\overline{a} \cap \overline{c} = \emptyset$.

13. Let A be the set of points different from the origin in the Euclidean plane. For $p, q \in A$, define $p \sim q$ if $p = q$ or the line through the distinct points p and q passes through the origin.

    (a) Prove that $\sim$ defines an equivalence relation on A.

    (b) Find the equivalence classes of $\sim$.

14. Show that $(\mathcal{P}(\mathsf{Z}), \subseteq)$ is a partially ordered set.

15. Let $A = \{1, 2, 4, 6, 8\}$ and, for $a, b \in A$, define $a \preceq b$ if and only if $\frac{b}{a}$ is an integer.

    (a) Prove that $\preceq$ defines a partial order on A.

    (b) Draw the Hasse diagram for $\preceq$.

    (c) List all minimum, minimal, maximum and maximal elements.

    (d) Is $(A, \preceq)$ totally ordered? Explain.

16. Let $(A, \preceq)$ be a poset and $a, b \in A$. Can a and b have two least upper bounds? Explain.

3

Functions

3.1 Domain, Range, One-to-One, Onto

Next in importance to the primitive notion of "set" is the idea of "function." This is a term which nowadays is introduced informally quite early in the mathematics curriculum. It's often defined as a rule which associates to each element of a set A (usually a set of real numbers) an element of another set B (often R as well). The function keys on a calculator are so named because when a number x is entered into a calculator and a "function key" f is pressed, another number $f(x)$ appears.[1] The essential characteristic of a function is that the value which it associates with a given element is uniquely determined by that element. If we enter 4 into a calculator and then press some function key, we are presented with exactly one number, not a choice of several numbers: $4 \boxed{\sqrt{}} = 2$, for instance, not ± 2.

Here is the formal definition of "function."

3.1.1 DEFINITION ▶ A *function* from a set A to a set B is a binary relation f from A to B with the property that, for every $a \in A$, there is exactly one $b \in B$ such that $(a, b) \in f$. (In some areas of mathematics, *map* is a commonly employed synonym for *function*.)

In this definition, the idea that a function should associate with each element $a \in A$ a unique element $b \in B$ is captured by viewing a function f as a subset of $A \times B$ with the special property that for each $a \in A$, there is just one pair (a, b) in f having first coordinate a. If $A = \{1, 2, 3\}$, $B = \{x, y\}$ and f is

$$f = \{(1, x), (2, y), (3, x)\},$$

then f is just the rule which associates x with 1, y with 2, and x with 3.

---

[1] Here we assume that we haven't entered a number like -1 and then pressed the square root key!

There are two key points about the definition of "function." First, **every** a in A must be the first coordinate of an ordered pair in the function. Again with $A = \{1, 2, 3\}$ and $B = \{x, y\}$, the set

$$g = \{(1, x), (3, y)\}$$

is not a function from A to B because g contains no ordered pair with first coordinate 2. Second, each element of A must be the first coordinate of **exactly one** ordered pair. With the same A and B,

$$h = \{(1, x), (2, x), (3, y), (2, y)\}$$

is not a function because 2 is the first coordinate of two ordered pairs.

PROBLEM 1. Suppose A is the set of surnames of people listed in the Salt Lake City telephone directory. Is it likely that

$$f = \{(a, n) \mid a \text{ is on page } n\}$$

is a function from A to the set of natural numbers? Comment.

Solution. By definition of f, each element of A is the first coordinate of a pair in f, so the key question here is whether or not each $a \in A$ determines a unique $n \in \mathsf{N}$. Is each surname on a unique page in the telephone directory? This is not very likely since some surnames (Smith?) are undoubtedly listed on a number of different pages. It is unlikely that f is a function. ∎

When f is a function from A to B, the element $b \in B$, which is uniquely determined by the element $a \in A$, is denoted $f(a)$ and called the *image* of a. Thus,

(1)
$$\boxed{(a, b) \in f \text{ if and only if } b = f(a).}$$

3.1.2 FUNCTION NOTATION ▶

It is customary to write $f : A \to B$ to mean that f is a function from A to B and to write $f : a \mapsto b$ to mean that $f(a) = b$. (Note the differences between the symbols $\to$ between sets, as in $A \to B$, and $\mapsto$ between elements, as in $a \mapsto b$.)

Thus, the function $f = \{(1, x), (2, y), (3, x)\}$ could also be described by

$$f : \quad \begin{array}{c} 1 \mapsto x \\ 2 \mapsto y \\ 3 \mapsto x \end{array}$$

although the description as ordered pairs seems simpler.

Sometimes (often in calculus), a function is sufficiently nice that it is possible to write down a precise formula showing how $f(x)$ is determined by x; for example, $f(x) = x^3$, $f(x) = 3x - 7$, $f(x) = \ln(x)$. When we talk about "the function $f(x) = x^2$," we are really talking about that function $f : \mathsf{R} \to \mathsf{R}$ which associates with any $x \in \mathsf{R}$, its square x^2; that is to say, $x \mapsto x^2$. As a binary relation, $f = \{(x, x^2) \mid x \in \mathsf{R}\}$. When this function is graphed as usual in the xy-plane, we are, in actual fact, making a picture of the ordered pairs which comprise it.

3.1.3 DEFINITIONS ▶

Let $f : A \to B$ be a function from A to B.

- The *domain* of f, written dom f, is the set A.
- The *target* of f is the set B.
- The *range* or *image* of f, written rng f, is

$$\text{rng } f = \{b \in B \mid (a, b) \in f \text{ for some } a \in A\}$$
$$= \{b \in B \mid b = f(a) \text{ for some } a \in A\}.$$

- The function is *onto* or *surjective* if its range is the target, rng $f = B$; that is, every $b \in B$ is of the form $b = f(a)$ for some $a \in A$; equivalently,

> For any $b \in B$, the equation $b = f(x)$ has a solution $x \in A$.

- It is *one-to-one* (1–1) or *injective* if and only if different elements of A have different images: in symbols, $a_1 \neq a_2 \to f(a_1) \neq f(a_2)$; equivalently (taking the contrapositive),

> If $f(a_1) = f(a_2)$, then $a_1 = a_2$.

- It is a *bijection* or *bijective function* if it is both one-to-one and onto.

Some Discrete Examples

EXAMPLE 2   Suppose $A = \{1, 2, 3, 4\}$, $B = \{x, y, z\}$ and

$$f = \{(1, x), (2, y), (3, z), (4, y)\}.$$

Then f is a function $A \to B$ with domain A and target B. Since rng $f = \{x, y, z\} = B$, f is onto. Since $f(2) = f(4)$ $(= y)$ but $2 \neq 4$, f is not one-to-one. [In fact, there can exist no one-to-one function $A \to B$. Why not? See Exercise 25(a).]   ▲

EXAMPLE 3   Suppose $A = \{1, 2, 3\}$, $B = \{x, y, z, w\}$ and

$$f = \{(1, w), (2, y), (3, x)\}.$$

Then $f : A \to B$ is a function with domain A and range $\{w, y, x\}$. Since rng $f \neq B$, f is not onto. [No function $A \to B$ can be onto. Why not? See Exercise 25(b).] This function is one-to-one because $f(1)$, $f(2)$, and $f(3)$ are all different: If $f(a_1) = f(a_2)$, then $a_1 = a_2$.   ▲

EXAMPLE 4   Suppose $A = \{1, 2, 3\}$, $B = \{x, y, z\}$,

$$f = \{(1, z), (2, y), (3, y)\} \quad \text{and} \quad g = \{(1, z), (2, y), (3, x)\}.$$

Then f and g are functions from A to B. The domain of f is A and dom $g = A$ too. The range of f is $\{z, y\}$, which is a proper subset of B, so f is not onto. On the other hand, g is onto because rng $g = \{z, y, x\} = B$. This function is also one-to-one because $g(1)$, $g(2)$, and $g(3)$ are all different: If $g(a_1) = g(a_2)$, then $a_1 = a_2$. Notice that f is not one-to-one: $f(2) = f(3)$ $(= y)$, yet $2 \neq 3$.   ▲

EXAMPLE 5   Let $f: \mathsf{Z} \to \mathsf{Z}$ be defined by $f(x) = 2x - 3$. Then $\operatorname{dom} f = \mathsf{Z}$. To find $\operatorname{rng} f$, note that

$$b \in \operatorname{rng} f \leftrightarrow b = 2a - 3 \qquad \text{for some integer } a$$
$$\leftrightarrow b = 2(a-2) + 1 \qquad \text{for some integer } a$$

and this occurs if and only if b is odd. Thus, the range of f is the set of odd integers. Since $\operatorname{rng} f \neq \mathsf{Z}$, f is not onto. It is one-to-one, however: If $f(x_1) = f(x_2)$, then $2x_1 - 3 = 2x_2 - 3$ and $x_1 = x_2$.   ▲

EXAMPLE 6   Let $f: \mathsf{N} \to \mathsf{N}$ be defined by $f(x) = 2x - 3$. This might look like a perfectly good function, as in the last example, but actually there is a difficulty. If we try to calculate $f(1)$, we obtain $f(1) = 2(1) - 3 = -1$ and $-1 \notin \mathsf{N}$. Hence, no function has been defined.   ▲

PROBLEM 7.   Define $f: \mathsf{Z} \to \mathsf{Z}$ by $f(x) = x^2 - 5x + 5$. Determine whether or not f is one-to-one and/or onto.

Solution.   To determine whether or not f is one-to-one, we consider the possibility that $f(x_1) = f(x_2)$. In this case, $x_1^2 - 5x_1 + 5 = x_2^2 - 5x_2 + 5$, so $x_1^2 - x_2^2 = 5x_1 - 5x_2$ and $(x_1 - x_2)(x_1 + x_2) = 5(x_1 - x_2)$. This equation indeed has solutions with $x_1 \neq x_2$: Any x_1, x_2 satisfying $x_1 + x_2 = 5$ will do, for instance, $x_1 = 2$, $x_2 = 3$. Since $f(2) = f(3) = -1$, we conclude that f is not one-to-one.

   Is f onto? Recalling that the graph of $f(x) = x^2 - 5x + 5$, $x \in \mathsf{R}$, is a parabola with vertex $(\frac{5}{2}, -\frac{5}{4})$, clearly any integer less than -1 is not in the range of f. Alternatively, it is easy to see that 0 is not in the range of f because $x^2 - 5x + 5 = 0$ has no integer solutions (by the quadratic formula). Either argument shows that f is not onto.   ∎

PROBLEM 8.   Define $f: \mathsf{Z} \to \mathsf{Z}$ by $f(x) = 3x^3 - x$. Determine whether or not f is one-to-one and/or onto.

Solution.   Suppose $f(x_1) = f(x_2)$ for $x_1, x_2 \in \mathsf{Z}$. Then $3x_1^3 - x_1 = 3x_2^3 - x_2$, so $3(x_1^3 - x_2^3) = x_1 - x_2$ and

$$3(x_1 - x_2)(x_1^2 + x_1 x_2 + x_2^2) = x_1 - x_2.$$

If $x_1 \neq x_2$, we must have $x_1^2 + x_1 x_2 + x_2^2 = \frac{1}{3}$, which is impossible since x_1 and x_2 are integers. Thus, $x_1 = x_2$ and f is one-to-one.

   Is f onto? If yes, then the equation $b = f(x) = 3x^3 - x$ has a solution in Z for every integer b. This seems unlikely and, after a moment's thought, it occurs to us that the integer $b = 1$, for example, cannot be written this way: $1 = 3x^3 - x$ for some integer x implies $x(3x^2 - 1) = 1$. But the only pairs of integers whose product is 1 are the pairs $1, 1$ and $-1, -1$. So here, we would require $x = 3x^2 - 1 = 1$ or $x = 3x^2 - 1 = -1$, neither of which is possible. The integer $b = 1$ is a counterexample to the assertion that f is onto, so f is not onto.   ∎

Pause 1   Define $g: \mathsf{Z} \to \mathsf{Z}$ by $g(x) = 2x^2 + 7x$. Is g onto? Is g one-to-one?

Functions of a Real Variable

EXAMPLE 9   Let $f: \mathsf{R} \to \mathsf{R}$ be defined by $f(x) = 2x - 3$. The domain of f is R and $\operatorname{rng} f = \mathsf{R}$ since any real number y can be expressed $y = 2x - 3$ (for $x = \frac{1}{2}(y + 3)$). Graphically, this means that any horizontal line intersects the graph of $y = 2x - 3$. (See Fig 3.1.) Since $\operatorname{rng} f = \mathsf{R}$, f is onto. It is also one-to-one (the argument in Example 5 can be used again), so being onto and one-to-one, it is a bijection from R to R. ▲

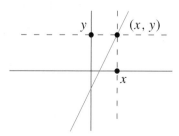

Figure 3.1   The graph of $y = 2x - 3$.

EXAMPLE 10   Let $g: \mathsf{R} \to \mathsf{R}$ be defined by $g(x) = x^2$. The domain of g is R; the range of g is the set of nonnegative real numbers. Since this is a proper subset of R, g is not onto. Neither is g one-to-one since $g(3) = g(-3)$, but $3 \neq -3$. ▲

EXAMPLE 11   Define $h: [0, \infty) \to \mathsf{R}$ by $h(x) = x^2$. This function is identical to the function g of the preceding example except for its domain. By *restricting the domain* of g to the nonnegative reals we have produced a function h which is one-to-one since $h(x_1) = h(x_2)$ implies $x_1^2 = x_2^2$ and hence $x_1 = \pm x_2$. Since $x_1 \geq 0$ and $x_2 \geq 0$, we must have $x_1 = x_2$. ▲

EXAMPLE 12   Let $f: \mathsf{R} \to \mathsf{R}$ be defined by $f(x) = 3x^3 - x$. Students of calculus should be able to plot the graph of f and see immediately that f is onto, but not one-to-one. (See Fig 3.2.) ▲

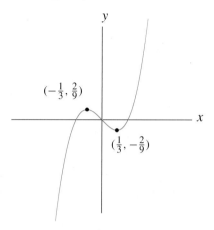

Figure 3.2   The graph of $y = 3x^3 - x$.

Contrast this last example with Problem 8 on page 74. There we saw that the function with the same rule as the one here but with Z as domain was one-to-one but not onto. A function is more than a rule; domain and range are critical too.

The absolute value of a number x, denoted $|x|$, is defined by

$$|x| = \begin{cases} x & \text{if } x \geq 0 \\ -x & \text{if } x < 0. \end{cases}$$

The absolute value function is a function with domain R and range $[0, \infty) = \{y \in R \mid y \geq 0\}$. It is not one-to-one because, for example, $|2| = |-2|$.

For any real number x, the *floor* of x, written $\lfloor x \rfloor$, is the greatest integer less than or equal to x, that is, the unique integer $\lfloor x \rfloor$ satisfying

$$x - 1 < \lfloor x \rfloor \leq x.$$

The *ceiling* of x, written $\lceil x \rceil$, is the least integer greater than or equal to x, that is, the unique integer $\lceil x \rceil$ satisfying

$$x \leq \lceil x \rceil < x + 1.$$

For instance, $\lfloor 2.01 \rfloor = 2$, $\lfloor 1.99 \rfloor = 1$, $\lfloor 15 \rfloor = 15$, $\lfloor -2.01 \rfloor = -3$, and $\lceil 2.01 \rceil = 3$, $\lceil 1.99 \rceil = 2$, $\lceil 15 \rceil = 15$, $\lceil -2.01 \rceil = -2$.

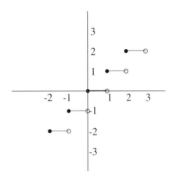

Figure 3.3   The graph of the floor function $y = \lfloor x \rfloor$.

By the *floor function*, we mean the function $f : R \to R$ defined by $f(x) = \lfloor x \rfloor$. (At one time, the floor function was commonly called the *greatest integer function*.) Similarly, the *ceiling function* is the function $g : R \to R$ defined by $g(x) = \lceil x \rceil$. The domain of these functions is R and both have range Z. The graph of the floor function is shown in Fig 3.3.

Functions whose graphs resemble the floor function are encountered frequently. In January 2001, the Canadian post office set rates for mail within Canada as shown in Table 3.1. The graph of this postal function would certainly resemble the floor function, with jumps at 30, 50, 100, and 200 gms.

The Identity Function

The special function we now define looks innocuous and it is, but it arises in so many situations that it is helpful to give it a name.

Table 3.1   2001 Canadian postal rates.

Weight w (gms)	Cost
$w \le 30$	$0.47
$30 < w \le 50$	$0.75
$50 < w \le 100$	$0.94
$100 < w \le 200$	$1.55
$200 < w \le 500$	$2.05

3.1.7 DEFINITION ▶ For any set A, the *identity function on A* is the function $\iota_A \colon A \to A$ defined by $\iota_A(a) = a$ for all $a \in A$. In terms of ordered pairs,

$$\iota_A = \{(a, a) \mid a \in A\}.$$

When there is no possibility of confusion about A, we will often write ι, rather than ι_A. (The Greek symbol ι is pronounced "yōta", so that "ι_A" is read "yota sub A."

The graph of the identity function on R is the familiar line with equation $y = x$. The identity function on a set A is indeed a function $A \to A$ since, for any $a \in A$, there is precisely one pair of the form $(a, y) \in \iota$, namely, the pair (a, a).

Pause 2 Prove that the identity function on a set A is one-to-one and onto.

Answers to Pauses

1. Since the graph of $g(x) = 2x^2 + 7x$, $x \in$ R, is a parabola, g is not onto. To determine whether or not g is one-to-one, we must see if $g(x_1) = g(x_2)$ implies that x_1 must equal x_2. Suppose $g(x_1) = g(x_2)$. Then $2x_1^2 + 7x_1 = 2x_2^2 + 7x_2$ and so $2(x_1 - x_2)(x_1 + x_2) = 7(x_2 - x_1)$. If $x_1 \ne x_2$, then $x_1 + x_2 = -\frac{7}{2}$, which is impossible for $x_1, x_2 \in$ Z. Thus, $x_1 = x_2$; g is one-to-one.
2. If $\iota(a_1) = \iota(a_2)$, then $a_1 = a_2$ (since $\iota(a_1) = a_1$ and $\iota(a_2) = a_2$), so ι is one-to-one. To show it is onto, we have to show that the equation $a = \iota(x)$ has a solution for any a. It is evident that $x = a$ works.

EXERCISES

The symbol [BB] means that an answer can be found in the Back of the Book.

1. Determine whether or not each of the following relations is a function with domain $\{1, 2, 3, 4\}$. For any relation that is not a function, explain why it isn't.

   (a) [BB] $f = \{(1, 1), (2, 1), (3, 1), (4, 1), (3, 3)\}$
   (b) $f = \{(1, 2), (2, 3), (4, 2)\}$
   (c) [BB] $f = \{(1, 1), (2, 1), (3, 1), (4, 1)\}$
   (d) $f = \{(1, 1), (1, 2), (1, 3), (1, 4)\}$
   (e) $f = \{(1, 4), (2, 3), (3, 2), (4, 1)\}$

2. Suppose A is the set of students currently registered at the University of Calgary, B is the set of professors at the University of Calgary, and C is the set of courses currently being offered at the University of Calgary. Under what conditions is each of the following a function?

(a) [BB] $\{(a, b) \mid a$ is taking a course from $b\}$

(b) $\{(a, c) \mid a$'s first class each week is in $c\}$

(c) $\{(a, c) \mid a$ has a class in c on Friday afternoon$\}$

Explain your answers.

3. [BB] Suppose A and B are nonempty sets. Can $A \times B$ ever be a function $A \to B$? Explain.

4. Give an example of a function $N \to N$ which is

(a) [BB] one-to-one but not onto;

(b) onto but not one-to-one;

(c) neither one-to-one nor onto;

(d) both one-to-one and onto.

5. Let X be the set of all countries in the British Commonwealth and Y be the set of all people who live in these countries.

(a) [BB] Show that Prime Minister: $X \to Y$ and Domicile: $Y \to X$ are functions.

(b) Show that Prime Minister is one-to-one but not onto.

(c) Show that Domicile is onto but not one-to-one.

6. [BB] Give examples of some common functions whose graphs resemble the graph of the floor function.

7. Let $S = \{1, 2, 3, 4, 5\}$ and define $f: S \to Z$ by

$$f(x) = \begin{cases} x^2 + 1 & \text{if } x \text{ is even} \\ 2x - 5 & \text{if } x \text{ is odd.} \end{cases}$$

Express f as a subset of $S \times Z$. Is f one-to-one?

8. The addition and multiplication of real numbers are functions add, mult: $R \times R \to R$, where

$$\text{add}(x, y) = x + y; \quad \text{mult}(x, y) = xy.$$

(a) [BB] Is add one-to-one? Is it onto?

(b) Is mult one-to-one? Is it onto?

Explain your answers.

9. [BB] Define $g: Z \to B$ by $g(x) = |x| + 1$. Determine (with reasons) whether or not g is one-to-one and whether or not it is onto in each of the following cases.

(a) $B = Z$

(b) $B = N$

10. Define $f: A \to A$ by $f(x) = 3x + 5$. Determine (with reasons) whether or not f is one-to-one and whether or not f is onto in each of the following cases.

(a) [BB] $A = Q$

(b) $A = N$

11. Define $h: A \to A$ by $h(x) = x^2 + 2$. Determine (with reasons) whether or not h is one-to-one and whether or not h is onto in each of the following cases.

(a) $A = Z$

(b) $A = N$

12. Define $g: A \to A$ by $g(x) = 3x^2 + 14x - 51$. Determine (with reasons) whether or not g is one-to-one and whether or not g is onto in each of the following cases.

(a) $A = Z$

(b) $A = R$

13. Define $f: A \to B$ by $f(x) = x^2 + 14x - 51$. Determine (with reasons) whether or not f is one-to-one and whether or not it is onto in each of the following cases.

(a) [BB] $A = N$, $B = \{b \in Z \mid b \geq -100\}$

(b) $A = Z$, B as in (a)

(c) $A = R$, $B = \{b \in R \mid b \geq -100\}$

14. Find the domain and range of each of the given functions of a real variable. In each case, determine whether or not the function is one-to-one and whether or not it is onto.

(a) [BB] $f: R \to R$ defined by $f(x) = x^3$.

(b) $g: R \to R$ defined by $g(x) = x|x|$.

(c) $\beta: (\frac{4}{3}, \infty) \to R$ defined by $\beta(x) = \log_2(3x - 4)$.

(d) $f: R \to R$ defined by $f(x) = 2^{x-1} + 3$.

15. (a) [BB] Define $f: R \to R$ by $f(x) = 3x^3 + x$. Graph f in order to determine whether or not f is one-to-one and/or onto.

(b) [BB] Define $f: Z \to Z$ by $f(x) = 3x^3 + x$. Determine (with reasons) whether or not f is one-to-one and/or onto.

16. (a) Define $g: R \to R$ by $g(x) = x^3 - x + 1$. Graph g in order to determine whether or not g is one-to-one and/or onto.

(b) Define $g: Z \to Z$ by $g(x) = x^3 - x + 1$. Determine (with reasons) whether or not g is one-to-one and/or onto.

(c) Repeat (b) for the function $g: N \to N$ defined by $g(x) = x^3 - x + 1$.

17. Determine whether or not each of the following defines a one-to-one and/or an onto function. Either give a proof or exhibit a counterexample to justify every answer.

(a) [BB] $f(n, m) = 2n + 3m$; $f: N \times N \to N$

(b) [BB] $f(n, m) = 2n + 3m$; $f: Z \times Z \to Z$

(c) $f(n, m) = 14n + 22m$; $f: N \times N \to N$

(d) $f(n, m) = 89n + 246m$; $f: Z \times Z \to Z$ [Hint: $1 = (-17)(246) + 47(89)$.]

(e) $f(n, m) = n^2 + m^2 + 1$; $f: Z \times Z \to N$

(f) $f(n, m) = \lfloor \frac{n}{m} \rfloor + 1$; $f: N \times N \to N$

18. For each of the following, find the largest subset A of R such that the given formula for $f(x)$ defines a function f with domain A. Give the range of f in each case.

(a) [BB] $f(x) = \dfrac{1}{x-3}$

(b) $f(x) = \dfrac{1}{\sqrt{1-x}}$

19. In each of the following cases, explain why the given function is not one-to-one. Then restrict the domain of the function to as large a set as possible so as to make a one-to-one function.

(a) $f = \{(a, \alpha), (b, \beta), (c, \gamma), (d, \alpha)\}$

(b) [BB] $f: \text{R} \to \text{R}$ defined by $f(x) = -4x^2 + 12x - 9$

(c) $f: \text{R} \to \text{R}$ defined by $f(x) = \sin x$

20. Let S be a set containing the number 5. Let $A = \{f: S \to S\}$ be the set of all functions $S \to S$. For $f, g \in A$, define $f \sim g$ if $f(5) = g(5)$.

(a) Prove that $\sim$ defines an equivalence relation on A.

(b) Find the equivalence class of $f = \{(5, a), (a, b), (b, b)\}$ in the case $S = \{5, a, b\}$.

21. Let A be a set and let $f: A \to A$ be a function. For $x, y \in A$, define $x \sim y$ if $f(x) = f(y)$.

(a) Prove that $\sim$ is an equivalence relation on A.

(b) For $A = \text{R}$ and $f(x) = \lfloor x \rfloor$, find the equivalence classes of 0, $\frac{7}{5}$, and $-\frac{3}{4}$.

(c) Suppose $A = \{1, 2, 3, 4, 5, 6\}$ and

$$f = \{(1, 2), (2, 1), (3, 1), (4, 5), (5, 6), (6, 1)\}.$$

Find all equivalence classes.

22. [BB] Let $X = \{a, b\}$ and $Y = \{1, 2, 3\}$.

(a) List all the functions from X to Y and Y to X.

(b) List all the one-to-one functions from X to Y and Y to X.

(c) List all the onto functions from X to Y and Y to X.

23. Let $X = \{a, b, c\}$ and $Y = \{1, 2, 3, 4\}$.

(a) How many one-to-one functions are there from $X \to Y$ and $Y \to X$? In each case, list all the functions.

(b) How many onto functions are there $X \to Y$ and $Y \to X$? In each case, list all the functions.

24. [BB] Given sets X and Y with $|X| = m$ and $|Y| = n$, guess a general formula for the number of functions $X \to Y$. (A table showing the possible values for each possible pair $m, n \in \{1, 2, 3, 4\}$ may be helpful.)

25. Suppose A and B are sets, A containing n elements and B containing m elements.

(a) If $n > m$, prove that no function $A \to B$ can be one-to-one.

(b) If $m > n$, prove that no function $A \to B$ can be onto.

26. (a) [BB ($\to$)] Suppose that A and B are sets each containing the same (finite) number of elements and that $f: A \to B$ is a function. Prove that f is one-to-one if and only if f is onto.

(b) Give an example of a one-to-one function $\text{N} \to \text{N}$ which is not onto. Does this contradict (a)? Explain.

(c) Give an example of an onto function $\text{N} \to \text{N}$ which is not one-to-one. Does this contradict (a)? Explain.

27. [BB] Let $f: \text{R} \to \text{R}$ be defined by $f(x) = x + \lfloor x \rfloor$.

(a) Graph this function.

(b) Find the domain and range of f.

28. Define $s: \text{R} \to \text{R}$ by $s(x) = x - \lfloor x \rfloor$. Is s one-to-one? Is it onto? Explain.

3.2 INVERSES AND COMPOSITION

The Inverse of a Function

Suppose that f is a one-to-one onto function from A to B. Given any $b \in B$, there exists $a \in A$ such that $f(a) = b$ (because f is onto) and only one such a (because f is one-to-one). Thus, for each $b \in B$, there is precisely one pair of the form $(a, b) \in f$. It follows that the set $\{(b, a) \mid (a, b) \in f\}$, obtained by reversing the ordered pairs of f, is a function from B to A (since each element of B occurs precisely once as the first coordinate of an ordered pair).

EXAMPLE 13 If $A = \{1, 2, 3, 4\}$ and $B = \{x, y, z, t\}$, then

$$f = \{(1, x), (2, y), (3, z), (4, t)\}$$

is a one-to-one onto function from A to B and, reversing its pairs, we obtain a function $B \to A$: $\{(x, 1), (y, 2), (z, 3), (t, 4)\}$. ▲

3.2.1 DEFINITION ▶ A function $f: A \to B$ *has an inverse* if and only if the set obtained by reversing the ordered pairs of f is a function $B \to A$. If $f: A \to B$ has an inverse, the function

$$\boxed{f^{-1} = \{(b, a) \mid (a, b) \in f\}}$$

is called the *inverse* of f.

We pronounce f^{-1}, "f inverse," terminology which should not be confused with $\frac{1}{f}$: f^{-1} is simply the name of a certain function, the inverse of f.[2]

If $f: A \to B$ has an inverse $f^{-1}: B \to A$, then f^{-1} also has an inverse because reversing the pairs of f^{-1} gives a function, namely f: thus, $(f^{-1})^{-1} = f$.

EXAMPLE 14 If $A = \{1, 2, 3, 4\}$ and $B = \{x, y, z, t\}$, and

$$f = \{(1, x), (2, y), (3, z), (4, t)\}$$

then

$$f^{-1} = \{(x, 1), (y, 2), (z, 3), (t, 4)\}$$

and $(f^{-1})^{-1} = \{(1, x), (2, y), (3, z), (4, t)\} = f$. ▲

At the beginning of this discussion, we saw that any one-to-one onto function has an inverse. On the other hand, suppose that a function $f: A \to B$ has an inverse $f^{-1}: B \to A$. Then f must be onto for, given $b \in B$, there is some pair $(b, a) \in f^{-1}$ (since dom $f^{-1} = B$), and so the pair (a, b) is in f. Moreover, f must be one-to-one.

Pause 3 Why?

So we have the following proposition.

3.2.2 PROPOSITION ▶ A function $f: A \to B$ has an inverse $B \to A$ if and only if f is one-to-one and onto.

For any function g, remember that $(x, y) \in g$ if and only if $g(x) = y$; in particular, $(b, a) \in f^{-1}$ if and only if $a = f^{-1}(b)$. Thus,

$$a = f^{-1}(b) \leftrightarrow (b, a) \in f^{-1} \leftrightarrow (a, b) \in f \leftrightarrow f(a) = b.$$

---

[2]Readers may wonder whether or not it is ever possible for f^{-1} and $\frac{1}{f}$ to coincide. Indeed it is, and this observation led to an article by R. Cheng, A. Dasgupta, B. R. Ebanks, L. F. Larson, and R. B. McFadden, "When Does $f^{-1} = \frac{1}{f}$?," *American Mathematical Monthly* **105** (1998), no. 8, 704–717.

The equivalence of the first and last equations here is very important:

(2)
$$a = f^{-1}(b) \text{ if and only if } f(a) = b.$$

For example, if for some function f, $\pi = f^{-1}(-7)$, then we can conclude that $f(\pi) = -7$. If $f(4) = 2$, then $4 = f^{-1}(2)$.

The solution to the equation $2x = 5$ is $x = \frac{5}{2} = 2^{-1} \cdot 5$. Generally, to solve the equation $ax = b$, we ask if $a \neq 0$, and if this is the case, we multiply each side of the equation by a^{-1}, obtaining $x = a^{-1}b = \frac{b}{a}$. Since all real numbers except 0 have a multiplicative inverse, checking that $a \neq 0$ is just checking that a has an inverse.

Look again at statement (2). We solve the equation $f(x) = y$ for x in the same way we solve $ax = b$ for x. We first ask if f has an inverse, and if it does, apply f^{-1} to each side of the equation, obtaining $x = f^{-1}(y)$.

The "application" of f^{-1} to each side of the equation $y = f(x)$ is very much like multiplying each side by f^{-1}. "Multiplying by f^{-1}" may sound foolish, but there is a context (called *group theory*) in which it makes good sense. Our intent here is just to provide a good way to remember the fundamental relationship expressed in (2).

EXAMPLE 15 If $f: \mathsf{R} \to \mathsf{R}$ is defined by $f(x) = 2x - 3$, then f is one-to-one and onto, so an inverse function exists. According to (2), if $y = f^{-1}(x)$, then $x = f(y) = 2y - 3$. Thus, $y = \frac{1}{2}(x + 3) = f^{-1}(x)$. ▲

EXAMPLE 16 Let $A = \{x \in \mathsf{R} \mid x \leq 0\}$, $B = \{x \in \mathsf{R} \mid x \geq 0\}$ and define $f: A \to B$ by $f(x) = x^2$. This is just the squaring function with domain restricted so that it is one-to-one as well as onto. Since f is one-to-one and onto, it has an inverse. To obtain $f^{-1}(x)$, let $y = f^{-1}(x)$, deduce [by the relationship expressed in (2)] that $f(y) = x$ and so $y^2 = x$. Solving for y, we get $y = \pm\sqrt{x}$. Since $x = f(y)$, $y \in A$, so $y \leq 0$. Thus, $y = -\sqrt{x}$; $f^{-1}(x) = -\sqrt{x}$. ▲

EXAMPLE 17 Denoting the positive real numbers by R^+, let $f: \mathsf{R} \to \mathsf{R}^+$ be defined by $f(x) = 3^x$. Since f is one-to-one and onto, it has an inverse. To find $f^{-1}(x)$, let $y = f^{-1}(x)$ and write down its equivalent form, $x = f(y) = 3^y$. Solving for y, we get $y = \log_3 x$. So $f^{-1}(x) = \log_3 x$. ▲

PROBLEM 18. Let $A = \{x \mid x \neq \frac{1}{2}\}$ and define $f: A \to \mathsf{R}$ by $f(x) = \dfrac{4x}{2x - 1}$.

Is f one-to-one? Find rng f. Explain why $f: A \to$ rng f has an inverse. Find dom f^{-1}, rng f^{-1}, and a formula for $f^{-1}(x)$.

Solution. Suppose $f(a_1) = f(a_2)$. Then $\dfrac{4a_1}{2a_1 - 1} = \dfrac{4a_2}{2a_2 - 1}$, so $8a_1a_2 - 4a_1 = 8a_1a_2 - 4a_2$, hence $a_1 = a_2$. Thus f is one-to-one.

Next,

$$y \in \text{rng } f \leftrightarrow y = f(x) \quad \text{for some } x \in A$$

$$\leftrightarrow \text{there is an } x \in A \text{ such that } y = \frac{4x}{2x-1}$$

$$\leftrightarrow \text{there is an } x \in A \text{ such that } 2xy - y = 4x$$

$$\leftrightarrow \text{there is an } x \in A \text{ such that } x(2y-4) = y.$$

If $y = 2$, the equation $x(2y-4) = y$ becomes $0 = 2$ and no x exists. On the other hand, if $y \neq 2$, then $2y-4 \neq 0$ and so, dividing by $2y-4$, we obtain $x = \frac{y}{2y-4}$. (It is easy to see that such x is never $\frac{1}{2}$; that is, $x \in A$.) Thus $y \in \text{rng } f$ if and only if $y \neq 2$. So $\text{rng } f = B = \{y \in \mathbb{R} \mid y \neq 2\}$.

Since $f \colon A \to B$ is one-to-one and onto, it has an inverse $f^{-1} \colon B \to A$. Also, $\text{dom } f^{-1} = \text{rng } f = B$ and $\text{rng } f^{-1} = \text{dom } f = A$. To find $f^{-1}(x)$, set $y = f^{-1}(x)$. Then

$$x = f(y) = \frac{4y}{2y-1}$$

and, solving for y, we get $y = \frac{x}{2x-4} = f^{-1}(x)$. ∎

The following important property of f^{-1} will be used in Section 3.3.

3.2.3 PROPOSITION ▶ If $f \colon A \to B$ is one-to-one and onto, then $f^{-1} \colon B \to A$ is also one-to-one and onto.

Pause 4 Prove Proposition 3.2.3.

Composition of Functions

3.2.4 DEFINITION ▶ If $f \colon A \to B$ and $g \colon B \to C$ are functions, then the *composition of g and f* is the function $g \circ f \colon A \to C$ defined by $(g \circ f)(a) = g(f(a))$ for all $a \in A$.

EXAMPLE 19 If $A = \{a, b, c\}$, $B = \{x, y\}$, and $C = \{u, v, w\}$, and if $f \colon A \to B$ and $g \colon B \to C$ are the functions

$$f = \{(a, x), (b, y), (c, x)\}, \quad g = \{(x, u), (y, w)\},$$

then

$$(g \circ f)(a) = g(f(a)) = g(x) = u,$$

$$(g \circ f)(b) = g(f(b)) = g(y) = w,$$

$$(g \circ f)(c) = g(f(c)) = g(x) = u$$

and so $g \circ f = \{(a, u), (b, w), (c, u)\}$. ▲

It is seldom the case that $g \circ f = f \circ g$. In this last example, for instance, $f \circ g$ is not even defined. We must be careful then to distinguish between "the composition of g and f" and "the composition of f and g."

EXAMPLE 20 If f and g are the functions $\mathsf{R} \to \mathsf{R}$ defined by

$$f(x) = 2x - 3, \quad g(x) = x^2 + 1,$$

then both $g \circ f$ and $f \circ g$ are defined and we have

$$(g \circ f)(x) = g(f(x)) = g(2x - 3) = (2x - 3)^2 + 1$$

and

$$(f \circ g)(x) = f(g(x)) = f(x^2 + 1) = 2(x^2 + 1) - 3. \quad \blacktriangle$$

EXAMPLE 21 In the definition of $g \circ f$, it is required that rng $f \subseteq B = \operatorname{dom} g$. If $f: \mathsf{R} \to \mathsf{R}$ and $g: \mathsf{R} \setminus \{1\} \to \mathsf{R}$ are the functions defined by

$$f(x) = 2x - 3 \quad \text{and} \quad g(x) = \frac{x}{x - 1},$$

then $g \circ f$ is not defined because rng $f = \mathsf{R} \nsubseteq \operatorname{dom} g$. On the other hand, $f \circ g$ is defined and

$$(f \circ g)(x) = 2\left(\frac{x}{x - 1}\right) - 3. \quad \blacktriangle$$

EXAMPLE 22 If $f: A \to A$ is any function and ι denotes the identity function on A, then $f \circ \iota$ is a function $A \to A$. Also, for any $a \in A$, $(f \circ \iota)(a) = f(\iota(a)) = f(a)$. Similarly, $(\iota \circ f)(a) = f(a)$ for any $a \in A$. Thus, $f \circ \iota = f = \iota \circ f$. It is for just this reason that ι is called the identity function on A; it behaves with respect to $\circ$ the way the number 1 behaves with respect to multiplication. $\quad \blacktriangle$

Under what conditions are two functions f and g equal? Since a function is a set and two sets are equal if and only if they contain the same elements, it must be that $f = g$ if and only if it is the case that

$$(a, b) \in f \text{ if and only if } (a, b) \in g.$$

Suppose $f = g$ and $a \in \operatorname{dom} f$. Then $(a, b) \in f$ for some $b = f(a)$, so $(a, b) \in g$. Thus, $a \in \operatorname{dom} g$ and $b = g(a)$. This shows that $\operatorname{dom} f \subseteq \operatorname{dom} g$ and, for every $a \in \operatorname{dom} f$, $f(a) = g(a)$. Reversing the roles of f and g, we see similarly that $\operatorname{dom} g \subseteq \operatorname{dom} f$ and $g(a) = f(a)$ for every $a \in \operatorname{dom} g$. We are led to the following necessary and sufficient condition for the equality of functions.

3.2.5 EQUALITY OF FUNCTIONS ▶

Functions f and g are equal if and only if they have the same domain, same target, and $f(a) = g(a)$ for every a in the common domain.

Notice how we proved that $f \circ \iota = f$ in Example 22. We noted that each of the functions $f \circ \iota$ and f has domain A and target A and proved that $(f \circ \iota)(a) = f(a)$ for all $a \in A$.

Pause 5 Suppose $f: A \to B$ is a function. Explain why $\iota_B \circ f = f$.

One of the fundamental properties of composition of functions is the content of the next proposition.

3.2.6 PROPOSITION ▶ Composition of functions is an associative operation.

Proof We must prove that $(f \circ g) \circ h = f \circ (g \circ h)$ whenever each of the two functions—$(f \circ g) \circ h$ and $f \circ (g \circ h)$—is defined. Thus, we assume that for certain sets A, B, C, and D, h is a function $A \to B$, g is a function $B \to C$, and f is a function $C \to D$. A direct proof is suggested.

Since the domain of $(f \circ g) \circ h$ is the domain of $f \circ (g \circ h)$ (namely, the set A), we have only to prove that $((f \circ g) \circ h)(a) = (f \circ (g \circ h))(a)$ for any $a \in A$. For this, we have

$$((f \circ g) \circ h)(a) = (f \circ g)(h(a)) = f(g(h(a)))$$

and

$$(f \circ (g \circ h))(a) = f((g \circ h)(a)) = f(g(h(a)))$$

as desired. ∎

If $f: A \to B$ has an inverse $f^{-1}: B \to A$, then, recalling (2),

$$f^{-1}(b) = a \text{ if and only if } b = f(a).$$

So for any $a \in A$,

$$a = f^{-1}(b) = f^{-1}(f(a)) = f^{-1} \circ f(a).$$

In other words, the composition $f^{-1} \circ f = \iota_A$, the identity function on A. Similarly, for any element $b \in B$,

$$b = f(a) = f(f^{-1}(b)) = f \circ f^{-1}(b).$$

Thus, the composition $f \circ f^{-1} = \iota_B$ is the identity function on B. We summarize.

3.2.7 PROPOSITION ▶ Functions $f: A \to B$ and $g: B \to A$ are inverses if and only if $g \circ f = \iota_A$ and $f \circ g = \iota_B$; that is, if and only if

$$g(f(a)) = a \text{ and } f(g(b)) = b \quad \text{for all } a \in A \text{ and all } b \in B.$$

PROBLEM 23. Show that the functions $f: \mathsf{R} \to (1, \infty)$ and $g: (1, \infty) \to \mathsf{R}$ defined by

$$f(x) = 3^{2x} + 1, \qquad g(x) = \tfrac{1}{2}\log_3(x - 1)$$

are inverses.

Solution. For any $x \in \mathsf{R}$,

$$(g \circ f)(x) = g(f(x)) = g(3^{2x} + 1)$$

$$= \tfrac{1}{2}(\log_3[(3^{2x} + 1) - 1])$$

$$= \tfrac{1}{2}(\log_3 3^{2x}) = \tfrac{1}{2}2x = x$$

and for any $x \in (1, \infty)$,

$$(f \circ g)(x) = f(g(x)) = f(\tfrac{1}{2} \log_3(x - 1))$$

$$= 3^{2(\frac{1}{2} \log_3(x-1))} + 1$$

$$= 3^{\log_3(x-1)} + 1 = (x - 1) + 1 = x.$$

By Proposition 3.2.7, f and g are inverses. ∎

We have noted the similarity in the solutions to the equations $ax = b$ and $f(x) = y$. If a (or f) has an inverse, we multiply by this inverse and obtain $x = a^{-1}b$ (or $x = f^{-1}(y)$). It is interesting also to observe that the connection between a and its inverse a^{-1} (their product is 1) is strikingly similar to the connection between f and its inverse f^{-1} (their composition is the identity function).

Answers to Pauses

3. Suppose $f(a_1) = f(a_2)$ for some $a_1, a_2 \in A$. Let $b = f(a_1) \ (= f(a_2))$. Then (a_1, b) and (a_2, b) are both in f, so (b, a_1) and (b, a_2) are both in f^{-1}. But f^{-1} is a function and so each element of B occurs just once as the first coordinate of an ordered pair in f^{-1}. It follows that $a_1 = a_2$; f is one-to-one.

4. To show that f^{-1} is one-to-one, assume $f^{-1}(b_1) = f^{-1}(b_2)$. Setting $a_1 = f^{-1}(b_1)$ and $a_2 = f^{-1}(b_2)$, we have $a_1 = a_2$. Therefore, $b_1 = f(a_1) = f(a_2) = b_2$. To show that f^{-1} is onto, let $a \in A$. Then $f(a) = b \in B$, so $a = f^{-1}(b)$ as desired.

5. Since $f: A \to B$ and $\iota_B: B \to B$, the composition $\iota_B \circ f$ is a function $A \to B$. Thus, $\operatorname{dom} \iota_B \circ f = A = \operatorname{dom} f$ and target $f = $ target $\iota_B \circ f = B$. Also, for any $a \in A$, $\iota_B \circ f(a) = \iota_B(f(a)) = f(a)$, because $f(a) \in B$ and $\iota_B(b) = b$ for all $b \in B$. Thus, the functions $\iota_B \circ f$ and f are equal.

EXERCISES

The symbol [BB] means that an answer can be found in the Back of the Book.

1. [BB] Let $A = \{1, 2, 3, 4, 5\}$. Find the inverse of each of the following functions $f: A \to A$.

   (a) $f = \{(1, 2), (2, 3), (3, 4), (4, 5), (5, 1)\}$
   (b) $f = \{(1, 2), (2, 4), (3, 3), (4, 1), (5, 5)\}$

2. Graph each of the following functions and find each inverse. Specify the domain and range of each inverse.

   (a) [BB] $f: \mathsf{R} \to \mathsf{R}$ given by $f(x) = 3x + 5$
   (b) $f: \mathsf{R} \to \mathsf{R}$ given by $f(x) = x^3 - 2$
   (c) $\beta: (\tfrac{4}{3}, \infty) \to \mathsf{R}$ given by $\beta(x) = \log_2(3x - 4)$
   (d) $g: \mathsf{R} \to \mathsf{R}$ given by $g(x) = x|x|$

3. Show that each of the following functions $f: A \to \mathsf{R}$ is one-to-one. Find the range of each function and a suitable inverse.

   (a) [BB] $A = \{x \in \mathsf{R} \mid x \neq 4\}$, $f(x) = 1 + \dfrac{1}{x - 4}$

   (b) $A = \{x \in \mathsf{R} \mid x \neq -1\}$, $f(x) = 5 - \dfrac{1}{1 + x}$

   (c) $A = \{x \in \mathsf{R} \mid x \neq -\tfrac{1}{2}\}$, $f(x) = \dfrac{3x}{2x + 1}$

   (d) $A = \{x \in \mathsf{R} \mid x \neq -3\}$, $f(x) = \dfrac{x - 3}{x + 3}$

4. Define $f: \mathsf{Z} \to \mathsf{N}$ by $f(x) = \begin{cases} 2|x| & \text{if } x < 0 \\ 2x + 1 & \text{if } x \geq 0. \end{cases}$

   (a) Show that f has an inverse.
   (b) Find $f^{-1}(2586)$.

5. Suppose A is the set of all married people, mother: $A \to A$ is the function which assigns to each married person his/her mother, and father and spouse have similar meanings. Give sensible interpretations of each of the following:

   (a) [BB] mother $\circ$ mother
   (b) mother $\circ$ father
   (c) father $\circ$ mother
   (d) mother $\circ$ spouse
   (e) spouse $\circ$ mother
   (f) father $\circ$ spouse
   (g) spouse $\circ$ spouse
   (h) (spouse $\circ$ father) $\circ$ mother
   (i) spouse $\circ$ (father $\circ$ mother)

6. Let $S = \{1, 2, 3, 4, 5\}$ and $T = \{1, 2, 3, 8, 9\}$ and define functions $f: S \to T$ and $g: S \to S$ by

   $$f = \{(1, 8), (3, 9), (4, 3), (2, 1), (5, 2)\} \text{ and}$$

   $$g = \{(1, 2), (3, 1), (2, 2), (4, 3), (5, 2)\}.$$

   (a) [BB] Find $f \circ g$ or explain why $f \circ g$ is not defined. Repeat for $g \circ f$, $f \circ f$, and $g \circ g$.
   (b) Which of f, g are one-to-one? Which are onto? Explain.
   (c) Find f^{-1} if it exists. If it doesn't, explain why not.
   (d) Find g^{-1} if it exists. If it doesn't, explain why not.

7. Let $S = \{1, 2, 3, 4\}$ and define functions $f, g: S \to S$ by

   $$f = \{(1, 3), (2, 2), (3, 4), (4, 1)\} \text{ and}$$

   $$g = \{(1, 4), (2, 3), (3, 1), (4, 2)\}.$$

   Find

   (a) [BB] $g^{-1} \circ f \circ g$
   (b) $f \circ g^{-1} \circ g$
   (c) $g \circ f \circ g^{-1}$
   (d) $g \circ g^{-1} \circ f$
   (e) $f^{-1} \circ g^{-1} \circ f \circ g$

8. [BB] Let f, g and $h: R \to R$ be defined by

   $$f(x) = x + 2, \quad g(x) = \frac{1}{x^2 + 1}, \quad h(x) = 3.$$

   Compute $g \circ f(x)$, $f \circ g(x)$, $h \circ g \circ f(x)$, $g \circ h \circ f(x)$, $g \circ f^{-1} \circ f(x)$, and $f^{-1} \circ g \circ f(x)$.

9. Let A be the set $A = \{x \in R \mid x > 0\}$ and define $f, g, h: A \to R$ by

   $$f(x) = \frac{x}{x + 1}, \quad g(x) = \frac{1}{x}, \quad h(x) = x + 1.$$

   Find $g \circ f(x)$, $f \circ g(x)$, $h \circ g \circ f(x)$, and $f \circ g \circ h(x)$.

10. [BB] Let $f: R \to R$ be a function, let c be a real number, and define $g: R \to R$ by $g(x) = x - c$. Explain how the graphs of f and $g \circ f$ are related.

11. Let $f: R \to R$ be a function, let c be a real number, and define $g: R \to R$ by $g(x) = x - c$. Explain how the graphs of f and $f \circ g$ are related.

12. Let $f: R \to R$ be a function and let $g: R \to R$ be defined by $g(x) = -|x|$.

    (a) [BB] How are the graphs of f and $f \circ g$ related?
    (b) How are the graphs of f and $g \circ f$ related?

13. Let A denote the set $R \setminus \{0, 1\}$. Let ι denote the identity function on A and define the functions $f, g, h, s, r: A \to A$ by

    $$f(x) = 1 - \frac{1}{x}, \quad g(x) = \frac{1}{1 - x}, \quad h(x) = \frac{1}{x},$$

    $$r(x) = \frac{x}{x - 1}, \quad s(x) = 1 - x.$$

    (a) Show that $f \circ g = \iota$ and $g \circ r = s$. Complete the table, thereby showing that the composition of any two of the given functions is one of the given five, or the identity.[3]

$\circ$	ι	f	g	h	r	s
ι						
f			ι			
g					s	
h						
r						
s						

    (b) Which of the given six functions have inverses? Find (and identify) any inverses which exist.

14. Let $A = \{1, 2, 3\}$ and define $f_1, f_2, f_3, f_4, f_5: A \to A$ as follows:

    $$f_1 = \{(1, 1), (2, 3), (3, 2)\}$$

    $$f_2 = \{(1, 3), (2, 2), (3, 1)\}$$

    $$f_3 = \{(1, 2), (2, 1), (3, 3)\}$$

    $$f_4 = \{(1, 2), (2, 3), (3, 1)\}$$

    $$f_5 = \{(1, 3), (2, 1), (3, 2)\}.$$

---

[3]The table you will construct in this exercise is the multiplication table for an important mathematical object known as a *group*. This particular group is the smallest one which is not commutative.

(a) Show that each composite function $f_i \circ f_j$ is one of the given functions, or the identity, by completing a table like the following one.[4] For example, if $f_2 \circ f_3 = f_4$, the entry in row f_2, column f_3, should be f_4.

$\circ$	ι	f_1	f_2	f_3	f_4	f_5
ι						
f_1						
f_2						
f_3						
f_4						
f_5						

(b) Find the inverses of those of the six functions in (a) which have inverses.

15. Let $S = \{1, 2, 3, 4, 5\}$ and let $f, g, h: S \to S$ be the functions defined by

$$f = \{(1, 2), (2, 1), (3, 4), (4, 5), (5, 3)\}$$

$$g = \{(1, 3), (2, 5), (3, 1), (4, 2), (5, 4)\}$$

$$h = \{(1, 2), (2, 2), (3, 4), (4, 3), (5, 1)\}.$$

(a) [BB] Find $f \circ g$ and $g \circ f$. Are these functions equal?

(b) Explain why f and g have inverses but h does not. Find f^{-1} and g^{-1}.

(c) Show that $(f \circ g)^{-1} = g^{-1} \circ f^{-1} \neq f^{-1} \circ g^{-1}$.

16. [BB] Let $A = \{x \in \mathsf{R} \mid x \neq 2\}$ and $B = \{x \in \mathsf{R} \mid x \neq 1\}$. Define $f: A \to B$ and $g: B \to A$ by

$$f(x) = \frac{x}{x - 2}, \qquad g(x) = \frac{2x}{x - 1}.$$

(a) Find $(f \circ g)(x)$.

(b) Are f and g inverses? Explain.

17. Let A be a subset of R and suppose $f: A \to A$ is a function with the property that $f^{-1}(x) = \dfrac{1}{f(x)}$ for all $x \in A$.

(a) Show that $0 \notin A$.

(b) Show that $f^4 = \iota_A$, the identity function on A. (By f^4, we mean $f \circ f \circ f \circ f$.)

18. Suppose $f: A \to B$ and $g: B \to C$ are functions.

(a) [BB] If $g \circ f$ is one-to-one and f is onto, show that g is one-to-one.

(b) If $g \circ f$ is onto and g is one-to-one, show that f is onto.

19. (a) [BB] Prove that the composition of one-to-one functions is a one-to-one function.

(b) Show, by an example, that the converse of (a) is not true.

(c) Show that if $g \circ f$ is one-to-one, then f must be one-to-one.

20. (a) Prove that the composition of onto functions is onto.

(b) [BB] Show, by an example, that the converse of (a) is not true.

(c) Show that if $g \circ f$ is onto, then g must be onto.

21. [BB] Is the composition of two bijective functions bijective? Explain.

22. Define $f: \mathsf{Z} \to \mathsf{Z}$ by

$$f(n) = \begin{cases} n - 2 & n \geq 1000 \\ f(f(n + 4)) & n < 1000. \end{cases}$$

(a) Find the values of $f(1000)$, $f(999)$, $f(998)$, $f(997)$, and $f(996)$.

(b) Guess a formula for $f(n)$.

(c) Guess the range of f.

23. Show that the function $f: \mathsf{R} \to \mathsf{R}$ defined by $f(x) = \dfrac{x}{\sqrt{x^2 + 2}}$ is one-to-one.

Find rng f and a suitable inverse.

24. Let $t: \mathsf{R} \to \mathsf{R}$ be the function defined by $t(x) = 2\lfloor x \rfloor - x$.

(a) Graph t and use this to decide whether or not t has an inverse.

(b) [BB] Prove that t is one-to-one without using the graph of t.

(c) Prove that t is onto without using the graph of t.

(d) Find a formula for $t^{-1}(x)$.

3.3 ONE-TO-ONE CORRESPONDENCE AND THE CARDINALITY OF A SET

In this section we think about the size of a set, or, more correctly, its "cardinality." We show that our natural instincts about the relative sizes of finite sets can be

---

[4]As with Exercise 13, this table also describes the multiplication for the smallest group which is not commutative.

extended to allow the comparison of infinite sets too, though sometimes with surprising results. It turns out, for example, that the natural numbers and the set of all rational numbers have the same "size," whereas both these sets are "smaller" than the set of real numbers.

In discussions about cardinality, it is common to employ the term *one-to-one correspondence* rather than its synonym, *one-to-one onto function*. Either of the statements

"*A* and *B* are in one-to-one correspondence" or

"There is a one-to-one correspondence between *A* and *B*"

means that there is a one-to-one onto function *f* from *A* to *B*. Recall from Section 3.2 that this implies there is also a one-to-one onto function from *B* to *A*, namely, f^{-1}.

3.3.1   DEFINITIONS ▶   A *finite* set is a set which is either empty or in one-to-one correspondence with the set $\{1, 2, 3, \ldots, n\}$ of the first *n* natural numbers, for some $n \in \mathsf{N}$. A set which is not finite is called *infinite*.

If *A* is a finite set and $A \neq \emptyset$, then, for some natural number *n*, there exists a one-to-one onto function $f : \{1, 2, 3, \ldots, n\} \to A$. Letting $f(i) = a_i$, this means that $A = \{a_1, a_2, \ldots, a_n\}$.

If *A* is a finite set, the *cardinality* of *A* is the number of elements in *A*; this is denoted $|A|$. Thus, $|\emptyset| = 0$ and, if $A = \{a_1, a_2, \ldots, a_n\}$, then $|A| = n$.

EXAMPLES 24
- $|\{a, b, x\}| = 3$,
- the letters of the English alphabet comprise a set of cardinality 26,
- $|\{x \in \mathsf{R} \mid x^2 + 1 = 0\}| = 0$. ▲

How might we determine if two finite sets contain the same number of elements? We could count the elements in each set, but if the sets were large, this method would be slow and highly unreliable. Confronted with two enormous pails of jelly beans and asked to determine if the number of jelly beans in each pail is the same, the best strategy would be to pair the beans in the two pails. Remove one jelly bean from each pail and lay the two aside; remove another jelly bean from each pail, lay these aside, and so on. If the last jelly beans in each pail are removed together, then certainly we would know that the numbers in each pail were the same.

If the numbers are the same, the set of pairs of beans removed from the pails defines a one-to-one onto function from the jelly beans in one pail to those in the other. Different beans in the first pail are paired with different beans in the second (one-to-one) and every bean in the second pail is paired with some bean in the first (onto).

If two finite sets *A* and *B* have the same cardinality—we write $|A| = |B|$— then there is a one-to-one onto function from *A* to *B*. Conversely, if there is a one-to-one onto function from *A* to *B*, then $|A| = |B|$. This idea allows us to extend the notion of "same size".

3.3.2 DEFINITION ▶ Sets A and B have the *same cardinality* and we write $|A| = |B|$, if and only if there is a *one-to-one correspondence* between them; that is, if and only if there exists a one-to-one onto function from A to B (or from B to A).

EXAMPLES 25
- $a \mapsto x$, $b \mapsto y$ is a one-to-one correspondence between $\{a, b\}$ and $\{x, y\}$; hence, $|\{a, b\}| = |\{x, y\}| (= 2)$.
- The function $f : \mathsf{N} \longrightarrow \mathsf{N} \cup \{0\}$ defined by $f(n) = n - 1$ is a one-to-one correspondence between N and $\mathsf{N} \cup \{0\}$; so $|\mathsf{N}| = |\mathsf{N} \cup \{0\}|$.
- The function $f : \mathsf{Z} \longrightarrow 2\mathsf{Z}$ defined by $f(n) = 2n$ is a one-to-one correspondence between the set Z of integers and the set $2\mathsf{Z}$ of even integers; thus, Z and $2\mathsf{Z}$ have the same cardinality. ▲

Many people find the second and third examples here surprising and perhaps mildly disconcerting; nevertheless, they must be accepted. Remember that our definition of "same cardinality" coincides with what we know to be the case with finite sets.

We can prove that $\{a, b\}$ and $\{x, y\}$ have the same cardinality (without using the word "two") by pairing the elements in the two sets

$$a \mapsto x, \quad b \mapsto y$$

and this is exactly how we argue, for instance, that Z and $2\mathsf{Z}$ have the same cardinality. We pair 0 with 0, 1 with 2, -1 with -2, 2 with 4, -2 with -4, n with $2n$.

Many readers of this book will be familiar with the graph of the function $\mathsf{R} \to \mathsf{R}$ defined by $f(x) = 2^x$. This is a one-to-one function and its range is the set R^+ of positive reals. We conclude that the real numbers and the positive real numbers have the same cardinality. In fact, the real numbers have the same cardinality as any interval, for example, the interval $(0, 1)$. (Exercise 12(c) shows that any open interval has the same cardinality as $(0, 1)$.)

PROBLEM 26. Show that the set R^+ of positive real numbers has the same cardinality as the open interval $(0, 1) = \{x \in \mathsf{R} \mid 0 < x < 1\}$.

Solution. Let $f : (0, 1) \to \mathsf{R}^+$ be defined by

$$f(x) = \frac{1}{x} - 1.$$

We claim that f establishes a one-to-one correspondence between $(0, 1)$ and R^+.

To show that f is onto, we have to show that any $y \in \mathsf{R}^+$ is $f(x)$ for some $x \in (0, 1)$. But

$$y = \frac{1}{x} - 1 \text{ implies } x = \frac{1}{1 + y}$$

which is in $(0, 1)$ since $y > 0$. Therefore,

$$y \in \mathsf{R}^+ \text{ implies } y = f\left(\frac{1}{1 + y}\right)$$

so f is indeed onto. Also, f is one-to-one because

$$f(x_1) = f(x_2) \rightarrow \frac{1}{x_1} - 1 = \frac{1}{x_2} - 1$$

$$\rightarrow \frac{1}{x_1} = \frac{1}{x_2}$$

$$\rightarrow x_1 = x_2. \qquad \blacksquare$$

Just as the notion of cardinality partitions finite sets into classes (each class consisting of sets with the same number of elements), so cardinality also partitions infinite sets. In this book, we consider just two classes of infinite sets.

3.3.3 DEFINITIONS ▶ A set A is *countably infinite* if and only if $|A| = |\mathbf{N}|$ and *countable* if and only if it is either finite or countably infinite. A set which is not countable is *uncountable*.

The symbol $\aleph_0$ (pronounced "aleph naught") has traditionally been used to denote the cardinality of the natural numbers. Thus, a countably infinite set has cardinality $\aleph_0$. As the name suggests, **countably** infinite sets are those whose elements can be listed in a systematic and definite way, because to list them is to rank them as first, second, third, and so on, and this ranking establishes the required one-to-one correspondence with $\mathbf{N}$.

PROBLEM 27. Show that $|\mathbf{Z}| = \aleph_0$.

Solution. The set of integers is infinite. To show they are countably infinite, we list them: $0, 1, -1, 2, -2, 3, -3, \ldots$. This list is just $f(1), f(2), f(3), \ldots$ where $f: \mathbf{N} \rightarrow \mathbf{Z}$ is defined by

$$f(n) = \begin{cases} \frac{1}{2}n & \text{if } n \text{ is even} \\ -\frac{1}{2}(n-1) & \text{if } n \text{ is odd,} \end{cases}$$

which is certainly both one-to-one and onto. $\qquad \blacksquare$

Pause 6 Can you express this function f with a single equation?

When we say list "in a systematic and definite way" we mean that when one person terminates the list (with three dots), another should be able to continue. In a good list, it should be possible to determine the position of any element. It is not acceptable to "list" the integers as $0, 1, 2, 3, \ldots, -1, -2, -3, \ldots$ because -1 does not have a position: 0 is the first number, 1 is the second, but in what position is -1? Similarly, it is not acceptable to "list" the integers as $\ldots, -3, -2, -1, 0, 1, 2, 3, \ldots$ because no integer has a position. What's the first element in the list? In fact, this is not a list at all. On the other hand, listing the integers this way—$0, 1, -1, 2, -2, 3, -3, \ldots$—assigns every integer a definite position.

Pause 7 In the list $0, 1, -1, 2, -2, 3, -3, \ldots$, -1 is the third number and 3 is the sixth. In what position is 1003? What number is in position 1003?

The following argument establishes the remarkable fact that the set $N \times N = \{(m, n) \mid m, n \in N\}$ is countable; it has the same cardinality as N.

PROBLEM 28. Show that $|N \times N| = |N|$.

Solution. The elements of $N \times N$ can be listed by the scheme illustrated in Fig 3.4. The arrows indicate the order in which the elements of $N \times N$ should be listed—$(1, 1), (2, 1), (1, 2), (1, 3), (2, 2), \ldots$. Wherever the arrows terminate, there is no difficulty in continuing, so each ordered pair acquires a definite position. ∎

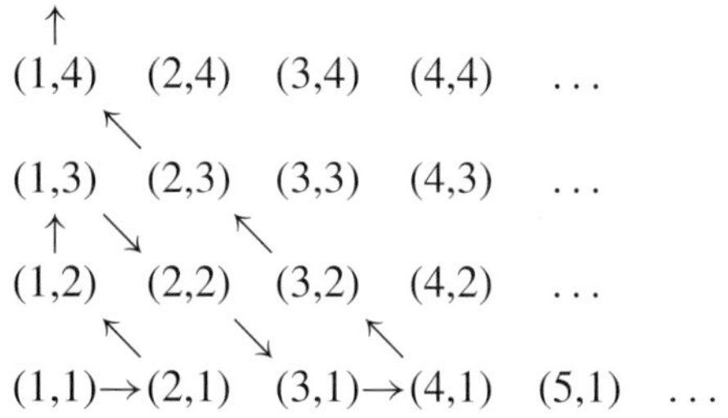

$\uparrow$
$(1,4) \quad (2,4) \quad (3,4) \quad (4,4) \quad \ldots$
$\nwarrow$
$(1,3) \quad (2,3) \quad (3,3) \quad (4,3) \quad \ldots$
$\uparrow \searrow \quad \nwarrow$
$(1,2) \quad (2,2) \quad (3,2) \quad (4,2) \quad \ldots$
$\nwarrow \quad \searrow \quad \nwarrow$
$(1,1) \rightarrow (2,1) \quad (3,1) \rightarrow (4,1) \quad (5,1) \quad \ldots$

Figure 3.4   The elements of $N \times N$ can be systematically listed by starting at $(1, 1)$ and following the arrows.

In Fig 3.4 suppose we write the positive rational $\frac{m}{n}$ at the point (m, n). Then every rational number appears in the picture, actually many times over, since, for instance, $\frac{2}{3}$ appears beside the points $(2, 3), (4, 6), (6, 9), (40, 60)$, and so on. By ignoring the repetitions of a given fraction, the set of positive rationals can be enumerated. We conclude that the set of positive rationals is a countable set.

Pause 8 The scheme which we have just described for listing the positive rationals begins $1, 2, \frac{1}{2}, \frac{1}{3}$ (ignore the number $\frac{2}{2}$, which has already been listed), $3, 4, \frac{3}{2}, \frac{2}{3}, \frac{1}{4}$. Continue the list in order to determine the position of $\frac{3}{4}$.

Not so long ago, there came to light another way to establish a one-to-one correspondence between the positive rationals and N. For a very neat and direct approach to an interesting problem, see Exercise 35 of Section 4.3.

This result is astonishing when you consider how much "bigger" than the natural numbers the set of positive rationals "seems." It is nonetheless true that these infinite sets have exactly the same cardinality. Actually, more is true. Since we are able to list the positive rational numbers—$a_1, a_2, a_3, \ldots$—it becomes easy to list **all** the rationals, simply by intertwining the positive and negative rationals, like this—$0, a_1, -a_1, a_2, -a_2, a_3, \ldots$. What do we conclude? The entire set Q of rational numbers is countable!

Pause 9 In the enumeration of the rationals just described, in what position is $\frac{3}{4}$?

At this point, we might be tempted to conjecture that every set is countable, but such is indeed not the case! We present another old and well-known argument which shows that the set of real numbers is not countable. Actually it shows that the set $(0, 1) = \{x \in \mathsf{R} \mid 0 < x < 1\}$ of real numbers between 0 and 1 is not countable. It then follows that R cannot be countable because if it were, so also would the subset $(0, 1)$ because a subset of a countable set is countable. (See the remarks at the end of this section.)

PROBLEM 29. Show that $(0, 1)$ is uncountable.

Solution. The argument we present is a splendid example of a proof by contradiction.

Suppose that there does exist a list $a_1, a_2, a_3, \ldots$ of all the real numbers between 0 and 1. Write each of these numbers in decimal form, agreeing to write $0.200\ldots$ rather than $0.199\ldots$, for example, so that the same number does not appear twice. Our list would look like this:

$$a_1 = 0.a_{11}a_{12}a_{13}a_{14} \cdots$$

$$a_2 = 0.a_{21}a_{22}a_{23}a_{24} \cdots$$

$$a_3 = 0.a_{31}a_{32}a_{33}a_{34} \cdots$$

$$\vdots$$

and remember, every real number between 0 and 1 is supposed to be here. We can, however, write down a number which is guaranteed not to be in the list. Look at a_{11}. If $a_{11} = 1$, let $b_1 = 2$; otherwise, let $b_1 = 1$. Next look at a_{22}. If $a_{22} = 1$, let $b_2 = 2$; otherwise, let $b_2 = 1$. Then look at a_{33}, in general continuing down the diagonal of the following square array

$$a_{11} \ a_{12} \ a_{13} \cdots$$
$$a_{21} \ a_{22} \ a_{23} \cdots$$
$$a_{31} \ a_{32} \ a_{33} \cdots$$
$$\vdots \quad \vdots \quad \vdots$$

and defining

$$b_i = \begin{cases} 2 & \text{if } a_{ii} = 1 \\ 1 & \text{if } a_{ii} \neq 1. \end{cases}$$

Thus, b_i is always different from a_{ii}. Consider the number $b = 0.b_1b_2b_3\ldots$. Since each b_i is 1 or 2, $b \neq 0.000\ldots = 0$ and $b \neq 0.999\ldots = 1$. Thus, b is in the interval $(0, 1)$, so it must be a_i, for some i. But $b \neq a_1$ since b differs from a_1 in the first decimal place, $b \neq a_2$ since b differs from a_2 in the second decimal place and, generally, $b \neq a_i$ since b differs from a_i in the ith decimal place. The hypothesis that the reals in $(0, 1)$ are countable has led us to a contradiction, so the hypothesis is false. The real numbers are not countable. ∎

The "diagonal" argument just sketched, due to the great German mathematician Georg Cantor (1845–1918), should be read several times. It is ingenious and certainly not the sort of thing most of us would think up by ourselves. Here's the important thing to remember, though. Many infinite sets are countable—the natural numbers, the integers, the rationals—but there are some sets which are much bigger (the real numbers for instance). Infinite sets come in different "sizes," just as do finite sets.

At this point, some readers may wonder if there are any subsets of the real numbers whose cardinality lies between that of N and R. This is a very good question! In work for which he was awarded a Fields Medal[5] in 1966, Paul Cohen showed that the usual axioms of set theory do not give enough information to settle it. As a consequence, mathematicians sometimes add the *Continuum Hypothesis*, first formulated by Cantor, to these basic axioms.

3.3.4 CONTINUUM HYPOTHESIS ▶

There is no set A with $\aleph_0 < |A| < |R|$.

3.3.5 REMARK ▶

This is the first time we have used the inequality symbol $<$ with infinite sets. While we have defined the notion of "same cardinality", we have not discussed the possibility of one infinite set being "smaller" than another. For any two sets A and B, we define $|A| \leq |B|$ if and only if there is a one-to-one function $A \to B$, and $|A| < |B|$ if $|A| \leq |B|$ but $|A| \neq |B|$. It is not hard to see that this agrees with intuition on finite sets, so it is a reasonable definition. The problem, however, is to prove that if $|A| \leq |B|$ and $|B| \leq |A|$, then $|A| = |B|$. This is easy if A and B are finite, but the deep "Schröder-Bernstein" theorem for infinite sets.

We conclude with two remarks which you are encouraged to remember and use freely in the exercises which follow.

1. A subset of a countable set is countable.

Why? In this section, we showed that the set of positive rationals is countable by taking the listing of all fractions suggested by Fig 3.4 and omitting repetitions. In a similar way, if B is a subset of a countable set A, the elements of B can be enumerated by taking a listing of A and omitting those elements which are not in B.

2. The concept of "same cardinality" is an equivalence relation on sets, in particular, it is transitive. (See Exercise 10.)

It follows that if we want to prove that a certain set A is countable, for instance, it is sufficient to show that A has the same cardinality as some other set which is known to be countable. Similarly, if we want to prove that a set is uncountable, it is enough to show that it has the same cardinality as a set which we know is uncountable.

Answers to Pauses

6. $f(n) = \frac{1}{4}[1 + (-1)^n(2n - 1)]$
7. In the listing of the integers, natural number k comes in position $2k$ and integer $-k$ in position $2k + 1$. Thus, 1003 comes in position 2006 and the number in position 1003 is -501.

---

[5] A Fields Medal is the highest honor which can be bestowed upon a mathematician. There is no Nobel Prize in mathematics.

8. The listing $1, 2, \frac{1}{2}, \frac{1}{3}, 3, 4, \frac{3}{2}, \frac{2}{3}, \frac{1}{4}, \frac{1}{5}, 5, 6, \frac{5}{2}, \frac{4}{3}, \frac{3}{4}, \ldots$.
   Thus, $\frac{3}{4}$ is in position 15.

9. The rationals are listed $0, 1, -1, 2, -2, \frac{1}{2}, -\frac{1}{2}, \ldots$. The number $\frac{3}{4}$ is 30th in the list.

EXERCISES

The symbol [BB] means that an answer can be found in the Back of the Book.

1. [BB] An enormous crowd covers the field of a base-ball stadium for a rock concert. Suggest an easy way to determine whether or not the size of the crowd exceeds the number of available seats in the stadium.

2. [BB] At first glance, the perfect squares—1, 4, 9, 16 25, ... —seem to be a very sparse subset of the natural numbers. Galileo, however, argued that there were as many perfect squares as there are natural numbers. What was his reasoning likely to have been?

3. Find a one-to-one correspondence between each of the following pairs of sets:
   (a) $\{x, y, \{a, b, c\}\}$ and $\{14, -3, t\}$
   (b) $2\mathsf{Z}$ and $17\mathsf{Z}$
   (c) [BB] $\mathsf{N} \times \mathsf{N}$ and $\{a + bi \in \mathsf{C} \mid a, b \in \mathsf{N}\}$
   (d) N and $\{\frac{m}{n} \mid m \in \mathsf{N}, n = 1, 2\}$

4. Find the inverse of the function $f \colon \mathsf{N} \to \mathsf{Z}$ defined in Problem 27.

5. (a) Suppose $f \colon \mathsf{N} \to \mathsf{Z}$ is a one-to-one onto function. Prove that the function $g \colon \mathsf{N} \times \mathsf{N} \to \mathsf{N} \times \mathsf{Z}$ defined by $g(m, n) = (m, f(n))$ is one-to-one and onto.
   (b) Find a one-to-one correspondence between $\mathsf{N} \times \mathsf{N}$ and $\mathsf{N} \times \mathsf{Z}$.

6. [BB] True or false and explain: If $A \subsetneqq B$, then $|A|$ and $|B|$ do not have the same cardinality.

7. Suppose S is a set and for $A, B \in \mathcal{P}(S)$, we define $A \preceq B$ to mean $|A| \leq |B|$. Is this relation a partial order on $\mathcal{P}(\mathsf{S})$? Explain.

8. Show that for any sets A and B, $|A \times B| = |B \times A|$.

9. [BB] Let X, Y and Z be sets.
   (a) True or false: $(X \times Y) \times Z = X \times (Y \times Z)$. Provide a proof or give a counterexample.
   (b) Find a one-to-one correspondence $(X \times Y) \times Z \to X \times (Y \times Z)$.

10. Prove that the notion of same cardinality is an equivalence relation on the family of all sets. Explain, with reference to Definition 3.3.2.

11. (a) [BB] Find a one-to-one correspondence between the intervals $(1, \infty)$ and $(3, \infty)$. What do you conclude about the cardinalities of $(1, \infty)$ and $(3, \infty)$? [Recall that $(a, \infty) = \{x \mid x > a\}$. See Definition 2.2.3.]
    (b) Let a and b be real numbers. Find a one-to-one correspondence between (a, ∞) and (b, ∞).

12. (a) [BB] Prove that the intervals $(0, 1)$ and $(1, 2)$ have the same cardinality.
    (b) Prove that $(0, 1)$ and $(4, 6)$ have the same cardinality.
    (c) Prove that any two open intervals (a, b) and (c, d) have the same cardinality.

13. (a) [BB] Prove that the intervals $(0, 1)$ and $(10, \infty)$ have the same cardinality.
    (b) Prove that the intervals $(2, 5)$ and $(10, \infty)$ have the same cardinality.
    (c) Prove that the intervals (a, b) and (c, ∞) have the same cardinality. Assume $a < b$.

14. Let a and b be real numbers with $a < b$. Show that the set R^+ of positive real numbers has the same cardinality as the open interval $(a, b) = \{x \in \mathsf{R} \mid a < x < b\}$.

15. [BB] Prove that the function defined by $f(x) = 3^x$ establishes a one-to-one correspondence between the real numbers R and the positive real numbers R^+. What can you conclude about cardinality?

16. [BB] Prove that R and the interval (a, b) have the same cardinality. Assume $a < b$.

17. Prove that each of the following sets is countable by listing its elements in a systematic and definite way. (Show at least the first dozen terms of your lists.)

(a) all positive and negative integer powers of 2

(b) those natural numbers which leave a remainder of 1 when divided by 3

(c) $N \times \{1, 2, 3\}$

(d) [BB] those positive rational numbers $\frac{m}{n}$ with n odd

(e) $N \times Z$

(f) $Z \times Z$

18. Determine, with justification, whether each of the following sets is finite, countably infinite, or uncountable:

   (a) [BB] $\{x \in R \mid 1 < x < 2\}$

   (b) $\{x \in Q \mid 1 < x < 2\}$

   (c) $\{\frac{m}{n} \mid m, n \in N, m < 100, 5 < n < 105\}$

   (d) $\{\frac{m}{n} \mid m, n \in Z, m < 100, 5 < n < 105\}$

   (e) [BB] $\{a + bi \in C \mid a, b \in N\}$

   (f) $\{(a, b) \in Q \times Q \mid a + b = 1\}$

   (g) $\{(a, b) \in R \times R \mid b = \sqrt{1 - a^2}\}$

19. Which of the following sets is finite, which is countably infinite, which is uncountable? Explain your answers.

   (a) the set of grains of sand on a beach

   (b) $\{3^n \mid n \in Z\}$

   (c) the set of words in the English language

   (d) the set of sentences in the English language

20. Give an example of each of the following or explain why no example exists.

   (a) [BB] an infinite collection of finite sets (no two the same) whose union is finite

   (b) a finite collection of finite sets whose union is infinite

   (c) [BB] an infinite collection of infinite sets whose union is finite

21. [BB] Let S_1 and S_2 denote spheres of radii 1 and 100, respectively. Prove that the points on the surface of S_1 and those on the surface of S_2 are sets with the same cardinality.

22. [BB] Let S be an infinite set and let x be an element not in S. Prove that S and $S \cup \{x\}$ are sets of the same cardinality. (You may assume that S contains a countably infinite subset.)

23. Prove that the points of a plane and the points of a sphere are sets of the same cardinality.

24. [BB] Suppose A is a finite set and B is a countably infinite set and that $A \cap B = \emptyset$. Show that $A \cup B$ is countably infinite.

25. **(a)** Show that if A and B are countable sets, then so is $A \times B$.

   (b) Show that the set of all polynomials of degree at most one with integer coefficients is countable.

26. Prove that the function defined by $f(x) = \dfrac{x - \frac{1}{2}}{x(x - 1)}$ defines a one-to-one correspondence from $(0, 1)$ to R.

27. Prove that for any set X, the cardinalities of X and $\mathcal{P}(X)$ are different.

28. Let S be the set of all real numbers in the interval $(0, 1)$ whose decimal expansions are infinite and contain only 3 and 4—for example, $0.343434\ldots$ and $0.333\ldots$, but not $0.34 = 0.34000\ldots$. Prove that S is uncountable.

29. Let S be the set of all real numbers in the interval $(0, 1)$ whose decimal expansions involve only 0 and 1. Prove that S is uncountable.

30. What is the Fields Medal and after whom was it named? Write a short note about this prize and its founder.

REVIEW EXERCISES FOR CHAPTER 3

1. Let $f = \{(1, 2), (2, 3), (3, 4), (4, 1)\}$ and $g = \{(1, 3), (2, 1), (3, 4), (4, 2), (5, 1)\}$. Find f^{-1} and $g \circ f$. Is g one-to-one? Explain.

2. Let $f : Z \to Z$ be defined by $f(m) = 3|m| + 1$. Is f one-to-one? Is f onto? Give reasons for your answers.

3. Define $f : Z \to Z$ by $f(x) = 2x^3 + x$. Show that f is one-to-one but not onto.

4. Can a function be a reflexive relation? Explain.

5. Is it possible for a function $f : R \to R$ to be symmetric as a relation? Give an example or explain why the answer is no.

6. Is it possible for a function $f : R \to R$ to be transitive as a relation? Give an example or explain why the answer is no.

7. Answer true or false and justify your answers.

   (a) For real numbers x and y, $|x| = y^2 \to x = y|y|$.

   (b) For real numbers x and y, $x = y|y| \to |x| = y^2$.

   (c) For real numbers x and y, $|x| = y^2 \leftrightarrow x = y|y|$.

8. Find subsets A and B of R, with A as large as possible, such that $f: A \to B$ defined by $f(x) = x^2 + 4x - 7$ is one-to-one and onto.

9. Suppose $f: A \to B$ and $g: B \to C$ are functions. If $g \circ f$ is one-to-one and f is onto, show that g is one-to-one.

10. Let $|A| = n$ and $|B| = m$ for $n, m \in \mathsf{N}$. Use the definition of *cardinality of a finite set* to show that if $A \cap B = \emptyset$, then $|A \cup B| = n + m$.

11. Show that the open intervals $(1, 3)$ and $(0, \infty)$ have the same cardinality.

12. (a) Let S be the set of all real numbers in the interval $(0, 1)$ whose decimal expansions contain only 0's, 2's and 7's. Prove that S is uncountable.

    (b) Let S' be the elements of S (defined in (a)) whose decimal expansions contain only finitely many 2's and 7's. What is the cardinality of S'?

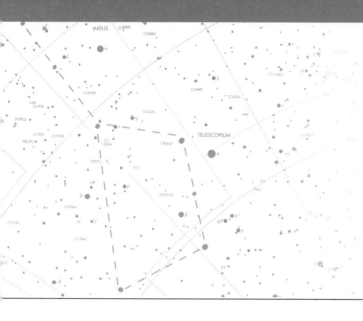

The Integers

4.1 THE DIVISION ALGORITHM

What are some of the important properties of the real numbers? Most of us learned early in life that they have a definite order. Later, we met the symbol $\leq$, whose properties are best summarized by saying that it is a partial order; that is, the binary relation $\leq$ is

reflexive: $a \leq a$ for all $a \in \mathsf{R}$,
antisymmetric: if $a \leq b$ and $b \leq a$ for $a, b \in \mathsf{R}$, then $a = b$, and
transitive: if $a \leq b$ and $b \leq c$, for $a, b, c \in \mathsf{R}$, then $a \leq c$.

(See Section 2.5.)

Readers should ensure that they agree with the following inequalities:

$$-15 \leq -8 \leq -\frac{1}{3} \leq -0.3 \leq -\frac{1}{4} \leq -0.25 \leq 0 \leq 1 \leq 3.14 \leq \pi \leq \frac{22}{7} \leq 4.$$

Which of the $\leq$ signs in the preceding list could be changed to the strict inequality $<$? The answer is all of them, except for $-\frac{1}{4} \leq -0.25$. Since $-\frac{1}{4} = -0.25$, the strict inequality $-\frac{1}{4} < -0.25$ is not true. Is it clear why each of the remaining $\leq$ signs could be replaced by $<$ signs?

Pause 1   Why is $-\frac{1}{3} < -0.3$? Why is $3.14 < \pi < \frac{22}{7}$?

We can add two real numbers a and b and obtain their sum $a + b$, or we can multiply two real numbers a and b and obtain their product $a \cdot b$ (usually written ab, without the centered dot). These operations of addition and multiplication satisfy a number of important properties, the last three of which relate to order.

4.1.1 PROPERTIES OF $+$ AND $\cdot$ ▶

Let a, b, and c be real numbers. Then

1. **(closure)** $a + b$ and ab are both real numbers.
2. **(commutativity)** $a + b = b + a$ and $ab = ba$.

3. (**associativity**) $(a + b) + c = a + (b + c)$ and $(ab)c = a(bc)$.

4. (**identities**) $a + 0 = a$ and $a \cdot 1 = a$.

5. (**distributivity**) $a(b + c) = ab + ac$ and $(a + b)c = ac + bc$.

6. (**additive inverse**) $a + (-a) = 0$.

7. (**multiplicative inverse**) $a(\frac{1}{a}) = 1$ if $a \neq 0$.

8. $a \leq b$ implies $a + c \leq b + c$.

9. $a \leq b$ and $c \geq 0$ implies $ac \leq bc$.

10. $a \leq b$ and $c \leq 0$ implies $ac \geq bc$.

A third well-known operation, *subtraction*, is defined in terms of addition by the rule

$$a - b = a + (-b).$$

It is not unusual for a set of real numbers to have no smallest element; for example, there is no smallest element in the set $\{\frac{1}{2}, \frac{1}{3}, \frac{1}{4}, \frac{1}{5}, \ldots\}$. Similarly, there is no smallest positive number. On the other hand, this sort of thing does not occur with sets of natural numbers, according to the *Well-Ordering Principle*.

4.1.2   WELL-ORDERING PRINCIPLE ▶

Any nonempty set of natural numbers has a smallest element.

In the rest of this section, we concentrate on the integers. All the properties of paragraph 4.1.1 hold for the integers, in most instances because they hold for all real numbers. One exception is the closure property, which simply says that the sum and product of integers are integers. We also say that the integers are *closed* under addition and under multiplication. Not all sets of real numbers are closed under these operations; for instance, the set of odd integers is not closed under addition since, for example, $1 + 3 = 4$, which is not odd.

Pause 2   Is the set of negative integers closed under multiplication? Is the set of natural numbers closed under multiplication? Is the set of natural numbers closed under subtraction?

When some of us were school children, division of 58 by 17 was performed with a configuration like this,

$$
\begin{array}{r}
3 \\
17\overline{)58} \\
51 \\
\hline
7
\end{array}
$$

which served to illustrate that the answer is "3 remainder 7." When 58 is divided by 17, the *quotient* is 3 and the *remainder* is 7; equivalently,

$$58 = 3(17) + 7.$$

It is also true that $58 = 2(17) + 24$, but we were taught that the remainder must be less than the divisor. We now prove that this sort of division is always possible and, moreover, the quotient and the remainder are unique.

4.1.3 THEOREM ▶ Given natural numbers a and b, there are unique nonnegative integers q and r, with $0 \le r < b$, such that $a = qb + r$.

Proof Consider the sequence of nonnegative multiples of b; that is, $0 \cdot b = 0$, $1 \cdot b = b$, $2 \cdot b = 2b$, $3 \cdot b = 3b, \ldots$. The first term in this increasing sequence of numbers is 0, which is less than a since a is a natural number. On the other hand, some term is bigger than a [for example, $(2a)b > a$ because $2b > 1$], so, by the Well-Ordering Principle, the set of multiples of b which exceed a has a smallest element, say $(q + 1)b$. So we have $qb \le a < (q + 1)b$. (See Fig 4.1.) Set $r = a - qb$. Since $qb \le a$, we have $r \ge 0$. Since $(q + 1)b > a$, we have $r < b$. Hence, $0 \le r < b$, and we have found q and r as required.

Figure 4.1

To see that q and r are unique, assume that a can be expressed in the given form in two ways; that is, suppose that $a = q_1 b + r_1$ and $a = q_2 b + r_2$ with $0 \le r_1 < b$ and $0 \le r_2 < b$. Then $(q_1 - q_2)b = r_2 - r_1$. Now $(q_1 - q_2)b$ is an integral multiple of b, while $-b < r_2 - r_1 < b$. The only possibility is that this multiple is 0, and so $q_1 = q_2$, $r_1 = r_2$ as desired. ∎

4.1.4 DEFINITIONS ▶ If a and b are natural numbers and $a = qb + r$ for nonnegative integers q and r with $0 \le r < b$, the integer q is called the *quotient* and the integer r is called the *remainder* when a is divided by b.

When a and b are both positive and a is divided by b, it is useful to note that the quotient is the integer part of the number displayed by a calculator used to divide a by b. (The remainder is $a - qb$.)

PROBLEM 1. Find the quotient q and the remainder r and write $a = qb + r$ when $a = 19$ is divided by $b = 7$. Find the quotient and remainder when 589,621 is divided by 7893 and when 11,109,999,999 is divided by 1111.

Solution. When 19 is divided by 7, the quotient is 2 and the remainder is 5. We have $19 = 2(7) + 5$.

Dividing 589,621 by 7893, one author's calculator displayed 74.701761. Thus, $q = 74$ and $r = 589{,}621 - 74(7893) = 5539$.

For the last part of the question, we have to be a little resourceful since one of the numbers here is too large for the author's unsophisticated calculator! Instead, we note that

$$11{,}109{,}999{,}999 = 11{,}110{,}000{,}000 - 1$$

$$= 1111(10{,}000{,}000) - 1$$

$$= 1111(9,999,999 + 1) - 1$$
$$= 1111(9,999,999) + 1110.$$

So the quotient is 9,999,999 and the remainder is 1110.   ∎

With only slight modifications, Theorem 4.1.3 can be generalized to situations where a and b are integers, not just natural numbers. Note, however, that in order to preserve the uniqueness of q, the possibility $b = 0$ must be excluded.[1] (Also, in this case, the condition $0 \leq r < b$ would be impossible to meet.)

4.1.5 THE DIVISION ALGORITHM ▶

Let $a, b \in \mathbb{Z}$, $b \neq 0$. Then there exist unique integers q and r, with $0 \leq r < |b|$, such that $a = qb + r$.

Proof In Theorem 4.1.3, we proved this in the case where a and b are both positive. If $a = 0$, then $q = 0$, $r = 0$ gives a solution. We will consider the other cases individually.

Case 1: $b > 0$ and $a < 0$.
Since $-a > 0$, we can apply Theorem 4.1.3 to $-a$ and b obtaining q and r, with $0 \leq r < b$, such that $-a = qb + r$. Therefore, $a = (-q)b - r$. If $r = 0$, $a = (-q)b$, while if $r > 0$, $a = (-q-1)b + (b-r)$ with $0 < b - r < b = |b|$. In either case, we have expressed a in the desired form.

Case 2: $b < 0$ and either $a > 0$ or $a < 0$.
Here $-b > 0$, so Theorem 4.1.3 and Case 1 tell us that there exist integers q and r, $0 \leq r < -b = |b|$, such that $a = q(-b) + r = (-q)b + r$. Again we have expressed a in the desired form.

The proof of uniqueness follows as in Theorem 4.1.3.   ∎

Chapter 8 of this book is devoted to the subject of *algorithms*. Here, we simply remark that *algorithm* means "definite procedure" and that the theorem known as the *Division Algorithm* takes its name from the fact that there is a definite procedure for determining q and r, given a and b. For $a, b > 0$, $q = \lfloor \frac{a}{b} \rfloor$ is the integer part of $\frac{a}{b}$ and $r = a - qb$, as illustrated in Problem 1. For possibly negative a and/or b, the procedure is given in Proposition 4.1.6.

EXAMPLE 2 Verify that $a = qb + r$ in each of the following cases and, in the process, notice that q is not always what we might expect.

a	b	q	r
58	17	3	7
58	-17	-3	7
-58	17	-4	10
-58	-17	4	10

▲

When a is written in the form $a = qb + r$, with $0 \leq r < |b|$, we shall show that q is the floor or ceiling of $\frac{a}{b}$ according as b is positive or negative. (The *floor*

---

[1] Division by 0 always causes problems!

and *ceiling* functions were defined in paragraph 3.1.6.) In the preceding table, for instance, when $a = -58$, $b = 17$, $\lfloor -58/17 \rfloor = \lfloor -3.41\ldots \rfloor = -4$, the value recorded for q. When $a = -58$, $b = -17$, $\lceil -58/-17 \rceil = \lceil 3.41 \rceil = 4$, again the recorded value of q.

4.1.6 PROPOSITION ▶   If $a = qb + r$, with $0 \le r < |b|$, then

$$q = \begin{cases} \lfloor \frac{a}{b} \rfloor \text{ if } b > 0 \\ \lceil \frac{a}{b} \rceil \text{ if } b < 0. \end{cases}$$

Proof   We consider the case $b > 0$ and leave the other possibility to the exercises. By definition, $x - 1 < \lfloor x \rfloor \le x$ for any x. Let $x = \frac{a}{b}$ and $k = \lfloor x \rfloor$. Then

$$\frac{a}{b} - 1 < k \le \frac{a}{b}.$$

Multiplying by the positive number b (and using Property 9 of paragraph 4.1.1), we obtain $a - b < kb \le a$ and then, multiplying by -1,

$$-a \le -kb < -a + b$$

(by Property 10). Adding a gives $0 \le a - kb < b$ so, letting $r = a - kb$, we have $a = kb + r$ with $0 \le r < |b|$. By uniqueness of q, $q = k = \lfloor \frac{a}{b} \rfloor$, as asserted. ∎

EXAMPLE 3   With $a = -1027$ and $b = 38$, we have

$$\left\lfloor \frac{a}{b} \right\rfloor = \left\lfloor -\frac{1027}{38} \right\rfloor = \lfloor -27.026\ldots \rfloor = -28 = q.$$

It is straightforward now to determine r: $r = a - qb = -1027 + 1064 = 37$. Thus, $-1027 = -28(38) + 37$. With $a = 1{,}234{,}567$ and $b = -103$, we have $\lceil \frac{a}{b} \rceil = \lceil -11{,}986.087\ldots \rceil = -11{,}986 = q$. As before, we note that $r = a - qb = 1{,}234{,}567 - (-11{,}986)(-103) = 9$. Thus, $1{,}234{,}567 = -103(-11{,}986) + 9$. ▲

Representing Natural Numbers in Various Bases

As an application of some of the ideas we have presented so far, we discuss briefly the representation of natural numbers in bases other than the familiar one.

     Unless they have reason to suspect otherwise, most people expect that the numbers they encounter in their day-to-day lives are presented in *base* 10. The integer 2159, for example, is assumed to mean

$$2 \cdot 10^3 + 1 \cdot 10^2 + 5 \cdot 10 + 9.$$

The digits 9, 5, 1, and 2 are called, respectively, the *units*, *tens*, *hundreds*, and *thousands* digits of this integer. If we lived in a base 12 world, however, we would interpret 2159 as

$$2 \cdot 12^3 + 1 \cdot 12^2 + 5 \cdot 12 + 9.$$

When we represent a number in a base other than 10, we shall employ notation such as $(2159)_{12}$. Thus,

$$(2159)_{12} = 2 \cdot 12^3 + 1 \cdot 12^2 + 5 \cdot 12 + 9.$$

4.1.7 DEFINITION ▶ Given a fixed natural number $b > 1$, the *base b representation* of a natural number N is the expression $(a_{n-1}a_{n-2} \ldots a_0)_b$, where $a_0, a_1, \ldots, a_{n-1}$ are those integers $0 \le a_i < b$ which satisfy $N = a_{n-1}b^{n-1} + a_{n-2}b^{n-2} + \cdots + a_1 b + a_0$. Thus,

$$(a_{n-1}a_{n-2} \ldots a_0)_b = a_{n-1}b^{n-1} + a_{n-2}b^{n-2} + \cdots + a_1 b + a_0.$$

The base 5 representation of 117 is 432, written $117 = (432)_5$, because $117 = 4(5^2) + 3(5) + 2$. In base 3, $117 = (11,100)_3$ because $117 = 1(3^4) + 1(3^3) + 1(3^2)$. While 10 is the most familiar base, nowadays bases 2, 8, and 16 are also common. Base 2 is also known as *binary*, base 8 as *octal*, and base 16 as *hexadecimal*. The basic hexadecimal digits are 0–9, A, B, C, D, E, and F. For example, $10 = A_{16}$, $11 = B_{16}$ and $(4C)_{16} = 4 \cdot 16 + 12 = 76$.

To convert from base b to base 10 is easy: Use the definition of $(\ldots)_b$ in Definition 4.1.7. What about converting from base 10 to base b?

Suppose that $N = (a_{n-1}a_{n-2} \ldots a_0)_b$ is a number given in base b. Then

$$N = a_{n-1}b^{n-1} + a_{n-2}b^{n-2} + \cdots + a_1 b + a_0$$

$$= (a_{n-1}b^{n-2} + a_{n-2}b^{n-3} + \cdots + a_1)b + a_0 = q_0 b + a_0$$

where $q_0 = a_{n-1}b^{n-2} + a_{n-2}b^{n-3} + \cdots + a_1$. Since $0 \le a_0 < b$, when N is divided by b, the remainder is a_0 and the quotient is q_0. Similarly,

$$q_0 = (a_{n-1}b^{n-3} + a_{n-2}b^{n-4} + \cdots + a_2)b + a_1 = q_1 b + a_1$$

with $q_1 = a_{n-1}b^{n-3} + a_{n-2}b^{n-4} + \cdots + a_2$, so that when q_0 is divided by b, the remainder is a_1 and the quotient is q_1. In a similar way, we see that a_2 is the remainder when q_1 is divided by b. Thus, the digits in the representation of N in base b are, from right to left, the remainders when first N, and then successive quotients, are divided by b.

PROBLEM 4. Find the binary, octal, and hexadecimal representations of the number 2159.

Solution. We have $2159 = 1079(2) + 1$, $1079 = 539(2) + 1$, $539 = 269(2) + 1$, $269 = 134(2) + 1$, $134 = 67(2) + 0$, $67 = 33(2) + 1$, $33 = 16(2) + 1$, $16 = 8(2) + 0$, $8 = 4(2) + 0$, $4 = 2(2) + 0$, $2 = 2(1) + 0$ and $1 = 0(2) + 1$. The sequence of remainders is 1, 1, 1, 1, 0, 1, 1, 0, 0, 0, 0, 1, so, writing these in reverse order, $2159 = (100,001,101,111)_2$.

The base 8 and base 16 representations can be obtained by similar means, or from the base 2 representation. Having determined that $2159 = 2^{11} + 2^6 + 2^5 + 2^3 + 2^2 + 2 + 1$, we have

$$2159 = 2^2(2^3)^3 + (2^3)^2 + (2^2)2^3 + 2^3 + 7$$

$$= 4(8^3) + 1(8^2) + 5(8) + 7 = (4157)_8$$

and

$$2159 = 2^{11} + 2^6 + 2^5 + 2^3 + 2^2 + 2 + 1$$
$$= 2^3(2^4)^2 + 2^2(2^4) + 2(2^4) + 15$$
$$= 8(16)^2 + 6(16) + 15 = (86F)_{16}. \qquad \blacksquare$$

PROBLEM 5.   Convert 21,469 to octal and to hexadecimal notation.

Solution.   We have $21,469 = 2683(8) + 5$, $2683 = 335(8) + 3$, $335 = 41(8) + 7$, $41 = 5(8) + 1$ and $5 = 0(8) + 5$. Thus, $21,469 = (51,735)_8$. Similarly, $21,469 = 1341(16) + 13$, $1341 = 83(16) + 13$, $83 = 5(16) + 3$ and $5 = 0(16) + 5$, so $21,469 = (53DD)_{16}$. $\qquad \blacksquare$

Answers to Pauses

1. $\frac{1}{3}$ is the same as the decimal $.333\ldots$. The three dots "$\ldots$" mean that the 3's continue indefinitely. We often write $\frac{1}{3} = 0.\dot{3}$ and say that $\frac{1}{3}$ equals "point 3 repeated." Since $0.\dot{3}$ is larger than 0.3, its **negative**, $-0.\dot{3}$, is less than -0.3. The number π is perhaps the most fascinating number of all. Its decimal expansion begins 3.14159, so it is definitely bigger than 3.14. As a decimal, $\frac{22}{7} = 3.\overline{142857}$, where the line indicates that the sequence of numbers underneath is repeated indefinitely; thus,

$$\frac{22}{7} = 3.\underbrace{142857}\,\underbrace{142857}\,\underbrace{142857}\,142857\ldots.$$

Thus, we see that $\pi < \frac{22}{7}$. Unlike $\frac{1}{3}$ and $\frac{22}{7}$, the decimal expansion of π continues indefinitely, without any kind of pattern.

2. The set of negative integers is not closed under multiplication since, for example, $(-2)(-3) = 6$ is not a negative integer. The set of natural numbers is closed under multiplication since the product of natural numbers is a natural number. The natural numbers are not closed under subtraction since, for example, $4 - 10 = -6$ is not a natural number.

EXERCISES

The symbol [BB] means that an answer can be found in the Back of the Book.

1. [BB] Use the properties given in paragraph 4.1.1 to derive a second distributive law: $(a + b)c = ac + bc$ for any real numbers a, b, c.

2. True or false? If false, give a counterexample.

   (a) [BB] Subtraction is a closed operation on the real numbers.

   (b) Subtraction of real numbers is commutative.

   (c) Subtraction of real numbers is associative.

3. Show (by means of a counterexample) that Property 9 of paragraph 4.1.1 does not hold if $c < 0$.

4. [BB] Find the quotient q and the remainder r as defined by the division algorithm, 4.1.5, in each of the following cases.

   (a) $a = 500$; $b = 17$
   (b) $a = -500$; $b = 17$
   (c) $a = 500$; $b = -17$
   (d) $a = -500$; $b = -17$

5. Find q and r as defined by the division algorithm, 4.1.5, in each of the following cases.

   (a) $a = 5286$; $b = 19$
   (b) $a = -5286$; $b = 19$

(c) $a = 5286; b = -19$

(d) $a = -5286; b = -19$

(e) $a = 19; b = 5286$

(f) $a = -19; b = 5286$

6. Find integers q and r, with $0 \leq r < |b|$, such that $a = bq + r$ in each of the following cases.

(a) $a = 12{,}345; b = -39$

(b) $a = -27{,}361; b = -977$

(c) $a = -102{,}497; b = -1473$

(d) $a = 98{,}764; b = 4789$

(e) $a = -41{,}391; b = -755$

(f) $a = 555{,}555{,}123; b = 111{,}111{,}111$

(g) $a = 81{,}538{,}416{,}000; b = 38{,}754$

7. [BB] Fix a natural number $n > 1$ and define $f : \mathbb{Z} \to \mathbb{Z}$ by setting $f(a)$ equal to the quotient when a is divided by n; that is, $f(a) = q$ where $a = qn + r$ with $0 \leq r < n$.

(a) Find the domain and range of f.

(b) Is f one-to-one?

(c) Is f onto?

Explain your answers.

8. Suppose $n > 1$ is a natural number and $f : \mathbb{Z} \to \mathbb{N} \cup \{0\}$ is that function which associates with each $a \in \mathbb{Z}$ its remainder upon division by n; thus, if $a = qn + r$ with $0 \leq r < n$, then $f(a) = r$.

(a) Find the domain and range of f.

(b) Is f one-to-one?

(c) Is f onto?

Explain your answers.

9. [BB] Complete the proof of Proposition 4.1.6 by showing that if $a = qb + r$ with $0 \leq r < |b|$ and $b < 0$, then $q = \lceil a/b \rceil$.

10. Find the binary, octal, and hexadecimal representations for each of the following integers (given in base 10).

(a) [BB] 4034

(b) 57,483

(c) 185,178

11. (a) Suppose the natural number N is $(a_{n-1} \ldots a_0)_b$ (in base b). Prove that $n - 1 = \lfloor \log_b N \rfloor$ (and, hence, that N has $1 + \lfloor \log_b N \rfloor$ digits in base b).

(b) How many digits does 7^{254} have in its base 10 representation?

(c) How many digits does the number 319^{566} have in its base 10 representation?

4.2 DIVISIBILITY AND THE EUCLIDEAN ALGORITHM

When we say that one integer *divides* another, we mean "divides exactly," that is, with a remainder of 0.

4.2.1 DEFINITION ▶ Given integers a and b with $b \neq 0$, we say that b is a *divisor* or a *factor* of a and that *a is divisible by b* if and only if $a = qb$ for some integer q. We write $b \mid a$ to signify that a is divisible by b and say "b divides a."

For example, 3 is a divisor of 18, -7 is a divisor of 35, $16 \mid -64$, $-4 \nmid 38$. (As always, a slash through a symbol negates the meaning of that symbol. Just as $\neq$ means "not equal" and $\notin$ means "does not belong to," so $\nmid$ means "does not divide.")

Note that $1 \mid n$ for any integer n because $n = n \cdot 1$.

Pause 3 Show that $n \mid 0$ for any integer $n \neq 0$.

PROBLEM 6. Given three consecutive integers $a, a + 1, a + 2$, prove that one of them is divisible by 3.

Solution. By the division algorithm, we can write $a = 3q + r$ with $0 \leq r < 3$. Since r is an integer, we must have $r = 0$, $r = 1$ or $r = 2$. If $r = 0$, then $a = 3q$ is divisible by 3. If $r = 1$, then $a = 3q + 1$, so $a + 2 = 3q + 3$ is divisible by 3. If $r = 2$, then $a = 3q + 2$, so $a + 1 = 3q + 3$ is divisible by 3. ∎

The following proposition is elementary but very important; it is used (often implicitly) all the time.

4.2.2 PROPOSITION ▶ Suppose a, b, and c are integers such that $c \mid a$ and $c \mid b$. Then $c \mid (xa + yb)$ for any integers x and y.

Proof Since $c \mid a$, we know that $a = q_1 c$ for some integer q_1. Since $c \mid b$, we also have $b = q_2 c$ for some integer q_2. Thus, $xa + yb = xq_1 c + yq_2 c = (q_1 x + q_2 y)c$. Since $q_1 x + q_2 y$ is an integer, $c \mid (xa + yb)$ as required. ∎

The most important properties of "divides" are summarized in the next proposition.

4.2.3 PROPOSITION ▶ The binary relation $\mathcal{R}$ on N defined by $(a, b) \in \mathcal{R}$ if and only if $a \mid b$ is a partial order.

Proof We have to show that the relation is reflexive, antisymmetric, and transitive.

Reflexive: For any $a \in$ N, $a \mid a$ because $a = 1 \cdot a$.

Antisymmetric: Suppose $a, b \in$ N are such that $a \mid b$ and $b \mid a$. Then $b = q_1 a$ for some natural number q_1 and $a = q_2 b$ for some natural number q_2. Thus, $a = q_2(q_1 a) = (q_1 q_2)a$. Since $a \neq 0$, $q_1 q_2 = 1$, and, since q_1 and q_2 are natural numbers, we must have $q_1 = q_2 = 1$; thus, $a = b$.

Transitive: If $a, b, c \in$ N are such that $a \mid b$ and $b \mid c$, then $b = q_1 a$ and $c = q_2 b$ for some natural numbers q_1 and q_2. Thus, $c = q_2 b = q_2(q_1 a) = (q_1 q_2)a$, with $q_1 q_2$ an integer. So $a \mid c$. ∎

The proposition says that $($N$, \mid)$ is a partially ordered set. In fact, $(A, \mid)$ is a poset for any set A of natural numbers. At this point, we encourage you to review Section 2.5 and, in particular, the terminology associated with posets.

EXAMPLE 7 Let $A = \{1, 2, 3, 4, 5, 6\}$. In the poset $(A, \mid)$, the element 4 is maximal because there is no $a \in A$ satisfying $4 \mid a$ except $a = 4$. This element is not a maximum however, since, and for example, $5 \nmid 4$. Similarly, 5 and 6 are maximal elements which are not maximum. ▲

Pause 4 Draw the Hasse diagram for this poset and find all minimal, minimum, maximal, and maximum elements.

The Greatest Common Divisor

4.2.4 DEFINITION ▶ Let a and b be integers not both of which are 0. An integer g is the *greatest common divisor (gcd)* of a and b if and only if g is the largest common divisor of a and b; that is, if and only if

1. $g \mid a$, $g \mid b$ and,
2. if c is any integer such that $c \mid a$ and $c \mid b$, then $c \leq g$.

We write $g = \gcd(a, b)$ to signify that g is the greatest common divisor of a and b.

EXAMPLES 8
- the greatest common divisor of 15 and 6 is 3,
- $\gcd(-24, 18) = 6$,
- $\gcd(756, 210) = 42$,
- $\gcd(-756, 210) = 42$,
- $\gcd(-756, -210) = 42$. ▲

Pause 5 If a and b are integers such that $a \mid b$, what is the greatest common divisor of a and b?

Pause 6 Suppose a is a nonzero integer. What is $\gcd(a, 0)$?

It follows almost immediately from the definition that two integers, not both 0, have exactly one greatest common divisor. (See Exercise 14.) Note also that since 1 is a common divisor of any two integers, the greatest common divisor of two integers is always positive.

The seventh book of Euclid's *Elements* (300 B.C.) describes a procedure now known as the Euclidean algorithm for finding the gcd of two integers a and b. It is based on the fact that when $a = qb + r$, then $\gcd(a, b) = \gcd(b, r)$. For instance, since $\underline{58} = 3(\underline{17}) + \underline{7}$, it must be that $\gcd(58, 17) = \gcd(17, 7)$. Also, since $\underline{75} = 3(\underline{21}) + \underline{12}$, it must be that $\gcd(75, 21) = \gcd(21, 12)$.

4.2.5 LEMMA ▶ If $a = qb + r$ for integers a, b, q, and r, then $\gcd(a, b) = \gcd(b, r)$.

Proof If $a = b = 0$, then $a = qb + r$ says $r = 0$. Similarly, if $b = r = 0$, then $a = 0$. In either case, the result is true since neither $\gcd(a, b)$ nor $\gcd(b, r)$ is defined. Thus, it remains to consider the case that $g_1 = \gcd(a, b)$ and $g_2 = \gcd(b, r)$ are both well defined integers. First, $g_2 \mid b$ and $g_2 \mid r$, so $g_2 \mid (qb + r)$; that is, $g_2 \mid a$. Thus, g_2 is a common divisor of a and b and, since g_1 is the greatest common divisor of a and b, we have $g_2 \leq g_1$.

On the other hand, since $g_1 \mid a$ and $g_1 \mid b$, we know that $g_1 \mid (a - qb)$; that is, $g_1 \mid r$. As a common divisor of b and r, it cannot exceed the gcd of these numbers. Thus, $g_1 \leq g_2$, so $g_1 = g_2$ as desired. ∎

The Euclidean algorithm involves nothing more than a repeated application of this lemma.

4.2.6 EUCLIDEAN ALGORITHM ▶ Let a and b be natural numbers with $b < a$. To find the greatest common divisor of a and b, write

$$a = q_1 b + r_1, \quad \text{with } 0 \leq r_1 < b.$$

If $r_1 \neq 0$, write $b = q_2 r_1 + r_2$, with $0 \leq r_2 < r_1$.
If $r_2 \neq 0$, write $r_1 = q_3 r_2 + r_3$, with $0 \leq r_3 < r_2$.
If $r_3 \neq 0$, write $r_2 = q_4 r_3 + r_4$, with $0 \leq r_4 < r_3$.

Continue this process until some remainder $r_{k+1} = 0$. Then the greatest common divisor of a and b is r_k, the last nonzero remainder.

Why does this process work, and why must some remainder be 0? The answer to the second question is that either $r_1 = 0$ or the set of nonzero remainders is a nonempty set of natural numbers and hence has a smallest element, r_k, by the Well-Ordering Principle. So the next remainder, being smaller than r_k, must be 0. It follows by Lemma 4.2.5 that r_k is the gcd of a and b because

$$\gcd(a, b) = \gcd(b, r_1) = \gcd(r_1, r_2)$$

$$= \gcd(r_2, r_3) = \cdots = \gcd(r_k, r_{k+1}) = \gcd(r_k, 0) = r_k.$$

PROBLEM 9. Find the greatest common divisor of 630 and 196.

Solution. We have $630 = 3(196) + 42$
$$196 = 4(42) + 28$$
$$42 = 1(28) + 14$$
$$28 = 2(14) + 0.$$

The last nonzero remainder is 14, so this is gcd(630, 196). ∎

Let us write down these same equations in a different way, expressing each of the remainders—$r_1 = 42, r_2 = 28, r_3 = 14$—in terms of $a = 630$ and $b = 196$. We have

$$42 = 630 - 3(196) \qquad\qquad = a - 3b$$
$$28 = 196 - 4(42)$$
(1) $$= b - 4r_1 = b - 4(a - 3b) \qquad = -4a + 13b$$
$$14 = 42 - 28$$
$$= r_1 - r_2 = (a - 3b) - (-4a + 13b) = 5a - 16b.$$

Recording only the remainders and the coefficients a and b, these equations can be neatly coded by the array

$$\begin{matrix} 42 & 1 & -3 \\ 28 & -4 & 13 \\ 14 & 5 & -16. \end{matrix}$$

Adding to the top of this array the rows

$$\begin{matrix} 630 & 1 & 0 \\ 196 & 0 & 1, \end{matrix}$$

which correspond to the equations $630 = 1a + 0b$ and $196 = 0a + 1b$, respectively, we obtain the array

$$\begin{matrix} 630 & 1 & 0 \\ 196 & 0 & 1 \\ 42 & 1 & -3 \\ 28 & -4 & 13 \\ 14 & 5 & -16, \end{matrix}$$

which is easy to remember. Each row after the first two is of the form $x - qy$, where x and y are the two rows preceding it and q is the quotient when the first number in x is divided by the first number in y. When 630 is divided by 196, the quotient is 3, so the third row is

$$(630\ 1\ 0) - 3(196\ 0\ 1) = 42\ 1\ -3.$$

When 196 is divided by 42, the quotient is 4, so the fourth row is

$$(196 \ 0 \ 1) - 4(42 \ 1 \ -3) = 28 \ -4 \ 13.$$

When 42 is divided by 28, the quotient is 1; so the last row is

$$(42 \ 1 \ -3) - (28 \ -4 \ 13) = 14 \ 5 \ -16.$$

Since the last remainder, 14, divides the previous remainder, 28, the next remainder is 0. So gcd(630, 196) = 14, the last nonzero remainder.

Here is another example of this procedure.

PROBLEM 10.   Find gcd(1800, 756).

Solution.

$$
\begin{array}{rrr}
1800 & 1 & 0 \\
756 & 0 & 1 \\
288 & 1 & -2 \\
180 & -2 & 5 \\
108 & 3 & -7 \\
72 & -5 & 12 \\
36 & 8 & -19 \\
\end{array}
$$

Since the last nonzero remainder is 36, gcd(1800, 756) = 36.   ∎

4.2.7   DEFINITION ▶   Nonzero integers are *relatively prime* if and only if their greatest common divisor is 1; in other words, if and only if 1 is the only positive integer which divides both the given integers.

Pause 7   Show that 17,369 and 5472 are relatively prime.

There is a very important property of the greatest common divisor which we have so far overlooked. As seen most clearly in equations (1), each remainder which arises in the course of applying the algorithm is an *integral linear combination* of the given two numbers a and b; that is, each remainder can be written in the form $ma + nb$ for integers m and n. In particular, the greatest common divisor of a and b, being the last nonzero remainder, is an integral linear combination of a and b. For example, as we saw in (1), gcd(630, 196) = 14 = $5a - 16b = 5(630) - 16(196)$. Similarly, the calculations of Problem 10 show that gcd(1800, 756) = 36 = 8(1800) − 19(756).

Pause 8   In Pause 7 we asked you to show that the integers 17,369 and 5472 are relatively prime. Find integers m and n so that $17{,}369m + 5472n = 1$.

4.2.8   REMARK ▶   Since, by definition, the greatest common divisor of two integers a and b, which are not both 0, is a positive number, it is clear that $\gcd(a, b) = \gcd(|a|, |b|)$. For example, gcd(15, −36) = gcd(15, 36) = 3 and gcd(−18, −99) = gcd(18, 99) = 9. Since $|a|$ and $|b|$ are natural numbers, the Euclidean algorithm in fact can be used to find the greatest common divisor of *any* pair of nonzero integers a and b.

We have been illustrating through examples a general fact: The greatest common divisor of two natural numbers a and b is a linear combination of a and b. If

$a = 0$ and $b \geq 1$, then $\gcd(a, b) = b = 0(a) + 1(b)$ is also a linear combination of a and b. Using our remark, it is clear that if $\gcd(|a|, |b|)$ is a linear combination of $|a|$ and $|b|$, then so is $\gcd(a, b)$. For example, since $\gcd(1800, 756) = 36 = 8(1800) - 19(756)$, $\gcd(1800, -756) = 36 = 8(1800) + 19(-756)$. The following theorem, the proof of which follows by noting that each remainder in the Euclidean algorithm is a linear combination of the previous two, describes the most important property of the greatest common divisor. (See also Exercise 24.)

4.2.9 THEOREM ▶ The greatest common divisor of integers a and b is an integral linear combination of them; that is, if $g = \gcd(a, b)$, then there are integers m and n such that $g = ma + nb$.

The corollaries which follow illustrate ways in which this theorem is used.

4.2.10 COROLLARY ▶ Suppose a, b and x are integers such that $a \mid bx$. If a and b are relatively prime, then $a \mid x$.

Proof We know that there are integers m and n so that $ma + nb = 1$ [because $\gcd(a, b) = 1$]. Multiplying by x, we obtain $max + nbx = x$. But $a \mid max$ and $a \mid nbx$ because $nbx = n(bx)$ and we are given that $a \mid bx$. Thus, a divides the sum $max + nbx = x$. ∎

4.2.11 COROLLARY ▶ The greatest common divisor of nonzero integers a and b is divisible by every common divisor of a and b.

Proof Let $g = \gcd(a, b)$. By Theorem 4.2.9, $g = ma + nb$ for integers m and n. Thus, by Proposition 4.2.2, if c is a common divisor of a and b, then $c \mid g$. ∎

PROBLEM 11. Suppose a, b and c are three nonzero integers with a and c relatively prime. Show that $\gcd(a, bc) = \gcd(a, b)$.

Solution. Let $g_1 = \gcd(a, bc)$ and $g_2 = \gcd(a, b)$. Since $g_2 \mid b$, we know that $g_2 \mid bc$. Since also $g_2 \mid a$, we have $g_2 \leq g_1$. On the other hand (and just as in Corollary 4.2.10), since $\gcd(a, c) = 1$, there are integers m and n such that $ma + nc = 1$. Multiplying by b, we obtain $mab + nbc = b$. Since $g_1 \mid a$ and $g_1 \mid bc$, it must also be the case that $g_1 \mid b$. Then, since $g_1 \mid a$, $g_1 \leq g_2$. Therefore, $g_1 = g_2$. ∎

The Least Common Multiple

In paragraph 2.5.6, the greatest lower bound of two elements a and b in a partially ordered set $(A, \preceq)$ was defined to be an element $g = a \wedge b \in A$ with the properties

1. $g \preceq a$, $g \preceq b$, and
2. if $c \preceq a$ and $c \preceq b$ for some $c \in A$, then $c \preceq g$.

Corollary 4.2.11 therefore shows that every pair of natural numbers has a glb in the poset $(\mathsf{N}, \mid)$, namely, their greatest common divisor: For $a, b \in \mathsf{N}$,

$$a \wedge b = \gcd(a, b).$$

It is also true that every pair of natural numbers has a least upper bound.

4.2.12 DEFINITION ▶ If a and b are nonzero integers, we say that ℓ is the *least common multiple (lcm)* of a and b and write $\ell = \mathrm{lcm}(a, b)$ if and only if ℓ is a positive integer satisfying

1. $a \mid \ell$, $b \mid \ell$ and,
2. if m is any positive integer such that $a \mid m$ and $b \mid m$, then $\ell \leq m$.

EXAMPLES 12
- The least common multiple of 4 and 14 is 28.
- $\mathrm{lcm}(-6, 21) = 42$.
- $\mathrm{lcm}(-5, -25) = 25$. ▲

Since $|ab|$ is a common multiple of a and b, the least common multiple always exists and does not exceed $|ab|$. Remember also that a least common multiple is always positive (by definition).

In the exercises (see also Exercise 29 of Section 4.3), we ask you to derive the formula

(2)
$$\gcd(a, b)\,\mathrm{lcm}(a, b) = |ab|,$$

which holds for any nonzero integers a and b and gives a quick way to compute least common multiples.

EXAMPLES 13
- Since $\gcd(6, 21) = 3$, it follows that $\mathrm{lcm}(6, 21) = \frac{6(21)}{3} = \frac{126}{3} = 42$.
- Since $\gcd(630, -196) = 14$ (Problem 9), it follows that $\mathrm{lcm}(630, -196) = \frac{630(196)}{14} = 8820$. ▲

Just as the greatest common divisor of a and b is divisible by all common divisors of a and b, we can show that the least common multiple of a and b is a divisor of all common multiples of a and b (see Exercise 27). Thus, the least common multiple of natural numbers a and b is their least upper bound in the poset $(\mathsf{N}, \mid)$ (see again paragraph 2.5.6). For $a, b \in \mathsf{N}$,

$$a \vee b = \mathrm{lcm}(a, b).$$

Remembering that a lattice is a partially ordered set in which every two elements have a greatest lower bound and a least upper bound, the following proposition is immediate.

4.2.13 PROPOSITION ▶ The poset $(\mathsf{N}, \mid)$ is a lattice.

The Lattice of Divisors of a Natural Number

Implicit in the definition of lattice is the fact that the greatest lower bound and least upper bound of every pair of elements should lie in the set. The poset $(\{2, 3, 4\}, \mid)$, for instance, is not a lattice because while, for example, 2 and 3 have a glb in N, this element is not in $\{2, 3, 4\}$.

We have seen that $(\mathsf{N}, \mid)$ is a lattice. Also, for many subsets A of N, $(A, \mid)$ is a lattice. For example, if n is any natural number and $A = \{d \in \mathsf{N} \mid d \mid n\}$ is the set of positive divisors of n, then $(A, \mid)$ is a lattice. (See Exercise 33.)

EXAMPLE 14 With $A = \{1, 2, 3, 5, 6, 10, 15, 30\}$, the set of positive divisors of $n = 30$, $(A, |)$ is a lattice whose Hasse diagram is shown in Fig 4.2. Notice that every pair of elements of A has a glb and a lub. For every pair of elements $a, b \in A$, $a \wedge b$ is the unique element below (and connected with lines to) both a and b and $a \vee b$ is the unique element above (and connected with lines to) both a and b. ▲

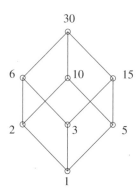

Figure 4.2 The Hasse diagram for $(A, |)$, where A is the set of divisors of 30. Compare with Fig 2.4, p. 65.

Answers to Pauses

3. For any integer n, we have $0 = qn$ for the integer $q = 0$.

4. Since $1 \mid a$ for all $a \in A$, no $a \neq 1$ can be minimal. The only minimal element is 1, and it is a minimum. The elements 4, 5, and 6 are maximal. For instance, 4 is maximal because there is no $a \in A$ such that $4 \mid a$, except $a = 4$. There are no maximum elements; for example, 4 is not maximum because $a \mid 4$ for all $a \in A$ is not true.

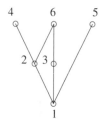

5. The largest divisor of a is $|a|$. Since $|a|$ is a common divisor of a and b, this must be their gcd.

6. Since a is a common divisor of a and 0, the greatest common divisor of these numbers is $|a|$.

7.

17,369	1	0
5472	0	1
953	1	-3
707	-5	16
246	6	-19
215	-17	54
31	23	-73
29	-155	492
2	178	-565
1	-2647	8402

The last nonzero remainder—1 in this case—is the gcd. So 17,369 and 5472 are relatively prime.

8. $1 = -2647(17,369) + 8402(5472)$.

EXERCISES

The symbol [BB] means that an answer can be found in the Back of the Book.

1. [BB] We have seen in this section that $(N, |)$ is a partially ordered set.

(a) Is it totally ordered?

(b) Does it have a maximum? A minimum?

Explain your answers.

2. Consider the partially ordered set $(\{2, 4, 6, 8\}, |)$.

(a) Explain why every pair of elements in this poset has a greatest lower bound.

(b) Does every pair of elements have a least upper bound?

(c) Is the poset a lattice?

Explain your answers.

3. Draw the Hasse diagrams for each of the following partially ordered sets.

(a) [BB] $(\{2, 3, 4, 5, 6, 7\}, |)$

(b) $(\{1, 2, 3, 4, 5, 6, 7, 8, 9, 10\}, |)$

4. List all minimal, minimum, maximal, and maximum elements for each of the posets in Exercise 3.

5. [BB] Let n be a natural number. Given n consecutive integers, $a, a + 1, a + 2, \ldots, a + n - 1$, show that one of them is divisible by n.

6. [BB] Prove that $n^2 - 2$ (n an integer) is never divisible by 4.

7. Given that a and x are integers, $a > 1$, $a \,|\, (11x + 3)$, and $a \,|\, (55x + 52)$, find a.

8. [BB] Suppose a and b are integers with the same remainder upon division by some natural number n. Prove that $n \,|\, (a - b)$.

9. True or false? In each case, justify your answer with a proof or a counterexample (all variables represent integers).

(a) [BB] if $a \,|\, b$ and $b \,|\, -c$, then $a \,|\, c$.

(b) if $a \,|\, b$ and $c \,|\, b$, then $ac \,|\, b$.

(c) if $a \,|\, b$ and $a \,|\, c$, then $a \,|\, bc$.

(d) if $a \,|\, b$ and $c \,|\, d$, then $ac \,|\, bd$.

(e) if $a \,|\, b$ and $c \,|\, \dfrac{b}{a}$, then $c \,|\, b$ and $a \,|\, \dfrac{b}{c}$.

10. Suppose a and b are relatively prime integers and c is an integer such that $a \,|\, c$ and $b \,|\, c$. Prove that $ab \,|\, c$.

11. In each of the following cases, find the greatest common divisor of a and b and express it in the form $ma + nb$ for suitable integers m and n.

(a) [BB] $a = 93$, $b = 119$

(b) [BB] $a = -93$, $b = 119$

(c) [BB] $a = -93$, $b = -119$

(d) $a = 1575$, $b = 231$

(e) $a = 1575$, $b = -231$

(f) $a = -1575$, $b = -231$

(g) $a = -3719$, $b = 8416$

(h) $a = 100{,}996$, $b = 20{,}048$

(i) $a = 28{,}844$, $b = -15{,}712$

(j) $a = 12{,}345$, $b = 54{,}321$

12. Which of the pairs of integers in Exercise 11 are relatively prime?

13. [BB] If a and b are relatively prime integers, prove that $\gcd(a + b, a - b) = 1$ or 2.

14. [BB] Prove that integers a and b can have at most one greatest common divisor.

15. Prove that for integers a and b, $\gcd(a, a + b) = \gcd(a, b)$. [*Hint*: Mimic the proof of Lemma 4.2.5.]

16. (a) [BB] Find a pair of integers x and y such that $17{,}369x + 5472y = 4$. (See Pause 8.)

(b) Find integers x and y such that $154x + 260y = 4$.

(c) [BB] Show that no integers x, y satisfy $154x + 260y = 3$.

(d) Show that no integers x, y satisfy $196x + 260y = 14$.

17. (a) [BB] Given integers d, x and y, suppose there exist integers m and n such that $d = mx + ny$. Prove that $\gcd(x, y) \,|\, d$.

(b) Is the converse of (a) true? If $\gcd(x, y) \,|\, d$, need there exist integers m and n such that $d = mx + ny$?

18. If $k \in N$, prove that $\gcd(3k + 2, 5k + 3) = 1$.

19. [BB] Let a, b, $c \in N$. Prove that $\gcd(ac, bc) = c(\gcd(a, b))$.

20. If $a \in N$, prove that $\gcd(a, a + 2) = \begin{cases} 1 \text{ if } a \text{ is odd} \\ 2 \text{ if } a \text{ is even.} \end{cases}$

21. [BB] Prove that $\gcd(n, n + 1) = 1$ for any $n \in N$. Find integers x and y such that $nx + (n + 1)y = 1$.

22. Prove that if a, b, and c are natural numbers, $\gcd(a, c) = 1$ and $b \mid c$, then $\gcd(a, b) = 1$.

23. Let n and s be positive integers and suppose k is the least positive integer such that $n \mid ks$. Prove that $k = \dfrac{n}{\gcd(n, s)}$.

24. [BB] Use the Well-Ordering Principle to prove Theorem 4.2.9. [*Hint*: Consider the set of positive linear combinations of a and b.]

25. [BB] Find $\mathrm{lcm}(63, 273)$ and $\mathrm{lcm}(56, 200)$. [*Hint*: An easy method takes advantage of formula (2) of this section.]

26. [BB; (a)] Find the lcm of each of the pairs of integers given in Exercise 11.

27. Let a and b be nonzero integers with $\mathrm{lcm}(a, b) = \ell$. Let m be any common multiple of a and b. Prove that $\ell \mid m$.

28. [BB] Prove that $\gcd(a, b) \mid \mathrm{lcm}(a, b)$ for any nonzero integers a and b.

29. Establish formula (2) of this section by proving that the least common multiple of nonzero integers a and b is $|ab|/\gcd(a, b)$.

30. (a) [BB] Let x be an integer expressed in base 10. Suppose the sum of the digits of x is divisible by 3. Prove that x is divisible by 3.

(b) [BB] Prove that if an integer x is divisible by 3, then, when written in base 10, the sum of the digits of x is divisible by 3.

(c) Repeat (a) and (b) for the integer 9.

4.2.14 DEFINITION ▼

The greatest common divisor of n integers $a_1, a_2, \ldots, a_n$, not all zero, is a number g which is a common divisor of these integers (that is, $g \mid a_1, g \mid a_2, \ldots, g \mid a_n$) and which is the largest of all such common divisors (that is, if $c \mid a_1, c \mid a_2, \ldots, c \mid a_n$, then $c \le g$). It is denoted $\gcd(a_1, \ldots, a_n)$.

31. (a) Suppose a, b, c are nonzero integers. Show that $\gcd(a, b, c) = \gcd(\gcd(a, b), c)$.

(b) Show that the gcd of nonzero integers a, b, c is an integral linear combination of them.

(c) Find $\gcd(105, 231, 165)$ and express this as an integral linear combination of the given three integers.

(d) Answer (c) for the integers 6279, 8580, and 2873.

(e) Answer (c) for the integers 5577, 18,837, and 25,740.

32. Suppose that $(A_1, \preceq_1)$ and $(A_2, \preceq_2)$ are partial orders.

(a) Show that the definition
$$(x_1, x_2) \preceq (y_1, y_2) \leftrightarrow x_1 \preceq_1 y_1 \text{ and } x_2 \preceq_2 y_2$$
for $(x_1, x_2), (y_1, y_2) \in A_1 \times A_2$ makes $A_1 \times A_2$ a partially ordered set.

(b) Let $A_1 = A_2 = \{2, 3, 4\}$. Assign to A_1 the partial order $\le$ and to A_2 the partial order $|$. Partially order $A_1 \times A_2$ as in part (a). Show all relationships of the form $(x_1, x_2) \not\succeq (y_1, y_2)$.

(c) Draw the Hasse diagram for the partial order in (b).

(d) Find any maximal, minimal, maximum, and minimum elements which may exist in the partial order of (b).

(e) With A_1 and A_2 as in part (b), find those glbs and lubs which exist for each of the following pairs of elements.

　i. $(2, 2)$, $(3, 3)$
　ii. $(4, 2)$, $(3, 4)$
　iii. $(3, 2)$, $(2, 4)$
　iv. $(3, 2)$, $(3, 4)$

(f) Show, by example, that if $(A_1, \preceq_1)$ and $(A_2, \preceq_2)$ are total orders, then $(A_1 \times A_2, \preceq)$ need not be a total order.

33. (a) Let n be a natural number, $n > 1$. Let $A = \{a \in \mathbb{N} \mid a \mid n\}$. Prove that $(A, |)$ is a lattice.

(b) [BB] Draw the Hasse diagram associated with the poset given in (a) for $n = 12$.

(c) Draw the Hasse diagram associated with the poset given in (a) for $n = 36$.

(d) Draw the Hasse diagram associated with the poset given in (a) for $n = 90$.

34. (a) [BB] Let g be the greatest common divisor of integers m and n, not both 0. Show that there are infinitely many pairs a, b of integers such that $g = am + bn$.

(b) Suppose m and n are relatively prime integers each greater than 1.

　i. Show that there exist unique integers a, b with $0 < b < m$ such that $am + bn = 1$.

　ii. Show that there exist unique integers a, b with $0 < a < n$ and $0 < b < m$ such that $am = bn + 1$.

4.3 PRIME NUMBERS

In Section 4.2, we saw what it means to say that one integer "divides" another. The basic building blocks of divisibility are the famous "prime numbers". We examine some of the properties of prime numbers in this section.

4.3.1 DEFINITION ▶ A natural number $p \geq 2$ is called *prime* if and only if the only natural numbers which divide p are 1 and p. A natural number $n > 1$ which is not prime is called *composite*. Thus, $n > 1$ is composite if $n = ab$ where a and b are natural numbers with $1 < a, b < n$.

Thus, 2, 3, and 5 are primes while 4, 6, and 10 are composite. Note that the natural number 1 is neither prime nor composite. It is interesting to note that exactly one quarter of the numbers between 1 and 100 (inclusive) are prime. These primes are shown in Table 4.1.

Table 4.1 The primes less than 100.

2	3	5	7	11
13	17	19	23	29
31	37	41	43	47
53	59	61	67	71
73	79	83	89	97

You may have needed more than a few seconds to verify that some of the entries in Table 4.1 are, in fact, prime. Such verification gets increasingly more difficult as the numbers get bigger. One of the great challenges of mathematics and an active area of research today is that of finding efficient algorithms for checking whether or not large integers are primes. By the end of 1988, the largest known prime number was $2^{216,091} - 1$. This was discovered in 1985 by computer scientists working at Geosciences Corporation, Houston, who required just three hours on a Cray X-MP computer to verify that this number was, indeed, prime. Not everyone was impressed. A vice president of Chevron Oil was apparently quoted as saying, "The results are interesting, if true, but they are certainly not going to help me find oil."

The discovery of new primes is viewed as both a test of humankind's ingenuity and the reliability of new computers, so, when a new prime is discovered, it is widely publicized. Early in 1992, David Slowinski and Paul Gage of Cray Research Inc. announced that $2^{756,839} - 1$ is prime. Then in 1994 the same people announced the primality of $2^{859,433} - 1$, a number with a mere 258,716 digits. On November 13, 1996 Joel Armengaud, a 29-year-old computer programmer working in France, discovered that $2^{1,398,269} - 1$ is prime. A year later, Gordon Spence of Hampshire, England, with a Pentium PC running a program downloaded over the World Wide Web,[2] found that $2^{2,976,221} - 1$ is prime. This number, with 895,932 digits, would fill a 450-page book all by itself! New primes are currently being discovered at such a rate that it is difficult for the authors of textbooks to keep pace. In June 1999, using the "Lucas-Lehmer Test," Nayan Hajratwala discovered that $2^{6,972,593} - 1$, a number with 2,098,960 digits, was prime, and a year later this number had not been beaten. The search for the next largest prime presents a challenge that will never end because it has been known since the time of Euclid that the number of primes is infinite. To prove this fact, we first need a short lemma.

4.3.2 LEMMA ▶ Given any natural number $n > 1$, there exists a prime p such that $p \mid n$.

Proof We give a proof by contradiction. Thus, we suppose the lemma is false. In this case, the set of natural numbers greater than 1 which are not divisible by any prime is not empty. By the Well-Ordering Principle, the set contains a smallest element m. Since $m \mid m$, but m is not divisible by any prime, m itself cannot be prime, so m is divisible by some natural number a, with $1 < a < m$. By minimality of m, a must be divisible by a prime p. Since $p \mid a$ and $a \mid m$, we have $p \mid m$, a contradiction. ∎

Now we can easily show that there are an infinite number of primes. The wonderfully simple argument we present has been known since 300 B.C. and is commonly attributed to Euclid since it appears in Euclid's *Elements*. It is a model of a proof by contradiction.

4.3.3 THEOREM ▶ There are infinitely many primes.

Proof If the theorem is not true, there are just a finite number of primes $p_1, p_2, \ldots, p_t$. Let $n = (p_1 p_2 \cdots p_t) + 1$. By Lemma 4.3.2, n is divisible by some prime and hence by some p_i. Since $p_1 p_2 \cdots p_t$ is also divisible by p_i, the number $n - (p_1 p_2 \cdots p_t)$ must also be divisible by p_i. Thus, 1 is divisible by p_i, a contradiction. ∎

Pause 9 Although the number of primes is infinite, the number of natural numbers which are not prime is infinite too. Find an easy way to see this.

We now return to the question of determining whether or not a given integer is prime. Several elementary observations can be made at the outset. Any even integer is divisible by 2 and so is not prime (unless it equals 2). Similarly, any integer larger than 5 whose units digit is 0 or 5 is divisible by 5 and hence not prime. If n is not prime, Lemma 4.3.2 tells us that n is divisible by some prime. So, to verify that n is not prime, it is enough just to test the **prime** numbers less than n (rather than **all** numbers less than n) when searching for a divisor of n.

Here is another fact which decreases considerably the amount of testing that must be done when checking for divisors of an integer.

---

[2]The World Wide Web is full of information about large primes. For instance, there is lots of up-to-the-minute news in "The Prime Pages:" http://www.utm.edu/research/primes/.

4.3.4 LEMMA ▶ If a natural number $n > 1$ is not prime, then n is divisible by some prime number $p \le \sqrt{n}$.

Proof Since n is not prime, n can be factored $n = ab$ with $1 < a \le b < n$. Then $a \le \sqrt{n}$ since otherwise, $ab > \sqrt{n}\sqrt{n} = n$, a contradiction. As a natural number greater than 1, it follows that a is divisible by some prime p. Since $p \mid a$ and $a \mid n$, $p \mid n$ by transitivity. Also, since $a \le \sqrt{n}$ and $p \mid a$, we must have $p \le \sqrt{n}$. Thus, p is the desired prime factor of n. ∎

As a consequence of Lemma 4.3.4, when testing a natural number n for primality, we need only consider the possibility of a prime divisor less than or equal to the square root of n. For example, to verify that 97 is prime, we need only check the prime numbers less than or equal to $\sqrt{97}$. Since none of 2, 3, 5, 7 is a divisor of 97, we are assured that 97 is a prime number. It is apparent that Lemma 4.3.4 reduces dramatically the work involved in a primality check.

One simple procedure which goes back over 2000 years for finding all the primes less than or equal to some given integer is named after the Greek Eratosthenes (ca. 200 B.C.), chief librarian of the great library at Alexandria and a contemporary of Archimedes. Reportedly, Eratosthenes was the first person to estimate the circumference of the earth.

Pause 10 Who was Archimedes?

4.3.5 THE SIEVE OF ERATOSTHENES ▶ To find all the prime numbers less than or equal to a given natural number n,

- list all the integers from 2 to n,
- circle 2 and then cross out all multiples of 2 in the list,
- circle 3, the first number not yet crossed out or circled, and then cross out all multiples of 3,
- circle 5, the first number not yet crossed out or circled, and cross out all multiples of 5,
- at the general stage, circle the first number which is neither crossed out nor circled and cross out all its multiples,
- continue until all numbers less than or equal to $\sqrt{n}$ have been circled or crossed out.

When the procedure is finished, those integers which are not crossed out are the primes not exceeding n.

EXAMPLE 15 As an example of this procedure, we verify that the list of primes $p < 100$ given in Table 4.1 is correct. All the integers from 2 to 100 are listed in Figure 4.3. Initially, 2 was circled and then all even integers were crossed out with a single stroke. At the second stage, 3 was circled and all multiples of 3 not yet crossed out were crossed out with two strokes. Then 5 was circled and all multiples of 5 not yet crossed out were crossed out with three strokes. Finally, 7 was circled and all multiples of 7 not previously crossed out were crossed out with four strokes. The primes less than 100 are those which are not crossed out. ▲

$$\begin{array}{ccccccccccc}
② & ③ & \cancel{4} & ⑤ & \cancel{6} & ⑦ & \cancel{8} & \cancel{9} & \cancel{10} & 11 \\
\cancel{12} & 13 & \cancel{14} & \cancel{15} & \cancel{16} & 17 & \cancel{18} & 19 & \cancel{20} & \cancel{21} \\
\cancel{22} & 23 & \cancel{24} & \cancel{25} & \cancel{26} & \cancel{27} & \cancel{28} & 29 & \cancel{30} & 31 \\
\cancel{32} & \cancel{33} & \cancel{34} & \cancel{35} & 36 & 37 & 38 & \cancel{39} & \cancel{40} & 41 \\
\cancel{42} & 43 & \cancel{44} & \cancel{45} & 46 & 47 & 48 & \cancel{49} & \cancel{50} & \cancel{51} \\
\cancel{52} & 53 & 54 & \cancel{55} & 56 & \cancel{57} & 58 & 59 & \cancel{60} & 61 \\
\cancel{62} & \cancel{63} & 64 & \cancel{65} & 66 & 67 & 68 & \cancel{69} & \cancel{70} & 71 \\
\cancel{72} & 73 & 74 & \cancel{75} & 76 & \cancel{77} & 78 & 79 & \cancel{80} & \cancel{81} \\
\cancel{82} & 83 & 84 & \cancel{85} & 86 & \cancel{87} & 88 & 89 & \cancel{90} & \cancel{91} \\
\cancel{92} & \cancel{93} & 94 & \cancel{95} & 96 & 97 & 98 & \cancel{99} & \cancel{100} &
\end{array}$$

Figure 4.3 The Sieve of Eratosthenes used to determine the primes $p \leq 100$.

The prime numbers are considered "building blocks" because of the following important theorem, which generalizes observations such as $8 = 2 \cdot 2 \cdot 2$, $15 = 3 \cdot 5$ and $60 = 2 \cdot 2 \cdot 3 \cdot 5$.

4.3.6 THEOREM ▶ Every natural number $n \geq 2$ can be written $n = p_1 p_2 \cdots p_r$ as the product of prime numbers $p_1, p_2, \ldots, p_r$ or, by grouping equal primes, in the form $n = q_1^{\alpha_1} q_2^{\alpha_2} \cdots q_s^{\alpha_s}$ as the product of powers of distinct primes $q_1, q_2, \ldots, q_s$.

Proof If the result is false, then the set of integers $n \geq 2$ which cannot be written as the product of primes is not empty and so, by the Well-Ordering Principle, contains a smallest element m. This number cannot be prime, so $m = ab$ with $1 < a, b < m$. By minimality of m, each of a and b is the product of primes, hence so is m, a contradiction. ∎

EXAMPLE 16
$$\begin{aligned}
100 &= 2 \cdot 2 \cdot 5 \cdot 5 = 2^2 5^2 \\
1176 &= 2 \cdot 2 \cdot 2 \cdot 3 \cdot 7 \cdot 7 = 2^3 \cdot 3 \cdot 7^2 \\
21340 &= 2 \cdot 2 \cdot 5 \cdot 11 \cdot 97 = 2^2 \cdot 5 \cdot 11 \cdot 97.
\end{aligned}$$ ▲

Theorem 4.3.6 can be strengthened. There is, in fact, just one way to express a natural number as the product of primes, but in order to prove this, we need a preliminary result which is actually a special case of something we proved earlier (see Corollary 4.2.10). Notice that $3 \mid 12$ and regardless of how 12 is factored—$12 = 2(6) = 3(4)$—3 always divides one of the factors. The next proposition describes the most important property of a prime number.

4.3.7 PROPOSITION ▶ Suppose a and b are integers and p is a prime such that $p \mid ab$. Then $p \mid a$ or $p \mid b$.

This indeed follows from Corollary 4.2.10 because, if $p \nmid a$, then p and a are relatively prime.

Now suppose a prime p divides the product $a_1 a_2 \cdots a_k$ of k integers. By Proposition 4.3.7, $p \mid a_1$ or $p \mid a_2 \cdots a_k$. In the latter case, applying the proposition again, we see that $p \mid a_2$ or $p \mid a_3 \cdots a_k$. In this way, we obtain the following corollary.

4.3.8 COROLLARY ▶ If a prime p divides the product $a_1 a_2 \cdots a_k$ of integers, then p divides one of the a_i.

Now to prove that a natural number $n > 1$ can be factored in just one way as the product of primes, assume that n can be factored in two different ways:

$$(3) \qquad\qquad n = p_1 p_2 \cdots p_k = q_1 q_2 \cdots q_\ell$$

with all the p_i and q_i primes. After canceling equal factors on each side of this equation, we either have an equation which says that the product of certain primes is 1 (an absurdity) or another equation like (3), where none of the primes p_i is among the primes q_j. Then, since $p_1 \mid p_1 p_2 \cdots p_k$, it would follow that $p_1 \mid q_1 q_2 \cdots q_\ell$. By Corollary 4.3.8, $p_1 \mid q_j$ for one of the primes q_j. Since both p_1 and q_j are primes, this forces $p_1 = q_j$, a contradiction.

4.3.9 THE FUNDAMENTAL THEOREM OF ARITHMETIC ▶

The fact that every integer $n > 1$ can be written uniquely as the product of prime numbers is called the "Fundamental Theorem of Arithmetic." Because of its importance, we state it again.

Every integer $n \geq 2$ can be written in the form

$$(4) \qquad\qquad n = p_1 p_2 \cdots p_r$$

for unique primes $p_1, p_2, \ldots, p_r$; equivalently, every integer $n \geq 2$ can be written

$$(5) \qquad n = q_1^{\alpha_1} q_2^{\alpha_2} \cdots q_s^{\alpha_s} \qquad \text{(the } prime\ decomposition \text{ of } n)$$

as the product of powers of distinct prime numbers $q_1, q_2, \ldots, q_s$. These primes and the exponents $\alpha_1, \alpha_2, \ldots, \alpha_s$ are unique.

As noted, the decomposition in (5) is called the *prime decomposition* of n. For example, the prime decomposition of 120 is $2^3 \cdot 3 \cdot 5$.

4.3.10 DEFINITION ▶ The *prime factors* or *prime divisors* of an integer $n \geq 2$ are the prime numbers which divide n. The *multiplicity* of a prime divisor p of n is the largest number α such that $p^\alpha \mid n$.

Thus, the prime factors of an integer n are the primes p_i or q_i given in (4) and (5) and the multiplicities of $q_1, \ldots, q_s$ are the exponents $\alpha_1, \ldots, \alpha_s$, respectively. The prime factors of 14 are 2 and 7 and each has multiplicity 1. The prime factors of 120 are 2, 3, and 5; 2 has multiplicity 3, while 3 and 5 each have multiplicity 1.

PROBLEM 17. Find the prime decomposition of the greatest common divisor of nonzero integers a and b.

Solution. By the Fundamental Theorem of Arithmetic, a and b can be expressed in the form

$$a = \pm p_1^{\alpha_1} p_2^{\alpha_2} \cdots p_r^{\alpha_r}, \quad b = \pm p_1^{\beta_1} p_2^{\beta_2} \cdots p_r^{\beta_r}$$

for certain primes $p_1, p_2, \ldots, p_r$ and integers $\alpha_1, \alpha_2, \ldots, \alpha_r, \beta_1, \beta_2, \ldots, \beta_r$. (By allowing the possibility that some of the α_i or β_i are 0, we can assume that the same r primes occur in the decompositions of both a and b.) We claim that the greatest common divisor of a and b is

$$g = p_1^{\min(\alpha_1,\beta_1)} p_2^{\min(\alpha_2,\beta_2)} \cdots p_r^{\min(\alpha_r,\beta_r)},$$

where $\min(\alpha_i, \beta_i)$ denotes the smaller of the two nonnegative integers α_i and β_i.

Note that $g \mid a$ since the exponent $\min(\alpha_i, \beta_i)$ does not exceed α_i, the exponent of the prime p_i in a. Similarly, $g \mid b$. Next, assume that $c \mid a$ and $c \mid b$. Since any prime which divides c also divides a and b, the only primes dividing c must be among $p_1, p_2, \ldots, p_r$. Thus, $c = p_1^{\gamma_1} p_2^{\gamma_2} \cdots p_r^{\gamma_r}$ for some integers γ_i. Since $c \mid a$, however, we must have $\gamma_i \le \alpha_i$ for each i and, similarly, $\gamma_i \le \beta_i$ for each i because $c \mid b$. Hence, $\gamma_i \le \min(\alpha_i, \beta_i)$ for each i, so $c \mid g$ and the result follows. ∎

Prime numbers have held a special fascination for humankind ever since the Greeks realized there were infinitely many of them. They are so familiar, yet there are many questions concerning them which are easy to state but hard to answer. Many remain unsolved today. In the rest of this section, we briefly survey some of what is known and some of what is unknown about primes, but few details and no proofs will be given. For more information, the interested reader might consult Silverman's book,[3] for example.

Mersenne Primes

While it is very difficult to determine whether or not a given large integer is prime, the problem may be easier for special classes of integers. For example, integers of the form $7n$, $n \in \mathbb{N}$, are never prime if $n > 1$ since they are obviously divisible by 7. What about integers of the form $n^2 - 1$? Well, since $n^2 - 1 = (n-1)(n+1)$, these cannot be prime either as long as $n > 2$. Father Marin Mersenne (1588–1648) was interested in integers of the form $2^n - 1$. He showed that these could only be prime if n itself was prime. (See Exercise 17.) Then he noted that it wasn't sufficient just for n to be prime since $2^{11} - 1 = 2047 = 23 \cdot 89$ is not prime. He conjectured that $2^p - 1$ is prime if p is any of the primes 2, 3, 5, 7, 13, 17, 19, 31, 67, 127, 257 and composite for the other primes $p \le 257$. Unfortunately, he was wrong on several counts. For example, $2^{61} - 1$ is prime while $2^{257} - 1$ is not. (Resist the urge to factor!) By the end of 1997, just 35 primes of the form $2^p - 1$ had been found. Interestingly, all the large primes which have been discovered in recent years are in this list. In Mersenne's honor, primes of the form $2^p - 1$ are called *Mersenne primes*. While Mersenne primes continue to be discovered, the following problem is still unresolved.

4.3.11 OPEN PROBLEM ▶

Are there infinitely many Mersenne primes?

Fermat Primes

Another interesting class of prime numbers is the set of so-called *Fermat primes*, these being prime numbers of the form $2^{2^n} + 1$. For $n = 0, 1, 2, 3, 4$, indeed $2^{2^n} + 1$

---

[3] Joseph H. Silverman, *A Friendly Introduction to Number Theory*, second edition (Prentice Hall, 2001).

is prime and it was a guess of the seventeenth-century lawyer Pierre de Fermat (1601–1665), perhaps the most famous amateur mathematician of all time, that $2^{2^n} + 1$ is prime for all $n \geq 0$. In 1732, however, the great Swiss mathematician Léonhard Euler (1707–1783) showed that $2^{2^5} + 1$ is not prime—it is divisible by 641—and, to this day, no further Fermat primes have been discovered.

When a Cray supercomputer tires of hunting for Mersenne primes, it turns to a search for Fermat primes! In a number of cases, complete factorizations of $2^{2^n} + 1$ are known; in other cases, one or two prime factors are known. There are Fermat *numbers*—numbers of the form $2^{2^n} + 1$—known to be composite, though no one knows a single factor! As of 1995, the smallest Fermat number whose primality was unsettled was $2^{2^{24}} + 1$. More details, together with an indication as to how testing huge numbers for primality is used to check the reliability of some of the world's largest computers, can be found in an article by Jeff Young and Duncan A. Buell.[4]

| *Pause 11* | What are the first five Fermat primes? |

4.3.12 OPEN PROBLEM ▶ Are there more than five Fermat primes?

How Many Primes Are There?

We have discussed Euclid's observation that there are infinitely many primes (Theorem 4.3.3), so what does the question just posed really mean?

There is great interest in the proportion of natural numbers which are prime. Students who have enjoyed advanced calculus will know that the series

$$\frac{1}{1} + \frac{1}{2} + \frac{1}{3} + \frac{1}{4} + \cdots$$

diverges (the partial sums increase without bound as more and more terms are added), while the series

$$\frac{1}{1^2} + \frac{1}{2^2} + \frac{1}{3^2} + \frac{1}{4^2} + \cdots$$

converges (to $\pi^2/6$ in fact).[5] Notice that the second sum here is part of the first. So the fact that the second sum is finite while the first is infinite shows that the sequence of perfect squares $1^2, 2^2, 3^2, \ldots$ forms only a tiny part of the sequence of all natural numbers $1, 2, 3, \ldots$.

Which of these sequences does the set of primes most resemble? Does the sum of reciprocals of the primes

$$\frac{1}{2} + \frac{1}{3} + \frac{1}{5} + \frac{1}{7} + \cdots$$

converge or diverge? As a matter of fact, it diverges, so, in some sense, there really are a "lot" of prime numbers, many more than there are perfect squares, for example.

---

[4]"The Twentieth Fermat Number is Composite," *Mathematics of Computation* **50** (1988), 261–263.

[5]The reader who has not studied advanced calculus can relax! The ideas to which we make reference here are not critical to an overall appreciation of the mysteries of the primes.

A sophisticated result concerning the number of primes is the important "Prime Number Theorem," which was proven independently by Jacques Hadamard and Charles-Jean de la Vallée-Pousin in 1896. It gives an approximation to the function $\pi(x)$, which is the number of primes $p \leq x$. For example, $\pi(5) = 3$ since there are three primes $p \leq 5$. We have earlier noted that there are 25 primes $p \leq 100$; thus, $\pi(100) = 25$.

4.3.13 PRIME NUMBER THEOREM ▶ Let $\pi(x)$ denote the number of primes $p \leq x$. Then

$$\lim_{x \to \infty} \frac{\pi(x)}{x / \ln x} = 1; \quad \text{equivalently,} \quad \pi(x) \sim \frac{x}{\ln x}.$$

Students uncertain about limits might wish a translation! First of all, we say "is asymptotic to" for $\sim$. So the second statement of the theorem reads "$\pi(x)$ is asymptotic to $x / \ln x$," from which we infer that $\pi(x)$ is approximately equal to $x / \ln x$ for large x, the approximation getting better and better as x grows. Setting $x = 100$, Theorem 4.3.13 asserts that the number of primes $p \leq 100$ is roughly $100 / \ln 100 \approx 21.715$. Note that

$$\frac{\pi(100)}{100 / \ln 100} \approx \frac{25}{21.715} \approx 1.151.$$

[The symbol $\approx$ means "approximately."] Setting $x = 1,000,000$, the theorem says that the number of primes under 1 million is roughly $1,000,000 / \ln 1,000,000 \approx 72,382$. In fact, there are 78,498 such primes. Note that

$$\frac{\pi(1,000,000)}{1,000,000 / \ln 1,000,000} \approx \frac{78,498}{72,382} \approx 1.084.$$

As x gets larger, the fraction $\frac{\pi(x)}{x / \ln x}$ gets closer and closer to 1.

On June 23, 1993, at a meeting at the Isaac Newton Institute in Cambridge, England, Andrew Wiles of Princeton University announced a proof of "Fermat's Last Theorem," arguably the most famous open mathematics problem of all time.

4.3.14 FERMAT'S LAST THEOREM ▶ For any integer $n > 2$, the equation $a^n + b^n = c^n$ has no nonzero integer solutions a, b, c.

Notice that it is sufficient to prove this theorem just for the case that n is a prime. For example, if we knew that $a^3 + b^3 = c^3$ had no integral solutions, then neither would $A^{3n} + B^{3n} = C^{3n}$. If the latter had a solution, so would the former, with $a = A^n$, $b = B^n$, $c = C^n$.

When we first learned the Theorem of Pythagoras for right-angled triangles, we discovered that there are many triples a, b, c of integers which satisfy $a^2 + b^2 = c^2$; for example, $3, 4, 5$ and $5, 12, 13$, but are there triples of integers a, b, c satisfying $a^3 + b^3 = c^3$ or $a^7 + b^7 = c^7$ or $a^n + b^n = c^n$ for any values of n except $n = 2$?

Pierre de Fermat was notorious for scribbling ideas in the margins of whatever he was reading. In 1637, he wrote in the margin of Diophantus's book *Arithmetic* that he had found a "truly wonderful" proof that $a^n + b^n = c^n$ had no solutions in the positive integers for $n > 2$, but that there was insufficient space to write it down. Truly wonderful it must have been because for over 350 years, mathematicians were unable to find a proof, though countless many tried!

Amateur and professional mathematicians alike devoted years and even lifetimes to working on Fermat's Last Theorem. The theorem owes its name, by the way, to the fact that it is the last of the many conjectures made by Fermat during his lifetime to have resisted resolution. By now, most of Fermat's unproven suggested theorems have been settled (and found to be true).

In 1983, Gerd Faltings proved that, for each $n > 2$, the equation $a^n + b^n = c^n$ could have at most a finite number of solutions. While this was a remarkable achievement, it was a long way from showing that this finite number was zero. Then, in 1985, Kenneth Ribet of Berkeley showed that Fermat's Last Theorem was a consequence of a conjecture first proposed by Yutaka Taniyama in 1955 and clarified by Goro Shimura in the 1960s. It was a proof of the Shimura-Taniyama conjecture which Andrew Wiles announced on June 23, 1993, a truly historic day in the world of mathematics. The months following this announcement were extremely exciting as mathematicians all over the world attempted to understand Wiles's proof. Not unexpectedly in such a complex and lengthy argument, a few flaws were found. By the end of 1994, however, Wiles and one of his former graduate students, Richard Taylor, had resolved the remaining issues to the satisfaction of all.

Among the many exciting accounts of the history of Fermat's Last Theorem and of Wiles's work, we draw special attention to an article by Barry Cipra, "Fermat's Theorem—at Last!", which was the leading article in *What's Happening in the Mathematical Sciences*, Vol. **3** (1995–1996), published by the American Mathematical Society. Faltings himself wrote "The Proof of Fermat's Last Theorem by R. Taylor and A. Wiles" for the *Notices of the American Mathematical Society*, Vol. 42 (1995), No. 7. In fact, many excellent accounts have been written. We cite just a few. There is one entitled "The Marvelous Proof" by Fernando Q. Gouvêa, which appeared in the *American Mathematical Monthly*, Vol. 101 (1994), and others by Ram Murty, *Notes of the Canadian Mathematical Society*, Vol. 25 (September 1993) and by Keith Devlin, Fernando Gouvêa, and Andrew Granville, *Focus*, Vol. 13, Mathematical Association of America (August 1993).

More Open Problems

So far, we have only peeked into the Pandora's box of fascinating but unanswered problems concerning prime numbers. Here are a few more.

There are intriguing questions concerning prime "gaps," the distances between consecutive primes. On the one hand, there are arbitrarily long gaps in the list of primes. One way to see this is to observe that if the first $n + 1$ primes are $p_1, p_2, \ldots, p_{n+1}$, then all the numbers between $p_1 p_2 \cdots p_n + 2$ and $p_1 p_2 \cdots p_n + p_{n+1} - 1$ are composite. (See Exercise 26.) On the other hand, there are also very small gaps in the primes, for example, gaps of length two such as between 3 and 5 or 11 and 13.

Integers p and $p + 2$ which are both prime are called *twin primes*. For example, 3 and 5, 5 and 7, 11 and 13, 41 and 43 are twin primes. So are $4,648,619,711,505 \times 2^{60,000} \pm 1$, a 1999 discovery of Heinz-Georg Wassing, Antal Jarai, and Karl-Heinz Indlekofer. As of mid-2000, these numbers, each with 18,075 digits, remained the largest known pair of twin primes. It is possible, though not likely, that there are no more. Unlike prime numbers, it is not known whether or there are infinitely many twin primes. One tantalizing result, proved by Viggo Brun in 1921 using a variation of the Sieve of Eratosthenes, is that the sum of the reciprocals of just the twin primes converges. Having noted that the sum of the reciprocals of all the primes diverges, Brun's result is evidence that the number of twin primes is "small." On the other hand, there are 224,376,048 twin prime pairs less than 100 billion and this seems like a pretty big number. Is the total number finite? No one knows.

4.3.15 THE TWIN PRIME CONJECTURE ▶

There are infinitely many twin primes.

Observe that $4 = 2+2$, $6 = 3+3$, $8 = 5+3$, $28 = 17+11$, and $96 = 43+53$. Every even integer **appears** to be the sum of two primes. Is this really right? The conjecture that it is, first made by Christian Goldbach in 1742 in a letter to Euler, has proven remarkably resistant to solution. In 1937, I. M. Vinogradov showed that every sufficiently large integer can be written as the sum of at most **four** primes. *Sufficiently large* means that there is a positive integer n_0 such that every integer larger than n_0 satisfies the condition. In 1966, J. Chen showed that every sufficiently large **even** integer can be written as $x + y$ where x is prime and y is either prime or the product of two primes. So we seem to be close to a solution of Goldbach's conjecture, though the last step is often the hardest.

4.3.16 THE GOLDBACH CONJECTURE ▶

Every even integer greater than 2 is the sum of two primes.

Answers to Pauses

9. For instance, numbers of the form $7n$ are not prime, and there are infinitely many of these.
10. Archimedes was a Greek scientist of the third century B.C. perhaps best known for the "Principle of Archimedes." This states that the weight of the fluid displaced by a floating object is equal to the weight of the object itself.
11. The first five Fermat primes are $2^{(2^0)} + 1 = 2^1 + 1 = 3$, $2^{(2^1)} + 1 = 2^2 + 1 = 5$, $2^{(2^2)} + 1 = 2^4 + 1 = 17$, $2^{(2^3)} + 1 = 2^8 + 1 = 257$, and $2^{(2^4)} + 1 = 2^{16} + 1 = 65537$.

EXERCISES

The symbol [BB] means that an answer can be found in the Back of the Book.

1. Determine whether or not each of the following integers is a prime.

   (a) [BB] 157
   (b) [BB] 9831
   (c) 9833
   (d) 55,551,111
   (e) $2^{216,090} - 1$

2. Can Lemma 4.3.4 be strengthened? In other words, does there exist a number $r < \sqrt{n}$ such that if n is not prime, then n has a prime factor $p \le r$?

3. (a) [BB] Suppose p is the smallest prime factor of an integer n and $p > \sqrt{n/p}$. Prove that n/p is prime.

(b) [BB] Express 16,773,121 as the product of primes given that 433 is this number's smallest prime factor.

4. Find the prime decomposition of each of the following natural numbers.

   (a) [BB] 856
   (b) 2323
   (c) 6647
   (d) 9970
   (e) [BB] $(2^8 - 1)^{20}$
   (f) 55,551,111

5. [BB] Use the Fundamental Theorem of Arithmetic to prove that for no natural number n does the integer 14^n terminate in 0.

6. Let $A = \{\frac{m}{n} \in \mathbb{Q} \mid 3 \nmid n\}$.

   (a) List five different elements of A at most one of which is an integer.
   (b) [BB] Prove that A is closed under addition.
   (c) Prove that A is closed under multiplication.

7. Let A be any subset of $\mathbb{Z} \setminus \{0\}$ and, for $a, b \in A$, define $a \sim b$ if ab is a perfect square (that is, the square of an integer). Show that $\sim$ defines an equivalence relation on A.

8. True or false? Explain your answers.

   (a) For all $n \in \mathbb{N}$, $n > 1$, there exists a prime p such that $p \mid n$.
   (b) [BB] There exists a prime p such that $p \mid n$ for all $n \in \mathbb{N}$, $n > 1$.

   (The position of the universal quantifier makes a world of difference!)

9. Define $f: \mathbb{N} \setminus \{1\} \to \mathbb{N}$ by setting $f(n)$ equal to the largest prime divisor of n.

   (a) Find the range of f.
   (b) Is f one-to-one?
   (c) Is f onto?
   (d) Why did we not express f as a function $\mathbb{N} \to \mathbb{N}$?

   Explain your answers.

10. Determine whether or not each of the following functions $\mathbb{N} \times \mathbb{N} \to \mathbb{N}$ is one-to-one. Explain your answers.

    (a) [BB] $f(n, m) = 2^m 6^n$
    (b) $f(n, m) = 36^m 6^n$

11. **(a)** [BB] Find $\pi(10)$ and approximate values of $10/\ln 10$ and $\frac{\pi(10)}{10/\ln 10}$ (three decimal place accuracy).

(b) Find $\pi(50)$ and approximate values of $50/\ln 50$ and $\frac{\pi(50)}{50/\ln 50}$.

(c) Find $\pi(95)$ and approximate values of $95/\ln 95$ and $\frac{\pi(95)}{95/\ln 95}$.

12. **(a)** Use the Sieve of Eratosthenes to list all the primes less than 200. Find $\pi(200)$ and the values of $200/\ln 200$ and $\frac{\pi(200)}{200/\ln 200}$ (to three decimal places).

    (b) Use the Sieve of Eratosthenes to list all the primes less than 500. Find $\pi(500)$ and the values of $500/\ln 500$ and $\frac{\pi(500)}{500/\ln 500}$ (to three decimal places).

13. Estimate the number of primes less than 5000, less than 50,000, less than 500,000, and less than 5,000,000.

14. [BB] Let p and q be distinct primes and n a natural number. If $p \mid n$ and $q \mid n$, why must pq divide n?

15. [BB] Given distinct positive integers a and b, show that there exists an integer $n \geq 0$ such that $a + n$ and $b + n$ are relatively prime.

16. For any natural number n, let $d(n)$ denote the number of positive divisors of n. For example, $d(4) = 3$ because 4 has three positive divisors; namely, 1, 2, and 4.

    (a) Describe those natural numbers n for which $d(n) = 2$.
    (b) [BB] Describe those natural numbers n for which $d(n) = 3$.
    (c) Describe those natural numbers n for which $d(n) = 5$.

17. **(a)** [BB] Is $2^{15} - 1$ prime? Explain your answer.
    (b) Is $2^{91} - 1$ prime? Explain your answer.
    (c) Show that if $2^n - 1$ is prime, then necessarily n is prime.
    (d) Is the converse of (c) true? (If n is prime, need $2^n - 1$ be prime?)

18. **(a)** [BB] Show that $2^6 + 1$ is not prime.
    (b) Show that $2^{20} + 1$ is not prime.
    (c) Show that if $2^n + 1$ is prime, then necessarily n is a power of 2.

19. **(a)** Show that the sum of two odd prime numbers is never prime.
    (b) Is (a) true if the word odd is deleted?

20. [BB] Show that the sum of two consecutive primes is never twice a prime.

21. If n is an odd integer, show that $x^2 - y^2 = 2n$ has no integer solutions.

22. [BB] True or false: $\{n \in \mathbb{N} \mid n > 2 \text{ and } a^n + b^n = c^n \text{ for some } a, b, c \in \mathbb{N}\} = \emptyset$?

23. [BB] Suppose that a, b, and c are integers each two of which are relatively prime. Prove that $\gcd(ab, bc, ac) = 1$.

24. Let a, b, and c be integers each relatively prime to another integer n. Prove that the product abc is relatively prime to n.

25. Given that p is prime, $\gcd(a, p^2) = p$ and $\gcd(b, p^3) = p^2$, find

(a) [BB] $\gcd(ab, p^4)$

(b) $\gcd(a + b, p^4)$

26. Let p_1, p_2, $\ldots$, p_{n+1} denote the first $n + 1$ primes (in order). Prove that every number between $p_1 p_2 \cdots p_n + 2$ and $p_1 p_2 \cdots p_n + p_{n+1} - 1$ (inclusive) is composite. How does this show that there are gaps of arbitrary length in the sequence of primes?

27. If the greatest common divisor of integers a and b is the prime p, what are the possible values of

(a) [BB] $\gcd(a^2, b)$?

(b) $\gcd(a^3, b)$?

(c) $\gcd(a^2, b^3)$?

28. [BB] Let a and b be natural numbers with $\gcd(a, b) = 1$ and ab a *perfect square*; that is, $ab = x^2$ for some natural number x. Prove that a and b are also perfect squares.

29. Let a and b be natural numbers.

(a) Find the prime decomposition of $\operatorname{lcm}(a, b)$ in terms of the prime decompositions of a and b and prove your answer. (See Problem 17, p. 118.)

(b) Use (a) to prove formula (2) of Section 4.2 in the case $a, b > 0$: $\gcd(a, b) \operatorname{lcm}(a, b) = ab$.

30. [BB] Prove that an integer which is both a square (a^2 for some a) and a cube (b^3 for some b) is also a sixth power.

31. Let a and b be integers. Let p be a prime. Answer true or false and explain:

(a) [BB] If $p \mid a^{11}$, then $p \mid a$.

(b) If $p \mid a$ and $p \mid (a^2 + b^2)$, then $p \mid b$.

(c) If $p \mid (a^9 + a^{17})$, then $p \mid a$.

32. [BB] Show that there are infinitely many triples of integers a, b, c which satisfy $a^2 + b^2 = c^2$.

33. (a) [BB] Prove that every odd positive integer of the form $3n + 2$, $n \in \mathbb{N}$, has a prime factor of the same form. What happens if the word odd is omitted?

(b) Repeat (a) for positive integers of the form $4n + 3$.

(c) Repeat (a) for positive integers of the form $6n + 5$.

(d) Prove that there are infinitely many primes of the form $6n + 5$.

34. Suppose p and $p + 2$ are twin primes and $p > 3$. Prove that $6 \mid (p + 1)$.

35. Let $\mathbb{Q}^+$ denote the set of positive rational numbers. If $\frac{m}{n}$ is in $\mathbb{Q}^+$, then

- either $m = 1$ or $m = p_1^{e_1} p_2^{e_2} \cdots p_k^{e_k}$ is the product of powers of distinct primes p_1, p_2, $\ldots$, p_k;
- either $n = 1$ or $n = q_1^{f_1} q_2^{f_2} \cdots q_\ell^{f_\ell}$ is the product of powers of distinct primes q_1, q_2, $\ldots$, q_ℓ

and, in the case $m \neq 1$ and $n \neq 1$, we may assume that no p_i equals any q_j. Now define $f: \mathbb{Q}^+ \to \mathbb{N}$ by

$$f(1) = 1$$

$$f\left(\frac{m}{n}\right) = \begin{cases} p_1^{2e_1} p_2^{2e_2} \cdots p_k^{2e_k} & \text{if } n = 1, m \neq 1 \\ q_1^{2f_1-1} q_2^{2f_2-1} \cdots q_\ell^{2f_\ell-1} & \text{if } m = 1, n \neq 1 \\ p_1^{2e_1} \cdots p_k^{2e_k} q_1^{2f_1-1} \cdots q_\ell^{2f_\ell-1} & \text{if } m \neq 1, n \neq 1 \end{cases}$$

(a) Find $f(8)$ [BB], $f(\frac{1}{8})$, $f(100)$ and $f(\frac{40}{63})$.

(b) Find x such that $f(x) = 1{,}000{,}000$ [BB], t such that $f(t) = 10{,}000{,}000$ and s such that such that $f(s) = 365{,}040$.

(c) Show that f is a bijection.

(This exercise, which gives a direct proof that the positive rationals are countable, is due to Yoram Sagher. See the *American Mathematical Monthly*, Vol. 96 (1989), no. 9, p. 823.)

36. For positive integers a and b, define $a \sim b$ if there exist integers $n \geq 1$ and $m \geq 1$ such that $a^m = b^n$.

(a) Prove that $\sim$ defines an equivalence relation on $\mathbb{N}$.

(b) Find $\overline{3}$, $\overline{4}$, and $\overline{144}$.

(c) Find the equivalence class of $a \in \mathbb{N}$. [*Hint*: Write $a = p_1^{\alpha_1} p_2^{\alpha_2} \cdots p_k^{\alpha_k}$.]

37. (a) Write a computer program which implements the Sieve of Eratosthenes as described in the text.

(b) Use your program to enumerate all primes less than 1000.

(c) What is $\pi(1000)$? Compare this with approximate values (three decimal place accuracy) of $1000/\ln 1000$ and $\frac{\pi(1000)}{1000/\ln 1000}$.

4.4 CONGRUENCE

If it is 11 p.m. in Los Angeles, what time is it in Toronto? If he or she were aware that Los Angeles is three time zones west of Toronto, a person might well respond (correctly) "2 a.m." The process by which the time in Toronto was obtained is sometimes called *clock arithmetic*; more properly, it is *addition modulo* 12: $11 + 3 = 14 = 2$ (modulo 12). It is based upon the idea of *congruence*, the subject of this section.

4.4.1 DEFINITION ▶ Let $n > 1$ be a fixed natural number. Given integers a and b, we say that a is *congruent to b modulo n* (or a is *congruent to b mod n* for short) and we write $a \equiv b \pmod{n}$, if and only if $n \mid (a - b)$. The number n is called the *modulus* of the congruence.

EXAMPLES 18 $3 \equiv 17 \pmod{7}$ because $3 - 17 = -14$ is divisible by 7; $-2 \equiv 13 \pmod{3}$ because $-2 - 13 = -15$ is divisible by 3; $60 \equiv 10 \pmod{25}$; $-4 \equiv -49 \pmod{9}$. ▲

As a binary relation on Z, congruence is

reflexive: $a \equiv a \pmod{n}$ for any integer a,
symmetric: if $a \equiv b \pmod{n}$, then $b \equiv a \pmod{n}$ and
transitive: if $a \equiv b \pmod{n}$ and $b \equiv c \pmod{n}$, then $a \equiv c \pmod{n}$.

Congruence is reflexive because $a - a = 0$ is divisible by n. It is symmetric because if $n \mid (a - b)$, then $n \mid (b - a)$ because $b - a = -(a - b)$. It is transitive because if both $a - b$ and $b - c$ are divisible by n, then so is their sum, which is $a - c$. Thus, for any $n > 1$, congruence mod n is an equivalence relation on Z.

We urge you to review the basic features of equivalence relations which were discussed in Section 2.4. Recall, for instance, that an equivalence relation partitions the underlying set into subsets called equivalence classes, the equivalence class of an element being the set of those elements to which it is equivalent. The equivalence classes of congruence mod n are called *congruence classes*.

4.4.2 DEFINITION ▶ The *congruence class mod n* of an integer a is the set of all integers to which a is congruent mod n. It is denoted $\overline{a}$. Thus,

$$\overline{a} = \{b \in \mathsf{Z} \mid a \equiv b \pmod{n}\}.$$

[Since congruence mod n is symmetric, we do not have to remember whether to write $a \equiv b \pmod{n}$ or $b \equiv a \pmod{n}$ in this definition. You can't have one without the other!]

EXAMPLE 19 Let $n = 5$. Since $-8 - 17 = -25$ is divisible by 5, we have $-8 \equiv 17 \pmod{5}$. Thus, -8 belongs to the congruence class of 17; in symbols, $-8 \in \overline{17}$. Notice also that $17 \in \overline{-8}$. In fact, you should check that $\overline{-8} = \overline{17}$. (See Proposition 4.4.3.) ▲

Let us find all congruence classes of integers mod 5. To begin,

$$\overline{0} = \{b \in \mathsf{Z} \mid b \equiv 0 \pmod{5}\}$$

$$= \{b \in \mathsf{Z} \mid 5 \mid (b - 0)\}$$

$$= \{b \in \mathsf{Z} \mid b = 5k \text{ for some integer } k\}$$

and

$$\overline{1} = \{b \in \mathsf{Z} \mid b \equiv 1 \pmod 5\}$$

$$= \{b \in \mathsf{Z} \mid 5 \mid (b - 1)\}$$

$$= \{b \in \mathsf{Z} \mid b - 1 = 5k \text{ for some integer } k\}$$

$$= \{b \in \mathsf{Z} \mid b = 5k + 1 \text{ for some integer } k\}.$$

In Section 2.4, we introduced the notation $5\mathsf{Z}$ and $5\mathsf{Z} + 1$ for these sets $\overline{0}$ and $\overline{1}$, respectively. Continuing, we find that

$$\overline{2} = \{b \in \mathsf{Z} \mid b = 5k + 2 \text{ for some } k \in \mathsf{Z}\} = 5\mathsf{Z} + 2$$

$$\overline{3} = \{b \in \mathsf{Z} \mid b = 5k + 3 \text{ for some } k \in \mathsf{Z}\} = 5\mathsf{Z} + 3$$

and

$$\overline{4} = \{b \in \mathsf{Z} \mid b = 5k + 4 \text{ for some } k \in \mathsf{Z}\} = 5\mathsf{Z} + 4.$$

It is useful to observe that the five congruence classes determined so far *partition* the integers in the sense that they are pairwise disjoint and their union is Z.

Why are they pairwise disjoint? If $a \in (5\mathsf{Z} + r) \cap (5\mathsf{Z} + s)$ for $0 \le r, s \le 4$, then $a = 5k + r = 5\ell + s$ for some integers k and ℓ and so $r - s = 5(\ell - k)$ would be a multiple of 5. For r, s between 0 and 4 this can only happen if $r = s$.

Why is their union Z? For any $a \in \mathsf{Z}$, by the Division Algorithm (Theorem 4.1.5), we can write $a = 5q + r$ with q, r integers and $0 \le r < 5$. Since $r \in \{0, 1, 2, 3, 4\}$, the integer a is in $5\mathsf{Z}$, $5\mathsf{Z} + 1$, $5\mathsf{Z} + 2$, $5\mathsf{Z} + 3$, or $5\mathsf{Z} + 4$.

Since congruence classes are equivalence classes and, in general, any two equivalence classes are disjoint or equal (Proposition 2.4.4), it follows that for any integer n, $\overline{n}$ must be one of $\overline{0}$, $\overline{1}$, $\overline{2}$, $\overline{3}$, or $\overline{4}$. With which of these classes, for example, does $\overline{5}$ coincide? Observe that $5 \in \overline{5}$, since $5 \equiv 5 \pmod 5$. Also, as noted previously, $5 \in \overline{0}$. Since the classes $\overline{0}$ and $\overline{5}$ are not disjoint, they are equal: $\overline{5} = \overline{0}$.

Similarly, since 6 belongs to both $\overline{6}$ and $\overline{1}$, we have $\overline{6} = \overline{1}$. You should also see that $\overline{7} = \overline{2}$, $\overline{8} = \overline{3}$, $\overline{-14} = \overline{1}$, and so on.

Our next few results are almost immediate consequences of the theory of equivalence relations.

4.4.3 PROPOSITION ▶ Let a, b, and n be integers with $n > 1$. Then the following statements are equivalent.

(1) $n \mid (a - b)$.

(2) $a \equiv b \pmod n$.

(3) $a \in \overline{b}$.

(4) $b \in \bar{a}$.

(5) $\bar{a} = \bar{b}$.

Proof   Each of the implications $(1) \rightarrow (2) \rightarrow (3)$ is a direct consequence of definitions and $(3) \rightarrow (4)$ follows from the symmetry of congruence. We may therefore complete the proof by establishing $(4) \rightarrow (5)$ and $(5) \rightarrow (1)$. To see that $(4) \rightarrow (5)$, let $b \in \bar{a}$. Then $b \equiv a \pmod{n}$ so $\bar{a} = \bar{b}$ by Proposition 2.4.3. To see that $(5) \rightarrow (1)$, suppose that $\bar{a} = \bar{b}$. Then $a \in \bar{b}$ because $a \in \bar{a}$, so $a \equiv b \pmod{n}$ and $n \mid (a - b)$.   ∎

Of all the equivalences established in Proposition 4.4.3, we emphasize perhaps the most important one.

4.4.4 COROLLARY ▶   For integers a, b, and n with $n > 1$,

$$a \equiv b \pmod{n} \text{ if and only if } \bar{a} = \bar{b}.$$

The next proposition generalizes the special case $n = 5$, which we have already investigated.

4.4.5 PROPOSITION ▶   Any integer is congruent mod n to its remainder upon division by n. Thus, there are n congruence classes of integers mod n corresponding to each of the n possible remainders

$$\begin{aligned}
\bar{0} &= n\mathsf{Z} \\
\bar{1} &= n\mathsf{Z} + 1 \\
\bar{2} &= n\mathsf{Z} + 2 \\
&\vdots \\
\overline{n-1} &= n\mathsf{Z} + (n - 1).
\end{aligned}$$

These congruence classes partition Z; that is, they are disjoint sets whose union is the set of all integers.

Proof   Suppose a is an integer. The remainder when a is divided by n is the number r, $0 \leq r < n$, obtained when we write $a = qn + r$ according to the Division Algorithm (Theorem 4.1.5). Since $a - r = qn$ is divisible by n, we obtain $a \equiv r \pmod{n}$, as claimed.

Since $0 \leq r < n$, the integer r is one of $0, 1, 2, \ldots, n - 1$. Thus, a belongs to one of the specified classes. It remains only to show that these classes are disjoint. For this, note that if $r_1, r_2 \in \{0, 1, \ldots, n - 1\}$, then $r_1 \not\equiv r_2 \pmod{n}$. Thus, $r_1 \in \bar{r_1}$, but $r_1 \notin \bar{r_2}$, so the congruence classes $\bar{r_1}$ and $\bar{r_2}$ are not equal. Since congruence classes are equivalence classes, $\bar{r_1} \cap \bar{r_2} = \emptyset$, by Proposition 2.4.4. Thus, the classes $\bar{0}, \ldots, \overline{n-1}$ are indeed disjoint.   ∎

EXAMPLE 20   Suppose $n = 32$. There are 32 congruence classes of integers mod 32; namely,

$$\bar{0} = 32\mathsf{Z}, \quad \bar{1} = 32\mathsf{Z} + 1, \quad \ldots, \quad \overline{31} = 32\mathsf{Z} + 31.$$

To determine the class to which a specific integer belongs, say 3958, we use the Division Algorithm to write $3958 = 123(32) + 22$.[6] Thus, 22 is the remainder, so $3958 \equiv 22 \pmod{32}$ by Proposition 4.4.5 and $3958 \in \overline{22}$. ▲

Pause 12 Find an integer r, $0 \le r < 18$, such that $3958 \equiv r \pmod{18}$. Do the same for -3958.

Proposition 4.4.5 says that every integer is congruent modulo n to one of the n integers $0, 1, 2, \ldots, n - 1$. Therefore, when working mod n, it is customary to assume that all integers are between 0 and $n - 1$ (inclusive) and to replace any integer a outside this range by its remainder upon division by n. This remainder is called "a (mod n)" and the process of replacing a by a (mod n) is called *reduction modulo n*.

4.4.6 DEFINITION ▶ If $n > 1$ is a natural number and a is any integer, a (mod n) is the remainder r, $0 \le r < n$, obtained when a is divided by n.

EXAMPLES 21 -17 (mod 5) $= 3$, 28 (mod 6) $= 4$ and -30 (mod 9) $= 6$.
The integer 29 reduced modulo 6 is 5. ▲

Just as for equations, to "solve" a congruence or a system of congruences involving one or more unknowns means to find all possible values of the unknowns which make the congruences true, always respecting the convention given in Definition 4.4.6. Without this convention, any even number is a solution to the congruence $2x \equiv 0 \pmod 4$. With the convention, however, we give only $x = 0$ and $x = 2$ as solutions.

PROBLEM 22. Solve each of the following congruences if possible. If no solution exists, explain why not.

(a) $3x \equiv 1 \pmod 5$
(b) $3x \equiv 1 \pmod 6$
(c) $3x \equiv 3 \pmod 6$

Solution. Simple congruences like these with small moduli are probably best solved by trying all possible values of x mod n.

(a)
$$\begin{aligned}
&\text{If } x = 0, \ 3x = 0 \equiv 0 \quad (\text{mod } 5). \\
&\text{If } x = 1, \ 3x = 3 \equiv 3 \quad (\text{mod } 5). \\
&\text{If } x = 2, \ 3x = 6 \equiv 1 \quad (\text{mod } 5). \\
&\text{If } x = 3, \ 3x = 9 \equiv 4 \quad (\text{mod } 5). \\
&\text{If } x = 4, \ 3x = 12 \equiv 2 \quad (\text{mod } 5).
\end{aligned}$$

Since the modulus is 5, we want x in the range $0 \le x < 5$. Thus, the only solution to the congruence is $x = 2$.

---

[6]Remember the easy way to see this. Use a calculator to compute $\frac{3958}{32} = 123.68\ldots$. The integer part of this number (the floor of $\frac{3958}{32}$) is the quotient given by the Division Algorithm. Refer to Proposition 4.1.6.

(b) There is no solution to this congruence because the values of $3x \pmod 6$ are just 0 and 3:

$$3(0) = 0, \quad 3(1) = 3, \quad 3(2) = 6 \equiv 0,$$

$$3(3) = 9 \equiv 3, \quad 3(4) = 12 \equiv 0, \quad 3(5) = 15 \equiv 3.$$

(c) The calculations in part (b) show that $3x \equiv 3 \pmod 6$ has solutions $x = 1$, $x = 3$, and $x = 5$. ∎

Suppose we want to find the sum $1017 + 2876 \pmod 7$. This can be accomplished in two ways. We could evaluate $1017 + 2876 = 3893 \equiv 1 \pmod 7$, but equally, we could reduce the integers 1017 and 2876 modulo 7 first like this:

$$1017 \equiv 2 \pmod 7$$

$$2876 \equiv 6 \pmod 7$$

$$1017 + 2876 \equiv 2 + 6 = 8 \equiv 1 \pmod 7.$$

The second approach is particularly useful when forming products since it keeps the numbers involved small. Observe:

$$(1017)(2876) \equiv (2)(6) = 12 \equiv 5 \pmod 7.$$

Computing powers of an integer modulo a natural number n is often just a mental (rather than calculator) exercise if we continually work mod n. For example,

$$(1017)^2 \equiv 2^2 = 4 \pmod 7$$

$$(1017)^3 = (1017)^2(1017) \equiv 4(2) = 8 \equiv 1 \pmod 7$$

$$(1017)^4 = (1017)^3(1017) \equiv 1(2) = 2 \pmod 7$$

$$(1017)^5 = (1017)^4(1017) \equiv 2(2) = 4 \pmod 7$$

and so on. Here is an easy way to compute $(1017)^{12}$:

$$(1017)^{12} = ((1017)^4)^3 \equiv 2^3 = 8 \equiv 1 \pmod 7.$$

The following proposition describes the general principles which guarantee that the sorts of calculations we have been performing are valid.

4.4.7 PROPOSITION ▶ If $a \equiv x \pmod n$ and $b \equiv y \pmod n$, then

(a) $a + b \equiv x + y \pmod n$ and
(b) $ab \equiv xy \pmod n$.

Proof Direct proofs of each part are suggested, and these we provide.

(a) We have to check that $(a + b) - (x + y)$ is divisible by n. This difference is $(a - x) + (b - y)$, which is the sum of two numbers each divisible by n, so the difference is itself divisible by n.

(b) We have to check that $ab - xy$ is divisible by n. Subtracting and adding ay, we notice that

$$ab - xy = ab - ay + ay - xy = a(b - y) + (a - x)y,$$

each term on the right again being divisible by n. Thus, $ab - xy$ is divisible by n. ∎

PROBLEM 23. Suppose a and b are integers and $3 \mid (a^2 + b^2)$. Show that $3 \mid a$ and $3 \mid b$.

Solution. We wish to prove that $a \equiv 0$ (mod 3) and $b \equiv 0$ (mod 3). If this is false, then $a \equiv 1$ or 2 (mod 3) and so, by part (b) of Proposition 4.4.7, $a^2 \equiv 1$ or 4 (mod 3). Since $4 \equiv 1$ (mod 3), we must have $a^2 \equiv 1$ (mod 3). Similarly, $b^2 \equiv 1$ (mod 3) and so, by (a) of the proposition, $a^2 + b^2 \equiv 1 + 1 = 2$ (mod 3), a contradiction. ∎

While addition and multiplication of congruences behave as we would hope, we must be **exceedingly** careful when dividing each side of a congruence. Dividing each side of the congruence $30 \equiv 12$ (mod 9) by 3, for example, produces the false statement $10 \equiv 4$ (mod 9). In general, we can only divide a congruence by an integer which is **relatively prime** to the modulus, as asserted by the next proposition. (We leave its proof to the exercises.)

4.4.8 PROPOSITION ▶ If $ac \equiv bc$ (mod n) and $\gcd(c, n) = 1$, then $a \equiv b$ (mod n).

For example, given $28 \equiv 10$ (mod 3), we can divide by 2 and obtain $14 \equiv 5$ (mod 3) because $\gcd(2, 3) = 1$. The linear congruences $2x \equiv 1$ (mod 7) and $6x \equiv 3$ (mod 7) have the same solutions since we can divide each side of the second congruence by 3, this being relatively prime to 7. On the other hand, multiplying a congruence by a number not relatively prime to the modulus changes the congruence. For example, the solutions to $2x \equiv 1$ (mod 9) and $6x \equiv 3$ (mod 9) are different!

Pause 13 Solve $2x \equiv 1$ (mod 9) and $6x \equiv 3$ (mod 9).

PROBLEM 24. Solve each of the following pairs of congruences, if possible. If no solution exists, explain why not.

   (a) $2x + 3y \equiv 1$ (mod 6)
        $x + 3y \equiv 4$ (mod 6)

   (b) $2x + 3y \equiv 1$ (mod 6)
        $x + 3y \equiv 5$ (mod 6)

Solution. We solve simple systems of linear congruences by the ad hoc methods with which we first solved systems of linear equations, by adding and subtracting and occasionally multiplying, though only by numbers relatively prime to the modulus (for a reason illustrated by Pause 13).

   (a) Adding the two congruences gives $3x + 6y \equiv 5$ (mod 6). Since $6y \equiv 0$ (mod 6), we have $3x \equiv 5$ (mod 6). This congruence has no solution because the values of $3x$ (mod 6) are 0 and 3. Thus, no x, y satisfy the given pair of congruences.

(b) This time, adding the two congruences gives $3x \equiv 0$ (mod 6) and hence $x \equiv 0$, $x \equiv 2$ or $x \equiv 4$ (mod 6). If $x \equiv 0$, then the second congruence says $3y \equiv 5$ (mod 6) to which there is no solution. If $x \equiv 2$, the second congruence says $2 + 3y \equiv 5$, hence $3y \equiv 3$ (mod 6), so $y \equiv 1$ (mod 6), $y \equiv 3$ (mod 6) or $y \equiv 5$ (mod 6). If $x \equiv 4$, the second congruence reads $4 + 3y \equiv 5$, so $3y \equiv 1$ (mod 6) and there is no solution. The pair of congruences has three solutions: $x \equiv 2$ (mod 6) and $y \equiv 1$ (mod 6), $y \equiv 3$ (mod 6) or $y \equiv 5$ (mod 6). ∎

Every integer a has an *additive inverse* modulo n, that is, there exists an integer x satisfying $a + x \equiv 0$ (mod n); for example, take $x = -a$ or $x = n - a$ to respect the convention in 4.4.6 above. The existence of additive inverses means that congruences of the form $a + x \equiv b$ (mod n) always have a solution obtained by adding the additive inverse of a to each side: $x = b + (-a) = b - a$. On the other hand, not every congruence of the form $ax \equiv b$ (mod n) has a solution.

The congruence $3x \equiv b$ (mod 6) may not have a solution (for example, when $b = 1$). Contrast this with the congruence $3x \equiv b$ (mod 7), which has a solution for any b since each of the integers $0, 1, 2, 3, 4, 5, 6$ is $3x$ (mod 7) for some x:

$$3(0) = 0, \quad 3(1) = 3, \quad 3(2) = 6, \quad 3(3) = 9 \equiv 2,$$

$$3(4) = 12 \equiv 5, \quad 3(5) = 15 \equiv 1, \quad 3(6) = 18 \equiv 4 \quad (\text{mod } 7).$$

A solution to $3x \equiv 1$ (mod 7), for instance, is $x = 5$. The difference between $3x \equiv b$ (mod 6) and $3x \equiv b$ (mod 7) is the modulus. In the second case, 3 is relatively prime to the modulus, 7, whereas in the first case, it is not.

4.4.9 PROPOSITION ▶

Let $n > 1$ be a natural number and let a be an integer with $\gcd(a, n) = 1$.

(a) There exists an integer s such that $sa \equiv 1$ (mod n). [We call s a multiplicative *inverse* of a (mod n).]

(b) For any integer b, the congruence $ax \equiv b$ (mod n) has a solution.

(c) The solution to $ax \equiv b$ (mod n) is unique mod n in the sense that if $ax_1 \equiv b$ (mod n) and $ax_2 \equiv b$ (mod n), then $x_1 \equiv x_2$ (mod n).

Proof

(a) By Theorem 4.2.9, the greatest common divisor of two integers is a linear combination of them. Here, since $\gcd(a, n) = 1$, we know there are integers s and t such that $sa + tn = 1$. Since $sa - 1$ is divisible by n, $sa \equiv 1$ (mod n).

(b) By part (a), we know that a has an inverse s (mod n). Multiplying the congruence $ax \equiv b$ by s, we have $sax \equiv sb$ (mod n) and so $x \equiv sb$. It is straightforward to verify that $x = sb$ is indeed a solution to $ax \equiv b$ (mod n) since $a(sb) = (as)b \equiv 1(b) \equiv b$ (mod n).

(c) Uniqueness follows directly from Proposition 4.4.8. Since $\gcd(a, n) = 1$, if $ax_1 \equiv ax_2$ (mod n), then $x_1 \equiv x_2$ (mod n). ∎

It is useful to observe that our proof [in part (b)] that $ax \equiv b$ (mod n) has a solution which essentially repeats the steps by which we solve the **equation**

$ax = b$ in real numbers. To solve $ax = b$, with a, b real numbers and $a \neq 0$, we multiply both sides by the (multiplicative) inverse of a to obtain $x = a^{-1}b$. We solve the congruence $ax \equiv b \pmod{n}$ exactly the same way. The only problem is to find the inverse of $a \pmod{n}$, if this exists.

PROBLEM 25. Solve the congruence $20x \equiv 101 \pmod{637}$.

Solution. We use the method described in the proof of Proposition 4.4.9. Write $-7(637) + 223(20) = 1$ and obtain $223(20) \equiv 1 \pmod{637}$. Thus, 223 is the inverse of 20 (mod 637). Multiplying each side of the given congruence by 223 gives $x \equiv 223(101) = 22{,}523 \equiv 228 \pmod{637}$. Thus, $x = 228$ provides a solution and one which is unique mod 637: Any other solution is congruent mod 637 to 228. ∎

PROBLEM 26. Part (a) of Proposition 4.4.9 says that if $\gcd(a, n) = 1$, then a has an inverse mod n. Show that the converse is also true: If a has an inverse mod n, then $\gcd(a, n) = 1$.

Solution. If a has an inverse mod n, there is an integer s such that $sa \equiv 1 \pmod{n}$, so $sa - 1 = qn$ for some integer q. Writing this as $1 = sa - qn$, it follows that any natural number which divides both n and a must divide 1 and hence be 1. So $\gcd(a, n) = 1$. ∎

As another application of Proposition 4.4.8, we derive *Fermat's Little Theorem*.

Pause 14 What was Fermat's "big" theorem?

4.4.10 FERMAT'S LITTLE THEOREM ▶ If p is a prime and $p \nmid c$, then $c^{p-1} \equiv 1 \pmod{p}$.

Proof We have $\gcd(c, p) = 1$. Thus, by Proposition 4.4.8, no two of the integers $c, 2c, \ldots, (p-1)c$ are congruent mod p. The same proposition also shows that none of the elements $c, 2c, \ldots, (p-1)c$ is 0 (mod p). Thus, modulo p, the $p-1$ integers $c, 2c, \ldots, (p-1)c$ are precisely $1, 2, \ldots, p-1$, in some order. Thus,

$$c \cdot 2c \cdot 3c \cdots (p-1)c \equiv 1 \cdot 2 \cdot 3 \cdots (p-1) \pmod{p}.$$

Letting $x = 1 \cdot 2 \cdot 3 \cdots (p-1)$,[7] this equation reads

$$xc^{p-1} \equiv x \pmod{p}$$

and, since $\gcd(p, x) = 1$, we can cancel x and obtain $c^{p-1} \equiv 1$, as required. ∎

Because of Fermat's Little Theorem, none of the congruences $2^2 \equiv 1 \pmod{3}$, $4^6 \equiv 1 \pmod{7}$, $9^{10} \equiv 1 \pmod{11}$ are surprises. Similarly, since 13,331 is prime, Fermat's Little Theorem shows $4^{13{,}330} \equiv 1 \pmod{13{,}331}$, $4^{13{,}331} \equiv 4 \pmod{13{,}331}$, and $4^{13{,}332} \equiv 16 \pmod{13{,}331}$.

---

[7]The product of all the integers from 1 to n inclusive is commonly denoted $n!$—read "n factorial"—so we could also write $(p-1)!$ instead of x. See Definition 5.1.2.

Explain the last two congruences here.

12. An integer a is congruent (mod 18) to its remainder when it is divided by 18. Since $\frac{3958}{18} = 219.\dot{8}$, $\lfloor\frac{3958}{18}\rfloor = 219$ and we find $3958 = 219(18) + 16$. The remainder is 16, so $3958 \equiv 16$ (mod 18). Similarly, $\lfloor -\frac{3958}{18}\rfloor = -220$, so $-3958 = -220(18) + 2$ and $-3958 \equiv 2$ (mod 18).

13. $2x \equiv 1$ (mod 9) implies $x = 5$, but $6x \equiv 3$ (mod 9) implies $x = 2$, $x = 5$ or $x = 8$.

14. Theorem 4.3.14.

15. $4^{13,331} = 4(4^{13,330}) \equiv 4(1) = 4$ (mod 13,331); $4^{13,332} = 4(4)(4^{13,330}) \equiv 16(1) = 16$ (mod 13,331), using Fermat's Little Theorem in each case to deduce that $4^{13,330} \equiv 1$ (mod 13,331).

EXERCISES

The symbol [BB] means that an answer can be found in the Back of the Book.

1. [BB] This question concerns congruence mod 7.

(a) List three positive and three negative integers in $\overline{5}$ and in $\overline{-3}$.

(b) What is the general form of an integer in $\overline{5}$ and of an integer in $\overline{-3}$?

2. This question concerns congruence mod 13.

(a) List four positive and five negative integers in $\overline{3}$ and in $\overline{-2}$.

(b) What is the general form of an integer in $\overline{3}$ and of an integer in $\overline{-2}$?

3. Find a (mod n) in each of the following cases.

(a) [BB] $a = 1286$, $n = 39$

(b) $a = 43,197$, $n = 333$

(c) [BB] $a = -545,608$, $n = 51$

(d) $a = -125,617$, $n = 315$

(e) $a = 11,111,111,111$, $n = 1111$

4. True or false? Give a reason for each answer.

(a) [BB] $\overline{2} = \overline{18}$ (mod 10)

(b) $7 \in \overline{-13}$ (mod 5)

(c) [BB] $-8 \equiv 44$ (mod 13)

(d) With respect to congruence mod 29, $\overline{17} \cap \overline{423} = \emptyset$

(e) $-18 \notin \overline{400}$ (mod 19)

5. List all congruence classes giving the most usual and one other name for each.

(a) [BB] congruence mod 3

(b) congruence mod 5

(c) congruence mod 8

6. Carry out each of the indicated calculations, giving the answer mod n.

(a) [BB] $21,758,623 + 17,123,055$, $n = 6$

(b) $(21,758,623)(17,123,055)$, $n = 6$

(c) $(17,123)^{50}$, $n = 6$

(d) 10^4, 10^8, 10^{12}, 10^{20}, 10^{24}, $n = 7$

(e) [BB] 2, 2^2, 2^3, $\ldots$, 2^{10}, $n = 11$

(f) 4, 4^2, 4^3, 4^4, 4^5, 4^6, 4^7, 4^8, 4^9, 4^{10}, $n = 11$

7. Find $a + b$ (mod n), ab (mod n) and $(a+b)^2$ (mod n) in each of the following situations:

(a) $a = 4003$, $b = -127$, $n = 85$

(b) $a = 17,891$, $b = 14,485$, $n = 143$

(c) $a = -389,221$, $b = 123,450$, $n = 10,000$

8. [BB] If $a \in \mathbb{Z}$ and $a \not\equiv 0$ (mod 7), show that $a \equiv 5^k$ (mod 7) for some integer k.

9. Find all integers x, $0 \le x < n$, satisfying each of the following congruences mod n. If no such x exists, explain why not.

(a) [BB] $3x \equiv 4$ (mod n), $n = 6$

(b) $4x \equiv 2$ (mod n), $n = 6$

(c) $4x \equiv 3$ (mod n), $n = 7$

(d) $4x \equiv 3$ (mod n), $n = 6$

(e) [BB] $2x \equiv 18$ (mod n), $n = 50$

(f) [BB] $5x \equiv 1$ (mod n), $n = 11$

(g) $5x \equiv 5$ (mod n), $n = 25$

(h) $4x \equiv 301$ (mod n), $n = 592$

(i) $65x \equiv 27$ (mod n), $n = 169$

(j) $4x \equiv 320$ (mod n), $n = 592$

(k) [BB] $16x \equiv 301 \pmod{n}$, $n = 595$

(l) $79x \equiv 15 \pmod{n}$, $n = 722$

(m) $155x \equiv 1185 \pmod{n}$, $n = 1404$

10. (a) Given integers a, b, c, d, x, and a prime p, suppose $(ax+b)(cx+d) \equiv 0 \pmod{p}$. Prove that $ax+b \equiv 0 \pmod{p}$ or $cx + d \equiv 0 \pmod{p}$.

(b) Find all integers x, $0 \le x < n$, which satisfy each of the following congruences. If no x exists, explain why not.

  i. [BB] $x^2 \equiv 4 \pmod{n}$, $n = 13$

  ii. $(2x + 1)(3x + 4) \equiv 0 \pmod{n}$, $n = 17$

  iii. $3x^2 + 14x - 5 \equiv 0 \pmod{n}$, $n = 97$

  iv. [BB] $x^2 \equiv 2 \pmod{n}$, $n = 6$

  v. $x^2 \equiv -2 \pmod{n}$, $n = 6$

  vi. $4x^2 + 3x + 7 \equiv 0 \pmod{n}$, $n = 5$

11. Find all integers x and y, $0 \le x, y < n$, which satisfy each of the following pairs of congruences. If no x, y exist, explain why not.

(a) [BB] $2x + y \equiv 1 \pmod{n}$ $n = 6$
$x + 3y \equiv 3 \pmod{n}$

(b) [BB] $x + 5y \equiv 3 \pmod{n}$ $n = 9$
$4x + 5y \equiv 1 \pmod{n}$

(c) $x + 5y \equiv 3 \pmod{n}$ $n = 8$
$4x + 5y \equiv 1 \pmod{n}$

(d) $7x + 2y \equiv 3 \pmod{n}$ $n = 15$
$9x + 4y \equiv 6 \pmod{n}$

(e) $3x + 5y \equiv 14 \pmod{n}$ $n = 28$
$5x + 9y \equiv 6 \pmod{n}$

12. [BB] Prove Proposition 4.4.8.

13. [BB] If $a \equiv b \pmod{n}$, show that $\gcd(a, n) = \gcd(b, n)$.

14. Let a and n be natural numbers with $n > 1$. Let r and s be integers and suppose $r \equiv s \pmod{n}$. True or false? Explain.

(a) [BB] $a^r \equiv a^s \pmod{n}$

(b) $r^a \equiv s^a \pmod{n}$

15. Suppose a and b are integers, $n > 1$ is a natural number, and $a \equiv b \pmod{n}$. True or false? In each case, prove or give a counterexample.

(a) [BB] $3a \equiv b^2 \pmod{n}$

(b) $a^2 \equiv b^2 \pmod{n}$

(c) $a^2 \equiv b^3 \pmod{n}$

(d) $a^2 \equiv b^2 \pmod{n^2}$

16. (a) [BB] Use the result of Problem 23 to prove that $\sqrt{2}$ is irrational; that is, show that it is impossible to write $\sqrt{2} = \frac{a}{b}$ for integers a, b. [*Hint:* Without loss of generality, assume that a and b have no common factors, other than ± 1. Why is there no loss of generality?]

(b) Let n be a natural number, $n \equiv 2 \pmod{3}$. Use Problem 23 to prove that $\sqrt{n}$ is irrational.

17. (a) Find all integers x, $0 \le x < n$, which satisfy each of the following congruences.

  i. $x^2 \equiv 1 \pmod{n}$, $n = 5$

  ii. $x^2 \equiv 1 \pmod{n}$, $n = 7$

  iii. $x^2 \equiv 1 \pmod{n}$, $n = 13$

(b) [BB] Suppose $p \ne 2$ is a prime. Find all integers x, $0 \le x < p$, such that $x^2 \equiv 1 \pmod{p}$.

18. (a) Find all integers x, $0 \le x < n$, which satisfy each of the following congruences.

  i. [BB] $x^2 \equiv 1 \pmod{n}$, $n = 5^2$

  ii. $x^2 \equiv 1 \pmod{n}$, $n = 5^3$

  iii. $x^2 \equiv 1 \pmod{n}$, $n = 7^2$

(b) Suppose $p \ne 2$ is a prime and k is any natural number. Find all integers x, $0 \le x < p^k$, which satisfy $x^2 \equiv 1 \pmod{p^k}$.

19. (a) Find all integers x, $0 \le x < n$, which satisfy each of the following congruences.

  i. $x^2 \equiv 1 \pmod{n}$, $n = 2$

  ii. $x^2 \equiv 1 \pmod{n}$, $n = 2^2$

  iii. $x^2 \equiv 1 \pmod{n}$, $n = 2^3$

  iv. $x^2 \equiv 1 \pmod{n}$, $n = 2^4$

(b) Let k be a natural number. Find all integers x, $0 \le x < 2^k$, which satisfy $x^2 \equiv 1 \pmod{2^k}$.

20. In each of the following, the given integer p is a prime.

(a) [BB] Find 18^{8970}, 18^{8971} and 18^{8972}, each mod p, $p = 8971$.

(b) Find $53^{20,592}$, $53^{20,593}$, $53^{20,594}$, each mod p, $p = 20,593$.

(c) Find 3^{508}, 3^{509}, 3^{512}, each mod p, $p = 509$.

(d) Find $6^{10,588}$, $6^{10,589}$, $6^{10,594}$, each mod p, $p = 10,589$.

(e) Find 8^{4056}, 8^{4058}, 8^{4060}, each mod p, $p = 4057$.

(f) Find 2^{3948} and $2^{3941} \pmod{p}$, $p = 3943$.

21. [BB] Show that $x^{97} - x + 1 \equiv 0 \pmod{97}$ has no solutions.

22. Let A be the set of congruence classes of integers modulo some natural number n. For $\bar{a}, \bar{b} \in A$, define $\bar{a} \preceq \bar{b}$ if

$ab \equiv a^2 \pmod{n}$. Prove or disprove that $\preceq$ is a partial order in each of the following cases.

(a) $n = p$ is a prime.

(b) $n = pq$ is the product of two distinct primes.

(c) n is divisible by the square of a prime. [*Hint*: It might be helpful first to consider the case $n = 12$.]

4.5   APPLICATIONS OF CONGRUENCE
International Standard Book Numbers

Since 1968, most published books have been assigned a ten-digit number called the *International Standard Book Number*—abbreviated ISBN— which identifies the country of publication, the publisher, and the book itself. In fact, all relevant information is stored in the first nine digits; the tenth digit is a *check digit* whose sole purpose is to give us confidence that the first nine digits are correct.

When a university department wishes to place an order for textbooks for a forthcoming semester, it is common for each faculty member to send to some administrative person the ISBNs of the books he or she will use and then for a single list of all desired ISBNs to be produced and sent to the bookstore. There lists from many departments are collected, grouped in various ways, and sent to publishers. It is not hard to see that there are many opportunities for numbers to be copied incorrectly. Thus, some easy way to encourage accuracy is essential. If the digits of an ISBN are denoted $a_1, a_2, \ldots, a_{10}$, with the first nine in the range 0–9, then a_{10} is chosen in the range 0–10 so that

(6) $$a_1 + 2a_2 + 3a_3 + \cdots + 9a_9 + 10a_{10} \equiv 0 \pmod{11}.$$

If a_{10} happens to be 10, it is recorded as an X.

For example, if the first nine digits are 0-936031-03, the tenth digit is chosen so that

$$1(0) + 2(9) + 3(3) + 4(6) + 5(0)$$
$$+ 6(3) + 7(1) + 8(0) + 9(3) + 10a_{10} \equiv 0 \pmod{11};$$

thus, $103 + 10a_{10} \equiv 0 \pmod{11}$, so $a_{10} = 4$. The ISBN would be recorded as 0-936031-03-4. If this number were copied to another list with an error in the fourth digit, say as 0-935031-03-4, a computer could easily check that

$$1(0) + 2(9) + 3(3) + 4(\underline{5}) + 5(0) + 6(3) + 7(1) + 8(0) + 9(3) + 10(4)$$
$$= 139 \equiv 7 \not\equiv 0 \pmod{11}$$

and hence the existence of a mistake would come to light. Note that it would be necessary to check previous lists to determine the precise error. Obtaining 7 instead of 0 as the result of our calculation provides no clue as to which digit was in error. For instance, the number 0-936031-53-4 also differs from the correct ISBN in one digit and for it we again have

$$1(0) + 2(9) + 3(3) + 4(6) + 5(0) + 6(3)$$
$$+ 7(1) + 8(5) + 9(3) + 10(4) = 183 \equiv 7 \pmod{11}.$$

We can show that the test given in (6) always detects errors in single digits (see Exercise 5), although it is quite possible for errors in two digits to cancel each other and so to go undetected.

Pause 16

Change two digits in 0-936031-03-4 so that $a_1 + 2a_2 + 3a_3 + \cdots + 9a_9 + 10a_{10} \equiv 0$ (mod 11) (and hence the changes would not be detected by our test).

On the other hand, when a number is copied, a common error is for consecutive digits to be transposed: We might dial the phone number 754-3628 instead of 754-3268, for instance.

PROBLEM 27. Show that the test in (6) detects transpositions of consecutive digits.

Solution. If $a_1 a_2 \cdots a_{10}$ is a correct ISBN number, then

$$a = a_1 + 2a_2 + \cdots + ia_i + (i + 1)a_{i+1} + \cdots + 10a_{10} \equiv 0 \pmod{11}.$$

Suppose that the digits a_i and a_{i+1} are transposed. We claim that the miscopied number $a_1 a_2 \cdots a_{i+1} a_i \cdots a_{10}$ does not satisfy (6). To see this, let

$$b = a_1 + 2a_2 + \cdots + ia_{i+1} + (i + 1)a_i + \cdots + 10a_{10}$$

and note that $a - b = i(a_i - a_{i+1}) + (i + 1)(a_{i+1} - a_i) = a_{i+1} - a_i$. Since $0 \le a_i, a_{i+1} \le 9$, the difference $a_{i+1} - a_i$ cannot be 0 (mod 11) (unless $a_i = a_{i+1}$, in which case there was no transposition error) and so $a \not\equiv b$ (mod 11) and $b \not\equiv 0$ (mod 11). The miscopied number does not pass the test. ∎

We conclude this section by observing that the test given in (6) can be abbreviated by writing it in the form $w \cdot a \equiv 0$ (mod 11), where a is the *vector* $(a_1, a_2, \ldots, a_{10})$, w the *weight vector* $(1, 2, \ldots, 10)$ and · denotes *dot product*. For readers unfamiliar with the concepts of vector and dot product, a *vector* is an n-tuple of numbers, $(a_1, a_2, \ldots, a_n)$ (n can be any natural number), and the dot product of a $= (a_1, a_2, \ldots, a_n)$ and b $= (b_1, b_2, \ldots, b_n)$ is the number

$$a \cdot b = a_1 b_1 + a_2 b_2 + \cdots + a_n b_n.$$

These notions are not critical to an understanding of this section but useful, perhaps, to students with some linear algebra background.

Universal Product Codes

Check digits permeate today's society. Besides forming part of an ISBN, they are attached to identification numbers of airline tickets, money orders, credit cards, bank accounts, drivers' licenses, and most items found in stores.

Figure 4.4 shows the kind of *universal product code* (UPC) with which North American consumers are familiar. Most goods for sale today can be identified uniquely by a special number called a *universal product number*. A universal product **code** is a way to represent a universal product **number** as a pattern of black and white stripes of various thicknesses. For obvious reasons, a universal product code is also called a *bar code*. (Later we shall see how this code works.)

Figure 4.4   A universal product code.

The universal product numbers we want to discuss are 12-digit numbers of the form x-xxxxx-xxxxx-x, where each x stands for a single digit between 0 and 9; for example, 0-12345-67890-1. The last digit is a check digit which serves to affirm (with some degree of confidence) the accuracy of the preceding eleven. For example, the universal product number shown in Fig 4.4 is 0-64200-11589-6 and 6 is the check digit. It is not unusual for a scanning device to misread a bar code. The use of a check digit, therefore, provides a way to alert the cashier that the number must be read again.

The check digit is determined by the rule

(7)   3(sum of digits in odd positions)

$$+ \text{(sum of digits in even positions)} \equiv 0 \pmod{10}.$$

For the number encoded by the bar code in Fig 4.4, for instance, we have

$$\text{sum of digits in odd positions} = 0 + 4 + 0 + 1 + 5 + 9 = 19$$

$$\text{sum of digits in even positions} = 6 + 2 + 0 + 1 + 8 + 6 = 23$$

and $3(19) + 23 = 80 \equiv 0 \pmod{10}$.

As with ISBNs, the rule for determining the check digit of a universal product number can be phrased in the language of vectors. If a_1–$a_2 \cdots a_6$–$a_7 \cdots a_{11}$–a_{12} is a universal product number and a denotes the vector $(a_1, a_2, \ldots, a_{12})$, then (7) just says $w \cdot a \equiv 0 \pmod{10}$ for a certain weight vector w.

Pause 17   What is w?

Finally, we consider the nature of a bar **code** itself. Notice that the bar code in Fig 4.4 begins and ends with a black-white-black sequence of thin stripes and is separated into distinct halves by a white-black-white-black-white sequence of thin stripes. In each half of the code, each of the digits 0–9 corresponds to a sequence of four stripes of varying thicknesses. In the **left** half, the pattern is white-black-white-black; in the **right** half, it is black-white-black-white.

Using 0 to denote a thin white stripe and 00, 000, 0000 to denote increasingly thicker white stripes and, similarly, using 1 to denote a thin black stripe and 11, 111, 1111 to denote increasingly thicker black stripes, Table 4.2 shows the right and left stripe sequences to which each digit in the range 0–9 corresponds. For example, on the left, 6 is encoded 0101111, that is, by

thin white (0) $---$ thin black (1) $---$ thin white

(0) $---$ thickest black (1111).

Table 4.2   The pattern of stripes used to encode each of the digits 0–9 on the left and right sides of a bar code.

	Encoding	
Digit	On left side	On right side
0	0001101	1110010
1	0011001	1100110
2	0010011	1101100
3	0111101	1000010
4	0100011	1011100
5	0110001	1001110
6	0101111	1010000
7	0111011	1000100
8	0110111	1001000
9	0001011	1110100

Note that the stripe sequence which encodes a digit on the right can be obtained from the stripe sequence on the left by interchanging black and white. (Also note that there is no connection between the sequences used to encode a digit and the base 2 representation of that digit.)

Only for convenience of the consumer is the universal product number sometimes written beneath the sequence of stripes which encodes it. The scanning device used by the cashier to read the code is, of course, only able to detect light and dark stripes.

How does such a device know whether a bar code has been passed over it from left to right or from right to left? Suppose the first four stripes read by the scanner are 0011101. At first thought, we might assume that this was the code for a digit on the left (after all, it begins with a white stripe), but it could also be the code for a digit on the right read backward (since it ends with a black stripe). Notice, however, that there is **no** digit whose left code is 0011101; thus, the scanner has read from right to left and the **last** digit in the product number has been coded 1011100 (the digits of 0011101 in reverse); the last digit in the product number must be 4. Since the sequence of stripes used to encode a digit, when taken in reverse order, is not the sequence of any other encoding, the scanner always knows the direction in which it is reading the code.

Pause 18   There is a simpler way to tell whether the code is being read left to right or right to left. Can you find it? (Study the sum of the digits in the codes given in Table 4.2.)

We refer readers who wish to learn more about check digits to a very interesting article by Joseph A. Gallian which appeared in 1991 in an issue of *The College Mathematics Journal*.[8]

The Chinese Remainder Theorem

The notion of congruence is so basic and useful that some people just have to spread the word. In the lunch room the other day, a discrete math student (DMS) boasted to his friend as follows:

> **DMS:** Calculators are neat, but there are lots of ways to work with numbers without having to use one. The trick is to keep the numbers small by working with remainders.
> **Friend:** What do you mean?
> **DMS:** I'm thinking of a three digit number. When I divide it by 12, I get a remainder of 4.
> **Friend:** So what?
> **DMS:** When I divide the same number by 25, I get a remainder of 15.
> **Friend:** So what?
> **DMS:** What was my number? I'll give you a hint. It's bigger than 200 but smaller than 500.
> **Friend (10 minutes later):** I give up.
> **DMS:** It's easy. I'll show you how to do it. (*See Exercise 19.*)

We conclude this section by considering certain systems of simultaneous congruences. The system

$$x \equiv 1 \pmod{4}$$
$$x \equiv 0 \pmod{30},$$

for example, has no solution because the first congruence says x is odd while the second says x is even. This situation occurs precisely because the moduli, 4 and 30, are not relatively prime.

Suppose m and n are relatively prime natural numbers. Then, for any integers a and b, the pair of congruences

$$(8) \qquad \begin{aligned} x &\equiv a \pmod{m} \\ x &\equiv b \pmod{n} \end{aligned}$$

has a solution. One way to see this is to note that there are integers s and t such that $sm + tn = 1$, by Theorem 4.2.9. Thus, $x = a(tn) + b(sm)$ is a solution to (8) since $tn \equiv 1 \pmod{m}$ and $sm \equiv 1 \pmod{n}$. Moreover, x is unique mod mn in the sense that, if x' is another solution, then $x \equiv x' \pmod{mn}$. For this, observe that if x and x' both satisfy (8), then $x - x'$ is divisible by both m and n and hence by mn. (See Exercise 10 of Section 4.2.)

PROBLEM 28.   Solve the system $x \equiv 2 \pmod{4}$
$x \equiv 6 \pmod{7}$.

[8]Joe Gallian, "The Mathematics of Identification Numbers," *The College Mathematics Journal* **22**, No. 3 (May 1991).

Solution. We write $1 = (-1)(7) + 2(4)$ and obtain $x = 2(-1)(7) + 6(2)(4) = 34$. Thus, mod 28, $34 \equiv 6$ is the unique solution to the given pair of congruences. ∎

We have been investigating a special case of the *Chinese Remainder Theorem*, named after the country where it was first discovered (circa A.D. 350).

4.5.1 CHINESE REMAINDER THEOREM ▶

Suppose $m_1, m_2, \ldots, m_t$ are pairwise relatively prime integers, that is, any two of them are relatively prime. Then, for any integers $a_1, a_2, \ldots, a_t$, the system of congruences

$$x \equiv a_1 \quad (\text{mod } m_1)$$

$$x \equiv a_2 \quad (\text{mod } m_2)$$

$$\vdots$$

$$x \equiv a_t \quad (\text{mod } m_t)$$

has a solution which is unique modulo the product $m_1 m_2 \cdots m_t$.

The proof, which is just an extension of the $t = 2$ case already discussed and a straightforward exercise in mathematical induction, is left to the exercises of Section 5.1.

The Chinese Remainder Theorem has many practical consequences, two of which we now describe. Each provides a convincing example of how "pure" mathematics can have very relevant applications.

Determining Numbers by Their Remainders

Let n be a natural number. By the Fundamental Theorem of Arithmetic (Theorem 4.3.9), n can be written $n = p_1^{\alpha_1} p_2^{\alpha_2} \cdots p_t^{\alpha_t}$, where $p_1, p_2, \ldots, p_t$ are unique distinct primes and the exponents $\alpha_i > 0$. Let $a > 1$ be some integer and let a_i be the remainder when a is divided by $p_i^{\alpha_i}$. By Proposition 4.4.5, $a \equiv a_i$ (mod $p_i^{\alpha_i}$); in other words,

$$a \equiv a_1 \quad (\text{mod } p_1^{\alpha_1})$$

$$a \equiv a_2 \quad (\text{mod } p_2^{\alpha_2})$$

$$\vdots$$

$$a \equiv a_t \quad (\text{mod } p_t^{\alpha_t}).$$

Since the numbers $p_1^{\alpha_1}, \ldots, p_t^{\alpha_t}$ are relatively prime, the Chinese Remainder Theorem says that a is unique mod n. Thus, if $0 < a < n$, then the integer a itself is uniquely determined. This idea is crucial for the manipulation of "large" numbers in computers since it implies that in order to store a number a larger than a given computer's capacity, it is sufficient to store the set of remainders which a leaves upon division by a set of prime powers. We illustrate.

PROBLEM 29. Let $n = 90 = 2 \cdot 5 \cdot 9$. Suppose a is an integer, $0 < a < 90$, which has remainders 1, 3, and 4 upon division by the prime powers 2, 5, and 9, respectively. We claim that a has been identified uniquely. Why?

Solution.   We are given that a is a solution to the system

$$x \equiv 1 \quad (\text{mod } 2)$$
$$x \equiv 3 \quad (\text{mod } 5)$$
$$x \equiv 4 \quad (\text{mod } 9).$$

The Chinese Remainder Theorem says that x is unique mod 90. Thus, x is congruent to exactly one a in the range $0 < a < 90$.

To find a, we solve the first two congruences by the method proposed earlier. We write $1(5) - 2(2) = 1$ and obtain $x \equiv 1(5) - 3(2)(2) = -7 \equiv 3 \pmod{10}$. Solving this congruence and the third one, we write $1(10) - 1(9) = 1$ and determine $a \equiv 4(10) - 3(9) = 13 \pmod{90}$. Thus, $a = 13 + 90t$ for some t. Since $0 < a < 90$, we must have $a = 13$. ∎

Suppose that a and b are integers and that $m_1, \ldots, m_t$ are pairwise relatively prime. (Typically, the m_i are powers of distinct prime numbers.) Let a_i and b_i be the remainders when a and b are divided by m_i, respectively. Since $a \equiv a_i$ $(\text{mod } m_i)$ and $b \equiv b_i$ $(\text{mod } m_i)$, we know that $ab \equiv a_i b_i$ $(\text{mod } m_i)$ by Proposition 4.4.7. The Chinese Remainder Theorem says that ab is determined uniquely, mod mn, by the remainders $a_1 b_1, \ldots, a_t b_t$. This idea can be exploited when a and b are (very) large to find the product ab in shorter time than is possible by conventional methods. Again, we illustrate with an example.

EXAMPLE 30   Let $a = 64$ and $b = 79$. We have

$$\begin{array}{llll}
a \equiv 0 & (\text{mod } 4) & \qquad b \equiv 3 & (\text{mod } 4) \\
a \equiv 1 & (\text{mod } 9) & \quad\text{and}\quad b \equiv 7 & (\text{mod } 9) \\
a \equiv 14 & (\text{mod } 25) & \qquad b \equiv 4 & (\text{mod } 25).
\end{array}$$

Thus, ab is a solution to the set of congruences

$$x \equiv 0(3) \equiv 0 \quad (\text{mod } 4)$$
$$x \equiv 1(7) \equiv 7 \quad (\text{mod } 9)$$
$$x \equiv 14(4) = 56 \equiv 6 \quad (\text{mod } 25).$$

We solve the first two congruences by writing $1 = 9 - 2(4)$ and obtaining $x \equiv 0(9) - 7(2)(4) = -56 \equiv 16 \pmod{36}$. Then $-9(36) + 13(25) = 1$ gives $x \equiv 6(-9)(36) + 16(13)(25) = 3256 \equiv 556 \pmod{900}$ as a solution to all three. Thus, $ab = 556 + 900t$ for some integer t. Since crude estimates give $4200 < ab < 5600$, we could determine $ab = 556 + 900(5) = 5056$, this being the only value of $556 + 900t$ in this range. ▲

Cryptography

Cryptography (also known as "cryptology") is the study of ways in which messages can be coded so that a third party, intercepting the code, will have great difficulty recovering the original text. Such coding is of importance not just to the military. Today, when so much information is transmitted electronically, secrecy is of paramount concern to us all. As a second application of the Chinese Remainder Theorem (and of other ideas introduced in this section), we discuss the *RSA Algorithm*, a method of encoding a message discovered in 1977 by Ronald Rivest, Adi Shamir, and Leonard Adleman.

By assigning to the letters of the alphabet the numbers 01, 02, ... , 26, to a space the number 27 and to various punctuation marks numbers beyond 27, it is apparent that any message determines a number M. "HELLO," for instance, would determine the number $M = 0805121215$. Here is how the RSA Algorithm converts M to another number E.

Choose two different primes p and q and a natural number s relatively prime to both $p - 1$ and $q - 1$. Let $r = pq$. Let E be the remainder when M^s is divided by r and use E to encode M; that is, transmit E instead of M. Since $E \equiv M^s$ (mod r), we have $E \equiv M^s$ (mod p) and $E \equiv M^s$ (mod q).

Using the RSA Algorithm, and publicizing the values of r and s, anybody can send you an encoded message which you can decode by the following scheme. Since $\gcd(s, p - 1) = 1$, there exist integers a and x such that $as + x(p - 1) = 1$, so $M = M^{as+x(p-1)} = (M^s)^a(M^{p-1})^x \equiv E^a$ (mod p) (by Fermat's Little Theorem). Similarly, $M \equiv E^b$ (mod q) for an integer b which you can determine because you know s and q. Since $M \equiv E^a$ (mod p) and $M \equiv E^b$ (mod q), by the Chinese Remainder Theorem, M is uniquely determined modulo r and, if $M < r$, then M is a uniquely determined integer. The success of this decoding procedure lies in knowing the factorization $r = pq$. Finding the prime factors of a large integer is practically impossible if r is, say, a 100-digit number. Thus, we have described a procedure which can be public knowledge but which ensures that messages transmitted to you are secure.

PROBLEM 31. Suppose $r = 17 \cdot 59 = 1003$ and $s = 3$. A secret agent wishes to send you the message "GO." In this case, $M = 715$. The agent calculates $M^s = (715)^3$ and, since $(715)^3 \equiv 579$ (mod 1003), she transmits the number $E = 579$. You receive this number and must decode it. How is this done?

Solution. First, note that $\gcd(3, 16) = 1$ and $11(3) + (-2)(16) = 1$, so $a = 11$. Also, $39(3) + (-2)(58) = 1$, so $b = 39$. (It is always possible to choose a and b positive. See Exercise 21.) Then $E^a = (579)^{11} \equiv 1^{11} \equiv 1$ (mod 17), while $E^b = (579)^{39} \equiv (48)^{39} \equiv 48(3)^{19} \equiv 48(27)22^4 \equiv 7$ (mod 59). The problem is then reduced to solving the pair of congruences

(9)
$$x \equiv 1 \pmod{17}$$
$$x \equiv 7 \pmod{59}.$$

Using the methods described earlier, we obtain $x \equiv 715$ (mod 1003), which was the word transmitted. ∎

Pause 19 What "methods described earlier"? Show how to get 715 as the solution.

As a final remark, we note that, in practice, we do not convert a message to a single enormous number M and then encode M. Instead we divide the message into blocks of characters of a fixed length, convert each block to a number, and then encode these numbers. For example, the message

THE PROJECT IS DELAYED TWO MONTHS

might be divided into blocks of length four,

THEP ROJE CTIS DELA YEDT WOMO NTHS

and then into numbers

20080516 18151005 03200919 04051201 25050420 23151315 14200819

each of which could be coded as described previously.

Readers interested in learning more about cryptography might consult the excellent little book *Cryptology*, by Albrecht Beutelspacher, published by the Mathematical Association of America in 1994.

Answers to Pauses

16. We have seen that changing the fourth digit from 6 to 5 gives a number with $a_1 + 2a_2 + \cdots + 9a_9 + 10a_{10} \equiv 7 \pmod{11}$, so if we also change the first digit from 0 to 4, we should get $a_1 + 2a_2 + \cdots + 9a_9 + 10a_{10} \equiv 0 \pmod{11}$, as desired. It is easy to check that the incorrect ISBN 4-935031-03-4, which has errors in two digits, would pass our test.

17. $w = (3, 1, 3, 1, 3, 1, 3, 1, 3, 1, 3, 1)$.

18. The sum of the digits which code a number on the left is odd, whereas the sum of the digits which code a number on the right is even. If the scanner at first reads 0011101, for example, the sum of digits being even implies that the number is on the right.

19.
$$\begin{array}{rrr} 59 & 1 & 0 \\ 17 & 0 & 1 \\ 8 & 1 & -3 \\ 1 & -2 & 7 \end{array}$$

So $1 = -2(59) + 7(17)$ and $1(-2)(59) + 7(7)(17) = 715$ is a solution.

EXERCISES

The symbol [BB] means that an answer can be found in the Back of the Book.

1. Which of the following are valid ISBNs and which are not? Explain.

    (a) [BB] 0-123456-78-9
    (b) [BB] 0-432091-05-5
    (c) 1-667132-42-6
    (d) 9-412883-19-6
    (e) 3-492166-27-X

2. Find the digit A so that each of the following numbers is a valid ISBN number.

    (a) [BB] 3-416109-27-A
    (b) A-461228-37-4
    (c) 9-123A45-51-X
    (d) [BB] 2-729188-A6-2
    (e) 8-9A5398-28-4

3. Is there an ISBN number of the form A-315266-78-2? Explain.

4. [BB] Show that the test given in (6) to detect an error in an ISBN is equivalent to the test $a_1 + 2a_2 + \cdots + 9a_9 \equiv a_{10} \pmod{11}$.

5. Show that an error in any single digit of an ISBN will always be detected by the test in (6).

6. (a) [BB] Change the third and sixth digits of the ISBN 2-429783-29-0 so that the resulting number is not a valid ISBN.

  (b) Change the third and sixth digits of the ISBN in (a) so that the resulting number is still valid.

  (c) Show that any two digits of an ISBN can always be changed so that the errors would not be detected by the test in (6).

7. Consider the following alternative check for correctness of an ISBN number. Instead of (6), the check digit a_{10} is determined by the rule $a_1 + a_2 + \cdots + a_9 + a_{10} \equiv 0$ (mod 11).

  (a) [BB] Express this rule in the form $w \cdot a \equiv 0$ (mod 11) for some vectors w and a.

  (b) Show that this rule detects an error in a single digit.

  (c) Show that this rule does not detect a transposition of two (different) digits.

8. Try to identify the universal product numbers defined by each of the bar codes in Fig 4.5.

Figure 4.5   Bar codes for Exercise 8.

9. Which of the following are valid universal product numbers and which are not? Explain.

  (a) [BB] 0-12345-67890-1

  (b) 1-66326-73551-5

  (c) [BB] 2-52998-17394-9

  (d) 9-53889-22687-3

  (e) 8-41285-19384-2

10. In each of the following cases, find the digit x so that the given number is a valid universal product number.

  (a) [BB] 0-12x89-29109-4

  (b) 1-29347-49x26-8

  (c) 4-29217-10258-x

  (d) 5-91057-x9332-2

  (e) x-79154-91937-6

11. [BB] This exercise concerns the test labeled (7) on page 138.

  (a) Show that the test will detect a single error in an even position digit of a universal product number.

  (b) Show by example that the test will not necessarily detect two errors in even positions.

12. Repeat both parts of Exercise 11 for odd positions.

13. **(a)** [BB] Give an example of a valid universal product number which differs from the valid number 1-23456-98732-6 in the fourth and fifth positions.

(b) [BB] Give an example of an invalid product number which differs from the valid number 1-23456-98732-6 in the fourth and fifth positions.

14. Will the test in (7) detect the transposition of two (different) adjacent digits? Explain.

15. A certain state proposes to determine the check digit a_{12} of a 12-digit driver's license number $a_1a_2\cdots a_{12}$ by the rule

$$12a_1 + 11a_2 + 10a_3 + \cdots + 2a_{11} + a_{12} \equiv 0 \pmod{10}.$$

(a) Express this test in the form $w \cdot a \equiv 0 \pmod{10}$ for vectors w and a.

(b) [BB] Does this test detect an error in a single digit?

(c) Does this test detect the error resulting from the transposition of consecutive digits?

(d) Does it detect transposition errors in general?

Explain your answers.

16. Consider the general test

$$w \cdot a \equiv 0 \pmod{n}$$

for correctness of a k-digit number $a_1a_2\ldots a_k$, $0 \le a_i \le 9$, where $a = (a_1, a_2, \ldots, a_k)$, $w = (w_1, w_2, \ldots, w_k)$ is a weight vector and $n > 1$.

(a) [BB] If this is to be a sensible test, we should have $n > 9$. Why?

(b) Assume $n > 9$. If w_i is relatively prime to n for all i, show that the test will detect single-digit errors.

(c) Assume $n > 9$. Show that the test will detect the error resulting from transposition of a_i and a_j provided $w_i - w_j$ is relatively prime to n.

17. In each case, find the smallest nonnegative integer x which satisfies the given system of congruences.

(a) [BB] $x \equiv 3 \pmod 5$
$x \equiv 4 \pmod 7$

(b) $x \equiv 1 \pmod 4$
$x \equiv 8 \pmod 9$

(c) $x \equiv 3 \pmod 5$
$x \equiv 7 \pmod 8$

(d) [BB] $x \equiv 6 \pmod 8$
$x \equiv 17 \pmod{25}$

(e) $x \equiv 3 \pmod{1917}$
$x \equiv 75 \pmod{385}$

(f) $x \equiv 1003 \pmod{17,369}$
$x \equiv 2974 \pmod{5472}$

(g) $x \equiv 1 \pmod 4$
$x \equiv 8 \pmod 9$
$x \equiv 10 \pmod{25}$

18. For each of the following, find the smallest positive integer which has the given sequence of remainders when divided by the prime powers 4, 3, and 25, respectively.

(a) [BB] 1, 2, 3

(b) 2, 0, 6

(c) 3, 1, 17

(d) [BB] 0, 2, 10

(e) 3, 2, 24

19. [BB] Reflect on the conversation between discrete math student and friend on page 140. What was the number?

20. In each of the following cases, find a positive integer x such that $ab \equiv x$ modulo a suitable integer. Assuming that $ab < 50,000$, find ab itself, if possible, and explain your reasoning.

(a) [BB] $a \equiv 3, b \equiv 3 \pmod 4$
$a \equiv 0, b \equiv 8 \pmod 9$
$a \equiv 4, b \equiv 18 \pmod{25}$

(b) $a \equiv 7, b \equiv 7 \pmod 8$
$a \equiv 9, b \equiv 8 \pmod{27}$
$a \equiv 29, b \equiv 18 \pmod{125}$

(c) $a \equiv 1, b \equiv 2 \pmod 8$
$a \equiv 8, b \equiv 1 \pmod{27}$
$a \equiv 55, b \equiv 82 \pmod{125}$

(d) $a \equiv 1, b \equiv 6 \pmod 8$
$a \equiv 2, b \equiv 6 \pmod{27}$
$a \equiv 12, b \equiv 97 \pmod{125}$

(e) [BB] $a \equiv 1, b \equiv 3 \pmod 4$
$a \equiv 3, b \equiv 4 \pmod{25}$
$a \equiv 10, b \equiv 79 \pmod{343}$

(f) $a \equiv 3, b \equiv 1 \pmod 4$
$a \equiv 17, b \equiv 6 \pmod{25}$
$a \equiv 78, b \equiv 122 \pmod{343}$

(g) $a \equiv 3, b \equiv 4 \pmod 5$
$a \equiv 36, b \equiv 42 \pmod{49}$
$a \equiv 7, b \equiv 8 \pmod{11}$
$a \equiv 1, b \equiv 3 \pmod{13}$

(h) $a \equiv 2, b \equiv 4 \pmod 5$
$a \equiv 14, b \equiv 36 \pmod{49}$
$a \equiv 1, b \equiv 10 \pmod{11}$
$a \equiv 11, b \equiv 7 \pmod{13}$

21. [BB] Suppose m and n are relatively prime positive integers. Show that there exist integers s and t with $s > 0$ such that $sm + tn = 1$. [*Hint:* It might be helpful first to consider some specific examples. For instance, can you find s and t, $s > 0$, such that $7s + 22t = 1$?]

The remaining exercises are based upon the method for encoding messages described at the end of this section.

22. Suppose $p = 17$, $q = 23$, and $s = 5$. How would you encode each of the following "messages"?

   (a) [BB] X

   (b) HELP

   (c) [BB] AIR

   (d) BYE

   (e) NOW

23. Suppose $p = 5$, $q = 7$, and $s = 5$. Decode each of the following encoded "messages."

   (a) [BB] 31

   (b) 24

   (c) [BB] 7

   (d) 11

   (e) 23

24. Suppose $p = 17$, $q = 59$, and $s = 3$.

   (a) [BB] If you receive $E = 456$, what is the message?

   (b) If you receive $E = 926$, what is the message?

REVIEW EXERCISES FOR CHAPTER 4

1. Find the quotient and remainder when 11,109,999,999,997 is divided by 1111.

2. **(a)** Convert $(1100101)_2$ to base 10.

   (b) Convert 32,145 to octal.

3. **(a)** The sum of the digits of the number 8215 is $8 + 2 + 1 + 5 = 16 \equiv 7 \pmod 9$. Observe also that $8215 = 9(912) + 7 \equiv 7 \pmod 9$. Does this hold for any number? Explain.

   (b) Suppose we want to compute the product 8215×3567 modulo 9. Replacing these numbers by those obtained by adding their digits and reducing modulo 9 gives $16 \times 21 \equiv 7 \times 3 = 21 \equiv 3 \pmod 9$. Is it true that $8215 \times 3567 \equiv 3 \pmod 9$? Explain.

4. Find gcd$(2700, -504)$ and express it as an integral linear combination of the given integers.

5. Suppose x, a and b are integers such that $x \mid ab$. If x and a are relatively prime, prove that $x \mid b$.

6. For any $k \in \mathbb{N}$, prove that gcd$(4k + 3, 7k + 5) = 1$.

7. **(a)** Give Euclid's proof that there are infinitely many primes.

   (b) State the Fundamental Theorem of Arithmetic.

8. Is $2^{119} - 1$ prime? What about $3^{109} - 1$? What about $4^{109} - 1$? Explain your answers.

9. True or false (and justify):

   (a) If $b \mid a$ and $\dfrac{a}{b} \mid c$, then $bc \mid a$.

   (b) If $b \mid a$ and $\dfrac{a}{b} \mid c$, then $a \mid bc$.

10. **(a)** Compute $3^{80} \pmod 7$.

   (b) Find all integers x such that $5x \equiv 1 \pmod{100}$. Briefly explain your answer.

11. Find all integers x and y, $0 \le x, y < 10$, which satisfy

$$x + 5y \equiv 5 \pmod{10}$$

$$5x + 3y \equiv 1 \pmod{10}.$$

12. **(a)** Find the digit A so that A-253228-46-7 is a valid ISBN number.

   (b) Find the digit x so that 3-25814-39x75-6 is a valid universal product code.

13. Find the smallest nonnegative integer which satisfies

$$x \equiv 5 \pmod{341}$$

$$x \equiv 11 \pmod{189}.$$

5

Induction and Recursion

5.1 MATHEMATICAL INDUCTION

One of the most basic methods of proof is *mathematical induction*, which is a way to establish the truth of a statement about all the natural numbers or, sometimes, all sufficiently large integers. Mathematical induction is important in every area of mathematics. In addition to the examples presented in this section, other proofs by mathematical induction appear elsewhere in this book in a variety of different contexts. In the index (see "induction"), we draw attention to some theorems whose proofs serve as especially good models of the technique.

PROBLEM 1. A certain store sells envelopes in packages of five and packages of twelve and you want to buy n envelopes. Prove that for every $n \geq 44$, this store can fill an order for exactly n envelopes (assuming an unlimited supply of each type of envelope package).

Solution. If you want to purchase 44 envelopes, you can buy two packages of twelve and four packages of five. If you want to purchase 45 envelopes, you can buy nine packages of five. If you want to purchase 46 envelopes, pick up three packages of twelve and two packages of five. If you want to buy 47 envelopes, get one package of twelve and seven packages of five and, if you want 48 envelopes, purchase four packages of twelve.

The obvious difficulty with this way of attacking the problem is that it never ends. Even supposing that we continued laboriously to answer the question for n as big as 153, say, could we be sure of a solution for $n = 154$? What is needed is a general, not an ad hoc, way to continue; that is, if it is possible to fill an order for exactly k envelopes at this store, we would like to be able to deduce that the store can also fill an order for $k + 1$ envelopes. Then, knowing that we can purchase exactly 44 envelopes and knowing that we can always continue, we could deduce that we can purchase exactly 45 envelopes. Knowing this, and knowing that we can always continue, we would know that we can purchase exactly 46 envelopes. And so on.

149

Suppose—just suppose—that it is possible to buy exactly k envelopes at this store, where $k \geq 44$. If this purchase requires seven packages of five, then exchanging these for three packages of twelve fills an order of exactly $k + 1$ envelopes. On the other hand, if k envelopes are purchased without including seven packages of five, then the order for k envelopes included at most 30 envelopes in packages of five and so, since $k \geq 44$, at least two packages of twelve must have been required. Exchanging these for five packages of five then fills exactly an order for $k + 1$ envelopes. We conclude that any order for $n \geq 44$ envelopes can be filled exactly. ∎

This example demonstrates the key ingredients of a proof by mathematical induction. Asked to prove something about all the integers greater than or equal to a particular given integer—for instance, that any order of $n \geq 44$ envelopes can be filled with packages of five and twelve—we first establish truth for the first integer—for example, $n = 44$—and then show how the truth of the statement for $n = k$ enables us to deduce truth for $n = k + 1$.

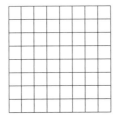

Figure 5.1   An 8×8 board.

PROBLEM 2.   Chess is a game played on an 8×8 grid, that is, a board consisting of eight rows of eight small squares. (See Fig 5.1.) Suppose our board is *defective* in the sense that one of its squares is missing. Given a box of L-shaped *trominos* like this, ⌐, each of which covers exactly three squares of a chess board, is it possible to tile the board without overlapping or going off the board? Show, in fact, that it is possible to tile any $2^n \times 2^n$ defective board.

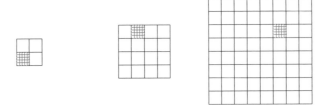

Figure 5.2   Three defective boards.

Solution.   We begin our solution by thinking of some easier situations. Figure 5.2 shows defective 2×2, 4×4, and 8×8 boards (with the missing square highlighted in each case). Certainly the 2×2 board can be tiled because its shape is exactly that of a single tromino. A little experimentation would show how to tile the

4×4 board. Rather than proceeding case by case, however, we use the idea suggested by our first example and attempt to understand how a solution for one particular board can be used to obtain a solution for the next bigger board. Suppose then that we know how to tile any $2^k \times 2^k$ defective board. How might we tile a defective board of the next size, $2^{k+1} \times 2^{k+1}$? The idea is to realize that a $2^{k+1} \times 2^{k+1}$ board can be divided into four boards, each of size $2^k \times 2^k$, as shown in Fig 5.3.

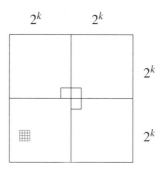

Figure 5.3

One of these smaller boards contains the missing square and so is defective. Now place a tromino at the center so as to cover squares in each of the three remaining smaller boards. Each of the boards is now "defective" and so, by assumption, can be tiled with trominos. So we have tiled the larger board! ∎

The two examples discussed so far made assertions about infinitely many consecutive integers. In each case, we adopted the following strategy.

- Verify that there is a solution for the smallest integer.
- Show how a solution for one integer leads to a solution for the next.

We now give a formal statement of the principle which has been at work, the *Principle of Mathematical Induction.*

5.1.1 THE PRINCIPLE OF MATHEMATICAL INDUCTION ▶

Given a statement $\mathcal{P}$ concerning the integer n, suppose

1. $\mathcal{P}$ is true for some particular integer n_0;
2. if $\mathcal{P}$ is true for some particular integer $k \geq n_0$, then it is true for the next integer $k + 1$.

Then $\mathcal{P}$ is true for all integers $n \geq n_0$.

In Step 2, the assumption that $\mathcal{P}$ is true for some particular integer is known as the *induction hypothesis.*

In our first example, we had to prove that any order of n envelopes, $n \geq 44$, could be filled with packages of five and of twelve: n_0 was 44 and the induction hypothesis was the assumption that there was a way to purchase k envelopes with packages of five and twelve. In the second example, we had to demonstrate that any defective board of size $2^n \times 2^n$, $n \geq 1$, could be covered in a certain way; n_0 was 1 and the induction hypothesis was the assumption that we could properly cover a $2^k \times 2^k$ board.

Our next example is suggested by the following pattern. Notice that

$$1 = \quad 1 = 1^2$$
$$1 + 3 = \quad 4 = 2^2$$
$$1 + 3 + 5 = \quad 9 = 3^2$$
$$1 + 3 + 5 + 7 = 16 = 4^2$$
$$1 + 3 + 5 + 7 + 9 = 25 = 5^2.$$

The first odd integer is 1^2; the sum of the first two odd integers is 2^2; the sum of the first three odd integers is 3^2 and so on. It appears as if the sum of the first n odd integers might always be n^2. The picture in Fig 5.4 adds force to this possibility.

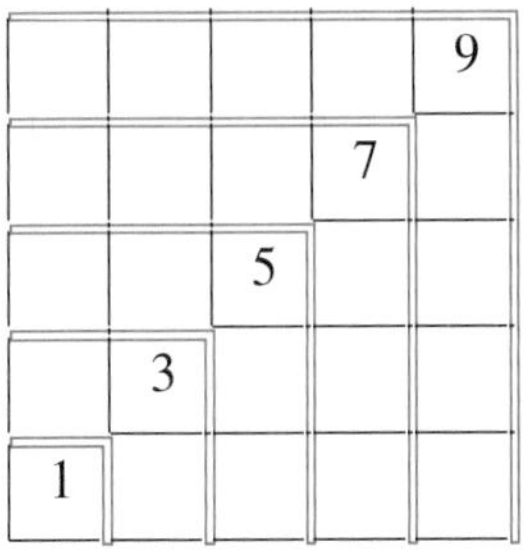

Figure 5.4   The sum of the first n odd integers is n^2.

PROBLEM 3.   Prove that for any integer $n \geq 1$, the sum of the odd integers from 1 to $2n - 1$ is n^2.

Before solving this problem, we remark that the sum in question is often written

(1) $$1 + 3 + 5 + \cdots + (2n - 1),$$

where the first three terms here—$1 + 3 + 5$—are present just to indicate that odd numbers are being added, beginning with 1, and the last term, $2n - 1$, describes the last term and gives a formula for the general term: The second odd number is $2(2) - 1$; the third odd number is $2(3) - 1$, the kth odd number is $2k - 1$. Do not infer from this expression that each of the three numbers 1, 3, 5 is always present. For example, when $n = 2$, $2n - 1 = 3$ and so the sum in (1) is, by definition, $1 + 3$.

We can also describe the sum (1) with *sigma* notation,

(2) $$1 + 3 + 5 + \cdots + (2n - 1) = \sum_{i=1}^{n} (2i - 1),$$

so named because the capital Greek letter, $\sum$, used to denote summation, is pronounced "sigma." The letter i is called the *index of summation*; the $i = 1$ at the bottom and the n at the top mean that the summation starts with $i = 1$ and continues with $i = 2, 3$ and so on, until $i = n$. Thus, the first term in the sum is $2i - 1$ with $i = 1$; that is, $2(1) - 1 = 1$. The second term is $2i - 1$ with $i = 2$; that is, $2(2) - 1 = 3$. Summing continues until $i = n$: The last term is $2n - 1$.

Pause 1   Write $\sum_{i=1}^{4}(2i - 1)$ without using sigma notation and evaluate this sum.

Problem 3 asks us to prove that, for all integers $n \geq 1$,

(3) $$1 + 3 + 5 + \cdots + (2n - 1) = n^2$$

or, equivalently, that

$$\sum_{i=1}^{n}(2i - 1) = n^2.$$

Solution. In this problem, $n_0 = 1$. When $n = 1$, $1 + 3 + 5 + \cdots + (2n - 1)$ means "the sum of the odd integers from 1 to $2(1) - 1 = 1$." Thus, the sum is just 1. Since 1^2 is also 1, the statement is true for $n = 1$. Now suppose that k is an integer, $k \geq 1$, and the statement is true for $n = k$; in other words, suppose

$$1 + 3 + 5 + \cdots + (2k - 1) = k^2 \qquad \text{induction hypothesis.}$$

We must show that the statement is true for the next integer, $n = k + 1$; namely, we must show that

$$1 + 3 + 5 + \cdots + (2(k + 1) - 1) = (k + 1)^2.$$

Since $2(k + 1) - 1 = 2k + 1$, we have to show

$$1 + 3 + 5 + \cdots + (2k + 1) = (k + 1)^2.$$

$$!!!! \text{——— } \textbf{WAIT A SECOND} \text{ ———}!!!!$$

The sum on the left is the sum of the odd integers from 1 to $2k + 1$; this is the sum of the odd integers from 1 to $2k - 1$, plus the next odd integer, $2k + 1$:

$$1 + 3 + 5 + \cdots + (2k + 1) = [1 + 3 + 5 + \cdots + (2k - 1)] + (2k + 1).$$

By the induction hypothesis, we know that

$$1 + 3 + 5 + \cdots + (2k + 1)$$
$$= 1 + 3 + 5 + \cdots + (2k - 1) + (2k + 1) = k^2 + (2k + 1).$$

Since $k^2 + (2k + 1) = (k + 1)^2$, this is the result we wanted. By the Principle of Mathematical Induction, statement (3) is true for all integers $n \geq 1$. ∎

Why did we "wait a second" in the preceding argument? It has been the authors' experience that students sometimes confuse their statement of what is to be proven when $n = k + 1$ with the start of their actual proof. Consequently, we strongly recommend the following approach to a proof by mathematical induction:

- verify the statement for $n = n_0$;
- write down the induction hypothesis (the statement for $n = k$) in the form "Now suppose that ... " and be explicit about what is being assumed;
- write down what is to be proven (the statement for $n = k + 1$) in the form "We must show that ... " again being very explicit about what is to be shown; and finally (after waiting a second);

- give a convincing argument as to why the statement for $n = k + 1$ is true (and make sure this argument uses the induction hypothesis).

We continue with several examples which you should take as models for proofs by mathematical induction.

PROBLEM 4. Prove that for any natural number $n \geq 1$,

$$1^2 + 2^2 + 3^2 + \cdots + n^2 = \frac{n(n + 1)(2n + 1)}{6}.$$

Solution. When $n = 1$, the sum of the integers from 1^2 to 1^2 is 1. Also

$$\frac{1(1 + 1)(2 \cdot 1 + 1)}{6} = 1,$$

so the statement is true for $n = 1$. Now suppose that $k \geq 1$ and the statement is true for $n = k$; that is, suppose that

$$1^2 + 2^2 + 3^2 + \cdots + k^2 = \frac{k(k + 1)(2k + 1)}{6}.$$

We have to show that the statement is true for $n = k + 1$; that is, we have to show that

$$1^2 + 2^2 + 3^2 + \cdots + (k + 1)^2$$

$$= \frac{(k + 1)[(k + 1) + 1][2(k + 1) + 1]}{6} = \frac{(k + 1)(k + 2)(2k + 3)}{6}.$$

!!!!———— **WAIT A SECOND** ————!!!!

Observe that

$$1^2 + 2^2 + 3^2 + \cdots + (k + 1)^2 = (1^2 + 2^2 + 3^2 + \cdots + k^2) + (k + 1)^2$$

$$= \frac{k(k + 1)(2k + 1)}{6} + (k + 1)^2$$

$$= \frac{k(k + 1)(2k + 1) + 6(k + 1)^2}{6}$$

$$= \frac{(k + 1)[k(2k + 1) + 6(k + 1)]}{6}$$

$$= \frac{(k + 1)[2k^2 + 7k + 6]}{6}$$

$$= \frac{(k + 1)(k + 2)(2k + 3)}{6}.$$

which is just what we wanted. By the Principle of Mathematical Induction, the statement is true for all integers $n \geq 1$. ∎

PROBLEM 5. Prove that for any integer $n \geq 1$, $2^{2n} - 1$ is divisible by 3.

Solution. When $n = 1$, $2^{2(1)} - 1 = 2^2 - 1 = 4 - 1 = 3$ is divisible by 3. Now suppose that $k \geq 1$ and the statement is true for $n = k$; that is, suppose that $2^{2k} - 1$ is divisible by 3. We must prove that the statement is true for $n = k + 1$; that is, we must prove that $2^{2(k+1)} - 1$ is divisible by 3. The key to what follows is the fact that we must somehow involve the induction hypothesis. Observe that $2^{2(k+1)} - 1 = 2^{2k+2} - 1 = 4(2^{2k}) - 1$. This is helpful since it introduces 2^{2k}. By the induction hypothesis, $2^{2k} - 1 = 3t$ for some integer t, so $2^{2k} = 3t + 1$. Now it's smooth sailing. We have

$$2^{2(k+1)} - 1 = 4(2^{2k}) - 1 = 4(3t + 1) - 1 = 12t + 4 - 1 = 12t + 3 = 3(4t + 1).$$

Thus, $2^{2(k+1)} - 1$ is divisible by 3, as required. By the Principle of Mathematical Induction, $2^{2n} - 1$ is divisible by 3 for all integers $n \geq 1$. ∎

Pause 2 Prove that $3^{2n} - 1$ is divisible by 8 for every $n \geq 1$.

Let n be a given positive integer. It is convenient to have some notation for the product of all the integers between 1 and n since this sort of product occurs frequently in statistical and in counting problems. (See Chapter 7.)

5.1.2 DEFINITION ▶ Define $0! = 1$ and, for any integer $n \geq 1$, define

$$n! = n(n - 1)(n - 2) \cdots (3)(2)(1).$$

The symbol $n!$ is read "n factorial." The first few factorials are $0! = 1$, $1! = 1$, $2! = 2 \cdot 1 = 2$, $3! = 3(2)(1) = 6$, $4! = 4(3)(2)(1) = 24$. It is useful to notice that $4! = 4(3!)$, $5! = 5(4!)$, and so on. Thus, if we know that $8! = 40{,}320$, then it is easy to deduce that $9! = 9(40{,}320) = 362{,}880$. Factorials grow very quickly. James Stirling (1730) provided an important estimate for the size of $n!$ when n is large.

5.1.3 STIRLING'S APPROXIMATION ▶

$$\lim_{n \to \infty} \frac{n!}{\sqrt{2\pi n}(n/e)^n} = 1; \quad \text{equivalently,} \quad n! \sim \sqrt{2\pi n}\left(\frac{n}{e}\right)^n.$$

In this formula, $e = 2.71828\ldots$ denotes the base of the natural logarithms. Remember from our discussion of the Prime Number Theorem (Theorem 4.3.13) that we read "is asymptotic to" at the symbol $\sim$. Thus, Stirling's formula says that $n!$ is asymptotic to $\sqrt{2\pi n}(n/e)^n$, meaning that $n!$ is approximately equal to $\sqrt{2\pi n}(n/e)^n$, for large n. For example, $15! \approx \sqrt{30\pi}(15/e)^{15} \approx 1.3 \times 10^{12}$, which, by most people's standards, is indeed a large number.

Our next problem provides another indication of the size of $n!$, albeit a rather crude one. For example, it says that $15! > 2^{15} = 32{,}768$ and $30! > 2^{30} \approx 10^9$.

PROBLEM 6. Prove that $n! > 2^n$ for all $n \geq 4$.

Solution. In this problem, $n_0 = 4$ and certainly $4! = 24 > 16 = 2^4$. Thus, the statement is true for n_0. Now suppose that $k \geq 4$ and the statement is true for

$n = k$. Thus, we suppose that $k! > 2^k$. We must prove that the statement is true for $n = k + 1$; that is, we must prove that $(k + 1)! > 2^{k+1}$. Now

$$(k + 1)! = (k + 1)k! > (k + 1)2^k$$

using the induction hypothesis. Since $k \geq 4$, certainly $k + 1 > 2$, so $(k + 1)2^k > 2 \cdot 2^k = 2^{k+1}$. We conclude that $(k + 1)! > 2^{k+1}$ as desired. By the Principle of Mathematical Induction, we conclude that $n! > 2^n$ for all integers $n \geq 4$.   ∎

Pause 3   What was the induction hypothesis in this problem?

Pause 4   Why did the induction in this example start at $n = 4$ instead of some smaller integer?

The Principle of Mathematical Induction is one of the most powerful tools of mathematics. With it, we can prove many interesting things, but if it is not applied correctly, we can also prove some interesting things which are not true!

PROBLEM 7.   What is the flaw in the following argument, which purports to show that

$$2 + 4 + 6 + \cdots + 2n = (n - 1)(n + 2)$$

for all positive integers n?

"Assume that $2 + 4 + 6 + \cdots + 2k = (k - 1)(k + 2)$ for some integer k. Then

$$2 + 4 + 6 + \cdots + 2(k + 1) = (2 + 4 + 6 + \cdots + 2k) + 2(k + 1)$$
$$= (k - 1)(k + 2) + 2(k + 1)$$

(by the induction hypothesis)

$$= k^2 + k - 2 + 2k + 2$$
$$= k^2 + 3k$$
$$= k(k + 3)$$
$$= [(k + 1) - 1][(k + 1) + 2],$$

which is the given statement for $n = k + 1$. It follows, by the Principle of Mathematical Induction, that the statement is true for all positive integers n."

Solution.   The inductive step, as given, is correct, but we neglected to check the case $n = 1$, for which the statement is most definitely false.   ∎

There is another form of the Principle of Mathematical Induction, called the *strong form*, because, at first glance, it appears to be more powerful than the principle stated previously. The two forms are completely equivalent, however: The collection of statements which can be proven true using one form is exactly

5.1.4 PRINCIPLE OF MATHEMATICAL INDUCTION (STRONG FORM) ▶

the collection which can be proven true using the other. It just so happens that in certain problems, the strong form is more convenient than the other.

Given a statement $\mathcal{P}$ concerning the integer n, suppose

1. $\mathcal{P}$ is true for some integer n_0;
2. if $k > n_0$ is any integer and $\mathcal{P}$ is true for all integers ℓ in the range $n_0 \leq \ell < k$, then it is true also for k.

Then $\mathcal{P}$ is true for all integers $n \geq n_0$.

The two forms of the Principle of Mathematical Induction differ only in the statement of the induction hypothesis (the assumption in the second step). Previously, we assumed the truth of the statement for just one particular integer and we had to prove it true for the next largest integer. In the strong form of mathematical induction, we assume the truth of the statement for **all** integers less than some integer and prove that the statement is true for that integer. When we first encounter mathematical induction, it seems to be the weak form which is always used; problems requiring the strong form are seldom encountered. In the analysis of finite structures, however, the strong form is employed extensively. We want to acquire knowledge about structures of a certain size from knowledge about similar structures of smaller size. Recall that part of the Fundamental Theorem of Arithmetic states that every natural number greater than 1 is the product of primes. The strong form of mathematical induction affords a very straightforward proof of this result.

PROBLEM 8. Use the strong form of mathematical induction to prove that every natural number $n \geq 2$ is either prime or the product of prime numbers. (See 4.3.9, the Fundamental Theorem of Arithmetic.)

Solution. The theorem is a statement about all integers $n \geq 2$. The first such integer, $n_0 = 2$, is prime, so the assertion of the theorem is true. Now let $k > 2$ and suppose that the assertion is true for all positive integers ℓ, $2 \leq \ell < k$; in other words, suppose that every integer ℓ in the interval $2 \leq \ell < k$ is either prime or the product of primes. We must prove that k has this same property. If k is prime, there is nothing more to do. On the other hand, if k is not prime, then k can be factored $k = ab$, where a and b are integers satisfying $2 \leq a, b < k$. By the induction hypothesis, each of a and b is either prime or the product of primes. Thus, k is the product of primes, as required. By the Principle of Mathematical Induction, we conclude that every $n \geq 2$ is prime or the product of primes. ∎

Our next problem demonstrates another common error in "proofs" by mathematical induction.

PROBLEM 9. Canada has a two-dollar coin known colloquially as the "toonie." What is wrong with the following argument, which purports to prove that any debt of $n > 1$ Canadian dollars can be repaid (exactly) with only toonies?

Here $n_0 = 2$. We begin by noting that any two-dollar debt can be repaid with a single toonie. Thus, the assertion is true for $n = 2$.

Now let $k \geq 2$ and suppose that the assertion is true for all ℓ, $2 \leq \ell < k$. We must prove that the assertion is true for $n = k$. For this, we apply the induction hypothesis to $k - 2$ and see that a $(k - 2)$-dollar debt can be repaid with toonies. Adding one more toonie allows us to repay k dollars with only toonies, as required. By the Principle of Mathematical Induction, any debt of $n > 1$ dollars can be repaid with toonies.

Solution. The problem here is that the latter part of the argument does not work if $k = 3$.

The induction hypothesis—that the assertion is true for all ℓ, $2 \leq \ell < k$— was applied to $\ell = k - 2$. If $k = 3$, however, then $k - 2 = 1$ and the induction hypothesis cannot be applied. ∎

We conclude with a brief discussion about the equivalence of the two Principles of Mathematical Induction and the Well-Ordering Principle.

Mathematical Induction and Well Ordering

Recall that the Well-Ordering Principle (4.1.2) says that any nonempty set of natural numbers has a smallest element. This can be proved using the weak form of the Principle of Mathematical Induction. Here is the argument.

A set containing just one element has a smallest member, the element itself, so the Well-Ordering Principle is true for sets of size $n_0 = 1$. Now suppose it is true for sets of size k; that is, assume that any set of k natural numbers has a smallest member. Given a set S of $k + 1$ numbers, remove one element a. The remaining k numbers have a smallest element, say b, and the smaller of a and b is the smallest element of S. This proves that any finite set of natural numbers has a smallest element. We leave to the reader (Exercise 17) the extension of this result to arbitrary subsets of N.

Conversely, we may use the Well-Ordering Principle to prove the Principle of Mathematical Induction (weak form). For suppose that $\mathcal{P}$ is a statement involving the integer n which we wish to establish for all integers greater than or equal to some given integer n_0. Assume

1. $\mathcal{P}$ is true for $n = n_0$, and
2. if $\mathcal{P}$ is true for an integer $k \geq n_0$, then it is also true for $k + 1$.

How does the Principle of Well-Ordering show that $\mathcal{P}$ is true for all $n \geq n_0$? For convenience we assume that $n_0 \geq 1$. (The case $n_0 < 0$ can be handled with a slight variation of the argument we present.)

If $\mathcal{P}$ is not true for all $n \geq n_0$, then the set S of natural numbers $n \geq n_0$ for which $\mathcal{P}$ is false is not empty. By the Well-Ordering Principle, S has a smallest element a. Now $a \neq n_0$ because we have established that $\mathcal{P}$ is true for $n = n_0$. Thus, $a > n_0$, so $a - 1 \geq n_0$. Also, $a - 1 < a$. By minimality of a, $\mathcal{P}$ is true for $k = a - 1$. By assumption 2, $\mathcal{P}$ is true for $k + 1 = a$, a contradiction. We are forced to conclude that our starting assumption is false: $\mathcal{P}$ must be true for all $n \geq n_0$.

The preceding paragraphs show that the Principles of Well-Ordering and Mathematical Induction (weak form) are equivalent. With minor variations in the reasoning, we can prove that the Principles of Well Ordering and Mathematical

Induction (strong form) are equivalent. It follows, therefore, that the three principles are logically equivalent.

1. $\sum_{i=1}^{4}(2i-1) = [2(1)-1]+[2(2)-1]+[2(3)-1]+[2(4)-1] = 1+3+5+7 = 16.$

2. When $n = 1$, $3^{2(1)}-1 = 3^2-1 = 9-1 = 8$ is divisible by 8. Now suppose that $k \geq 1$ and the statement is true for $n = k$; that is, suppose that $3^{2k}-1$ is divisible by 8. Thus, $3^{2k}-1 = 8t$ for some integer t and so $3^{2k} = 8t+1$. We have

$$3^{2(k+1)}-1 = 3^{2k+2}-1 = 9(3^{2k})-1$$

$$= 9(8t+1)-1 = 72t+9-1 = 72t+8 = 8(9t+1).$$

Thus, $3^{2(k+1)}-1$ is divisible by 8, as required. By the Principle of Mathematical Induction, $3^{2n}-1$ is divisible by 8 for all integers $n \geq 1$.

3. The induction hypothesis was that $k! > 2^k$ for some particular integer $k \geq 4$.

4. The statement $n! > 2^n$ is not true for $n < 4$; for example, $3! = 6$, whereas $2^3 = 8$.

EXERCISES

The symbol [BB] means that an answer can be found in the Back of the Book.

1. Write each of the following sums without using $\sum$ and evaluate.

(a) [BB] $\sum_{i=1}^{5} i^2$

(b) [BB] $\sum_{i=1}^{4} 2^i$

(c) [BB] $\sum_{t=1}^{1} \sin \pi t$

(d) $\sum_{j=0}^{2} 3^{j+2}$

(e) $\sum_{k=-1}^{4}(2k^2-k+1)$

(f) $\sum_{k=0}^{n}(-1)^k$

2. List the elements of each of the following sets:

(a) $\{\sum_{i=0}^{n}(-1)^i \mid n = 0, 1, 2, 3\}$

(b) $\{\sum_{i=1}^{n} 2^i \mid n \in \mathbb{N}, 1 \leq n \leq 5\}$

3. Prove that it is possible to fill an order for $n \geq 32$ pounds of fish given bottomless wheelbarrows full of 5-pound and 9-pound fish.

4. Use mathematical induction to prove the truth of each of the following assertions for all $n \geq 1$.

(a) [BB] n^2+n is divisible by 2.

(b) n^3+2n is divisible by 3.

(c) [BB] $n^3+(n+1)^3+(n+2)^3$ is divisible by 9.

(d) 5^n-1 is divisible by 4.

(e) 8^n-3^n is divisible by 5.

(f) $5^{2n}-2^{5n}$ is divisible by 7.

(g) $10^{n+1}+10^n+1$ is divisible by 3.

(h) a^n-b^n is divisible by $a-b$ for any integers a, b with $a-b \neq 0$.

5. (a) [BB] Prove by mathematical induction that

$$1+2+3+\cdots+n = \frac{n(n+1)}{2}$$

for any natural number n.

(b) Prove by mathematical induction that

$$1^3+2^3+\cdots+n^3 = \frac{n^2(n+1)^2}{4}$$

for any natural number n.

(c) Use the results of (a) and (b) to establish that

$$(1+2+3+\cdots+n)^2 = 1^3+2^3+\cdots+n^3$$

for all $n \geq 1$.

6. Use mathematical induction to establish the truth of each of the following statements for all $n \geq 1$.

(a) [BB] $1+2+2^2+2^3+\cdots+2^n = 2^{n+1}-1$

(b) [BB] $1^2-2^2+3^2-4^2+\cdots+(-1)^{n-1}n^2 = (-1)^{n-1}\dfrac{n(n+1)}{2}$

(c) $1^2 + 3^2 + 5^2 + \cdots + (2n-1)^2 = \dfrac{n(2n-1)(2n+1)}{3}$

(d) $1 \cdot 2 \cdot 3 + 2 \cdot 3 \cdot 4 + 3 \cdot 4 \cdot 5 + \cdots + n(n+1)(n+2) = \dfrac{n(n+1)(n+2)(n+3)}{4}$

(e) $\dfrac{1}{1 \cdot 2} + \dfrac{1}{2 \cdot 3} + \dfrac{1}{3 \cdot 4} + \cdots + \dfrac{1}{n(n+1)} = \dfrac{n}{n+1}$

7. [BB; (a)] Rewrite each of the sums in Exercise 6 using $\sum$ notation.

8. Use mathematical induction to establish each of the following formulas.

(a) [BB] $\displaystyle\sum_{i=1}^{n}(i+1)2^i = n2^{n+1}$

(b) $\displaystyle\sum_{i=1}^{n}\dfrac{i^2}{(2i-1)(2i+1)} = \dfrac{n(n+1)}{2(2n+1)}$

(c) $\displaystyle\sum_{i=1}^{n}(2i-1)(2i) = \dfrac{n(n+1)(4n-1)}{3}$

9. Use mathematical induction to establish each of the following inequalities.

(a) [BB] $2^n > n^2$, for $n \geq 5$.

(b) $(1 + \frac{1}{2})^n \geq 1 + \frac{n}{2}$, for $n \in \mathbb{N}$.

(c) For any $p \in \mathbb{R}$, $p > -1$, $(1+p)^n \geq 1 + np$ for all $n \in \mathbb{N}$.

(d) For any integer $n \geq 2$, $\frac{1}{n+1} + \frac{1}{n+2} + \frac{1}{n+3} + \cdots + \frac{1}{2n} > \frac{13}{24}$.

(e) $\displaystyle\sum_{i=1}^{n}\dfrac{1}{\sqrt{i}} > \sqrt{n}$ for $n \geq 2$.

10. Suppose $c, x_1, x_2, \ldots, x_n, y_1, y_2, \ldots, y_n$ are $2n + 1$ given numbers. Prove each of the following assertions by mathematical induction.

(a) [BB] $\displaystyle\sum_{i=1}^{n}(x_i + y_i) = \sum_{i=1}^{n}x_i + \sum_{i=1}^{n}y_i$ for $n \geq 1$.

(b) $\displaystyle\sum_{i=1}^{n}cx_i = c\sum_{i=1}^{n}x_i$ for $n \geq 1$.

(c) $\displaystyle\sum_{i=2}^{n}(x_i - x_{i-1}) = x_n - x_1$ for $n \geq 2$.

11. [BB] Find the fault in the following "proof" that in any group of n people, everybody is the same age.

> Suppose $n = 1$. If a group consists of just one person, everybody is the same age. Suppose that in any group of k people, everyone is the same age. Let $G = \{a_1, a_2, \ldots,$

$a_{k+1}\}$ be a group of $k + 1$ people. Since each of the groups $\{a_1, a_2, \ldots, a_k\}$ and $\{a_2, a_3, \ldots, a_{k+1}\}$ consists of k people, everybody in each group has the same age, by the induction hypothesis. Since a_2 is in each group, it follows that all $k + 1$ people a_1, $a_2, \ldots, a_{k+1}$ have the same age.

12. [BB] Find the fault in the following "proof" by mathematical induction that

$$1 + 2 + 3 + \cdots + n = \dfrac{(2n+1)^2}{8}$$

for all natural numbers n.

> If $1 + 2 + 3 + \cdots + k = \dfrac{(2k+1)^2}{8}$, then
>
> $1 + 2 + 3 + \cdots + (k+1)$
>
> $= (1 + 2 + 3 + \cdots + k) + (k+1)$
>
> $= \dfrac{(2k+1)^2}{8} + (k+1)$
>
> $= \dfrac{4k^2 + 4k + 1 + 8k + 8}{8}$
>
> $= \dfrac{4k^2 + 12k + 9}{8}$
>
> $= \dfrac{(2k+3)^2}{8} = \dfrac{[2(k+1)+1]^2}{8}$
>
> and so truth for k implies truth for $k + 1$.

13. What is wrong with the following "proof" that any order for $n \geq 10$ pounds of fish can be filled with only 5-pound fish?

> We use the strong form of mathematical induction. Here $n_0 = 10$. Since an order for 10 pounds of fish can be filled with two 5-pound fish, the assertion is true for $n = 10$. Now let $k > 10$ be an integer and suppose that any order for ℓ pounds of fish, $10 \leq \ell < k$, can be filled with only 5-pound fish. We must prove that an order for k pounds can be similarly filled. But by the induction hypothesis, we can fill an order for $k - 5$ pounds of fish, so, adding one more 5-pounder, we can fill the order for k pounds. By the Principle of Mathematical Induction, we conclude that the assertion is true for all $n \geq 10$.

14. One of several differences between the Canadian and American games of football is that in Canada, a team can score a single point without first having scored a touchdown. So it is clear that any score is possible in the Canadian game. Is this so in the American game? Indeed this seems to be the case, even assuming (this is not true!) that in the United States, points can be scored only three at a time (with a field goal) or seven at a time (with a converted touchdown). Here is an argument.

> Assume that k points can be achieved with multiples of 3 or 7. Here's how to reach $k+1$ points. If k points are achieved with at least two field goals, subtracting these and adding a touchdown gives $k + 1$ points. On the other hand, if the k points are achieved with at least two touchdowns, subtracting these and adding five field goals also gives $k + 1$ points.

Does this argument show that any score is possible in American football? Can it be used to show something about the nature of possible scores?

15. [BB] It is tempting to think that if a statement involving the natural number n is true for many consecutive values of n, it must be true for all n. In this connection, the following example due to Euler is illustrative. Let $f(n) = n^2 + n + 41$.

(a) Convince yourself (perhaps with a computer algebra package like Maple or Mathematica) that $f(n)$ is prime for $n = 1, 2, 3, \ldots 39$ but that $f(40)$ is not prime.

(b) Show that for any n of the form $n = k^2 + 40$, $f(n)$ is not prime.

16. [BB] Prove that a set with n elements, $n \geq 0$, contains 2^n subsets.

17. [BB] Suppose that any nonempty finite set of natural numbers has a smallest element. Prove that **any** nonempty set of natural numbers has a smallest element.

18. (a) Prove that for any integer $n \geq 1$, any set of n positive real numbers has a smallest element.

(b) Prove that the result of (a) is not true for infinite sets of positive real numbers in general but that it is true for some infinite sets.

(c) What is the name of the principle which asserts that any nonempty set of natural numbers has a smallest element?

19. [BB] Let $n \geq 1$ and let $A, B_1, B_2, \ldots, B_n$ be sets. Generalize the result of Exercise 18, Section 2.2 by proving that $A \cup \left(\bigcap_{i=1}^{n} B_i \right) = \bigcap_{i=1}^{n} (A \cup B_i)$ for all $n \geq 1$.

20. Prove that $A \cap \left(\bigcup_{i=1}^{n} B_i \right) = \bigcup_{i=1}^{n} (A \cap B_i)$ for any sets $A, B_1, B_2, \ldots, B_n$.

21. In Section 2.2, we defined the symmetric difference of two sets. More generally, the symmetric difference of $n \geq 3$ sets $A_1, \ldots, A_n$ can be defined inductively as follows:

$$A_1 \oplus \cdots \oplus A_n = (A_1 \oplus \cdots \oplus A_{n-1}) \oplus A_n.$$

Prove that for any $n \geq 2$, $A_1 \oplus A_2 \oplus \cdots \oplus A_n$ consists of those elements in an odd number of the sets $A_1, \ldots, A_n$.

22. Prove the Chinese Remainder Theorem, 4.5.1, by mathematical induction.

23. [BB] For $n \geq 3$, the greatest common divisor of n nonzero integers $a_1, a_2, \ldots, a_n$ can be defined inductively by

$$\gcd(a_1, \ldots, a_n) = \gcd(a_1, \gcd(a_2, \ldots, a_n)).$$

Prove that $\gcd(a_1, a_2, \ldots, a_n)$ is an integral *linear combination* of $a_1, a_2, \ldots, a_n$ for all $n \geq 2$; that is, prove that there exist integers $s_1, \ldots, s_n$ such that $\gcd(a_1, \ldots, a_n) = s_1 a_1 + s_2 a_2 + \cdots + s_n a_n$.

24. The definition of the greatest common divisor of $n \geq 3$ integers given in Exercise 23 differs from that given in the exercises to Section 4.2. Suppose $a_1, \ldots, a_n$ are nonzero integers. Show that $\gcd(a_1, \ldots, a_n)$, as defined in Exercise 23, satisfies the properties given in Definition 4.2.14.

25. Suppose n and $m_1, m_2, \ldots, m_t$ are natural numbers, and that the m_i are pairwise relatively prime. Suppose each m_i divides n. Prove that the product $m_1 m_2 \cdots m_t$ divides n. [*Hint:* Induction on t and Exercise 10 of Section 4.2.]

26. Define $f : \mathbb{Z} \to \mathbb{Z}$ by

$$f(n) = \begin{cases} n - 2 & n \geq 1000 \\ f(f(n + 4)) & n < 1000. \end{cases}$$

(a) Find the values of $f(1000 - n)$ for $n = 0, 1, 2, 3, 4, 5$.

(b) Guess a formula for $f(1000 - n)$ valid for $n \geq 0$ and prove your answer.

(c) Find $f(5)$ and $f(20)$.

(d) What is the range of f?

5.1.5 DEFINITION ▼

A set A of integers is called an *ideal* if and only if

(i) $0 \in A$,

(ii) if $a \in A$, then also $-a \in A$, and

(iii) if $a, b \in A$, then $a + b \in A$.

27. For any integer $n \geq 0$, recall that $n\mathsf{Z} = \{kn \mid k \in \mathsf{Z}\}$ denotes the set of multiples of n.

    (a) Prove that $n\mathsf{Z}$ is an ideal of the integers.

    (b) Let A be any ideal of Z. Prove that $A = n\mathsf{Z}$ for some $n \geq 0$ by establishing each of the following statements.

    i. If A contains only one element, then A is of the desired form.

    Now assume that A contains more than one element.

    ii. Show that A contains a positive number.

    iii. Show that A contains a smallest positive number n.

    iv. $n\mathsf{Z} \subseteq A$, where n is the integer found in iii.

    v. $A \subseteq n\mathsf{Z}$. [*Hint*: The division algorithm, 4.1.5.]

28. [BB] Prove that for every integer $n \geq 2$, the number of lines obtained by joining n distinct points in the plane, no three of which are collinear, is $\frac{1}{2}n(n-1)$.

29. An n-sided polygon (commonly shortened to n-gon) is a closed planar figure bounded by n straight sides no two of which intersect unless they are adjacent, in which case they intersect just at a vertex. Thus, a 3-gon is just a triangle, a 4-gon is a quadrilateral, a 5-gon is a pentagon, and so on. An n-gon is *convex* if the line joining any pair of nonadjacent vertices lies entirely within the figure. A rectangle, for example, is convex. Prove that the sum of the interior angles of a convex n-gon is $(n-2)180°$ for all $n \geq 3$.

30. Suppose a rectangle is subdivided into regions by means of straight lines each extending from one border of the rectangle to another. Prove that the regions of the "map" so obtained can be colored with just two colors in such a way that bordering "countries" have different colors.

31. (a) [BB] Given an equal arm balance capable of determining only the relative weights of two quantities and eight coins, all of equal weight except possibly one which is lighter, explain how to determine if there is a light coin and how to identify it in just two weighings.

    (b) Given an equal arm balance as in (a) and $3^n - 1$ coins, $n \geq 1$, all of equal weight except possibly one which is lighter, show how to determine if there is a light coin and how to identify it in at most n weighings.

32. True or false? In each case, give a proof or provide a counterexample which disproves the given statement.

    (a) [BB] $5^n + n + 1$ is divisible by 7 for all $n \geq 1$

    (b) $\sum_{k=0}^{n}(k+1) = n(n+3)/2$ for all $n \geq 1$

    (c) If $n \geq 2$, $\gcd\left(\dfrac{(n+2)!}{3}, \dfrac{(n+3)!}{2}\right) = \dfrac{(n+2)!}{6}$

    (d) (For students of calculus) $n^{15} \geq 2^n$ for all $n \geq 1$

33. Let n be any integer greater than 1. Show that the following argument is valid. (See Section 1.5.)

$$p_1 \rightarrow p_2$$
$$p_2 \rightarrow p_3$$
$$\vdots$$
$$\frac{p_{n-1} \rightarrow p_n}{p_1 \rightarrow p_n}$$

34. (For students who have completed a course in differential calculus) State and prove (by mathematical induction) a formula for $\frac{d}{dx}x^n$ which holds for all $n \geq 1$.

5.1.6 NOTATION ▼

The product of n elements $a_1, a_2, \ldots, a_n$ is denoted $\prod\limits_{r=1}^{n} a_r$.

35. Let x be a real number, $x \neq \pm 1$. Prove that

$$\prod_{r=1}^{n}(1 + x^{2^r}) = \frac{1 - x^{2^{r+1}}}{1 - x^2}$$

for any integer $n \geq 1$.

36. [BB] (For students of calculus) The condition $x^2 \neq 1$ is necessary in Exercise 35 since otherwise we would have a denominator of 0 on the right. However,

$$\lim_{x \to 1} \frac{1 - x^{2^{r+1}}}{1 - x^2} \quad \text{and} \quad \lim_{x \to -1} \frac{1 - x^{2^{r+1}}}{1 - x^2}$$

both exist. Use the result of Exercise 35 to find these limits.

37. Find an expression for $\prod\limits_{r=2}^{n} \dfrac{2r-1}{2r-3}$ valid for $n \geq 2$ and prove by mathematical induction that your answer is correct.

38. For any $n \geq 0$, let $F_n = 2^{2^n} + 1$. (Those numbers F_n which are prime are called *Fermat primes*. See Section 4.3.)

(a) Prove that $\displaystyle\prod_{r=0}^{n-1} F_r = F_n - 2$ for all $n \geq 1$.

(b) Prove that $\gcd(F_m, F_n) = 1$ for any positive integers m and n with $m \neq n$.

(c) Use the result of (b) to give another proof (different from Euclid's) that there are infinitely many primes.

39. For a given natural number n, prove that the set of all polynomials of degree at most n with integer coefficients is countable. [*Hint*: Let P_n denote the set of all polynomials of degree at most n with integer coefficients. The result for P_1 was Exercise 25(b) in Section 3.3.]

40. (a) Prove that the strong form of the Principle of Mathematical Induction implies the Well-Ordering Principle.

(b) Prove that the Well-Ordering Principle implies the strong form of the Principle of Mathematical Induction. (Assume $n_0 \geq 1$.)

41. In this section, we have studied two formulations of the Principle of Mathematical Induction.

(a) Use either of these to establish the following (peculiar?) third formulation.

Suppose $\mathcal{P}(n)$ is a statement about the natural number n such that

1. $\mathcal{P}(1)$ is true;
2. For any $k \geq 1$, $\mathcal{P}(k)$ true implies $\mathcal{P}(2k)$ true; and,
3. For any $k \geq 2$, $\mathcal{P}(k)$ true implies $\mathcal{P}(k-1)$ true.

(b) Prove that for any two nonnegative numbers x and y, $\dfrac{x+y}{2} \geq \sqrt{xy}$.

(c) Use the Principle of Mathematical Induction in the form given in part (a) to generalize the result of part (b), thus establishing the *arithmetic mean–geometric mean inequality*: For any $n \geq 1$ and any n nonnegative real numbers $a_1, a_2, \ldots, a_n$,

$$\frac{a_1 + a_2 + \cdots + a_n}{n} \geq \sqrt[n]{a_1 a_2 \cdots a_n}.$$

42. Let m and n be relatively prime integers each greater than 1. Assume you have an unlimited supply of m- and n-cent stamps. Using only these stamps, show that

(a) it is not possible to purchase a selection of stamps worth precisely $mn - m - n$ cents;

(b) for any $r > mn - m - n$, it is possible to purchase a selection of stamps worth exactly r cents. [*Hint*: There exist integers a and b, with $0 < a < n$ and $0 < b < m$ such that $bn = am - 1$ and hence such that $(n-a)m = (m-b)n - 1$. See Exercise 34 of Section 4.2. Now try downward induction!]

5.2 RECURSIVELY DEFINED SEQUENCES

Suppose n is a natural number. How should we define 2^n? We could write

$$2^n = \underbrace{2 \cdot 2 \cdot 2 \cdots 2}_{n \text{ 2's}}$$

or

(4) $$2^1 = 2 \quad \text{and, for } k \geq 1, \quad 2^{k+1} = 2 \cdot 2^k.$$

The latter statement is an example of a *recursive definition*. It explicitly defines 2^n when $n = 1$ and then, assuming 2^n has been defined for $n = k$, defines it for $n = k+1$. By the Principle of Mathematical Induction, we know that 2^n has been defined for all integers $n \geq 1$.

Another expression that is most naturally defined recursively is $n!$, which was introduced in Section 5.1. If we write

$$0! = 1 \quad \text{and, for } k \geq 0, \quad (k+1)! = (k+1)k!,$$

then it follows by the Principle of Mathematical Induction that $n!$ has been defined for every $n \geq 0$.

Sequences of numbers are often defined recursively. A *sequence* is a function whose domain is some infinite set of integers (often $\mathbb{N}$) and whose range is a set of real numbers. Since its domain is countable, we can and usually do describe a sequence by simply listing its range. The sequence which is the function $f : \mathsf{N} \to \mathsf{R}$ defined by $f(n) = n^2$, for instance, is generally described by the list $1, 4, 9, 16, \ldots$, the idea being to write down enough numbers from the start of the list that the rest can be inferred. The numbers in the list (the range of the function) are called the *terms* of the sequence. Sometimes we start counting at 0 (if the function has domain $\mathsf{N} \cup \{0\}$) so that the terms are denoted $a_0, a_1, a_2, \ldots$.

The sequence $2, 4, 8, 16, \ldots$ can be defined recursively like this:

$$(5) \qquad\qquad a_1 = 2 \quad \text{and, for } k \geq 1, \quad a_{k+1} = 2a_k.$$

By this, we understand that $a_1 = 2$ and then, setting $k = 1$ in the second part of the definition, that $a_2 = 2a_1 = 2(2) = 4$. With $k = 2$, $a_3 = 2a_2 = 2(4) = 8$; with $k = 3$, $a_4 = 2a_3 = 2(8) = 16$ and so on. Evidently, (5) defines the sequence we had in mind. Again, the definition is recursive because each term in the sequence beyond the first is defined in terms of the previous term.

The equation $a_{k+1} = 2a_k$ in (5), which defines one member of the sequence in terms of a previous one, is called a *recurrence relation*. The equation $a_1 = 2$ is called an *initial condition*.

There are other possible recursive definitions which describe the same sequence as (5). For example, we could write

$$a_0 = 2 \quad \text{and, for } k \geq 0, \quad a_{k+1} = 2a_k,$$

or we could say

$$a_1 = 2 \quad \text{and, for } k \geq 2, \quad a_k = 2a_{k-1}.$$

[Verify that these definitions give the same sequence, $2, 4, 8, 16, \ldots$.]

Sometimes, after computing a few terms of a sequence which has been defined recursively, we can guess an explicit formula for a_n. In (5), for instance, $a_n = 2^n$. We say that $a_n = 2^n$ is the *solution* to the recurrence relation. Our goal in this section and the next is to gain some skill at solving recurrence relations.

PROBLEM 10. Write down the first six terms of the sequence defined by $a_1 = 1$, $a_{k+1} = 3a_k + 1$ for $k \geq 1$. Guess a formula for a_n and prove that your formula is correct.

Solution. The first six terms are

$$\begin{aligned} a_1 &= 1 \\ a_2 &= 3a_1 + 1 = 3(1) + 1 = 4 \\ a_3 &= 3a_2 + 1 = 3(4) + 1 = 13 \\ a_4 &= 40 \\ a_5 &= 121 \\ a_6 &= 364. \end{aligned}$$

Since there is multiplication by 3 at each step, we might suspect that 3^n is involved in the answer. After trial and error, we guess that $a_n = \frac{1}{2}(3^n - 1)$ and verify this by mathematical induction.

When $n = 1$, the formula gives $\frac{1}{2}(3^1 - 1) = 1$, which is indeed a_1, the first term in the sequence.

Now assume that $k \geq 1$ and that $a_k = \frac{1}{2}(3^k - 1)$. We wish to prove that $a_{k+1} = \frac{1}{2}(3^{k+1} - 1)$. We have

$$a_{k+1} = 3a_k + 1 = 3\tfrac{1}{2}(3^k - 1) + 1$$

using the induction hypothesis. Hence,

$$a_{k+1} = \tfrac{1}{2}3^{k+1} - \tfrac{3}{2} + 1 = \tfrac{1}{2}(3^{k+1} - 1)$$

as required. By the Principle of Mathematical Induction, our guess is correct. ∎

PROBLEM 11. A sequence is defined recursively by $a_0 = 1$, $a_1 = 4$ and $a_n = 4a_{n-1} - 4a_{n-2}$ for $n \geq 2$. Find the first six terms of this sequence. Guess a formula for a_n and establish the validity of your guess.

Solution. Here there are two initial conditions—$a_0 = 1$, $a_1 = 4$. Also, the recurrence relation, $a_n = 4a_{n-1} - 4a_{n-2}$, defines the general term as a function of two previous terms. The first six terms of the sequence are

$$a_0 = 1$$

$$a_1 = 4$$

$$a_2 = 4a_1 - 4a_0 = 4(4) - 4(1) = 12$$

$$a_3 = 4a_2 - 4a_1 = 4(12) - 4(4) = 32$$

$$a_4 = 4a_3 - 4a_2 = 4(32) - 4(12) = 80$$

$$a_5 = 4a_4 - 4a_3 = 4(80) - 4(32) = 192.$$

Finding a general formula for a_n requires some ingenuity. Let us carefully examine some of the first six terms. We note that $a_3 = 32 = 24 + 8 = 3(8) + 8 = 4(8)$ and $a_4 = 80 = 64 + 16 = 4(16) + 16 = 5(16)$ and $a_5 = 192 = 6(32)$. We guess that $a_n = (n+1)2^n$. To prove this, we use the strong form of mathematical induction (with $n_0 = 0$).

When $n = 0$, we have $(0 + 1)2^0 = 1(1) = 1$, in agreement with the given value for a_0. When $n = 1$, $(1 + 1)2^1 = 4 = a_1$. Now that the formula has been verified for $k = 0$ and $k = 1$, we may assume that $k > 1$ and that $a_n = (n+1)2^n$ for all n in the interval $0 \leq n < k$. We wish to prove the formula is valid for $n = k$; that is, we wish to prove that $a_k = (k+1)2^k$. Since $k \geq 2$, we know that $a_k = 4a_{k-1} - 4a_{k-2}$. Applying the induction hypothesis to $k - 1$ and to $k - 2$ (each of which is in the range $0 \leq n < k$), we have $a_{k-1} = k2^{k-1}$ and $a_{k-2} = (k-1)2^{k-2}$. Thus,

$$a_k = 4(k2^{k-1}) - 4(k-1)2^{k-2} = 2k2^k - k2^k + 2^k = k2^k + 2^k = (k+1)2^k$$

as required. By the Principle of Mathematical Induction, the formula is valid for all $n \geq 0$. ∎

In effect, our method of verifying the formula $a_n = (n+1)2^n$ in Problem 11 amounts simply to checking that it satisfies both initial conditions and also the given recurrence relation. We prefer the more formal approach of mathematical induction since it emphasizes this important concept and avoids pitfalls associated with working on both sides of an equation at once.

As previously mentioned, there is nothing unique about a recursive definition. The sequence in the last example can also be defined by

$$a_0 = 1, a_1 = 4 \quad \text{and, for } n \geq 1, \quad a_{n+1} = 4a_n - 4a_{n-1}.$$

In this case, we again obtain $a_n = (n + 1)2^n$. We could also say

$$a_1 = 1, a_2 = 4 \quad \text{and, for } n \geq 1, \quad a_{n+2} = 4a_{n+1} - 4a_n$$

but then, labeling the first term a_1 instead of a_0 would give $a_n = n2^{n-1}$. Other variants are also possible.

Some Special Sequences

Suppose you have $50 in an old shoe box and acquire a paper route which nets you $14 a week. Assuming all this money goes into your shoe box on a weekly basis (and you never borrow from it), after your first week delivering papers, your shoe box will contain $64; after two weeks, $78; after three weeks, $92 and so on. A sequence of numbers like $50, 64, 78, 92, \ldots$, where each term is determined by adding the same fixed number to the previous one, is called an *arithmetic sequence*. The fixed number is called the *common difference* of the sequence (because the difference of successive terms is constant throughout the sequence).

EXAMPLES 12
- $50, 64, 78, 92, \ldots$ is an arithmetic sequence with common difference 14;
- $-17, -12, -7, -2, 3, 8, \ldots$ is an arithmetic sequence with common difference 5;
- $103, 99, 95, 91, \ldots$ is an arithmetic sequence with common difference -4. ▲

5.2.1 DEFINITION ▶ The *arithmetic sequence* with first term a and *common difference* d is the sequence defined by

$$a_1 = a \quad \text{and, for } k \geq 1, \quad a_{k+1} = a_k + d.$$

The general arithmetic sequence thus takes the form

$$a, a + d, a + 2d, a + 3d, \ldots$$

and it is easy to see that, for $n \geq 1$, the nth term of the sequence is

(6)
$$\boxed{a_n = a + (n - 1)d.}$$

We leave a formal proof to the exercises and also a proof of the fact that the sum of n terms of the arithmetic sequence with first term a and common difference d is

(7)
$$\boxed{S = \frac{n}{2}[2a + (n - 1)d].}$$

EXAMPLE 13 The first 100 terms of the arithmetic sequence $-17, -12, -7, -2, 3, \ldots$ have the sum

$$S = \tfrac{100}{2}[2(-17) + 99(5)] = 50(-34 + 495) = 23{,}050.$$

The 100th term of this sequence is $a_{100} = -17 + 99(5) = 478$ by (6). The number 2038 occurs as the 412th term, as we see by solving $-17 + (n-1)5 = 2038$. ▲

Many people with paper routes deposit their earnings in a bank account which pays interest instead of into a shoe box, which does not. Fifty dollars in a bank account which pays 1% interest per month accumulates to $50 + (.01 \times 50) = 50(1 + .01) = 50(1.01)$ dollars after one month. After another month, the original investment will have accumulated to what it was at the start of the month plus 1% of this amount; that is,

$$50(1.01) + .01(50)(1.01) = 50(1.01)(1 + .01) = 50(1.01)^2.$$

After three months, the accumulation is $50(1.01)^3$ dollars; after twelve months, it is $50(1.01)^{12}$ dollars ($\approx \$56.34$). A sequence of numbers such as

$$50, 50(1.01), 50(1.01)^2, \ldots$$

in which each term is determined by multiplying the previous term by a fixed number is called a *geometric sequence*. The fixed number is called the *common ratio*.

EXAMPLES 14
- $50, 50(1.01), 50(1.01)^2, \ldots$ is a geometric sequence with common ratio 1.01;
- $3, -6, 12, -24, \ldots$ is a geometric sequence with common ratio -2;
- $9, 3, 1, \tfrac{1}{3}, \ldots$ is a geometric sequence with common ratio $\tfrac{1}{3}$. ▲

5.2.2 DEFINITION ▶ The *geometric sequence* with first term a and *common ratio* r is the sequence defined by

$$a_1 = a \quad \text{and, for } k \geq 1, \quad a_{k+1} = ra_k.$$

The general geometric sequence thus has the form

$$a, ar, ar^2, ar^3, ar^4, \ldots$$

the nth term being $a_n = ar^{n-1}$. This is straightforward to prove, as is the following formula for the sum S of n terms, provided $r \neq 1$.

(8)
$$S = \frac{a(1 - r^n)}{1 - r}$$

EXAMPLE 15 The sum of 29 terms of the geometric sequence with $a = 8^{12}$ and $r = -\tfrac{1}{2}$ is

$$S = 8^{12}\,\frac{1 - (-\tfrac{1}{2})^{29}}{1 - (-\tfrac{1}{2})} = 2^{36}\,\frac{1 + (\tfrac{1}{2})^{29}}{\tfrac{3}{2}} = \frac{2^{36} + 2^7}{\tfrac{3}{2}} = \tfrac{1}{3}(2^{37} + 2^8). \qquad ▲$$

Pause 5   What is the 30th term of the geometric sequence just described?

Leonardo Fibonacci,[1] also known as Leonardo of Pisa, was one of the brightest mathematicians of the Middle Ages. His writings in arithmetic and algebra were standard authorities for centuries and are largely responsible for the introduction into Europe of the Arabic numerals 0, 1, ... , 9 we use today. Fibonacci was fond of problems, his most famous of which is concerned with rabbits!

Suppose that newborn rabbits start producing offspring by the end of their second month of life and that after this point, they produce a pair a month (one male, one female). Assuming just one pair of rabbits initially, how many pairs of rabbits, Fibonacci asked, will be alive after one year? The sequence which gives the number of pairs at the end of successive months is the famous *Fibonacci sequence*.

After one month, there is still only one pair of rabbits in existence, but after a further month, this pair is joined by its offspring; thus, after two months, there are two pairs of rabbits. At the end of any month, the number of pairs of rabbits is the number alive at the end of the previous month plus the number of pairs alive two months ago, since each pair alive two months ago produced one pair of offspring. We obtain the sequence 1, 1, 2, 3, 5, 8, 13, ... , which is defined recursively as follows.

5.2.3 THE FIBONACCI SEQUENCE ▶

$$f_1 = 1, f_2 = 1 \quad \text{and, for } k \geq 2, \quad f_{k+1} = f_k + f_{k-1}$$

Pause 6   Think of the Fibonacci sequence as a function fib: $\mathbb{N} \to \mathbb{N}$. List eight elements of this function as ordered pairs. Is fib a one-to-one function? Is it onto?

Pause 7   What is the answer to Fibonacci's question?

Although we have found an explicit formula for the nth term of most of the sequences discussed so far, there are many sequences for which such a formula is difficult or impossible to obtain. (This is one reason why recursive definitions are important.) Is there a specific formula for the nth term of the Fibonacci sequence? As a matter of fact, there is, though it is certainly not one which many people would discover by themselves. We show in the next section (see Problem 20) that the nth term of the Fibonacci sequence is the closest integer to the number

(9) $$\frac{1}{\sqrt{5}}\left(\frac{1 + \sqrt{5}}{2}\right)^n.$$

The first few values of $\frac{1}{\sqrt{5}}\left(\frac{1+\sqrt{5}}{2}\right)^n$ are approximately 0.72361, 1.17082, 1.89443, 3.06525, 4.95967, 8.02492, 12.98460, and 21.00952; the integers closest to these numbers are the first eight terms of the Fibonacci sequence.

Finally, note that we must be careful with sequences **apparently** defined recursively, since some recursive definitions do not define actual sequences! Consider, for example,

$$a_1 = 1 \quad \text{and, for } k > 1, \quad a_k = \begin{cases} 1 + a_{k/2} & \text{if } k \text{ is even} \\ 1 + a_{3k-1} & \text{if } k \text{ is odd.} \end{cases}$$

---

[1] Fibonacci was born around the year 1180 and died in 1228. "Fibonacci" is a contraction of "Filius Bonaccii," Latin for "son of Bonaccio."

What happens if we try to write down the first few terms of this sequence?

$$a_1 = 1$$
$$a_2 = 1 + a_1 = 1 + 1 = 2$$
$$a_3 = 1 + a_8 = 1 + (1 + a_4) = 2 + a_4 = 2 + (1 + a_2) = 3 + a_2 = 5$$
$$a_4 = 1 + a_2 = 1 + 2 = 3,$$

but then

$$a_5 = 1 + a_{14} = 1 + (1 + a_7) = 2 + a_7$$
$$= 2 + (1 + a_{20}) = 3 + a_{20} = 3 + (1 + a_{10})$$
$$= 4 + a_{10} = 4 + (1 + a_5) = 5 + a_5$$

and, to our dismay, we have reached the absurdity $5 = 0$. Obviously, no sequence has been defined.

Pause 8 Compute the first six terms of the sequence "defined" as follows:

$$a_1 = 1 \quad \text{and, for } k > 1, \quad a_k = \begin{cases} 1 + a_{k/2} & \text{if } k \text{ is even} \\ 1 + a_{3k+1} & \text{if } k \text{ is odd.} \end{cases}$$

Answers to Pauses

5. Here $n = 30$, $a = 8^{12} = (2^3)^{12} = 2^{36}$ and $r = -(2^{-1})$, so

$$a_{30} = 2^{36}[-(2^{-1})]^{29} = -2^{36}2^{-29} = -2^7 = -128.$$

6. The most obvious eight pairs are $(1, 1)$, $(2, 1)$, $(3, 2)$, $(4, 3)$, $(5, 5)$, $(6, 8)$, $(7, 13)$, and $(8, 21)$. The Fibonacci function is not one-to-one because $(1, 1) \in$ fib and $(2, 1) \in$ fib but $1 \neq 2$. It's not onto since, for example, 4 is not in the range.

7. The number of rabbits in existence after twelve months is the **thirteenth** term of the Fibonacci sequence, 233.

8. $a_1 = 1$; $a_2 = 1 + a_1 = 1 + 1 = 2$; $a_3 = 1 + a_{10} = 1 + 1 + a_5 = 2 + 1 + a_{16} = 3 + a_{16}$. Now $a_{16} = 1 + a_8 = 1 + 1 + a_4 = 1 + 1 + 1 + a_2 = 3 + 2 = 5$, so that $a_3 = 3 + 5 = 8$. Continuing, $a_4 = 1 + a_2 = 3$; $a_5 = 1 + a_{16} = 1 + 5 = 6$; $a_6 = 1 + a_3 = 1 + 8 = 9$. The first six terms are $1, 2, 8, 3, 6, 9$.

EXERCISES

The symbol [BB] means that an answer can be found in the Back of the Book.

1. Give recursive definitions of each of the following sequences:

(a) [BB] $1, 5, 5^2, 5^3, 5^4, \ldots$

(b) $5, 3, 1, -1, -3, \ldots$

(c) $4, 1, 3, -2, 5, -7, 12, -19, 31, \ldots$

(d) $1, 2, 0, 3, -1, 4, -2, \ldots$

2. (a) [BB] Find the first seven terms of the sequence $\{a_n\}$ defined by

$a_1 = 16$, and for $k \geq 1$,

$$a_{k+1} = \begin{cases} 1 & \text{if } a_k = 1 \\ \frac{1}{2}a_k & \text{if } a_k \text{ is even} \\ \frac{1}{2}(a_k - 1) & \text{if } a_k \neq 1 \text{ is odd.} \end{cases}$$

(b) Repeat part (a) with $a_1 = 17$.

(c) Repeat part (a) with $a_1 = 18$.

(d) Repeat part (a) with $a_1 = 100$.

3. Let $a_1, a_2, a_3, \ldots$ be the sequence defined by $a_1 = 1$, $a_{k+1} = 3a_k$ for $k \geq 1$. Prove that $a_n = 3^{n-1}$ for all $n \geq 1$.

4. [BB] Suppose $a_1, a_2, a_3, \ldots$ is a sequence of integers such that $a_1 = 0$ and, for $n > 1$, $a_n = n^3 + a_{n-1}$. Prove that $a_n = \frac{(n-1)(n+2)(n^2+n+2)}{4}$ for every integer $n \geq 1$.

5. Define the sequence $a_1, a_2, a_3, \ldots$ by $a_1 = 0$, $a_2 = \frac{1}{2}$ and $a_{k+2} = \frac{1}{2}(a_k + a_{k+1})$ for $k \geq 1$. Find the first seven terms of this sequence. Prove that $a_n = \frac{1}{3}\left(1 - (-\frac{1}{2})^{n-1}\right)$ for every $n \geq 1$.

6. [BB] Let $a_1, a_2, a_3, \ldots$ be the sequence defined by $a_1 = 1$ and, for $n > 1$, $a_n = 2a_{n-1} + 1$. Write down the first six terms of this sequence. Guess a formula for a_n and prove that your guess is correct.

7. Let $a_1, a_2, a_3, \ldots$ be the sequence defined by $a_1 = \frac{3}{2}$ and $a_n = 5a_{n-1} - 1$ for $n \geq 2$. Write down the first six terms of this sequence. Guess a formula for a_n and prove that your guess is correct.

8. Suppose $a_0, a_1, a_2, \ldots$ is a sequence such that $a_0 = a_1 = 1$ and for $n \geq 1$, $a_{n+1} = n(a_n + a_{n-1})$.

   (a) Find a_2, a_3, a_4 and a_5.
   (b) Guess a formula for a_n, valid for $n \geq 0$, and use mathematical induction to prove that your guess is correct.

9. [BB] Consider the sequence defined by $a_1 = 1$, $a_{n+1} = (n+1)^2 - a_n$ for $n \geq 1$. Find the first six terms. Guess a general formula for a_n and prove that your answer is correct.

10. Let $a_1, a_2, a_3, \ldots$ be the sequence defined by $a_1 = 1$, $a_{k+1} = (k+1)a_k$ for $k \geq 1$. Find a formula for a_n and prove that your formula is correct.

11. [BB] Suppose $a_1, a_2, a_3, \ldots$ is a sequence of integers such that $a_1 = 0$, $a_2 = 1$ and for $n > 2$, $a_n = 4a_{n-2}$. Guess a formula for a_n and prove that your guess is correct.

12. A sequence is defined recursively by $a_0 = 2$, $a_1 = 3$ and $a_n = 3a_{n-1} - 2a_{n-2}$ for $n \geq 2$.

   (a) Find the first five terms of this sequence.
   (b) Guess a formula for a_n.
   (c) Verify that your guess in (b) is correct.
   (d) Find a formula for a_n which involves only one preceding term.

13. Let $a_1, a_2, a_3, \ldots$ be the sequence defined by $a_1 = 1$, $a_2 = 0$ and for $n > 2$, $a_n = 4a_{n-1} - 4a_{n-2}$. Prove that $a_n = 2^n(1 - \frac{n}{2})$ for all $n \geq 1$.

14. Let $a_1, a_2, a_3, \ldots$ be the sequence defined by
$$a_1 = 1, \quad \text{and for } k \geq 1, \quad a_{k+1} = k^2 a_k.$$
Find the first six terms of this sequence. Guess a general formula for a_n and prove your answer by mathematical induction.

15. Consider the arithmetic sequence with first term 2 and common difference 3.

   (a) [BB] Find the first ten terms and the 123rd term of this sequence.
   (b) [BB] Does 752 belong to this sequence? If so, what is the number of the term where it appears?
   (c) Repeat (b) for 1023 and 4127.
   (d) [BB] Find the sum of the first 75 terms of this sequence.

16. [BB] Consider the arithmetic sequence with first term 7 and common difference $-\frac{1}{2}$.

   (a) Find the 17th and 92nd terms.
   (b) Find the sum of the first 38 terms.

17. An arithmetic sequence begins 116, 109, 102.

   (a) Find the 300th term of this sequence.
   (b) Determine whether or not -480 belongs to this sequence. If it does, what is its term number?
   (c) Find the sum of the first 300 terms of the sequence.

18. The arithmetic sequence $4, 15, 26, 37, \ldots$ begins with a *perfect square*, that is, an integer of the form k^2, where k is also an integer. Find the next three perfect squares in this sequence.

19. Establish formulas (6) [BB] and (7) for the nth term and the sum of the first n terms of the arithmetic sequence with first term a and common difference d.

20. [BB] Consider the geometric sequence with first term 59,049 and common ratio $-\frac{1}{3}$.

   (a) Find the first ten terms and the 33rd term of this sequence.
   (b) Find the sum of the first 12 terms.

21. Consider the geometric sequence which begins $-3072, 1536, -768$.

   (a) Find the 13th and 20th terms of this sequence.
   (b) Find the sum of the first nine terms.

22. If the first term of a geometric sequence is 48 and the sixth term is $-\frac{3}{2}$, find the sum of the first ten terms.

23. (a) [BB] Find, to four decimal places, the 129th term of the geometric sequence which begins $-0.00001240, 0.00001364$.
   (b) [BB] Find the approximate sum of the first 129 terms of the sequence in (a).

24. [BB] Verify formula (8) for the sum of n terms of a geometric sequence with first term a and common ratio $r \neq 1$.

25. Consider the sequence defined recursively by $a_1 = 1$ and for $n > 1$, $a_n = \sum_{i=1}^{n-1} a_i$. Write down the first six terms of this sequence, guess a formula for a_n valid for $n \geq 2$, and prove your answer.

26. (a) Find the sum of 18 terms of the geometric sequence with first term 7/1024 and common ratio 8.

(b) [BB] Suppose $|r| < 1$. Explain why the sum of the first n terms of the geometric sequence with first term a and common ratio r is approximately $\frac{a}{1-r}$.

(c) [BB] Approximate $\sum_{k=0}^{100} \frac{3}{2^k}$.

(d) Find the approximate sum of the first 1 million terms of the geometric sequence which begins 144, 48, 16.

27. (a) Find the 19th and 100th terms of the geometric sequence which has first term 98,415 and common ratio $\frac{1}{3}$.

(b) Find the sum of the first 15 terms of the sequence in (a).

(c) Find the approximate sum of the first 10,000 terms of the sequence in (a).

28. Given that each sum below is the sum of part of an arithmetic or geometric sequence, find each sum.

(a) [BB] $75 + 71 + 67 + 63 + \cdots + (-61)$

(b) $75 + 15 + 3 + \frac{3}{5} + \cdots + \frac{3}{5^7}$

(c) $-52 - 41 - 30 - 19 + \cdots + 949$

(d) $1 - \frac{1}{2} + \frac{1}{4} - \frac{1}{8} + \cdots + \frac{1}{2^{60}}$

(e) $2 + 6 + 18 + 54 + \cdots + 354{,}294$

29. [BB] Is it possible for an arithmetic sequence to be also a geometric sequence? Explain your answer.

30. A bank account pays interest at the rate of $100i\%$ a year. Assume an initial balance of P, which accumulates to s_n after n years.

(a) Find a recursive definition for s_n.

(b) Find a formula for s_n.

31. [BB] Maurice borrows $1000 at an interest rate of 15%, compounded annually.

(a) How much does Maurice owe after two years?

(b) In how many years will the debt grow to $2000?

32. On January 1, 2001, you have $50 in a savings account, which pays interest at the rate of 1% per month. At the end of January and at the end of each month thereafter, you deposit $56 to this account. Assuming no

withdrawals, what will be your balance on January 1, 2002?

33. On June 1, you win $1 million in a lottery and immediately acquire numerous "friends," one of whom offers you the deal of a lifetime. In return for the million, she'll pay you a cent today, two cents tomorrow, four cents the next day, eight cents the next, and so on, stopping with the last payment on June 21.

(a) Assuming you take this deal, how much money will you receive on June 21?

(b) Should you take the deal? Explain.

(c) Would you take the deal if payments continued for the entire month of June?

34. Define a sequence $\{a_n\}$ recursively as follows:

$$a_0 = 0, \quad \text{and for } n > 0, \quad a_n = a_{\lfloor n/5 \rfloor} + a_{\lfloor 3n/5 \rfloor} + n.$$

Prove that $a_n \leq 20n$ for all $n \geq 0$. (Recall that $\lfloor x \rfloor$ denotes the floor of the real number x. See paragraph 3.1.6.)

35. The number $\Phi = \dfrac{1 + \sqrt{5}}{2} \approx 1.618$ is known as the *golden mean*. It has many remarkable properties. For instance, the geometric sequence $1, \Phi, \Phi^2, \ldots$ satisfies the Fibonacci recurrence relation $a_{n+1} = a_n + a_{n-1}$. Establish this fact.

36. [BB] Suppose we think of the Fibonacci sequence as going backward as well as forward. What seven terms precede $1, 1, 2, 3, 5, 8, \ldots$? How is f_{-n} related to f_n?

37. [BB] Let $\{f_n\}$ denote the Fibonacci sequence. Prove that $f_{n+1}f_n = \sum_{i=1}^{n} f_i^2$ for all $n \geq 1$.

38. Represent the Fibonacci sequence by $f_1 = f_2 = 1$, $f_n = f_{n-1} + f_{n-2}$ for $n > 2$.

(a) Verify the formula $f_1 + f_2 + f_3 + \cdots + f_n = f_{n+2} - 1$ for $n = 4, 5, 6$.

(b) Prove that the formula in (a) is valid for all $n \geq 1$.

39. Show that, for $n \geq 2$, the nth term of the Fibonacci sequence is less than $(7/4)^{n-1}$. [Use the definition of the Fibonacci sequence, not the approximation to f_n given in equation (9).]

40. [BB] What is wrong with the following argument, which purports to prove that all the Fibonacci numbers after the first two are even?

> Let f_n denote the nth term of the Fibonacci sequence. We prove that f_n is even for all $n \geq 3$ using the strong form of the Principle

of Mathematical Induction. The Fibonacci sequence begins 1, 1, 2. Certainly $f_3 = 2$ is even and so the assertion is true for $n_0 = 3$. Now let $k > 3$ be an integer and assume that the assertion is true for all n, $3 \le n < k$; that is, assume that f_n is even for all $n < k$. We wish to show that the assertion is true for $n = k$; we wish to show that f_k is even. But $f_k = f_{k-1} + f_{k-2}$. Applying the induction hypothesis to $k - 1$ and to $k - 2$, we conclude that each of f_{k-1} and f_{k-2} is even, hence, so is the sum. By the Principle of Mathematical Induction, f_n is even for all $n \ge 3$.

41. Let $f_1 = f_2 = 1$, $f_k = f_{k-1} + f_{k-2}$ for $k > 2$ be the Fibonacci sequence. Which terms of this sequence are even? Prove your answer.

42. For $n \ge 1$, let a_n denote the number of ways to express n as the sum of natural numbers, taking order into account. For example, $3 = 3 = 1 + 1 + 1 = 2 + 1 = 1 + 2$, so $a_3 = 4$.

   (a) [BB] Find the first five terms of the sequence $\{a_n\}$.

   (b) Guess and then establish a formula for a_n.

43. For $n \ge 1$, let b_n denote the number of ways to express n as the sum of 1's and 2's, taking order into account. Thus, $b_4 = 5$ because $4 = 1 + 1 + 1 + 1 = 2 + 2 = 2 + 1 + 1 = 1 + 1 + 2 = 1 + 2 + 1$.

   (a) Find the first five terms of the sequence $\{b_n\}$.

   (b) Find a recursive definition for b_n and identify this sequence.

44. (a) [BB] Let a_n be the number of ways of forming a line of n people distinguished only by sex. For example, there are four possible lines of two people—MM, MF, FM, FF—so $a_2 = 4$. Find a recurrence relation satisfied by a_n and identify the sequence $a_1, a_2, a_3, \ldots$.

   (b) Let a_n be the number of ways in which a line of n people can be formed such that no two males are standing beside each other. For example, $a_3 = 5$ because there are five ways to form lines of three people with no two males beside each other; namely, FFF, MFF, FMF, FFM, MFM. Find a recurrence relation satisfied by a_n and identify the sequence $a_1, a_2, a_3, \ldots$.

45. Define the Fibonacci sequence by $f_1 = f_2 = 1$, $f_{n+1} = f_n + f_{n-1}$ for $n \ge 2$.

   (a) Prove that $\gcd(f_{n+1}, f_n) = 1$ for all $n \ge 1$.

   (b) Prove that $f_n = f_{n-m+1}f_m + f_{n-m}f_{m-1}$ for any positive integers n and m with $n > m > 1$.

   (c) Prove that for any positive integers n and m, the greatest common divisor of f_n and f_m is $f_{\gcd(n,m)}$.

46. Suppose u_n and v_n are sequences defined recursively by

$$u_1 = 0, v_1 = 1, \text{ and, for } n \ge 1,$$

$$u_{n+1} = \tfrac{1}{2}(u_n + v_n), v_{n+1} = \tfrac{1}{4}(u_n + 3v_n).$$

   (a) Prove that $v_n - u_n = \frac{1}{4^{n-1}}$ for $n \ge 1$.

   (b) Prove that u_n is an increasing sequence; that is, $u_{n+1} > u_n$ for all $n \ge 1$.

   (c) Prove that v_n is a decreasing sequence; that is, $v_{n+1} < v_n$ for all $n \ge 1$.

   (d) Prove that $u_n = \frac{2}{3} - \frac{1}{6} \cdot \frac{1}{4^{n-2}}$ for all $n \ge 1$.

(This problem is taken from a Portuguese examination designed to test the level of mathematical knowledge of graduating high school students. It was reprinted in *Focus*, the newsletter of the Mathematical Association of America **13**, no. 3, June 1993, p. 13.)

5.2.4 DEFINITION ▼

The *powers of a function* $f: A \to A$ are defined recursively by

$$f^1 = f \quad \text{and, for } n > 1, \quad f^n = f \circ f^{n-1}.$$

47. Suppose $f: \mathbb{N} \to \mathbb{N}$ is defined by

$$f(m) = \begin{cases} \frac{2m}{3} & \text{if } m \equiv 0 \pmod 3 \\ \frac{4m-1}{3} & \text{if } m \equiv 1 \pmod 3 \\ \frac{4m+1}{3} & \text{if } m \equiv 2 \pmod 3. \end{cases}$$

   (a) Prove that f is one-to-one and onto.

   (b) [BB] Find the first ten terms of the sequence $f^n(1)$.

   (c) Find the sequence $f^n(2)$, $n \ge 1$.

   (d) Find the sequence $f^n(4)$, $n \ge 1$.

   (e) Find the first ten terms of the sequence $f^n(8)$, $n \ge 1$.[2]

---

[2]It is unknown whether or not the terms of this sequence ever repeat.

48. Define $g : \mathbb{N} \to \mathbb{N}$ by

$$g(m) = \begin{cases} \frac{m}{2} & \text{if } m \text{ is even} \\ \frac{3m+1}{2} & \text{if } m \text{ is odd.} \end{cases}$$

This function is known as the "$3m + 1$" function. It is suspected that for **any** starting number m, the sequence

$g(m), g^2(m), g^3(m), \ldots$ eventually terminates with 1. Verify this assertion for each of the five integers $m = 341, 96, 104, 336,$ and 133.[3]

49. (For students who have had a course in linear algebra) Give a recursive definition of the *determinant* of an $n \times n$ matrix, for $n \geq 1$.

5.3 SOLVING RECURRENCE RELATIONS; THE CHARACTERISTIC POLYNOMIAL

Recursively defined sequences were introduced in the previous section. Given a particular recurrence relation and certain initial conditions, you were encouraged to guess a formula for the nth term and prove that a guess was correct. Guessing is an important tool in mathematics and a skill which can be sharpened through practice, but we now confess that there is a definite procedure for solving most of the recurrence relations we have encountered so far.

In this section, we describe a procedure for solving recurrence relations of the form

(10) $$a_n = r a_{n-1} + s a_{n-2} + f(n)$$

where r and s are constants and $f(n)$ is some function of n. Such a recurrence relation is called a *second order linear recurrence relation with constant coefficients*. If $f(n) = 0$, the relation is called *homogeneous*. *Second order* refers to the fact that the recurrence relation (10) defines a_n as a function of the two terms preceding it, *linear* to the fact that the terms of the sequence appear by themselves, to the first power, and with just constant coefficients. You should consult more specialized books in combinatorics for a general treatment of constant coefficient recurrence relations where a_n is a function of any number of terms of the form $c a_{n-i}$, $c \in \mathbb{R}$.[4]

EXAMPLES 16 Here are some second order linear recurrence relations with constant coefficients.

- $a_n = a_{n-1} + a_{n-2}$, the recurrence relation which appears in the definition of the Fibonacci sequence. This is homogeneous with $r = s = 1$. Notice that we have modified slightly the definition $a_{n+1} = a_n + a_{n-1}$ given in paragraph 5.2.3 so that it is readily seen to be of the type we are considering here.
- $a_n = 5a_{n-1} - 6a_{n-2} + n$. Here $r = 5$, $s = -6$, $f(n) = n$.
- $a_n = 3a_{n-1}$. This is homogeneous with $r = 3$, $s = 0$. ▲

EXAMPLES 17 Consider the following two recurrence relations.

- $a_n = 5a_{n-1} - 3a_{n-3}$
- $a_n = a_{n-1} a_{n-2} + n^2$

---

[3]While this conjecture has been established for all integers $m < 2^{40} \approx 10^{12}$, it is unknown whether it holds for all integers! This problem has attracted the interest of many people, some of whom have offered a sizeable monetary reward for its solution! We refer the interested reader to the excellent article, "The $3x + 1$ Problem and Its Generalizations," by Jeffrey C. Lagarias, *American Mathematical Monthly* **92** (1985) no. 1, 1–23.

[4]See, for example, Alan Tucker, *Applied Combinatorics* Wiley, New York: (1980).

Neither is of interest to us in this section. The first is not second order while the second is not linear. ▲

With the homogeneous recurrence relation $a_n = ra_{n-1} + sa_{n-2}$, which can be rewritten in the form

$$a_n - ra_{n-1} - sa_{n-2} = 0,$$

we associate the quadratic polynomial

$$x^2 - rx - s,$$

which is called the *characteristic polynomial* of the recurrence relation. Its roots are called the *characteristic roots* of the recurrence relation. For example, the recurrence relation $a_n = 5a_{n-1} - 6a_{n-2}$ has characteristic polynomial $x^2 - 5x + 6$ and characteristic roots 2 and 3.

The following theorem, whose proof is left to the exercises, shows how to solve any second order linear **homogeneous** recurrence relation with constant coefficients.

5.3.1 THEOREM ▶ Let x_1 and x_2 be the roots of the polynomial $x^2 - rx - s$. Then the solution of the recurrence relation $a_n = ra_{n-1} + sa_{n-2}$, $n \geq 2$, is

$$a_n = \begin{cases} c_1 x_1^n + c_2 x_2^n & \text{if } x_1 \neq x_2 \\ c_1 x^n + c_2 n x^n & \text{if } x_1 = x_2 = x. \end{cases}$$

In each case, c_1 and c_2 are constants determined by initial conditions.

PROBLEM 18. Solve the recurrence relation $a_n = 5a_{n-1} - 6a_{n-2}$, $n \geq 2$, given $a_0 = 1$, $a_1 = 4$.

Solution. The characteristic polynomial, $x^2 - 5x + 6$, has distinct roots $x_1 = 2, x_2 = 3$. Theorem 5.3.1 tells us that the solution is $a_n = c_1(2^n) + c_2(3^n)$. Since $a_0 = 1$, we must have $c_1(2^0) + c_2(3^0) = 1$ and, since $a_1 = 4$, we have $c_1(2^1) + c_2(3^1) = 4$. Therefore,

$$\begin{aligned} c_1 + c_2 &= 1 \\ 2c_1 + 3c_2 &= 4. \end{aligned}$$

Solving, we have $c_1 = -1$, $c_2 = 2$, so the solution is $a_n = -2^n + 2(3^n)$. (You are encouraged to verify that this formula is correct.) ∎

PROBLEM 19. Solve the recurrence $a_n = 4a_{n-1} - 4a_{n-2}$, $n \geq 2$, with initial conditions $a_0 = 1$, $a_1 = 4$. (We solved this by guesswork and ingenuity in Problem 11.)

Solution. The characteristic polynomial, $x^2 - 4x + 4$, has the repeated root $x = 2$. Hence, the solution is $a_n = c_1(2^n) + c_2 n(2^n)$. The initial conditions yield $c_1 = 1$, $2c_1 + 2c_2 = 4$, so $c_2 = 1$. Thus, $a_n = 2^n + n(2^n) = (n+1)2^n$. ∎

PROBLEM 20. Find a formula for the nth term of the Fibonacci sequence.

Solution. To simplify the algebra which follows, we take for initial conditions $a_0 = a_1 = 1$ rather than $a_1 = a_2 = 1$. Hence, the recurrence relation to be solved is $a_n = a_{n-1} + a_{n-2}$, $n \geq 2$, and we must remember that the nth term will be

a_{n-1}. The characteristic polynomial, $x^2 - x - 1$, has distinct roots $\frac{1 \pm \sqrt{5}}{2}$. Hence, the solution to our recurrence relation is

$$a_n = c_1 \left(\frac{1+\sqrt{5}}{2} \right)^n + c_2 \left(\frac{1-\sqrt{5}}{2} \right)^n.$$

The initial conditions give

$$c_1 + c_2 = 1$$

$$c_1 \left(\frac{1+\sqrt{5}}{2} \right) + c_2 \left(\frac{1-\sqrt{5}}{2} \right) = 1$$

yielding $c_1 = \frac{1}{\sqrt{5}} \left(\frac{1+\sqrt{5}}{2} \right)$ and $c_2 = -\frac{1}{\sqrt{5}} \left(\frac{1-\sqrt{5}}{2} \right)$. Thus, the solution is

$$a_n = \frac{1}{\sqrt{5}} \left(\frac{1+\sqrt{5}}{2} \right) \left(\frac{1+\sqrt{5}}{2} \right)^n - \frac{1}{\sqrt{5}} \left(\frac{1-\sqrt{5}}{2} \right) \left(\frac{1-\sqrt{5}}{2} \right)^n$$

$$= \frac{1}{\sqrt{5}} \left(\frac{1+\sqrt{5}}{2} \right)^{n+1} - \frac{1}{\sqrt{5}} \left(\frac{1-\sqrt{5}}{2} \right)^{n+1}.$$

The nth term of the Fibonacci sequence is

$$a_{n-1} = \frac{1}{\sqrt{5}} \left(\frac{1+\sqrt{5}}{2} \right)^n - \frac{1}{\sqrt{5}} \left(\frac{1-\sqrt{5}}{2} \right)^n.$$

In Exercise 18, we ask you to use this result to obtain the simpler formula (9) on p. 168. ∎

Pause 9 Why is the real number $\frac{1}{\sqrt{5}} \left(\frac{1+\sqrt{5}}{2} \right)^n - \frac{1}{\sqrt{5}} \left(\frac{1-\sqrt{5}}{2} \right)^n$ an integer for any $n \geq 1$?

We now turn our attention to the general second order recurrence relation $a_n = r a_{n-1} + s a_{n-2} + f(n)$ and show that the solution is closely related to the corresponding homogeneous recurrence relation.

Suppose we could find one specific solution p_n to the given recurrence relation. (Such a function p_n is called a *particular solution*.) Thus, $p_n = r p_{n-1} + s p_{n-2} + f(n)$. Suppose t_n were another solution. Then we would also have $t_n = r t_{n-1} + s t_{n-1} + f(n)$ and, subtracting,

$$t_n - p_n = r(t_{n-1} - p_{n-1}) + s(t_{n-2} - p_{n-2}).$$

This equation shows that $t_n - p_n$ satisfies the homogeneous recurrence relation $a_n = r a_{n-1} + s a_{n-2}$. Setting $t_n - p_n = q_n$, we have $t_n = p_n + q_n$ where p_n is a particular solution to the given recurrence relation and q_n satisfies the associated homogeneous recurrence relation. This is the content of our next theorem.

5.3.2 THEOREM ▶ Let p_n be any particular solution to the recurrence relation $a_n = r a_{n-1} + s a_{n-2} + f(n)$ ignoring initial conditions. Let q_n be the solution to the homogeneous recurrence $a_n = r a_{n-1} + s a_{n-2}$ given by Theorem 5.3.1, again ignoring initial conditions. Then $p_n + q_n$ is the solution to the recurrence relation $a_n = r a_{n-1} + s a_{n-2} + f(n)$. The initial conditions determine the constants in q_n.

As we have seen, the result of this theorem seems plausible. In the exercises, you will be invited to supply a proof. The main point of the theorem is that once some particular solution to the recurrence has been found, the problem is reduced to the homogeneous case, which we have already considered. Finding a particular solution can be difficult. As we shall see, a useful trick is to try a formula for p_n

which is of the same type as $f(n)$. For example, if $f(n)$ is a linear function, try a linear function for p_n.

PROBLEM 21. Solve the recurrence relation $a_n = -3a_{n-1} + n$, $n \geq 1$, where $a_0 = 1$.

Solution. Since $f(n) = n$ is linear, we try a linear function for p_n; that is, we set $p_n = a + bn$ and attempt to determine a and b. Putting this expression for p_n in the given recurrence relation, we obtain

$$a + bn = -3[a + b(n-1)] + n = -3a + 3b + (1 - 3b)n.$$

This equation will hold if $a = -3a + 3b$ and $b = 1 - 3b$, which is the same as $a = \frac{3}{16}$, $b = \frac{1}{4}$. We conclude that $p_n = \frac{3}{16} + \frac{1}{4}n$ is a particular solution to the recurrence, ignoring initial conditions.

The corresponding homogeneous recurrence relation in this case is $a_n = -3a_{n-1}$, whose characteristic polynomial is $x^2 + 3x$. The characteristic roots are -3 and 0, so the solution to the homogeneous recurrence relation is

$$q_n = c_1(-3)^n + c_2(0^n) = c_1(-3)^n.$$

Thus,

$$p_n + q_n = \frac{3}{16} + \frac{1}{4}n + c_1(-3)^n.$$

Since $a_0 = 1$, $\frac{3}{16} + \frac{1}{4}(0) + c_1(-3)^0 = 1$; that is, $\frac{3}{16} + c_1 = 1$. Thus, $c_1 = \frac{13}{16}$ and the solution is $a_n = \frac{3}{16} + \frac{1}{4}n + \frac{13}{16}(-3)^n$. ∎

PROBLEM 22. Solve $a_n = 2a_{n-1} + 3a_{n-2} + 5^n$, $n \geq 2$, given $a_0 = -2$, $a_1 = 1$.

Solution. In looking for a particular solution, it seems reasonable to try $p_n = a(5^n)$. Substituting into the recurrence relation, we get

$$a(5^n) = 2a(5^{n-1}) + 3a(5^{n-2}) + 5^n$$

$$25a = 10a + 3a + 25$$

$$12a = 25,$$

so $a = \frac{25}{12}$ and $p_n = \frac{25}{12}(5^n)$ is a particular solution.

Next we solve the homogeneous recurrence relation $a_n = 2a_{n-1} + 3a_{n-2}$. The characteristic polynomial, $x^2 - 2x - 3$, has distinct roots -1 and 3, so the solution is $q_n = c_1(-1)^n + c_2(3^n)$. By Theorem 5.3.2, the given recurrence relation has the solution

$$p_n + q_n = \frac{25}{12}(5^n) + c_1(-1)^n + c_2(3^n).$$

The initial conditions give

$$a_0 = -2 = \frac{25}{12} + c_1 + c_2$$

$$a_1 = 1 = \frac{25}{12}(5) - c_1 + 3c_2,$$

hence, $c_1 = -\frac{17}{24}$, $c_2 = -\frac{27}{8}$. Our solution is $a_n = \frac{25}{12}(5^n) - \frac{17}{24}(-1)^n - \frac{27}{8}(3^n)$. ∎

Answers to Pauses

9. All the terms of the Fibonacci sequence are integers!

EXERCISES

The symbol [BB] means that an answer can be found in the Back of the Book.

1. [BB] Solve the recurrence relation $a_n = a_{n-1} + 6a_{n-2}$, $n \geq 2$, given $a_0 = 1$, $a_1 = 3$.

2. Solve the recurrence relation $a_n = -6a_{n-1} + 7a_{n-2}$, $n \geq 2$, given $a_0 = 32$, $a_1 = -17$.

3. [BB] Solve the recurrence relation $a_n = 6a_{n-1} - 9a_{n-2}$, $n \geq 2$, given $a_0 = -5$, $a_1 = 3$.

4. Solve the recurrence relation $a_{n+1} = 7a_n - 10a_{n-1}$, $n \geq 2$, given $a_1 = 10$, $a_2 = 29$.

5. [BB] Solve the recurrence relation $a_n = -8a_{n-1} - a_{n-2}$, $n \geq 2$, given $a_0 = 0$, $a_1 = 1$.

6. Solve the recurrence relation $a_n = -5a_{n-1} + 6a_{n-2}$, $n \geq 2$, given $a_0 = 5$, $a_1 = 19$.

7. [BB] Solve the recurrence relation $a_{n+1} = 2a_n + 3a_{n-1}$, $n \geq 1$, given $a_0 = 0$, $a_1 = 8$.

8. Solve the recurrence relation $a_n = 2a_{n-1} - a_{n-2}$, $n \geq 2$, given $a_0 = 40$, $a_1 = 37$.

9. Solve the recurrence relation $9a_n = 6a_{n-1} - a_{n-2}$, $n \geq 2$, given $a_0 = 3$, $a_1 = -1$.

10. (a) [BB] Solve the recurrence relation $a_n = -2a_{n-1} + 15a_{n-2}$, $n \geq 2$, given $a_0 = 1$, $a_1 = -1$.
    (b) [BB] Solve the recurrence relation $a_n = -2a_{n-1} + 15a_{n-2} + 24$, $n \geq 2$, given $a_0 = 1$, $a_1 = -1$.

11. (a) Solve the recurrence relation $a_{n+1} = -8a_n - 16a_{n-1}$, $n \geq 1$, given $a_0 = 5$, $a_1 = 17$.
    (b) Solve the recurrence relation $a_{n+1} = -8a_n - 16a_{n-1} + 5$, $n \geq 1$, given $a_0 = 2$, $a_1 = -1$.

12. Solve the recurrence relation $a_n = 4a_{n-1} - 4a_{n-2} + n$, $n \geq 2$, given $a_0 = 5$, $a_1 = 9$.

13. (a) [BB] Solve the recurrence relation $a_n = 4a_{n-1}$, $n \geq 1$, given $a_0 = 1$.
    (b) [BB] Solve the recurrence relation $a_n = 4a_{n-1} + 8^n$, $n \geq 1$, given $a_0 = 1$.
    (c) [BB] Verify that your answer to (b) is correct.

14. (a) Solve the recurrence relation $a_n = 5a_{n-1} - 6a_{n-2}$, $n \geq 2$, given $a_0 = 2$, $a_1 = 11$.
    (b) Solve the recurrence relation $a_n = 5a_{n-1} - 6a_{n-2} + 3n$, $n \geq 2$, given $a_0 = 2$, $a_1 = 14$.
    (c) Verify that your answer to (b) is correct.

15. (a) Solve the recurrence relation $a_n = -6a_{n-1} - 9a_{n-2}$, $n \geq 2$, given $a_0 = 1$, $a_1 = -4$.

    (b) Solve the recurrence relation $a_n = -6a_{n-1} - 9a_{n-2} + n^2 + 3n$, $n \geq 2$, given $a_0 = \frac{179}{128}$, $a_1 = -\frac{21}{128}$.
    (c) Verify that your answer to (b) is correct.

16. (a) [BB] Solve the recurrence relation $a_n = 4a_{n-1} - 9$, $n \geq 1$, given $a_0 = 4$.
    (b) [BB] Solve the recurrence relation $a_n = 4a_{n-1} + 3n2^n$, $n \geq 1$, given $a_0 = 4$.

17. Solve the recurrence relation $a_n = 5a_{n-1} - 2a_{n-2} + 3n^2$, $n \geq 2$, given $a_0 = 0$, $a_1 = 3$.

18. [BB] Using the result of Problem 20, show that the nth term of the Fibonacci sequence is the integer closest to $\frac{1}{\sqrt{5}}(\frac{1+\sqrt{5}}{2})^n$. This result was stated without proof in the text. [*Hint:* Show that $\left|\frac{1}{\sqrt{5}}(\frac{1-\sqrt{5}}{2})\right| < \frac{1}{2}$.]

19. Let a_n denote the number of n-digit numbers, each of whose digits is 1, 2, 3, or 4 and in which the number of 1's is even.

    (a) Find a recurrence relation for a_n.
    (b) Find an explicit formula for a_n.

20. The Towers of Hanoi is a popular puzzle. It consists of three pegs and a number of discs of differing diameters, each with a hole in the center. The discs initially sit on one of the pegs in order of decreasing diameter (smallest at top, largest at bottom, as in Fig 5.5), thus forming a triangular tower. The object is to move the tower to one of the other pegs by transferring the discs to any peg one at a time in such a way that no disc is ever placed upon a smaller one.

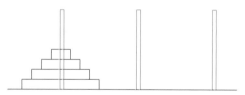

Figure 5.5 The Towers of Hanoi.

    (a) Solve the puzzle when there are $n = 2$ discs and show your moves by completing a little table like that below. [The pegs are labeled A, B, C, and we have used an asterisk (∗) to denote an empty peg. The disks are numbered in order of increasing size, thus disk 1 is the smallest.]

	A	B	C
Initial position	1,2	*	*
Move 1	??	??	??
Move 2	??	??	??
etc.			

Also solve the puzzle, with a similar table, when $n = 3$. How many moves are required in each case?

(b) Give a recurrence relation for a_n, the number of moves required to transfer n discs from one peg to another.

(c) Find an explicit formula for a_n.

(d) Suppose we can move a disc a second. Estimate the time required to transfer the discs if $n = 8$, $n = 16$, $n = 32$, and $n = 64$.

21. Suppose we modify the traditional rules for the Towers of Hanoi as described in the preceding question by requiring that one move discs only to an **adjacent** peg. Answer all four parts of the previous question for this new version of the puzzle.

22. [BB] Let $a_n = ra_{n-1} + sa_{n-2}$, $n \geq 2$, be a second order homogeneous recurrence relation with constant coefficients.

(a) If x is a root of the characteristic polynomial and c is any constant, show that $a_n = cx^n$ satisfies the given recurrence relation for $n \geq 2$.

(b) If p_n and q_n both satisfy the given recurrence for $n \geq 2$, show that $a_n = p_n + q_n$ also satisfies the recurrence for $n \geq 2$.

(c) Using (a) and (b), prove Theorem 5.3.1 for the case where the characteristic polynomial has distinct roots.

23. Let $a_n = ra_{n-1} + sa_{n-2}$, $n \geq 2$, be a second order homogeneous recurrence relation with constant coefficients and assume that its characteristic polynomial has just one (repeated) nonzero root.

(a) [BB] If x is the characteristic root, show that $r = 2x$ and $s = -x^2$.

(b) Conclude from (a) that if c is any constant, then $a_n = cnx^n$ is a solution of the recurrence relation.

(c) Use (b) and the previous exercise to prove Theorem 5.3.1 for this case.

24. Let $a_n = ra_{n-1} + sa_{n-2} + f(n)$, $n \geq 2$, be a second order recurrence relation with constant coefficients.

(a) Show that if p_n satisfies this recurrence relation for $n \geq 2$ and q_n satisfies the associated homogeneous recurrence relation $a_n = ra_{n-1} + sa_{n-2}$ for $n \geq 2$, then $p_n + q_n$ satisfies the given relation for $n \geq 2$.

(b) Complete the proof of Theorem 5.3.2.

5.4 Solving Recurrence Relations; Generating Functions

The brief introduction to generating functions which we give in this section belies the importance of this concept in combinatorial mathematics. Our purpose is just to give some indication as to how generating functions can be used to solve recurrence relations.

Roughly speaking, a generating function is a polynomial that "goes on forever," that is, an expression of the form

$$f(x) = a_0 + a_1 x + a_2 x^2 + a_3 x^3 + \cdots + a_n x^n + \cdots .$$

Unlike the usual polynomial, in which the coefficients a_i are all zero after a certain point, a generating function usually has infinitely many nonzero terms. There is an obvious correspondence between generating functions and sequences $a_0, a_1, a_2, \ldots$; namely,

$$a_0 + a_1 x + a_2 x^2 + a_3 x^3 + \cdots \quad \longleftrightarrow \quad a_0, a_1, a_2, a_3, \ldots .$$

5.4.1 DEFINITION ▶ The *generating function* of a sequence $a_0, a_1, a_2, \ldots$ is the expression $f(x) = a_0 + a_1 x + a_2 x^2 + \cdots .$

EXAMPLES 23   The generating function of the sequence $1, 2, 3, \ldots$ of natural numbers is $f(x) = 1 + 2x + 3x^2 + \cdots$, while the generating function of the arithmetic sequence $1, 4, 7, 10, \ldots$ is $f(x) = 1 + 4x + 7x^2 + 10x^3 + \cdots$. ▲

Two generating functions can be added and multiplied term by term just like polynomials. If $f(x) = a_0 + a_1 x + a_2 x^2 + \cdots$ and $g(x) = b_0 + b_1 x + b_2 x^2 + \cdots$, then

$$f(x) + g(x) = (a_0 + b_0) + (a_1 + b_1)x + (a_2 + b_2)x^2 + \cdots$$

$$f(x)g(x) = (a_0 b_0) + (a_1 b_0 + a_0 b_1)x + (a_0 b_2 + a_1 b_1 + a_2 b_0)x^2 + \cdots .$$

Note that while generating functions have infinitely many terms, the definitions of addition and multiplication involve no infinite sums; for example, the coefficient of x^n in the product $f(x)g(x)$ is the finite sum

$$a_0 b_n + a_1 b_{n-1} + a_2 b_{n-2} + \cdots + a_n b_0 .$$

PROBLEM 24.   If

$$f(x) = 1 + x + x^2 + \cdots + x^n + \cdots$$

and

$$g(x) = 1 - x + x^2 - x^3 + \cdots + (-1)^n x^n + \cdots ,$$

find $f(x) + g(x)$ and $f(x)g(x)$.

Solution. $f(x) + g(x) = (1 + x + x^2 + \cdots + x^n + \cdots)$
$$+ (1 - x + x^2 - x^3 + \cdots + (-1)^n x^n + \cdots)$$
$$= (1 + 1) + (1 - 1)x + (1 + 1)x^2$$
$$+ \cdots + (1 + (-1)^n)x^n + \cdots$$
$$= 2 + 2x^2 + 2x^4 + \cdots$$

$$f(x)g(x) = (1 + x + x^2 + \cdots + x^n + \cdots)$$
$$\cdot (1 - x + x^2 - x^3 + \cdots + (-1)^n x^n + \cdots)$$
$$= 1 + [1(-1) + 1(1)]x + [1(1) + 1(-1) + 1(1)]x^2 + \cdots$$
$$= 1 + x^2 + x^4 + x^6 + \cdots$$ ∎

Students who have studied calculus for more than one year should notice the obvious similarity between generating functions and power series and will be comfortable with the fact that generating functions often can be expressed as the quotient of polynomials. An important example is

(11) $$\frac{1}{1-x} = 1 + x + x^2 + x^3 + \cdots ,$$

which shows that $\dfrac{1}{1-x}$ is the generating function of the sequence $1, 1, 1, \ldots$.

Pause 10   Suppose a is a real number. Show that $\dfrac{1}{1-ax}$ is the generating function for a certain geometric sequence.

In combinatorics, the fundamental distinction between power series and generating functions is that, whereas power series in calculus are **functions** $\mathbb{R} \to \mathbb{R}$ with radii of convergence, generating functions are purely formal objects which will never be "evaluated" at a specific real number x. Thus, we do not worry about what the infinite sum $a_0 + a_1 x + a_2 x^2 + \cdots$ means. Whereas it is a topic of interest in calculus to prove formula (11), for instance, and to discover for which real numbers this formula is valid, for us the proof is a routine application of the definition of multiplication of generating functions.

$$(1-x)(1 + x + x^2 + \cdots + x^n + \cdots)$$
$$= 1 + [1(1) - 1(1)]x + [1(1) - 1(1)]x^2 + \cdots + [1(1) - 1(1)]x^n + \cdots$$
$$= 1 + 0x + 0x^2 + \cdots + 0x^n + \cdots$$
$$= 1$$

Another very useful formula for us is

(12) $$\dfrac{1}{(1-x)^2} = 1 + 2x + 3x^2 + 4x^3 + \cdots + (n+1)x^n + \cdots,$$

which says that $\dfrac{1}{(1-x)^2}$ is the generating function of the sequence of natural numbers.

Pause 11   Prove formula (12).

Suppose $f(x)$ is the generating function of the sequence $0, 1, 2, 3, \ldots$; that is, $f(x) = 0 + 1x + 2x^2 + 3x^3 + 4x^4 + \cdots$. Then

$$f(x) = x + 2x^2 + 3x^3 + 4x^4 + \cdots$$
$$= x(1 + 2x + 3x^2 + 4x^3 + \cdots) = x\dfrac{1}{(1-x)^2}$$

by (12), so $f(x)$ takes the simpler form $f(x) = \dfrac{x}{(1-x)^2}$.

We now present a few examples which show how generating functions can be used to solve recurrence relations.

PROBLEM 25.   Solve the recurrence relation $a_n = 3a_{n-1}$, $n \geq 1$, given $a_0 = 1$.

Solution.   Consider the generating function $f(x) = a_0 + a_1 x + a_2 x^2 + \cdots + a_n x^n + \cdots$ of the sequence $a_0, a_1, a_2, \ldots$. Multiplying by $3x$ and writing the product $3xf(x)$ below $f(x)$ so that terms involving x^n match, we obtain

$$f(x) = a_0 + \quad a_1 x + \quad a_2 x^2 + \quad \cdots + \quad a_n x^n + \cdots$$
$$3xf(x) = \qquad 3a_0 x + 3a_1 x^2 + \quad \cdots + 3a_{n-1}x^n + \cdots.$$

Subtracting gives

$$f(x) - 3xf(x)$$

$$= a_0 + (a_1 - 3a_0)x + (a_2 - 3a_1)x^2 + \cdots + (a_n - 3a_{n-1})x^n + \cdots .$$

Since $a_0 = 1$, $a_1 = 3a_0$ and, in general, $a_n = 3a_{n-1}$, this says that $(1 - 3x)f(x) = 1$. Thus, $f(x) = \dfrac{1}{1 - 3x}$ and, using (11),

$$f(x) = 1 + 3x + (3x)^2 + \cdots + (3x)^n + \cdots$$

$$= 1 + 3x + 9x^2 + \cdots + 3^n x^n + \cdots .$$

We conclude that a_n, which is the coefficient of x^n in $f(x)$, must equal 3^n. So we have $a_n = 3^n$ as the solution to our recurrence relation. ∎

PROBLEM 26.   Solve the recurrence relation $a_n = 2a_{n-1} - a_{n-2}$, $n \geq 2$, given $a_0 = 3$, $a_1 = -2$.

Solution.   Letting $f(x)$ be the generating function of the sequence in question, we have

$$\begin{aligned}
f(x) &= a_0 + a_1 x + a_2 x^2 + \cdots + a_n x^n + \cdots \\
2xf(x) &= \quad\quad 2a_0 x + 2a_1 x^2 + \cdots + 2a_{n-1} x^n + \cdots \\
x^2 f(x) &= \quad\quad\quad\quad\quad a_0 x^2 + \cdots + a_{n-2} x^n + \cdots .
\end{aligned}$$

Therefore,

$$f(x) - 2xf(x) + x^2 f(x) = a_0 + (a_1 - 2a_0)x + (a_2 - 2a_1 + a_0)x^2 + \cdots$$

$$+ (a_n - 2a_{n-1} + a_{n-2})x^n + \cdots$$

$$= 3 - 8x$$

since $a_0 = 3$, $a_1 = -2$ and $a_n - 2a_{n-1} + a_{n-2} = 0$ for $n \geq 2$.
    So $(1 - 2x + x^2)f(x) = 3 - 8x$, $(1 - x)^2 f(x) = 3 - 8x$ and

$$f(x) = \frac{1}{(1 - x)^2} (3 - 8x)$$

$$= (1 + 2x + 3x^2 + \cdots + (n + 1)x^n + \cdots)(3 - 8x) \quad\quad \text{by (12)}$$

$$= 3 - 2x - 7x^2 - 12x^3 + \cdots + [3(n + 1) - 8n]x^n + \cdots$$

$$= 3 - 2x - 7x^2 - 12x^3 + \cdots + (-5n + 3)x^n + \cdots$$

and $a_n = 3 - 5n$ is the desired solution. [Verify that this recurrence can also be solved by Theorem 5.3.1 of Section 5.3.] ∎

PROBLEM 27.   Solve the recurrence $a_n = -3a_{n-1} + 10a_{n-2}$, $n \geq 2$, given $a_0 = 1$, $a_1 = 4$.

Solution.   Letting $f(x)$ be the generating function of the sequence in question, we have

$$f(x) = a_0 + a_1x + a_2x^2 + \cdots + a_nx^n + \cdots$$
$$3xf(x) = 3a_0x + 3a_1x^2 + \cdots + 3a_{n-1}x^n + \cdots$$
$$10x^2f(x) = 10a_0x^2 + \cdots + 10a_{n-2}x^n + \cdots .$$

Therefore,

$$f(x) + 3xf(x) - 10x^2f(x) = a_0 + (a_1 + 3a_0)x + (a_2 + 3a_1 - 10a_0)x^2$$
$$+ \cdots + (a_n + 3a_{n-1} - 10a_{n-2})x^n + \cdots$$
$$= 1 + 7x$$

since $a_0 = 1$, $a_1 = 4$ and $a_n + 3a_{n-1} - 10a_{n-2} = 0$ for $n \geq 2$. So

$$(1 + 3x - 10x^2)f(x) = 1 + 7x$$

and

$$f(x) = \frac{1 + 7x}{1 + 3x - 10x^2} = \frac{1 + 7x}{(1 + 5x)(1 - 2x)} .$$

At this point, it is useful to recall the method of *partial fractions*. We set

$$\frac{1}{(1 + 5x)(1 - 2x)} = \frac{A}{1 + 5x} + \frac{B}{1 - 2x} = \frac{A(1 - 2x) + B(1 + 5x)}{(1 + 5x)(1 - 2x)} .$$

Equating numerators, $1 = (A + B) + (-2A + 5B)x$, so $A + B = 1$, $-2A + 5B = 0$. Solving for A and B, we get $A = \frac{5}{7}$, $B = \frac{2}{7}$. Therefore,

$$\frac{1}{(1 + 5x)(1 - 2x)} = \frac{5}{7}\left(\frac{1}{1 + 5x}\right) + \frac{2}{7}\left(\frac{1}{1 - 2x}\right)$$

and so $f(x) = \dfrac{1 + 7x}{(1 + 5x)(1 - 2x)}$

$$= \frac{5}{7}\left(\frac{1}{1 + 5x}\right)(1 + 7x) + \frac{2}{7}\left(\frac{1}{1 - 2x}\right)(1 + 7x)$$

$$= \tfrac{5}{7}(1 + (-5x) + (-5x)^2 + \cdots)(1 + 7x)$$

$$+ \tfrac{2}{7}(1 + 2x + (2x)^2 + \cdots)(1 + 7x)$$

$$= \tfrac{5}{7}(1 - 5x + 25x^2 + \cdots + (-5)^nx^n + \cdots)(1 + 7x)$$

$$+ \tfrac{2}{7}(1 + 2x + 4x^2 + \cdots + 2^nx^n + \cdots)(1 + 7x)$$

$$= \tfrac{5}{7}(1 + 2x - 10x^2 + \cdots + [(-5)^n + 7(-5)^{n-1}]x^n + \cdots)$$

$$+ \tfrac{2}{7}(1 + 9x + 18x^2 + \cdots + [2^n + 7(2^{n-1})]x^n + \cdots)$$

$$= \tfrac{5}{7}(1 + 2x - 10x^2 + \cdots + 2(-5)^{n-1}x^n + \cdots)$$

$$+ \tfrac{2}{7}(1 + 9x + 18x^2 + \cdots + 9(2^{n-1})x^n + \cdots)$$

$$= 1 + 4x - 2x^2 + \cdots + (-\tfrac{2}{7}(-5)^n + \tfrac{9}{7}(2^n))x^n + \cdots$$

and hence $a_n = -\frac{2}{7}(-5)^n + \frac{9}{7}(2^n)$ is the desired solution. (Again, we suggest that you verify that this recurrence can also be solved by the methods of Section 5.3.) ∎

PROBLEM 28. Solve the recurrence relation $a_n = -a_{n-1} + 2n - 3$, $n \geq 1$, given $a_0 = 1$.

Solution. Let $f(x)$ be the generating function of the sequence $a_0, a_1, a_2, \ldots$. Then

$$f(x) = a_0 + a_1 x + a_2 x^2 + \cdots + a_n x^n + \cdots$$
$$xf(x) = \quad\quad a_0 x + a_1 x^2 + \cdots + a_{n-1} x^n + \cdots .$$

Therefore,

$$f(x) + xf(x) = a_0 + (a_1 + a_0)x + (a_2 + a_1)x^2 + \cdots + (a_n + a_{n-1})x^n + \cdots .$$

We are given that $a_0 = 1$ and $a_n + a_{n-1} = 2n - 3$. Thus $a_1 + a_0 = 2(1) - 3 = -1$, $a_2 + a_1 = 2(2) - 3 = 1$, and so on. We obtain $(1 + x)f(x) = 1 - x + x^2 + \cdots + (2n - 3)x^n + \cdots$ and so

$$f(x) = \frac{1}{1+x}(1 - x + x^2 + \cdots + (2n - 3)x^n + \cdots)$$

$$= (1 + (-x) + (-x)^2 + \cdots + (-x)^n + \cdots)$$
$$\cdot (1 - x + x^2 + \cdots + (2n - 3)x^n + \cdots)$$

$$= (1 - x + x^2 - x^3 + \cdots + (-x)^n + \cdots)$$
$$\cdot (1 - x + x^2 + \cdots + (2n - 3)x^n + \cdots)$$

$$= 1 - 2x + 3x^2 + \cdots$$
$$+ [(2n - 3) - (2n - 5) + \cdots + (-1)^{n-1}(-1) + (-1)^n]x^n + \cdots .$$

Now a_n is the coefficient of x^n, the term in the square brackets:

$$a_n = \left\{ \sum_{k=0}^{n-1} (-1)^k [2n - (2k + 3)] \right\} + (-1)^n$$

$$= (2n - 3) \sum_{k=0}^{n-1} (-1)^k - 2 \sum_{k=0}^{n-1} (-1)^k k + (-1)^n.$$

We leave it to you to verify that

$$\sum_{k=0}^{n-1} (-1)^k k = \begin{cases} -\frac{n}{2} & \text{if } n \text{ is even} \\ \frac{n-1}{2} & \text{if } n \text{ is odd.} \end{cases}$$

So, if n is even, $a_n = 0 - 2(-\frac{n}{2}) + 1 = n + 1$, while if n is odd, $a_n = (2n - 3) - 2(\frac{n-1}{2}) - 1 = 2n - 3 - n + 1 - 1 = n - 3$. The solution is

$$a_n = \begin{cases} n + 1 & \text{if } n \text{ is even} \\ n - 3 & \text{if } n \text{ is odd.} \end{cases}$$

Note that this solution could also be written as $a_n = 2(-1)^n + n - 1$, which is the answer one would obtain by the methods of Section 5.3.   ∎

10. Replacing x by ax in (11), we see that

$$\frac{1}{1 - ax} = 1 + (ax) + (ax)^2 + (ax)^3 + \cdots = 1 + ax + a^2x^2 + a^3x^3 + \cdots .$$

From this, we see that $\dfrac{1}{1 - ax}$ is the generating function for the sequence $1, a, a^2, a^3, \ldots$, which is the geometric sequence with first term 1 and common ratio a.

11. $(1 - x)^2(1 + 2x + 3x^2 + \cdots + (n + 1)x^n + \cdots)$

$$= (1 - 2x + x^2)(1 + 2x + 3x^2 + \cdots + (n + 1)x^n + \cdots)$$

$$= 1 + [1(2) - 2(1)]x + [1(3) - 2(2) + 1(1)]x^2 + \cdots$$

$$+ [1(n + 1) - 2(n) + 1(n - 1)]x^n + \cdots = 1$$

since $n + 1 - 2n + n - 1 = 0$.

EXERCISES

The symbol [BB] means that an answer can be found in the Back of the Book.

1. What sequence is associated with each of the following generating functions?

   (a) [BB] $(2 - 3x)^2$

   (b) $\dfrac{x^4}{1 - x}$

   (c) [BB] $\dfrac{1}{(1 + 3x)^2}$

   (d) $\dfrac{1}{(1 - x^3)^2}$

   (e) $\dfrac{x^2}{(1 + x)^2} + \dfrac{1}{1 + 5x}$

2. Express the generating function of each of the following sequences as a polynomial or as the quotient of polynomials.

   (a) [BB] $1, 2, 5, 0, 0, \ldots$

   (b) $0, 1, 4, 1, 0, 0, \ldots$

   (c) $1, 2, 4, 8, 16, \ldots$

   (d) $1, -1, 1, -1, \ldots$

   (e) $3, 3, 3, \ldots$

   (f) [BB] $1, 0, 1, 0, \ldots$

   (g) $1, -2, 3, -4, \ldots$

3. [BB] Using the method of generating functions, solve the recurrence relation $a_n = 2a_{n-1}, n \geq 1$, given $a_0 = 1$. Compare this solution to the sequence defined by (5), p. 164.

4. Use the method of generating functions to solve the recurrence relation $a_n = 3a_{n-1} + 1, n \geq 1$, given $a_0 = 1$. Compare your solution with the sequence given in Problem 10, p. 164.

5. [BB] Use the method of generating functions to solve the recurrence relation $a_n = 5a_{n-1} - 6a_{n-2}$, given $a_0 = 1$, $a_1 = 4$. Verify your answer by comparing with Problem 18.

6. Use the method of generating functions to solve the recurrence relation $a_n = 4a_{n-1} - 4a_{n-2}, n \geq 2$, given $a_0 = 1$, $a_1 = 4$. Compare your solution with the sequence given in Problem 19, p. 174.

7. [BB] Use generating functions to find a formula for a_n given $a_0 = 5$ and $a_n = a_{n-1} + 2^n$ for $n \geq 1$.

8. (a) Use the method of the characteristic polynomial (as in Section 5.3) to solve the recurrence relation $a_n = 4a_{n-1} - 4a_{n-2} + 4^n, n \geq 2$, with the initial conditions $a_0 = 2$, $a_1 = 8$.

(b) Solve the recurrence in (a) by means of generating functions.

(c) Which of the preceding methods do you prefer in this case?

9. Solve each of the following using generating functions. In each case, use the methods of Section 5.3 to verify your answer.

(a) [BB] $a_n = -5a_{n-1}$, $n \geq 1$, given $a_0 = 2$.

(b) $a_n = -5a_{n-1} + 3$, $n \geq 1$, given $a_0 = 2$.

10. Solve each of the following using generating functions. Verify your answer by the method of Section 5.3.

(a) $a_n = 4a_{n-1} - 3a_{n-2}$, $n \geq 2$, given $a_0 = 2$, $a_1 = 5$.

(b) $a_n = -10a_{n-1} - 25a_{n-2}$, $n \geq 2$, given $a_0 = 1$, $a_1 = 25$.

11. [BB] Using generating functions, solve the recurrence relation $a_n = 2a_{n-1} + a_{n-2} - 2a_{n-3}$, $n \geq 3$, given $a_0 = 1$, $a_1 = 3$, $a_2 = 6$. (Note that this recurrence is not second order and so cannot be solved by the methods of Section 5.3.)

12. [BB] Use the method of generating functions to solve the recurrence relation $a_n = a_{n-1} + a_{n-2} - a_{n-3}$, $n \geq 3$, given $a_0 = 2$, $a_1 = -1$, $a_2 = 3$.

[Hint : $\dfrac{1}{(ax+b)^2(cx+d)} = \dfrac{Ax+B}{(ax+b)^2} + \dfrac{C}{cx+d}$.]

13. The Pell sequence is defined by $p_0 = 1$, $p_1 = 2$ and $p_n = 2p_{n-1} + p_{n-2}$ for $n \geq 2$.

(a) Use the characteristic polynomial to solve this recurrence relation.

(b) Show that p_n is the integer closest to $\left(\frac{2+\sqrt{2}}{4}\right)(1 + \sqrt{2})^n$.

(c) Find the generating function of the Pell sequence, finding explicitly its first four terms.

Remarkably, there exist closed form solutions for p_n. For example,

$$p_n = \sum_{\substack{i,j,k \geq 0 \\ i+j+2k=n}} \frac{(i+j+k)!}{i!j!k!}.$$

See the *American Mathematical Monthly* **107** (2000), no. 4, p. 370 for three verifications of this formula.

14. This question concerns the Fibonacci sequence defined by the recurrence relation $a_n = a_{n-1} + a_{n-2}$, where $a_0 = a_1 = 1$.

(a) Suppose $f(x)$ is the generating function of the Fibonacci sequence. Show that $f(x) = \dfrac{1}{1 - x - x^2}$.

(b) Find α and β such that $1 - x - x^2 = (1 - \alpha x)(1 - \beta x)$.

(c) Find A and B, in terms of α and β, such that

$$\frac{1}{1-x-x^2} = \frac{A}{1-\alpha x} + \frac{B}{1-\beta x}.$$

(d) Use the results of the previous parts to obtain a formula for a_n and compare your answer with that found in Problem 20, p. 174. Which method for finding a_n do you prefer?

REVIEW EXERCISES FOR CHAPTER 5

1. Using mathematical induction, show that $\displaystyle\sum_{i=1}^{n} 3^{i-1} = \dfrac{3^n - 1}{2}$ for all $n \geq 1$.

2. Using mathematical induction, show that $\left(1 - \dfrac{1}{2}\right)^n \geq 1 - \dfrac{n}{2}$ for all $n \geq 1$.

3. Use mathematical induction to prove that $10^{n+2} + 10^n + 1$ is divisible by 3 for all $n \geq 1$.

4. Give a recursive definition of each of the following sequences:

(a) $1, 5, 29, 173, 1037, \ldots$

(b) $3, 5, 13, 85, 3613, \ldots$

(c) $1, 9, 36, 100, 225, \ldots$

5. Let a_n be defined recursively by $a_1 = 0$, $a_2 = \frac{1}{3}$ and, for $k \geq 1$, $a_{k+2} = \frac{1}{2}(a_k + a_{k+1})$. Prove that $a_n = \frac{2}{9}(1 - (-\frac{1}{2})^{n-1})$ for all integers $n \geq 1$.

6. Consider the arithmetic sequence which begins $5, 9, 13$.

(a) Find the 32nd and 100th terms of this sequence.

(b) Does 125 belong to the sequence? If so, where does it occur?

(c) Repeat (b) for the numbers 429 and 1000.

(d) Find the sum of the first 18 terms.

7. The first two terms of a sequence are 6 and 2.

(a) If the sequence is arithmetic, find the 27th term and the sum of the first 30 terms.

(b) If the sequence is geometric, find an expression for the 27th term and the sum of the first 30 terms.

8. (a) Define the Fibonacci sequence.

(b) Is it possible for three successive terms of the Fibonacci sequence to be odd?

(c) Is it possible for two successive terms in the Fibonacci sequence to be even?

Justify your answers.

9. Solve the recurrence relation $a_n = 5a_{n-1} - 4a_{n-2}$, $n \geq 2$, given that $a_0 = -3$ and $a_1 = 6$. Use the characteristic

polynomial as described in Section 5.3.

10. Solve Exercise 9 using the method of generating functions described in Section 5.4.

11. Solve $a_n = 4a_{n-1} + 5a_{n-2} + 3^n$, $n \geq 2$, given $a_0 = 4$, $a_1 = -1$.

12. (For students of calculus) Let $f_1, f_2, f_3, \ldots$ denote the Fibonacci sequence as defined in 5.2.3. Evaluate $\sum_{1}^{\infty} \dfrac{f_k}{100^k}$ exactly. Then approximate this sum to 19 decimal places and admire its beauty. What do you notice?

6

Principles of Counting

6.1 THE PRINCIPLE OF INCLUSION–EXCLUSION

Glenys is thinking about registering for a word processing course. Of the 100 people who have registered so far, she discovers that three quarters are men and 80% own personal digital assistants (PDAs).

(a) Find a formula for the number of women in the course who do not have PDAs. How large might this number be? How small?

(b) How many of the men registered in this course could conceivably own PDAs?

The object of this chapter is to illustrate some basic principles of counting. We consider first the number of elements in various combinations of finite sets such as union, intersection, and difference. As in Section 3.3, the number of elements in a finite set S will be denoted $|S|$.

To analyze Glenys's questions, we introduce the sets U of all registrants, M of male registrants, and P of those registrants who own a PDA. We are given that $|U| = 100$, $|M| = 75$, and $|P| = 80$. The set of women who do not have a PDA is $M^c \cap P^c$ and it is the size of this set with which part (a) is concerned. By one of the laws of De Morgan,

$$M^c \cap P^c = (M \cup P)^c$$

and so

$$|M^c \cap P^c| = |(M \cup P)^c| = 100 - |M \cup P|.$$

How big is $M \cup P$? Adding the number of men and the number of registrants who have PDAs counts twice the men with PDAs (see Fig 6.1), so

$$|M \cup P| = |M| + |P| - |M \cap P| = 75 + 80 - |M \cap P| = 155 - |M \cap P|.$$

Therefore,

$$|M^c \cap P^c| = 100 - (155 - |M \cap P|) = |M \cap P| - 55,$$

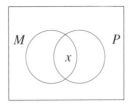

Figure 6.1 $|M \cup P| = |M| + |P| - |M \cap P|$ because Region x gets counted twice in the sum $|M| + |P|$.

which is a formula for the number of women without PDAs. Since $M \cap P$ is a subset of M, a set of 75 people, $|M \cap P| \leq |M| = 75$ and so $M^c \cap P^c \leq 75 - 55 = 20$. Conceivably, $|M^c \cap P^c|$ could be 0 (if $|M \cap P| = 55$). So the number of women without PDAs is between 0 and 20 (inclusive).

Part (b) asks about $M \cap P$. Observing that $|M \cap P| - 55 \geq 0$ (because it is the number of women without PDAs), we see that $|M \cap P| \geq 55$; that is, at least 55 men must own PDAs. As noted previously, the upper bound for $|M \cap P|$ is 75 and it is possible for this number to be realized.

Suppose, in fact, that $|M \cap P| = 72$. Then

$$|M \cup P| = |M| + |P| - |M \cap P| = 75 + 80 - 72 = 83.$$

There are 83 people in the class who are either men or own a PDA.

How many men registered in the class would not have PDAs? Surely, $75 - 72 = 3$:

$$|M \setminus P| = |M| - |M \cap P| = 75 - 72 = 3.$$

How many of the owners of PDAs would be women? Surely, $80 - 72 = 8$:

$$|P \setminus M| = |P| - |P \cap M| = 80 - 72 = 8.$$

How many of those registered are either men without PDAs or women with PDAs? This question, which asks for $|M \oplus P|$, can be answered in several ways. It is the number of men without PDAs plus the number of women with a PDA; that is,

$$|M \oplus P| = |M \setminus P| + |P \setminus M| = 3 + 8 = 11.$$

It is also the number of people who are men or owners of a PDA but not both:

$$|M \oplus P| = |M \cup P| - |M \cap P| = 83 - 72 = 11.$$

We could also legitimately argue that this number is

$$|M \oplus P| = |M| + |P| - 2|M \cap P| = 75 + 80 - 2(72) = 155 - 144 = 11$$

because adding the number of men and the number of people with PDAs counts twice the "unwanted" individuals, the men with PDAs.

We summarize the basic principles of counting, most of which we have just illustrated.

6.1.1 PROPOSITION ▶ Let A and B be subsets of a finite universal set U. Then

(a) $|A \cup B| = |A| + |B| - |A \cap B|$

(b) $|A \cap B| \leq \min\{|A|, |B|\}$, the minimum of $|A|$ and $|B|$

(c) $|A \setminus B| = |A| - |A \cap B| \geq |A| - |B|$

(d) $|A^c| = |U| - |A|$

(e) $|A \oplus B| = |A \cup B| - |A \cap B| = |A| + |B| - 2|A \cap B| = |A \setminus B| + |B \setminus A|$

(f) $|A \times B| = |A| \times |B|$

Proof (a) If $A = \emptyset$, then $A \cap B = \emptyset$, and $|A| = |A \cap B| = 0$ so the result holds because each side of (a) equals $|B|$. Similarly, the result holds if $B = \emptyset$, so we assume henceforth that neither A nor B is empty.

Suppose $A \cap B = \emptyset$. Let $A = \{a_1, a_2, \ldots, a_r\}$ and $B = \{b_1, b_2, \ldots, b_s\}$. Then

$$A \cup B = \{a_1, a_2, \ldots, a_r, b_1, b_2, \ldots, b_s\}$$

and, since there is no repetition among the elements listed here,

$$|A \cup B| = r + s = |A| + |B| = |A| + |B| - |A \cap B|$$

and the desired formula is true.

On the other hand, if $A \cap B \neq \emptyset$ and we let

$$A \setminus B = \{a_1, a_2, \ldots, a_r\}$$
$$B \setminus A = \{b_1, b_2, \ldots, b_s\}$$
$$A \cap B = \{x_1, x_2, \ldots, x_t\}$$

then

$$A = \{a_1, a_2, \ldots, a_r, x_1, x_2, \ldots, x_t\} \quad \text{and}$$
$$B = \{b_1, b_2, \ldots, b_s, x_1, x_2, \ldots, x_t\},$$

so

$$A \cup B = \{a_1, \ldots, a_r, b_1, \ldots, b_s, x_1, \ldots, x_t\}$$

with no repetition among the elements listed in any of these sets. Thus,

$$|A| + |B| - |A \cap B| = (r + t) + (s + t) - t = r + s + t = |A \cup B|$$

and the formula holds in this case as well.

We omit the proofs of (b)–(e) since, as with the proof of part (a), these follow in a straightforward way from the definitions of $\cup$, $\cap$, $\setminus$, $\oplus$, and set complement. The proof of (f) is left to the exercises. ∎

The formula for the number of elements in the union of two sets—part (a) of Proposition 6.1.1—is the simplest case of the *Principle of Inclusion-Exclusion*, which gives the general formula for the number of elements in the union of any finite collection of finite sets.

Suppose, for instance, we have three finite sets A, B, and C. Then

$$|A \cup B \cup C| = |A \overset{\downarrow}{\cup} (B \cup C)|$$

$$= |A| + |B \cup C| - |A \overset{\Downarrow}{\cap} (B \cup C)|$$

$$= |A| + |B \overset{\downarrow}{\cup} C| - |(A \cap B) \overset{\downarrow}{\cup} (A \cap C)|$$

$$= |A| + \big[|B| + |C| - |B \cap C|\big]$$
$$\quad - \big[|A \cap B| + |A \cap C| - |(A \cap B) \cap (A \cap C)|\big]$$

$$= |A| + |B| + |C| - |B \cap C| - |A \cap B| - |A \cap C| + |A \cap B \cap C|,$$

where, at the spots marked with single arrows, we used the fact that $|X \cup Y| = |X| + |Y| - |X \cap Y|$ and, at the last step, that $(A \cap B) \cap (A \cap C) = A \cap B \cap C$.

Pause 1   What happened at the double arrow?

Here is the formula for the cardinality of the union of four finite sets:

$$|A \cup B \cup C \cup D|$$

$$= |A| + |B| + |C| + |D|$$
$$\quad - |A \cap B| - |A \cap C| - |A \cap D| - |B \cap C| - |B \cap D| - |C \cap D|$$
$$\quad + |A \cap B \cap C| + |A \cap B \cap D| + |A \cap C \cap D| + |B \cap C \cap D|$$
$$\quad - |A \cap B \cap C \cap D|.$$

The general pattern should be evident. Add the cardinalities of each of the sets, subtract the cardinalities of the intersections of all pairs of sets, add the cardinalities of all intersections of the sets taken three at a time, subtract the cardinalities of all intersections of the sets taken four at a time, and so on.

6.1.2 PRINCIPLE OF INCLUSION– EXCLUSION ▶

Given a finite number of finite sets, $A_1, A_2, \ldots, A_n$, the number of elements in the union $A_1 \cup A_2 \cup \cdots \cup A_n$ is

$$|A_1 \cup A_2 \cup \cdots \cup A_n| = \sum_i |A_i| - \sum_{i<j} |A_i \cap A_j| + \sum_{i<j<k} |A_i \cap A_j \cap A_k|$$

$$- \cdots + (-1)^{n+1} |A_1 \cap A_2 \cap \cdots \cap A_n|,$$

where the first sum is over all i, the second sum is over all pairs i, j with $i < j$, the third sum is over all triples i, j, k with $i < j < k$, and so forth.

PROBLEM 1.   Of 30 personal computers (PCs) owned by faculty members in a certain university department, 20 run Windows, eight have 21 inch monitors, 25 have CD-ROM drives, 20 have at least two of these features, and six have all three.
  (a) How many PCs have at least one of these features?
  (b) How many have none of these features?
  (c) How many have exactly one feature?

Solution. Let W be the set of PCs running under Windows, M the set of PCs with 21 inch monitors, and C the set with CD-ROM drives. We are given that $|W| = 20$, $|M| = 8$, $|C| = 25$,

$$|(W \cap M) \cup (W \cap C) \cup (M \cap C)| = 20,$$

and $|W \cap M \cap C| = 6$. By the Principle of Inclusion–Exclusion,

$$\begin{aligned}
20 &= |(W \cap M) \cup (W \cap C) \cup (M \cap C)| \\
&= |W \cap M| + |W \cap C| + |M \cap C| - |(W \cap M) \cap (W \cap C)| \\
&\quad - |(W \cap M) \cap (M \cap C)| - |(W \cap C) \cap (M \cap C)| \\
&\quad + |(W \cap M) \cap (W \cap C) \cap (M \cap C)|.
\end{aligned}$$

Since each of the last four terms here is $|W \cap M \cap C|$, we obtain

$$20 = |W \cap M| + |W \cap C| + |M \cap C| - 2|W \cap M \cap C|;$$

therefore,

$$|W \cap C| + |W \cap M| + |M \cap C| = 20 + 2(6) = 32.$$

(a) The number of PCs with at least one feature is

$$\begin{aligned}
|W \cup M \cup C| &= |W| + |M| + |C| \\
&\quad - |W \cap M| - |W \cap C| - |M \cap C| + |W \cap M \cap C| \\
&= 20 + 8 + 25 - \big(|W \cap M| + |W \cap C| + |M \cap C|\big) + 6 \\
&= 59 - 32 = 27.
\end{aligned}$$

(b) It follows that $30 - 27 = 3$ PCs have none of the specified features.

(c) Since the number of computers with exactly one feature is the number with at least one feature less the number with at least two, the number with exactly one is $27 - 20 = 7$. ∎

Pause 2 Continuing the preceding problem, how many faculty have computers with exactly two of the three features described?

Some readers may notice that the problems discussed so far could also have been solved by Venn diagrams, which are an acceptable approach wherever possible. When more than three sets are involved, however, Venn diagrams can no longer be used. Thus, our solutions, which avoid Venn diagrams, illustrate techniques which can be applied in general.

PROBLEM 2. Suppose 18 of the 30 personal computers in Problem 1 have Pentium III processors, including ten of those running Windows, all of those with 21 inch monitors, and 15 of those with CD-ROM drives. Suppose also that every computer has at least one of the four features now specified. How many have at least three features?

Solution.   Let P be the set of PCs with Pentium III processors. Then, with W, M, and C as before, the question asks for the number

$$n = |(W \cap M \cap C) \cup (W \cap M \cap P) \cup (W \cap C \cap P) \cup (M \cap C \cap P)|$$
$$= |W \cap M \cap C| + |W \cap M \cap P| + |W \cap C \cap P| + |M \cap C \cap P|$$
$$- 6|W \cap M \cap C \cap P| + 4|W \cap M \cap C \cap P| - |W \cap M \cap C \cap P|$$
$$= |W \cap M \cap C| + |W \cap M \cap P| + |W \cap C \cap P| + |M \cap C \cap P|$$
$$- 3|W \cap M \cap C \cap P|.$$

We are given that

$$30 = |W \cup M \cup C \cup P|$$
$$= |W| + |M| + |C| + |P|$$
$$- |W \cap M| - |W \cap C| - |W \cap P| - |M \cap C| - |M \cap P| - |C \cap P|$$
$$+ |W \cap M \cap C| + |W \cap M \cap P| + |W \cap C \cap P| + |M \cap C \cap P|$$
$$- |W \cap M \cap C \cap P|$$
$$= |W| + |M| + |C| + |P|$$
$$- |W \cap M| - |W \cap C| - |W \cap P| - |M \cap C| - |M \cap P| - |C \cap P|$$
$$+ n + 2|W \cap M \cap C \cap P|$$

because

$$|W \cap M \cap C| + |W \cap M \cap P| + |W \cap C \cap P| + |M \cap C \cap P|$$
$$= n + 3|W \cap M \cap C \cap P|.$$

In Problem 1, we learned that $|W \cap C| + |W \cap M| + |M \cap C| = 32$ and were told that $|W \cap M \cap C| = 6$. Here, since $M \subseteq P$, we have $W \cap M \cap C \cap P = W \cap M \cap C$ and so $|W \cap M \cap C \cap P| = 6$. Therefore,

$$30 = 20 + 8 + 25 + 18 - 32 - 10 - 8 - 15 + n + 2(6),$$

so $n = 30 - 71 + 65 - 12 = 12$.   ∎

PROBLEM 3.   How many integers between 1 and 300 (inclusive) are

(a) divisible by at least one of 3, 5, 7?

(b) divisible by 3 and by 5 but not by 7?

(c) divisible by 5 but by neither 3 nor 7?

Solution.   Let A, B, and C be the sets of those integers between 1 and 300 which are divisible by 3, by 5, and by 7, respectively; thus,

$$A = \{n \mid 1 \le n \le 300, 3 \mid n\}$$

$$B = \{n \mid 1 \le n \le 300, 5 \mid n\}$$

$$C = \{n \mid 1 \le n \le 300, 7 \mid n\}.$$

(a) To be divisible by either 3 or 5 or 7 is to be in at least one of the sets A, B, or C. Thus, part (a) asks us to find $|A \cup B \cup C|$. To determine this number, we need $|A|$, $|B|$, $|C|$, $|A \cap B|$, $|A \cap C|$, $|B \cap C|$ and $|A \cap B \cap C|$. The elements of A are 3, 6, 9, 12, ..., 300, so $|A| = 100$. Note that this number is $\lfloor \frac{300}{3} \rfloor$, where, as in paragraph 3.1.6, $\lfloor x \rfloor$ denotes the floor of the real number x.

In general, for natural numbers a and b, the number of positive integers less than or equal to a and divisible by b is $\lfloor \frac{a}{b} \rfloor$. (See Exercise 18.) Thus, we also have

$$|B| = \left\lfloor \frac{300}{5} \right\rfloor = 60 \quad \text{and} \quad |C| = \left\lfloor \frac{300}{7} \right\rfloor = 42.$$

Next we have to find $A \cap B$, the set of integers between 1 and 300 which are divisible by both 3 and 5. Since 3 and 5 are relatively prime numbers, any number divisible by each of them must be divisible by their product. (See Exercise 10 of Section 4.2.) Therefore, $A \cap B$ is just the set of integers between 1 and 300 which are divisible by 15 and, similarly, $A \cap C$, $B \cap C$, and $A \cap B \cap C$ are the sets of integers between 1 and 300 which are divisible by 21, 35, and 105, respectively. So we have

$$|A \cap B| = \left\lfloor \frac{300}{15} \right\rfloor = 20, \quad |A \cap C| = \left\lfloor \frac{300}{21} \right\rfloor = 14,$$

$$|B \cap C| = \left\lfloor \frac{300}{35} \right\rfloor = 8, \quad |A \cap B \cap C| = \left\lfloor \frac{300}{105} \right\rfloor = 2$$

and hence, $|A \cup B \cup C| = 100 + 60 + 42 - 20 - 14 - 8 + 2 = 162$.

(b) The numbers divisible by 3 and by 5, but not by 7, are precisely those numbers in $(A \cap B) \setminus C$, a set of cardinality $|A \cap B| - |A \cap B \cap C| = 20 - 2 = 18$.

(c) The numbers divisible by 5 but by neither 3 nor 7 are those in $B \setminus (A \cup C)$, a set of cardinality $|B| - |B \cap (A \cup C)|$. Since

$$B \cap (A \cup C) = (B \cap A) \cup (B \cap C),$$

the Principle of Inclusion–Exclusion gives

$$|B \cap (A \cup C)| = |B \cap A| + |B \cap C| - |(B \cap A) \cap (B \cap C)|$$

$$= |B \cap A| + |B \cap C| - |B \cap A \cap C|$$

because $(B \cap A) \cap (B \cap C) = B \cap A \cap C$. Therefore, $|B \cap (A \cup C)| = 20 + 8 - 2 = 26$ and the number we seek is $|B| - 26 = 60 - 26 = 34$. ∎

Pause 3   In Problem 3, we characterized the set of integers divisible by both 3 and 5 as those divisible by 15. In general, for any natural numbers a and b, how can we characterize the set of natural numbers divisible by both a and b?

| Pause 4 | Suppose the Sieve of Eratosthenes (4.3.5) is used to enumerate all primes between 1 and 100. How many integers will remain after the first five steps of the procedure? |

Answers to Pauses

1. See equation (3) on p. 44.
2. The number of computers with exactly two of the features is the number with at least two less the number with exactly three; that is, $20 - 6 = 14$.
3. This set is just the set of integers divisible by the *least common multiple* of a and b since $a \mid n$, $b \mid n$ if and only if $\text{lcm}(a, b) \mid n$. (See Section 4.2.)
4. The integers which have been crossed out after five steps are those which are divisible by, but not equal to, at least one of 2, 3, 5, 7, 11. Letting A, B, C, D, and E be the sets of integers between 2 and 100 which are divisible by 2, by 3, by 5, by 7, and by 11, respectively, the set of numbers divisible by at least one of these has cardinality

$$|A \cup B \cup C \cup D \cup E| = (50 + 33 + 20 + 14 + 9)$$
$$- (16 + 10 + 7 + 4 + 6 + 4 + 3 + 2 + 1 + 1)$$
$$+ (3 + 2 + 1 + 1) = 79.$$

(There are no terms arising from intersections of four or five sets, since these intersections are empty.) These 79 numbers, with the exception of 2, 3, 5, 7, 11, are those which have been crossed out after five steps. So 25 numbers remain. (Remember that the list of numbers used with the Sieve of Eratosthenes starts with 2.)

EXERCISES

The symbol [BB] means that an answer can be found in the Back of the Book.

1. [BB] In a group of 15 pizza experts, ten like Canadian bacon, seven like anchovies, and six like both.

   (a) How many people like at least one of these toppings?
   (b) How many like Canadian bacon but not anchovies?
   (c) How many like exactly one of the two toppings?
   (d) How many like neither?

2. Multiple personality disorder (MPD) is a condition in which different personalities exist within one person and at various times control that person's behavior. In a recent survey of people with MPD, it was reported that "98% had been emotionally abused, 89% had been physically abused, and **most** had experienced both types of abuse." Make this statement more precise.

3. Among the 30 students registered for a course in discrete mathematics, 15 people know the JAVA programming language, 12 know HTML, and five know both of these languages.

   (a) How many students know at least one of JAVA or HTML?
   (b) How many students know only JAVA?
   (c) How many know only HTML?
   (d) How many know exactly one of the languages, JAVA and HTML?
   (e) How many students know neither JAVA nor HTML?

4. [BB] In a recent survey of college graduates, it was found that 200 had undergraduate degrees in arts, 95 had undergraduate degrees in science, and 120 had graduate degrees. Fifty-five of those with undergraduate arts degrees had also a graduate degree, 40 of those with science degrees had a graduate degree, 25 people had undergraduate degrees in both arts and science, and five people had undergraduate degrees in arts and science and also a graduate degree.

   (a) How many people had at least one of the types of degrees mentioned?

(b) How many people had an undergraduate degree in science but no other degree?

5. The owner of a corner store stocks popsicles, gum, and candy bars. After school one day, he is swamped by an influx of 15 school children. They are in and out of his store in minutes. Later, the clerk reports that ten children purchased popsicles, seven purchased gum, twelve purchased candy bars, five purchased popsicles and gum, six purchased popsicles and candy bars, and two purchased gum and candy bars. The owner is very upset. Why?

6. **(a)** In a group of 82 students, 59 are taking English, 46 are taking mathematics, and 12 are taking neither of these subjects. How many are taking both English and math?

   (b) In a group of 97 students, the number taking English is twice the number taking math. Fifty-three students take exactly one of these subjects and 15 are taking neither course. How many students are taking math? How many are taking English?

7. [BB] Of the 2300 delegates at a political convention, 1542 voted in favor of a motion to decrease the deficit, 569 voted in favor of a motion dealing with environmental issues, and 1197 voted in favor of a motion not to increase taxes. Of those voting in favor of the motion concerning environmental issues, 327 also voted to decrease the deficit, and 92 voted not to increase taxes. Eight hundred and thirty-nine people voted to decrease the deficit while also voting against increasing taxes, but of these 839, only 50 voted also in favor of the motion dealing with the environment.

   (a) How many delegates did not vote in favor of any of the three motions?

   (b) How many of those who voted against increasing taxes voted in favor of neither of the other two motions?

8. Seven members of a group of nineteen people dislike the New Democratic Party (NDP), ten dislike the Liberals, eleven dislike the Conservatives, and six dislike the Canadian Alliance. Five of the group dislike both the Liberals and the New Democratic Party, five dislike both the NDP and the Conservatives, six dislike the Liberals and Conservatives, three dislike the New Democratic and Canadian Alliance parties, four dislike the Liberals and the Alliance, and five dislike the Conservatives and the Alliance. Three people dislike the Conservatives, Liberals, and the NDP, while two dislike the Liberals, NDP, and Alliance; three dislike the Conservatives, New

Democrats, and Alliance; and four dislike the Conservatives, Liberals, and Alliance. Two people dislike all four parties. How many members of the group like all four parties?

9. The owner of a convenience store reports that of 890 people who bought bottled fruit juice in a recent week,

   - 750 bought orange juice
   - 400 bought apple juice
   - 100 bought grapefruit juice
   - 50 bought citrus punch
   - 328 bought orange juice and apple juice
   - 25 bought orange juice and grapefruit juice
   - 12 bought orange juice and citrus punch
   - 35 bought apple juice and grapefruit juice
   - 8 bought apple juice and citrus punch
   - 33 bought grapefruit juice and citrus punch
   - 4 bought orange juice, apple juice, and citrus punch
   - 17 bought orange juice, apple juice, and grapefruit juice
   - 2 bought citrus punch, apple juice, and grapefruit juice
   - 9 bought orange juice, grapefruit juice and citrus punch.

   Determine the numbers of people who bought

   (a) [BB] all four kinds of juice
   (b) grapefruit juice, but nothing else
   (c) exactly two kinds of juice
   (d) more than two kinds of juice

10. Suppose U is a set containing 75 elements and A_1, A_2, A_3, A_4 are subsets of U with the following properties:

    - each subset contains 28 elements;
    - the intersection of any two of the subsets contains 12 elements;
    - the intersection of any three of the subsets contains 5 elements;
    - the intersection of all four subsets contains 1 element.

    (a) [BB] How many elements belong to none of the four subsets?

    (b) How many elements belong to exactly two of the four subsets?

11. [BB] How many integers between 1 and 500 are

    (a) divisible by 3 or 5?
    (b) divisible by 3 but not by 5 or 6?

12. **(a)** How many integers less than 500 are relatively prime to 500?

    (b) How many integers less than 9975 are relatively prime to 9975?

13. [BB] How many integers between 1 and 250 are divisible by at least one of the three integers 4, 6, and 15?

14. (a) [BB] How many integers between 1 and 1000 (inclusive) are not divisible by 2, 3, 5, or 7?
    (b) How many integers between 1 and 1000 (exclusive) are not divisible by 2, 3, 5, or 7?

15. Find the number of integers between 1 and 10,000 inclusive which are
    (a) divisible by at least one of $3, 5, 7, 11$;
    (b) divisible by 3 and 5, but not by either 7 or 11;
    (c) divisible by exactly three of $3, 5, 7, 11$;
    (d) divisible by at most three of $3, 5, 7, 11$.

16. How many integers between 1 and 10^6 (inclusive) are neither perfect squares nor perfect cubes? (See Exercise 30, Section 4.3.)

17. [BB] How many primes are less than 200? Explain your answer.

18. [BB] Let a and b be natural numbers. Show that the number of positive integers less than or equal to a and divisible by b is $\lfloor \frac{a}{b} \rfloor$.

19. Suppose A and B are finite sets with $|A \cup B| < |A| + |B|$. Show that A and B have an element in common.

20. Suppose A and B are finite sets. Prove that $|A \times B| = |A| \times |B|$.

21. Suppose A and B are subsets of a universal set U. Find a formula for $|A \cup B^c|$ and prove that it is correct.

22. Let A, B and C be sets. Prove that
    (a) [BB] $|(A \oplus B) \cap C| = |A \cap C| + |B \cap C| - 2|A \cap B \cap C|$
    (b) $|A \oplus B \oplus C| = |A| + |B| + |C| - 2|A \cap B| - 2|A \cap C| - 2|B \cap C| + 4|A \cap B \cap C|$

23. Prove the Principle of Inclusion–Exclusion by mathematical induction. (For this, the result of Exercise 20, Section 5.1 will be helpful.)

6.2 THE ADDITION AND MULTIPLICATION RULES

The Principle of Inclusion–Exclusion gives a formula for the number of elements in the union of a finite number of finite sets. If the sets are pairwise disjoint (that is, if the intersection of any pair of sets is empty), then the formula takes a particularly simple form.

6.2.1 PRINCIPLE OF INCLUSION–EXCLUSION (DISJOINT SETS) ▶

Given n pairwise disjoint finite sets, $A_1, A_2, \ldots, A_n$, then

$$|A_1 \cup A_2 \cup \cdots \cup A_n| = \sum_{i=1}^{n} |A_i|.$$

In this section, we use the general term *event* to mean the outcome of any process or experiment, for example, the course a student selects to complete his or her degree. Events are *mutually exclusive* if no two of them can occur together; for example, if a student selects one course to complete his or her degree, then Mathematics 2320 and Statistics 2500 are mutually exclusive.

Notice that if n sets $A_1, A_2, \ldots, A_n$ correspond to events with the elements of A_i representing the ways in which the corresponding event can occur, then the events are mutually exclusive if and only if the A_i are pairwise disjoint. Thus, the Principle of Inclusion–Exclusion for disjoint sets translates into the following basic principle of counting.

6.2.2 ADDITION RULE ▶

The number of ways in which precisely one of a collection of mutually exclusive events can occur is the sum of the numbers of ways in which each event can occur.

For example, a student who needs one course to complete her degree decides to take either computer science or statistics and makes lists of the courses for which she is eligible. The computer science courses might comprise the set $A_1 = \{CS2602, CS2700, CS2721, CS2800, CS2500\}$ and the statistics courses the set $A_2 = \{S2510, S2511, S3500\}$. There are $5 + 3 = 8$ ways in which this student can register for a course because the two sets here are disjoint: The events "register for CS" and "register for Statistics" are mutually exclusive.

It is Friday night and Ursula has been invited to two parties but feels more inclined to go to a movie; there are six new movies in town. Assuming she is not permitted two engagements in the same evening (that is, assuming that the events "party" and "movie" are mutually exclusive), there are $2 + 6 = 8$ possible ways to spend her evening.

In the two examples just given, the individual events A_i were clearly specified. By contrast, the addition rule is probably most often applied in situations where this is not the case. A problem asking for the number of ways in which a certain event can occur can sometimes be analyzed by partitioning the event into mutually exclusive subevents (cases), precisely one of which must occur, and then applying the addition rule. It is for the problem solver to find a convenient set of subevents.

PROBLEM 4.   In how many ways can you get a total of six when rolling two dice?

Solution.   The event "get a six" is the union of the mutually exclusive subevents.

- A_1: "two 3's"
- A_2: "a 2 and a 4"
- A_3: "a 1 and a 5"

Event A_1 can occur in one way, A_2 can occur in two ways (depending on which die lands 4), and A_3 can occur in two ways, so the number of ways to get a six is $1 + 2 + 2 = 5$.   ∎

In Section 6.1, we observed that if A and B are finite sets, then the Cartesian product $A \times B$ contains $|A| \times |B|$ elements. More generally, if $A_1, A_2, \ldots, A_n$ are finite sets,

$$|A_1 \times A_2 \times \cdots \times A_n| = \prod_{i=1}^{n} |A_i|.$$

As before, thinking of A_i as the set of ways a certain event can occur, we are led to another basic principle of counting.

6.2.3 MULTIPLICATION RULE ▶

The number of ways in which a sequence of events can occur is the product of the numbers of ways in which each individual event can occur.

If Ursula's parents had a change of mind and permitted her to go to a party after the show, there would be $6 \times 2 = 12$ ways in which she could spend her Friday night, as shown in Fig 6.2. The figure also illustrates why this number is the cardinality of the Cartesian product Movie × Party, where Movie $= \{1, 2, 3, 4, 5, 6\}$ and Party $= \{1, 2\}$.

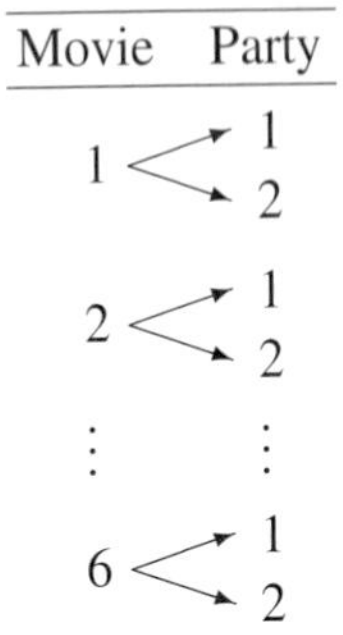

Figure 6.2 There are $6 \times 2 = 12$ ways to choose one of six movies and then one of two parties.

Suppose that there are five computer science courses and three statistics courses in which a student is eligible to enrol and that this student wishes to take a course in each subject. There are $5 \times 3 = 15$ ways in which the choice can be made.

If Jack, Maurice, and Tom each roll two dice, then the number of ways in which Jack can get a total of six, Maurice a total of four, and Tom a total of three is $5 \times 3 \times 2 = 30$ since there are five ways to get a total of six, as shown before, three ways to get a total of four (two 2's, a 1 and a 3, or a 3 and a 1), and two ways to get a total of three.

PROBLEM 5. How many numbers in the range 1000–9999 do not have any repeated digits?

Solution. Imagine enumerating all numbers of the desired type in the spirit of Fig 6.2. There are nine choices for the first digit (any of 1–9). Once this has been chosen, there remain still nine choices for the second (the chosen first digit cannot be repeated but 0 can now be used). There are now eight choices for the third digit and seven for the fourth. Altogether, there are $9 \times 9 \times 8 \times 7 = 4536$ possible numbers. ∎

PROBLEM 6. License plates in the Canadian province of Ontario consist of four letters followed by three of the digits 0–9 (not necessarily distinct). How many different license plates can be made in Ontario?

Solution. There are 26 ways in which the first letter can be chosen, 26 ways in which the second can be chosen, and similarly for the third and fourth. By the multiplication rule, the number of ways in which the three letters can be chosen is $26 \times 26 \times 26 \times 26 = 26^4$. By the same reasoning there are 10^3 ways in which the final three digits of an Ontario license plate can be selected and, all in all, $26^4 \times 10^3 = 456,976,000$ different license plates which can be manufactured by the government of Ontario under its current system. ∎

Pause 5 How many different license plates can be made by a government which permits four letters followed by three digits or three digits followed by four letters?

PROBLEM 7. Darlene wishes to fly from Portland, Maine, to Portland, Oregon. There are two possible routes she can take, namely, Portland–Detroit–Portland and Portland–Chicago–Portland. Two flights from Portland to Detroit each connect with three flights from Detroit to Portland, while two flights from Portland to Chicago each connect with four flights from Chicago to Portland.

    (a) Find the total number of different flight sequences from Portland, Maine, to Portland, Oregon, via Detroit.

    (b) Repeat (a) if "via Detroit" is omitted.

Solution.

    (a) The multiplication rule applies. There are $2 \times 3 = 6$ possibilities.

    (b) By (a), there are six routes via Detroit. Also, there are $2 \times 4 = 8$ routes via Chicago. Using the addition rule, the number of routes via Detroit or Chicago is $6 + 8 = 14$. ∎

To determine the number of ways in which some event can occur, it is often helpful to break the event into mutually exclusive subevents or cases, one of which must occur. By counting the number of ways in which each case can occur and adding these numbers, the addition rule gives the total number of ways in which the entire event can occur. Such a subtle use of the addition rule is probably its most important virtue.

PROBLEM 8. How many even numbers in the range 100–999 have no repeated digits?

Solution. The question is equivalent to asking for the number of ways in which one can write down an even number in the range 100–999 without repeating digits. This event can be partitioned into two mutually exclusive cases.

    Case 1: The number ends in 0.
    In this case, there are nine choices for the first digit (1–9) and then eight for the second (since 0 and the first digit must be excluded). So there are $9 \times 8 = 72$ numbers of this type.

    Case 2: The number does not end in 0.
    Now there are four choices for the final digit (2, 4, 6, and 8), then eight choices for the first digit (0 and the last digit are excluded), and eight choices for the second digit (the first and last digits are excluded). There are $4 \times 8 \times 8 = 256$ numbers of this type.

By the addition rule, there are $72 + 256 = 328$ even numbers in the range 100–999 with no repeated digits. ∎

Pause 6 In Case 2 of Problem 8, suppose we choose the last digit, then the **middle**, and finally the first digit. By considering two subcases,

    2a: the middle digit is 0

    2b: the middle digit is not 0

show again that there are 256 even numbers in the range 100–999 with no repeated digits and last digit different from 0.

Pause 7   Answer Problem 8 again by considering the four cases:

1. first two digits are even,
2. first two digits are odd,
3. first digit even, second digit odd,
4. first digit odd, second digit even.

These last two Pauses illustrate an important point. There are usually several ways to approach a combinatorial problem!

PROBLEM 9.   A typesetter (long ago) has before him 26 trays, one for each letter of the alphabet. Each tray contains ten copies of the same letter. In how many ways can he form a three letter "word" which requires at most two different letters? By "word," we mean any sequence of three letters—xpt, for example—not necessarily a real word from the dictionary. Two "ways" are different unless they use the identical pieces of type.

Solution.   The event "at most two different letters" is comprised of two mutually exclusive cases:

Case 1: The first two letters are the same.
Here, the third letter can be arbitrary; that is, any of the 258 letters which remain after the first two are set can be used. So the number of ways in which this case can occur is $260 \times 9 \times (260 - 2) = 603,720$.

Case 2: The first two letters are different.
In this case, the third letter must match one of the first two, so it must be one of the 18 letters remaining in the two trays used for the first two letters. The number of ways in which this case occurs is $260 \times 250 \times 18 = 1,170,000$.

By the addition rule, the number of ways to form a word using at most two different letters is $603,720 + 1,170,000 = 1,773,720$.   ∎

With many counting problems, there are several different ways of arriving at the same answer. Another approach to the previous problem is to count the total number of ways of forming words and to subtract from this the number of cases in which all three letters are different. (Recall that $|A^c| = |U| - |A|$.) There are $260 \times 259 \times 258 = 17,373,720$ possible words of which $260 \times 250 \times 240 = 15,600,000$ consist of three different letters. So the number of ways of forming words in which at most two different letters are used is $17,373,720 - 15,600,000 = 1,773,720$ as before.

PROBLEM 10.   Continuing Problem 9, determine the number of ways of forming words which use **exactly** two different letters.

Solution.   We subtract from the 1,773,720 cases requiring at most two different letters the number of ways of forming words which use just one letter. So the answer is $1,773,720 - (260 \times 9 \times 8) = 1,755,000$.   ∎

It is instructive to observe that this number could also be obtained by consideration of three mutually exclusive cases:

Case 1: The first two letters are the same but the third is different.

Case 2: The first and third letters are the same but the second is different.

Case 3: The second and third letters are the same but different from the first.

By this method, we obtain $(260 \times 9 \times 250) + (260 \times 250 \times 9) + (250 \times 260 \times 9) = 3(260 \times 250 \times 9) = 1,755,000$, as before.

Pause 8 In Problems 9 and 10, we counted the number of ways of forming certain three letter words, not the number of different words which can be formed. How many different "words" can be formed in each of these problems?

6.2.4 PROPOSITION ▶ A set of cardinality n contains 2^n subsets (including the empty set and the entire set itself).

Proof There are several ways to prove this fundamental result. We present one here which uses the ideas of this section.

Given n objects $a_1, a_2, \ldots, a_n$, each subset corresponds to a sequence of choices. Is a_1 in the subset? Is a_2 in the subset? Finally, is a_n in the subset? There are two answers to the first question, two for the second, and so on. In all, there are

$$\underbrace{2 \times 2 \times \cdots \times 2}_{n \text{ factors}} = 2^n$$

ways in which all n choices can be made. Thus, there are 2^n subsets. ∎

Answers to Pauses

5. As in Problem 6, the number of different plates with the letters first is $26^4 \times 10^3 = 456,976,000$. Similarly, the number of different plates with the digits first is also $10^3 \times 26^4 = 456,976,000$. By the addition rule, the number of plates altogether is $456,976,000 + 456,976,000 = 913,952,000$.

6. **Case 2a:** The middle digit is 0.
   Here, there are four choices for the last digit, one choice for the middle digit, and eight choices for the first (which can be any of 1–9 except that chosen for the last digit). There are 32 such numbers.

   Case 2b: The middle digit is not 0.
   In this case, there are again four choices for the last digit, there are eight choices for the middle (which is not 0 and not the last digit), and just seven choices for the first digit since neither of the digits used for the last or middle position is eligible. There are $4 \times 8 \times 7 = 224$ such numbers.
   As before, there are $32 + 224 = 256$ even numbers in Case 2.

7. **Case 1:** The first digit can be chosen in four ways (2, 4, 6, or 8), then the second digit in four ways (avoid repeating the first), and the third in three. There are $4 \times 4 \times 3 = 48$ ways for Case 1 to occur.

   Case 2: The first digit can be chosen in five ways, the second in four, and the last in five. There are $5 \times 4 \times 5 = 100$ ways for Case 2 to occur.

Case 3: There are four choices for the first digit, five for the second, and four for the third (just avoid repeating the first). There are $4 \times 5 \times 4 = 80$ ways for Case 3 to occur.

Case 4: There are five choices for the first digit, five choices for the second, and four for the third (avoid repeating the second). There are $5 \times 5 \times 4 = 100$ ways for Case 4 to occur.

By the addition rule, the number we seek is $48 + 100 + 80 + 100 = 328$, as before.

8. In Problem 9, there are 26^2 different words which have the same first two letters and $(26)(25)(2)$ which have the first two letters different. So the answer is $26^2 + (26)(25)(2) = 1976$.

In Problem 10, there are $1976 - 26 = 1950$ words which use exactly two different letters.

EXERCISES

The symbol [BB] means that an answer can be found in the Back of the Book.

1. **(a)** [BB] Roberta needs two courses to complete her university degree. There are three sections of the first course which meet in the morning and four afternoon sections. In how many different ways can she select the two courses she needs?

   (b) [BB] Suppose Roberta needs just one course to complete her degree. In how many ways can she select this course?

2. A building supplies store carries metal, wood, and plastic moldings. Metal and wood molding comes in two different colors. Plastic molding comes in six different colors.

   (a) How many choices of molding does this store offer?

   (b) If each kind and each color of molding comes in four different lengths, how many choices does the consumer have in the purchase of one piece of molding?

3. [BB] In how many of the three-digit numbers 000–999 are all the digits different?

4. **(a)** How many numbers in the range 100–999 have no repeated digits?

   (b) Without using the result of Problem 8, determine how many **odd** numbers in the range 100–999 have no repeated digits.

   (c) Answer (b) again, this time making use of (a) and Problem 8.

5. **(a)** [BB] Some license plates in California consist of one of the digits 1–9, followed by three (not necessarily distinct) letters and then three of the digits 0–9 (not necessarily distinct). How many possible license plates can be produced by this method?

   (b) [BB] Other California license plates consist of one of the digits 1–9 followed by three letters and then three of the digits 0–9. (The same digit or letter can be used more than once.) How many license plates of this type can be made?

   (c) [BB] What is the maximum number of license plates in California that can be made assuming plates have one of the two types described in (a) and (b)?

6. In Mark Salas, the 1991 Detroit Tigers had probably the only *palindromic* player in major league baseball (certainly, the only palindromic catcher). A *palindrome* is a word that reads the same forward and backward, like SALAS.

   (a) How many five-letter palindromes (not necessarily real words) can be made from the letters of the English alphabet?

   (b) How many eight-letter palindromes are possible?

   (c) How many "words" not exceeding eight letters in length are palindromes?

   (d) One of the most famous palindromes of all time is one which might have been uttered by Napoleon (had his native tongue been English): ABLE WAS I ERE I SAW ELBA. How many palindromes (not necessarily of real words) are this of this length?

7. [BB] From a group of 13 men, 6 women, 2 boys, and 4 girls,

   (a) In how many ways can a man, a woman, a boy, and a girl be selected?

   (b) In how many ways can a man or a girl be selected?

   (c) In how many ways can one person be selected?

8. In how many ways can one draw from a standard deck of 52 playing cards

   (a) a heart or a spade?

   (b) an ace or a king?

   (c) a card numbered 2 through 10?

   (d) a card numbered 2 through 10 or a king?

9. [BB] Using only the digits 1, 3, 4, and 7,

   (a) how many two-digit numbers can be formed?

   (b) how many three-digit numbers can be formed?

   (c) how many two- or three-digit numbers can be formed?

10. How many possible telephone numbers consist of seven digits, the first two in the range 2–9 (inclusive), the third in the range 1–9 (inclusive), and each of the last four in the range 0–9 (inclusive)?

11. [BB] A company produces combination locks, the combinations consisting of three different numbers in the range 0–59 (inclusive) which must be dialed in order. How many different combinations are possible?

12. In how many ways can two adjacent squares be selected from an 8 × 8 chess board?

13. [BB] New parents wish to give their baby one, two, or three different names. In how many ways can the baby be named if the parents will choose from a book containing 500 names?

14. There are three different roads from Cupids to Harbour Grace and five different roads from Harbour Grace to Heart's Desire.

   (a) How many different routes are there from Cupids to Heart's Desire via Harbour Grace?

   (b) How many different round trips are there from Cupids to Heart's Desire and back, passing through Harbour Grace each way?

   (c) Repeat (b) if you don't want to drive on any road more than once.

15. How many three-digit numbers contain the digits 2 and 5 but none of the digits 0, 3, 7?

16. You are dealt four cards from a standard deck of 52 playing cards. In how many ways can you get

   (a) [BB] four of a kind (four 2's, or four kings, etc.)?

   (b) two (different) pairs?

   (c) four of a kind or two (different) pairs?

   (d) three, but not four, of a kind?

   (e) at least one pair?

17. Two dice are rolled.

   (a) In how many ways can a total of eight arise?

   (b) In how many ways can a total of seven arise?

   (c) In how many ways can a total of eight or seven arise?

   (d) In how many ways can one get "doubles" (the dice land with the same side up)?

18. [BB] Make a table which shows all the totals which are possible when two dice are rolled and the number of ways in which each total can occur.

19. (a) In how many ways can two dice land?

   (b) In how many ways can five dice?

   (c) In how many ways can n dice land?

   (d) If n dice are rolled, in how many ways can they land not all showing the same number?

20. A coin is tossed four times.

   (a) [BB] Make a list of all the possible outcomes. For example, the sequence HHTH, representing head, head, tail, head, is one possibility. How many possibilities are there altogether?

   (b) In how many ways can you get exactly one head?

   (c) In how many ways can you get exactly two heads?

   (d) In how many ways can you get exactly three heads?

   (e) In how many ways can you get at least one head?

21. (a) How many five-digit numbers can be formed using the digits 0–9 inclusive if repetitions are allowed? (Leading 0's are not allowed: 07392, for example, should not be considered a five-digit number.)

   (b) How many five-digit numbers can be formed if repetition is not allowed?

   (c) How many five-digit numbers have one or more repeated digits?

22. How many possible license plates can be manufactured if a license plate consists of three letters followed by three digits and

   (a) [BB] the digits must be distinct; the letters can be arbitrary?

   (b) the letters must be distinct; the digits can be arbitrary?

   (c) the digits and the letters must be distinct?

23. The complete menu from a local gourmet restaurant is shown in Table 6.1.

Appetizers	Beverages
cod au gratin	house wine
squid ribs	imported wine
snail's tails	soft drink
smoked seaweed	coffee
	tea
	milk

Entrée	Vegetables
flipper pie	spinach
cod tongues	fiddleheads
caplin fillets	cabbage
	turnip
	potatoes

(a) In how many ways can John plan a four-course meal, one course from each group?

(b) Suppose Mary wants two appetizers and one course from each other group. How many meals are possible?

(c) Mike wants a two-course meal. How many choices does he have?

(d) Suppose Rod wants at most one course from each group and at least one course from some group. How many meals are possible?

24. Suppose $n = p_1^{\alpha_1} p_2^{\alpha_2} \cdots p_r^{\alpha_r}$ is the decomposition of n into the product of distinct primes $p_1, p_2, \ldots, p_r$. How many (unordered) pairs $\{s, t\}$ of positive integers satisfy both $st = n$ and $\gcd(s, t) = 1$?

25. [BB] Let $A = \{a_1, a_2, \ldots, a_n\}$ be a set of n elements and $B = \{0, 1\}$.

(a) Show that there are 2^n functions from A to B.

(b) Show that there are $2^n - 2$ onto functions from A to B.

26. Let $A = \{a_1, a_2, \ldots, a_n\}$ be a set of n elements, $n \geq 1$, and $B = \{1, 2, 3\}$.

(a) Prove that there are 3^n functions $A \to B$.

(b) How many functions $A \to B$ are **not** onto?

(c) How many onto functions are there $A \to B$? Justify your answer.

6.3   THE PIGEON-HOLE PRINCIPLE

This section deals with a deceptively simple restatement of a certain fact about functions between finite sets, namely, that if A and B are finite sets with $|A| > |B|$, no function $f : A \to B$ can be one-to-one. For example, no function $\{1, 2, 3\} \to \{x, y\}$ can be one-to-one.

Suppose that B is a set of bird houses, A is a set of pigeons, and f is a function which assigns a bird house to each pigeon. The statement that f is not one-to-one is just the observation that at least two pigeons are confined to the same house. Thus, we obtain the so-called *Pigeon-Hole Principle*.

6.3.1   THE PIGEON-HOLE PRINCIPLE ▶

If n objects are put into m boxes and $n > m$, then at least one box contains two or more of the objects.

We said that this principle is "deceptively" simple. The deception often lies in recognizing that the principle can be applied. For example, within any group of 13 people, there must be two who have their birthdays in the same month. (The people are the objects and the months of the year are the boxes.)

PROBLEM 11.   Given five points inside a square whose sides have length 2, prove that two are within $\sqrt{2}$ of each other.

Solution.   Subdivide the square into four squares with sides of length 1, as shown. By the Pigeon-Hole Principle, at least two of the five chosen points must lie in, or on the boundary of, the same smaller square. But these points are at most $\sqrt{2}$ apart (the length of the diagonal of a smaller square).   ∎

We continue with some subtler applications of the Pigeon-Hole Principle.

PROBLEM 12.   Show that among $n + 1$ arbitrarily chosen integers, there must exist two whose difference is divisible by n.

Solution.   Denote the integers $a_1, a_2, \ldots, a_{n+1}$. As we know, the n congruence classes $\overline{0}, \overline{1}, \ldots, \overline{n-1}$ of integers mod n partition the integers into disjoint sets whose union is Z. (See Proposition 4.4.5.) Put each integer a_i into its own congruence class. By the Pigeon-Hole Principle, two integers, a_i and a_j will go to the same class $\overline{k}$. Since $a_i \in \overline{k}$, $\overline{a_i} = \overline{k}$ (Proposition 4.4.3) and, similarly, $\overline{a_j} = \overline{k}$. Thus, $\overline{a_i} = \overline{a_j}$, so $a_i \equiv a_j \pmod{n}$; that is, $n \mid (a_i - a_j)$, as desired.   ∎

The solution to this problem involves a property of the integers which is often used in Pigeon-Hole problems: If two numbers lie in the same congruence class of integers mod n, then their difference is divisible by n.

PROBLEM 13.   Prove that in any list of ten natural numbers, $a_1, a_2, \ldots, a_{10}$, there is a string of consecutive items of the list, $a_\ell, a_{\ell+1}, a_{\ell+2}, \ldots$, whose sum is divisible by 10. (We include the possibility that the "string" consists of just one number.)

Solution.   Consider the ten numbers

$$a_1, a_1 + a_2, \ldots, a_1 + a_2 + \cdots + a_{10}.$$

If any of these is divisible by 10, we have the desired conclusion; otherwise, each number lies in one of the nine congruence classes $\overline{1}, \overline{2}, \ldots, \overline{9}$ of integers mod 10. By the Pigeon-Hole Principle, two of them must lie in the same class and hence have a difference divisible by 10. Again we reach the desired conclusion because, if $s > t$, $(a_1 + a_2 + \cdots + a_s) - (a_1 + a_2 + \cdots + a_t) = a_{t+1} + a_{t+2} + \cdots + a_s$.   ∎

PROBLEM 14.   Martina has three weeks to prepare for a tennis tournament. She decides to play at least one set every day but not more than 36 sets in all. Show that there is a period of consecutive days during which she will play exactly 21 sets.

Solution.   Suppose Martina plays a_1 sets on day 1, a_2 sets on day 2, and so on, a_{21} sets on the last day of her preparation period. Consider the 21 natural numbers $a_1, a_1 + a_2, \ldots, a_1 + a_2 + \cdots + a_{21}$. Since each $a_i \geq 1$ and since the sum $a_1 + a_2 + \cdots + a_{21}$ is the total number of sets Martina will play, we have

$$1 \leq a_1 < a_1 + a_2 < \cdots < a_1 + a_2 + \cdots + a_{21} \leq 36.$$

Now the only natural number between 1 and 36 which is divisible by 21 is 21 itself. Therefore, each of these sums is either 21 or it belongs to one of the 20 nonzero congruence classes of integers mod 21. In the first case, we have $a_1 + a_2 + \cdots + a_i = 21$ for some i; so Martina would play precisely 21 sets on days 1 through i giving the desired result. In the second case, by the Pigeon-Hole Principle, two sums lie in the same congruence class and hence have difference divisible by 21. As in Problem 13, we obtain $a_{t+1} + a_{t+2} + \cdots + a_s$ divisible by 21 for some s and t. Since this is a number between 1 and 36, it must equal 21 and we conclude that Martina plays 21 games on days $t + 1$ through s. ∎

PROBLEM 15. Suppose Martina has 11 weeks to prepare for her tournament and that she intends to play at least one set a day and at most 132 practice sets in all. Draw again the conclusion that during some period of consecutive days, Martina will play precisely 21 sets.

Solution. The approach used before doesn't work this time (see Exercise 9), so we find a slightly different solution. We let b_i be the number of sets Martina plays on days 1 through i inclusive ($b_i = a_1 + a_2 + \cdots + a_i$ in the notation of Problem 14) and consider the 154 numbers

$$b_1, b_2, \ldots, b_{77}, b_1 + 21, b_2 + 21, \ldots, b_{77} + 21.$$

The largest number here, $b_{77} + 21$, is at most $132 + 21 = 153$. By the Pigeon-Hole Principle, we conclude that two are the same. Since $b_i > b_j$ if $i > j$, the only way for two to be equal is for $b_i = b_j + 21$ for some i and j with $i > j$. Therefore, $b_i - b_j = 21$ and Martina plays 21 sets on days $j + 1$ through i. ∎

We noted earlier that in any group of 13 people, at least two must have birthdays in the same month. The same must hold in any larger group, but surely something stronger is also true. If at most two people in a group of 30 had birthdays in the same month, we would account for at most 24 people. Thus, in any group of 30, there must be at least **three** with birthdays in the same month.

Recall the definition of the *ceiling function* given in 3.1.6. For any real number x, $\lceil x \rceil$ means the least integer which is greater than or equal to x. For example, $\lceil 3.5 \rceil = 4$, $\lceil 0.24 \rceil = 1$, $\lceil -2.9 \rceil = -2$ and $\lceil \frac{30}{12} \rceil = 3$.

6.3.2 PIGEON-HOLE PRINCIPLE (STRONG FORM) ▶ If n objects are put into m boxes and $n > m$, then some box must contain at least $\lceil \frac{n}{m} \rceil$ objects.

EXAMPLE 16 If there are 44 chairs positioned around five tables in a room, some table must have at least $\lceil \frac{44}{5} \rceil = 9$ chairs around it. ▲

To prove the strong form of the Pigeon-Hole Principle, we establish the truth of its contrapositive. Note that

$$\left\lceil \frac{n}{m} \right\rceil < \frac{n}{m} + 1 \quad \text{and hence} \quad \left\lceil \frac{n}{m} \right\rceil - 1 < \frac{n}{m}$$

because, for any real number x,

$$x \le \lceil x \rceil < x + 1.$$

Thus, if a box contains fewer than $\lceil \frac{n}{m} \rceil$ objects, then it contains at most $\lceil \frac{n}{m} \rceil - 1$ and so fewer than $\frac{n}{m}$ objects. If all m boxes are like this, we account for fewer than $m \times \frac{n}{m} = n$ objects.

PROBLEM 17.   In any group of six people, at least three must be mutual friends or at least three must be mutual strangers.

Solution.   Pretend you are one of the six people and put the other $n = 5$ people into $m = 2$ "boxes" labeled "my friends" and "strangers to me." Since $\lceil n/m \rceil = \lceil 5/2 \rceil = 3$, by the strong form of the Pigeon-Hole Principle, at least one of these boxes must contain three people. Suppose three people are your friends. If any two of these three are friends, then together with you, we have a set of three mutual friends. If no two of these three are friends, then these three people are mutual strangers. In either case, we have one of the desired conclusions. The remaining possibility, that three of the five people are strangers to you, leads again to the desired conclusion with an argument, which we omit, similar to the one already presented. ∎

Pause 9   Provide the details of this omitted argument.

Answers to Pauses

9. Either the three people in this group are mutual friends or there are two who are strangers. Since each of these two is a stranger to you, we have a set of three mutual strangers.

EXERCISES

The symbol [BB] means that an answer can be found in the Back of the Book.

1. [BB] Show that in any group of eight people, at least two have birthdays which fall on the same day of the week in any given year.

2. Write down any six natural numbers. Verify that there is a string of consecutive numbers in your list (possibly a "string" of just one number) whose sum is divisible by 6. Prove that this must always be the case.

3. In any list of n natural numbers, prove that there must always exist a string of consecutive numbers (possibly just one number in the string) whose sum is divisible by n.

4. [BB] In a group of 100 people, several will have their birthdays in the same month. At least how many must have birthdays in the same month? Why?

5. A standard deck of playing cards contains 52 cards divided into four suits (club, diamond, heart, spade)

of 13 denominations (Ace, 2, 3, ... , 10, Jack, Queen, King). How many cards of a single suit must be present in any set of n cards? How many cards of the same denomination? Explain.

6. (a) [BB] If 20 processors are interconnected and every processor is connected to at least one other, show that at least two processors are directly connected to the same number of processors.

   (b) [BB] Is the result of (a) still true without the assumption that every processor is connected to at least one other? Explain.

7. Thirty buses are to be used to transport 2000 refugees from Gander to St. John's, Newfoundland. Each bus has 80 seats. Assume one seat per passenger.

   (a) Prove that one of the buses will carry at least 67 passengers.

(b) Prove that one of the buses will have at least 14 empty seats.

8. In a gathering of 30 people, there are 104 different pairs of people who know each other.

(a) Show that some person must have at least seven acquaintances.

(b) Show that some person must have fewer than seven acquaintances.

9. [BB] Try to solve Problem 15 with the method used to solve Problem 14. What goes wrong?

10. [BB] Brad has five weeks to prepare for his driver's test. His mother volunteers to drive with him for either 15 minutes or a half hour every day until the test, but not for more than 15 hours in all. Show that during some period of consecutive days, Brad and his mother will drive for exactly eight and three quarter hours.

11. Linda has six weeks to prepare for an examination and at most 50 hours available to study. She plans to study at least an hour a day and a whole number of hours each day. Show that no matter how she schedules her study time, there is a period of consecutive days during which she will have studied exactly 33 hours.

12. George has promised to increase his vocabulary by learning the meanings of 90 new words during his summer holidays. Suppose he has 53 days in which to accomplish this task and will learn at least one new word a day. Show that during some span of consecutive days, George will learn precisely 15 new words.

13. [BB] In a 12-day period, a small business mailed 195 bills to customers. Show that during some period of three consecutive days, at least 49 bills were mailed.

14. The circumference of a roulette wheel is divided into 36 sectors to which the numbers $1, 2, 3, \ldots, 36$ are assigned in some arbitrary manner. Show that there must be three consecutive sectors whose assigned numbers add to at least 56.

15. How many seats in a large auditorium have to be occupied to be certain that at least three people seated have the same first and last initials?

16. [BB] Of any 26 points within a rectangle measuring 20 cm by 15 cm, show that at least two are within 5 cm of each other.

17. Of any five points chosen within an equilateral triangle whose sides have length 1, show that two are within a distance of $\frac{1}{2}$ of each other.

18. A cake is in the shape of a regular hexagon with each of its sides exactly 30 cm long. Seven flowers of icing adorn the top. Show that at least two flowers are not more than 30 cm apart.

19. Let $S = \{2, 3, 5, 7, 11, 13, 17, 19\}$ be the set of prime numbers less than 20. If A is a subset of S, we can form the sum and product of the elements of A. For example, if $A = \{7, 11, 13\}$, then the associated sum is $7 + 11 + 13 = 31$ and the associated product is $7(11)(13) = 1001$.

(a) Use the Pigeon-Hole Principle to show that there are four nonempty subsets of S with the same sum.

(b) Are there two nonempty subsets of S with the same product? Explain.

20. [BB] Given any positive integer n, show that some multiple of n is an integer whose representation in base 10 requires just 3's and 0's.
[*Hint*: Let $M_1 = 3$, $M_2 = 33$, $M_3 = 333, \ldots$. Then think about the remainders when each of these numbers is divided by n.]

21. Show that some multiple of 2002 consists of a string of 1's followed by a string of 0's.

22. [BB] Show that the decimal expansion of a rational number must, after some point, become periodic or stop. [*Hint*: Think about the remainders in the process of long division.]

23. One hundred and one numbers are chosen from the set of natural numbers $\{1, 2, 3, \ldots, 200\}$. Show that one must be a multiple of another. [*Hint*: Any natural number can be written in the form $2^k a$ with $k \geq 0$ and a odd.]

24. In a room where there are more than 50 people with ages between 1 and 100, show that

(a) [BB] Either two people have the same age or there are two people whose ages are consecutive integers.

(b) Either two people have the same age or one person's age is a multiple of another's.

(c) Some of the people shake hands. Show that at least two shook the same number of hands. ("No hands" is a possibility.)

(d) Some people shake hands. Show that among those who shook at least one hand, two people shook the same number of hands.

25. (a) Let A be a set of seven (distinct) natural numbers none of which exceeds 21. Prove that the sums of the elements in all the nonempty subsets of A are not distinct.

(b) Improve the result of (a) by showing that the result holds under the assumption that the integers of A do not exceed 23.

(c) Assume none of the elements of A exceeds 12. At least how many subsets of A must have the same sum?

26. (a) [BB] Show that in any group of ten people, there is either a set of three mutual strangers or a set of four mutual friends.

(b) Show that in any group of 20 people, there is either a set of four mutual strangers or a set of four mutual friends.

27. Suppose $a_1, a_2, \dots, a_{10}$ are ten integers between 1 and 100 (inclusive).

(a) Prove that there exist two subsets $\{a_{i_1}, a_{i_2}, \dots, a_{i_r}\}$ and $\{a_{j_1}, a_{j_2}, \dots, a_{j_s}\}$ with equal sums.

(b) Prove that there exist two **disjoint** subsets $\{a_{i_1}, a_{i_2}, \dots, a_{i_r}\}$ and $\{a_{j_1}, a_{j_2}, \dots, a_{j_s}\}$ with equal sums.

28. Prove the (first form of the) Pigeon-Hole Principle by mathematical induction on m, the number of boxes.

29. Given any 52 integers, show that there exist two whose sum or difference is divisible by 100. [*Hint*: $x_1 \pm x_i$.]

REVIEW EXERCISES FOR CHAPTER 6

1. Suppose A and B are nonempty finite sets and $|A \cup B| < |A| + |B|$. Show that $A \cap B \neq \emptyset$.

2. Using the Principle of Inclusion–Exclusion, find the number of integers between 1 and 2000 (inclusive) which are divisible by at least one of 2, 3, 5, 7.

3. John Sununu was once the governor of New Hampshire, and his name reminds one of the authors of a palindrome.

(a) What is a *palindrome* and what made the author think of this?

(b) How many seven-letter palindromes (not necessarily real words) begin with the letter "S" and contain at most three different letters?

(c) How many seven-letter palindromes (not necessarily real words) are there in all?

(d) Name a palindromic rock band of the late 1970s which is still popular today.

4. Two Math 2320 students are arguing about the number of palindromes with nine and ten letters. One claims there are more with ten letters than with nine; the other

says there are the same number in each case. Who is right? Explain.

5. Four sets A_1, A_2, A_3, A_4 have the property that $A_i \cap A_j \cap A_k = \emptyset$ whenever i, j, k are distinct. In addition, $|A_i \cap A_j| = 1$ whenever $i \neq j$ and $|A_i| = 5$ for all i. Find the number of elements that belong to at least one of the sets.

6. Seventy cars sit on a parking lot. Thirty have stereo systems, 30 have air conditioners and 40 have sun roofs. Thirty of the cars have at least two of these three options and 10 have all three. How many cars on the lot have at least one of these three options? How many have exactly one?

7. State the **strong form** of the Pigeon-Hole Principle.

8. Show that among 18 arbitrarily chosen integers, there must exist two whose difference is divisible by 17.

9. Use the Pigeon-Hole Principle and the definition of *infinite set* to prove that $\mathbb{Z}$ is infinite.

10. Show that, of any ten points chosen within an equilateral triangle whose sides have length 1, there are two whose distance apart is at most $\frac{1}{3}$.

Permutations and Combinations

7.1 PERMUTATIONS

The registrar of a school must schedule six examinations in an eight-day period and has promised the students not to schedule more than one examination per day. How many different schedules can be made? The answer to this question involves a straightforward application of the multiplication rule. The first exam can be given on any of the eight days, but once this day has been settled, the second exam can be given on only one of the remaining seven days. There are $8 \times 7 = 56$ pairs of days on which the first two examinations can be scheduled. Once the first two exams have been scheduled, there remain six days on which to schedule the third exam. Continuing to reason this way, we see that there are $8 \times 7 \times 6 \times 5 \times 4 \times 3 = 20,160$ possible schedules.

In how many ways can a blue, a white, and a red marble be put into 10 numbered boxes? If there is no limit on the number of marbles which can be put into a box, there are 10 possible boxes into which the blue marble can be placed and then 10 choices for the white one and 10 choices for the red one, giving $10^3 = 1000$ possibilities altogether. Suppose, on the other hand, we are not allowed to put more than one marble into a box. As before, there are 10 possible boxes for the blue marble, but now there are just 9 possible boxes for the white one and 8 for the red. Altogether, there are $10 \times 9 \times 8 = 720$ possibilities.

The solution to a counting problem often involves the product of consecutive integers. Remember that $n!$ denotes the product of all the natural numbers from 1 to n (inclusive):

$$\boxed{n! = n(n-1)(n-2) \cdots (3)(2)(1)}$$

(see Definition 5.1.2). The notation $P(n, r)$ denotes the product of the first r factors of $n!$.

7.1.1 DEFINITION ▶ For integers n and r, $n \geq 1$, $0 \leq r \leq n$, the symbol $P(n,r)$ is defined by $P(n,0) = 1$ and, for $r > 0$,

$$P(n,r) = \underbrace{n(n-1)(n-2)\cdots(n-r+1)}_{r \text{ factors}}.$$

For example, $P(6,2) = 6 \cdot 5 = 30$ (the first two factors of 6!), $P(7,3) = 7 \cdot 6 \cdot 5 = 210$ (the first three factors of 7!), $P(8,5) = 8 \cdot 7 \cdot 6 \cdot 5 \cdot 4 = 6720$, $P(12,0) = 1$, and so on.

Pause 1 Find $P(5,3)$, $P(4,4)$, and $P(7,2)$.

Among the combinatorial problems to which $P(n,r)$ is the answer is the following one, which we have already illustrated.

7.1.2 PROPOSITION ▶ Given natural numbers r and n with $r \leq n$, the number of ways to place r marbles of different colors into n numbered boxes, at most one marble to a box, is $P(n,r)$.

Notice that

$$\underbrace{n(n-1)(n-2)\cdots(n-r+1)}_{P(n,r)}\underbrace{(n-r)(n-r-1)\cdots(3)(2)(1)}_{(n-r)!} = n!.$$

Thus,

$$\boxed{P(n,r) = \frac{n!}{(n-r)!}}$$

a formula which holds also for $r = 0$ and $r = n$ because $P(n,0) = 1$ and $0! = 1$.

Some people have been known to protest that this formula is difficult to use because, for example, they are unable to evaluate numbers like $P(25,1) = \frac{25!}{24!}$ with their calculator, 25! being too large to compute. It is good for us all to notice that

$$\frac{25!}{24!} = \frac{25 \cdot \cancel{24} \cdot \cancel{23} \cdots \cancel{3} \cdot \cancel{2}}{\cancel{24} \cdot \cancel{23} \cdots \cancel{3} \cdot \cancel{2}} = 25.$$

Pause 2 Find $\frac{20!}{17!}$, $\frac{100!}{98!}$, and $P(7,0)$.

For us, the symbol $P(n,r)$ is primarily a notational device which makes it easy to write down the answers to certain combinatorial problems. We caution against trying to fit a problem involving a straightforward application of the multiplication rule into the context of marbles and boxes.

PROBLEM 1. How many pairs of dance partners can be selected from a group of 12 women and 20 men?

Solution. The first woman can be paired with any of 20 men, the second woman with any of the remaining 19 men, the third with any of the remaining 18, and so on. There are $20 \cdot 19 \cdot 18 \cdots 9 = P(20,12)$ possible couples. ∎

There is a second question, having to do with *permutations*, to which $P(n, r)$ is the answer and which explains the "P" in "$P(n, r)$."

7.1.3 DEFINITION ▶ A *permutation* of a set of distinct symbols is an arrangement of them in a line in some order.

EXAMPLES 2 *ab* and *ba* are permutations of the symbols a and b; 1642, 4126, and 6241 are permutations of the symbols 1, 2, 4, and 6. ▲

The permutations 1642, 4126, and 6241 are also examples of 4-*permutations* of the symbols 1, 2, 3, 4, 5, 6, that is, permutations of the symbols 1, 2, 3, 4, 5, 6 *taken four at a time*. Here are some 3-permutations of the 26 letters of the alphabet: *zxy, aqr, cat, how*.

7.1.4 DEFINITION ▶ For natural numbers r and n, $r \leq n$, an *r-permutation* of n symbols is a permutation of r of them, that is, an arrangement of r of the symbols in a line in some order.

As a direct consequence of the multiplication rule, we have the following.

7.1.5 PROPOSITION ▶ The number of permutations of n symbols is $n!$. The number of r-permutations of n symbols is $P(n, r)$.

For example, there are $3! = 6$ permutations of a, b, c, namely, *abc, acb, bac, bca, cab*, and *cba*. (There are three choices for the first symbol and then two for the second and one for the third.) There are $6 \cdot 5 \cdot 4 \cdot 3 = 360 = P(6, 4)$ ways in which four of the six creatures—man, woman, boy, girl, dog, cat—can walk in a line down a road, one after the other.

PROBLEM 3. There are $7! = 5040$ ways in which seven people can form a line. In how many ways can seven people form a circle?

Solution. A circle is determined by the order of the people to the right of any one of the individuals, say Eric. There are six possibilities for the person on Eric's right, then five possibilities for the next person, four for the next, and so on. The number of possible circles is $6! = 720$. ∎

Another way to obtain $6!$ is to relate the two problems, line and circle. Each circle determines seven lines, determined by asking the people to join hands and then breaking the circle at one of the seven people.

$$\text{No. of circles} \times 7 = \text{no. of lines} = 7!$$

$$\text{Therefore, no. of circles} = \frac{7!}{7} = 6!$$

PROBLEM 4. A man, a woman, a boy, a girl, a dog, and a cat are walking down a long and winding road one after the other.
(a) In how many ways can this happen?
(b) In how many ways can this happen if the dog comes first?

(c) In how many ways can this happen if the dog immediately follows the boy?

(d) In how many ways can this happen if the dog (and only the dog) is between the man and the boy?

Solution.

(a) There are $6! = 720$ ways for six creatures to form a line.

(b) If the dog comes first, the others can form $5!$ lines behind.

(c) If the dog immediately follows the boy, then the dog-boy pair should be thought of as a single object to be put into a line with four others. There are $5! = 120$ such lines.

(d) If the man, dog, and boy appear in this order, then thinking of man-dog-boy as a single object to be put into a line with three others, we see that there are $4!$ possible lines. Similarly, there are $4!$ lines in which the boy, dog, and man appear in this order. So, by the addition rule, there are $4! + 4! = 48$ lines in which the dog (and only the dog) is between the man and the boy. ∎

PROBLEM 5.   In how many ways can ten adults and five children stand in a line so that no two children are next to each other?

Solution.   Imagine a line of ten adults named $A, B, \ldots, J$,

$$\times D \times J \times H \times C \times I \times E \times B \times A \times G \times F \times,$$

the $\times$'s representing the 11 possible locations for the children. For each such line, the first child can be positioned in any of the 11 spots, the second child in any of the remaining 10, and so on. Hence, the children can be positioned in $11 \cdot 10 \cdot 9 \cdot 8 \cdot 7 = P(11, 5)$ ways. For each such positioning, there are $10!$ ways of ordering the adults $A, \ldots, J$, so, by the multiplication rule, the number of lines of adults and children is $10! P(11, 5)$. ∎

Pause 3   In how many ways can ten adults and five children stand in a circle so that no two children are next to each other?

PROBLEM 6.   In how many ways can the letters of the English alphabet be arranged so that there are exactly ten letters between a and z?

Solution.   There are $P(24, 10)$ arrangements of the letters of the alphabet (excluding a and z) taken ten at a time, and hence $2 \cdot P(24, 10)$ strings of 12 letters, each beginning and ending with an a and a z (either letter coming first in a string). For each of these strings, there are $15!$ ways to arrange the 14 remaining letters and the string. So there are altogether $2 \cdot P(24, 10) \cdot 15!$ arrangements of the desired type. ∎

Pause 4   The answer to Problem 6 is also $30(24!)$. Why?

Answers to Pauses

1. $P(5, 3) = 5 \cdot 4 \cdot 3 = 60$; $P(4, 4) = 4 \cdot 3 \cdot 2 \cdot 1 = 24$; $P(7, 2) = 7 \cdot 6 = 42$.

2. $\frac{20!}{17!} = 20 \cdot 19 \cdot 18 = 6840$; $\frac{100!}{98!} = 100 \cdot 99 = 9900$; $P(7, 0) = \frac{7!}{7!} = 1$.

3. Arrange the adults into a circle in one of 9! ways. There are then 10 locations for the first child, 9 for the second, 8 for the third, 7 for the fourth, and 6 for the fifth. The answer is $9!(10 \cdot 9 \cdot 8 \cdot 7 \cdot 6) = 9! \cdot P(10,5)$.

4. First we count the number of arrangements of the prescribed type in which a precedes z. For each of the 24! permutations of the letters b through y, there are 15 locations for a after which the position of z is fixed. Thus, there are 15(24!) arrangements with a preceding z. Similarly, there are 15(24!) arrangements with z preceding a giving, altogether, 2(15)(24!) arrangements.

EXERCISES

The symbol [BB] means that an answer can be found in the Back of the Book.

1. [BB] How many ways are there to distribute eight different books among thirteen people if no person is to receive more than one book?

2. Eight horses are entered in a race in which a first, second, and third prize will be awarded. Assuming no ties, how many different outcomes are possible?

3. [BB] A club has ten members. In how many ways can they choose a slate of four officers consisting of a president, vice president, secretary, and treasurer?

4. There are 30 people in a class learning about permutations. One after another, eight people gradually slip out the back door. In how many ways can this exodus occur?

5. [BB] Find the number of ways in which six children can ride a toboggan if one of the three girls must steer (and therefore sit at the back).

6. Four cats and five mice enter a race. In how many ways can they finish with a mouse placing first, second, and third?

7. (a) In how many ways can ten boys and four girls sit in a row?
   (b) [BB] In how many ways can they sit in a row if the boys are to sit together and the girls are to sit together?
   (c) In how many ways can they sit in a row if the girls are to sit together?
   (d) In how many ways can they sit in a row if **just** the girls are to sit together?

8. Answer all parts of Exercise 7 with "circle" instead of "row." (In every circle, people sit facing inward.)

9. [BB] In how many ways is it possible to sit seven knights at a round table if

(a) Friday, Saturday, and Sunday insist on sitting together?
(b) Wednesday refuses to sit next to Saturday or Sunday?

10. [BB] Let X and Y be sets with $|X| = m$ and $|Y| = n$.
    (a) How many bijective functions are there $X \to X$?
    (b) How many injective functions are there $X \to Y$?

11. How many permutations of the letters a, b, c, d, e, f, g contain neither the pattern bge nor the pattern eaf?

12. [BB] In how many ways can two couples, the Noseworthys and the Abbotts, form a line so that
    (a) the Noseworthys are beside each other?
    (b) the Noseworthys are not beside each other?
    (c) each couple is together?
    (d) the Noseworthys are beside each other but the Abbotts are not?
    (e) at least one couple is together?
    (f) exactly one couple is together?

13. Repeat Exercise 12, assuming, in each part, that a dog also forms part of the line.

14. Three couples, the Smiths, Joneses, and Murphys, are going to form a line.
    (a) [BB] In how many such lines will Mr. and Mrs. Jones be next to each other?
    (b) In how many such lines will Mr. and Mrs. Jones be next to each other and Mr. and Mrs. Murphy be next to each other?
    (c) In how many such lines will at least one couple be next to each other?

15. How many permutations of the letters a, b, c, d, e, f, g have either two or three letters between a and b?

16. (a) [BB] How many seven-digit numbers have no repeated digits?

(b) How many seven-digit numbers with no repeated digits contain a 3 but not a 6?

(Leading zeros are not permitted in either part of this question.)

17. **(a)** In how many numbers with seven distinct digits do only the digits 1–9 appear?

(b) [BB] How many of the numbers in (a) contain a 3 and a 6?

(c) In how many of the numbers in (a) do 3 and 6 occur consecutively (in either order)?

(d) [BB] How many of the numbers in (a) contain neither a 3 nor a 6?

(e) How many of the numbers in (a) contain a 3 but not a 6?

(f) In how many of the numbers in (a) do exactly one of the numbers 3, 6 appear?

(g) In how many of the numbers in (a) do neither of the consecutive pairs 36 and 63 appear?

18. Let X and Y be finite nonempty sets, $|X| = m$, $|Y| = n \leq m$. Let $f(n, m)$ denote the number of partitions of X into n subsets. Prove that the number of surjective functions $X \to Y$ is $n! f(n, m)$.

7.2 COMBINATIONS

There are 12 different kinds of drinks for sale in a store and a customer wants to buy two different kinds. How many choices does this person have? There are 12 women and 20 men in a club, five of whom will be chosen to organize a Christmas party. In how many ways can these five people be chosen? If a coin is tossed eight times, in how many ways can exactly five heads turn up? These are the sorts of questions which we answer in this section.

In the last section, we found that the number of ways to place three different colored marbles into ten numbered boxes, at most one to a box, is $P(10, 3) = 720$. Suppose the marbles have the same color. Are there still 720 possibilities? In fact, the number is much smaller. When there are red, blue, and green marbles, for each choice of boxes, there are $3! = 6$ different ways of assigning the marbles to them corresponding to the six permutations of "red," "blue," "green." Figure 7.1 illustrates the six ways the marbles can be assigned to boxes 3, 4, and 8, for instance. When the marbles have the same color, these six assignments are all the same.

BOX 1	BOX 2	BOX 3	BOX 4	$\cdots$	BOX 8	$\cdots$
rbg						
rgb		red	blue		green	
brg		red	green		blue	
bgr		blue	red		green	
grb		blue	green		red	
gbr		green	red		blue	
		green	blue		red	

Figure 7.1   There are $3! = 6$ ways to assign three different colored marbles to three boxes.

It is useful to notice that when the marbles have the same color, each configuration of marbles in boxes simply amounts to a choice of three boxes. Thus, the number of ways to put three identical marbles into ten boxes is just the number of ways to select three boxes out of ten. Furthermore, for a given three boxes, there are $3 \times 2 \times 1 = 3! = 6$ ways in which three marbles of different colors can be assigned to these boxes. By the multiplication rule, the number of ways of putting three marbles of different colors into ten boxes (at most one to a box) is the product of the number of ways of selecting three boxes and the number of ways then to assign the marbles to these boxes.

$$P(10, 3) = (\text{number of ways to select 3 boxes from 10}) \times 3!$$

To summarize,

Number of ways to select three boxes from ten

$$= \text{number of ways to put three identical marbles into ten boxes}$$

$$= \frac{P(10, 3)}{3!} = \frac{10!}{7!3!} = 120.$$

7.2.1 DEFINITION ▶ For integers r and n, $n \geq 0$ and $0 \leq r \leq n$, the *binomial coefficient* $\binom{n}{r}$ (read "n choose r") is defined by

$$\boxed{\binom{n}{r} = \frac{n!}{r!(n-r)!}.}$$

The reason for the name "binomial coefficient" will become clear in Section 7.5. The reason for saying "n choose r" is given in the following proposition.

7.2.2 PROPOSITION ▶ Let n and r be integers with $n \geq 0$ and $0 \leq r \leq n$. The number of ways to choose r objects from n is $\binom{n}{r}$.

Proof If $r = 0$, the result is true because there is just one way to choose 0 objects (do nothing!), while $\binom{n}{0} = \frac{n!}{0!(n-0)!} = 1$ because $0! = 1$. Thus, we may assume that $r \geq 1$ and hence $n \geq 1$. Let N be the number we are seeking; that is, there are N ways to choose r objects from the n given objects. Notice that for each way of choosing r objects, there are $r!$ ways to order them. By the multiplication rule, the number of r-permutations of n objects [which we know is $P(n, r)$] is the number of ways to choose r objects multiplied by $r!$, the number of ways to order the r objects.

$$P(n, r) = N \times r!$$

Therefore,

$$N = \frac{P(n, r)}{r!} = \frac{n!}{r!(n-r)!} = \binom{n}{r},$$

which is what we wanted to show. ∎

Pause 5 Explain why $\binom{n}{r} = \binom{n}{n-r}$.

PROBLEM 7. Wanda is going to toss a coin eight times. In how many ways can she get five heads and three tails?

Solution. Wanda might get a string of five heads followed by three tails (denote this possibility HHHHHTTT); or a string of three tails followed by five heads, TTTHHHHH; or the sequence HTHHTHTH; and so on. The number of such sequences is the number of ways of selecting five occasions (from the eight) on which the heads should arise or, equivalently, the number of ways of selecting the three occasions on which tails should come up. The answer is $\binom{8}{5} = \binom{8}{3} = 56$. ∎

As is so often the case in mathematics, in Proposition 7.2.2, we solved one problem (in how many ways r objects can be selected from n?) by relating it to another problem for which we knew the answer (how many r-permutations of n objects are there?). This is an extremely useful and important idea.

7.2.3 DEFINITIONS ▶ A *combination* of a set of objects is a subset of them. A subset of r objects is called an *r-combination* or a combination of the objects *taken r at a time*.

Thus,

7.2.4 COROLLARY ▶ The number of r-combinations of n objects is $\binom{n}{r}$.

EXAMPLES 8
- The number of 2-combinations of the digits $0, 1, \ldots, 9$ is
$$\binom{10}{2} = \frac{10!}{2!8!} = \frac{10 \cdot 9}{2} = 45.$$

- The number of combinations of letters of the alphabet taken six at a time is
$$\binom{26}{6} = \frac{26!}{6!20!} = \frac{26 \cdot 25 \cdot 24 \cdot 23 \cdot 22 \cdot 21}{6 \cdot 5 \cdot 4 \cdot 3 \cdot 2 \cdot 1} = 230{,}230.$$

- The number of ways to put three identical marbles into ten boxes is the number of ways of selecting three of ten boxes:
$$\binom{10}{3} = \frac{10!}{3!7!} = \frac{10 \cdot 9 \cdot 8}{3 \cdot 2} = 120.$$

- The number of ways to choose two kinds of drinks from a dozen different kinds is
$$\binom{12}{2} = \frac{12!}{2!10!} = \frac{12 \cdot 11}{2} = 66.$$

- The number of ways to choose five people from a group of 32 is
$$\binom{32}{5} = \frac{32!}{5!27!} = \frac{32 \cdot 31 \cdot 30 \cdot 29 \cdot 28}{5 \cdot 4 \cdot 3 \cdot 2 \cdot 1} = 201{,}376. \quad ▲$$

The distinction between permutation and combination is the distinction between order and selection. Box 1, box 2, box 4 is one combination of boxes; box 2, box 1, box 4 is the same combination, but a different permutation. A permutation takes order into account; a combination involves only selection.

There are $\binom{32}{5} = 201{,}376$ ways to choose five people from 32, but $5! \times 201{,}376 = P(32,5)$ lines of five people which can be formed from a group of 32, a much larger number, because order is important in a line. The line Ruby, Mary, Colin, Tom, Eileen is different from the line Mary, Tom, Eileen, Ruby, Colin, although these two selections of people are the same.

Pause 6   Mr. Hiscock has ten children but his car holds only five people (including driver). When he goes to the circus, in how many ways can he select four children to accompany him?

PROBLEM 9.   In how many ways can 20 students out of a class of 32 be chosen to attend class on a late Thursday afternoon (and take notes for the others) if

(a) Paul refuses to go to class?

(b) Michelle insists on going?

(c) Jim and Michelle insist on going?

(d) either Jim or Michelle (or both) go to class?

(e) just one of Jim and Michelle attend?

(f) Paul and Michelle refuse to attend class together?

Solution.

(a) The answer is $\binom{31}{20} = 84{,}672{,}315$ since, in effect, it is necessary to select 20 students from the 31 students excluding Paul.

(b) Now the number of possibilities is $\binom{31}{19} = 141{,}120{,}525$ since 19 students must be chosen from 31.

(c) The answer is $\binom{30}{18} = 86{,}493{,}225$, it being necessary to choose the remaining 18 students from a group of 30.

(d) Let J be the set of classes of 20 which contain Jim and M the set of classes of 20 which contain Michelle. The question asks for $|J \cup M|$. Using the Principle of Inclusion–Exclusion, we obtain $|J \cup M| = |J| + |M| - |J \cap M| = \binom{31}{19} + \binom{31}{19} - \binom{30}{18} = 195{,}747{,}825$.

An alternative method of obtaining this answer would be to count separately the cases—Jim (but not Michelle) goes to class, Michelle (but not Jim) goes to class, both Jim and Michelle go to class—and add the results. (This follows because $|J \cup M| = |J \setminus M| + |M \setminus J| + |J \cap M|$.) We obtain $\binom{30}{19} + \binom{30}{19} + \binom{30}{18} = 2(54{,}627{,}300) + 86{,}493{,}225 = 195{,}747{,}825$, as before. Yet another method would be to subtract from the total number of possible classes of 20 the number containing neither Jim nor Michelle. This gives $\binom{32}{20} - \binom{30}{20} = 225{,}792{,}840 - 30{,}045{,}015 = 195{,}747{,}825$.

(e) Using the formula $|J \oplus M| = |J| + |M| - 2|J \cap M|$, we obtain $\binom{31}{19} + \binom{31}{19} - 2\binom{30}{18} = 109{,}254{,}600$.

(f) The number of classes containing Paul and Michelle is $\binom{30}{18}$ by part (c), so the number which do not contain both is $\binom{32}{20} - \binom{30}{18} = 139{,}299{,}615$.

Alternatively, observe that this part asks for $|(P \cap M)^c| = |P^c \cup M^c|$ (by one of the laws of De Morgan). So the answer is $|P^c| + |M^c| - |P^c \cap M^c| = \binom{31}{20} + \binom{31}{20} - \binom{30}{20} = 139,299,615.$ ∎

PROBLEM 10.   How many committees of five people can be chosen from 20 men and 12 women

(a) if exactly three men must be on each committee?

(b) if at least four women must be on each committee?

Solution.

(a) We must choose three men from 20 and then two women from 12. The answer is $\binom{20}{3}\binom{12}{2} = 1140(66) = 75,240.$

(b) We calculate the cases of four women and five women separately and add the results (using the addition rule). The answer is $\binom{12}{4}\binom{20}{1} + \binom{12}{5}\binom{20}{0} = 495(20) + 792 = 10,692.$ ∎

The combinatorial symbols we have presented in Sections 7.1 and 7.2 are summarized, together with their interpretations, in Table 7.1.

Table 7.1   Some combinatorial symbols and their uses.

$n!$	• number of permutations of n distinct symbols
$P(n, r) = \dfrac{n!}{(n - r)!}$	• number of ways to put r different colored marbles into n numbered boxes, at most one to a box
	• number of permutations of n distinct symbols used r at a time
$\dbinom{n}{r} = \dfrac{n!}{r!(n - r)!}$	• number of ways to put r identical marbles into n boxes, at most one to a box
	• number of combinations of n distinct symbols used r at a time
	• number of ways to choose r objects from n

Answers to Pauses

5. Suppose we have n white marbles and we wish to paint r of them black. Choosing the r marbles is equivalent to choosing the $n - r$ marbles which are to remain white. Thus, each choice of r marbles from n corresponds to a choice of the remaining $n - r$, so the numbers of choices are the same. By Proposition 7.2.2, these are $\binom{n}{r}$ and $\binom{n}{n-r}$, respectively.

6. The question involves choosing, not order. There are $\binom{10}{4} = \frac{10!}{4!6!} = 210$ different ways.

EXERCISES

The symbol [BB] means that an answer can be found in the Back of the Book.

1. [BB] A group of people is comprised of six from Nebraska, seven from Idaho, and eight from Louisiana.

(a) In how many ways can a committee of six be formed with two people from each state?

(b) In how many ways can a committee of seven be formed with at least two people from each state?

2. In how many ways can 12 players be divided into two teams of six for a game of street hockey?

3. [BB] How many 12-digit 0–1 strings contain precisely five 1's?

4. How many different signals, each consisting of seven flags arranged in a column, can be formed from three identical red flags and four identical blue flags?

5. [BB] In how many ways can Tom, Billie, and Peter share 15 salmon of different sizes

(a) if each takes five?

(b) if the youngest boy takes seven salmon and the others each take four?

(c) if one boy takes seven salmon and the others each take four?

6. [BB] How many subsets of a set of ten apples contain at most three apples?

7. In a popular lottery known as "Lotto 6/49," a player marks a card with six different numbers from the integers 1–49 and wins if his or her numbers match six randomly selected such numbers.

(a) In how many ways can a player complete a game card?

(b) How many cards should you complete in order to have at least one chance in a million of winning?

(c) In how many ways can a player complete a game card so that no number matches any of the six selected?

(d) In how many ways can a player complete a game card so that at least one number matches one of the six selected?

8. (a) [BB] How many five-card hands dealt from a standard deck of 52 playing cards are all of the same suit?

(b) How many five-card hands contain exactly two aces?

9. A newcomers' club of 30 people wants to choose an executive board consisting of president, secretary, treasurer, and two other officers. In how many ways can this be accomplished?

10. [BB] A coin is tossed ten times and the sequence of heads and tails observed.

(a) How many different sequences are possible?

(b) In how many of these sequences are there exactly four heads?

11. [BB] An urn contains 15 red numbered balls and ten white numbered balls. A sample of five balls is selected.

(a) How many different samples are possible?

(b) How many samples contain all red balls?

(c) How many samples contain three red balls and two white balls?

12. A group of eight scientists is composed of five mathematicians and three geologists.

(a) In how many ways can five people be chosen to visit an oil rig?

(b) Suppose the five people chosen to visit the rig must be comprised of three mathematicians and two geologists. Now in how many ways can the group be chosen?

13. [BB] From a hundred used cars sitting on a lot, 20 are to be selected for a test designed to check certain safety requirements. These cars will then be put back onto the lot and, again, 20 will be selected for a test designed to check antipollution standards.

(a) In how many ways can the first selection be made?

(b) In how many ways can the second selection be made?

(c) In how many ways can both selections be made?

(d) In how many ways can both selections be made if exactly five cars are to undergo both tests?

14. In how many ways can a team of six be chosen from 20 players so as to

(a) include both the strongest and the weakest player?

(b) include the strongest but exclude the weakest player?

(c) exclude both the strongest and weakest player?

15. A woman has nine close friends.

(a) In how many ways can she invite six of these to dinner?

(b) Repeat (a) if two of her friends are divorced (from each other) and will not attend together.

(c) Repeat (a) if the friends consist of three single people and three married couples and, if a husband or wife is invited, the spouse must be invited too.

16. The Head of the Department of Mathematical Sciences at a certain university has 12 mathematicians, seven computer scientists, and three statisticians in his employ. He wishes to appoint some committees from among these 22 people.

(a) How many five-member committees can he appoint?

(b) How many five-member committees, each containing at least one statistician, can he appoint?

(c) A certain professor of mathematics, Dr. G, and a certain colleague, Dr. P, refuse to serve together on the same committee. How many five-member committees can be formed so as not to contain both Dr. G and Dr. P?

(d) How many five-member committees can be formed so that the number of mathematicians is greater than the number of computer scientists and the number of computer scientists is greater than the number of statisticians?

17. A student must answer exactly eight questions out of ten on a final examination.

(a) [BB] In how many ways can she choose the questions to answer?

(b) Repeat (a) if she must answer the first three questions.

(c) [BB] Repeat (a) if she must answer at least three of the last five questions and at most four of the first five.

(d) Repeat (a) if she must answer at least three of the last five questions.

18. Suppose U is a set of 52 elements which contains five subsets A_1, A_2, A_3, A_4, A_5 with the following properties:

- each set contains 23 elements;
- the intersection of any two of the sets contains ten elements;
- the intersection of any three of the sets contains four elements;
- the intersection of any four of the sets contains one element;
- the intersection of all the sets is empty.

How many elements belong to none of the five subsets?

19. Ten points in the plane, no three collinear, are given.

(a) [BB] How many different line segments are formed by joining pairs of these points?

(b) How many different triangles are formed by the line segments in (a)?

(c) If A is one of the ten points, how many of the triangles in (b) have A as a vertex?

20. (a) [BB] How many triangles are determined by the vertices of a regular 12-sided polygon?

(b) Repeat (a) if the sides of the polygon are not to be the sides of any triangle.

21. A *diagonal* of a polygon is a line joining two nonadjacent vertices.

(a) [BB] How many diagonals does an octagon have?

(b) *Loonie* is the colloquial name for the Canadian one dollar coin. It has the shape of a regular 11-sided polygon. How many diagonals does a loonie have?

22. (a) How many diagonals does a regular n-sided polygon have?

(b) Which regular n-sided polygon has three times as many diagonals as sides?

23. [BB] Prove that the product of any n consecutive natural numbers is divisible by $n!$.

24. (a) Use Definition 7.2.1 to prove that $\binom{n}{k}\binom{n-k}{\ell} = \binom{n}{\ell}\binom{n-\ell}{k}$, where n, k, and ℓ are natural numbers with $k + \ell \le n$.

(b) Establish the identity in part (a) without appealing to Definition 7.2.1. [*Hint*: In how many ways can one choose two teams, one of size k and the other of size ℓ, from a group of n people?]

25. (a) Let k and n be natural numbers with $k < n$. Use the definition of $\binom{n}{r}$ given in 7.2.1 to prove that $\binom{n}{k} = \binom{n-1}{k-1} + \binom{n-1}{k}$.

(b) Establish the identity in (a) **without** appealing to any definition.

26. (a) Use Definition 7.2.1 to prove that $\binom{2n}{2} = 2\binom{n}{2} + n^2$ for any natural number n.

(b) Establish the identity in (a) without using the definition.

7.3 REPETITIONS

In the previous two sections we counted the number of ways of putting r marbles into n boxes with at most one marble to a box in two cases: The marbles have different colors and the marbles are all the same color. Suppose we allow any number of marbles in a box. If the marbles are all colored differently, then there are n choices for the first marble, n for the second, and so forth. There are, altogether, $n \times n \times \cdots \times n = n^r$ possibilities. When the marbles are all the same color, however, we expect far fewer possibilities.

Suppose we want to place three white marbles into ten boxes and we are allowed to put as many marbles into a box as we like. There are three mutually exclusive ways in which this can be accomplished: Each marble goes into a different box [in one of $\binom{10}{3} = 120$ ways], two marbles go into the same box but the third goes into its own box (in one of $10 \times 9 = 90$ ways), or all the marbles go into the same box (in one of 10 ways). By the addition rule, the number of ways in all is $120 + 90 + 10 = 220$. We note in passing that $220 = \binom{10+3-1}{3}$.

In how many ways can ten marbles, all of the same color, be put into three boxes? While this problem is obviously similar to the previous, it becomes quickly apparent that a case-by-case examination of the possibilities is very difficult this time. There are three ways in which the marbles all go into the same box, but in how many ways can they be placed into exactly two boxes, or into exactly three? The question seems quite complicated, so we consider it from a new perspective.

One way to put the marbles into the boxes is to put three marbles into box 1, five into box 2, and two into box 3. This situation can be represented by the 12-digit string 000100000100, which has strings of three, five, and two 0's separated by two 1's. There is a one-to-one correspondence between ways of putting the marbles into the boxes and 12-digit strings consisting of two 1's and ten 0's. The string 001000000100 would represent the situation of two marbles in box 1, six in box 2, and two in box 3. The string 100000010000 would indicate that there are no marbles in box 1, six in box 2, and four in box 3. To count the number of ways of putting ten marbles in three boxes is, therefore, just to count the number of 12-digit 0–1 strings which contain precisely two 1's. This latter number is easy to find! There are $\binom{12}{2}$ such numbers corresponding to the number of ways of selecting two positions for the 1's from 12 positions. Therefore, the number of ways to put ten identical marbles into three boxes, any number to a box, is $\binom{12}{2} = 66 = \binom{3+10-1}{10}$. We have seen two instances of the next proposition.

7.3.1 PROPOSITION ▶ The number of ways to put r identical marbles into n boxes is $\binom{n+r-1}{r}$.

PROBLEM 11. Doughnuts come in 30 different varieties and Catherine wants to buy a dozen. How many choices does she have?

Solution. Imagine that the 30 varieties are in $n = 30$ boxes labeled "chocolate white," "Boston creme," "peanut crunch," and so on. Catherine can indicate her choice by dropping $r = 12$ (identical) marbles into the boxes. So there are $\binom{30+12-1}{12} = 7,898,654,920$ possibilities. ∎

Pause 7	David wants to buy 30 doughnuts and finds just 12 varieties available. In how many ways can he make his selection?

The basic facts about marbles and boxes which we have discussed so far are summarized in Table 7.2.

Table 7.2   The number of ways to put r marbles into n numbered boxes.

	Same Color	All Different Colors
At most one to a box	$\binom{n}{r}$	$P(n, r)$
Any number in a box	$\binom{n + r - 1}{r}$	n^r

In Section 7.1, we counted the number of ways to put three marbles of different colors into ten boxes, at most one marble to a box; and, in Section 7.2, we considered the same problem for identical marbles. What about the intermediate possibility that two marbles have the same color, say red, and the third is a different color, perhaps blue? In how many ways can these three marbles be put into ten boxes, at most one to a box? There are $\binom{10}{2}$ ways in which to pick boxes for the red marbles and then $\binom{8}{1}$ ways to choose the box for the blue marble. By the multiplication rule, the number of ways to put two reds and one blue marble into ten boxes, at most one to a box, is

$$\binom{10}{2} \times \binom{8}{1} = \frac{10!}{2!8!} \times \frac{8!}{7!1!} = \frac{10!}{2!1!7!} = 360.$$

The number of ways to put two red, four blue, and three green marbles into ten boxes, at most one to a box, is $\binom{10}{2}$ (choose boxes for the reds) $\times \binom{8}{4}$ (choose boxes for the blues) $\times \binom{4}{3}$ (choose boxes for the greens); that is,

$$\frac{10!}{2!8!} \times \frac{8!}{4!4!} \times \frac{4!}{3!1!} = \frac{10!}{2!4!3!1!}.$$

PROBLEM 12.   Suppose there are ten players to be assigned to three teams, the Xtreme, the Maniax, and the Enforcers. The Xtreme and the Maniax are to receive four players each and the Enforcers are to receive two. In how many ways can this be done?

Solution.   The assignment of players is accomplished by choosing four players from ten for the Xtreme, then choosing four players from the remaining six for the Maniax, and assigning the remaining two players to the Enforcers. The number of possible teams is

$$\binom{10}{4} \times \binom{6}{4} = 210(15) = 3150. \qquad \blacksquare$$

Pause 8 In how many ways can 14 men be divided into six named teams, two with three players and four with two? In how many ways can 14 men be divided into two unnamed teams of three and four teams of two?

In how many ways can the letters of the word *easy* be rearranged? The question just asks for the number of permutations of four different letters: The answer is 4! = 24. In how many ways can the letters of the word *ease* be rearranged? This is a slightly different problem because of the repeated *e*: When the first and last letters of *easy* are interchanged, we get two different arrangements of the four letters, but when the first and last letters of *ease* are interchanged we get the same word. To see how to count the ways in which the four letters of *ease* can be arranged, we imagine the list of all these arrangements.

s e a e

a e e s

e e s a

⋮

Pretending for a moment that the two *e*'s are different (say one is a capital *E*), then each "word" in this list will produce two different arrangements of the letters *e a s E*. (See Fig 7.2.)

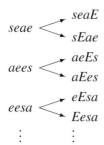

Figure 7.2

The list on the right contains the 4! = 24 arrangements of the four letters *e a s E*, so the list on the left contains half as many. There are $\frac{4!}{2!} = 12$ ways in which the letters of *ease* can be arranged.

PROBLEM 13. In how many ways can the letters of the word *attention* be rearranged?

Solution. The word *attention* has nine letters, three of one kind, two of another, and four other different letters. The number of rearrangements of this word is $\frac{9!}{3!2!} = 30{,}240$. ∎

This answer is obtained, as before, by imagining the list of arrangements and imagining how many arrangements could be formed if the letters were all different. If the two *n*'s were different, each rearrangement would produce two more, giving a second list twice as long. If the three *t*'s were different, each "word" in this second list would yield 3! = 6 more. For instance, replacing the three t's by t, T, and τ, then *tanetoeti* yields

taneToeτi, taneτoeTi, Tanetoeτi, Taneτoeti, τanetoiTi, τaneToiti

corresponding to the 3! permutations of t, T, τ.

We would obtain a third list 3! times as long as the second and $3! \times 2$ times as long as the first. Since this third list consists of all the permutations of the nine symbols a, t, T, e, n, τ, i, o, N, it contains 9! "words," so the original list contained $\frac{9!}{3!2!}$ "words."

Pause 9 In how many ways can the letters of the word *REARRANGE* be rearranged?

Answers to Pauses

7. As in Problem 11, we imagine the 12 varieties in boxes. Each of David's possible decisions can be indicated by dropping 30 identical marbles into these boxes. There are $\binom{12+30-1}{30} = 3{,}159{,}461{,}968$ possibilities ($r = 30$, $n = 12$).

8. If the teams are named, the answer is

$$\binom{14}{3} \times \binom{11}{3} \times \binom{8}{2} \times \binom{6}{2} \times \binom{4}{2} = 364(165)(28)(15)(6) = 151{,}351{,}200.$$

This number is $2!4! = 48$ times the number of divisions into unnamed teams of sizes 3, 3, 2, 2, 2, and 2 because, for each division into unnamed teams, there are two ways to name the teams of three and 4! ways to name the teams of two. The answer in the case of unnamed teams is $\frac{151,351,200}{48} = 3{,}153{,}150$.

9. There are $\frac{9!}{3!2!2!} = 15{,}120$ rearrangements of the letters of the word *REARRANGE*.

EXERCISES

The symbol [BB] means that an answer can be found in the Back of the Book.

1. [BB] In how many ways can 30 identical dolls be placed on seven different shelves?

2. There are 15 questions on a multiple-choice exam and five possible answers to each question.

   (a) In how many ways can the exam be answered?
   (b) In how many ways can the exam be answered with exactly eight answers correct?

3. [BB] How many different outcomes are possible when five dice are rolled? (Two 6's, two 5's, and a 1 is one outcome.)

4. [BB] How many different collections of ten coins can be made from pennies, nickels, dimes, and quarters?

5. A florist sells roses in five different colors.

   (a) How many bunches of a half dozen roses can be formed?

   (b) How many bunches of a half dozen can be formed if each bunch must contain at least one rose of each color?

6. There are 20 varieties of chocolates available and Linda wants to buy eight chocolates.

   (a) [BB] How many choices does she have?
   (b) How many choices does she have if her boyfriend insists that at least one chocolate should have a cherry center?
   (c) How many choices does she have if there remain only two chocolates with caramel centers? (At least 20 chocolates of all other varieties are available.)

7. (a) In how many ways can five different mathematics books, three different physics books, and four different chemistry books be arranged on a shelf?

(b) Repeat (a) if all books of the same subject are to be together.

(c) [BB] Repeat (a) if three of the five mathematics books are the same.

(d) Repeat (c) if, in addition, all the physics books are the same.

8. [BB] In how many ways can 12 people form four groups of three if

(a) the groups have names?

(b) the groups are unnamed?

9. In how many ways can 18 different books be given to Tara, Danny, Shannon, and Mike so that one person has six books, one has two books, and the other two people have five books each?

10. (a) Twenty basketball players are going to be drafted by the professional basketball teams in Philadelphia, Boston, Miami, and Toronto such that each team drafts five players. In how many ways can this be accomplished?

(b) In how many ways can 20 players be divided into four unnamed teams of five players each?

11. Suppose there are 12 baseball players to be drafted by San Diego, Houston, and Toronto. If San Diego is to get six players, Houston four players, and Toronto two, in how many ways can this be accomplished?

12. (a) [BB] In how many ways can ten red balls, ten white balls, and ten blue balls be placed in 60 different boxes, at most one ball to a box?

(b) In how many ways can ten red balls, ten white balls, and ten blue balls be placed in 60 different boxes?

13. In how many ways can two white rooks, two black bishops, eight black pawns, and eight white pawns be placed on a prescribed 20 squares of a chess board?

14. [BB] A committee wishes to award one scholarship of $10,000, two scholarships of $5000, and five scholarships of $1000. The list of potential award winners has been narrowed to 13 possibilities. In how many ways can the scholarships be awarded?

15. Find the number of arrangements of the letters of each of the following words:

(a) [BB] *SCIENTIFIC*

(b) *SASKATOON*

(c) *PICCININI*

(d) *CINCINNATI*

16. [BB] A department store in downtown Victoria, British Columbia, has 30 flags to hang along its roof line to celebrate Queen Victoria's birthday. If there are ten red flags, five white flags, seven yellow flags, and eight blue flags, how many ways can the flags be displayed in a row?

17. In a university residence there are five single rooms, five doubles, and five rooms which hold three students each. In how many ways can 30 students be assigned to the 15 rooms? (All rooms are numbered.)

18. (a) Show that there is a one-to-one correspondence between the number of ways to put ten identical marbles into three boxes and the number of ordered triples (x, y, z) of nonnegative integers which satisfy $x + y + z = 10$.

(b) How many triples (x, y, z) of nonnegative integers satisfy $x + y + z = 10$?

(c) How many 5-tuples (x, y, z, u, v) of nonnegative integers satisfy $x + y + z + u + v = 19$?

19. (a) [BB] Show that there is a one-to-one correspondence between the solutions to

$$(*) \qquad x_1 + x_2 + x_3 + x_4 = 21$$

with x_1, x_2, x_3, x_4 nonnegative integers, $x_1 \geq 8$, and solutions to

$$(**) \qquad x_1 + x_2 + x_3 + x_4 = 13$$

with x_1, x_2, x_3, x_4 nonnegative integers. How many solutions are there?

(b) Show that there is a one-to-one correspondence between the solutions to (*) with $x_1 \geq 8$, $x_2 \geq 8$, and solutions to

$$(***) \qquad x_1 + x_2 + x_3 + x_4 = 5$$

with x_1, x_2, x_3, x_4 nonnegative integers. How many solutions are there?

(c) How many solutions are there to (*) with all variables nonnegative integers not exceeding 7? [*Hint*: Let A_i be the set of solutions to (*) with $x_i \geq 8$.]

(d) How many solutions (x_1, x_2, x_3, x_4) are there to the equation

$$x_1 + x_2 + x_3 + x_4 = 35$$

with all x_i integers, $0 \leq x_i \leq 10$?

20. How many integer solutions are there to the equation $x + y + z + w = 20$ subject to $x \geq 1$, $y \geq 2$, $z \geq 3$, and $w \geq 4$?

7.4 DERANGEMENTS

There are four houses on a street. A mischievous postman arrives with one letter addressed to each house. In how many ways can he deliver the letters, one to each house, so that no letter arrives at the correct house? The person in charge of coats at the arts center loses all claim checks and returns coats randomly to the patrons. What are the chances that nobody gets their own coat?

7.4.1 DEFINITION ▶ A *derangement* of n distinct symbols which have some natural order is a permutation in which no symbol is in its correct position. The number of derangements of n symbols is denoted D_n.

EXAMPLES 14
- There is just one derangement of the symbols 1, 2 (namely, 21), so $D_2 = 1$.
- There are two derangements of 1, 2, 3 (312 and 231), so $D_3 = 2$.
- There are nine derangements of 1, 2, 3, 4,

$$
\begin{array}{ccc}
2341 & 2413 & 2143 \\
3142 & 3412 & 3421 \\
4123 & 4312 & 4321
\end{array}
$$

so $D_4 = 9$. There are nine ways our mischievous postman can deliver four letters, each to the wrong address. ▲

Finding a general formula for D_n involves a nice application of the Principle of Inclusion–Exclusion, binomial coefficients, and a famous irrational number. We consider again D_4, the number of derangements of 1, 2, 3, 4.

Let A_1 be the set of permutations of 1, 2, 3, 4 in which the number 1 is in the first position. Let A_2 be the set of permutations of 1, 2, 3, 4 in which the number 2 is in the second position. Define A_3 and A_4 similarly. The set of permutations in which at least one of the four numbers is left in its natural position is then $A_1 \cup A_2 \cup A_3 \cup A_4$ and the complement of this set consists precisely of the derangements of 1, 2, 3, 4. Since there are $4! = 24$ permutations altogether, the number of derangements of four symbols is

$$
D_4 = 4! - |A_1 \cup A_2 \cup A_3 \cup A_4|.
$$

How many elements are in the union $A_1 \cup A_2 \cup A_3 \cup A_4$? By the Inclusion-Exclusion Principle,

$$
(1) \quad |A_1 \cup A_2 \cup A_3 \cup A_4| = \sum_i |A_i| - \sum_{i<j} |A_i \cap A_j| + \sum_{i<j<k} |A_i \cap A_j \cap A_k|
$$
$$
- |A_1 \cap A_2 \cap A_3 \cap A_4|.
$$

Now $|A_1| = 3!$ (1 is in the first position, numbers 2, 3, and 4 go to any of the next three positions). Similarly, $|A_i| = 3!$ for any i.

The set $A_1 \cap A_2$ contains those permutations of 1, 2, 3, 4 in which 1 and 2 are in the correct positions. There are just two such permutations, 1234 and 1243. Thus, $|A_i \cap A_j| = 2!$ for each of the $6 = \binom{4}{2}$ terms in the second sum on the right of equation 1.

Similar reasoning shows that each of the $4 = \binom{4}{3}$ terms $|A_i \cap A_j \cap A_k|$ equals 1, as does the last term. Therefore,

$$|A_1 \cup A_2 \cup A_3 \cup A_4| = 4(3!) - \binom{4}{2}(2!) + \binom{4}{3}(1!) - 1$$

and hence,

$$D_4 = 4! - \left(4(3!) - \binom{4}{2}2! + \binom{4}{3}1! - 1\right)$$

$$= 4! - 4! + \frac{4!}{2!2!}2! - \frac{4!}{3!} + 1$$

$$= 4! - 4! + \frac{4!}{2!} - \frac{4!}{3!} + \frac{4!}{4!}$$

$$= 4!\left(1 - \frac{1}{1!} + \frac{1}{2!} - \frac{1}{3!} + \frac{1}{4!}\right)$$

$$= 24\left(1 - 1 + \frac{1}{2} - \frac{1}{6} + \frac{1}{24}\right)$$

$$= 24 - 24 + 12 - 4 + 1 = 9.$$

To find a general formula for D_n, for general n, we mimic this calculation of D_4.

7.4.2 PROPOSITION ▶ The number of derangements of $n \geq 1$ ordered symbols is

$$D_n = n!\left(1 - \frac{1}{1!} + \frac{1}{2!} - \frac{1}{3!} + \cdots + (-1)^n \frac{1}{n!}\right).$$

For example,

$$D_5 = 5!\left(1 - \frac{1}{1!} + \frac{1}{2!} - \frac{1}{3!} + \frac{1}{4!} - \frac{1}{5!}\right)$$

$$= 5! - 5! + 5 \cdot 4 \cdot 3 - 5 \cdot 4 + 5 - 1 = 60 - 20 + 5 - 1 = 44.$$

Some readers may have encountered the number e, the base of the natural logarithm, in a calculus course. The *Taylor expansion* for e^x is the formula

(2) $$e^x = 1 + \frac{x}{1!} + \frac{x^2}{2!} + \frac{x^3}{3!} + \cdots ,$$

which, among its uses, allows us to approximate various powers of e. To approximate e itself, for instance, we evaluate the first several terms of the Taylor expansion for e^x with $x = 1$:

$$e = e^1 \approx 1 + \frac{1}{1!} + \frac{1^2}{2!} + \frac{1^3}{3!} + \frac{1^4}{4!} = 1 + 1 + \frac{1}{2} + \frac{1}{6} + \frac{1}{24} \approx 2.708.$$

Better approximations can be obtained by including more terms of the Taylor expansion. To approximate $\sqrt{e}$, we have

$$\sqrt{e} = e^{1/2} = e^{.5} \approx 1 + \frac{.5}{1!} + \frac{(.5)^2}{2!} + \frac{(.5)^3}{3!}$$

$$= 1 + .5 + .125 + .0208\dot{3} \approx 1.646.$$

With $x = -1$, we have

$$\frac{1}{e} = e^{-1} \approx 1 - \frac{1}{1!} + \frac{1}{2!} - \frac{1}{3!} + \cdots + (-1)^n \frac{1}{n!}$$

for any $n \geq 1$.

Comparison with the formula for D_n given in Proposition 7.4.2 shows that $D_n \approx n!e^{-1} = \frac{n!}{e}$.[1] Equivalently, $\frac{D_n}{n!} \approx \frac{1}{e} \approx 0.368$, an observation which shows that the derangements of n symbols comprise about 36.8% of all the permutations of n symbols. This percentage can also be viewed as the probability that a randomly selected permutation of n symbols will be a derangement. It is curious that for any $n > 5$, this probability is 36.8% to one decimal place. The chance that a postman, delivering letters randomly, will deliver six letters to six incorrect addresses is just about the same as the chance that he will deliver 50 letters to incorrect addresses.

Pause 10   Express $\frac{D_4}{4!}$ as a decimal and compare with $\frac{1}{e}$. Do the same for D_5.

Answers to Pauses   **10.** $\frac{D_4}{4!} = \frac{9}{24} = 0.375$. Even for $n = 4$, $\frac{D_n}{n!}$ is remarkably close to $\frac{1}{e}$. $\frac{D_5}{5!} = \frac{44}{120} = .3\dot{6}$. In fact, for $n > 5$, $\frac{D_n}{n!} = .368\ldots$ agrees with $\frac{1}{e}$ to three decimal places.

EXERCISES

The symbol [BB] means that an answer can be found in the Back of the Book.

1. [BB] Find D_6, D_7, and D_8.

2. A simple code is made by permuting the letters of the alphabet such that every letter is replaced by a different letter. How many different codes can be made in this way?

3. [BB] Eleven books are arranged on a shelf in alphabetical order by author name. In how many ways can your little sister rearrange these books so that no book is in its original position?

4. Fifty students take an exam. For the purposes of grading, the teacher asks the students to exchange papers so that no one marks his or her own paper. In how many ways can this be accomplished?

---

[1] Students who have studied Taylor polynomials should quickly observe that the error in approximating D_n by $\frac{n!}{e}$ is less than $\frac{1}{n+1}$.

5. Your letter carrier has a drinking problem. When she delivers letters to the seven houses in your block, she delivers completely at random. Suppose she has seven letters, one addressed to each house, and she delivers one letter to each house. In how many ways can this be accomplished if

   (a) no letter arrives at the right house?
   (b) [BB] at least one letter arrives at the right house?
   (c) all letters arrive at the right house?

6. Twenty people check their hats at a theater. In how many ways can their hats be returned so that

   (a) no one receives his or her own hat?
   (b) at least one person receives his or her own hat?
   (c) [BB] exactly one person receives his or her own hat?
   (d) at least two people receive their own hats?
   (e) at most two people receive their own hats?

7. (a) [BB] In how many ways can the integers 1 through 9 be permuted such that no odd integer will be in its natural position?
   (b) In how many ways can the integers 1 through 9 be permuted such that no even integer is in its natural position?

(c) In how many ways can the integers 1 through 9 be permuted such that exactly four of the nine integers are in their natural positions?

8. Without any calculation, prove that

$$n! = D_n + \binom{n}{1}D_{n-1} + \binom{n}{2}D_{n-2}$$

$$+ \cdots + \binom{n}{n-1}D_1 + 1.$$

9. Prove Proposition 7.4.2.

10. (a) [BB] Prove that $D_n \equiv (-1)^n \pmod{n}$.
    (b) [BB] Prove that D_n is even if and only if n is odd.

11. (a) [BB] Prove that $D_n = (n-1)(D_{n-1} + D_{n-2})$ for all $n \geq 3$.
    (b) Prove that $D_n = nD_{n-1} + (-1)^n$ for all $n \geq 2$.
    (c) Use (b) to find D_n and $\frac{D_n}{n!}$ to six decimal places for $n = 1, 2, 3, \ldots, 10$. What is the approximation to $\frac{1}{e}$ to six decimal places?

12. While the concept of a derangement of zero objects is rather dubious, the formula for D_n given by Proposition 7.4.2 does make sense for $n = 0$ and gives $D_0 = 1$. Taking $D_0 = 1$, show that the generating function for the sequence $\frac{D_0}{0!}, \frac{D_1}{1!}, \frac{D_2}{2!}, \ldots$ is $\frac{e^{-x}}{1-x}$.

7.5 THE BINOMIAL THEOREM

The arrangement of integers in Fig 7.3 is named after Blaise Pascal (1623–1662) because of the many applications in combinatorics and probability which Pascal found for it.[2] Each row begins and ends with a 1; each other number in a row sits between two consecutive numbers in the row above it and is the sum of these two. For example, $10 = 4 + 6$, $6 = 1 + 5$ and $21 = 15 + 6$, as shown in the figure.

This famous triangle has many fascinating properties. For example, if we number the rows starting at zero and interpret each row as the "digits" of a single number written in base 10,[3] then row k of the triangle is 11^k. For example, row zero is $1 = 11^0$, row one is $11 = 11^1$, row two is $121 = 11^2$, row three is $1331 = 11^3$, and row five is

$$1(10^5) + 5(10^4) + 10(10^3) + 10(10^2) + 5(10^1) + 1(10^0) = 11^5,$$

and so on.

---

[2] The triangle itself was known in India as early as the third century B.C.
[3] In this context, by the "digits" of a number, we mean the coefficients required to write the number as a linear combination of powers of 10.

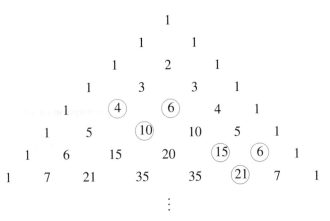

Figure 7.3   Pascal's triangle.

If, beginning at each initial 1, we follow the diagonal pattern suggested by the figure to the right, adding alternate numbers along the way (at the positions denoted •), we discover the terms of the Fibonacci sequence. For example, after two 1's, we obtain the sums $1 + 1 = 2$, $1 + 2 = 3$, $1 + 3 + 1 = 5$ (see Fig 7.4), $1 + 4 + 3 = 8$, $1 + 5 + 6 + 1 = 13$, and so on.

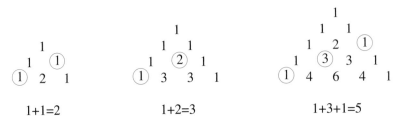

Figure 7.4   The Fibonacci sequence, hidden within Pascal's triangle.

Our major interest in Pascal's triangle concerns its connection with the expansion of $(x + y)^n$, for the numbers in row n of the triangle give the requisite coefficients. As shown in the following equations, which we number starting with 0, the coefficients needed to expand $(x + y)^0$ are in row 0. In row 1, we see the coefficients needed to expand $(x + y)^1$; in row 2, the coefficients needed to expand $(x + y)^2$; and so forth.

$$(x + y)^0 = \underline{1}$$

$$(x + y)^1 = x + y = \underline{1}x + \underline{1}y$$

$$(x + y)^2 = x^2 + 2xy + y^2 = \underline{1}x^2 + \underline{2}xy + \underline{1}y^2$$

$$(x + y)^3 = x^3 + 3x^2y + 3xy^2 + y^3 = \underline{1}x^3 + \underline{3}x^2y + \underline{3}xy^2 + \underline{1}y^3$$

$$(x + y)^4 = x^4 + 4x^3y + 6x^2y^2 + 4xy^3 + y^4$$

$$= \underline{1}x^4 + \underline{4}x^3y + \underline{6}x^2y^2 + \underline{4}xy^3 + \underline{1}y^4$$

How might we obtain the expansion of $(x + y)^{11}$? We could continue to build Pascal's triangle as far as row 11, but it would be better to observe that the numbers in Pascal's triangle are just the values of the binomial coefficients $\binom{n}{k}$. Figure 7.5 shows Pascal's triangle again, but in terms of these combinatorial symbols.

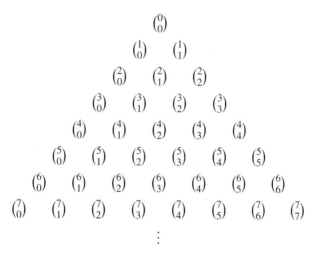

Figure 7.5   A useful presentation of Pascal's triangle.

Note that $\binom{n}{0} = \binom{n}{n} = \frac{n!}{n!0!} = 1$ because $0! = 1$, so each row begins and ends with a 1 as it should. The reader is encouraged to evaluate all the symbols in this second presentation of Pascal's triangle in order to verify that they do indeed agree with the numbers in Fig 7.3. A proof that this correspondence extends to further rows of the triangle depends upon the identity

(3)
$$\binom{n}{k} = \binom{n-1}{k-1} + \binom{n-1}{k}$$

(the proof of which was Exercise 25, Section 7.2) and is left to the exercises. The advantage in identifying the entries of the triangle as the symbols $\binom{n}{r}$ should be apparent: Any particular row can be obtained without first constructing all the rows above it. Row 11, for instance, which gives the coefficients needed to expand $(x + y)^{11}$, is

$$\binom{11}{0}, \quad \binom{11}{1}, \quad \binom{11}{2}, \quad \binom{11}{3}, \quad \cdots, \quad \binom{11}{10}, \quad \binom{11}{11}$$

and these numbers are, respectively, the coefficients of the terms

$$x^{11}, \quad x^{10}y, \quad x^9y^2, \quad x^8y^3, \quad \cdots, \quad xy^{10}, \quad y^{11}.$$

Notice that each of these terms is of the form $x^i y^j$, that the sum $i + j$ of the exponents is always 11, and that the exponent of x decreases one at a time from 11 to 0 as the exponent of y increases from 0 to 11.

7.5.1 THE BINOMIAL THEOREM ▶

For any x and y and any natural number n,

$$(x + y)^n = \sum_{k=0}^{n} \binom{n}{k} x^{n-k} y^k$$

$$= x^n + \binom{n}{1} x^{n-1} y + \binom{n}{2} x^{n-2} y^2 + \cdots + \binom{n}{n-1} xy^{n-1} + y^n.$$

It is, of course, from this theorem that binomial coefficients take their name.

While the Binomial Theorem can be proven by mathematical induction, there is a different argument which explains why the binomial coefficients $\binom{n}{k}$ are involved. Computing $(x + y)^2$ from first principles, we obtain

$$(x + y)^2 = (x + y)(x + y)$$

$$= x(x + y) + y(x + y)$$

$$= xx + xy + yx + yy = x^2 + 2xy + y^2.$$

Computing $(x + y)^3$ the same way, we have

$$(x + y)^3 = (x + y)(x + y)(x + y)$$

$$= (x + y)(xx + xy + yx + yy)$$

$$= xxx + xxy + xyx + xyy + yxx + yxy + yyx + yyy$$

$$= x^3 + 3x^2 y + 3xy^2 + y^3.$$

In the third line of this expansion, each of the $8 = 2^3$ terms xxx, xxy, and so on, corresponds to the selection of an x or a y from each of the original three factors $x + y$. Since xxx arises only when an x is selected from each factor, there is just one x^3 term in the final expansion. The coefficient of $x^2 y$ in the final expansion is 3 because there are three terms involving two x's and one y; namely, xxy, xyx, and yxx. The reason why there are three terms is that there are three ways to select one y (and hence two x's) from the three factors $x + y$.

In the full expansion of $(x + y)^n$, before any simplification and before like terms are collected, there are 2^n terms. Each such term is a string of x's and y's, n symbols in all, and corresponds to a selection of x or y from each of the n factors $x + y$. Any term containing k y's (and hence $n - k$ x's) simplifies to $x^{n-k} y^k$. The eventual coefficient of $x^{n-k} y^k$ is the number of terms which simplify to $x^{n-k} y^k$, this number being $\binom{n}{k}$ because there are $\binom{n}{k}$ ways to select k factors for the y's.

We conclude this section with a sampling of problems which can be solved using the Binomial Theorem.

PROBLEM 15. Using the Binomial Theorem, expand $(3x^2+2y)^5$ and simplify.

Solution.

$$(3x^2 + 2y)^5 = (3x^2)^5 + \binom{5}{1}(3x^2)^4(2y) + \binom{5}{2}(3x^2)^3(2y)^2$$

$$+ \binom{5}{3}(3x^2)^2(2y)^3 + \binom{5}{4}(3x^2)(2y)^4 + (2y)^5$$

$$= 243x^{10} + 5(81x^8)(2y) + 10(27x^6)(4y^2) + 10(9x^4)(8y^3)$$

$$+ 5(3x^2)(16y^4) + 32y^5$$

$$= 243x^{10} + 810x^8y + 1080x^6y^2 + 720x^4y^3 + 240x^2y^4 + 32y^5 \quad \blacksquare$$

PROBLEM 16. Using the Binomial Theorem, expand $\left(x - \dfrac{4}{x}\right)^6$ and simplify.

Solution.

$$\left(x - \frac{4}{x}\right)^6 = x^6 + \binom{6}{1}x^5\left(-\frac{4}{x}\right) + \binom{6}{2}x^4\left(-\frac{4}{x}\right)^2 + \binom{6}{3}x^3\left(-\frac{4}{x}\right)^3$$

$$+ \binom{6}{4}x^2\left(-\frac{4}{x}\right)^4 + \binom{6}{5}x\left(-\frac{4}{x}\right)^5 + \left(-\frac{4}{x}\right)^6$$

$$= x^6 + 6x^5\left(-\frac{4}{x}\right) + 15x^4\left(\frac{16}{x^2}\right) + 20x^3\left(-\frac{64}{x^3}\right) + 15x^2\left(\frac{256}{x^4}\right)$$

$$+ 6x\left(-\frac{1024}{x^5}\right) + \frac{4096}{x^6}$$

$$= x^6 - 24x^4 + 240x^2 - 1280 + \frac{3840}{x^2} - \frac{6144}{x^4} + \frac{4096}{x^6} \quad \blacksquare$$

PROBLEM 17. Find the coefficient of x^{16} in the expansion of $\left(2x^2 - \dfrac{x}{2}\right)^{12}$.

Solution. The general term in the expansion of this expression by the Binomial Theorem is

$$\binom{12}{k}(2x^2)^{12-k}\left(-\frac{x}{2}\right)^k = \binom{12}{k}2^{12-k}\left(-\frac{1}{2}\right)^k x^{24-k}.$$

We want $24 - k = 16$; thus, $k = 8$. The coefficient is $\binom{12}{8}2^4\left(-\frac{1}{2}\right)^8 = \frac{1}{16}\binom{12}{8} = \frac{495}{16}$. $\quad\blacksquare$

PROBLEM 18. Prove that $\binom{n}{0} + \binom{n}{1} + \binom{n}{2} + \cdots + \binom{n}{n} = 2^n$ for all natural numbers n.

Solution. Consider $(x + y)^n = \sum_{k=0}^{n} \binom{n}{k} x^{n-k} y^k$. Setting $x = y = 1$, we obtain

$$(1+1)^n = \sum_{k=0}^{n} \binom{n}{k} 1^{n-k} 1^k; \text{ that is, } 2^n = \sum_{k=0}^{n} \binom{n}{k}, \text{ as desired.} \quad \blacksquare$$

EXERCISES

The symbol [BB] means that an answer can be found in the Back of the Book.

1. [BB] For each of the following, expand using the Binomial Theorem and simplify.

    (a) $(x + y)^6$
    (b) $(2x + 3y)^6$

2. Use the Binomial Theorem to expand $(a+4b)^5$. Simplify your answer.

3. For each of the following, expand using the Binomial Theorem and simplify.

    (a) $(2x^3 - y^2)^8$
    (b) [BB] $(2x^3 - x^2)^8$
    (c) $\left(2x^3 - \dfrac{1}{x^2}\right)^8$

4. [BB] Find the fourth term in the binomial expansion of $(x^3 - 2y^2)^{12}$.

5. Consider the binomial expansion of $(x + y)^{20}$.

    (a) What are the first three terms?
    (b) What are the last three terms?
    (c) [BB] What is the seventh term? The fifteenth term?
    (d) What is the coefficient of $x^{13} y^7$?

6. [BB] Consider the binomial expansion of $(2x - y)^{16}$.

    (a) How many terms are there altogether?
    (b) Is there a middle term or are there two middle terms? Find and explain.

7. Consider the binomial expansion of $(x + 3y)^{17}$.

    (a) How many terms are there altogether?
    (b) Is there a middle term or are there two middle terms? Find and explain.

8. [BB] Find the coefficient of $x^3 y^7$ in the binomial expansion of $(4x + 5y)^{10}$.

9. [BB] Find the coefficient of the term containing y^8 in the binomial expansion of $(x + 3y^2)^{17}$.

10. What is the coefficient of x^5 in the binomial expansion of $(x - 2x^{-2})^{20}$?

11. [BB] What is the coefficient of x^{27} in the binomial expansion of $\left(\dfrac{3}{x} + x^2\right)^{18}$?

12. Find the coefficient of x^{25} in the binomial expansion of $\left(2x - \dfrac{3}{x^2}\right)^{58}$.

13. Prove that $4^n > n^4$ for all integers $n \geq 5$.

14. Sequences of integers, $\{x_n\}, \{y_n\}, n \geq 0$, are defined by $(1 + \sqrt{2})^n = x_n + y_n \sqrt{2}$.

    (a) [BB] Find the first three terms and the fifth term of each sequence.
    (b) Find formulas for x_n and y_n.
    (c) Without using the result of (b), establish the recursion formulas

    $$x_{n+1} = x_n + 2y_n$$
    $$y_{n+1} = x_n + y_n$$

    for $n \geq 0$.

15. (For students who have studied calculus) In Problem 8 on p. 14, we proved that $\sqrt{2}$ is irrational, that is, not the quotient of integers. In this exercise, we give an alternative proof using the binomial theorem.

    (a) If n is an integer, show that there exist integers x_n and y_n, depending on n, such that $(\sqrt{2} - 1)^n = x_n + y_n \sqrt{2}$.
    (b) Explain why $\lim_{n \to \infty} (\sqrt{2} - 1)^n = 0$.
    (c) Suppose that $\sqrt{2}$ is rational and show that (a) and (b) lead to a contradiction.

16. [BB] At the beginning of this section, we said that the numbers in row n of Pascal's triangle, interpreted as the "digits" of a single number in base 10, give the integer 11^n. Prove this.

17. Suppose that we construct Pascal's triangle as described at the beginning of this section. Prove that the entries of row n are $\binom{n}{0}$, $\binom{n}{1}$, $\binom{n}{2}$, ..., $\binom{n}{n}$. [*Hint:* Identity (3).]

18. Prove the Binomial Theorem by mathematical induction. [*Hint:* Identity (3) will be useful.]

19. (a) Show that $\sum_{k=2}^{6} \binom{k}{2} = \binom{7}{3}$ and interpret this result with reference to Pascal's triangle.

(b) Show that $\sum_{k=r}^{n} \binom{k}{r} = \binom{n+1}{r+1}$ for any r and n, $1 \leq r \leq n$, and interpret with reference to Pascal's triangle.

20. [BB] Show that $\binom{n}{0} - \binom{n}{1} + \binom{n}{2} - \cdots + (-1)^n \binom{n}{n} = 0$ for all natural numbers n.

21. (a) By direct calculation, show that $3^4 = 1 + 2\binom{4}{1} + 4\binom{4}{2} + 8\binom{4}{3} + 16$.

(b) Find a simple expression for $\binom{n}{0} + 2\binom{n}{1} + 4\binom{n}{2} + \cdots + 2^n \binom{n}{n}$ valid for all natural numbers n and prove your answer.

22. (a) [BB] Find a simple expression for $\sum_{k=1}^{n} k\binom{n}{k}$ and prove your answer.

(b) Find a simple expression for $\sum_{k=1}^{n} k^2 \binom{n}{k}$ and prove your answer.

[*Hint:* In both of these questions, to begin, it may prove helpful to evaluate the sum for small values of n.]

23. Consider the Fibonacci sequence $a_0, a_1, a_2, \ldots$, where $a_0 = a_1 = 1$ and, for $n \geq 1$, $a_{n+1} = a_n + a_{n-1}$. Express a_n for $n \geq 1$ as the sum of certain binomial coefficients and prove your answer. [*Hint:* See remarks at the beginning of this section.]

REVIEW EXERCISES FOR CHAPTER 7

1. In how many ways can seven boys and six girls stand in a row if the girls are to stand together but the boys must not stand together?

2. Do Exercise 1 again with "circle" instead of "row."

3. How many permutations of the letters a, b, c, d, e, f contain at least one of the patterns aeb and bcf?

4. Find an expression for the number of five-card poker hands which contain exactly three kings.

5. A committee of seven is to be chosen from eight men and nine women.

(a) How many such committees contain at least six women?

(b) How many such committees contain either Bob or Alice but not both?

6. A Middle East peace conference will be attended by five Arab countries, four Western countries, and Israel. Each delegation will sit together at a round table subject only to the requirement that no Arab delegation be seated next to the Israeli delegation. How many seating arrangements are possible?

7. Frank wants to buy 12 muffins and finds seven different types available. In how many ways can he make his selection?

8. In how many ways can the letters of the word *NUNAVUT* be rearranged?

9. Find the coefficient of x^5 in the binomial expansion of $(x - 2x^{-2})^{20}$. *Do not simplify.*

10. Find the coefficient of x^{-6} in $\left(16x^2 - \dfrac{1}{2x}\right)^{12}$. Simplify your answer.

11. (a) Give a verbal argument for the truth of the identity $\binom{n}{k} = \binom{n-1}{k-1} + \binom{n-1}{k}$.

(b) If $n \geq k+2$ and $k \geq 2$, show that $\binom{n}{k} - \binom{n-2}{k} - \binom{n-2}{k-2}$ is even.

12. Find a simple expression for $3^n - \binom{n}{1}3^{n-1} + \cdots + (-1)^k \binom{n}{k}3^{n-k} + \cdots + (-1)^n$.

8

Algorithms

8.1 WHAT IS AN ALGORITHM?

Computers! Computers! Computers! Are they the bane of our existence or the most important invention since the discovery of the wheel? Certainly there is no denying that computers have changed almost every aspect of our lives. They are used by stockbrokers to track the performance of stocks and to provide guidance as to which should be bought and sold. Much of the blame for the October 1987 North American stock market crash was attributed to computers all giving the same "sell" advice at the same time. Nowadays, virtually every company puts personal computers on the desks of their employees, in the belief that computers both increase productivity and cut revenue losses. Upon discovery of a potential flaw in some part, a car manufacturer is able almost instantly to produce a list of the names and addresses of all the people across North America who have purchased a car containing that particular part. The list of computer applications is literally endless. But what is a computer?

A computer is fundamentally just a box containing wires and chips (these being tiny circuits built onto small boards) all cleverly put together so that the machine can follow instructions given to it. The box of wires and chips, and peripheral items such as monitor and keyboard, are called *hardware*; the instructions fed to the machine are called *software*. Computer programming involves writing software that allows a machine to perform various tasks which, by hand, would be tedious or so time consuming as to be essentially impossible to do. Machines work incredibly quickly, never get tired, and are excellent at following orders; however, they will only perform as well as the instructions presented to them. One of the oldest adages in the computer business is "Garbage in—garbage out!".

There are two parts to a computer program. There is the process or sequence of steps which is necessary to complete the given task and the translation of this process into a language which the computer can understand. In this chapter, we explore the first of these ideas, asking the reader to think about the process or *algorithm* by which familiar tasks are accomplished.

The word "algorithm" evolved from the older word "algorism" which is a corruption of the surname of a ninth-century Persian, Abu Ja'far Muḥammad ibn Mûsâ al-Khwârizmî, who wrote an important book setting forth rules for performing arithmetic with the arabic numerals $1, 2, 3 \ldots$ we use today.[1] While the older term "algorism" referred primarily to arithmetical rules, "algorithm" today is used in a more general sense as a virtual synonym for "procedure." Its ingredients are an input, an output, and a sequence of precise steps for converting the input to the output. In Chapter 4, we studied the Euclidean algorithm which, upon input of positive integers a and b, outputs the greatest common divisor of a and b through a precise series of steps.

As children, after learning the addition rules for single-digit numbers, we were taught a procedure for adding numbers with any number of digits; upon input of two arbitrary integers, the procedure outputs their sum.

$$
\begin{array}{cccc}
1 & 2 & 3 & 4 \\
5 & 6_1 & 7_1 & 8 \\
\hline
6 & 9 & 1 & 2
\end{array}
$$

We memorized multiplication tables for single-digit numbers and then were taught a procedure which, upon input of two arbitrary numbers, outputs their product.

Folklore has it that Russian peasants have an instinctive and flawless ability to multiply and divide by the number 2 and are able to use this successfully to multiply any two numbers. For example, they would compute 211×453 by the scheme shown in Fig 8.1. Each number on the left is the quotient when the number above it is divided by 2;[2] the numbers on the right are the product of 2 and the number above. After reaching 1 in the left column, the rows containing even numbers on the left are crossed out. The desired product is then the sum of the numbers which remain uncrossed on the right!

$$
\begin{array}{rcr}
211 & \times & 453 \\
105 & & 906 \\
\cancel{52} & & \cancel{1812} \\
\cancel{26} & & \cancel{3624} \\
13 & & 7248 \\
\cancel{6} & & \cancel{14496} \\
3 & & 28992 \\
1 & & 57984 \\
\hline
& & 95583
\end{array}
$$

Figure 8.1   The Russian peasant method of multiplication.

Before the days of pocket calculators, school children were taught a procedure for finding the square root of a number. To find the square root of 2, for instance, a student was taught to produce the pattern shown in Fig 8.2. First, we find the

---

[1]Interestingly, the word "algebra" derives from the Latin title of this book, *Ludus algebrae et almucgrabalaeque*.

[2]Remember that when a natural number a is divided by another natural number b, the *quotient* is the integer part of the fraction $\frac{a}{b}$. See Section 4.1.

largest integer whose square is less than 2 and write this number, 1, in the two places shown. The product of 1 and 1 is 1 and $2 - 1 = 1$. Write down 1 followed by two 0's. Next, we double the number on top, giving 2, and search for the largest integer x such that $x \times 2x$ does not exceed 100. We have $x = 4$, $4 \times 24 = 96$, and $100 - 96 = 4$, as shown. Follow this 4 by another two 0's, double the number on top, giving 28, and search for the largest single digit x such that $x \times 28x$ does not exceed 400. We have $x = 1$, $1 \times 281 = 281$, and $400 - 281 = 119$, and so forth. Whether or not you understand this scheme is much less important today than it was formerly. We mention it here not just as a historical curiosity, but as another example of an arithmetic algorithm, one which allows the computation of square roots to any degree of accuracy.

```
            1.   4     1     4     2
          | 2.  0  0 | 0  0 | 0  0 | 0  0
      1   | 1
     24   | 1   0  0
          |     9  6
    281   |        4  0  0
          |        2  8  1
   2824   |        1  1  9  0  0
          |        1  1  2  9  6
  28282   |              6  0  4  0  0
          |              5  6  5  6  4
          |              3  8  3  6
```

Figure 8.2   Part of an algorithm for determining $\sqrt{2}$.

In a course on Euclidean geometry, we learn a vast number of geometric algorithms usually called *constructions*. To bisect an angle with vertex A, for instance, we draw an arc with center A. If B and C are the points where this arc meets the arms of the angle, we then draw arcs with centers B and C and some suitably large radius (for example, the length of AB). If these arcs meet at the point P, then AP is the bisector of the angle at A. (See Fig 8.3.)

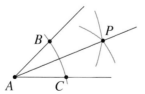

Figure 8.3   Bisection of the angle at A.

In high school and college, students learn algorithms for multiplying polynomials, for multiplying matrices, for solving systems of linear equations, for differentiating polynomials; the list goes on. In the rest of this chapter, we discuss important algorithms the reader may not have seen before and their relative strengths and weaknesses. Since this is not a text in computer programming, our description of algorithms is mostly in ordinary English, subject only to the require-ments of clarity and precision: It should always be clear to a programmer how to implement any of our algorithms in his or her favorite language.

PROBLEM 1. Describe an algorithm whose input is a list $a_1, a_2, \ldots, a_n$ of integers and whose output is their sum.

Solution. This question requires us to confront what an expression like $5 + 17 + 3 + 6 + 4$ really means. How do the basic rules for addition of two numbers extend to permit the addition of several numbers? Most people would calculate $5 + 17 + 3 + 6 + 4$ by adding from the left, like this:

$$\{[(5 + 17) + 3] + 6\} + 4 = [(22 + 3) + 6] + 4 = (25 + 6) + 4 = 31 + 4 = 35.$$

This approach is the idea behind our algorithm.

To find the sum of n integers $a_1, a_2, \ldots, a_n$,

Step 1. set $S = 0$;

Step 2. for $i = 1$ to n, replace S by $S + a_i$;

Step 3. output S.

The value of S output at Step 3 is the desired sum. ∎

Step 2 calls for a *loop*, a feature of every programming language. Whenever we write "for $i = 1$ to n" as part of an algorithm, we mean

- set $i = 1$ and execute the indicated statement (or statements);
- increase the value of i by 1 and, if $i \leq n$, execute the indicated statement(s);
- continue to increase the value of i by 1 and, as long as $i \leq n$, execute the indicated statement(s);
- if the stage $i > n$ is reached, skip to the next step.

The variable i which is used in this process is called a *counter*. For example, with a_1, a_2, a_3, a_4, a_5, respectively, equal to 5, 17, 3, 6, 4, the preceding algorithm begins by setting $S = 0$. At Step 2, the counter i is set equal to 1 and S is replaced by $S + a_1 = 0 + 5 = 5$. Then i is set equal to 2 and S is replaced by $S + a_2 = 5 + 17 = 22$. Continuing, S is replaced by $S + a_3 = 22 + 3 = 25$, then by $S + a_4 = 25 + 6 = 31$ and, finally, by $S + a_5 = 31 + 4 = 35$. Then i is set equal to 6, at which point the algorithm skips to Step 3 and outputs $S = 35$, the sum of $a_1, \ldots, a_5$.

PROBLEM 2. (Polynomial evaluation) Describe an algorithm which, upon input of $n+1$ integers $a_0, a_1, a_2, \ldots, a_n$ and an integer x, outputs the integer $a_0 + a_1 x + a_2 x^2 + \cdots + a_n x^n$.

Solution. Perhaps the most obvious approach to this problem is first to compute the powers of x from x to x^n and then to evaluate the expression $a_0 + a_1 x + a_2 x^2 + \cdots + a_n x^n$ term by term.

Given integers $a_0, a_1, \ldots, a_n$ and an integer x, to compute the number $a_0 + a_1x + a_2x^2 + \cdots + a_nx^n$,

Step 1. set $P = 1$;

Step 2. for $i = 1$ to n, replace P by Px, call this element x_i and store (note that $x_i = x^i$);

Step 3. set $S = a_0$;

Step 4. for $i = 1$ to n, replace S by $S + a_ix_i$;

Step 5. output S.

The value of S output at Step 5 is the desired integer $a_0 + a_1x + a_2x^2 + \cdots + a_nx^n$.

∎

EXAMPLE 3

To evaluate the polynomial $f(x) = -1 + 2x + 4x^2 - 3x^3$ at $x = 5$, the algorithm takes as input $a_0 = -1$, $a_1 = 2$, $a_2 = 4$, $a_3 = -3$ and $x = 5$ and proceeds as follows.

Step 1. $P = 1$;

Step 2. $i = 1$: replace P by $Px = 1(5) = 5$ (P is now 5) and let $x_1 = 5$;
  $i = 2$: replace P by $Px = 5(5) = 25$ (now $P = 25$) and let $x_2 = 25$;
  $i = 3$: replace P by $Px = 25(5) = 125$ (now $P = 125$) and let $x_3 = 125$.
  Since $i = n = 3$, Step 2 is complete.

Step 3. $S = a_0 = -1$;

Step 4. $i = 1$: replace S by $S + a_1x_1 = -1 + 2(5) = 9$ (now $S = 9$);
  $i = 2$: replace S by $S + a_2x_2 = 9 + 4(25) = 109$ (now $S = 109$);
  $i = 3$: replace S by $S + a_3x_3 = 109 - 3(125) = -266$ (now $S = -266$).
  Since $i = n = 3$, Step 4 is complete.

Step 5. Output $S = -266$. ▲

There is a much more efficient way to evaluate a polynomial, more efficient in the sense that it requires fewer arithmetical operations and fewer storage requirements. It is based upon the observation, due to William Horner, an eighteenth-century English school headmaster, that

$$a_0 + a_1x + a_2x^2 + \cdots + a_nx^n = a_0 + (a_1 + a_2x + a_3x^2 + \cdots + a_nx^{n-1})x$$

$$= a_0 + (a_1 + (a_2 + a_3x + \cdots + a_nx^{n-2})x)x$$

$$\vdots$$

$$= a_0 + (a_1 + (a_2 + \cdots + (a_{n-1} + a_nx)x)\cdots)x.$$

8.1.1 HORNER'S ALGORITHM ▶

Given integers $a_0, a_1, \ldots, a_n$ and an integer x, to evaluate the expression $a_0 + a_1x + a_2x^2 + \cdots + a_nx^n$,

Step 1. set $S = a_n$;

Step 2. for $i = 1$ to n, replace S by $a_{n-i} + Sx$;

Step 3. output S.

The final value of S is the desired number $a_0 + a_1x + a_2x^2 + \cdots + a_nx^n$.

EXAMPLE 4 We apply Horner's method to the same polynomial considered in Example 3, $f(x) = -1 + 2x + 4x^2 - 3x^3$, and again with $x = 5$.

Step 1. set $S = a_3 = -3$.

Step 2. $i = 1$: replace S by $a_2 + Sx = 4 - 3(5) = -11$ (now $S = -11$);
$i = 2$: replace S by $a_1 + Sx = 2 - 11(5) = -53$ (now $S = -53$);
$i = 3$: replace S by $a_0 + Sx = -1 - 53(5) = -266$ (now $S = -266$).
Since $i = n = 3$, Step 2 is complete.

Step 3. output $S = -266$. ▲

Pause 1 Show the successive values of S when Horner's method is used to evaluate the polynomial $f(x) = 4x^3 - 2x + 1$ at $x = -2$.

Many of the uses to which modern-day computers are put do not involve arithmetical calculations but take advantage of a machine's ability to perform certain logical operations. Of these, the most fundamental is its ability to compare, that is, to decide if two elements are equal. The next two sections are devoted to such important nonarithmetical procedures as searching, sorting, and enumerating permutations. We conclude this section with one example of an algorithm in which comparison is the basic operation.

PROBLEM 5. Given a "small" list, it is easy for a person to note the distinct items it contains. The distinct items in abc, dbc, abc, xbc, dbc are, of course, abc, dbc, and xbc. To record the distinct items in a list of a thousand or more items would be an almost intolerable task for most people, but it is a simple matter for a computer. Describe an algorithm which, upon input of a list $a_1, a_2, \ldots, a_n$, outputs the distinct items in this list.

Solution. The idea is straightforward. We compare each element in the list with those preceding it and, if it is different from all its predecessors, output it.

Step 1. output a_1;

Step 2. if $n = 1$, stop;
else for $i = 2$ to n,
if a_i does not equal any of $a_1, a_2, \ldots, a_{i-1}$, output a_i.

The algorithm outputs the distinct items among $a_1, a_2, \ldots, a_n$. ∎

Answers to Pauses

1. We have $n = 3$, $a_0 = 1$, $a_1 = -2$, $a_2 = 0$ and $a_3 = 4$. The first value of S is $S = 4$. Then $S = 0 + 4(-2) = -8$, $S = -2 - 8(-2) = 14$ and, finally, $S = 1 + 14(-2) = -27$.

EXERCISES

The symbol [BB] means that an answer can be found in the Back of the Book.

In the first three exercises, ruler *means an unmarked straight edge and* compass *means an instrument which allows the describing of circles or arcs thereof.*

1. [BB] Describe a procedure for finding the mid-point of a line segment, with only ruler and compass, and explain why your procedure works.

2. Given a line ℓ in the plane and a point P not on ℓ, describe a ruler and compass procedure for constructing the line through P parallel to ℓ, and explain why your procedure works.

3. Given a point P and a line ℓ, describe a ruler and compass procedure for constructing the line through P perpendicular to ℓ in each of the following cases. In each case, explain why your procedure works.

   (a) [BB] P is not on ℓ.
   (b) P is on ℓ.

4. Find each of the following products by the Russian peasant method.

   (a) 168×413
   (b) 461×973
   (c) 141×141

5. [BB] Explain why the Russian peasant method works.

6. Find the square roots of 3, 12, and 153, to four decimal places of accuracy, by the method used to compute $\sqrt{2}$ in the text and illustrated in Fig 8.2.

7. [BB] Describe an algorithm which, upon input of n real numbers, outputs their average.

8. [BB] Describe an algorithm which, upon input of n real numbers, outputs the maximum of these numbers.

9. Describe an algorithm which, upon input of n real numbers, $a_1, a_2, \ldots, a_n$, and another number, x, determines how many items in the list are equal to x.

10. Describe an algorithm which, given n real numbers, $a_1, a_2, \ldots, a_n$, outputs the number of a_i which lie in the range 85–90, inclusive.

11. [BB] Given an ordered list $a_1 \leq a_2 \leq \cdots \leq a_n$ of real numbers and a real number x, describe an algorithm which inserts x into its correct position in the list and outputs the ordered list of $n + 1$ numbers.

12. Describe an algorithm which, upon input of integers a and b and a natural number n, outputs all solutions of $ax \equiv b \pmod{n}$ in the range $0 \leq x < n$, if there are solutions, and otherwise outputs the words *no solution*.

13. [BB] Let n be a given natural number. Find an algorithm for writing an integer a in the form

    $$a = a_n n! + a_{n-1}(n-1)! + \cdots + a_2 2! + a_1$$

    with $0 \leq a_i \leq i$ for $i = 1, 2, \ldots, n - 1$. [*Hint*: Modify the procedure for converting from base 10 to base b described in Section 4.1.]

14. Describe an algorithm which, upon input of a number n given in base 10, outputs the digits of n (from right to left).

15. [BB] Describe an algorithm which, upon input of n distinct symbols $a_1, a_2, \ldots, a_n$, outputs all the subsets of $\{a_1, a_2, \ldots, a_n\}$. [*Hint*: One way to do this is to recognize that the subsets of $\{a_1, a_2, \ldots, a_n\}$ are in one-to-one correspondence with the binary representations of the numbers between 0 and $2^n - 1$.]

16. For each polynomial $f(x)$ and each value of x, list in order the successive values S which occur in the calculation of $f(x)$ by

    i. the algorithm described in Problem 2;
    ii. Horner's algorithm.

    Show your calculations.

    (a) [BB] $f(x) = 2x^2 - 3x + 1$; $x = 2$
    (b) $f(x) = 3x^2 + 1$; $x = 5$
    (c) $f(x) = -4x^3 + 6x^2 + 5x - 4$; $x = -1$
    (d) $f(x) = 17x^5 - 40x^3 + 16x - 7$; $x = 3$

17. [BB] Explain why Horner's algorithm works.

18. Describe an algorithm which upon input of a natural number n outputs the set of primes in the range 1–n. [*Hint*: Use the Sieve of Eratosthenes and take advantage of Lemma 4.3.4.]

19. Cite examples of five algorithms not mentioned in this text which you learned **after** leaving high school.

8.2 COMPLEXITY

Suppose that two mathematicians working in the research department of a large corporation are given a problem to solve. Each of these people has a solid background in discrete mathematics and so has no difficulty in finding a suitable algorithm, but the two algorithms obtained are very different. When implemented on a computer, one of the algorithms took far less time to produce an answer than the other. The head of the research department is pleased that an efficient algorithm was discovered, but also concerned because of the huge amount of computer time wasted running the inefficient algorithm. Is it possible to estimate the amount of time an algorithm will require before actually implementing it? The answer is yes, and we shall begin to see why in this section.

Just like a person, a computer requires a certain amount of time to carry out an arithmetical operation, a multiplication or an addition, for instance. So we can estimate the time an algorithm requires by calculating the number of arithmetical operations it involves. This is at best only approximate, of course, because different operations may well require different amounts of time.

EXAMPLE 6   In order to determine the distance between two points $(x_1, x_2, \ldots, x_n)$ and $(y_1, y_2, \ldots, y_n)$ in n-dimensional Euclidean space, we must calculate the number

$$\sqrt{(x_1 - y_1)^2 + (x_2 - y_2)^2 + \cdots + (x_n - y_n)^2}.$$

If $n = 2$, this requires two subtractions, two squarings, one addition, and one square root, six operations in all. (While we might expect multiplication and especially the square root operation to be more time consuming than addition and subtraction, we ignore this potential complication for now.) If $n = 3$, nine operations are required; if $n = 4$, 12 operations are required, and so on. In general, the number of operations in the n-dimensional case is $3n$. As is typical, the number of operations depends on the size of the input, in this case, the $2n$ numbers $x_1, \ldots, x_n, y_1, \ldots, y_n$.   ▲

There are other measures of the efficiency of algorithms in addition to time. For instance, different algorithms require different amounts of space to hold numbers in memory for later use; the less space required the better. We shall, however, concentrate entirely upon time estimates of efficiency, as measured by operation counts.

Often, and particularly in the graph theoretical algorithms which we will study in later chapters, it is uncertain exactly how many operations will be needed. Our approach will be to seek upper bound, also called *worst-case*, estimates of the number of operations which we know will never be exceeded.

In general, for a given algorithm, we try to find the *complexity function* $f: \mathbb{N} \to \mathbb{N}$, where, for some measure n of the size of the input, $f(n)$ is an upper bound for the number of operations required to carry out the algorithm. In the distance algorithm described in Example 6, it is logical to let n be the dimension of the Euclidean space; then $f(n) = 3n$. As we shall see, it is unusual to have such an exact count for the number of operations.

We have noted that estimating the running time of an algorithm by estimating the number of operations required is an inherently inexact process because it cannot be expected that all operations—addition, multiplication, square root—require equal amounts of time. There is yet another difficulty. The time required to perform the basic operations of arithmetic depends on the size of the integers involved. Clearly it takes more time to add or multiply two 50-digit numbers than it does to add or multiply two three-digit numbers. The following problem indicates that it is possible to account for this additional complication.

PROBLEM 7. Find the complexity function for adding two n-digit integers if the basic operation is addition of single-digit integers.

Solution. Suppose the integers to be added are $a = (a_{n-1}a_{n-2} \ldots a_1a_0)_{10}$ and $b = (b_{n-1}b_{n-2} \ldots b_1b_0)_{10}$, expressed in base 10 using the notation of Section 4.1. The units digit of a is a_0 and the units digit of b is b_0; the tens digit of a is a_1 and the tens digit of b is b_1, and so forth. The units digit of $a + b$ is obtained by adding a_0 and b_0, a single operation. To obtain the tens digit, we add a_1 and b_1; then, perhaps, we add 1, depending on whether or not there is a carry from the previous step. Hence, at most two single-digit additions (two operations) are required for the tens digit of $a + b$. Similarly, at most two operations are required for each digit of $a + b$ after the units digit. An upper bound for the number of operations is $f(n) = 1 + 2(n - 1) = 2n - 1$. ∎

In this problem, and typical of most complexity problems, we were unable to obtain an exact count for the number of single-digit additions required. Remember that for us, complexity is measured in worst-case terms. The addition of two n-digit numbers requires at most $2n - 1$ single-digit additions.

Because of the approximate nature of operation counts, it is useful to have some notation by which we can easily suggest the size of a function.

8.2.1 DEFINITION ▶ Let f and g be functions $\mathsf{N} \to \mathsf{R}$. We say that f is *Big Oh* of g and write $f = \mathcal{O}(g)$ if there is an integer n_0 and a positive real number c such that $|f(n)| \leq c|g(n)|$ for all $n \geq n_0$.[3]

8.2.2 REMARK ▶ 1. Instead of saying "There exists an integer n_0 such that $|f(n)| \leq c|g(n)|$ for all $n \geq n_0$, we often say simply "$|f(n)| \leq c|g(n)|$ for *sufficiently large n*."

2. If $f, g \colon \mathsf{N} \to \mathsf{R}$ are functions which count operations or, more generally, as long as $f(n)$ and $g(n)$ are positive for sufficiently large n, then the absolute value symbols around $f(n)$ and $g(n)$ in Definition 8.2.1 are not necessary. Whenever we apply this definition without absolute value symbols, it will be because the functions in question are either positive or positive for large n.

EXAMPLE 8 Let $f(n) = 15n^3$ and $g(n) = n^3$. With $n_0 = 1$ and $c = 15$, we see that $f = \mathcal{O}(g)$. ▲

Pause 2 It is also true that $g = \mathcal{O}(f)$. Why?

---

[3]There is also a *Little Oh* notation, which we do not introduce since $f(n) = o(g(n))$ is the same as $f(n) \prec g(n)$, notation which we prefer. See Definition 8.2.4.

EXAMPLE 9 Let $f(n) = n + 1$ and $g(n) = n^2$. If $n \geq 1$, $f(n) \leq n + n = 2n \leq 2n^2$. Taking $n_0 = 1$ and $c = 2$, we see that $f = \mathcal{O}(g)$. On the other hand, $g \neq \mathcal{O}(f)$, for suppose that $n^2 \leq c(n + 1)$ eventually, that is, for all n greater than or equal to some number n_0. Dividing by n, we have $n \leq c(1 + \frac{1}{n}) \leq 2c$ for all $n \geq n_0$. This, however, is impossible because $2c$ is a constant. ▲

Two fundamental properties of Big Oh are summarized in the next proposition.

8.2.3 PROPOSITION ▶ Let f, g, f_1, g_1 be functions $\mathsf{N} \to \mathsf{R}$.

(a) If $f = \mathcal{O}(g)$, then $f + g = \mathcal{O}(g)$.
(b) If $f = \mathcal{O}(f_1)$ and $g = \mathcal{O}(g_1)$, then $fg = \mathcal{O}(f_1 g_1)$.

Proof We prove each part directly from the definition and, in part (a), use the *triangle inequality* which states that $|a + b| \leq |a| + |b|$ for any real numbers a and b.

(a) There exists a positive constant c and an integer n_0 such that $|f(n)| \leq c|g(n)|$ for all $n \geq n_0$. So, for all $n \geq n_0$, $|f(n) + g(n)| \leq |f(n)| + |g(n)| \leq c|g(n)| + |g(n)| = (c + 1)|g(n)|$. With n_0 and $c + 1$ in Definition 8.2.1, we see that $f + g = \mathcal{O}(g)$.

(b) There exists a constant c and positive integer n_0 such that $|f(n)| \leq c|f_1(n)|$ for all $n \geq n_0$. There exists a constant d and positive integer n_1 such that $|g(n)| \leq d|g_1(n)|$ for all $n \geq n_1$. Let $n^* = \max\{n_0, n_1\}$. Then, if $n \geq n^*$, $|f(n)g(n)| = |f(n)||g(n)| \leq c|f_1(n)| \, d|g_1(n)| = cd|f_1(n)g_1(n)|$. Replacing the constants n_0 and c in Definition 8.2.1 with n^* and cd, respectively, we see that $fg = \mathcal{O}(f_1 g_1)$. ∎

EXAMPLE 10 We showed in Example 9 that $n + 1 = \mathcal{O}(n^2)$. Thus, by part (a) of Proposition 8.2.3, we know also that $n^2 + n + 1 = \mathcal{O}(n^2)$. Since $n + 1 = \mathcal{O}(n)$, part (b) tells us that $(n + 1)(n^2 + n + 1) = \mathcal{O}(n^3)$. ▲

Big Oh gives us a way to compare the relative sizes of functions.

8.2.4 DEFINITIONS ▶ If f and g are functions $\mathsf{N} \to \mathsf{R}$, we say that f has *smaller order* than g and write $f \prec g$ if and only if $f = \mathcal{O}(g)$ but $g \neq \mathcal{O}(f)$. If $f = \mathcal{O}(g)$ and $g = \mathcal{O}(f)$, then we say that f and g have the *same order* and write $f \asymp g$.

EXAMPLES 11
- Example 9 shows that $n + 1 \prec n^2$; thus, $n + 1$ has smaller order than n^2.
- Example 8 and the Pause which follows show that $15n^3 \asymp n^3$: $15n^3$ and n^3 have the same order. ▲

PROBLEM 12. Show that $n!$ has smaller order than n^n, that is, $n! \prec n^n$.

Solution. First note that $n! = n(n - 1)(n - 2) \cdots (3)(2) \leq n \cdot n \cdot n \cdots n \cdot n = n^n$ so, with $n_0 = c = 1$ in Definition 8.2.1, it follows that $n! = \mathcal{O}(n^n)$. On the other hand, $n^n \neq \mathcal{O}(n!)$, which we establish by contradiction. If $n^n = \mathcal{O}(n!)$, then, for some constant c, $n^n \leq cn!$, so $\frac{n^n}{n!} \leq c$ for all sufficiently large n. This is impossible because

$$\frac{n^n}{n!} = \frac{n}{n}\frac{n}{n-1}\cdots\frac{n}{1} > n$$

for $n > 2$. ∎

8.2.5 REMARK ▶ Suppose that $\lim_{n\to\infty} f(n)$ and $\lim_{n\to\infty} g(n)$ both exist, are positive, and $\lim_{n\to\infty} g(n) \neq 0$. Students of calculus may notice that if $f = \mathcal{O}(g)$, then $\frac{f(n)}{g(n)} \leq c$ for all sufficiently large n, and thus, $\lim_{n\to\infty}\frac{f(n)}{g(n)} \leq c$. In particular, if $\lim_{n\to\infty}\frac{f(n)}{g(n)} = \infty$, then $f \neq \mathcal{O}(g)$.

Students familiar with limits should appreciate the next proposition. We leave its proof to the exercises.

8.2.6 PROPOSITION ▶ Let f and g be functions $\mathsf{N} \to \mathsf{R}$.

(a) If $\displaystyle\lim_{n\to\infty} \frac{f(n)}{g(n)} = 0$, then $f \prec g$.

(b) If $\displaystyle\lim_{n\to\infty} \frac{f(n)}{g(n)} = \infty$, then $g \prec f$.

(c) If $\displaystyle\lim_{n\to\infty} \frac{f(n)}{g(n)} = L$ for some number $L \neq 0$, then $f \asymp g$.

EXAMPLES 13
- Since $\displaystyle\lim_{n\to\infty} \frac{n+1}{n^2} = 0$, $n+1 \prec n^2$, as earlier noted.
- Since $\displaystyle\lim_{n\to\infty} \frac{2n^2 - 3n + 6}{n^2} = 2$, $2n^2 - 3n + 6 \asymp n^2$. ▲

We have seen that $15n^3 \asymp n^3$ and $2n^2 - 3n + 6 \asymp n^2$. These examples illustrate a general principle whose proof we first describe in a specific case.

PROBLEM 14. Let $f(n) = 5n^2 - 6n + 3$ and $g(n) = n^2$. Show that f and g have the same order.

Solution. For $n \geq 1$, we have $5n^2 - 6n + 3 \leq 5n^2 + 6n + 3 \leq 5n^2 + 6n^2 + 3n^2 = 14n^2$ so, with $c = 14$ and $n_0 = 1$ in Definition 8.2.1, we see that $f = \mathcal{O}(g)$. On the other hand,

$$f(n) = 5n^2 - 6n + 3 = n^2\left(5 - \frac{6}{n} + \frac{3}{n^2}\right) \geq n^2\left(5 - \frac{6}{n} - \frac{3}{n^2}\right).$$

Since

$$\lim_{n\to\infty} \frac{6}{n} = \lim_{n\to\infty} \frac{3}{n^2} = 0,$$

there exists an integer n_0 such that both

$$\frac{6}{n} < \frac{5}{4} \quad \text{and} \quad \frac{3}{n^2} < \frac{5}{4}$$

for all $n \geq n_0$. (Evidently, $n_0 = 5$ will suffice.) It follows that

$$-\frac{6}{n} > -\frac{5}{4} \quad \text{and} \quad -\frac{3}{n^2} > -\frac{5}{4}$$

so that

$$f(n) = 5n^2 - 6n + 3 \geq n^2(5 - \frac{5}{4} - \frac{5}{4}) = n^2(5 - \frac{5}{2}) = \frac{5}{2}n^2.$$

Thus $g(n) = n^2 \leq \frac{2}{5}f(n)$ so, with $c = \frac{2}{5}$ in Definition 8.2.1, we see that $g = \mathcal{O}(f)$. Thus, $f \asymp g$.   ∎

Here is the general result.

8.2.7 PROPOSITION ▶

A polynomial has the same order as its highest power; that is, if $f(n) = a_t n^t + a_{t-1}n^{t-1} + \cdots + a_0$ is a polynomial of degree t, then $f(n) \asymp n^t$.

Proof

This follows immediately from part (c) of Proposition 8.2.6, but we provide here a direct proof which does not require any knowledge of calculus. Our proof mimics the solution to Problem 14.

For any natural number n, the triangle inequality gives

$$|f(n)| \leq |a_t n^t| + |a_{t-1}n^{t-1}| + \cdots + |a_1 n| + |a_0|$$

$$= |a_t|n^t + |a_{t-1}|n^{t-1} + \cdots + |a_1|n + |a_0|$$

$$\leq |a_t|n^t + |a_{t-1}|n^t + \cdots + |a_1|n^t + |a_0|n^t$$

$$= (|a_t| + |a_{t-1}| + \cdots + |a_1| + |a_0|)n^t.$$

Thus, $|f(n)| \leq cn^t$ for any $n \geq n_0 = 1$ with $c = |a_t| + |a_{t-1}| + \cdots + |a_1| + |a_0|$. By Definition 8.2.1, we see that $f(n) = \mathcal{O}(n^t)$. Now we prove that $n^t = \mathcal{O}(f(n))$.

Since $f \asymp -f$ for any $f: \mathsf{N} \to \mathsf{R}$ (see Exercise 15 at the end of this section), we may assume that $a_t > 0$. We have

$$f(n) = n^t\left(a_t + \frac{a_{t-1}}{n} + \frac{a_{t-2}}{n^2} + \cdots + \frac{a_0}{n^t}\right)$$

$$\geq n^t\left(a_t - \frac{|a_{t-1}|}{n} - \frac{|a_{t-2}|}{n^2} \cdots - \frac{|a_0|}{n^t}\right).$$

As n gets large, the numbers

$$\frac{|a_{t-1}|}{n}, \quad \frac{|a_{t-2}|}{n^2}, \quad \cdots, \quad \frac{|a_0|}{n^t}$$

approach 0, so all of them are eventually smaller than the positive number $\frac{a_t}{2t}$. Thus, there is an n_0 such that for any $n \geq n_0$,

$$\frac{|a_{t-1}|}{n} < \frac{a_t}{2t}, \quad \frac{|a_{t-2}|}{n^2} < \frac{a_t}{2t}, \quad \cdots, \quad \frac{|a_0|}{n^t} < \frac{a_t}{2t}.$$

Therefore,

$$-\frac{|a_{t-1}|}{n} > -\frac{a_t}{2t}, \quad -\frac{|a_{t-2}|}{n^2} > -\frac{a_t}{2t}, \quad \cdots, \quad -\frac{|a_0|}{n^t} > -\frac{a_t}{2t}$$

and

$$f(n) \geq n^t \left(a_t - \frac{a_t}{2t} - \frac{a_t}{2t} - \cdots - \frac{a_t}{2t} \right) = n^t \left(a_t - t\frac{a_t}{2t} \right) = n^t \left(\frac{a_t}{2} \right).$$

Thus, $n^t \leq \frac{2}{a_t} f(n)$ for all $n \geq n_0$ and, with $c = \frac{2}{a_t}$ in Definition 8.2.1, we obtain $n^t = \mathcal{O}(f(n))$, as required. ∎

In the exercises, you are asked to verify that the relation *same order* defines an equivalence relation on the class of functions $\mathsf{N} \to \mathsf{R}$; that is, the relation is

reflexive: $f \asymp f$ for all f;
symmetric: if $f \asymp g$, then $g \asymp f$; and
transitive: if $f \asymp g$ and $g \asymp h$, then $f \asymp h$.

(See Section 2.4.)

EXAMPLE 15   Since $5n^2 - 6n + 3 \asymp n^2$ and $2n^2 - 3n + 6 \asymp n^2$, it follows that $5n^2 - 6n + 3$ and $2n^2 - 3n + 6$ have the same order, something which is, perhaps, not so easy to see directly from the definition of Big Oh. In fact, transitivity and Proposition 8.2.7 imply that any two polynomials of the same degree have the same order.   ▲

There are also some natural connections between the binary relations $\asymp$ and $\prec$. For example, if $f \prec g$ and $g \asymp h$, then $f \prec h$. We have seen that $n + 1 \prec n^2$ and $n^2 \asymp 5n^2 + 6n + 3$, so we may conclude that $n + 1 \prec 5n^2 + 6n + 3$.

Proposition 8.2.7 says that any polynomial has the same order as its highest power. The next proposition shows that different powers of n never have the same order.

8.2.8 PROPOSITION ▶   Suppose a and b are real numbers and $1 < a < b$. Then $n^a \prec n^b$.

Proof   Take $n_0 = 1$ and $c = 1$ in Definition 8.2.1 to obtain $n^a = \mathcal{O}(n^b)$. To see that the reverse is not true, suppose $n^b < cn^a$ for some constant c and all sufficiently large n. Dividing by n^a, we have $n^{b-a} < c$. The left-hand side is a positive power of n and, hence, for large n, it is bigger than c. Thus, n^b is not $\mathcal{O}(n^a)$.   ∎

Students of calculus should note that Proposition 8.2.8 follows directly from part (b) of Proposition 8.2.6 because

$$\lim_{n \to \infty} \frac{n^b}{n^a} = \infty.$$

As a particular case of Proposition 8.2.8, we see that $1 \prec n^a$ for any positive power n^a of n. Thus, we begin to see that the idea of order permits the definition of a hierarchy among functions. For instance, we now know that

$$1 \prec \sqrt{n} \prec n \prec n\sqrt{n} \prec n^2 \prec n^{2.01} \prec \cdots .$$

A straightforward induction argument shows that $n^2 < 2^n$ for $n \geq 5$ (see Exercise 9 of Section 5.1). It follows that $2^n \neq \mathcal{O}(n)$; otherwise, we would have $2^n < cn$ and, hence, $n^2 < 2^n < cn$ for some constant c and sufficiently large n.

This implies $n < c$ which, for large n, is not true. Thus, $2^n \neq \mathcal{O}(n)$. Since $n = \mathcal{O}(2^n)$, we conclude that $n \prec 2^n$.

In a similar fashion, we can prove that $n^3 < 2^n$ for $n \geq 10$ and, hence, that $n^2 \prec 2^n$. (See Exercise 14 at the end of this section.) In fact,

$$n^a \prec b^n$$

for any real numbers a and b with $b > 1$. This follows directly from Proposition 8.2.6 since

$$\lim_{n \to \infty} \frac{b^n}{n^a} = \infty$$

for any real constants a, b with $b > 1$. (For students familiar with l'Hôpital's Rule, this result will be elementary.)

Our examination of the relative orders of common functions continues.

8.2.9 PROPOSITION ▶ $\log_b n \prec n$ for any real number b, $b > 1$.

Proof We apply $\log_b$ to the inequality $n < b^n$, which is true for sufficiently large n, obtaining $\log_b n < n$. With $c = 1$ in Definition 8.2.1, we see that $\log_b n = \mathcal{O}(n)$. Now $n \neq \mathcal{O}(\log_b n)$ since $n \leq c \log_b n$ implies $b^n \leq n^c$ (applying the function $x \mapsto b^x$) which, for large n, is not true. Thus, $\log_b n \prec n$. ∎

In the Exercises, you are asked to show that any two logarithm functions (with bases larger than 1) have the same order; that is,

$$\log_a n \asymp \log_b n$$

for any $a, b > 1$. For this reason, we now frequently omit mention of the base of a logarithm because most assertions about order which involve logarithms are true for any base $b > 1$.

Propositions 8.2.8 and 8.2.9, together with the facts that

$$n! \prec n^n$$

(Problem 12) and

$$b^n \prec n!$$

(Exercise 13 at the end of this section), allow our hierarchy of functions to be extended considerably. For $a, b > 1$, we have

$$\boxed{1 \prec \log n \prec n \prec n^a \prec b^n \prec n! \prec n^n.}$$

Figure 8.4 shows the graphs of the natural logarithm and of the functions defined by $f(x) = x$, $f(x) = x^2$, and $f(x) = 3^x$ and provides dramatic evidence of the relative orders and rates of growth of these functions. In particular, note the very slow growth of the logarithm and the very rapid growth of the exponential.

How does all of this relate to efficiency of algorithms? Suppose $\mathcal{A}$ and $\mathcal{B}$ are algorithms which accomplish the same task. If the problem to which these algorithms is applied is "small," they may work so quickly on even a microcomputer that differences in performance are not noticeable. Few of us can detect the

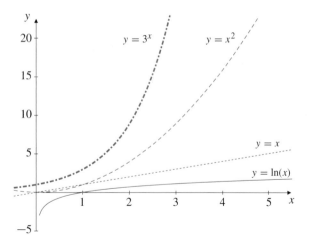

Figure 8.4   Some common complexity functions.

difference in a running time of 10^{-6} seconds and one that is 100 times as fast! As the problem grows in size, however, differences in running time become very significant. Suppose $\mathcal{A}$ has complexity $\mathcal{O}(n)$ while $\mathcal{B}$ is $\mathcal{O}(n^3)$. With $n = 5000$ and a computer which performs a million operations per second, algorithm $\mathcal{A}$ requires on the order of $5000 \times 10^{-6} = 0.005$ seconds while $\mathcal{B}$ requires about $5000^3 \times 10^{-6} = 125{,}000$ seconds, or about a day and a half! Some algorithms may involve so many operations that they are simply impractical to implement. We become very concerned, therefore, with the relative sizes of complexity functions when n is large. This is precisely what the concept of order is designed to measure.

If complexity functions f and g, which correspond respectively to algorithms $\mathcal{A}$ and $\mathcal{B}$, have the same order ($f \asymp g$), then $\mathcal{A}$ and $\mathcal{B}$ are considered to be very similar with respect to efficiency, whereas, if f has smaller order than g, then $\mathcal{A}$ is more efficient than $\mathcal{B}$. An algorithm whose complexity function has the same order as n is said to be *linear* or to run in *linear time*. An algorithm with complexity function the order of n^2 is called *quadratic*. More generally, algorithms with complexity functions of order n^a for some positive real number a are *polynomial* or said to run in *polynomial time*. Algorithms with complexity functions of order $\log n$ are *logarithmic* and algorithms with complexity of order a^n for some $a > 0$ are *exponential*.

We have seen that an algorithm which runs in logarithmic time is more efficient than a linear algorithm, which, in turn, is more efficient than a quadratic algorithm. Any polynomial algorithm is more efficient than one which is exponential.

Here a word of caution is in order. Saying that an exponential algorithm $\mathcal{A}$ is less efficient than a polynomial algorithm $\mathcal{B}$ which runs in polynomial time means only that when n is **sufficiently large**, algorithm $\mathcal{A}$ takes more time. For small values of n, the differences may be inconsequential; indeed, $\mathcal{A}$ may even be faster! For example, if $f(n) = 100n^3$ and $g(n) = 2^n$, then $f \prec g$, but we must have $n \geq 20$ for the polynomial algorithm to be faster. Similarly, $f(n) = n^2$ is "better" than $g(n) = \frac{n^3}{100}$ because, once n becomes sufficiently large, an algorithm

Table 8.1 Approximate time required to run an algorithm requiring $f(n)$ operations assuming a machine speed of 1 million operations per second.

$f(n)$ ⟍ n	10	100	5000
25	25×10^{-6} s	25×10^{-6} s	25×10^{-6} s
$\log_2 n$	3.3×10^{-6} s	6.6×10^{-6} s	1.2×10^{-5} s
n	10^{-5} s	10^{-4} s	5×10^{-3} s
$n \log_2 n$	3.3×10^{-5} s	6.6×10^{-6} s	6.1×10^{-2} s
n^2	10^{-4} s	.01 s	25 s
$2n^2 + 5n + 100$	3.5×10^{-4} s	.021 s	50 s
$n^3/100$	10^{-5} s	.01 s	21 min
2^n	10^{-3} s	4.0×10^{16} yr	4.5×10^{1491} yr

with the complexity function f will take less time. This is clearly not the case for small values of n, however.

These remarks are illustrated in Table 8.1, which compares a number of complexity functions $f(n)$. The entries, which are only approximate, show the time required to run an algorithm with $f(n)$ operations on a machine performing one million operations each second (like a snail). The functions are ordered in decreasing order of efficiency *for sufficiently large n* (except for n^2 and $2n^2 + 5n + 100$, which have the same order). From this table, we see that an algorithm with $f(n) = 25$ operations runs much more quickly than an algorithm with $f(n) = n$ or $f(n) = \frac{n^3}{100}$, once n is big enough. While for n as large as 100, an algorithm with $f(n) = \frac{n^3}{100}$ runs more quickly than one with $f(n) = 2n^2 + 5n + 100$, it is apparent that once $n = 5000$, the situation is reversed. Note how quickly an algorithm with $\log_2 n$ operations runs and, conversely, how slowly one with 2^n operations runs (for large n). (When viewing the numbers in the last line of the table, it is interesting to bear in mind that the universe has been around for about 7×10^9 years!) Algorithms with complexity functions which are $\mathcal{O}(\log_2 n)$ are highly efficient whereas exponential algorithms are so slow as to be useless from a practical point of view, once n is any reasonable size at all.

In recent years, following the initiative of Stephen Cook of the University of Toronto, computer scientists have identified a collection of problems for which no polynomial time algorithm is currently known. These problems, which are equivalent in the sense that the existence of a polynomial time algorithm for one of them would imply polynomial time algorithms for them all, are called *NP-complete*. The *NP* stands for nondeterministic polynomial time. Some of the

most interesting NP-complete problems are related to graphs and will be identified in later chapters. (See, for instance, the "Traveling Salesman's Problem" in
Chapter 10 or the problem of coloring the vertices of a graph, which is discussed
in Chapter 14.)

We conclude this section with a discussion of the complexity of some algorithms which will already be familiar to the reader.

PROBLEM 16. Show that the number of single-digit additions required in the
addition of m n-digit numbers does not exceed $m(2n + m)$. Conclude that the
complexity of such addition is $\mathcal{O}(n^2)$ if $m \le n$ and $\mathcal{O}(m^2)$ otherwise.

Solution. This addition problem has the following form.

$$
\left.
\begin{aligned}
r &= (r_{n-1}r_{n-2}\ldots r_1 r_0)_{10} \\
&\;\;\vdots \\
c &= (c_{n-1}c_{n-2}\ldots c_1 c_0)_{10} \\
b &= (b_{n-1}b_{n-2}\ldots b_1 b_0)_{10} \\
a &= \underline{(a_{n-1}a_{n-2}\ldots a_1 a_0)_{10}}
\end{aligned}
\right\} m \text{ numbers}
$$

We use repeatedly the result of Problem 7 that the addition of two n-digit numbers
requires at most $2n-1$ single-digit additions. Thus, the addition of a and b requires
at most $2n - 1$ additions. The sum is a number with at most $n + 1$ digits. The
addition of this number and c (which we think of as an $n + 1$-digit number
beginning with 0) requires another $2(n + 1) - 1 = 2n + 1$ operations, at most,
and gives a number with at most $n + 2$ digits. Adding the next number requires
at most $2(n + 2) - 1 = 2n + 3$ operations. Continuing this way, the final addition
with r requires at most $2[n + (m - 2)] - 1 = 2n + 2m - 5$ operations. Altogether,
the addition requires at most

$$(2n - 1) + (2n + 1) + (2n + 3) + \cdots + (2n + 2m - 5)$$

operations. This is the sum of an arithmetic sequence with $m - 1$ terms, first term
$a = 2n - 1$ and common difference $d = 2$, so it equals

$$\tfrac{m-1}{2}[2(2n - 1) + (m - 2)2] = (m - 1)(2n - 1) + (m - 2)(m - 1)$$

$$= (m - 1)(2n + m - 3)$$

$$\le m(2n + m).$$

[See formula (7) on p. 166.] If $m \le n$, the number of operations is at most
$n(2n + n) = n(3n) = 3n^2$, so the addition has complexity $\mathcal{O}(n^2)$. If $n \le m$,
$m(2n + m) \le m(2m + m) = 3m^2$, so the complexity is $\mathcal{O}(m^2)$. ∎

PROBLEM 17. Show that the usual method of multiplying two n-digit numbers
has complexity $\mathcal{O}(n^2)$. Count as equivalent operations the addition and multiplication of two single-digit numbers.[4]

---

[4]There are faster methods of multiplying. In 1962, Anatolii Karatsuba found a method which is $\mathcal{O}(n^{1.58})$.
By December 1988, the fastest known method was $\mathcal{O}(n(\log n)(\log \log n))$. No algorithm better than $\mathcal{O}(n)$

Solution. In the multiplication of two n-digit numbers by the usual method, there are n multiplications of an n-digit number by single-digit numbers. Each such multiplication requires at most $2n - 1$ operations. (See Exercise 2.) Then it is necessary to add n numbers, each of length at most $2n$. Using the result of Problem 16, this requires at most an additional $n[2(2n) + n] = n(5n)$ operations for a total not exceeding

$$n(2n - 1) + n(5n) \asymp n^2,$$

by Proposition 8.2.7. ∎

PROBLEM 18. One straightforward way to compute a positive integral power of a real number x is to compute $x(x) = x^2$, then $x(x^2) = x^3$, then $x(x^3) = x^4$, and so forth.

(a) Find the complexity of this algorithm in terms of the number of multiplications.

(b) Find an algorithm which is more efficient than the one described and indicate why your algorithm is more efficient.

Solution.

(a) This algorithm requires $n - 1$ multiplications.

(b) A seemingly more efficient way to compute x^n is to compute the powers $x, x^2, (x^2)^2 = x^4, (x^4)^2 = x^8$, and so on, stopping upon calculation of x^{2^k}, where $2^{k-1} < n \le 2^k$. This requires k multiplications. Now consider how the Russian peasant method computes the product $n = n \times 1$.[5] On the left, n is divided by 2 a total of k times, while on the right, 1 is multiplied by 2 a total of k times. The product n is determined as the sum of certain numbers on the right, that is, as the sum of at most k powers of 2. Then x^n is the product of the corresponding powers of x. For example,

$$37 = 1 + 2^2 + 2^5, \quad \text{so } x^{37} = x^1 x^{2^2} x^{2^5}$$

and generally,

$$n = \sum 2^{k_i}, \quad \text{so } x^n = x^{\Sigma 2^{k_i}} = x^{2^{k_1}} x^{2^{k_2}} \cdots$$

for which at most a further k multiplications are required. Thus, after at most $2k$ multiplications, we can compute x^n. Applying $\log_2$ to the inequality $2^{k-1} < n$, we have $k - 1 < \log_2 n$ and so $k < 1 + \log_2 n$. Thus, the number of multiplications required this time is $\mathcal{O}(\log n)$. Since $\log n \prec n$, the proposed algorithm is more efficient than the one described in (a). ∎

---

is possible. We refer the interested reader to "Ramanujan, Modular Equations, and Approximations to Pi or How to Compute One Billion Digits of Pi," by J. M. Borwein, P. B. Borwein, and D. H. Bailey, *American Mathematical Monthly* **96** (1989), 201–219, and to the bibliography of this article.

[5] The Russian peasant method can be very useful!

PROBLEM 19. Find a Big Oh estimate of the complexity of the Euclidean algorithm in terms of the number of divisions. (Given integers a and b, a "division" of a by b means the determination of q and r, $0 \le r < b$, such that $a = bq + r$.)

Solution. Given positive integers a and b with $b < a$, recall that the Euclidean algorithm finds the greatest common divisor of a and b by the following sequence of divisions,

$$
\begin{aligned}
a &= q_1 b + r_1, & r_1 &< b, \\
b &= q_2 r_1 + r_2, & r_2 &< r_1, \\
r_1 &= q_3 r_2 + r_3, & r_3 &< r_2, \\
&\ \ \vdots \\
r_{k-2} &= q_k r_{k-1} + r_k, & r_k &< r_{k-1}, \\
r_{k-1} &= q_{k+1} r_k,
\end{aligned}
$$

the process terminating with the first zero remainder r_{k+1}. The gcd is then r_k, the last nonzero remainder. (See Section 4.2.) The number of divisions and, hence, the number of basic operations is $k + 1$. How big is k? Note that $q_{k+1} > 1$ because $r_k < r_{k-1}$ and, since the other quotients q_i are positive integers, we have $q_i \ge 1$ for $1 \le i \le k$. Therefore,

$$
\begin{aligned}
r_k &\ge 1 \\
r_{k-1} &\ge 2r_k \ge 2 \\
r_{k-2} &\ge r_{k-1} + r_k \ge 3 \\
r_{k-3} &\ge r_{k-2} + r_{k-1} \ge 3 + 2 = 5 \\
&\ \ \vdots \\
r_1 &\ge r_2 + r_3 \\
b &\ge r_1 + r_2.
\end{aligned}
$$

Consider the two sequences

$$
\begin{array}{cccccccc}
1, & r_k, & r_{k-1}, & r_{k-2}, & r_{k-3}, & \ldots, & r_1, & b \\
1, & 1, & 2, & 3, & 5, & \ldots. &
\end{array}
$$

Each term in the first sequence is at least as big as the corresponding term in the second sequence. In particular, b is at least as large as the $(k + 2)$th term in the second sequence. The second sequence, however, is the Fibonacci sequence (see 5.2.3), whose kth term is the nearest integer to $\frac{1}{\sqrt{5}} \left(\frac{1+\sqrt{5}}{2} \right)^k$. Thus,

$$
b \ge \frac{1}{\sqrt{5}} \left(\frac{1+\sqrt{5}}{2} \right)^{k+2} - 1 = \frac{1}{\sqrt{5}} \left(\frac{1+\sqrt{5}}{2} \right)^2 \left(\frac{1+\sqrt{5}}{2} \right)^k - 1,
$$

from which it follows that $\left(\frac{1+\sqrt{5}}{2} \right)^k \le Ab + B$, where A and B are positive constants independent of b. Since $\frac{1+\sqrt{5}}{2} = C$ is also a constant and $C^k \le Ab + B$, an application of the logarithm (to any base) gives $k \log C \le \log(Ab + B)$. Since

$\log(An + B) \asymp \log n$ for $A, B > 0$ (see Exercise 22), it follows that the number of operations required is $\mathcal{O}(\log b)$. The Euclidean algorithm is very efficient.   ∎

Answers to Pauses   **2.** Take $n_0 = 1$ and $c = 1$.

EXERCISES

The symbol [BB] means that an answer can be found in the Back of the Book.

1. [BB] Consider the distance algorithm described in Example 6. Assume that addition and subtraction require the same amount of time, that addition is twice as fast as multiplication, and that multiplication is ten times as fast as the square root operation. Find the complexity function.

2. Show that the number of operations required to multiply an n-digit number by a single-digit number does not exceed $2n - 1$. Count the addition and the multiplication of two single-digit numbers as equivalent operations.

3. [BB] Show that the number of operations required to divide an n-digit number by a single-digit number does not exceed $3n$. Count as equivalent the following operations:

- the multiplication of two single-digit numbers,
- the division of a single-digit number or a two-digit number by a single-digit number, provided the quotient has only a single digit (e.g., $37 \div 9$ but not $37 \div 3$), and
- the subtraction of two two-digit numbers provided the difference has only a single digit (e.g., $37 - 32$, but not $37 - 22$).

4. Consider the problem of finding the minimum number in a set of n real numbers. Find an algorithm for this and find its complexity function. Take *comparison* in the form "$x < y$?" as the basic operation.

5. Let x be a real number and n a positive integer. Consider the following two algorithms for calculating x^{2^n}.

ALGORITHM $\mathcal{A}$:

Step 1. Set $a = 1$.
Step 2. For $i = 1$ to 2^n, replace a by xa.
Step 3. Output a.

ALGORITHM $\mathcal{B}$:

Step 1. Set $a = x$.
Step 2. For $i = 1$ to n, replace a by a^2.
Step 3. Output a.

Find complexity functions for each of these algorithms and explain why $\mathcal{B}$ is more efficient. Assume that multiplication is the basic operation.

6. (a) [BB] Justify the statement made in Section 8.1 that Horner's method of polynomial evaluation requires fewer arithmetic operations than the more obvious method described in Problem 2. Assume that powers of x cannot be stored.

(b) Repeat part (a) assuming it is possible to store powers of x.

7. Use Definition 8.2.1 to show that $f = \mathcal{O}(g)$ in each of the following cases of functions $f, g: \mathsf{N} \to \mathsf{R}$.

(a) [BB] $f(n) = 5n$, $g(n) = n^3$
(b) $f(n) = 17n^4 + 8n^3 + 5n^2 + 6n + 1$, $g(n) = n^4$
(c) [BB] $f(n) = 8n^3 + 4n^2 + 5n + 1$, $g(n) = 3n^4 + 6n^2 + 8n + 2$
(d) $f(n) = 2^n$, $g(n) = 3^n$
(e) $f(n) = 2n^2 + 3n + 1$, $g(n) = n^2 + 1$
(f) $f(n) = 2n^2 - 3n + 5$, $g(n) = n^2 - 7n$
(g) $f(n) = \log_2 n^5$, $g(n) = n$

8. [BB; (a), (c)] For each part of Exercise 7 decide whether $f \prec g$ or $f \asymp g$. Use Propositions 8.2.7 and 8.2.8, if appropriate.

9. If $f, g, h: \mathsf{N} \to \mathsf{R}$, $f = \mathcal{O}(h)$ and $g = \mathcal{O}(h)$, show that $f + g = \mathcal{O}(h)$.

10. (a) Show that $\mathcal{O}$ defines a transitive operation on the class of functions $\mathsf{N} \to \mathsf{R}$.

(b) [BB] Prove that $\prec$ defines a transitive relation on the class of functions $\mathsf{N} \to \mathsf{R}$.

11. [BB] Let f, g and h be functions $N \to R$ and suppose that $f \asymp g$ and $g \prec h$. Prove that $f \prec h$.

12. [BB] Show that $a^n \prec b^n$ if $0 < a < b$.

13. (a) [BB] Show that $2^n \prec n!$.
  (b) Show that $b^n \prec n!$ for any $b > 1$.

14. (a) Prove that $n^3 < 2^n$ for all $n \geq 10$.
  (b) Use the result of part (a) to show that $n^2 \prec 2^n$.

15. [BB] Let $f : N \to R$ be a function and $0 \neq k \in R$. Show that $kf \asymp f$.

16. Let A be the set of all functions $N \to R$ and, for $f, g \in A$, define $f \prec g$ as in Definition 8.2.4.

  (a) Does $\prec$ define a partial order on A?
  (b) For $f, g \in A$, define $f \preceq g$ if $f \prec g$ or $f = g$. Is $(A, \preceq)$ a partially ordered set?

17. Show that $\asymp$, as defined in Definitions 8.2.4, defines an equivalence relation on the class of functions $N \to R$.

18. [BB] Use any of the results of this section to show that $n \log_2 n \prec n^2$.

19. Find a function in the list

$$n^2, 1, n^3, n \log n, \log n, n^4, 2^n, n^n, n!, n, \pi^n, n^5$$

which has the same order as

  (a) $f(n) = 3n^4 + 6n^2 - 8n - 5$
  (b) [BB] $f(n) = n \log n + n^2$
  (c) $f(n) = 5$
  (d) $f(n) = 2^n + 4n^n$
  (e) $f(n) = 3n! - 17n^4$
  (f) $f(n) = n + \log n$

Justify your answers appealing to any of the results in this section that prove helpful.

20. Repeat Exercise 19 for each of the following functions:

  (a) [BB] $f(n) = 48n^5 - 7n^4 + 6n^3 + 5n^2 - 2n - 7$
  (b) $f(n) = 2^n + 375n^{1990}$
  (c) $f(n) = \pi^n + n^\pi$

  (d) $f(n) = 3n + 5n \log n + 7n^2$
  (e) $f(n) = 2 + 4 + 6 + \cdots + 2n$
  (f) $f(n) = 2 + 4 + 8 + \cdots + 2^n$
  (g) $f(n) = \dfrac{(5n + 7)(n^2 \log_2 n)}{n^2 + 4n}$

21. [BB] Show that $\log_a n \asymp \log_b n$ for any real numbers $a, b > 1$.

22. Suppose A and B are positive real numbers. Show that $\log(An + B)$ and $\log n$ have the same order. (Use $\log_2$, for instance.)

23. [BB] Show that $\log n! = \mathcal{O}(n \log n)$.

24. (a) Find a Big Oh estimate of the complexity of the algorithm for writing an integer a in base 2 in terms of the number of divisions. (See Section 4.1.) As in Problem 19, by a "division", we mean the determination of q and r such that a given integer $b = 2q + r$.
  (b) Suppose a division counts as three basic operations. How does this affect your answer to (a)?

25. The Russian peasant method is used to multiply two n-digit numbers.

  (a) Find a reasonable upper bound for the number of rows required by this process, that is, for the number of multiplications and divisions by 2.
  (b) Find a reasonable upper bound for the number of basic operations required in the multiplying and dividing by 2.
  (c) Find a reasonable upper bound for the number of basic operations required by the final addition.
  (d) Show that the number of basic operations in all is $\mathcal{O}(n^2)$.

Count as basic operations

  • the addition of two single-digit numbers,
  • the multiplication of two single-digit numbers, and
  • the division of a 2-digit number by a single digit number.

26. (For students of calculus) Prove Proposition 8.2.6.

8.3 SEARCHING AND SORTING

Perhaps the two most important tasks required of modern computers are the searching and sorting of lists. Since they are required by so many other processes and often must be repeated many times, efficient algorithms to search and sort lists are of supreme importance.

A car manufacturer, realizing that a certain part is a potential safety hazard, needs to search the records of all cars sold in recent years for those which contain this part. A university wishes to implement a system whereby, each semester, students will register in order of decreasing grade point average (those with the highest grade point average register first). In order to determine the numbers of students with grade point averages in various ranges, student records will have to be searched.

How often is it required to sort a list alphabetically? Surely this is a key step in many tasks, from the creation of telephone directories to dictionaries. Although computers can work very quickly, given a large input and an inefficient algorithm, they can be slowed to a crawl easily.

In this section, we discuss several algorithms for searching and sorting and discuss the complexity of each. Our approach to complexity will again always be worst case. While we would hope that a sorting algorithm would not take long to alphabetize a list of names input already in alphabetical order, our concern is the length of time such an algorithm could take when the input is as disorganized as possible.

We begin with an algorithm which searches a list for one particular entry.

8.3.1 A LINEAR SEARCH ALGORITHM ▶

To search a list $a_1, a_2, \ldots, a_n$ for the element x,

> for $i = 1$ to n
> > if $x = a_i$, output "true" and set $i = 2n$;
> if $i \neq 2n$, output "false".

The algorithm proceeds in an obvious way. Setting $i = 2n$ is a little trick which stops the loop as soon as x has been found and, if x is not in the list, ensures that "false" is output at the end.

EXAMPLE 20   For $x = -2$ and an input list

$$
\begin{array}{cccc}
6 & 0 & -2 & 1 \\
a_1 & a_2 & a_3 & a_4
\end{array}
$$

the algorithm sets $i = 1$, determines that $x \neq a_1 = 6$, and sets $i = 2$. Since $x \neq a_2 = 0$, it sets $i = 3$. Since $x = a_3 = -2$, it outputs "true" and sets $i = 2n = 8$. Since i is no longer in the range from 1 to n, the loop stops. Since $i = 2n$, the algorithm knows that x was found and does not output "false".

On the other hand, for $x = 2$ and the same input the algorithm determines that $x \neq a_1 = 6$, $x \neq a_2 = 0$, $x \neq a_3 = -2$, and $x \neq a_4 = 1$. At the final step, $i = 4 \neq 2n$, so the algorithm outputs "false."   ▲

In the worst of circumstances, when the element x is not in the list a_1, $a_2, \ldots, a_n$, the loop is executed n times and the algorithm requires $n + 1$ comparisons. In most searching and sorting algorithms, there are few arithmetic calculations. The best basic operation with which to measure efficiency is a *comparison*, that is, a statement of the form $x = a$, $x \leq a$, $x < a$, $x \geq a$, or $x > a$. This linear search algorithm requires at most $n + 1$ comparisons, so it is $\mathcal{O}(n)$.

8.3.2 REMARK ▶

Big Oh notation is sufficiently imprecise that we can be relatively careless when determining the complexity of an algorithm. A more thoughtful approach to the linear search algorithm might have led us to count two comparisons for each value of i, for after checking if $x = a_i$, the algorithm must also check to see if the end of the list has been reached, that is, if $i = n$. Such analysis would lead us to complexity $2n$ rather than n, but, as we know, $n \asymp 2n$, so the Big Oh estimate of complexity remains $\mathcal{O}(n)$.

It is often the case that a list to be searched is in some natural order. When searching a dictionary for the meaning of *obfuscate*, for instance, it is helpful to know which way to turn if the dictionary is open at *trisect*. When a list is already in order, it stands to reason that there should be a procedure for searching which is more efficient than a linear search. One commonly used such procedure is called *binary search*. The Binary Search Algorithm determines which half of an ordered list would have to contain the sought-after element, then which half of that half, and so on. At each stage, the list to be searched is one half the length of the previous list.

8.3.3 A BINARY SEARCH ALGORITHM ▶

To search for an element x in an ordered list $a_1 \le a_2 \le \cdots \le a_n$, proceed as follows.

> while $n > 0$
>> if $n = 1$ then
>>> if $x = a_1$ output "true" and set $n = 0$;
>>> else output "false" and set $n = 0$;
>> else
>>> set $m = \lfloor \frac{n}{2} \rfloor$;
>>> if $x = a_m$, output "true" and set $n = 0$;
>>> else if $x < a_m$
>>>> set $n = m$ and replace the current list with $a_1, \ldots, a_m$;
>>>> else replace n with $n - m$ and the current list with $a_{m+1}, \ldots, a_n$.
> end while

This is the first time we have encountered the "while/end while" construct, which works pretty much as we would imagine. In this particular case, as long as there is a list to examine ("while $n > 0$"), the algorithm decides in which half of the list the sought-for element may reside and replaces the current list with this half.

EXAMPLE 21   With $x = 12$ and the input

$$
\begin{array}{cccccccccc}
5 & 6 & 7 & 10 & 11 & 12 & 15 & 17 & 19 & 20 \\
a_1 & a_2 & a_3 & a_4 & a_5 & a_6 & a_7 & a_8 & a_9 & a_{10}
\end{array}
$$

of length $n = 10 > 0$, the algorithm sets $m = \lfloor \frac{10}{2} \rfloor = 5$. Since $x = a_m = 11$ and $x < a_m$ are both false, the algorithm replaces n by $n - m = 5$ and replaces the original list with

$$
\begin{array}{ccccc}
12 & 15 & 17 & 19 & 20 \\
a_1 & a_2 & a_3 & a_4 & a_5.
\end{array}
$$

The integer n is still positive, so the algorithm executes the statements inside the "while ... end while" again. Since $n \ne 1$, it sets $m = \lfloor \frac{5}{2} \rfloor = 2$ and, because

$x < a_2 = 15$, it sets $n = 2$ and replaces the current list with

$$\begin{array}{cc} 12 & 15 \\ a_1 & a_2. \end{array}$$

Now n is still positive, so the algorithm executes the statements between "while" and "end while" again. Since $n \neq 1$, the algorithm sets $m = \lfloor \frac{2}{2} \rfloor = 1$. Since $x = a_1 = 12$, the algorithm outputs "true" and sets $n = 0$. The "while" statement is no longer true, so the algorithm terminates. ▲

To determine the complexity of binary search, first suppose that $n = 2^k$ is a power of 2. The algorithm proceeds by progressively halving the length of the list; hence, it terminates after executing the statements between "while" and "end while", in the worst case, a total of k times. This involves $2k + 2$ comparisons, including the two required when $n = 1$. If n is not a power of 2, find k such that $2^{k-1} < n \leq 2^k$ and increase the length of the input list by adding terms all equal to a_n until the extended list has length 2^k. This list, and hence the original, is searched after $2k + 2$ comparisons. Since $2^{k-1} < n$, we have $k - 1 < \log_2 n$, so $k < 1 + \log_2 n$ and $2k + 2 < 4 + 2\log_2 n \asymp \log_2 n$. The binary search algorithm is $\mathcal{O}(\log_2 n)$.

We now turn our attention to sorting algorithms. A *sorting algorithm* will put a list of numbers into increasing order or a list of words into alphabetical order. We begin with perhaps the simplest of all sorting algorithms, the *bubble sort*. The Bubble Sort Algorithm makes a number of passes through a list, each time interchanging any consecutive pair of elements which are not in proper order. On the first pass, the largest element is shuffled to the end. On the next pass, the next largest element is shuffled to second from the end. In general, after the kth pass the list looks like this:

$$\underbrace{a_1, a_2, \ldots, a_{n-k}}_{\text{unsorted}} < \underbrace{a_{n-k+1} < \cdots < a_{n-1} < a_n}_{\text{sorted}}$$

Gradually, the largest elements "bubble" to the end in order.

EXAMPLE 22 Suppose the input list is $6, 3, 1, 4, 9, 2$. Initially, the first two numbers are compared and (in this case) interchanged, giving the list $3, 6, 1, 4, 9, 2$. Then the second and third elements are compared and, in this case, interchanged, giving $3, 1, 6, 4, 9, 2$. Next, the third and fourth elements are compared and switched, giving $3, 1, 4, 6, 9, 2$. No interchange of the fourth and fifth numbers is required. The first pass is completed with an interchange of the fifth and last numbers, giving $3, 1, 4, 6, 2, 9$. Notice that the largest number, 9, has bubbled to the end.

The second pass through the list is like the first, but with one exception. Since the largest element is now at the end, it is not necessary to compare the second last element with the last. Here are the four steps of the second pass in our example. The numbers being compared at each stage are underlined.

$$\underline{3, 1}, 4, 6, 2, 9 \to 1, \underline{3, 4}, 6, 2, 9 \to 1, 3, \underline{4, 6}, 2, 9$$

$$\to 1, 3, 4, \underline{6, 2}, 9 \to 1, 3, 4, 2, 6, 9$$

Notice that the two largest numbers, 6 and 9, are now at the end of the list in correct order. The third pass is like the second, but one step shorter since it is not necessary to compare the fourth and fifth elements. Here are the three steps of the third pass.

$$\underline{1, 3}, 4, 2, 6, 9 \rightarrow 1, \underline{3, 4}, 2, 6, 9 \rightarrow 1, 3, \underline{4, 2}, 6, 9 \rightarrow 1, 3, 2, 4, 6, 9.$$

The fourth pass requires two comparisons,

$$\underline{1, 3}, 2, 4, 6, 9 \rightarrow 1, \underline{3, 2}, 4, 6, 9 \rightarrow 1, 2, 3, 4, 6, 9$$

and the fifth pass just one,

$$\underline{1, 2}, 3, 4, 6, 9 \rightarrow 1, 2, 3, 4, 6, 9. \qquad \blacktriangle$$

In the general description of the Bubble Sort Algorithm which follows, we use the term *swap* to denote the exchange of values of two variables. For instance, if $a = 4$ and $b = 2$ and we swap a and b, we mean that henceforth, $a = 2$ and $b = 4$. In the description of any sorting algorithm, we always assume that the elements to be sorted come from a totally ordered set, for example, the real numbers or the words of a language (with lexicographic ordering).

8.3.4 A BUBBLE SORT ALGORITHM ▶

To sort n elements $a_1, a_2, \ldots, a_n$ from least to greatest,
  for $i = n - 1$ down to 1,
    for $j = 1$ to i
      if $a_j > a_{j+1}$, swap a_j and a_{j+1}.

The algorithm consists of a loop inside a loop. In the "outer" loop, we set $i = n - 1$ and then execute the "inner" loop from $j = 1$ to $j = n - 1$. Then we drop i to $n - 2$ and execute the inner loop from $j = 1$ to $j = n - 2$, and so on until $i = 1$, at which point the inner loop is executed just once, with $j = 1$. The Bubble Sort Algorithm is popular with beginning programmers because it is short and simple to implement in any language. Let us make sure we understand it.

EXAMPLE 23 For the input list

$$\begin{array}{cccc} 4 & 0 & 3 & 2 \\ a_1 & a_2 & a_3 & a_4 \end{array}$$

we have $n = 4$. The algorithm sets $i = n - 1 = 3$ and runs the inner loop from $j = 1$ to 3. With $j = 1$, we have $a_1 > a_2$, so a_1 and a_2 are swapped, giving the list

$$\begin{array}{cccc} 0 & 4 & 3 & 2 \\ a_1 & a_2 & a_3 & a_4. \end{array}$$

With $j = 2$, we have $a_2 > a_3$, so a_2 and a_3 are swapped, giving

$$\begin{array}{cccc} 0 & 3 & 4 & 2 \\ a_1 & a_2 & a_3 & a_4. \end{array}$$

With $j = 3$, we have $a_3 > a_4$, so a_3 and a_4 are swapped, giving

$$\begin{array}{cccc} 0 & 3 & 2 & 4 \\ a_1 & a_2 & a_3 & a_4. \end{array}$$

Having completed the inner loop, the algorithm returns to the outer, drops i to $i = 2$, and executes the inner loop from $j = 1$ to $j = 2$. With $j = 1$, the inequality $a_1 > a_2$ is not true, so there is no swap. With $j = 2$, we have $a_2 > a_3$, so a_2 and a_3 are swapped, giving

$$\begin{array}{cccc} 0 & 2 & 3 & 4 \\ a_1 & a_2 & a_3 & a_4. \end{array}$$

Again, the algorithm returns to the outer loop, drops i to $i = 1$ and executes the inner loop just for $j = 1$. Since $a_1 > a_2$ is false, there is no swap. The algorithm now terminates.   ▲

What is the complexity of a bubble sort? With $i = n - 1$, the inner loop is executed $n - 1$ times and there are $n - 1$ comparisons. With $i = n - 2$, there are $n - 2$ comparisons, and so on. With $i = 1$, just one comparison is required. In all, the number of comparisons is

$$\sum_{i=1}^{n-1}(n - i) = \sum_{i=1}^{n-1} i = \frac{n(n - 1)}{2}$$

so that this algorithm is $\mathcal{O}(n^2)$. Note that this estimate of the complexity of the bubble sort is best possible: In any bubble sort, all $\frac{1}{2}n(n - 1)$ comparisons must be made before it is completed.

Many sorting algorithms do better than the bubble sort. On the other hand, it is known that no sorting algorithm has worst case complexity better than $\mathcal{O}(n \log n)$. With a view toward an eventual discussion of one $n \log n$ algorithm, we first introduce an algorithm for merging two sorted lists, an important task in its own right. Instructors with numerous examination booklets to put into alphabetical order have long since learned that speed can be improved by first dividing the booklets into smaller alphabetized groups and then merging the groups.

Our merging algorithm takes two ordered lists and produces a third ordered list comprised of the elements in the two given lists, in order.

EXAMPLE 24   Here is how our Merging Algorithm would sort the ordered lists

$$1, 3, 5, 7, 10, 11 \quad \text{and} \quad 2, 5, 6, 9.$$

We compare the first elements of each list. Since $1 \le 2$, we start our third list with 1, removing 1 from the first list. This leaves us with

$$3, 5, 7, 10, 11 \quad 2, 5, 6, 9 \quad \text{and} \quad 1.$$

We compare the first elements of the first two lists. Since $2 \le 3$, we place 2 after 1 in the third list and remove 2 from the second list. At this stage, we have the three lists

$$3, 5, 7, 10, 11 \quad 5, 6, 9 \quad \text{and} \quad 1, 2.$$

We continue to compare the first elements in the first two lists, appending the smaller to the third list while removing it from the list from which it came. If the first element in the first list equals the first element in the second, we append that element from the first list to the third (and remove it from the first). Here are the next six steps in our merge.

$$
\begin{array}{lll}
5, 7, 10, 11 & 5, 6, 9 & 1, 2, 3 \\
7, 10, 11 & 5, 6, 9 & 1, 2, 3, 5 \\
7, 10, 11 & 6, 9 & 1, 2, 3, 5, 5 \\
7, 10, 11 & 9 & 1, 2, 3, 5, 5, 6 \\
10, 11 & 9 & 1, 2, 3, 5, 5, 6, 7 \\
10, 11 & & 1, 2, 3, 5, 5, 6, 7, 9
\end{array}
$$

At this stage the second list is empty, so we simply append the remaining elements 10, 11 of the first list to the third, giving the final list 1, 2, 3, 5, 5, 6, 7, 9, 10, 11. ▲

8.3.5 A MERGING ALGORITHM ▶

To merge two given sorted lists

$$
\mathcal{L}_1 : a_1 \le a_2 \le \cdots \le a_s \quad \text{and} \quad \mathcal{L}_2 : b_1 \le b_2 \le \cdots \le b_t,
$$

of lengths s and t, into a single sorted list

$$
\mathcal{L}_3 : c_1 \le c_2 \le \cdots \le c_{s+t}
$$

of length $s + t$, proceed as follows.

Step 1. Set $\mathcal{L}_3$ equal to the empty list.

Step 2. If $\mathcal{L}_1$ is empty, set $\mathcal{L}_3 = \mathcal{L}_2$ and stop. If $\mathcal{L}_2$ is empty, set $\mathcal{L}_3 = \mathcal{L}_1$ and stop.

Step 3. Suppose $a_1 \le b_1$. Then append a_1 to $\mathcal{L}_3$ and, if this empties $\mathcal{L}_1$, append the remaining elements of $\mathcal{L}_2$ to $\mathcal{L}_3$ and stop. If $r > 0$ elements remain in $\mathcal{L}_1$, label them $a_1, a_2, \ldots, a_r$ in increasing order and repeat Step 3.

  Suppose $a_1 > b_1$. Then append b_1 to $\mathcal{L}_3$ and, if this empties $\mathcal{L}_2$, append the remaining elements of $\mathcal{L}_1$ to $\mathcal{L}_3$ and stop. If $r > 0$ elements remain in $\mathcal{L}_2$, label them $b_1, b_2, \ldots, b_r$ in increasing order and repeat Step 3.

EXAMPLE 25   We apply the merging algorithm to the lists

$$
\mathcal{L}_1 : \begin{array}{ccc} 3 & 5 & 8 \\ a_1 & a_2 & a_3 \end{array} \quad \text{and} \quad \mathcal{L}_2 : \begin{array}{ccc} 1 & 7 & 8 \\ b_1 & b_2 & b_3 \end{array}.
$$

Since neither list is empty, we proceed directly to Step 3. Since $3 = a_1 > b_1 = 1$, we append b_1 to the list $\mathcal{L}_3$, which was initially empty. Then we relabel the remaining elements 7 and 8 of $\mathcal{L}_2$ as b_1, b_2 respectively so that our lists are

$$
\mathcal{L}_1 : \begin{array}{ccc} 3 & 5 & 8 \\ a_1 & a_2 & a_3 \end{array} \quad \mathcal{L}_2 : \begin{array}{cc} 7 & 8 \\ b_1 & b_2 \end{array} \quad \text{and} \quad \mathcal{L}_3 : \begin{array}{c} 1 \\ c_1 \end{array}.
$$

We then repeat Step 3. Since $a_1 \le b_1$, we append $a_1 = 3$ to $\mathcal{L}_3$ and relabel the remaining elements 5 and 8 of $\mathcal{L}_1$ as a_1 and a_2, respectively. Our lists are now

$$
\mathcal{L}_1 : \begin{array}{cc} 5 & 8 \\ a_1 & a_2 \end{array} \quad \mathcal{L}_2 : \begin{array}{cc} 7 & 8 \\ b_1 & b_2 \end{array} \quad \text{and} \quad \mathcal{L}_3 : \begin{array}{cc} 1 & 3 \\ c_1 & c_2 \end{array}.
$$

We repeat Step 3. Since $a_1 \leq b_1$, we append $a_1 = 5$ to $\mathcal{L}_3$ and relabel the elements of $\mathcal{L}_1$, obtaining

$$\mathcal{L}_1: \begin{array}{c} 8 \\ a_1 \end{array} \qquad \mathcal{L}_2: \begin{array}{cc} 7 & 8 \\ b_1 & b_2 \end{array} \qquad \text{and} \qquad \mathcal{L}_3: \begin{array}{ccc} 1 & 3 & 5 \\ c_1 & c_2 & c_3. \end{array}$$

After the next repetition of Step 3 we have

$$\mathcal{L}_1: \begin{array}{c} 8 \\ a_1 \end{array} \qquad \mathcal{L}_2: \begin{array}{c} 8 \\ b_1 \end{array} \qquad \text{and} \qquad \mathcal{L}_3: \begin{array}{cccc} 1 & 3 & 5 & 7 \\ c_1 & c_2 & c_3 & c_4. \end{array}$$

Now, since $8 = a_1 \leq b_1 = 8$, we append a_1 to $\mathcal{L}_3$, giving

$$\mathcal{L}_1: \qquad \mathcal{L}_2: \begin{array}{c} 8 \\ b_1 \end{array} \qquad \text{and} \qquad \mathcal{L}_3: \begin{array}{ccccc} 1 & 3 & 5 & 7 & 8 \\ c_1 & c_2 & c_3 & c_4 & c_5. \end{array}$$

Since $\mathcal{L}_1$ is now empty, we append 8, the remaining element of $\mathcal{L}_2$, to $\mathcal{L}_3$, producing the final merged list:

$$\mathcal{L}_3: \begin{array}{cccccc} 1 & 3 & 5 & 7 & 8 & 8 \\ c_1 & c_2 & c_3 & c_4 & c_5 & c_6. \end{array}$$
▲

How many comparisons does the Merging Algorithm require to merge ordered lists $\mathcal{L}_1$ and $\mathcal{L}_2$ of lengths s and t into a single ordered list $\mathcal{L}_3$ of length $s + t$? Notice that while elements remain in $\mathcal{L}_1$ and $\mathcal{L}_2$, each element goes into $\mathcal{L}_3$ after one comparison. Eventually one list is empty and at least one element remains in the other; such remaining elements go to the end of $\mathcal{L}_3$ with no comparisons. It follows that, in all, at most $s + t - 1$ comparisons are needed.

PROBLEM 26. Find ordered lists of total length $s + t = 7$ which require exactly $7 - 1 = 6$ comparisons to merge.

Solution. The maximum number of comparisons, $s + t - 1$, occurs when just one number remains in one of the lists when the other is emptied. Here is such a situation.

$\mathcal{L}_1$	$\mathcal{L}_2$	$\mathcal{L}_3$	Comparisons Needed
1,3,5,7	2,4,6		
3,5,7	2,4,6	1	1
3,5,7	4,6	1,2	1
5,7	4,6	1,2,3	1
5,7	6	1,2,3,4	1
7	6	1,2,3,4,5	1
7		1,2,3,4,5,6	1
		1,2,3,4,5,6,7	0
		Total	6

∎

Pause 3 Find ordered lists of total length 7 which require just three comparisons to merge.

We are now able to describe an algorithm for sorting a list which is not only more efficient than the bubble sort, but, in fact, achieves the theoretical best possible worst case complexity. Here is the idea.

EXAMPLE 27   Suppose we are given the list

$$7, 11, 5, 9, 11, 4, 10, 15, 17, 3, 9, 6, 21, 1$$

of length 14. We group the elements into seven pairs and order each pair. This gives us seven pairs of ordered lists.

$$7, 11; \quad 5, 9; \quad 4, 11; \quad 10, 15; \quad 3, 17; \quad 6, 9; \quad 1, 21.$$

Next, we merge the first two ordered lists, merge the third and fourth ordered lists, and merge the fifth and sixth. The last ordered list—1, 21—is unchanged.

$$5, 7, 9, 11; \quad 4, 10, 11, 15; \quad 3, 6, 9, 17; \quad 1, 21.$$

At this point, we have four ordered lists. We merge the first two and the last two, giving

$$4, 5, 7, 9, 10, 11, 11, 15; \qquad 1, 3, 6, 9, 17, 21.$$

Finally, we merge the two lists that remain, obtaining the final ordered list

$$1, 3, 4, 5, 6, 7, 9, 9, 10, 11, 11, 15, 17, 21. \qquad \blacktriangle$$

In this example, it is useful to think of the given list as 14 ordered lists, each of length 1. After the first step, we were left with seven ordered lists of length 2; after the second, with four ordered lists; after the third, with two ordered lists, and after the fourth, with the desired single ordered list. Since the number of ordered lists is essentially halved at each stage, in general k steps will be needed to sort a list of n elements, where $2^{k-1} < n \le 2^k$. For example, $2^3 < 14 \le 2^4$, so four steps are required to sort a list of length 14.

Here is a general description of our *Merge Sort Algorithm*.

8.3.6   A MERGE SORT ALGORITHM ▶

To sort a list $a_1, a_2, \ldots, a_n$ into increasing order, proceed as follows.

Step 1. set $F = 0$.

Step 2. for $i = 1$ to n, let the list $\mathcal{L}_i$ be the single element a_i.

Step 3.   while $F = 0$
        if $n = 1$, set $F = 1$ and output $\mathcal{L}_1$;
        if $n = 2m$ is even
           for $i = 1$ to m
               • merge the sorted lists $\mathcal{L}_{2i-1}$ and $\mathcal{L}_{2i}$
                 and label the resulting sorted list $\mathcal{L}_i$;
               • set $n = m$.
        if $n = 2m + 1 > 1$ is odd
           for $i = 1$ to m
               • merge the sorted lists $\mathcal{L}_{2i-1}$ and $\mathcal{L}_{2i}$
                 and label the resulting sorted list $\mathcal{L}_i$;
               • change the label of the former list $\mathcal{L}_n$ to $\mathcal{L}_{m+1}$;
               • set $n = m + 1$.
     end while

The variable F in our algorithm is called a *flag*, a useful parameter to introduce in an algorithm to enable fast termination of a loop. Here, it enables us to get out of the "while" loop as soon as the merging is complete.

PROBLEM 28. What would the merge sort algorithm do with the list

$$\begin{array}{ccccccc} 2 & 9 & 1 & 4 & 6 & 5 & 3 \\ a_1 & a_2 & a_3 & a_4 & a_5 & a_6 & a_7 \end{array} ?$$

Solution. Initially $F = 0$ and lists $\mathcal{L}_1, \mathcal{L}_2, \ldots, \mathcal{L}_7$ are defined, each of length 1.

$$\mathcal{L}_1: 2 \quad \mathcal{L}_2: 9 \quad \mathcal{L}_3: 1 \quad \mathcal{L}_4: 4 \quad \mathcal{L}_5: 6 \quad \mathcal{L}_6: 5 \quad \mathcal{L}_7: 3.$$

At Step 3, since $F = 0$ and $n = 2(3) + 1$ is odd ($m = 3$), we form four new lists $\mathcal{L}_1, \mathcal{L}_2, \mathcal{L}_3, \mathcal{L}_4$ by merging the first six former lists in pairs into three and adding the seventh. We now have

$$\mathcal{L}_1: 2, 9 \quad \mathcal{L}_2: 1, 4 \quad \mathcal{L}_3: 5, 6 \quad \mathcal{L}_4: 3.$$

At this point, n is replaced by $m + 1 = 3 + 1 = 4$ and we repeat Step 3. Since $n = 2(2)$ is even ($m = 2$), we form two lists $\mathcal{L}_1$ and $\mathcal{L}_2$ by merging the former lists $\mathcal{L}_1, \mathcal{L}_2$ and $\mathcal{L}_3, \mathcal{L}_4$, respectively. At this point we have

$$\mathcal{L}_1: 1, 2, 4, 9 \quad \mathcal{L}_2: 3, 5, 6.$$

Now n is replaced by $m = 2$ and we repeat Step 3. Since $n = 2(1)$ is even ($m = 1$), the list $\mathcal{L}_1$ is formed by merging the former $\mathcal{L}_1$ and $\mathcal{L}_2$.

$$\mathcal{L}_1: 1, 2, 3, 4, 5, 6, 9$$

At this point, n is replaced by $m = 1$ and we repeat Step 3. Since $n = 1$, we set $F = 1$, output $\mathcal{L}_1$, the desired sorted list, and stop. ∎

What is the complexity of the Merge Sort Algorithm in terms of the number of comparisons which are required? As with binary search, it is helpful first to consider the case that the given list has length $n = 2^k$ because then Step 3 is executed exactly k times. Also, we enter Step 3 each time with an even number of lists and exit with an even number, until the last step.

Let us first ignore the at most three comparisons required to decide if $n = 1$, $n = 2m$, or $n = 2m + 1 > 1$ each time Step 3 is executed and just count comparisons required by the merging. Initially, we enter Step 3 with 2^k lists, each of length 1, which are merged in pairs. Each merge requires one comparison for a total of $2^{k-1} = 2^k - 2^{k-1}$ comparisons altogether. We pass through Step 3 a second time, entering with 2^{k-1} lists of length 2 which are merged in pairs. Each pair is merged with at most $2 + 2 - 1 = 4 - 1$ comparisons, for a total of $2^{k-2}(4 - 1) = 2^k - 2^{k-2}$. The third time we execute Step 3, we enter with 2^{k-2} lists of length 4 which are merged in pairs. Each pair is merged with at most $4 + 4 - 1 = 8 - 1$ comparisons, for a total of $2^{k-3}(8 - 1) = 2^k - 2^{k-3}$. Continuing this line of reasoning, we see that pass i through Step 3 requires at most $2^k - 2^{k-i}$ comparisons for the merging. The kth and final pass requires $2^k - 1$ comparisons, the result of merging two lists of length 2^{k-1}.

In total, the merging in the algorithm requires at most

$$(1) \qquad (2^k - 2^{k-1}) + (2^k - 2^{k-2}) + \cdots + (2^k - 4) + (2^k - 2) + (2^k - 1)$$

comparisons. There are k terms here. Since $1 + 2 + 4 + \cdots + 2^{k-1} = 2^k - 1$ (see Pause 4), the sum in equation (1) equals $k2^k - (2^k - 1) = k2^k - 2^k + 1$. In all, including the possible three checks to determine whether $k = 1$, $k = 2m$ or $k = 2m + 1 > 1$, the algorithm requires at most $3k + (k2^k - 2^k + 1) = k2^k - 2^k + 3k + 1$ comparisons.

In general, for a list of n elements, find k such that $2^{k-1} < n \le 2^k$. By extending the list to one of length 2^k (for example, by appending additional elements larger than any number in the list), we see that the given list can be sorted by sorting the larger list and then removing the additional large elements from the end of the list. By the preceding analysis, it follows that the given list can be sorted with at most $k2^k - 2^k + 3k + 1$ comparisons. Since $2^{k-1} < n$, $k < 1 + \log_2 n$ and $2^k < 2n$. So the number of comparisons is at most

$$k2^k - 2^k + 3k + 1 \le (1 + \log_2 n)2^k - 2^k + 3k + 1$$

$$= 2^k \log_2 n + 3k + 1$$

$$< 2n \log_2 n + 3 \log_2 n + 4.$$

Since $2n \log_2 n + 3 \log_2 n + 4 \sim n \log_2 n$, the number of comparisons is $\mathcal{O}(n \log n)$. Referring to Table 8.1 of Section 8.2, we note that this is a substantial improvement on $\mathcal{O}(n^2)$, the complexity of the bubble sort.

Pause 4 Explain why $1 + 2 + 4 + \cdots + 2^{k-1} = 2^k - 1$.

Answers to Pauses

3. We want ordered lists with the property that after three comparisons one is empty. Here is an example:

$\mathcal{L}_1$	$\mathcal{L}_2$	$\mathcal{L}_3$	Comparisons Needed
1,3	2,4,6,8,9		
3	2,4,6,8,9	1	1
3	4,6,8,9	1,2	1
	4,6,8,9	1,2,3	1
		1,2,3,4,6,8,9	0
		Total	3

4. The terms in this sum are those of a geometric sequence with $a = 1$, $r = 2$, $n = k$. The sum is $\frac{a(1 - r^n)}{1 - r}$. [See formula (8) on p. 167.]

EXERCISES

The symbol [BB] means that an answer can be found in the Back of the Book.

1. (a) [BB] Show the sequence of steps in using a binary search to find the number 2 in the list $1, 2, 3, 4, 5, 6, 7, 8, 9$. How many times is 2 compared with an element in the list? How many times would it be compared with an element in the list if we employed a linear search?

   (b) Repeat (a) if we are searching for the number 7.

2. (a) Describe a *ternary search* algorithm, which searches an ordered list for a given element by successively dividing the list into thirds and determining in which third the element must lie.

   (b) Show that the complexity of the algorithm found in (a) (in terms of comparisons) is $\mathcal{O}(\log_3 n)$.

3. (a) [BB] Describe an algorithm which, upon input of k numbers, each in the range 1–100 (inclusive), outputs the complement of this set with respect to $U = \{1, 2, \ldots, 100\}$.

   (b) Repeat (a) if 100 is replaced by n. Find a Big Oh estimate for the complexity of your algorithm, in terms of comparisons.

4. [BB] Show the sequence of steps involved in merging the sorted lists $1, 2, 3, 4, 5$ and $2, 4, 6, 8, 10$. How many comparisons are required?

5. Rewrite the Merging Algorithm using a flag to prevent the program from exiting during Step 1.

6. [BB] Sort the list $3, 1, 7, 2, 5, 4$ into increasing order

   (a) with a bubble sort;

   (b) with a merge sort.

   In each case, how many comparisons are needed? (For the merge sort, ignore comparisons required to check the size and parity of n at each iteration of Step 3.)

7. Repeat Exercise 6 for the list $10, 11, 15, 3, 18, 14, 7, 1$.

8. [BB] The median of a list of numbers is the middle number or the average of the middle two numbers, after the numbers have been listed from smallest to largest. Describe an algorithm for finding the median of an input list of n numbers. Find a Big Oh estimate of the complexity of your algorithm.

9. Suppose that the eight elements a, b, c, d, u, v, w, x are ordered

$$d < a < u < c < x < b < v < w.$$

   (a) [BB] Show the steps in a bubble sort applied to the list a, b, c, d, u, v, w, x. How many comparisons are required?

   (b) Repeat (a) using a merge sort.

10. [BB] Find an example of two ordered lists of lengths s and $t \geq 3$ which can be merged with

    (a) one comparison

    (b) t comparisons

    (c) exactly $s + t - 1$ comparisons. (For this part, assume also that $s \geq 3$.)

11. (a) [BB] Show the steps involved in the application of a bubble sort to the list c, a, e, b, d, where these letters have their natural alphabetical order.

    (b) Apply the same sequence of interchanges as required in part (a) to the list $1, 2, 3, 4, 5$; that is, if it is necessary to interchange the second and third elements at a certain stage in the bubble sort of (a), then interchange the second and third elements at the same stage in the "sort" of $1, 2, 3, 4, 5$. The final sequence is $2, 4, 1, 5, 3$, which describes the **order** in which the elements of c, a, e, b, d must be taken to list them in order.

    (c) Describe an algorithm whose input is a list $a_1, a_2, \ldots, a_n$ whose natural order is $a_{i_1}, a_{i_2}, \ldots, a_{i_n}$ and whose output is the sequence of indices $i_1, i_2, \ldots, i_n$ (in this order).

12. We explained in the text why the Merging Algorithm requires at most $s + t - 1$ comparisons to merge ordered lists of lengths s and t. For arbitrary s and t, do there exist ordered lists of these lengths for which precisely $s + t - 1$ comparisons are needed to merge? Explain.

13. The Binary Search Algorithm we have presented appears to be more efficient than a linear search, $\mathcal{O}(\log n)$ versus $\mathcal{O}(n)$, but the binary search assumes the input list is ordered. Suppose we modify the Binary Search Algorithm so that it accepts an unordered list as input by first using an efficient sorting algorithm and then the binary search as described in this section. Is this new algorithm still more efficient than a linear search? Explain.

14. [BB] What is the fewest number of comparisons required to merge ordered lists of lengths s and t? Explain your answer.

15. Ignoring repeated checks as to whether $n = 1$, $n = 2\ell$, or $n = 2\ell + 1 > 1$, we showed in the text that the number of comparisons in any merge sort of 2^k elements is at most $k2^k - 2^k + 1$. Give a specific example of a list of length 2^k, where precisely $k2^k - 2^k + 1$ comparisons are required.

16. [BB] In order to output the distinct items of a given list $a_1, a_2, \dots, a_n$, the following method is proposed. For each k, search the elements preceding a_k and, if a_k is not found, output a_k. Show that this algorithm can be accomplished with $\mathcal{O}(\log n!)$ comparisons.

17. Modify the Bubble Sort Algorithm to find an algorithm which reverses a sequence; that is, upon input

of $a_1, a_2, \dots, a_n$, the algorithm outputs the sequence $a_n, a_{n-1}, \dots, a_2, a_1$. How many swaps are required? Give a Big Oh estimate of complexity.

18. [BB] It is suggested that one way of searching for a number x in a given unordered list is first to sort the list using a merge sort and then to use a binary search algorithm. Is this more or less efficient than a linear search?

19. Given a list $a_1, a_2, \dots, a_n$, each item of which is 0 or 1, it is desired to put the list into increasing order. Would you use one of the sorting methods discussed in this section, or is there a better method? Explain.

8.4 ENUMERATION OF PERMUTATIONS AND COMBINATIONS

A truck driver for Deluxe Bakery has to deliver bread to many supermarkets scattered throughout a large metropolitan area. She has a map showing the precise locations of her delivery points. She knows how long it takes to drive from the bakery to each point and how long it takes to drive between any two delivery points. In what order should she deliver bread in order to minimize total traveling time? This problem, known as the *Traveling Salesman's Problem*, will be encountered again. Here, it serves to illustrate a situation where we would like to enumerate permutations. Assigning the supermarkets the numbers $1, 2, \dots, n$, the truck driver ideally would like to list all the permutations of $1, 2, \dots, n$ and, for each permutation, determine the time required for the route which visits supermarkets in the specified order.

To list all the permutations of a set of two or three or even four elements is not difficult. To enumerate all the permutations of larger sets becomes first tedious, and then ridiculous, at least by hand, because $n!$ grows so very rapidly with n: For instance, 15! is roughly 10^{12}.

To list all permutations of ten elements, say, it would seem sensible to enlist the support of a computer. For this, of course, a suitable algorithm is required, an algorithm which, upon input of a positive integer n, outputs a complete list of all the permutations of $1, 2, \dots, n$ without omission or repetition. We are confronted with two problems.

- How should the permutations be ordered?
- Given a permutation, how is the next one determined?

Lexicographic order—the way words are ordered in a dictionary—is one way to order permutations. In a dictionary, *terrible* precedes *terrific* because, reading from left to right, the first place where the words differ is at the sixth letter, and there b precedes f in the natural ordering of the letters of the alphabet. So we list the permutation *beadc* before *bedca*, but after *bcaed*. In a similar way, if the symbols are numbers instead of letters, 1357264 precedes 1357624 but follows 1347256.

With respect to lexicographic ordering, there is a procedure for finding the permutation which follows any given permutation.

8.4.1 PROPOSITION ▶

When the permutations of $1, 2, \ldots, n$ are ordered lexicographically, the permutation which follows a particular permutation π is obtained by

- reading the digits in π from right to left,
- noting the first consecutive pair xy where $x < y$,
- replacing x by the smallest of those digits to its right which are larger than x, and then
- writing down in increasing order the digits not yet used.

EXAMPLE 29

The permutation following 1374652 is 1375246 because, reading from right to left, the first consecutive pair xy with $x < y$ is 46, so we replace 4 by 5, the smallest digit to its right which is larger than 4, giving an initial string 1375, which is completed by writing down the unused digits, 2, 4, and 6, in increasing order. ▲

EXAMPLE 30

The permutations of $1, 2, 3, 4$ in lexicographic order are

1234	1243	1324	1342	1423	1432	2134	2143
2314	2341	2413	2431	3124	3142	3214	3241
3412	3421	4123	4132	4213	4231	4312	4321.

▲

8.4.2 ALGORITHM FOR ENUMERATING PERMUTATIONS ▶

To enumerate all permutations of $1, 2, \ldots, n$, the following algorithm can be used.

Given a natural number n, to enumerate the $n!$ permutations of $1, 2, \ldots, n$, proceed as follows.

Step 1. Set $t = 1$. Output Perm$(1) = 123 \ldots n$. If $n = 1$, stop.

Step 2. For $t = 1$ to $n! - 1$, given permutation Perm$(t) = \pi_1 \pi_2 \ldots \pi_n$, determine the next permutation Perm$(t + 1)$ as follows:

- find the largest j such that $\pi_j < \pi_{j+1}$;
- let $m = \min\{\pi_i \mid i > j, \pi_i > \pi_j\}$;
- let S be the complement of the set $\{\pi_1, \ldots, \pi_{j-1}, m\}$ with respect to $U = \{1, 2, \ldots, n\}$;
- sort the elements of S in increasing order, $b_1 < b_2 < \ldots < b_{n-j}$;
- output Perm$(t + 1) = \pi_1 \ldots \pi_{j-1} m b_1 \ldots b_{n-j}$.

With $n = 3$, for example, the permutations of $1, 2, 3$ are generated as shown in Table 8.2.

Any algorithm which involves $n!$ steps will take a long time to run, even for relatively small n. For example, at a million operations per second, 12! operations require almost eight minutes. In the Exercises, we ask you to show that the algorithm just described can be implemented with complexity $\mathcal{O}(n!n)$. (See Exercise 3 at the end of this section.)

Table 8.2 The permutations of $1, 2, 3$ as generated by Algorithm 8.4.2.

t	Perm(t)	j	m	S
1	$\begin{array}{ccc} 1 & 2 & 3 \\ \pi_1 \pi_2 \pi_3 \end{array}$	2	3	$\{\pi_1, m\}^c = \{1, 3\}^c = \{2\}$
2	$\begin{array}{ccc} 1 & 3 & 2 \\ \pi_1 \pi_2 \pi_3 \end{array}$	1	2	$\{m\}^c = \{2\}^c = \{1, 3\}$
3	$\begin{array}{ccc} 2 & 1 & 3 \\ \pi_1 \pi_2 \pi_3 \end{array}$	2	3	$\{\pi_1, m\}^c = \{2, 3\}^c = \{1\}$
4	$\begin{array}{ccc} 2 & 3 & 1 \\ \pi_1 \pi_2 \pi_3 \end{array}$	1	3	$\{m\}^c = \{3\}^c = \{1, 2\}$
5	$\begin{array}{ccc} 3 & 1 & 2 \\ \pi_1 \pi_2 \pi_3 \end{array}$	2	2	$\{\pi_1, m\}^c = \{3, 2\}^c = \{1\}$
6	$\begin{array}{ccc} 3 & 2 & 1 \end{array}$			

Deluxe Bakery is considering the introduction of bagels to its traditional product lines of bread and pastries and decides to select 6 of the supermarkets which carry its products for a trial. Towards this effort, it would be helpful for the marketing manager to review a list of all possible choices in order to choose the most useful 6. What is required then is a listing of all combinations of supermarkets taken 6 at a time.

As with permutations, combinations will be listed lexicographically; however, this time we must be careful to avoid listing both 142897 and 218479, for instance, since these represent the same combination. Suppose we agree to list elements in their natural (increasing) order, so that rather than 142897 or 218479, we will write 124789. Imagine that the combination $uvwxyz$ appears in our list. Then $u < v < w < x < y < z$ and so, if these symbols come from $\{1, 2, \ldots, 72\}$, it is apparent, for instance, that y cannot be 72 and x can be neither 71 nor 72. The largest possible value for z is 72, the largest for y is 71, the largest for x is 70, the largest for w is 69, the largest for v is 68, and the largest value for u is 67. Also, the smallest combination is 123456.

8.4.3 PROPOSITION ▶

When the combinations of $1, 2, \ldots, n$ taken r at a time are ordered lexicographically, the first combination is $123 \ldots r$ and the combination which immediately follows a given one is obtained by

- reading the combination from right to left until the first digit which can be increased is found,
- adding 1 to this digit (suppose this number is now k),
- leaving the digits before k as they were, but following k by $k + 1, k + 2, \ldots$ and so on until r digits in all have been written down.

In the lexicographic ordering of the combinations $1, 2, \ldots, 9$ taken six at a time, what combination follows 134589? Reading from right to left, the first digit which can be increased is 5. Increase this to 6 (so that the combination begins 1346) and complete the combination with 78, giving 134678. What follows 236789? The first digit which can be increased is 3. Increase this to 4 (the next combination begins 24) and complete with 5678, giving 245678.

In the lexicographic ordering of the combinations of $1, 2, 3, 4, 5, 6$ taken four at a time, after 1346 come 1356, 1456, 2345, 2346, 2356, and so on. The combinations of $1, 2, 3, 4, 5$ taken three at a time are

$$123, 124, 125, 134, 135, 145, 234, 235, 245, 345.$$

EXERCISES

The symbol [BB] means that an answer can be found in the Back of the Book.

1. **(a)** [BB] With a table like Table 8.2, show the steps involved in the enumeration of the first eight permutations of $1, 2, 3, 4$.
   (b) Repeat (a) for the second eight permutations of $1, 2, 3, 4$.
   (c) Repeat (a) for the last eight permutations of $1, 2, 3, 4$.

2. **(a)** Use the procedure outlined in this section to list the first 20 permutations of $1, 2, 3, 4, 5$.
   (b) [BB] In the lexicographic ordering of the permutations of $1, 2, 3, 4, 5$, what five permutations come after 42513?
   (c) In the lexicographic ordering of the permutations of $1, 2, 3, 4, 5$, what five permutations precede 42513?

3. **(a)** [BB] Show that the complexity of Algorithm 8.4.2 for enumerating permutations is $\mathcal{O}(n! n \log n)$, in terms of comparisons.
   (b) Improve the estimate in (a) by showing how to achieve $\mathcal{O}(n! n)$.

4. Let $\pi' = \text{Perm}(t + 1)$ be the permutation derived from $\text{Perm}(t) = \pi$ as in Proposition 8.4.1.
   (a) Prove that $\pi \prec \pi'$, where $\prec$ means precedes in lexicographic order.
   (b) Prove that if σ is a permutation satisfying $\pi \preceq \sigma \prec \pi'$, then $\sigma = \pi$. Thus, π' is the immediate successor of π relative to lexicographic ordering of permutations. (Here $\pi \preceq \sigma$ means $\pi \prec \sigma$ or $\pi = \sigma$.)

5. List, in lexicographic order, all combinations of $1, 2, 3, 4, 5, 6$
   (a) [BB] taken four elements at a time;
   (b) taken three elements at a time.

6. In the lexicographic ordering of all combinations of $1, 2, \ldots, 9$ taken five at a time, list, if possible, the three combinations which immediately precede and the three which immediately follow
   (a) [BB] 23469,
   (b) 13567,
   (c) 45789.

7. **(a)** List, in lexicographic order, the combinations of $1, 2, 3, 4, 5, 6, 7$ taken two at a time.
   (b) [BB] Use Proposition 8.4.3 and Algorithm 8.4.2 to describe an algorithm for enumerating the permutations of $1, 2, \ldots, n$ taken r at a time, not necessarily in lexicographic order.
   (c) Use parts (a) and (b) to list all permutations of $1, 2, 3, 4, 5, 6, 7$ taken two at a time.

8. **(a)** [BB] List, in lexicographic order, all combinations of $1, 2, \ldots, 8$ taken six elements at a time.
   (b) List, in lexicographic order, the first ten and the last ten combinations of $1, 2, \ldots, 8$ taken three at a time.
   (c) Use the list in (b) and the method proposed in Exercise 7(b) to enumerate the first ten and the last ten permutations of $1, 2, \ldots, 8$ taken three at a time.

9. [BB] Prove Proposition 8.4.3.

10. [BB] Suppose $a_1a_2a_3\ldots a_r$ appears in the lexicographic ordering of the combinations of the integers $1, 2, 3, \ldots, n$ taken r at a time, where $a_1 < a_2 < a_3 < \cdots < a_r$.

    (a) What is the biggest possible value for a_r? For a_1? For a_j?

(b) Under what conditions is this the first combination in the list?

(c) When is it last?

11. [BB] Describe an algorithm which enumerates all combinations of $1, 2, \ldots, n$ taken r at a time.

12. Suppose a set S has kn elements for natural numbers k and n. Describe an algorithm which will output all k element subsets of S.

REVIEW EXERCISES FOR CHAPTER 8

1. Describe how Horner's algorithm evaluates $f(x)$ when

    (a) $f(x) = 2x^3 - 4x + 1$ and $x = -3$
    (b) $f(x) = x^4 - 2x^3 + x^2 - 5x + 6$ and $x = 2$

2. Use the Russian peasant method to find

    (a) 149×712
    (b) 1018×72

3. Let $n > 1$ be an integer, let $S = \{1, 2, \ldots, n\}$ and let $\mathcal{R} = \{(a_1, b_1), (a_2, b_2), \ldots, (a_t, b_t)\}$ be a subset of $S \times S$. Describe algorithms which determine whether or not the relation $\mathcal{R}$ is

    (a) reflexive;
    (b) symmetric;
    (c) antisymmetric;
    (d) transitive;
    (e) a function.

In each case, find a reasonable estimate of the complexity of your algorithm.

4. Suppose we want an algorithm which, for an input of integers $a_1, a_2, \ldots, a_n$, outputs the largest and second largest integers. It is proposed to sort the list in decreasing order and to output the first two integers of the sorted list.

    (a) In terms of comparisons, what is the best complexity function for this algorithm?

    (b) Is there a better way to proceed? Explain and, if the answer is yes, describe a better algorithm.

5. Describe an algorithm which upon input of a list $a_1, a_2, \ldots, a_n$ of integers, outputs the largest and the smallest numbers of the list.

6. Describe an algorithm which implements the floor function; that is, upon input of a real number x, the output is $\lfloor x \rfloor$.

7. (a) Show that $1^k + 2^k + \cdots + n^k = \mathcal{O}(n^{k+1})$ for any $k \geq 1$.

    (b) Show that $1^2 + 2^2 + \cdots + n^2 \asymp n^3$.

8. Show that

    (a) $3n^3 - 5n^2 + 2n + 1 \asymp n^3$,
    (b) $n^4 - 2n^3 + 3n^2 - 5n + 7 \asymp n^4$.

Do not merely quote Proposition 8.2.7, but you may use ideas from its proof.

9. (a) (Requires a little knowledge of calculus) Show that $1 + \frac{1}{2} + \frac{1}{3} + \cdots + \frac{1}{n} = \mathcal{O}(\log n)$.

    (b) Show that the Sieve of Eratosthenes, used to find all primes less than or equal to a given integer n, can be implemented with an algorithm which is $\mathcal{O}(n \log n)$. Count comparisons and "crossing out" an integer as basic operations. (See Section 4.3 and especially paragraph 4.3.5.)

10. Show the sequence of steps in using a binary search to find the number 5 in the list $-8, -5, -1, 0, 1, 2, 4, 5, 8, 9, 12$. How many times is 5 compared with an element in the list? How many times would it be compared with an element in the list if we employed a linear search?

11. Show the sequence of steps involved in merging the sorted lists $-1, 3, 7, 10, 12, 15$ and $-4, -3, -1, 2$. How many comparisons are required?

12. Sort the list $9, -3, 1, 0, -4, 5, 3$ into increasing order

    (a) with a bubble sort;
    (b) with a merge sort.

In each case, how many comparisons are needed? (For the merge sort, ignore comparisons required to check the size and parity of n at each iteration of Step 3.)

13. In the lexicographic ordering of the permutations of $1, 2, 3, 4, 5, 6$, list the 20 permutations which follow 463152.

14. In the lexicographic ordering of all combinations of $1, 2, \ldots, 7$ taken four at a time, list, if possible, the three combinations which immediately precede and the three which immediately follow

    (a) 3467,
    (b) 1236,
    (c) 2347.

Graphs

9.1 A Gentle Introduction

There are many concrete, practical problems that can be simplified and solved by looking at them from a different point of view. The Königsberg Bridge Problem, which we will soon describe, was a long-standing problem until it was imaginatively solved in 1736 by the great Swiss mathematician Léonhard Euler (1707–1783). Beginning his scientific career shortly after the death of Sir Isaac Newton, Euler spent the last 17 years of his life blind, but still very active. Some of his mathematical contributions were to the theory of convergent sequences and to the calculus of variations. Much of what is taught today about quadratic equations, conic sections, and quadrics in Euclidean space is just as Euler himself laid out. Perhaps less known is that Euler was a superb designer of algorithms; he had the uncanny ability of making order out of chaos, of seeing simple routes through the most complicated situations. It is because of his imaginative solution to the Königsberg Bridge Problem that Euler is generally considered to be the father of modern-day graph theory.

In the eighteenth century, Königsberg was the capital of East Prussia.[1] The Pregel River flowed through town and split into two branches around Kneiphof island, which is labeled A in Fig 9.1. Seven bridges crossed the river, providing links among the four land masses labeled A, B, C, D in the figure. People wondered if it were possible to start on one of the land masses, walk over each of the seven bridges exactly once, and return to the starting point (without getting wet).

To find an abstract mathematical model of a concrete problem can be a difficult task requiring both ingenuity and experience. The primary aim of this chapter is to provide the reader with some of this experience by presenting several real-world

---

[1] An ice-free port on the southern coast of the Baltic Sea, Königsberg was completely destroyed during World War II. It was renamed Kaliningrad and transferred to the Soviet Union in 1945. Today, it is the westernmost city in Russia.

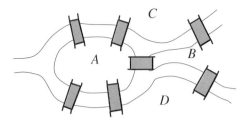

Figure 9.1   The bridges of Königsberg.

problems and showing how they can be formulated in mathematical terms. This process of translation into mathematics forces us to sift through all the details of the problem, deciding which ones are important and which are extraneous. (Those aspects of a "real" problem which make it interesting are sometimes quite irrelevant and serve primarily to create confusion!) Because the language of mathematics is precise and without ambiguity, problems which seem complicated when expressed in ordinary language often have surprisingly straightforward mathematical translations.

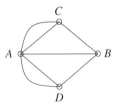

Figure 9.2   A graphical representation of the Königsberg Bridge Problem.

For the Königsberg Bridge Problem, Euler's idea was to realize that the physical layout of land, water, and bridges could be modeled by the graph shown in Fig 9.2. The land masses are represented by small circles (or vertices) and the bridges by lines (or edges) which can be straight or curved. By means of this graph, the physical problem is transformed into this mathematical one: Given the graph in Fig 9.2, is it possible to choose a vertex, then to proceed along the edges one after the other and return to the chosen vertex covering every edge exactly once? Euler was able to show that this was not possible. Can you? (We shall return to this problem and discuss its solution in Section 10.1, in the context of *Eulerian circuits*.)

The Three Houses—Three Utilities Problem is another physical situation which can be modeled by means of a graph. There are three houses, each of which is to be connected to each of three utilities—water, electricity, and telephone—by means of underground pipes. Is it possible to make these connections without any crossovers? Figure 9.3 shows how to describe this problem with a graph.

The houses and the utilities are represented by vertices and the pipes are the lines drawn between the vertices. When we discuss *planar graphs* in Chapter 14, we shall see that the answer to our question is no. Can you convince yourself of this now?

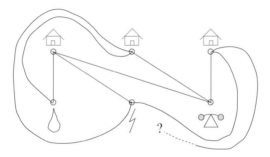

Figure 9.3   The Three Houses—Three Utilities Problem.

For a long time, there has been on the market a popular game called "Instant Insanity," which consists of four cubes, each of whose six faces is colored with one of four colors: red, blue, green, white. The object is to stack the cubes in such a way that each of the four colors appears on each side of the resulting column. Since there are over 40,000 possible ways to stack such cubes,[2] it is easy to understand how this game got its name! Certainly there ought to be a better way to solve this puzzle than trial and error. To illustrate the method we have in mind, we first need a way to picture a three-dimensional cube on paper. We do so as indicated in Fig 9.4, where we imagine that the cube has been opened and flattened, as if it were a small cardboard box.

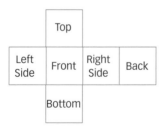

Figure 9.4   A way to picture a cube.

In Fig 9.5, we depict the four cubes of one version of "Instant Insanity" in this way and show a graph which records the pertinent information about them. The graph has four vertices, labeled R, B, G, and W, corresponding, respectively, to the four colors red, blue, green, and white. There is an edge labeled 3 joining vertices W and G because cube 3 has a white face and a green face on opposite sides of the cube. Cube 3 also has a white and red pair of opposite faces, so vertices W and R are joined with an edge labeled 3. For a similar reason, there is an edge labeled 3 joining vertices B and R. The circle labeled 1 at vertex R corresponds to the fact that cube 1 has a pair of opposite red faces.

Once again, a concrete physical problem has been represented by a graph. You do not have to visit Königsberg to simulate the Königsberg Bridge Problem; you do not need to build any houses to simulate the Three Houses—Three Utilities

---

[2]The exact number is $\frac{1}{2}(6^4 \times 4^3) = 41{,}472$.

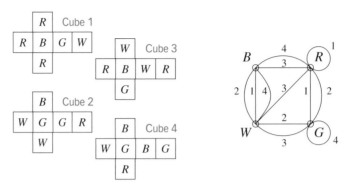

Figure 9.5 Four colored cubes and a graphical representation.

Problem; you do not even have to possess the actual cubes of "Instant Insanity" in order to play.

The graph in Fig 9.5 describes the four cubes of "Instant Insanity." Could it also be used to describe a stacking?

Figure 9.6

FRONT   BACK

W	Cube 4	B
R	Cube 3	W
G	Cube 2	W
B	Cube 1	W

Cube 1 in Fig 9.6 has its blue face at the front and white face at the back. This orientation can be shown in the graph by emphasizing the edge labeled 1 joining vertices B and W. Similarly, the green-white, red-white, and white-blue front-back faces can be shown by emphasizing three other edges in the graph. The stacking shown on the left of Fig 9.6 determines the *subgraph* on the right. This stacking does not solve the puzzle, however, because there are too many white faces at the back; moreover, any rotation of any of the cubes through 180° always results in more than one white face at the front or the back. This is more easily seen by observing that in the subgraph, there are too many edges joined to vertex W. On the other hand, there are exactly two edges in the subgraph (1 and 4) joined to vertex B, corresponding to the fact that there is just one blue face on the front (part of the B-W pair of faces of cube 1) and just one blue face at the back (part of the W-B pair of faces of cube 4).

When the cubes are stacked with their fronts and backs *correct* (that is, with four different colors on the front and four different colors on the back), the corresponding subgraph will

- contain all four vertices R, B, W, G;
- consist of four edges, one from each cube;
- have exactly two edges or one circle meeting at each vertex.

Figure 9.7 shows two subgraphs, each of which corresponds to a stacking of cubes in which the fronts and backs are correct.

Figure 9.7 Two acceptable subgraphs which are not edge disjoint.

Pause 1 Draw an arrangement of fronts and backs (like that on the left in Fig 9.6), which is represented by the subgraph on the left in Fig 9.7. All four colors should appear on the front and on the back.

As anyone who has tried this puzzle knows, getting the front and back of the column correct is easy. It is next to get the **sides** correct that provides the fun(?). Graphically, we require a second subgraph of the type described previously to represent a correct stacking of the sides. Moreover, since a given edge cannot represent both front-back and side-side at the same time, the second subgraph must be *edge disjoint* from the first: No edge can appear in both subgraphs. The subgraphs shown in Fig 9.7, for instance, are not edge disjoint; they have edges 3 and 4 in common. On the other hand, the subgraphs shown in Fig 9.8 are indeed edge disjoint and so provide us with a solution to "Instant Insanity," as shown in Fig 9.9.

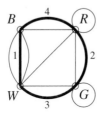
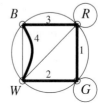

Figure 9.8 Two acceptable edge disjoint subgraphs.

	FRONT	BACK	RIGHT	LEFT
Cube 4	B	R	W	B
Cube 3	G	W	B	R
Cube 2	R	G	G	W
Cube 1	W	B	R	G

Figure 9.9 A solution to the game of "Instant Insanity."

9.1.1 REMARKS ▶

While our game of "Instant Insanity" had a unique solution, if we try to make other games by assigning colors to the faces of four cubes in other ways, it is not hard to find games where there are several winning configurations or where there is no solution at all. Also, whereas the subgraphs for our game had certain special properties, such as *connectedness*[3] and a lack of edge crossovers, in general, be on the lookout for other types of subgraphs, three of which are depicted in Fig 9.10.

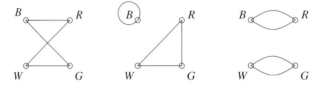

Figure 9.10   Possible subgraphs for other cube games.

Newspapers and magazines often contain mathematical teasers whose solutions can be helped by drawing simple graphs. We close this section with an example of one such puzzle and include another in the exercises for this section.

PROBLEM 1.   You and your buddy return home after a semester at college and are greeted at the airport by your mothers and your buddy's two sisters. Not uncharacteristically, there is a certain amount of hugging! Later, the other five people tell you the number of hugs they got and, curiously, these numbers are all different. Assume that you and your buddy did not hug each other, your mothers did not hug each other, and your buddy's sisters did not hug each other. Assume also that the same two people hugged at most once. How many people did you hug? How many people hugged your buddy?

Solution.   The conditions on who did not hug whom dictate that no person hugged more than four people. Since the other five hugged different numbers of people, this set of numbers must be $\{0, 1, 2, 3, 4\}$.

Next, it is important to realize that the person who hugged four others could not be your buddy. Why? The graph to the right should help. The four people other than you and your buddy are labeled A_1, A_2, B_1, B_2, and an edge indicates a hug. Think of A_1 and A_2 as sisters (or mothers), and B_1 and B_2 as mothers (or sisters).

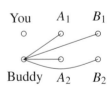

Remembering that you and your buddy did not hug, the graph shows that if your buddy had four hugs, it would be impossible for one of the group Buddy, A_1, A_2, B_1, B_2 to have reported no hugs. Thus, your buddy could not have had four hugs.

---

[3]A graph is connected if any two vertices are joined by a sequence of edges. We shall introduce the notion of connectedness formally in Section 10.1.

So the person who got four hugs is among the group A_1, A_2, B_1, B_2. Suppose A_1 got four hugs. (The argument which follows works in precisely the same way if we assume A_2 or B_1 or B_2 was the one with four hugs.) Since A_1 did not hug A_2, the hugging involving A_1 is as depicted in Fig 9.11(a).

Figure 9.11   (a)   (b)

Since one of the group A_1, A_2, B_1, B_2 received no hugs, it is apparent that this person was A_2. Consider again Fig 9.11(a). Since you didn't hug your buddy and B_1 didn't hug B_2, the only possible hugs not pictured there are between you or your buddy and B_1 or B_2.

Somebody received three hugs. Who could that be? If it were your buddy, he would have had to hug both B_1 and B_2, leaving nobody with only one hug. We conclude that either B_1 or B_2 received exactly three hugs. There is no loss of generality in assuming it was B_2. Thus, B_2 hugged A_1, you and your buddy. Your buddy now hugged at least two people (A_1 and B_2), so it follows that B_1 must be the person who had one hug. The final situation is shown in Fig 9.11(b). You and your buddy each hugged two people. ∎

Answers to Pauses

1. There are several possibilities for the fronts and backs of cubes which correspond to the subgraph shown in Fig 9.7. Here is one.

FRONT		BACK
B	Cube 4	R
G	Cube 3	W
R	Cube 2	G
W	Cube 1	B

EXERCISES

The symbol [BB] means that an answer can be found in the Back of the Book.

1. [BB] (Fictitious) A recently discovered map of the town of Königsberg shows that there was a ferry operating between the banks labeled C and D in Fig 9.1. Draw a graph in which the vertices are land masses and an edge between two vertices corresponds to a way to move between corresponding land masses.

2. [BB] Draw a configuration of two houses and two utilities, each house connected to each utility, but with no crossovers.

3. One of the owners of the houses in the Three Houses—Three Utilities Problem does not want a telephone. Is it now possible for the houses to be connected to utilities without crossovers? Draw a graph depicting the situation.

4. Find solutions, where possible, for the cube games pictured in Fig 9.12.

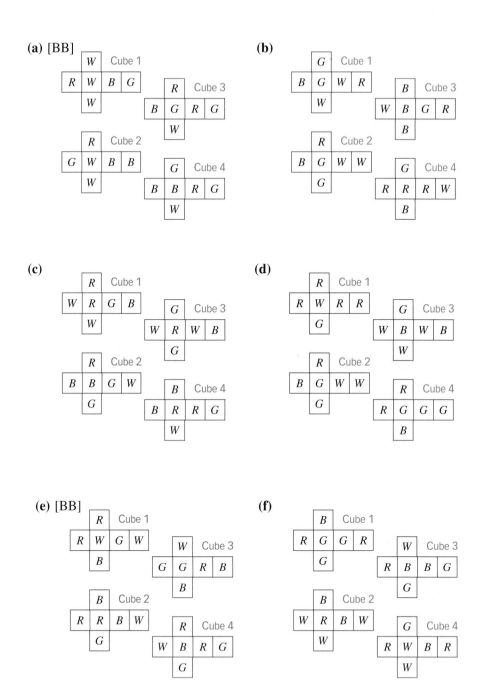

Figure 9.12 Cubes for Exercise 4.

(g)   **(h)**

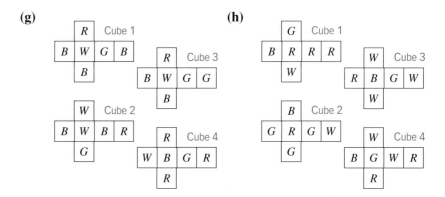

Figure 9.12   *Continued*

5. Suppose the vertices of a graph represent cities in a certain region and an edge joining two vertices indicates that there is a direct (nonstop) flight between those two cities. Geographers define the *beta index* of a graph as the ratio of the number of edges to the number of vertices, and view this number as a measure of connectivity of the region. Highly developed countries have high beta indices; poorly developed countries have low beta indices. Find the beta index of each of the following graphs.

(a) [BB]

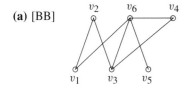

(b)

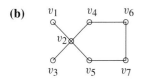

6. [BB] In the graph shown, the vertices represent the rooms of a one-story house and an edge between vertices means that the corresponding rooms have a wall in common. Draw a possible floor plan for this house.

7. You and a friend meet three other couples at a party and several handshakes take place. Nobody shakes hands with himself or herself, there are no handshakes within couples, and no one shakes hands with the same person more than once. The numbers of hands shaken by the other seven people (excluding you) are all different. How many hands did you shake? How many hands did your partner shake? Use a graph to aid your solution.

8. (a) A graph has six vertices every two of which are joined by an edge. Each edge is colored red or white. Show that the graph contains a monochromatic triangle.

    (b) Is the result of (a) true for a graph with five vertices? Explain.

9. [BB] A graph has six vertices every two of which are joined by an edge. Each vertex is colored red or white. Show that the graph contains at least **two** monochromatic triangles.

10. A graph has six vertices, every two of which are joined by an edge. Each edge is colored red or white. Show that the graph contains at least **two** monochromatic triangles.

9.2 DEFINITIONS AND BASIC PROPERTIES

Having used the term "graph" quite a bit already, it is time now to define the word properly and to introduce some of the basic terminology of graph theory.

A *graph* is a pair $(\mathcal{V}, \mathcal{E})$ of sets, $\mathcal{V}$ nonempty and each element of $\mathcal{E}$ a set of two distinct elements of $\mathcal{V}$. The elements of $\mathcal{V}$ are called *vertices*; the elements of $\mathcal{E}$ are called *edges*. Thus, if e is an edge, then $e = \{v, w\}$, where v and w are different elements of $\mathcal{V}$ called the *end vertices* or *ends* of e. (Colloquially, we often say that the edge $e = \{v, w\}$ *joins* vertices v and w.) We usually abandon set notation and refer to the edge vw; this is, of course, the same as the edge wv. The vertices v and w are said to be *incident* with the edge vw; the edge vw is *incident* with each vertex. Two vertices are *adjacent* if they are the end vertices of an edge; two edges are *adjacent* if they have a vertex in common. The number of edges incident with a vertex v is called the *degree* of that vertex and is denoted $\deg v$. If $\deg v$ is an even number, then v is said to be an *even vertex*; if $\deg v$ is an odd number, vertex v is *odd*. A vertex of degree 0 is said to be *isolated*. In this text, graphs will always be *finite*, meaning that the set of vertices and, hence, also the set of edges, are finite sets. Finally, when we say that $\mathcal{G}(\mathcal{V}, \mathcal{E})$ is a graph, we mean that $\mathcal{G}$ is a graph with vertex set $\mathcal{V}$ and edge set $\mathcal{E}$.

Usually we draw a picture of a graph rather than presenting it formally as sets of vertices and edges. For instance, the graph $\mathcal{G}$ with vertices

$$\mathcal{V} = \{v_1, v_2, v_3, v_4, v_5, v_6\}$$

and edges

$$\mathcal{E} = \{v_1 v_4, v_1 v_6, v_2 v_5, v_4 v_5, v_5 v_6\}$$

can be described as shown in Fig 9.13.

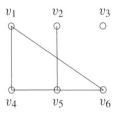

Figure 9.13   A picture of $\mathcal{G}(\mathcal{V}, \mathcal{E})$, where $\mathcal{V} = \{v_1, v_2, v_3, v_4, v_5, v_6\}$ and $\mathcal{E} = \{v_1 v_4, v_1 v_6, v_2 v_5, v_4 v_5, v_5 v_6\}$.

Many of the graphs presented in Section 9.1 are not graphs at all according to our definition! Many of them have *multiple edges*, that is, several edges incident with the same two vertices. Some of them have a *loop* at a vertex, that is, an edge which is incident with only one vertex. Since most of the graphs of interest to us will have neither multiple edges nor loops, we have opted for the definition of "graph" presented here. Nevertheless, it is convenient to have a term for more general types of graphs which do arise from time to time. A *pseudograph* is like a graph, but it may contain loops and/or multiple edges.

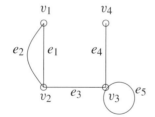

Figure 9.14

Figure 9.14 illustrates a pseudograph with four vertices and five edges. Vertices v_1 and v_2 are incident with edges e_1 and e_2; thus, e_1 and e_2 are multiple edges. Edge e_4 is incident with vertices v_3 and v_4. Edge e_5 is a loop because it is incident only with vertex v_3. Vertices v_2 and v_3 are adjacent while v_1 and v_3 are not. Edges e_1 and e_3 are adjacent while e_1 and e_4 are not. Vertex v_1 has degree 2, so v_1 is an even vertex. Counting the loop incident with v_3 twice (as it enters and leaves), v_3 is also an even vertex, with degree 4. Vertices v_2 and v_4 are odd: deg $v_2 = 3$ and deg $v_4 = 1$.

It is unfortunate that there is some lack of standardization of terminology in graph theory, especially in a subject which has more than its share of technical terms. Many words have almost "obvious" meanings, which are the same from book to book, but other terms are used differently by different authors. It is consequently vital when perusing books or articles on graph theory never to assume that you know the meanings of the graph theoretical terms employed. Always check the author's definitions carefully. For convenience, the definitions which we employ in this book are summarized in a glossary at the end. Readers may have also noticed on the inside covers a description of symbols and notation.

9.2.2 DEFINITION ▶ A graph $\mathcal{G}_1$ is a *subgraph* of another graph $\mathcal{G}$ if and only if the vertex and edge sets of $\mathcal{G}_1$ are, respectively, subsets of the vertex and edge sets of $\mathcal{G}$.

EXAMPLE 2 Each of the three graphs $\mathcal{G}_1$, $\mathcal{G}_2$, and $\mathcal{G}_3$ shown on the right of Fig 9.15 is a subgraph of the graph $\mathcal{G}$ on the left in this figure. (The subgraphs do not have to be drawn the same way they appear in the presentation of $\mathcal{G}$.) ▲

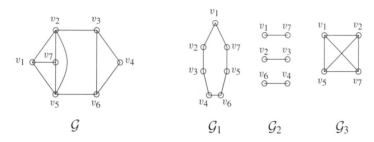

Figure 9.15 A graph and three subgraphs.

EXAMPLE 3 Figure 9.16 depicts a graph $\mathcal{G}$ and two subgraphs $\mathcal{G}_1$ and $\mathcal{G}_2$. The last graph in this figure, $\mathcal{G}_3$, is not a subgraph of $\mathcal{G}$ because the two vertices of degree 3 in $\mathcal{G}$ are not adjacent. ▲

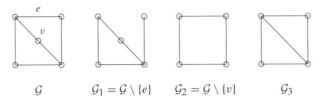

$$\mathcal{G} \qquad \mathcal{G}_1 = \mathcal{G} \setminus \{e\} \qquad \mathcal{G}_2 = \mathcal{G} \setminus \{v\} \qquad \mathcal{G}_3$$

Figure 9.16 A graph $\mathcal{G}$, two subgraphs, $\mathcal{G}_1$ and $\mathcal{G}_2$, and a graph $\mathcal{G}_3$ which is not a subgraph of $\mathcal{G}$.

A useful tool in graph theory is the deletion of an edge or a vertex from a graph. If e is an edge in a graph $\mathcal{G}$, we shall abuse notation and write $\mathcal{G} \setminus \{e\}$ to denote that subgraph of $\mathcal{G} = \mathcal{G}(\mathcal{V}, \mathcal{E})$ which has the same vertex set as $\mathcal{G}$ but whose edge set is $\mathcal{E} \setminus \{e\}$. For example, in Fig 9.16, the subgraph $\mathcal{G}_1$ is $\mathcal{G} \setminus \{e\}$. Similarly, $\mathcal{G} \setminus \{v\}$ denotes the graph $\mathcal{G}$ with the vertex v removed. The deletion of a vertex requires some care, since when a vertex is removed from a graph, all edges incident with that vertex must also be removed. Again with reference to Fig 9.16, subgraph $\mathcal{G} \setminus \{v\}$ is $\mathcal{G}_2$, not $\mathcal{G}_3$.

Each of the graphs pictured in Fig 9.17 has as many edges as possible (multiple edges and loops are not permitted in a "graph"). Such graphs have a name.

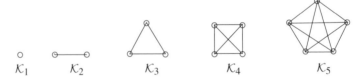

$$\mathcal{K}_1 \qquad \mathcal{K}_2 \qquad \mathcal{K}_3 \qquad \mathcal{K}_4 \qquad \mathcal{K}_5$$

Figure 9.17 The first five complete graphs.

9.2.3 DEFINITION ▶ For any positive integer n, the *complete graph on n vertices*, denoted $\mathcal{K}_n$, is that graph with n vertices every two of which are adjacent.

The graphs in Fig 9.18 are not complete, though they have other interesting properties. In each of them, no two top vertices are adjacent and no two bottom vertices are adjacent (they are examples of *bipartite* graphs), and in the two rightmost graphs every top vertex is adjacent to every bottom vertex (these two are *complete bipartite* graphs).

9.2.4 DEFINITIONS ▶ A *bipartite graph* is one whose vertices can be partitioned into two (disjoint) sets $\mathcal{V}_1$ and $\mathcal{V}_2$, called *bipartition sets*, in such a way that every edge joins a vertex in $\mathcal{V}_1$ and a vertex in $\mathcal{V}_2$. (In particular, there are no edges within $\mathcal{V}_1$ nor within $\mathcal{V}_2$.) A *complete bipartite graph* is a bipartite graph in which every vertex in $\mathcal{V}_1$ is joined to every vertex in $\mathcal{V}_2$. The complete bipartite graph on bipartition sets of m vertices and n vertices, respectively, is denoted $\mathcal{K}_{m,n}$.

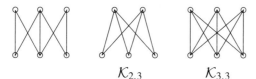

$\mathcal{K}_{2,3}$ $\mathcal{K}_{3,3}$

Figure 9.18   Three bipartite graphs, two of which are complete bipartite.

Figure 9.18 shows typical pictures of $\mathcal{K}_{2,3}$ and $\mathcal{K}_{3,3}$. The one vertex graph $\mathcal{K}_1$ shown in Fig 9.17 is also bipartite since bipartition sets are not required to be nonempty. It is helpful to note that a graph is bipartite if and only if its vertices can be colored with two colors such that every edge has ends of different colors.

EXAMPLE 4   Consider the graph on the left of Fig 9.19. Coloring vertex 1 red, vertex 2 white, and continuing to alternate these two colors through the vertices $3, \ldots, 8$ gives every edge different colored ends. So the graph is bipartite, the bipartition sets being the red vertices and the white vertices. By grouping these bipartition sets, the graph can be redrawn so that it more obviously appears bipartite, as shown on the right of Fig 9.19.   ▲

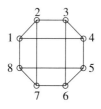 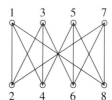

Figure 9.19   A graph and a way to show it is bipartite.

Pause 2   In the graph shown in Fig 9.19, what are the vertex sets $\{1, 3, 5, 7\}$ and $\{2, 4, 6, 8\}$ called? Is the graph complete? Is it complete bipartite?

It is not hard to show that a bipartite graph can contain no *triangles*.[4] (See Exercise 21.) For instance, graph $\mathcal{G}_3$ in Fig 9.16 is not bipartite because it contains a triangle (in fact, two triangles).

Pause 3   Look at the graphs accompanying Exercise 5 in Section 9.1. Are either of these bipartite? Explain.

Two very useful properties of graphs are described in the proposition and corollary which follow. With one proviso, the proofs we present are valid generally, so we state the propositions for pseudographs. The proviso is that a loop at a vertex shall add two to the degree of that vertex. [With reference to the pseudograph in Fig 9.14, recall that deg $v_3 = 4$.]

---

[4]A *triangle* in a graph is a set of three vertices with an edge joining each pair.

9.2.5 PROPOSITION ▶ **(Euler)** The sum of the degrees of the vertices of a pseudograph is an even number equal to twice the number of edges. In symbols, if $\mathcal{G}(\mathcal{V}, \mathcal{E})$ is a pseudograph, then

$$\sum_{v \in \mathcal{V}} \deg v = 2|\mathcal{E}|.$$

Proof Adding the degrees of all the vertices involves counting one for each edge incident with each vertex. How many times does an edge get counted? If it is not a loop, it is incident with two different vertices and so gets counted twice, once at each vertex. On the other hand, a loop at a vertex is also counted twice, by convention, in the degree of that vertex. ∎

EXAMPLE 5 The graph in Fig 9.19 has eight vertices each of degree 3. Since

$$\sum_{v \in \mathcal{V}} \deg v = 8(3) = 24 = 2|\mathcal{E}|,$$

it must have 12 edges, and it does. ▲

EXAMPLE 6 The pseudograph in Fig 9.14 has vertices of degrees 4, 3, 2, 1. Since $4+3+2+1 = 10$, the pseudograph must have five edges, and it does. (Note that a loop is **one** edge, but it adds **two** to the degree.) ▲

How many edges does $\mathcal{K}_{3,6}$ contain? This complete bipartite graph has six vertices of degree 3 and three of degree 6. Since

$$\sum_{v \in \mathcal{V}} \deg v = 6(3) + 3(6) = 36 = 2|\mathcal{E}|,$$

$\mathcal{K}_{3,6}$ has 18 edges.

9.2.6 DEFINITION ▶ Suppose that $d_1, d_2, \dots, d_n$ are the degrees of the vertices of a graph (or pseudograph) $\mathcal{G}$, ordered so that $d_1 \geq d_2 \geq \cdots \geq d_n$. Then $d_1, d_2, \dots, d_n$ is called the *degree sequence* of $\mathcal{G}$.

EXAMPLES 7 • The degree sequence of the pseudograph in Fig 9.14 is 4, 3, 2, 1.
• $\mathcal{K}_{2,3}$ has degree sequence 3, 3, 2, 2, 2. (See Fig 9.18.) ▲

Pause 4 Why can there not exist a graph whose degree sequence is 5, 4, 4, 3, 2, 1?

9.2.7 COROLLARY ▶ The number of odd vertices in a pseudograph is even.

Proof By Proposition 9.2.5, $\sum_{v \in \mathcal{V}} \deg v = 2|\mathcal{E}|$ is an even number. Since

$$\sum_{v \in \mathcal{V}} \deg v = \sum_{\substack{v \in \mathcal{V} \\ v \text{ even}}} \deg v + \sum_{\substack{v \in \mathcal{V} \\ v \text{ odd}}} \deg v$$

and the first sum on the right, being a sum of even numbers, is even, so also the second sum

$$\sum_{\substack{v \in \mathcal{V} \\ v \text{ odd}}} \deg v$$

must be even. Since the sum of an odd number of odd numbers is odd, the number of terms in the sum here, that is, the number of odd vertices, must be even. ∎

The pseudograph in Fig 9.14, for instance, has two odd vertices, v_2 and v_4. The complete bipartite graph $\mathcal{K}_{3,6}$ has six odd vertices (each of degree 3).

Pause 5 Find the degree of each vertex of the pseudograph $\mathcal{G}$ shown in Fig 9.20. What is the degree sequence for $\mathcal{G}$? Verify that the sum of the degrees of the vertices is an even number. Which vertices are even? Which are odd? Verify that the number of odd vertices is even.

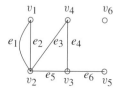

Figure 9.20

Answers to Pauses

2. The given vertex sets are the bipartition sets of the graph. The graph is not complete since, for example, vertices 1 and 3 are not incident with an edge; it isn't a complete bipartite graph either because, for example, vertices 1 and 6, which lie in different bipartition sets, are not incident with an edge.

3. Neither graph is bipartite. The graph in (a) contains the triangle $v_3 v_4 v_6$. While the graph in (b) does not contain a triangle, it contains the *5-cycle* $v_2 v_4 v_6 v_7 v_5 v_2$, which, like a triangle, causes problems. Try coloring the vertices of the graph in (b) with two colors—say red and white—so that the ends of every edge have different colors. If v_2 were colored white, then v_4 and v_5 would have to be red, so v_6 and v_7 would have to be white. Thus, the edge $v_6 v_7$ would have ends of the same color.

4. The sum of the degrees is 19, which is not an even number.

5. Vertices v_1, v_2, v_3, v_4, v_5, v_6 have degrees 2, 4, 3, 2, 1, and 0, respectively. The degree sequence of $\mathcal{G}$ is 4, 3, 2, 2, 1, 0. The sum of the degrees is $4+3+2+2+1+0 = 12$. The even vertices are v_1, v_2, v_4, and v_6; the odd vertices are v_3 and v_5. There are two odd vertices.

EXERCISES

The symbol [BB] means that an answer can be found in the Back of the Book.

1. [BB] Draw a graph with five vertices v_1, v_2, v_3, v_4, v_5 such that $\deg v_1 = 3$, v_2 is an odd vertex, $\deg v_3 = 2$, and v_4 and v_5 are adjacent.

2. Draw all possible graphs with three vertices v_1, v_2, v_3. How many edges are there in each graph? What is the degree sequence of each graph? Does this question make sense for pseudographs? Explain.

3. [BB] Give an example of a graph such that every vertex is adjacent to two vertices and every edge is adjacent to two edges.

4. (a) How many vertices and how many edges does the pseudograph contain? What is the degree sequence of this pseudograph?

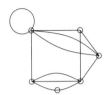

(b) Verify Proposition 9.2.5 and Corollary 9.2.7 for this pseudograph.

5. [BB] Draw a graph with five vertices and as many edges as possible. How many edges does your graph contain? What is the name of this graph and how is it denoted?

6. (a) What is the maximum degree of a vertex in a graph with n vertices?

(b) What is the maximum number of edges in a graph with n vertices?

(c) Given a natural number n, does there exist a graph with n vertices and the maximum possible number of edges?

7. Draw $\mathcal{K}_7$, $\mathcal{K}_{3,4}$, and $\mathcal{K}_{2,6}$.

8. Draw a graph with 64 vertices representing the squares of a chessboard. Connect two vertices with an edge if you can move legally between the corresponding squares with a single move of a knight. [The moves of a knight are L-shaped, two squares vertically (or horizontally) followed by one square horizontally (respectively, vertically).]

(a) Explain why this graph is bipartite.

(b) What are the degrees of the vertices?

9. Consider again the graphs accompanying Exercise 5 of Section 9.1, which we reproduce here.

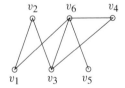 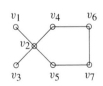

(a) [BB] For the graph on the left,

   i. Make a table which shows the least number of edges joining each pair of vertices in this graph. (Such a table displays the least number of stops required on air trips between cities in the region depicted by the graph.)

ii. Add up the numbers in each column of the table. Divide each column total by the degree of the corresponding vertex. These ratios are called *accessibility indices* since they measure the relative accessibility of the cities (by air). Which city is the most accessible? Which is the least accessible?

iii. Suppose a direct flight joining cities v_1 and v_3 is introduced. What is the new beta index of the graph? What are the new accessibility indices? Which city is most accessible now? Which city is now least accessible?

iv. Repeat part (iii), assuming a flight is introduced between cities v_2 and v_6 instead of between v_1 and v_3.

(b) Repeat the preceding questions for the graph on the right.

10. [BB] Verify Proposition 9.2.5 and Corollary 9.2.7 for the complete graph $\mathcal{K}_n$. What is the beta index of $\mathcal{K}_n$? (See Exercise 5, Section 9.1.)

11. Verify Proposition 9.2.5 and Corollary 9.2.7 for the complete bipartite graph $\mathcal{K}_{m,n}$. What is the beta index of $\mathcal{K}_{m,n}$?

12. [BB] At most social functions, there is a lot of handshaking. Prove that the number of people who shake the hands of an odd number of people is always even.

13. Which of the graphs is a subgraph of the graph in Fig 9.19?

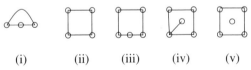

    (i)   (ii)   (iii)   (iv)   (v)

14. For each pair of graphs shown, discover whether or not the graph on the left is a subgraph of the one on the right. If it is not, explain why not. If it is, label the vertices of the subgraph, then use the same symbols to label the corresponding vertices of the graph on the right.

(a) [BB]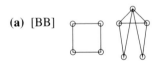

(b)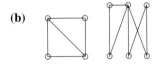

(c)

(d)
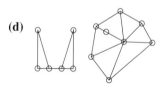

15. For each of the graphs shown, draw pictures of the subgraphs $\mathcal{G} \setminus \{e\}$, $\mathcal{G} \setminus \{v\}$, and $\mathcal{G} \setminus \{u\}$.

(a) [BB]
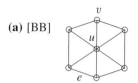
(b)
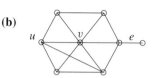

(c)
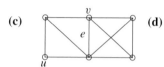
(d)
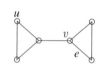

16. (a) [BB] What are the degrees of the vertices in the pseudograph?

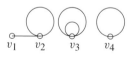

(b) [BB] Can there exist a graph with four vertices of degrees 1, 2, 3, and 4?

17. For each of the following sequences, determine if there exists a graph whose degree sequence is the one specified. In each case, either draw a graph, or explain why no graph exists.

(a) [BB] 4, 4, 4, 3, 2
(b) 100, 99, 98, ... , 3, 2, 2, 2
(c) [BB] 5, 5, 4, 3, 2, 1
(d) 1, 1, 1, 1, 1, 1
(e) 5, 4, 3, 2, 1
(f) 5, 4, 3, 2, 1, 1

(g) 6, 6, 4, 2, 2, 2, 1, 1

18. Does there exist a graph with five vertices, every vertex incident with at least one edge but no two edges adjacent? Explain.

19. (a) [BB] A graph has five vertices of degree 4 and two vertices of degree 2. How many edges does it have?
(b) A graph has degree sequence 5, 5, 4, 4, 3, 3, 3, 3. How many edges does it have?

20. Determine whether or not each of the graphs is bipartite. In each case, give the bipartition sets or explain why the graph is not bipartite.

(a) [BB]

(b)

(c) [BB]

(d)

(e)

(f)

21. [BB] Prove that a graph which contains a triangle cannot be bipartite.

22. (a) Must a subgraph of a bipartite graph be bipartite?
(b) Would your answer to (a) change if, in the definition of a bipartite graph, bipartition sets were required to be nonempty?

Explain your answers.

23. [BB] (Requires calculus) Prove that the number of edges in a bipartite graph with n vertices is at most $\frac{n^2}{4}$.

24. How many complete bipartite graphs have n vertices?

25. Let $\mathcal{V} = \{1, 2, 3, \ldots, n\}$.

(a) [BB] How many graphs are there with vertex set $\mathcal{V}$?
(b) How many of the graphs in (a) contain the triangle 123?
(c) [BB] What is the total number of triangles in all the graphs with vertex set $\mathcal{V}$?
(d) On average, how many triangles does a graph on n labeled vertices contain?

26. Suppose a graph has nine vertices each of degree 5 or 6. Prove that at least five vertices have degree 6 or at least six vertices have degree 5.

27. [BB] What is the largest possible number of vertices in a graph with 35 edges, all vertices having degree at least 3?

28. Let m and M denote the minimum and the maximum degrees of the vertices of a graph $\mathcal{G}$ with vertex set $\mathcal{V}$ and edge set $\mathcal{E}$. Show that

$$m \le \frac{2|\mathcal{E}|}{|\mathcal{V}|} \le M.$$

29. [BB] Suppose all vertices in a graph have odd degree k. Show that the total number of edges in $\mathcal{G}$ is a multiple of k.

30. [BB] Prove that in any graph with more than one vertex, there must exist two vertices of the same degree. [*Hint*: Pigeon-Hole Principle.]

31. Show that a set of nonnegative integers $\{d_1, d_2, \ldots, d_n\}$ is the set of degrees of some pseudograph if and only if $\sum_{i=1}^{n} d_i$ is even.

32. Suppose that $d_1, d_2, \ldots, d_n$ are the degrees of the vertices in some graph. Show that for any $t < n$,

$$\sum_{i=1}^{t} d_i \le t(t-1) + \sum_{i=t+1}^{n} \min\{t, d_i\}.$$

Remark: The given condition (holding for all $t < n$) and the requirement that $\sum_{i=1}^{n} d_i$ is even are sufficient as well as necessary for the existence of a graph with a prescribed set $d_1, \ldots, d_n$ of degrees.[5]

33. Can there exist a graph with 13 vertices, 31 edges, three vertices of degree 1, and seven vertices of degree 4? Explain.

9.3 ISOMORPHISM

It is important to know when two graphs are essentially the same and when they are essentially different. When we say graphs are "essentially the same," we mean that they differ only in the way they are labeled or drawn. There should be a one-to-one correspondence between the vertices of the graphs and a one-to-one correspondence between their edges such that corresponding vertices are incident with corresponding edges. The proper word for "essentially the same" is *isomorphic*.

There is a distinction between a graph and its picture. A graph is a set $\mathcal{V}$ and a set $\mathcal{E}$ of unordered pairs of elements of $\mathcal{V}$. A picture of it consists of dots and lines which can be drawn and arranged in many different ways. The bipartite graphs pictured in Fig 9.19, for instance, are isomorphic: The two pictures represent the same graph $\mathcal{G}(\mathcal{V}, \mathcal{E})$, where

$$\mathcal{V} = \{1, 2, 3, 4, 5, 6, 7, 8\}$$

and

$$\mathcal{E} = \{12, 14, 18, 23, 27, 34, 36, 45, 56, 58, 67, 78\}.$$

EXAMPLE 8   Graphs $\mathcal{G}_2$ and $\mathcal{G}_3$ in Fig 9.21 each consist of two edges incident with a common vertex. They are drawn differently, but the graphs are the same: $\mathcal{G}_2$ and $\mathcal{G}_3$ are isomorphic. The picture of $\mathcal{G}_1$ indicates that this graph has only one edge: $\mathcal{G}_1$ is different from $\mathcal{G}_2$ in an essential way; it is isomorphic to neither $\mathcal{G}_2$ nor $\mathcal{G}_3$.   ▲

---

[5]P. Erdös and T. Gallai, "Graphs with Prescribed Degrees of Vertices," *Matematikai Lapok* **11** (1960), 264–274.

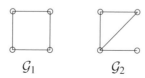

Figure 9.21 $\mathcal{G}_2$ and $\mathcal{G}_3$ are isomorphic, but neither is isomorphic to $\mathcal{G}_1$.

EXAMPLE 9   Figure 9.22 illustrates two graphs which are not isomorphic: Each graph consists of four vertices and four edges, but $\mathcal{G}_2$ contains a vertex of degree 1, while $\mathcal{G}_1$ has no such vertex.   ▲

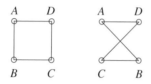

Figure 9.22   Two nonisomorphic graphs.

EXAMPLE 10   The graphs in Fig 9.23 are isomorphic and they have been labeled so as to show corresponding vertices. Either picture represents a graph with four vertices A, B, C, D, and four edges AB, BC, CD, and DA.   ▲

Figure 9.23   Two graphs labeled so as to show that they are isomorphic.

Pause 6   Show that the two graphs to the right are isomorphic by assigning the labels A, B, C, and D to appropriate vertices of the graph on the right. (It might help to think of the edges of $\mathcal{G}_2$ as pieces of string knotted at the vertices. How could $\mathcal{G}_2$ be rearranged to look like $\mathcal{G}_1$? Once you see this, it will be easy to do the required labeling.)

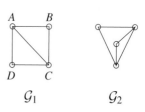

Here is the precise definition of the term "isomorphic."

9.3.1 DEFINITION ▶   Given graphs $\mathcal{G}_1 = \mathcal{G}_1(\mathcal{V}_1, \mathcal{E}_1)$ and $\mathcal{G}_2 = \mathcal{G}_2(\mathcal{V}_2, \mathcal{E}_2)$, we say that $\mathcal{G}_1$ is *isomorphic* to $\mathcal{G}_2$ and write $\mathcal{G}_1 \cong \mathcal{G}_2$ if there is a one-to-one function φ from $\mathcal{V}_1$ onto $\mathcal{V}_2$ such that

- if vw is an edge in $\mathcal{E}_1$, then $\varphi(v)\varphi(w)$ is an edge is $\mathcal{E}_2$, and
- every edge in $\mathcal{E}_2$ has the form $\varphi(v)\varphi(w)$ for some edge $vw \in \mathcal{E}_1$.

We call φ an *isomorphism* from $\mathcal{G}_1$ to $\mathcal{G}_2$ and, abusing notation, say that $\varphi \colon \mathcal{G}_1 \to \mathcal{G}_2$ is an isomorphism.

The definition of isomorphism is symmetric: If G_1 is isomorphic to G_2, then G_2 is isomorphic to G_1. In fact, if $\varphi\colon G_1 \to G_2$ is an isomorphism, then $\varphi^{-1}\colon G_2 \to G_1$ is an isomorphism. Thus, there is no ambiguity if we simply say that two graphs "are isomorphic."

We often say that graphs are isomorphic if and only if there is a bijection between their vertex sets which "preserves incidence relations." By this, we mean that a vertex v is incident with an edge e in G_1 if and only if $\varphi(v)$ is incident with $\varphi(e)$ in G_2. An isomorphism φ simply relabels vertices without changing any of the incidence relations. If $\varphi(v) = x$, think of x as the new label for v.

EXAMPLE 11

Consider the two graphs shown at the right. We encountered these earlier in Fig 9.23. Remembering our previous discussion, we see that if vertex u of G_2 is relabeled A, if vertex v is relabeled C, if w is relabeled D, and if x is relabeled B, then the pictures represent precisely the same graph. Having seen how to relabel the vertices of G_2 with the labels of G_1, it is easy to write down the isomorphism $G_2 \to G_1$ explicitly:

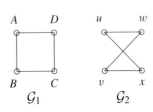

$$\varphi(u) = A, \quad \varphi(v) = C, \quad \varphi(w) = D, \quad \varphi(x) = B. \qquad \blacktriangle$$

Pause 7

Use your solution to Pause 6 to find an isomorphism $\varphi\colon G_2 \to G_1$ for the graphs G_1, G_2 shown in Fig 9.24.

Figure 9.24

The notion of isomorphism is exceedingly important in mathematics. If the definition looks complicated, the idea is very simple. Isomorphic objects, from a mathematical standpoint, are the same; they differ only in appearance. This idea is not new: $\frac{2}{4}$ and 0.5 look different, but they represent the same real number.

We have remarked that the definition of isomorphism is

> **symmetric:** $G_1 \cong G_2$ if and only if $G_2 \cong G_1$.

It is also

> **reflexive:** $G \cong G$ for any graph G

(because the identity map $G \to G$ is an isomorphism) and

> **transitive:** If $G_1 \cong G_2$ and $G_2 \cong G_3$, then $G_1 \cong G_3$

(because, if $\varphi_1\colon \mathcal{G}_1 \to \mathcal{G}_2$ and $\varphi_2\colon \mathcal{G}_2 \to \mathcal{G}_3$ are isomorphisms, then so is the composition $\varphi_2 \circ \varphi_1\colon \mathcal{G}_1 \to \mathcal{G}_3$). Thus, isomorphism is an equivalence relation on the set of all graphs.

The set of all graphs is therefore partitioned into disjoint equivalence classes called *isomorphism classes*. Any two graphs in the same equivalence class are isomorphic; two graphs in different equivalence classes are not isomorphic. The graphs pictured in Pause 6 belong to the same equivalence class; the graphs pictured in Fig 9.22 belong to different equivalence classes. When we casually remark that two graphs are "different," we really mean "in different isomorphism classes."

Usually, it is very difficult to prove that two graphs are isomorphic. In principle, we have to list all the one-to-one onto functions between vertex sets and, in each case, check whether or not the function preserves incidence relations. On the other hand, it is often easy to prove that graphs are not isomorphic. For instance, the graphs $\mathcal{G}_1$ and $\mathcal{G}_2$ in Fig 9.21 cannot possibly be isomorphic because $\mathcal{G}_2$ is *connected* in the sense that there is a sequence of adjacent edges between any two vertices, while $\mathcal{G}_1$ is not connected.

Since an isomorphism is a one-to-one onto function between vertex sets, isomorphic graphs have the same numbers of vertices. Many other properties are shared by isomorphic graphs.

Suppose $\varphi\colon \mathcal{G}_1 \to \mathcal{G}_2$ is an isomorphism from a graph $\mathcal{G}_1$ to another graph $\mathcal{G}_2$. If v is a vertex of degree k in $\mathcal{G}_1$ and if $v_1, v_2, \ldots, v_k$ are the vertices adjacent to v, then, in $\mathcal{G}_2$, $\varphi(v)$ is adjacent to the k vertices $\varphi(v_1), \varphi(v_2), \ldots, \varphi(v_k)$, but to no other vertex. Thus, the degree of $\varphi(v)$ is also k. It follows that isomorphic graphs have the same degree sequences and hence also the same numbers of edges, since the number of edges in a graph is one half the sum of the vertex degrees (Proposition 9.2.5).

9.3.2 PROPOSITION ▶   If $\mathcal{G}_1$ and $\mathcal{G}_2$ are isomorphic graphs, then $\mathcal{G}_1$ and $\mathcal{G}_2$ have the

- same number of vertices,
- same number of edges, and
- same degree sequences.

The graphs $\mathcal{G}_1$ and $\mathcal{G}_3$ shown in Fig 9.15 are not isomorphic because they have different numbers of vertices. The graphs shown in Fig 9.22 are not isomorphic because $\mathcal{G}_2$ contains a vertex of degree 1 while $\mathcal{G}_1$ does not.

Take care not to misinterpret Proposition 9.3.2. The proposition asserts three implications of the form

$$\text{``}\mathcal{G}_1 \cong \mathcal{G}_2 \to \ldots\text{''}$$

which must not be confused with double implications of the form

$$\text{``}\mathcal{G}_1 \cong \mathcal{G}_2 \leftrightarrow \ldots\text{''}$$

Proposition 9.3.2 says that **if** two graphs are isomorphic, **then** they must have certain properties. It does not say that two graphs with the properties listed are isomorphic. In fact, it is quite possible for two graphs which are not isomorphic

to have the same numbers of vertices and edges and the same degree sequences.
See Exercises 7 and 8.

Answers to Pauses

6. To the right, we show one of several
ways in which the vertices of $\mathcal{G}_2$ can be
labeled so that it becomes clear that the
graph represented by each picture is the
same.

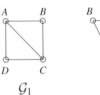

$\mathcal{G}_1$ $\mathcal{G}_2$

7. $\varphi(u) = B$, $\varphi(v) = C$, $\varphi(w) = A$, $\varphi(x) = D$. The isomorphism φ just relabels
vertices of $\mathcal{G}_2$.

EXERCISES

The symbol [BB] means that an answer can be found in the Back of the Book.

1. [BB] For each of the ten pairs of graphs which can be
obtained from those shown, either label the graphs so
as to exhibit an isomorphism or explain why the graphs
are not isomorphic.

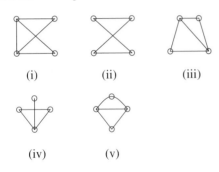

   (i)   (ii)   (iii)

   (iv)   (v)

2. (a) Draw a graph isomorphic to the one shown on the
left, but with no crossover of edges.

   (b) Same as (a) for the graph on the right.

3. (a) [BB] Draw all nonisomorphic graphs on $n = 3$ vertices. Give the degree sequence of each.

   (b) Repeat part (a) for $n = 4$.

4. For each pair of graphs shown,

   • if the graphs are not isomorphic, explain why not;

• if the graphs are isomorphic, exhibit an isomorphism
from one to the other and relabel the graph on the
right so as to show this isomorphism.

(a)

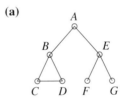

(b) [BB]

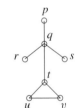

(c)

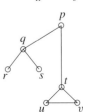

(d)

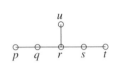

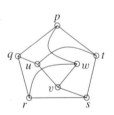

5. (a) [BB] Is the graph on the left isomorphic to $\mathcal{K}_{3,4}$? Explain.

(b) Is the graph on the right isomorphic to $\mathcal{K}_{4,4}$? Explain.

6. [BB] Explain why any graph is isomorphic to a subgraph of some complete graph.

9.3.3 DEFINITION ▼

Suppose $v_1, \ldots, v_n$ is a set of n vertices in a graph such that v_i and v_{i+1} are adjacent for $1 \le i \le n - 1$ and v_n and v_1 are also adjacent. Then the set of these n vertices and the n edges $v_1 v_2, v_2 v_3, \ldots v_{n-1} v_n, v_n v_1$ is called an *n-cycle*. A 3-cycle is often called a *triangle* and a 4-cycle a *quadrilateral*.

7. (a) [BB] Prove that two graphs which are isomorphic must contain the same number of triangles.

(b) Prove that, for any $n \ge 4$, two isomorphic graphs must contain the same number of *n*-cycles.

(c) How many edges are there in the graphs $\mathcal{G}_1$ and $\mathcal{G}_2$? How many vertices? What is the degree sequence of each graph? Are the graphs isomorphic? Explain.

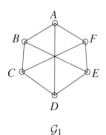

$\mathcal{G}_1$

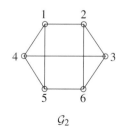

$\mathcal{G}_2$

8. Consider the following three graphs.

$\mathcal{G}_1$

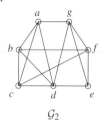

$\mathcal{G}_2$

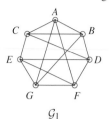

$\mathcal{G}_3$

(a) How many vertices and how many edges are there in each graph? What is the degree sequence of each graph? How many triangles are in each graph?

(b) For each pair of graphs, either exhibit an isomorphism between vertex sets or explain why the graphs are not isomorphic.

9. Show that the following graphs are not isomorphic.

10. (a) [BB] Suppose that graphs $\mathcal{G}$ and $\mathcal{H}$ have the same numbers of vertices and the same numbers of edges, and suppose that the degree of every vertex in $\mathcal{G}$ and in $\mathcal{H}$ is 2. Are $\mathcal{G}$ and $\mathcal{H}$ necessarily isomorphic? Explain.

(b) Suppose that graphs $\mathcal{G}$ and $\mathcal{H}$ have the same number of vertices and the same number of edges. Suppose that the degree sequences of $\mathcal{G}$ and $\mathcal{H}$ are the same and that neither graph contains a triangle. Are $\mathcal{G}$ and $\mathcal{H}$ necessarily isomorphic? Explain.

REVIEW EXERCISES FOR CHAPTER 9

1. In the Königsberg Bridge Problem, a tragic fire destroys the bridge from B to C and also one of the bridges from A to D. (See Fig 9.1.) Draw a graph representing the new situation. Show that it is now possible for someone to start on land mass B and walk over each of the bridges exactly once, returning to land mass B again.

2. (a) Draw a configuration of four houses and two utilities, each house connected to each utility, but with no crossovers.

(b) Let n be any positive integer. Motivated by 2(a), suggest a general result concerning n houses and 2 utilities. Draw a graph supporting your answer.

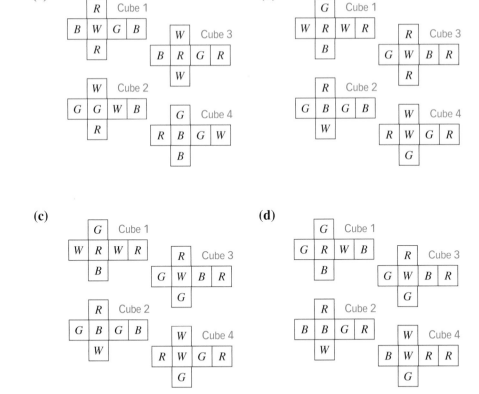

Figure 9.25 Cubes for Exercise 3.

3. Find solutions, where possible, for the cube games pictured in Fig 9.25.

4. (a) Draw a graph with six vertices at least three of which are odd and at least two of which are even.

   (b) Draw a graph with six vertices at most three of which are odd and at least two of which are even.

   (c) Is it possible to find a graph which satisfies the conditions in both 4(a) and 4(b) simultaneously? Explain your answer.

5. (a) Why is it not possible for a graph to have degree sequence 6, 6, 5, 5, 4, 4, 4, 4, 3?

   (b) Why is it not possible for a graph to have degree sequence 8, 8, 7, 6, 5, 4, 3, 2, 1?

6. Let $\mathcal{G}$ be a graph and let $\mathcal{H}$ be a subgraph of $\mathcal{G}$. Assume $\mathcal{H}$ contains at least three vertices.

(a) Is it possible for $\mathcal{G}$ to be bipartite and for $\mathcal{H}$ to be a complete graph?

(b) Is it possible for $\mathcal{G}$ to be a complete graph and for $\mathcal{H}$ to be bipartite?

Explain your answers.

7. Suppose a graph has 49 vertices, each of degree 4 or 5. Prove that at least 25 vertices have degree 4 or at least 26 vertices have degree 5.

8. Suppose $\mathcal{G}$ is a graph with n vertices, n edges and no vertices of degree 0 or 1. Prove that every vertex of $\mathcal{G}$ has degree 2.

9. For each pair of graphs shown,

   • if the graphs are not isomorphic, explain why not;
   • if the graphs are isomorphic, exhibit an isomorphism from one to the other.

(a)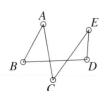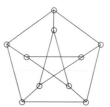

(b)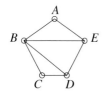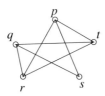

(c)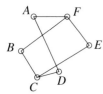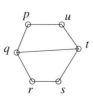

10. Determine whether or not the two graphs are isomorphic. (The one on the left is the "Petersen" graph, which we will encounter again in Chapter 10.)

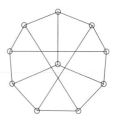

11. For each of the following cases, either explain why the two graphs are isomorphic or explain why they are not.

(a) $\mathcal{K}_4$ and $\mathcal{K}_{3,8}$

(b) $\mathcal{K}_{11}$ and $\mathcal{K}_{3,8}$

(c) $\mathcal{K}_{4,6}$ and $\mathcal{K}_{2,12}$

(d) $\mathcal{K}_{4,6}$ and $\mathcal{K}_{5,5}$

(e) $\mathcal{K}_{4,6}$ and $\mathcal{K}_{6,4}$

12. George is examining three graphs $\mathcal{G}_1$, $\mathcal{G}_2$, and $\mathcal{G}_3$. He gives correct arguments showing that $\mathcal{G}_1$ is not isomorphic to $\mathcal{G}_2$ and that $\mathcal{G}_2$ is not isomorphic to $\mathcal{G}_3$. Can he conclude that $\mathcal{G}_1$ is not isomorphic to $\mathcal{G}_3$? Explain.

13. Answer Exercise 12 again, assuming that George's correct arguments show that $\mathcal{G}_1$ is not isomorphic to $\mathcal{G}_2$ while $\mathcal{G}_2$ is isomorphic to $\mathcal{G}_3$.

14. Prove that $\mathcal{K}_{a,b} \cong \mathcal{K}_{c,d}$ if and only if $\{a, b\} = \{c, d\}$.

10

Paths and Circuits

10.1 EULERIAN CIRCUITS

Many real problems, when translated to questions about graphs, inquire about the possibility of walking through a graph in a particular way. Although our primary interest is in graphs, the definitions and results of this section are stated for pseudographs, since they apply equally, and with few additional complications, in the more general setting.

10.1.1 DEFINITIONS ▶ A *walk* in a pseudograph is an alternating sequence of vertices and edges, beginning and ending with a vertex, in which each vertex (except the last) is incident with the edge which follows and the last edge is incident with the edge which precedes it. The *length* of a walk is the number of edges in it. A walk is *closed* if the first vertex is the same as the last and otherwise *open*. A *trail* is a walk in which all edges are distinct; a *path* is a walk in which all vertices are distinct. A closed trail is called a *circuit*. A circuit in which the first vertex appears exactly twice (at the beginning and the end) and in which no other vertex appears more than once is a *cycle*. An *n-cycle* is a cycle with *n* vertices. It is *even* if *n* is even and *odd* if *n* is odd.

There are a lot of words here and, as always, we cannot emphasize too strongly the importance of coming to grips with their meanings. When thinking about the concepts of path and trail, it is perhaps helpful to note that a path is necessarily a trail: If all the vertices of a walk are different, then all the edges must be different too. As we show in the next example, the converse is not true: A trail need not be a path.

EXAMPLE 1 In the graph shown in Fig 10.1, $ABCEFCBD$ is a walk of length 7 which is neither a trail nor a path; $ABCEFCD$ is a trail, but not a path; $ABCEFCDBA$ is a closed walk which is not a circuit; $BCEFCDB$ is a circuit which is not a cycle; and $BCDB$ is a 3-cycle and hence an odd cycle. The closed walk $CEFCBDC$ is not a cycle because the first and last vertex appears a third time. ▲

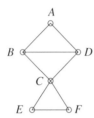

Figure 10.1

Pause 1 Verify each of the assertions just made.

While, strictly speaking, a walk should be specified by an alternating sequence of vertices and edges, it is often enough to specify only the vertices, as long as consecutive vertices are adjacent, as we have just done, or only the edges, as long as consecutive edges are adjacent.[1] An important type of circuit, known as an *Eulerian circuit*, is one which passes through every edge of a pseudograph.

10.1.2 DEFINITION ▶ An *Eulerian circuit* in a pseudograph is a circuit which contains every vertex and every edge. An *Eulerian pseudograph* is a pseudograph which contains an Eulerian circuit.

Note the reference to vertices in the definition of Eulerian circuit, which is crucial to the theory. See Exercise 5.

EXAMPLE 2 In the graph of Fig 10.1, the circuit $ABCEFCDA$ is not Eulerian because it does not contain the edge BD. As we shall soon see, this graph possesses no Eulerian circuit; it is not an Eulerian graph. ▲

EXAMPLE 3 Figure 10.2 further illustrates the difference between a circuit and an Eulerian circuit. The circuit $ABCDEFGHFA$ is not Eulerian since, while it encompasses all vertices, it omits four edges. The graph is Eulerian, however: $ABCDEFGHFADBEA$ is an Eulerian circuit. ▲

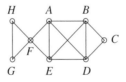

Figure 10.2 An Eulerian graph.

Eulerian circuits are named, of course, after Léonhard Euler, the solver of the Königsberg Bridge Problem, and their study is motivated by that problem. In Section 9.1, we saw that to follow the desired route over the bridges of Königsberg, you have to choose a vertex in the pseudograph of Fig 9.2 and find a walk which includes all the edges exactly once and leads back to the chosen vertex. With our present terminology, the Königsberg Bridge Problem asks if the pseudograph is Eulerian.

---

[1] In a **graph**, there is never a problem specifying just vertices or just edges.

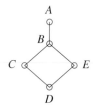

Figure 10.3   A graph which is not Eulerian.

In Theorem 10.1.4, we present the remarkably simple test which Euler found for the presence of an Eulerian circuit. First, since an Eulerian circuit provides a walk between every pair of vertices, any Eulerian pseudograph must be *connected*, in the following sense.

10.1.3 DEFINITION ▶   A pseudograph is *connected* if and only if there exists a walk between any two vertices.

As we ask you to show in Exercise 16, if a graph is connected, then there is actually a path between any two vertices, not just a walk.

An Eulerian graph must be more than just connected, however, as the graph in Fig 10.3 illustrates. The basic difficulty with this graph is that there is only one edge incident with A. Any circuit that begins at A cannot return to A without using this edge again, and any circuit which begins at a vertex other than A and attempts to include all edges, after using the edge BA, has to repeat it en route back to the starting vertex. The degrees of the vertices play a role in determining whether or not a graph is Eulerian. It is not hard to see that the degrees of the vertices of an Eulerian graph must be even.

In essence, we have already given the argument. In walking along an Eulerian circuit, every time we meet a vertex (other than the one where we started), either we leave on a loop and return immediately, never traversing that loop again, or we leave on an edge different from that by which we entered and traverse neither edge again. So the edges (other than loops) incident with any vertex in the middle of the circuit can be paired. So also can the edges incident with the first (and last) vertex since the edge by which we left it at the beginning can be paired with the edge by which we returned at the end. Thus, an Eulerian graph must not only be connected, but also have vertices of even degree. Conversely, a connected graph all of whose vertices are even must be Eulerian. To see why, it will be helpful to examine again the graph of Fig 10.2 (which is connected and has only even vertices) and to try to construct an Eulerian circuit with a strategy that might apply more generally.

We attempt to find an Eulerian circuit starting at A. To begin, we find **some** circuit which starts and ends at A, for instance, the circuit C_1: $ABCDEFA$. This circuit obviously is not Eulerian because it misses lots of edges. If we delete the edges of C_1 from the graph as well as vertex C, which is isolated after the edges of C_1 have been removed, we are left with the graph $\mathcal{G}_1$ on the left of Fig 10.4.

This graph $\mathcal{G}_1$ has a circuit C, FGH, which is connected to C_1 at vertex F. Thus, it can be used to enlarge C_1 as follows: Start at A, follow C_1 as far

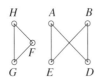

Figure 10.4

as F, then pass around C and complete C_1 to A. We obtain a second circuit C_2, $ABCDEFGHFA$, in the original graph, which is larger than C_1, but still not Eulerian because not all edges are yet included. As before, if the edges of C_2 are deleted (from the original graph), together with vertices F, G, and H which have become isolated, we are left with the graph on the right of Fig 10.4. This graph contains a circuit $ADBEA$, which makes contact with C_2 at A. Piecing together these circuits (follow C_2 to A, then $ADBEA$), we obtain a circuit C_3 which contains all the edges of the original graph, so C_3 is Eulerian. With the hindsight of this example, we are in a position to prove an important theorem.

10.1.4 THEOREM ▶ A pseudograph (with at least two vertices) is Eulerian if and only if it is connected and every vertex is even.

Proof ($\rightarrow$) We have already shown that an Eulerian pseudograph must be connected with each vertex even.

($\leftarrow$) For the converse, suppose that $\mathcal{G}$ is a connected pseudograph with all vertices of even degree. We must prove that $\mathcal{G}$ has an Eulerian circuit. Let v be any vertex of $\mathcal{G}$. If there are any loops incident with v, follow these first, one after the other without repetition. Then, since we are assuming that $\mathcal{G}$ has at least two vertices and since $\mathcal{G}$ is connected, there must be an edge vv_1 (with $v_1 \neq v$) incident with v. If there are loops incident with v_1, follow these one after the other without repetition. Then, since $\deg v_1$ is even and bigger than 0, there must be an edge v_1v_2 different from vv_1. Thus we have a trail from v to v_2 which we continue if possible. Each time we arrive at a vertex not encountered before, follow all the loops without repetition. Since the degree of each vertex is even, we can leave any vertex different from v on an edge not yet covered. Remembering that pseudographs in this book are always finite, we see that the process just described cannot continue indefinitely; eventually, we must return to v, having traced a circuit C_1. Notice that every vertex in C_1 is even since we entered and left on different edges each time it was encountered. At this point, it may happen that every edge has been covered; in other words, that C_1 is an Eulerian circuit, in which case we are done. If C_1 is not Eulerian, as in the preceding example, we delete from $\mathcal{G}$ all the edges of C_1 and all the vertices of $\mathcal{G}$ which are left isolated (that is, acquire degree 0) by this procedure. All vertices of the remaining graph $\mathcal{G}_1$ are even (since both $\mathcal{G}$ and C_1 have only even vertices) and of positive degree. Also, $\mathcal{G}_1$ and C_1 have a vertex u in common, because $\mathcal{G}$ is connected. (See Exercise 18.) Starting at u, and proceeding in $\mathcal{G}_1$ as we did in $\mathcal{G}$, we construct a circuit C in $\mathcal{G}_1$ which returns to u. Now combine C and C_1 by starting at v, moving along C_1 to u, then through C back to u, and then back to v on the remaining edges of C_1. We obtain a circuit C_2 in $\mathcal{G}$ which contains more edges than $\mathcal{C}_1$. If it contains all the edges of $\mathcal{G}$, it is Eulerian; otherwise, we repeat the process, obtaining a sequence of larger and larger circuits. Since our pseudograph is finite, the process must eventually stop, and it stops only with a circuit through all edges and vertices, that is, with an Eulerian circuit. ∎

Pause 2   Why is the graph shown in Fig 10.1 not Eulerian?

Not only does Theorem 10.1.4 give criteria for a pseudograph to be Eulerian, but its proof gives an algorithm for finding an Eulerian circuit when one exists. We have given one example. Here is another.

Pause 3   A power company's wires in a certain region follow the routes indicated in Fig 10.5. The vertices represent poles and the edges wires. After a severe storm, all the wires and poles must be inspected. Show that there is a round trip beginning at A, which allows a person to inspect each wire exactly once. Find such a trip.

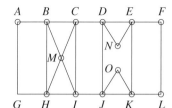

Figure 10.5

Sometimes, instead of finding an Eulerian circuit in a pseudograph, we want to find an Eulerian *trail* between two vertices, a trail which passes through every vertex and includes every edge. It is not hard to classify the pseudographs in which such a trail is possible, for adding one additional edge between the two vertices, the enlarged pseudograph is Eulerian. To see this, note that if the two vertices are u and v, then following an Eulerian trail from u to v and going back to u along the extra edge defines an Eulerian circuit. By Theorem 10.1.4, the enlarged pseudograph must be connected with all vertices even. So the original pseudograph must have been connected with all vertices even except u and v which are necessarily odd. (With the extra edge, they were even.) On the other hand, given a connected pseudograph with all vertices except u and v even, then certainly there is an Eulerian trail from u to v, for adding an extra edge between u and v produces an Eulerian pseudograph. Since an Eulerian circuit can begin at any vertex, imagine the one which begins by going from v to u along the added edge. Removing this added edge from the circuit then gives an Eulerian trail from u to v. We have established the following theorem.

10.1.5 THEOREM ▶   A pseudograph $\mathcal{G}$ possesses an Eulerian trail between two (different) vertices u and v if and only if $\mathcal{G}$ is connected and all vertices except u and v are even.

EXAMPLE 4   Consider the graph in Fig 10.6. Vertices A and B have degree 3, J and K have degree 2, and all the others have degree 4. Since the graph is connected and A and B are its only odd vertices, there exists an Eulerian trail from A to B. One Eulerian trail is $AGHFDGJHIFELIKLBCADCEB$. ▲

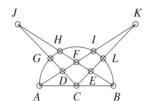

Figure 10.6

Answers to Pauses

1. $ABCEFCBD$ is not a trail (and hence not a path) because BC is a repeated edge; $ABCEFCD$ is a trail because all edges are distinct, but not a path because C is a repeated vertex; $ABCEFCDBA$ is a closed walk because its first and last vertices are the same, but not a circuit because edge AB is repeated; $BCEFCDB$ is a circuit because its edges are distinct and it begins and ends at vertex B, but it is not a cycle because its second and fifth vertices are the same. $BCDB$ is an odd cycle because it contains three edges. (Note that the circuit $CBDCEFC$ is not a cycle: Its first and fourth vertices are the same, but the fourth vertex is not last.)

2. The graph in Fig 10.1 is not Eulerian because not all vertices are even; for example, deg $B = 3$.

3. The desired round trip is just an Eulerian circuit in the graph of Fig 10.5. Such a circuit exists because the graph is connected and each vertex has even degree; vertices A, F, G, L, N, and O have degree 2; the rest have degree 4. In the search for an Eulerian circuit, an obvious circuit with which to begin is $ABCDEFLKJIHGA$. Joining this to the circuit $BMCIMHB$ at B gives A-$BMCIMHB$-$CDEFLKJIHGA$. Joining this to the circuit $DJOKEND$ at D gives the following routing for the desired inspection: A-$BMCIMHB$-C-$DJOKEND$-$EFLKJIHGA$.

EXERCISES

The symbol [BB] means that an answer can be found in the Back of the Book.

1. **(a)** [BB] Find a connected graph with as few vertices as possible which has precisely two vertices of odd degree.

   (b) Find a connected graph with as few vertices as possible which has precisely two vertices of even degree.

2. [BB] Answer the Königsberg Bridge Problem and explain.

3. In each case, explain why the graph is Eulerian and find an Eulerian circuit.

(a) [BB]

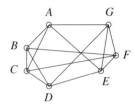

(b)

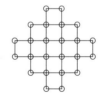

4. In each of the pseudographs shown, either describe an Eulerian circuit by numbering the edges or explain why no Eulerian circuit exists.

(a) [BB] **(b)**

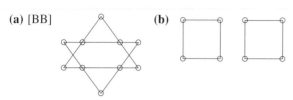

(c) **(d)**

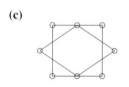

(e) **(f)**

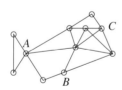
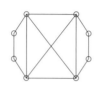

5. Suppose we modify the definition of Eulerian circuit by omitting the reference to vertices. Thus we propose that an Eulerian circuit be a circuit which contains every edge of a graph. Does Theorem 10.1.4 remain true? Explain.

6. (a) [BB] Is there an Eulerian trail from A to B? If yes, find one; if not, explain why not.

(b) Same question for A to C.

7. [BB] (Fictitious) A recently discovered map of the old town of Königsberg shows that there was a ferry operating between the areas labeled C and D in Fig 9.1.

(a) Is it possible to start on some land area, cross over each bridge exactly once, take the ferry exactly

once, and return to the starting point? Explain your answer.

(b) Is it possible to start on some land mass, walk over each bridge exactly once, take the ferry exactly once, and finish on some land mass (possibly different from the starting point)? Explain.

8. Euler's original article about the Königsberg Bridge Problem, which is dated 1736, presents a second similar problem with two islands, four rivers flowing around them, and 15 bridges connecting various land masses, as shown below.

(a) Is it possible to tour the region starting and finishing in the same area having walked over every bridge exactly once? Either describe such a tour or explain why none is possible.

(b) Is it possible to tour the region (with perhaps different starting and stopping points) having walked over every bridge exactly once? Either describe such a tour or explain why none is possible.

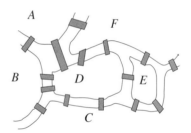

9. [BB] In Exercise 10 of Section 9.3, you were asked whether graphs $\mathcal{G}$ and $\mathcal{H}$ with the same numbers of vertices and edges and with every vertex in each graph of degree 2 need be isomorphic. Answer this question again, assuming in addition that the graphs are connected.

10. Suppose $\mathcal{G}_1$ and $\mathcal{G}_2$ are Eulerian graphs with no vertices in common. Let v_1 be a vertex in $\mathcal{G}_1$ and let v_2 be a vertex in $\mathcal{G}_2$. Join v_1 and v_2 with a single edge. What can be said about the resulting graph and why?

11. (a) [BB] For which values of $n > 1$, if any, is $\mathcal{K}_n$ Eulerian?

(b) [BB] For which values of $n > 1$, if any, does $\mathcal{K}_n$ possess an Eulerian trail? Explain.

12. (a) Find a necessary and sufficient condition on natural numbers m and n in order for $\mathcal{K}_{m,n}$ to be Eulerian. Prove your answer.

(b) Find a necessary and sufficient condition on natural numbers m and n in order for $\mathcal{K}_{m,n}$ to have an Eulerian trail. Assume $m \leq n$. Prove your answer.

13. Does there exist any sort of route in and around the figure which crosses every edge exactly once? Explain your answer.

14. [BB] Prove that any circuit in a graph must contain a cycle and that any circuit which is not a cycle contains at least two cycles.

15. [BB] Answer true or false and explain: Any closed walk in a graph contains a cycle.

16. Let u and v be distinct vertices in a graph $\mathcal{G}$. Prove that there is a walk from u to v if and only if there is a path from u to v.

17. [BB] For vertices u and v in a graph $\mathcal{G}$, define $u \sim v$ if $u = v$ or there exists a walk from u to v. Prove that $\sim$ defines an equivalence relation on the vertices of $\mathcal{G}$.

18. Complete some details in the proof of Theorem 10.1.4 by establishing the following. Suppose $\mathcal{C}_1(\mathcal{V}_1, \mathcal{E}_1)$ is a circuit in a connected graph $\mathcal{G}$ which does not contain all the edges of $\mathcal{G}$. Let $\mathcal{G}_1$ be that subgraph of $\mathcal{G}$ whose edge set is $\mathcal{E} \setminus \mathcal{E}_1$ and whose vertex set is $\mathcal{V}$ less those vertices of $\mathcal{V}_1$ which become isolated after the removal of the edges in $\mathcal{E}_1$. Prove that $\mathcal{G}_1$ and $\mathcal{C}_1$ have a vertex in common.

19. Suppose $\mathcal{G}_1$ and $\mathcal{G}_2$ are isomorphic graphs. Prove that either both $\mathcal{G}_1$ and $\mathcal{G}_2$ are connected or else neither is connected.

20. [BB] A graph $\mathcal{G}$ has 20 vertices. Any two distinct vertices x and y have the property that $\deg x + \deg y \geq 19$. Prove that $\mathcal{G}$ is connected.

21. (a) [BB] Let $\mathcal{G}$ be a connected graph with $n > 1$ vertices none of which has degree 1. Prove that $\mathcal{G}$ has at least n edges.

(b) Let $\mathcal{G}$ be a connected graph with n vertices. Prove that $\mathcal{G}$ has at least $n - 1$ edges.

22. Let $\mathcal{G}$ be a graph with n vertices and m edges, where $m > \frac{1}{2}(n - 1)(n - 2)$.

(a) Show that $\mathcal{G}$ does not have a vertex of degree 0.

(b) Show that $\mathcal{G}$ is connected.

10.1.6 DEFINITION ▼

A *component* of a graph is a maximal connected subgraph, that is, a connected subgraph which is properly contained in no other connected subgraph which has more vertices or more edges.

A graph with just one component is connected. Fig 10.7 shows graphs with two and with three components.

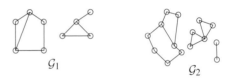

Figure 10.7 $\mathcal{G}_1$ has two components; $\mathcal{G}_2$ has three components.

23. Let $\mathcal{G}$ be a graph all of whose vertices have even degree. How can the Eulerian circuit algorithm described in Theorem 10.1.4 be modified to determine the number of components in $\mathcal{G}$?

24. [BB] Prove that isomorphic graphs have the same number of components.

25. Prove that a graph is bipartite if and only if it contains no odd cycles.

10.2 HAMILTONIAN CYCLES

An Eulerian circuit contains every edge of a graph exactly once. In this section, we discuss circuits which contain each vertex of a graph exactly once. Unlike in Section 10.1, the definitions and results of this section apply only to graphs, not to pseudographs.

10.2.1 DEFINITION ▶

A *Hamiltonian cycle* in a graph is a cycle which contains every vertex of the graph. A *Hamiltonian graph* is one with a Hamiltonian cycle.

Some authors define a *Hamiltonian circuit* as a circuit in which every vertex except the first and last appears exactly once. A circuit with no repeated vertices (except the first and the last) is a cycle, so a Hamiltonian circuit is a cycle. Thus the terms *Hamiltonian circuit* and *Hamiltonian cycle* are synonymous.

EXAMPLE 5   Graph $\mathcal{G}_1$ in Fig 10.8 is Hamiltonian: The cycle $ABCDEA$, for instance, is Hamiltonian. On the other hand, graph $\mathcal{G}_2$ is not Hamiltonian, but how can we convince ourselves of this fact?

Suppose that $\mathcal{G}_2$ has a cycle $\mathcal{H}$ which contains every vertex. Then $\mathcal{H}$ will contain A, which we note is a vertex of degree 2. Since we cannot enter and leave A on the same edge (edges of a cycle are distinct), it follows that both edges incident with A have to be part of $\mathcal{H}$. In particular, edge CA is in $\mathcal{H}$. The same argument applied to B shows that CB is part of $\mathcal{H}$ and, similarly, CD and CE are in $\mathcal{H}$: All four edges incident with C are part of $\mathcal{H}$. This situation is impossible, however: Since $\mathcal{H}$ is a cycle, vertex C can appear only once unless the cycle begins and ends at C. In either case, since $\mathcal{H}$ cannot use the same edge twice, exactly two edges incident with C can be part of $\mathcal{H}$.   ▲

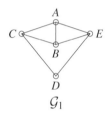
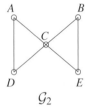

Figure 10.8 $\mathcal{G}_1$ is Hamiltonian; $\mathcal{G}_2$ is not.

While it is possible to decide precisely which graphs possess Eulerian circuits (Theorem 10.1.4), it is noteworthy that there has not yet been found a way to classify Hamiltonian graphs; in other words, there is no known theorem of the sort, "$\mathcal{G}$ is Hamiltonian if and only if ...". There are, however, some properties of cycles which are helpful in trying to find Hamiltonian cycles and which sometimes allow us to conclude that a particular graph is not Hamiltonian.

10.2.2 PROPERTIES OF CYCLES ▶   Suppose $\mathcal{H}$ is a cycle in a graph $\mathcal{G}$.

1. For each vertex v of $\mathcal{H}$, precisely two edges incident with v are in $\mathcal{H}$; hence, if $\mathcal{H}$ is a Hamiltonian cycle of $\mathcal{G}$ and a vertex v in $\mathcal{G}$ has degree 2, then both edges incident with v must be part of $\mathcal{H}$.

2. The only cycle contained in $\mathcal{H}$ is $\mathcal{H}$ itself. (We say that $\mathcal{H}$ contains no *proper cycles*.)

We have, in essence, already explained why each vertex of a cycle is incident with exactly two edges: Any vertex (except the first and last) appears exactly once in a cycle. If the cycle $\mathcal{H}$ is Hamiltonian, then every vertex is in $\mathcal{H}$ so both edges incident with any vertex of degree 2 must be in $\mathcal{H}$. This establishes Property 1.

Property 2 asserts that if C is a cycle contained in another cycle $\mathcal{H}$, then $C = \mathcal{H}$. We prove this by contradiction.

Suppose C is a cycle contained in $\mathcal{H}$ and $C \neq \mathcal{H}$. Then there is a vertex y in $\mathcal{H}$ which is not in C. Let x be any vertex in C. Since $\mathcal{H}$ contains both x and y, there is a path using edges of $\mathcal{H}$ from x to y. Thus, $\mathcal{H}$ contains some edge vw where vertex v is in C but w is not. (See Fig 10.9.) So $\mathcal{H}$ contains the two edges of C which are incident with v together with the edge vw. Altogether, there are three edges incident with v which must be part of $\mathcal{H}$. This contradicts Property 1. Thus $\mathcal{H}$ is not a cycle.

Figure 10.9

Pause 4   Answer true or false and explain: The graph shown at the right is not Hamiltonian because it contains the cycle $GHIG$.

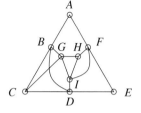

10.2.3 THE PETERSEN GRAPH ▶

As a deeper application of the properties of Hamiltonian graphs described in 10.2.2, we introduce a famous graph named after the Danish mathematician Julius Petersen (1839–1910).[2] We show that this graph, the one in Fig 10.10, is not Hamiltonian.

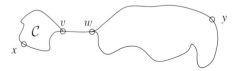

Figure 10.10   The Petersen graph is not Hamiltonian.

Suppose $\mathcal{H}$ is a Hamiltonian cycle. Then $\mathcal{H}$ must contain at least one of the five edges connecting the outer to the inner vertices. Since the graph is symmetric, there is no loss of generality in assuming that AF is part of $\mathcal{H}$. (Refer to Fig 10.11.) By Property 1, precisely one of the two edges FH and FI is in $\mathcal{H}$. Again, by symmetry, we may assume FH is part of the cycle while FI is not.

Since FI is not in $\mathcal{H}$, but two edges incident with I must be in $\mathcal{H}$ (Property 1), IG and ID are in $\mathcal{H}$. Now precisely one of the edges GB, GJ is in $\mathcal{H}$.

Suppose first that GB is in and GJ is out. Because precisely two edges incident with J are in $\mathcal{H}$ and JG is not, both JH and JE are part of $\mathcal{H}$. Thus,

[2]J. Petersen, "Die Theorie der regulären Graphen," *Acta Mathematica* **15** (1891), 193–220.

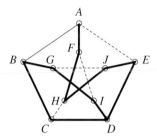

Figure 10.11

CH is out and both BC and CD are in $\mathcal{H}$. At this point, however, $\mathcal{H}$ contains the proper cycle $BCDIGB$, a contradiction. We conclude that GB cannot be part of $\mathcal{H}$ and hence that GJ is. An argument similar to the one just given now leads again to the false conclusion that $\mathcal{H}$ contains a proper cycle.

Pause 5   Give the details of this argument.

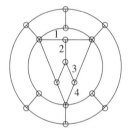

Figure 10.12   No Hamiltonian cycle contains 1234.

Pause 6   Look at the graph pictured in Fig 10.12. Show that no Hamiltonian cycle can contain edges $1, 2, 3, 4$.

    Hamiltonian graphs take their name from Sir William Rowan Hamilton (1805–1865), a contemporary and personal friend of William Wordsworth and Samuel Taylor Coleridge and indeed a man of many talents. By the age of 13, he had mastered one language for each year of his life, including Latin, Greek, Hebrew, Chinese, and Sanskrit! At 17, he had a firm grasp of calculus. He studied astronomy at Trinity College, Dublin, and later made important contributions to the study of optics. He is perhaps best known within the sphere of mathematics as the inventor of the *quaternions*, the first noncommutative *field* to be discovered. Hamilton's quaternions is an algebraic structure like the real numbers in which one can add, subtract, multiply, and divide (by any nonzero number), but in which multiplication is not commutative: There are quaternions a and b for which $ab \neq ba$. The reader may have encountered noncommutative systems before; for example, the set of all $n \times n$ matrices over the real numbers is not commutative. A basic difference between matrices and quaternions, however, and the thing that made Hamilton's discovery so remarkable, is the lack of divisibility in matrices. There are many nonzero matrices which are not invertible; on the other hand, every nonzero quaternion has an inverse.

Hamilton also invented a game which made use of a wooden regular *dodec-ahedron*, that is, a solid with 12 congruent faces, each of which is a regular pentagon. The vertices of the dodecahedron were labeled with the names of 20 cities of the world, and the object of the game was to find a route "around the world," along the edges of the solid, which passed through each city exactly once and led back to the city where the tour started.

Imagine that the pentagon on which the dodecahedron sits is stretched so that the solid collapses until it is flat. The result is a graph in the plane. (See Fig 10.13.) Hamilton's world tour is possible if and only if the graph contains what we now call a Hamiltonian cycle. The graph in fact is Hamiltonian; a Hamiltonian cycle is marked with the heavy lines in Fig 10.13.

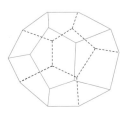 

Figure 10.13   A dodecahedron and its associated Hamiltonian graph.

There are some graphs in which a Hamiltonian cycle always exists (the complete graph on *n* vertices, for instance). Assigning the vertices of $\mathcal{K}_n$ the labels $v_1, v_2, \ldots, v_n$, then $v_1v_2\cdots v_nv_1$ is a cycle because we are assured of an edge between each pair of vertices. The cycle clearly passes through each vertex (except the first and the last) exactly once, so it is Hamiltonian. It would seem that a graph with "lots" of edges should have a good chance of being Hamiltonian. The following theorem, published by G. A. Dirac in 1952, provides further evidence in support of this idea.

10.2.4 THEOREM ▶ (**Dirac**[3]) If a graph $\mathcal{G}$ has $n \geq 3$ vertices and every vertex has degree at least $\frac{n}{2}$, then $\mathcal{G}$ is Hamiltonian.[4]

Proof Among all the possible paths in $\mathcal{G}$, suppose that $\mathcal{P}: v_1v_2\cdots v_t$ is longest in the sense that it uses the most vertices. Thus, there is no walk in $\mathcal{G}$ which uses more than t vertices without repeating some vertex. If some vertex w adjacent to v_1 is not in $\mathcal{P}$, then the walk $wv_1v_2\cdots v_t$ does not have repeated vertices and is longer than $\mathcal{P}$. Since this is contrary to the way $\mathcal{P}$ was chosen, every vertex adjacent to v_1 is in $\mathcal{P}$. (Similarly, every vertex adjacent to v_t is in $\mathcal{P}$.) Since $\deg v_1 \geq \frac{n}{2}$, $t \geq \frac{n}{2} + 1$, the "+1" counting v_1 itself. Since $n \geq 3$ and t is an integer, we conclude that $t \geq 3$ also.

---

[3]G. A. Dirac, "Some Theorems on Abstract Graphs," *Proceedings London Mathematical Society* **2** (1952), 69–81.

[4]There is a stronger version of this theorem, due to Oystein Ore, which says that a graph with $n \geq 3$ vertices is Hamiltonian as long as the sum of the degrees of any two nonadjacent vertices is at least n. See Exercise 14.

Claim: There is a pair of vertices v_k, v_{k+1} in $\mathcal{P}$ $(1 \le k < t)$ such that v_1 is adjacent to v_{k+1} and v_t is adjacent to v_k, as suggested in Fig 10.14.

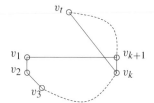

Figure 10.14

If this were not the case, then each of the vertices of $\mathcal{P}$ adjacent to v_1 would determine a vertex **not** adjacent to v_t (namely, its predecessor in $\mathcal{P}$). Since the vertices $v_2, \ldots, v_t$ are all different, there would be in $\mathcal{G}$ at least $\frac{n}{2}$ vertices not adjacent to v_t. These vertices, together with the vertices adjacent to v_t, account for at least n vertices in $\mathcal{G}$. Including v_t itself, we have found more than n vertices, which cannot be. This establishes the validity of our claim, from which it follows that $\mathcal{G}$ contains the cycle

$$\mathcal{C}: v_1 v_{k+1} v_{k+2} \cdots v_t v_k v_{k-1} \cdots v_1.$$

We show that $\mathcal{C}$ contains all vertices of $\mathcal{G}$ and hence is the desired Hamiltonian cycle. Remember that $\mathcal{C}$ contains at least $\frac{n}{2} + 1$ vertices, so there are less than $\frac{n}{2}$ vertices not in $\mathcal{C}$. Hence, any vertex w which is not in $\mathcal{C}$ must be adjacent to some vertex v_s of $\mathcal{C}$. Then w, v_s, and the remaining vertices of $\mathcal{C}$ in sequence would define a path longer than $\mathcal{P}$, contradicting the definition of $\mathcal{P}$. ∎

Pause 7

Show that Dirac's Theorem is false if $\frac{n}{2}$ is replaced by $\frac{n-1}{2}$ in its statement. [*Hint*: Examine those graphs presented in this section which are not Hamiltonian.]

Answers to Pauses

4. False! The graph is Hamiltonian: $AFEDIHGCBA$ is a Hamiltonian cycle. It is not the entire **graph** which must not contain a proper cycle, but any **Hamiltonian cycle**.

5. We are assuming that both AF and FH are in a Hamiltonian path $\mathcal{H}$. As before, FI is not in $\mathcal{H}$; therefore, both IG and ID are. Since edge GJ is also in $\mathcal{H}$ while GB is out, both BA and BC are in because two edges adjacent to B are part of $\mathcal{H}$ and BG is not in $\mathcal{H}$, as shown on the left. Using Property 2, we see that CH cannot be part of $\mathcal{H}$; otherwise $\mathcal{H}$ contains the proper cycle $ABCHFA$. So both CD and HJ are in. Now $\mathcal{H}$ contains the proper cycle $ABCDIGJHFA$, a contradiction.

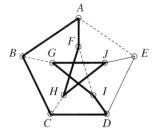
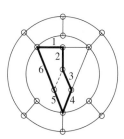

6. If a Hamiltonian cycle $\mathcal{H}$ were to contain edges 2 and 3 (see the graph on the right), then the third edge at the center could not be part of $\mathcal{H}$ (Property 1), so the edges labeled 5 and 6 would have to belong to $\mathcal{H}$ (Property 1 again). Now edges 1, 2, 3, 4, 5 and 6 form a proper cycle within $\mathcal{H}$. This contradicts Property 2.

7. The graph $\mathcal{G}_2$ in Fig 10.8, which is **not** Hamiltonian, has $n = 5$ vertices each of degree at least $2 = \frac{n-1}{2}$.

EXERCISES

The symbol [BB] means that an answer can be found in the Back of the Book.

1. [BB] Is the graph Hamiltonian? Is it Eulerian? Explain your answers.

2. [BB; (b),(d)] Determine whether or not each of the graphs of Exercise 4 of Section 10.1 is Hamiltonian. In each case, either label the edges with numbers so as to indicate a Hamiltonian cycle or explain why no such cycle exists.

10.2.5 DEFINITION ▼

A *Hamiltonian path* in a graph is a path which passes through every vertex exactly once.

3. Determine whether or not each of the graphs shown is Hamiltonian. Determine also whether or not each graph has a Hamiltonian path. In each case, either number the vertices so as to indicate a Hamiltonian cycle or path or explain why no such cycle or path exists.

(a) **(b)**

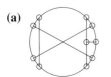

(c) **(d)**

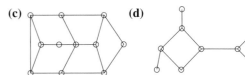

4. (a) [BB] Is the graph Hamiltonian?
(b) Is there a Hamiltonian path?
(c) [BB] Is it Eulerian?
(d) Is there an Eulerian trail?

Explain your answers.

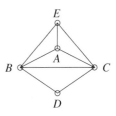

5. (a) (The Knight's Tour) Is it possible for a knight to tour a chessboard visiting every square exactly once and returning to its initial square? (See Exercise 8 of Section 9.2.)

(b) Is the sort of tour described in 5(a) possible on a 7×7 "chessboard"?

(For a complete classification of those $m \times n$ boards on which knight's tours are possible, the reader is directed to the interesting article by Allen J. Schwenk which appeared in the December 1991 issue of *Mathematics Magazine*.)

6. Is the graph of Fig 10.12 Hamiltonian? Display a Hamiltonian cycle or explain clearly why no such cycle exists.

7. The figure shows the floor plan of a single-story house with various doorways between rooms and other doorways leading outside.

$$O$$

(a) [BB] Is it possible to start outside, then to enter the house and walk through every room exactly once (without leaving the house), and finally to return outside? If yes, exhibit the route on a copy of the floor plan.

(b) Is there a route which starts outside and leads through every doorway in the house exactly once? (You are allowed to return outside or to reenter rooms as often as you want.)

8. [BB] In a group of $2n$ people, each person has at least n friends. Prove that the group can be seated in a circle, each person next to a friend.

9. (a) [BB] How many edges must a Hamiltonian cycle in $\mathcal{K}_n$ contain?

(b) How many Hamiltonian cycles does $\mathcal{K}_n$ have? (Begin all cycles at the same vertex.)

(c) [BB] What is the maximum number of *edge disjoint* Hamiltonian cycles in $\mathcal{K}_n$?[5] (Cycles are *edge disjoint* if no two of them have an edge in common.)

(d) Find all the Hamiltonian cycles in $\mathcal{K}_n$ for $n = 1, 2, 3, 4, 5$. In each case, exhibit a maximum number which are edge disjoint.

10. [BB] Draw a picture of a cube. By imagining that the bottom square is stretched until it is flat, draw a graph of the flattened cube. Is this graph Hamiltonian? If so, draw a Hamiltonian cycle. If not, explain why not.

11. The picture on the left is that of an icosahedron, a solid object whose faces consist of 20 congruent equilateral triangles. By stretching the base triangle and flattening, the icosahedron determines a graph in the plane (as shown on the right side of the figure). Find a Hamiltonian cycle in this graph.

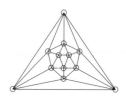

12. Make a model of a dodecahedron (Fig 10.13) and trace a Hamiltonian path along its edges. Do the same for the icosahedron. (*Suggestion*: Look for ideas in geometry texts, in their discussions of the Platonic solids. One such book is the wonderful work of H. S. M. Coxeter, *Introduction to Geometry*, Wiley, New York, 1961.)

13. (a) [BB] Suppose $\mathcal{G}$ is a graph with n vertices, each of which has degree $d \geq \frac{n-1}{2}$. Prove that $\mathcal{G}$ contains a Hamiltonian path. [*Hint*: Add an extra vertex to $\mathcal{G}$ which is adjacent to every vertex and use Dirac's Theorem.]

(b) Does the graph shown on the left have a Hamiltonian path? If so, find it. If it doesn't have one, explain why not.

(c) Repeat the previous question for the graph on the right.

(d) Does the converse of (a) hold; that is, if a graph has a Hamiltonian path, must the degree of every vertex be at least $\frac{n-1}{2}$? Explain your answer. What about the converse of Dirac's Theorem?

(e) [BB] Give an example of a graph which has a Hamiltonian path but no Hamiltonian cycle.

14. (Ore's Theorem) Suppose $\mathcal{G}$ is a graph with $n \geq 3$ vertices and that the sum of the degrees of any two nonadjacent vertices is at least n. Prove that $\mathcal{G}$ is Hamiltonian by starting with a path $\mathcal{P}: v_1 v_2 \cdots v_t$ of greatest length, as in the proof of Dirac's Theorem, and then considering separately the cases where

---

[5]In fact, any complete graph actually has this maximum number of edge disjoint cycles.

(a) [BB] v_1 and v_t are adjacent, and

(b) v_1 and v_t are not adjacent.

15. [BB] Suppose G is a graph with $n \geq 2$ vertices such that the sum of the degrees of any two nonadjacent vertices is at least $n - 1$. Prove that G has a Hamiltonian path.

16. Answer true or false and in each case either give a proof or provide a counterexample.

(a) A Hamiltonian graph contains no proper cycles.

(b) Every vertex in a Hamiltonian graph has degree 2.

(c) [BB] Every Eulerian graph is Hamiltonian.

(d) Every Hamiltonian graph is Eulerian.

17. Let G be a graph with at least three vertices.

(a) [BB] If there is a Hamiltonian path between any two vertices of G, need G contain a Hamiltonian cycle? Explain.

(b) If, at every vertex v in G, there is a Hamiltonian path which starts at v, need G contain a Hamiltonian cycle? Explain.

(c) Is it possible for there to exist an Eulerian trail between any two vertices of G? If so, need G contain an Eulerian circuit? Explain.

18. A connected graph G has 11 vertices and 53 edges. Show that G is Hamiltonian but not Eulerian.

19. Find a necessary and sufficient condition on m and n in order for $K_{m,n}$ to be Hamiltonian. Prove your answer.

10.3 THE ADJACENCY MATRIX

Graphs occur with increasing frequency in modern-day problems. While theoretically, any problem associated with a finite graph is solvable, in practice, the number of cases to consider is often so large and the time needed to deal with each case so great that an exhaustive search of all possibilities is impossible. Accordingly, the discovery of new graph-based algorithms and ways to improve efficiency are flourishing areas of mathematical research today. In order to write an algorithm which requires the input of a graph, we first must decide how to code the pertinent information which describes a graph. For this purpose, the *adjacency matrix* is commonly used.

10.3.1 DEFINITION ▶ Let G be a graph with n vertices labeled $v_1, v_2, \ldots, v_n$. For each i and j with $1 \leq i, j \leq n$, define

$$a_{ij} = \begin{cases} 1 & \text{if } v_i v_j \text{ is an edge} \\ 0 & \text{if } v_i v_j \text{ is not an edge.} \end{cases}$$

The *adjacency matrix* of G is the $n \times n$ matrix $A = [a_{ij}]$ whose (i, j) entry is a_{ij}.

EXAMPLE 6 Figure 10.15 shows a graph G and its adjacency matrix A. ▲

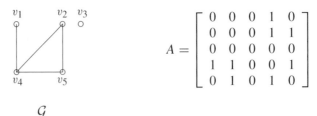

$$A = \begin{bmatrix} 0 & 0 & 0 & 1 & 0 \\ 0 & 0 & 0 & 1 & 1 \\ 0 & 0 & 0 & 0 & 0 \\ 1 & 1 & 0 & 0 & 1 \\ 0 & 1 & 0 & 1 & 0 \end{bmatrix}$$

G

Figure 10.15   A graph and its adjacency matrix.

EXAMPLE 7 The adjacency matrix of the complete bipartite graph on the sets $\mathcal{V}_1 = \{v_1, v_2\}$ and $\mathcal{V}_2 = \{v_3, v_4, v_5\}$ is

$$\begin{bmatrix} 0 & 0 & 1 & 1 & 1 \\ 0 & 0 & 1 & 1 & 1 \\ 1 & 1 & 0 & 0 & 0 \\ 1 & 1 & 0 & 0 & 0 \\ 1 & 1 & 0 & 0 & 0 \end{bmatrix}.$$ ▲

Since the adjacency matrix records all the incidence relations in a graph, it is not surprising that it can be used to give a lot of information about the graph. We list here some of the basic properties of an adjacency matrix. Some are obvious, others not quite so, and perhaps the most curious of all, due to Gustav Kirchhoff, we leave to Chapter 12.

10.3.2 PROPERTIES OF AN ADJACENCY MATRIX ▶

Let $\mathcal{G}$ be a graph with vertices $v_1, v_2, \ldots, v_n$ and let $A = [a_{ij}]$ be the adjacency matrix of $\mathcal{G}$.

1. The diagonal entries of A are all 0; that is, $a_{ii} = 0$ for $i = 1, \ldots, n$. This follows because an edge from vertex v_i to v_i is a loop and loops are not allowed in graphs. (Actually, we can define the adjacency matrix of a pseudograph and show that it has most of the properties given here, but we shall restrict our attention to graphs.)

2. The adjacency matrix is *symmetric*, that is, $a_{ij} = a_{ji}$ for all i, j.

   Conversely, given any symmetric matrix A which contains only 0's and 1's and only 0's on its diagonal, there exists a graph $\mathcal{G}$ whose adjacency matrix is A. Thus, there is a one-to-one correspondence between graphs and symmetric 0, 1 matrices with only 0's on the diagonal.

   The next three properties are less obvious than the first two. We will discuss them in some detail after a couple of Pauses.

3. $\deg v_i$ is the number of 1's in row i; this is also the number of 1's in column i since row i and column i are the same, by symmetry.

4. The (i, j) entry of A^2 is the number of different walks from v_i to v_j which include two edges; thus, the degree of v_i is the ith entry on the main diagonal of A^2.

5. In general, for any $n \geq 1$, the (i, j) entry of A^n is the number of walks from v_i to v_j which include n edges.

Pause 8 Can the 4×4 identity matrix $I = \begin{bmatrix} 1 & 0 & 0 & 0 \\ 0 & 1 & 0 & 0 \\ 0 & 0 & 1 & 0 \\ 0 & 0 & 0 & 1 \end{bmatrix}$ be the adjacency matrix of a graph?

Pause 9 Find a graph whose adjacency matrix is $A = \begin{bmatrix} 0 & 0 & 0 & 1 \\ 0 & 0 & 1 & 0 \\ 0 & 1 & 0 & 0 \\ 1 & 0 & 0 & 0 \end{bmatrix}.$

As mentioned, the powers of an adjacency matrix A evidently have special significance. In order to understand why the (i, j) entry of A^2 is the number of walks of length 2 between v_i and v_j in the graph corresponding to A, remember that this entry is the dot product of row i and column j of A. Since A contains only 0's and 1's, this dot product is just the number of coordinates in which row i and column j each have a 1. For example, if row i of A were the vector $[0, 0, 1, 0, 1, 1, 0]$ and column j the vector $[1, 1, 1, 0, 0, 1, 0]$, then the dot product (row i) $\cdot$ (column j) $= 2$, corresponding to the two coordinates (third and sixth) where each vector has a 1. How does it happen that a row and a column have a 1 in the same coordinate? In our example, row i and column j each have third coordinate 1; this corresponds to the fact that there is an edge in the graph between v_i and v_3 and an edge between v_j and v_3. There is a walk in the graph from v_i to v_j which uses two edges, namely, $v_i v_3 v_j$. The dot product of row i and column j is therefore the number of walks of length 2 from v_i to v_j.

In general, there is one walk of length 2 from a vertex to itself for each edge incident with that vertex. Thus, the degree of vertex v_i is the diagonal entry a_{ii} of the square of the adjacency matrix, as asserted in Property 4. The rest of Property 4 and Property 5 can be justified with similar arguments.

EXAMPLE 8 Referring to Fig 10.15, $A^2 = \begin{bmatrix} 1 & 1 & 0 & 0 & 1 \\ 1 & 2 & 0 & 1 & 1 \\ 0 & 0 & 0 & 0 & 0 \\ 0 & 1 & 0 & 3 & 1 \\ 1 & 1 & 0 & 1 & 2 \end{bmatrix}$.

The $(4, 2)$ entry of A^2 is 1 corresponding to the fact that there is precisely one walk of length 2 between v_4 and v_2 in the associated graph: $v_4 v_5 v_2$. The $(4, 4)$ entry of A^2 is 3; there are three walks of length 2 from v_4 back to v_4, one for each edge incident with v_4. ▲

Pause 10 The third power of the matrix A in Fig 10.15 is $A^3 = \begin{bmatrix} 0 & 1 & 0 & 3 & 1 \\ 1 & 2 & 0 & 4 & 3 \\ 0 & 0 & 0 & 0 & 0 \\ 3 & 4 & 0 & 2 & 4 \\ 1 & 3 & 0 & 4 & 2 \end{bmatrix}$.

The $(4, 5)$ entry of A^3 is 4. Thus, there are four walks of length 3 from v_4 to v_5 in the graph. What are they?

Since the adjacency matrix of a graph records the number of vertices and the adjacencies between them, the following theorem is straightforward.

10.3.3 THEOREM ▶ Two graphs are isomorphic if and only if their vertices can be labeled in such a way that the corresponding adjacency matrices are equal.

EXAMPLE 9 The graphs in Fig 10.16 have been labeled so that each has the adjacency matrix

$$\begin{bmatrix} 0 & 1 & 0 & 1 \\ 1 & 0 & 1 & 0 \\ 0 & 1 & 0 & 1 \\ 1 & 0 & 1 & 0 \end{bmatrix}.$$

(These are the graphs which appeared in Fig 9.23, with the labels A, B, C, D replaced by v_1, v_2, v_3, v_4, respectively.) As we earlier noted, these graphs are isomorphic. ▲

Figure 10.16   Two graphs labeled so as to show that they are isomorphic.

Suppose we are presented with two graphs which are already labeled. Can we tell from the adjacency matrices whether or not the graphs are isomorphic? Consider again the graphs of Fig 10.16, but labeled as in Fig 10.17.

Figure 10.17   Two isomorphic graphs.

The adjacency matrices A_1 and A_2 of $\mathcal{G}_1$ and $\mathcal{G}_2$, respectively, are

$$A_1 = \begin{bmatrix} 0 & 1 & 0 & 1 \\ 1 & 0 & 1 & 0 \\ 0 & 1 & 0 & 1 \\ 1 & 0 & 1 & 0 \end{bmatrix} \quad \text{and} \quad A_2 = \begin{bmatrix} 0 & 0 & 1 & 1 \\ 0 & 0 & 1 & 1 \\ 1 & 1 & 0 & 0 \\ 1 & 1 & 0 & 0 \end{bmatrix}.$$

Although these graphs are isomorphic, their matrices are not equal simply because of the way the graphs were presented to us, their labels already in place. Observe, however, that the following reassignment of labels to $\mathcal{G}_1$ defines an isomorphism $\mathcal{G}_1 \to \mathcal{G}_2$.

(1)
$$\begin{array}{ccc} v_1 & \to & u_1 \\ v_2 & \to & u_4 \\ v_3 & \to & u_2 \\ v_4 & \to & u_3 \end{array}$$

We use this isomorphism to obtain a *permutation matrix* P, that is, a matrix whose rows are the rows of the identity matrix, but not necessarily in their natural order. Precisely, let P be that 4×4 matrix with row 1 of the 4×4 identity matrix as its first row, row 2 of the identity as its fourth row, row 3 of the identity as its second row, and row 4 of the identity as its third row. In other words,

$$P = \begin{bmatrix} 1 & 0 & 0 & 0 \\ 0 & 0 & 1 & 0 \\ 0 & 0 & 0 & 1 \\ 0 & 1 & 0 & 0 \end{bmatrix}.$$

Notice how the isomorphism in (1) determines P. If the isomorphism maps v_i to u_j, we put row i of the identity matrix into row j of P. Just as the relabeling of the vertices of $\mathcal{G}_1$ given by (1) transforms $\mathcal{G}_1$ into $\mathcal{G}_2$, the matrix P transforms the adjacency matrix of $\mathcal{G}_1$ into the adjacency matrix of $\mathcal{G}_2$ in the sense that $PA_1 P^T = A_2$. The notation P^T means the *transpose* of the matrix P, that is, the matrix obtained from P by interchanging rows and columns.

$$PA_1 P^T = \begin{bmatrix} 1 & 0 & 0 & 0 \\ 0 & 0 & 1 & 0 \\ 0 & 0 & 0 & 1 \\ 0 & 1 & 0 & 0 \end{bmatrix} \begin{bmatrix} 0 & 1 & 0 & 1 \\ 1 & 0 & 1 & 0 \\ 0 & 1 & 0 & 1 \\ 1 & 0 & 1 & 0 \end{bmatrix} \begin{bmatrix} 1 & 0 & 0 & 0 \\ 0 & 0 & 0 & 1 \\ 0 & 1 & 0 & 0 \\ 0 & 0 & 1 & 0 \end{bmatrix}$$

$$= \begin{bmatrix} 0 & 1 & 0 & 1 \\ 0 & 1 & 0 & 1 \\ 1 & 0 & 1 & 0 \\ 1 & 0 & 1 & 0 \end{bmatrix} \begin{bmatrix} 1 & 0 & 0 & 0 \\ 0 & 0 & 0 & 1 \\ 0 & 1 & 0 & 0 \\ 0 & 0 & 1 & 0 \end{bmatrix}$$

$$= \begin{bmatrix} 0 & 0 & 1 & 1 \\ 0 & 0 & 1 & 1 \\ 1 & 1 & 0 & 0 \\ 1 & 1 & 0 & 0 \end{bmatrix} = A_2$$

Pause 11   Graph $\mathcal{G}_1$ shown at the right is the same as $\mathcal{G}$ in Fig 10.15, except that the labels v_1, v_2, v_3, v_4, v_5 have been replaced by u_1, u_4, u_3, u_2, u_5, respectively. Find the adjacency matrix A_1 of $\mathcal{G}_1$. Then find a permutation matrix such that $PAP^T = A_1$.

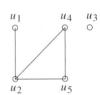

The proof of the following theorem, which we have been illustrating, is not especially interesting and will not be included. The key idea is that if $\mathcal{G}$ is a graph with adjacency matrix A, then an isomorphism from $\mathcal{G}$ to another graph just amounts to renumbering the vertices of $\mathcal{G}$. Renumbering vertices corresponds to permuting the rows (equivalently, computing PA) and permuting the columns of A (equivalently, computing AP^T), hence, changing A to the matrix PAP^T. See the Exercises for more details.

10.3.4 THEOREM ▶   Labeled graphs $\mathcal{G}_1$ and $\mathcal{G}_2$, with adjacency matrices A_1 and A_2, respectively, are isomorphic if and only if $A_2 = PA_1 P^T$ for some permutation matrix P.

Answers to Pauses   **8.** No, by Property 1.

9. Here is a graph with adjacency matrix $A = \begin{bmatrix} 0 & 0 & 0 & 1 \\ 0 & 0 & 1 & 0 \\ 0 & 1 & 0 & 0 \\ 1 & 0 & 0 & 0 \end{bmatrix}.$

10. There are four walks of length 3 from v_4 to v_5: $v_4v_5v_2v_5$, $v_4v_1v_4v_5$, $v_4v_5v_4v_5$, and $v_4v_2v_4v_5$.

$$
\textbf{11.}\ A_1 =
\begin{bmatrix}
0 & 1 & 0 & 0 & 0 \\
1 & 0 & 0 & 1 & 1 \\
0 & 0 & 0 & 0 & 0 \\
0 & 1 & 0 & 0 & 1 \\
0 & 1 & 0 & 1 & 0
\end{bmatrix}
\qquad
P =
\begin{bmatrix}
1 & 0 & 0 & 0 & 0 \\
0 & 0 & 0 & 1 & 0 \\
0 & 0 & 1 & 0 & 0 \\
0 & 1 & 0 & 0 & 0 \\
0 & 0 & 0 & 0 & 1
\end{bmatrix}
$$

The equation $PAP^T = A_1$ corresponds to the reassignment of labels to the graph $\mathcal{G}$, whose adjacency matrix is A. The matrix P is therefore the 5×5 matrix where row 1 of the identity is in row 1, where row 2 of the identity is in row 4, where row 3 of the identity is in row 3, where row 4 of the identity is in row 2, and row 5 of the identity is in row 5.

EXERCISES

The symbol [BB] means that an answer can be found in the Back of the Book.

1. Find the adjacency matrices of the graphs in Figs 10.1 [BB] and 10.3.

2. What is the adjacency matrix of K_n? Label the vertices of $K_{m,n}$ so that the adjacency matrix has an especially nice form.

3. **(a)** [BB] Let A be the adjacency matrix of the graph $\mathcal{G}_2$ shown in Fig 10.8. Determine the $(3, 5)$ entry of A^3 by inspection of the graph, that is, without writing down A explicitly. Determine the $(2, 2)$ entry of A^3 by similar means.

(b) Repeat part (a) for the graph $\mathcal{G}_1$ of Fig 10.8.

4. [BB] What is the significance of the total number of 1's in the adjacency matrix of a graph?

5. Let A be the adjacency matrix of a graph $\mathcal{G}$ whose vertex set is $\{v_1, \ldots, v_n\}$. Prove that the ith entry on the diagonal of A^3 equals twice the number of different triangles which contain vertex v_i.

6. Suppose that $\mathcal{G}$ is a graph with adjacency matrix A.

(a) [BB] Show that the number of walks of length 2 in $\mathcal{G}$ is the sum of the entries of the matrix A^2.

(b) Let d_i denote the degree of the vertex v_i in $\mathcal{G}$. Show that the sum of the entries of A^2 is also $\sum d_i^2$.

7. Find the adjacency matrices A_1 and A_2 of the graphs $\mathcal{G}_1$ and $\mathcal{G}_2$ shown. Find a permutation matrix P such that $A_2 = PA_1P^T$, thus proving that $\mathcal{G}_1$ and $\mathcal{G}_2$ are isomorphic. [*Hint:* See Pause 6 of Section 9.3.]

$\mathcal{G}_1$

$\mathcal{G}_2$

8. **(a)** [BB] Find the adjacency matrices A_1 and A_2 of the graphs $\mathcal{G}_1$ and $\mathcal{G}_2$ shown.

$\mathcal{G}_1$

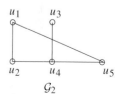

$\mathcal{G}_2$

(b) [BB] Explain why the function $\varphi \colon \mathcal{G}_1 \to \mathcal{G}_2$ defined by

$$\varphi(v_1) = u_4, \ \varphi(v_2) = u_5, \ \varphi(v_3) = u_1,$$
$$\varphi(v_4) = u_3, \ \varphi(v_5) = u_2$$

is an isomorphism.

(c) [BB] Find a permutation matrix P which corresponds to the isomorphism in (b) such that $PA_1P^T = A_2$.

9. Repeat Exercise 8 for the graphs $\mathcal{G}_1$ and $\mathcal{G}_2$ shown. For φ, take the function $\mathcal{G}_1 \to \mathcal{G}_2$ defined by

$$\varphi(v_1) = u_4, \ \varphi(v_2) = u_1, \ \varphi(v_3) = u_5,$$
$$\varphi(v_4) = u_6, \ \varphi(v_5) = u_3, \ \varphi(v_6) = u_2.$$

$$A_2 = \begin{bmatrix} 0 & 1 & 0 & 0 \\ 1 & 0 & 1 & 1 \\ 0 & 1 & 0 & 0 \\ 0 & 1 & 0 & 0 \end{bmatrix}$$

10. Repeat Exercise 8 for the graphs $\mathcal{G}_1$ and $\mathcal{G}_2$ shown. For φ, take the function $\mathcal{G}_1 \to \mathcal{G}_2$ defined by

$$\varphi(v_1) = u_1, \ \varphi(v_2) = u_2, \ \varphi(v_3) = u_6, \ \varphi(v_4) = u_8,$$
$$\varphi(v_5) = u_4, \ \varphi(v_6) = u_3, \ \varphi(v_7) = u_7, \ \varphi(v_8) = u_5.$$

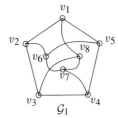

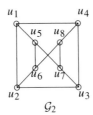

(b) $A_1 = \begin{bmatrix} 0 & 0 & 0 & 1 & 1 \\ 0 & 0 & 1 & 0 & 0 \\ 0 & 1 & 0 & 1 & 0 \\ 1 & 0 & 1 & 0 & 1 \\ 1 & 0 & 0 & 1 & 0 \end{bmatrix}$,

$A_2 = \begin{bmatrix} 0 & 1 & 1 & 0 & 0 \\ 1 & 0 & 0 & 1 & 1 \\ 1 & 0 & 0 & 0 & 0 \\ 0 & 1 & 0 & 0 & 1 \\ 0 & 1 & 0 & 1 & 0 \end{bmatrix}$

11. Let $A = \begin{bmatrix} a & b & c \\ p & q & r \\ x & y & z \end{bmatrix}$ and let $P = \begin{bmatrix} 0 & 1 & 0 \\ 0 & 0 & 1 \\ 1 & 0 & 0 \end{bmatrix}$.

(c) $A_1 = \begin{bmatrix} 0 & 1 & 0 & 1 & 0 & 1 \\ 1 & 0 & 1 & 0 & 1 & 0 \\ 0 & 1 & 0 & 1 & 0 & 1 \\ 1 & 0 & 1 & 0 & 1 & 0 \\ 0 & 1 & 0 & 1 & 0 & 1 \\ 1 & 0 & 1 & 0 & 1 & 0 \end{bmatrix}$,

Thus P is a permutation matrix whose rows are those of the 3×3 identity matrix in the order 2, 3, 1.

(a) [BB] Compute PA and compare with A.

(b) Compute AP^T and compare with A.

(c) Compute PAP^T and compare with A.

$A_2 = \begin{bmatrix} 0 & 1 & 0 & 1 & 1 & 0 \\ 1 & 0 & 1 & 0 & 0 & 1 \\ 0 & 1 & 0 & 1 & 0 & 1 \\ 1 & 0 & 1 & 0 & 1 & 0 \\ 1 & 0 & 0 & 1 & 0 & 1 \\ 0 & 1 & 1 & 0 & 1 & 0 \end{bmatrix}$

12. Let $A = \begin{bmatrix} a & b & c & d \\ p & q & r & s \\ x & y & z & w \\ \alpha & \beta & \gamma & \delta \end{bmatrix}$ and let P be the permu-

tation matrix $\begin{bmatrix} 0 & 0 & 1 & 0 \\ 1 & 0 & 0 & 0 \\ 0 & 0 & 0 & 1 \\ 0 & 1 & 0 & 0 \end{bmatrix}$.

Find PA, AP^T, and PAP^T without calculation but with explanations.

13. For each pair of matrices A_1, A_2 shown, decide whether or not there is a permutation matrix P with $A_2 = PA_1P^T$. Either find P or explain why no such P exists.

(a) [BB] $A_1 = \begin{bmatrix} 0 & 1 & 0 & 0 \\ 1 & 0 & 1 & 1 \\ 0 & 1 & 0 & 1 \\ 0 & 1 & 1 & 0 \end{bmatrix}$,

14. [BB] Let A be the adjacency matrix of a bipartite graph. Prove that the diagonal entries of A^{37} are all equal to 0.

15. Let A be the adjacency matrix of a graph $\mathcal{G}$.

(a) [BB] Find a necessary and sufficient condition for the matrix A^2 to be the adjacency matrix of some graph.

(b) Find a necessary and sufficient condition for A^3 to be an adjacency matrix.

16. Let A be the adjacency matrix of a graph $\mathcal{G}$ with at least two vertices. Prove that $\mathcal{G}$ is connected if and only if, for some natural number n, the matrix $B = A + A^2 + \cdots + A^n$ has no zero entries.

17. Suppose A_1 and A_2 are the adjacency matrices of isomorphic graphs $\mathcal{G}_1$ and $\mathcal{G}_2$, respectively. Show that A_1 and A_2 have the same characteristic polynomial. [*Hint:* First show that if P is a permutation matrix, then $P^T = P^{-1}$.]

18. [BB] Discuss other ways, besides the adjacency matrix, that a graph could be stored in a computer.

10.4   SHORTEST PATH ALGORITHMS

In this section, we consider graphs whose edges have numbers attached to them. Typically, the number associated with an edge is a unit of time, distance, cost, or "capacity" in some sense.

10.4.1   DEFINITION ▶   A *weighted graph* is a graph G in which each edge, e, is assigned a nonnegative real number, $w(e)$, called the *weight* of e. The *weight of a subgraph* of G (often a path or a trail) is the sum of the weights of the edges of the subgraph.

One famous problem concerning weighted graphs is known as the Traveling Salesman's Problem.

The Traveling Salesman's Problem

On a typical business trip, a traveling salesman visits various towns and cities. If he wants to avoid having to pass through the same community twice, he needs a Hamiltonian cycle through the map of towns and air routes. This map can be thought of as a graph in an obvious way. Assigning to each edge a weight equal to the distance between the cities at the ends, the traveling salesman's graph becomes a weighted graph. Among all Hamiltonian cycles (assuming there are any), what our salesman would like to find is one whose weight is a minimum, in order to economize on flying time and expenses. The problem of finding a minimum Hamiltonian cycle in a weighted (Hamiltonian) graph is called the *Traveling Salesman's Problem*. It can be solved by laboriously calculating all Hamiltonian cycles and then selecting the most economical. At the time of writing, it is unknown whether or not a more efficient algorithm is possible. In fact, the Traveling Salesman's Problem is many people's favorite example of an *NP-complete* problem: It is an open question as to whether or not there exists an efficient (polynomial time) algorithm for its solution.

Our aim in the rest of this section is to discuss a certain problem concerning weighted graphs for which, unlike the Traveling Salesman's Problem, complete and efficient solutions exist. A *shortest path* between two vertices in a weighted graph is a path of least weight. (In an unweighted graph, a shortest path means one with the fewest number of edges.) Numerous algorithms for finding shortest paths have been discovered. We will present two of them here and show how they apply to the graph of Fig 10.18, which we might view as a map, the vertices representing towns, the edges, roads, and the weights of the edges, distances.

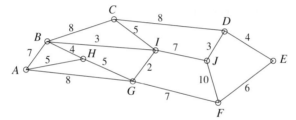

Figure 10.18

The first algorithm we present is due to Holland's Edsger Dijkstra (1930–), one of today's leading computer scientists and discoverer of a number of important graph algorithms. Dijkstra's algorithm[6] finds the shortest path from a specified vertex A to another specified vertex E. If continued indefinitely, it gives shortest paths from A to all other vertices in the graph. If no path exists, then this is identified by the algorithm. It proceeds by progressively assigning to each vertex v in the graph an ordered pair (x, d), where d is the shortest distance from A to v and xv is the last edge on the shortest path. Thus, if E is eventually labeled (w, t), the shortest path from A to E is t units long and the last edge of the shortest path is wE. The first coordinate of the label for w determines the second last edge of the shortest path, and by continuing to work backward, the entire path can be found. If vertex E never gets labeled, there is no path from A to E; the graph is not connected.

10.4.2 DIJKSTRA'S ALGORITHM ▶

To find a shortest path from vertex A to vertex E in a weighted graph, carry out the following procedure.

Step 1. Assign to A the label $(-, 0)$.

Step 2. Until E is labeled or no further labels can be assigned, do the following.

    (a) For each labeled vertex $u(x, d)$ and for each unlabeled vertex v adjacent to u, compute $d + w(e)$, where $e = uv$.

    (b) For each labeled vertex u and adjacent unlabeled vertex v giving minimum $d' = d + w(e)$, assign to v the label (u, d'). If a vertex can be labeled (x, d') for various vertices x, make any choice.

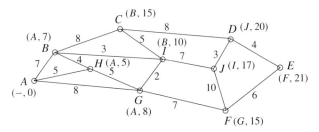

Figure 10.19   Dijkstra's shortest path algorithm applied to a weighted graph.

Here is how Dijkstra's algorithm works for the graph in Fig 10.18. (Refer to Fig 10.19.) First, give A the label $(-, 0)$. There are three edges incident with A with weights 7, 5, and 8. Since $d = 0$, vertex H gives the smallest value of $d + w(e)$, so H acquires the label $(A, 5)$. Now we repeat Step 2 for the two vertices labeled so far.

There are two unlabeled vertices adjacent to the already labeled vertex A. The numbers $d + w(e)$ are $0 + 7 = 7$ and $0 + 8 = 8$. There are also two unlabeled vertices adjacent to the other labeled vertex, H, and for these, the numbers $d + w(e)$ are

[6]E. W. Dijkstra, "A Note on Two Problems in Connection with Graphs," *Numerische Mathematik* **1** (1959), 269–271.

$5 + 4 = 9$ and $5 + 5 = 10$. The smallest $d + w(e)$ is 7, corresponding to the labeled vertex $u = A$ and the unlabeled $v = B$. Thus, B is labeled $(A, 7)$. Again we repeat Step 2.

Now there are three labeled vertices. There is one unlabeled vertex adjacent to the labeled vertex A, and for this, $d + w(e) = 0 + 8 = 8$. There is also just one unlabeled vertex adjacent to the labeled H, and here, $d + w(e) = 5 + 5 = 10$. There are two unlabeled vertices adjacent to the third labeled vertex B; for C, $d + w(e) = 7 + 8 = 15$ and for I, $d + w(e) = 10$. The smallest $d + w(e)$ is 8, corresponding to edge AG. So G acquires the label $(A, 8)$.

We repeat Step 2. There are four labeled vertices, A, B, H, and G, and the algorithm requires that we look at all unlabeled vertices adjacent to each of these. Since all vertices adjacent to A and to H have already been labeled, we have, in fact, only to look at B and G. There are two unlabeled vertices adjacent to B. For C, $d + w(e) = 7 + 8 = 15$; for I, $d + w(e) = 7 + 3 = 10$. There are two unlabeled vertices adjacent to G. For I, $d + w(e) = 8 + 2 = 10$; for F, $d + w(e) = 8 + 7 = 15$. The minimum $d + w(e)$ occurs with I and either edge BI or GI. We can therefore assign to I either the label $(B, 10)$ or the label $(G, 10)$; we opt for $(B, 10)$.

Repeating Step 2, we have only to look at vertices B, G and I. The only unlabeled vertex adjacent to B is C, for which $d + w(e) = 7 + 8 = 15$. The only unlabeled vertex adjacent to G is F, for which $d + w(e) = 8 + 7 = 15$. There are two unlabeled vertices adjacent to I. For C, $d + w(e) = 10 + 5 = 15$ and for J, $d + w(e) = 10 + 7 = 17$. Vertices C and F which tie for the minimum $d + w(e)$ are each labeled. Vertex F is labeled $(G, 15)$ and C can be labeled either $(B, 15)$ or $(I, 15)$; we choose $(B, 15)$.

Continuing in this way, the vertices of the graph acquire the labels shown in Fig 10.19. The shortest route from A to E has weight 21. A shortest path is $AGFE$, as we see by working backward from E. Since E was labeled last, the algorithm has actually found the length of a shortest route from A to any vertex. For instance, $ABIJ$ is a shortest path to J, of weight 17.

What is the complexity of Dijkstra's algorithm? We take as our basic operation an addition or comparison (which we weight equally). Each time we iterate Step 2, one or more new vertices acquire labels; the worst case occurs if only one new vertex is labeled each time. Assume, therefore, that at the kth iteration, k vertices have been labeled. Each of these vertices is conceivably adjacent to $n - k$ unlabeled vertices. For each such vertex there is one addition. After at most $k(n - k)$ additions in all, we must find the minimum of at most $k(n - k)$ numbers, a process requiring $k(n - k) - 1$ comparisons.[7] At worst, therefore, Dijkstra's algorithm requires

$$f(n) = \sum_{1}^{n-1} [\, 2k(n - k) - 1 \,] = \frac{n^3}{3} - \frac{4}{3}n + 1$$

operations, a function which is $\mathcal{O}(n^3)$. Note that $f(n)$ is also an upper bound for the total number of operations required if the algorithm runs until all vertices are labeled. [This was our assumption in computing $f(n)$.]

---

[7]Finding the minimum of t numbers requires $t - 1$ comparisons. (See Exercise 4 of Section 8.2.)

With a few minor modifications, the efficiency of Dijkstra's algorithm can be substantially increased. We present an improved version here and apply it, as with the former, to determine the weight of a shortest path from A to E in the graph of Fig 10.18.

In the new version, the starting point A is assigned a permanent label of 0, while all other vertices initially are assigned temporary labels of ∞. At each iteration, the temporary labels are decreased or left unchanged, and one additional vertex is assigned a permanent label, namely, the shortest distance from A to that particular vertex. This procedure continues until the required terminal point E acquires a permanent label, or until some iteration results in no temporary labels (including that of E) being changed. In the latter case, we can conclude that there

10.4.3 DIJKSTRA'S ALGORITHM (IMPROVED) ▶

is no path from A to E.

To find the length of a shortest path from vertex A to vertex E in a weighted graph, proceed as follows.

Step 1. Set $v_1 = A$ and assign to this vertex the permanent label 0. Assign every other vertex a temporary label of ∞, where ∞ is a symbol which, by definition, is deemed to be larger than any real number.

Step 2. Until E has been assigned a permanent label or no temporary labels are changed in (a) or (b), do the following.

    (a) Take the vertex v_i which most recently acquired a permanent label, say d. For each vertex v which is adjacent to v_i and has not yet received a permanent label, if $d + w(v_i v) < t$, the current temporary label of v, change the temporary label of v to $d + w(v_i v)$.

    (b) Take a vertex v which has a temporary label smallest among all temporary labels in the graph. Set $v_{i+1} = v$ and make its temporary label permanent. If there are several vertices v which tie for smallest temporary label, make any choice.

Here is how the algorithm works for the graph of Fig 10.18. We will describe the first few iterations in words, draw a figure summarizing the state we have reached, and then describe the remaining iterations. Readers might well wish to follow our description with their own diagrams.

At the start, $A = v_1$ is given the permanent label 0; all others are given temporary labels ∞. Next, H, B, and G have their temporary labels decreased to 5, 7, and 8, respectively, while all other temporary labels stay at ∞. Since 5 is the minimum of 5, 7, 8, we set $v_2 = H$ and make 5 its permanent label.

Now examine those vertices with temporary labels adjacent to v_2. For B, $5 + w(e) = 5 + 4 = 9$, but this is larger than its temporary label of 7, so no change occurs. For G, $5 + w(e) = 5 + 5 = 10$, again larger than the older temporary label of 8. Hence, no temporary labels are changed. The smallest temporary label in the graph at this point is 7, so this becomes the next permanent label and $v_3 = B$.

Next, we examine temporary vertices adjacent to v_3. This will assign to C a new temporary label of $7 + 8 = 15$ and, to I, a new temporary label of $7 + 3 = 10$.

The smallest temporary label is 8, on G; so $v_4 = G$ and this vertex acquires the permanent label 8. Figure 10.20 shows the present state of affairs. Permanent labels are circled; temporary labels are in parentheses.

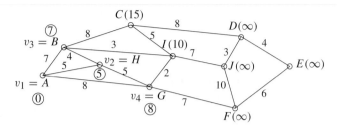

Figure 10.20   Dijkstra's algorithm, improved version.

Next, we focus on v_4. For I, we obtain $8 + 2 = 10$, the current label, so there is no change. For F, we get $8 + 7 = 15$, so F gets a temporary label of 15 and $v_5 = I$ is assigned the permanent label of 10. At the next iteration, we have a choice of C or F for v_6 and J gets a temporary label of 17. Assuming $v_6 = C$, then F will be v_7 and D will obtain a temporary label of 23.

At the next step, E gets a temporary label of 21 and $v_8 = J$. Then D gets its temporary label lowered to 20 and becomes v_9. Finally, $v_{10} = E$ with a permanent label of 21. Note that the permanent labels are exactly the second coordinates of the labels assigned to the vertices in Fig 10.19.

As stated, this algorithm gives the length of the shortest path but not the path itself. If at each choice of a permanently labeled vertex v_i, however, we also make note of the vertex v_ℓ from which its permanent label arose, using

(permanent label of v_ℓ) + (weight of edge $v_\ell v_i$) = permanent label of v_i,

then we can easily trace back the shortest path. Also—and the original algorithm has the same property—if allowed to continue until all vertices are labeled, the improved version will give the shortest distances from A to all other vertices.

Before introducing another shortest path algorithm, we justify the adjective *improved*, which we applied to the second version of Dijkstra's algorithm by showing that the complexity function for the second version indeed has smaller order than that for the original.

In the second version, exactly one vertex is given a permanent label at each iteration of Step 2. Assume v_k has just received a permanent label. There are, at worst, $n - k$ vertices adjacent to v_k. For each such vertex, one addition is required and then one comparison to determine whether or not to change a temporary label. This process requires $2(n - k)$ operations. Finally, to choose the smallest among $n - k$ temporary labels requires an additional $n - k - 1$ comparisons. In all, at most

$$f(n) = \sum_{k=1}^{n-1} [\, 2(n - k) + (n - k - 1) \,]$$

$$= \sum_{1}^{n-1} [\, 3(n-k) - 1\,]$$

$$= \frac{3}{2}n^2 - \frac{5}{2}n + 1 = \mathcal{O}(n^2)$$

operations are required. Recalling that the original version of Dijsktra's algorithm was $\mathcal{O}(n^3)$, we see that the second is indeed more efficient than the original. Again, we also observe that the complexity function is still $\mathcal{O}(n^2)$ even if it runs until all vertices get labeled.

If our goal is to find the shortest path between every pair of vertices in a weighted graph with n vertices, we can employ either version of Dijkstra's algorithm, letting it run until all vertices have acquired their final labels and repeating this procedure for each of the n possible starting points. The complexity functions increase by a factor of n to $\mathcal{O}(n^4)$ in the original version and to $\mathcal{O}(n^3)$ in the improved case. There is also an algorithm, due to R. W. Floyd[8] and S. Warshall,[9] which determines the shortest distances between all pairs of vertices in a graph. This algorithm is popular because it is so easy to describe.

10.4.4 THE FLOYD–WARSHALL ALGORITHM ▶

To find the shortest distances between all pairs of vertices in a weighted graph where the vertices are $v_1, v_2, \ldots, v_n$, carry out the following procedure.

Step 1. For $i = 1$ to n, set $d(i, i) = 0$. For $i \neq j$, if $v_i v_j$ is an edge, let $d(i, j)$ be the weight of this edge; otherwise, set $d(i, j) = \infty$.

Step 2. For $k = 1$ to n,

for $i, j = 1$ to n, let $d(i, j) = \min\{d(i, j), \ d(i, k) + d(k, j)\}$

The final value of $d(i, j)$ is the shortest distance from v_i to v_j.

Initially, the algorithm sets the shortest distance from v_i to v_j to be the length of edge $v_i v_j$, if this is an edge. After the first iteration of Step 2 ($k = 1$), this shortest distance has been replaced by the length of the path $v_i v_1 v_j$, if this is a path. In general, after stage k, the algorithm has determined the shortest distance from v_i to v_j via the vertices $v_1, v_2, \ldots, v_k$. This distance is the true shortest distance after $k = n$. In Fig 10.21, we show a graph, the initial values of $d(i, j)$, and the values of $d(i, j)$ after each change in k.

The Floyd–Warshall algorithm is very efficient from the point of view of storage since it can be implemented by just updating the matrix of distances with each change in k; there is no need to store different matrices. In many specific applications, it is faster than either version of Dijkstra's algorithm although, like the improved version, it too is $\mathcal{O}(n^3)$.

---

[8]R. W. Floyd, "Algorithm 97: Shortest Path," *Communications of the Association for Computing Machinery* **5** (1962), 345.

[9]S. Warshall, "A Theorem on Boolean Matrices," *Journal of the Association for Computing Machinery* **9** (1962), 11–12.

Initial values of $d(i, j)$

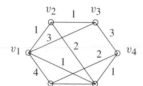

	v_1	v_2	v_3	v_4	v_5	v_6
v_1	0	1	3	∞	1	4
v_2	1	0	1	∞	2	∞
v_3	3	1	0	3	∞	∞
v_4	∞	∞	3	0	1	2
v_5	1	2	∞	1	0	2
v_6	4	∞	∞	2	2	0

After $k = 1$

0	1	3	∞	1	4
1	0	1	∞	2	5
3	1	0	3	4	7
∞	∞	3	0	1	2
1	2	4	1	0	2
4	5	7	2	2	0

$\longrightarrow$

After $k = 2$

0	1	2	∞	1	4
1	0	1	∞	2	5
2	1	0	3	3	6
∞	∞	3	0	1	2
1	2	3	1	0	2
4	5	6	2	2	0

$\longrightarrow$

After $k = 3$

0	1	2	5	1	4
1	0	1	4	2	5
2	1	0	3	3	6
5	4	3	0	1	2
1	2	3	1	0	2
4	5	6	2	2	0

After $k = 4$

0	1	2	5	1	4
1	0	1	4	2	5
2	1	0	3	3	5
5	4	3	0	1	2
1	2	3	1	0	2
4	5	5	2	2	0

$\longrightarrow$

After $k = 5$

0	1	2	2	1	3
1	0	1	3	2	4
2	1	0	3	3	5
2	3	3	0	1	2
1	2	3	1	0	2
3	4	5	2	2	0

$\longrightarrow$

After $k = 6$

0	1	2	2	1	3
1	0	1	3	2	4
2	1	0	3	3	5
2	3	3	0	1	2
1	2	3	1	0	2
3	4	5	2	2	0

Figure 10.21   An application of the Floyd–Warshall algorithm.

EXERCISES

The symbol [BB] means that an answer can be found in the Back of the Book.

1. The Traveling Salesman's Problem is that of finding the Hamiltonian cycle of least weight in a Hamiltonian graph. Is this what a traveling salesman necessarily wants to do? Discuss.

2. (a) [BB] Apply the original version of Dijkstra's algorithm to find the length of the shortest path from A to every other vertex in the figure. Show the final labels on all vertices. Also, find the shortest path from A to E.

(b) Apply the original version of Dijkstra's algorithm to find the length of the shortest path from H to every other vertex in the graph. Show the final labels on all vertices. Find the shortest path from H to D.

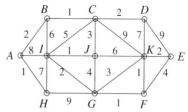

3. [BB; A only] Apply the improved version of Dijkstra's algorithm to find the length of a shortest path from A and from H to every other vertex in the graph of Exercise 2. In each case, exhibit an order in which permanent labels might be assigned.

4. Use the first form of Dijkstra's algorithm to find the shortest path from A to R (and its length) in the graph shown. Show the final labels on all vertices.

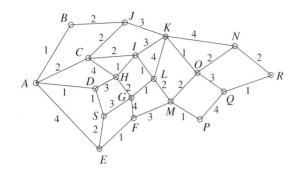

5. [BB] Use the improved version of Dijkstra's algorithm to find the length of a shortest path from A to R in the graph of Exercise 4. Also, exhibit an order in which permanent labels might be assigned.

6. Use the original form of Dijkstra's algorithm to find the shortest path from A to T in the graph shown. Label all vertices.

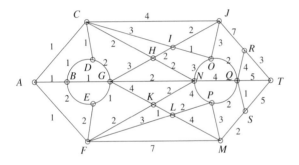

7. [BB] Use the improved version of Dijkstra's algorithm to find the shortest path from A to E (and its length). Label all vertices.

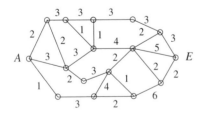

8. Use the improved version of Dijkstra's algorithm to find the shortest path from A to E in the graph shown. Show the final labels on all vertices.

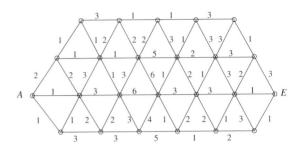

9. [BB] Could Dijkstra's algorithm (original version), employed to determine a shortest path from A to E in a weighted graph, terminate before E is labeled? Could the improved algorithm terminate before E acquires a permanent label? Explain.

10. (a) If weights were assigned to the edges of the graph shown in Exercise 1 of Section 10.2, the Traveling Salesman's Problem would not have a solution. Why not?

(b) Despite this observation, our salesman still has to complete the trip. Assigning weights as shown in the figure, and assuming the salesman starts at A and does not wish to travel along the same edge more than once, find a most efficient route for the trip.

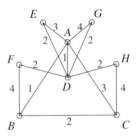

(c) Suppose the salesman is willing to cover the same edge more than once. Is the route found in (b) still the most efficient?

(d) Use the original form of Dijkstra's algorithm to find the shortest paths from E to each of the other vertices in the above graph. Label all vertices.

11. (a) [BB] How could any of the algorithms presented in this section be used to find the path requiring the fewest number of edges between two specified vertices in an *unweighted* graph?

(b) Use one of Dijkstra's algorithms to find the shortest path from A to B in the graph.

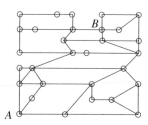

12. [BB] Suppose the improved version of Dijsktra's algorithm is used to find the shortest distances between all pairs of vertices in a graph on n vertices by being permitted to run until all vertices acquire their final labels. Why is this procedure $\mathcal{O}(n^3)$ (in terms of comparisons)?

13. Suppose the Floyd–Warshall algorithm is applied to the graph in Fig 10.18 (with vertices $A, \ldots, J$ relabeled $v_1, \ldots, v_{10}$, respectively).

   (a) What are the final values of $d(1, 1), d(1, 2), \ldots, d(1, 10)$?

   (b) [BB] Find the values of $d(1, 5), d(1, 6), d(3, 4),$ and $d(8, 5)$ after $k = 4$.

   (c) What is the initial value of $d(2, 5)$? How does this value change as k increases from 1 to 10?

14. The Floyd–Warshall algorithm is applied to the graph shown.

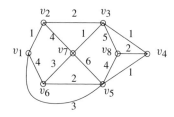

(a) Find the final values of $d(7, 1), d(7, 2), \ldots, d(7, 8)$.

(b) Find the values of $d(1, 2), d(3, 4), d(2, 5),$ and $d(8, 6)$ after $k = 4$.

(c) Find the values of $d(6, 8)$ at the start and as k varies from 1 to 8.

15. Inadvertently, David reverses the order of the k and the i, j loops in Step 2 of the Floyd–Warshall algorithm, implementing Step 2 in the form
For $i, j = 1$ to n,
   for $k = 1$ to n, let $d(i, j) = \min\{d(i, j), d(i, k) + d(k, j)\}$
Does this affect the algorithm? Explain.

16. [BB] Suppose the values of the $d(i, j)$ for two consecutive values of k are the same in an implementation of the Floyd–Warshall algorithm. Is it necessary to continue, that is, can it be assumed that the values will now remain constant?

17. Prove that the Floyd–Warshall algorithm works. Specifically, assume that $\mathcal{G}$ is a weighted graph with vertices $v_1, v_2, \ldots, v_n$ and assume that there is a path from v_i to v_j. Prove that when the algorithm terminates, $d(i, j)$ is the length of a shortest path from v_i to v_j.

18. Show that the Floyd–Warshall algorithm requires $\mathcal{O}(n^3)$ additions and comparisons.

✍ **19.** Discover what you can about Edsger Dijkstra and write a short note about him.

✍ **20.** Discover what you can about R. W. Floyd and write a short note about him.

REVIEW EXERCISES FOR CHAPTER 10

1. In the Königsberg Bridge Problem (see Fig 9.1), two new bridges are constructed, one joining A to C and the other B to D. Use Theorem 10.1.4 to show that the answer to the question is now yes.

2. One of the mayoralty candidates in Königsberg says that he can obtain a positive solution to the bridge problem by building only one new bridge. Is he telling the truth?

3. Suppose $\mathcal{G}_1$ and $\mathcal{G}_2$ are graphs with no vertices in common and assume that each graph possesses an Eulerian

trail. Show that it is possible to select vertices v and w of $\mathcal{G}_1$ and $\mathcal{G}_2$, respectively, such that if v and w are joined by a new edge, the resulting graph will possess an Eulerian trail.

4. True or false? Explain your answers in each case.

   (a) Every trail is a path.

   (b) Every open trail is a path.

   (c) If there is an open trail from vertex v to vertex w, then there is a path from v to w.

   (d) Every path is an open trail.

(e) If there is a path from vertex v to vertex w, then there is an open trail from v to w.

5. Let $\mathcal{G}$ be a connected graph with at least two vertices. Assume that every edge of $\mathcal{G}$ belongs to a unique cycle. Prove that $\mathcal{G}$ is Eulerian.

6. Is the graph Hamiltonian? Is it Eulerian? Explain your answers.

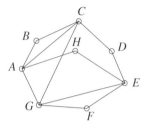

7. (a) How many edges must a Hamiltonian cycle in $\mathcal{K}_{n,n}$ contain?

(b) How many Hamiltonian cycles does $\mathcal{K}_{n,n}$ have? Assume $n \geq 2$.

(c) What is the maximum number of edge disjoint Hamiltonian cycles in $\mathcal{K}_{n,n}$?

(d) When $n = 4$, show that the maximum number of edge disjoint Hamiltonian cycles predicted in 7(c) is realized.

8. True or false? Explain your answers in each case.

(a) In a Hamiltonian graph, every edge belongs to some Hamiltonian cycle.

(b) In a Hamiltonian graph, every edge belongs to a cycle.

(c) Every Eulerian graph contains a subgraph which is Hamiltonian.

(d) Every Hamiltonian graph contains a subgraph which is Eulerian.

9. A connected graph $\mathcal{G}$ has 14 vertices and 88 edges. Show that $\mathcal{G}$ is Hamiltonian but not Eulerian.

10. The following questions refer to the graph $\mathcal{G}$ drawn on the left.

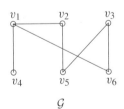

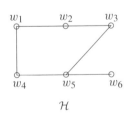

(a) Find the adjacency matrix A of $\mathcal{G}$.

(b) Without any calculation, determine the $(1, 5)$ entries of A^2, A^3, A^4, and A^5.

(c) Find an isomorphism $\varphi \colon \mathcal{G} \to \mathcal{H}$, where $\mathcal{H}$ is the graph on the right.

(d) If B is the adjacency matrix of $\mathcal{H}$, find a permutation matrix P such that $PAP^T = B$. Check your answer by writing down B and computing the product PAP^T.

11. Let $v_1, v_2, \ldots, v_8$ and $w_1, w_2, \ldots, w_{12}$ be the bipartition sets of the complete bipartite graph $\mathcal{K}_{8,12}$. Let A be the adjacency matrix of this graph, where the vertices are listed in the order $v_1, \ldots, v_8, w_1, \ldots, w_{12}$.

(a) What is the $(1, 5)$ entry of A?

(b) What is the $(8, 9)$ entry of A?

(c) What is the $(10, 12)$ entry of A^2?

(d) What is the $(5, 5)$ entry of A^2?

(e) What is the $(20, 6)$ entry of A^2?

(f) What is the $(5, 7)$ entry of A^4?

(g) What is the $(5, 7)$ entry of A^{15}?

12. Suppose $\mathcal{G}$ is a connected graph. Is it possible to determine from the adjacency matrix of $\mathcal{G}$ whether or not $\mathcal{G}$ is Eulerian? Explain.

13. Martha claims that a graph with adjacency matrix

$$A = \begin{bmatrix} 0 & 1 & 1 & 1 & 0 & 1 & 1 \\ 1 & 0 & 1 & 0 & 1 & 0 & 1 \\ 1 & 1 & 0 & 0 & 0 & 1 & 1 \\ 1 & 0 & 0 & 0 & 1 & 1 & 1 \\ 0 & 1 & 0 & 1 & 0 & 1 & 1 \\ 1 & 0 & 1 & 1 & 1 & 0 & 0 \\ 1 & 1 & 1 & 1 & 1 & 0 & 0 \end{bmatrix}$$

must be Hamiltonian. How can she be so sure?

14. If A is the adjacency matrix of a graph $\mathcal{G}$ and $A^2 = [b_{ij}]$, find $\frac{1}{2}\sum_i b_{ii}$.

15. Apply the first form of Dijkstra's algorithm to the following graph, showing the shortest distances from A to every other vertex. Exhibit an order in which a shortest path from A to E might be realized.

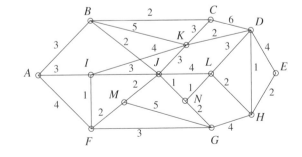

16. Apply the improved version of Dijkstra's algorithm to the graph of Exercise 15 to find the length of a shortest path from A to every other vertex.

17. Apply the original form of Dijkstra's algorithm to find the length of the shortest path from A to every other vertex. Show the final labels on all vertices. Also find the shortest path from A to H.

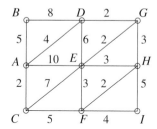

18. Apply the improved version of Dijkstra's algorithm to answer Exercise 17.

19. Apply the Floyd–Warshall algorithm, showing the initial values of $d(i, j)$ and the values of $d(i, j)$ at the end of every step.

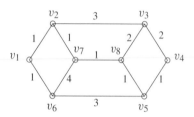

20. Apply the Floyd–Warshall algorithm to the graph in Exercise 17. Show the initial values of the $d(i, j)$, $i, j = 1, 2, \ldots, 9$; the values after $k = 1$, $k = 3$, and $k = 5$; and the final values. Explain how the algorithm can be used to find the shortest path from A to H.

21. An encyclopedia salesman, traveling by car, wishes to visit ten towns and return home without passing through the same town twice. He knows that one, but only one, of the towns has a direct connection to each of the other towns (that is, connections passing through no other towns). He also knows that there are a total of 39 such direct road connections between pairs of towns. Why is he confident that he will be able to find such a route?

Applications of Paths and Circuits

11.1 THE CHINESE POSTMAN PROBLEM

In this chapter, we present various applications of the ideas developed in Chapter 10. We use what we know about Eulerian graphs to solve the so-called *Chinese Postman Problem* and to reconstruct *RNA chains*. The latter subject requires the concept of a *digraph*, a graph in which the edges have directions. We show how Hamiltonian paths can be used in the study of *tournaments* and, finally, we will see how shortest path algorithms can be applied to *scheduling problems*.

A mail carrier who begins his route at the post office must deliver letters to each block in a certain part of town and then return to the post office. What is the least amount of walking the mail carrier must do? This problem, a version of which was first solved by H. E. Dudeney in 1917,[1] is today known as the *Chinese Postman Problem* because it was studied in most general form by the Chinese mathematician, Mei-ko Kwan (also known as Meigo Guan).[2] It is clear that the same problem is faced by delivery people of various kinds, by paper boys, by street repair crews, by snow plow operators, and so on.

We represent the mail carrier's problem by a weighted graph G with each vertex denoting a street corner and each edge a street of length the weight of that edge. If we assume that the post office is on a corner, then any route the postman follows corresponds to a closed walk in the graph which uses each edge at least once. If the graph G has an Eulerian circuit, then we have an optimal solution since every block will be walked exactly once. If not, then the postman will have to walk certain blocks more than once, but his aim is to plan these extra trips so that the total distance walked is as small as possible.

---

[1] See Angela Newing, "The Life and Work of H. E. Dudeney," *Mathematical Spectrum* **21** (1988/89), 37–44.

[2] M. K. Kwan, "Graphic Programming Using Odd or Even Points," *Chinese Math* **1** (1962), 273–277.

11.1.1 THE CHINESE POSTMAN PROBLEM ▶

Given a connected possibly weighted graph, find the shortest closed walk that covers every edge at least once.

The first step in the solution to this problem is the realization that the postman's walk will follow an Eulerian circuit in a pseudograph obtained from the given graph by the duplication of certain edges.

Suppose that the graph $\mathcal{G}$ shown in Fig 11.1 represents the streets which a postman must cover and suppose that all streets have the same length.

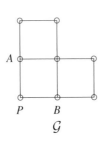

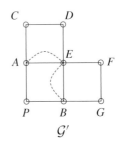

Figure 11.1

Since the postman's route covers all streets, it doesn't matter where the post office is, so assume it is at vertex P. Since $\mathcal{G}$ has vertices of odd degree (namely, A and B), $\mathcal{G}$ is not Eulerian, so the postman cannot service this area by walking each block exactly once. Our problem is to determine the least number of edges which should be duplicated in order to obtain an Eulerian pseudograph. Since all vertices of an Eulerian pseudograph must have even degree, we will have to add an extra copy (or extra copies) of certain of the edges incident with A and with B. If A and B were joined by an edge, then a single extra copy of that edge would certainly suffice. Since, however, A and B are not joined by an edge, we need at least two additional edges. The pseudograph $\mathcal{G}'$ shown on the right in Fig 11.1 (extra edges shown as dashed lines) is Eulerian and a solution to the Chinese Postman Problem. One Eulerian circuit is $PACDEAEBGFEBP$.

Unlike many graph theoretical problems, the Chinese Postman Problem has been solved: Procedures have been found for determining which edges of a graph must be duplicated in order to obtain the Eulerian pseudograph of least weight.[3]

EXAMPLE 1

We illustrate with reference to the graph $\mathcal{G}$ in Fig 11.2, where we again assume that all streets have the same length. There are four odd vertices in $\mathcal{G}$, labeled A, B, C, D in the figure. Consider all the partitions of these vertices into pairs and, for each partition, calculate the sum of the lengths of shortest paths between pairs. As the calculations in Table 11.1 show, the minimum sum of lengths is 5, and this is achieved with two partitions either of which yields a solution to the problem. If we choose, for instance, $\{A, B\}, \{C, D\}$, then a pseudograph of minimum weight is obtained by duplicating the edges along the shortest path from A to B and

[3]See J. Edmonds, "The Chinese Postman Problem," *Operations Research* **13** Suppl. 1 (1965), 353; J. Edmonds and E. L. Johnson, Matching, Euler Tours and the Chinese Postman, *Mathematical Programming* **5** (1973), 88–124; E. L. Lawler, *Combinatorial Optimization Networks and Matroids*, Holt, Rinehart and Winston, New York, 1975; and E. Minieka, *Optimization Algorithms for Networks and Graphs*, Marcel Dekker, New York, 1978.

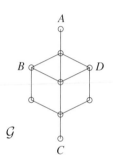

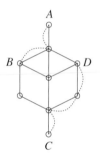

Figure 11.2

Table 11.1

Partition into pairs	Sum of lengths of shortest paths
$\{A, B\}, \{C, D\}$	$2 + 3 = 5$
$\{A, C\}, \{B, D\}$	$4 + 2 = 6$
$\{A, D\}, \{B, C\}$	$2 + 3 = 5$

along the shortest path from C to D, as shown on the right of Fig 11.2. As before, many people could have obtained our solution by inspection. Computers cannot see, however. Our intent has been to illustrate a concrete procedure which works in all cases and does not depend on a picture. ▲

11.1.2 THE CHINESE POSTMAN PROBLEM — AN ALGORITHM ▶

To find an Eulerian pseudograph of minimum weight by duplicating edges of a weighted connected graph $\mathcal{G}$,

Step 1. Find all the odd vertices in $\mathcal{G}$.

Step 2. For each partition of the odd vertices into pairs of vertices,

$$\{v_1, w_1\}, \{v_2, w_2\}, \ldots, \{v_m, w_m\},$$

find the length of a shortest path between each v_i and w_i and add these lengths.

Step 3. Take the partition for which the sum of lengths in Step 2 is minimum and, for each pair $\{v_i, w_i\}$ of vertices in this partition, duplicate the edges along a shortest path from v_i to w_i.

Pause 1 The second step of this algorithm assumes that the odd vertices of a graph can always be partitioned into pairs. Why should this be the case?

Pause 2 In the implementation of the algorithm, will it ever be necessary to duplicate an edge more than once? In other words, if an edge e has been duplicated for inclusion in the shortest path from v_i to w_i, will it perhaps be necessary later to duplicate it for inclusion in the shortest path from v_j to w_j?

In the Exercises, we ask you to show that the pseudograph produced by this algorithm is Eulerian and, furthermore, that given any Eulerian pseudograph obtained from $\mathcal{G}$ by duplicating certain edges, the odd vertices can be paired such that there are edge disjoint paths of new edges between them. From this, it follows that our algorithm indeed produces an Eulerian pseudograph of minimum weight.

Suppose we attach some weights to the edges of the graph we considered before, as shown in Fig 11.3. There are still four odd vertices and three partitions of these into pairs, but the lengths of shortest paths between pairs of odd vertices have changed.

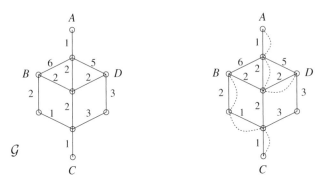

Figure 11.3

As the calculations in Table 11.2 show, the minimum sum of lengths is now 9, corresponding to the partition $\{A, D\}, \{B, C\}$. Duplicating the edges on a shortest path from A to D and from B to C, as shown on the right of Fig 11.3, we obtain an Eulerian pseudograph of minimum weight and a solution to the Chinese Postman Problem for the graph $\mathcal{G}$.

Table 11.2

Partition into pairs	Sum of lengths of shortest paths
$\{A, B\}, \{C, D\}$	$5 + 5 = 10$
$\{A, C\}, \{B, D\}$	$6 + 4 = 10$
$\{A, D\}, \{B, C\}$	$5 + 4 = 9$

Answers to Pauses

1. The number of odd vertices in a graph is even (Corollary 9.2.7).
2. No, it won't. To see why not, think about how the partition

$$\mathcal{P} = \{\{v_1, w_1\}, \{v_2, w_2\}, \dots, \{v_m, w_m\}\}$$

was determined. Denoting by ℓ_i the length of the shortest path between v_i and w_i, the sum, $L = \sum_i \ell_i$, is least among all similar sums arising from all partitions of the odd vertices into pairs.

If an edge e is duplicated because it is part of the shortest path between v_i and w_i, and it is later required again because it is part of the shortest path between v_j and w_j, then the partition obtained from $\mathcal{P}$ by replacing the pairs $\{v_i, w_i\}, \{v_j, w_j\}$ by the pairs $\{v_i, v_j\}, \{w_i, w_j\}$

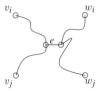

has sum of shortest path lengths less than or equal to $L - 2w(e)$, contradicting the minimality of L.

EXERCISES

The symbol [BB] means that an answer can be found in the Back of the Book.

1. Solve the Chinese Postman Problem for each of the graphs shown.

(a) [BB] 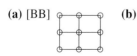   **(b)**

(c)   **(d)**

(e) 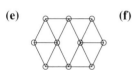   **(f)**

2. [BB; $\mathcal{K}_5$] Solve the Chinese Postman Problem for $\mathcal{K}_5$, $\mathcal{K}_6$, and $\mathcal{K}_{3,5}$.

3. [BB] Solve the Chinese Postman Problem.

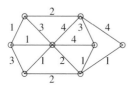

4. [BB] Solve the Chinese Postman Problem for each graph.

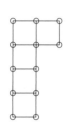   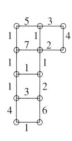

5. Solve the Chinese Postman Problem for each graph.

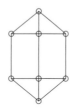   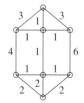

6. [BB] Suppose X and Y are any two vertices of a weighted connected graph $\mathcal{G}$. Explain why the shortest mail carrier's route from X to X has the same length as the shortest route from Y to Y.

7. Solve the Chinese Postman Problem for the unweighted graph shown.

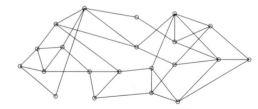

8. Solve the Chinese Postman Problem for the weighted graph.

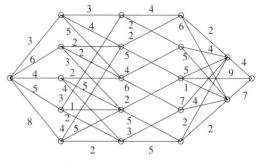

The remaining exercises concern the algorithm presented in this section for obtaining an Eulerian pseudograph of minimum weight from a given non-Eulerian connected graph $\mathcal{G}$ by duplicating certain edges.

9. [BB] Prove that the pseudograph $\mathcal{G}'$ which the algorithm finds is Eulerian.

10. Suppose $\mathcal{G}$ has n odd vertices. In how many ways can these be partitioned into $\frac{n}{2}$ pairs?

11. Suppose that $\mathcal{G}''$ is any Eulerian pseudograph obtained from $\mathcal{G}$ by duplicating certain edges.

    (a) Prove that each odd vertex of $\mathcal{G}$ is the initial vertex of a path consisting entirely of new edges (in $\mathcal{G}''$) and leading to another odd vertex.

    (b) Improve the result of (a) by showing that the odd vertices of $\mathcal{G}$ can, in fact, be partitioned into two equal sets—$\{v_1, v_2, \dots, v_m\}$ and $\{w_1, w_2, \dots, w_m\}$—with edge disjoint paths consisting entirely of new edges between each v_i and w_i.

    (c) Conclude that of all the pseudographs obtainable from $\mathcal{G}$ by duplicating edges, the algorithm yields an Eulerian pseudograph of minimum weight.

11.2 DIGRAPHS

Often, when graphs are used to model real-life situations, the edges represent lines of communication such as roads or pipes. In such situations, it is not unusual for the flow through an edge to have an associated direction.

11.2.1 DEFINITION ▶ A *digraph* is a pair $(\mathcal{V}, \mathcal{E})$ of sets, $\mathcal{V}$ nonempty and each element of $\mathcal{E}$ an ordered pair of distinct elements of $\mathcal{V}$. The elements of $\mathcal{V}$ are called vertices and the elements of $\mathcal{E}$ are called *arcs*.

Loosely speaking, a digraph is just a graph in which each edge has an orientation or direction assigned to it. (We use the words "orientation" and "direction" synonymously.) It can be pictured like a graph, with the orientation of an arc indicated by an arrow: We draw an arrow $u \longrightarrow v$ if $(u, v) \in \mathcal{E}$. Some digraphs are pictured in Fig 11.4.

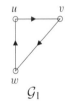
$\mathcal{G}_1$
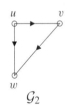
$\mathcal{G}_2$

$\mathcal{G}_3$

$\mathcal{G}_4$

Figure 11.4

As with most aspects of graph theory, there is a rather extensive terminology used for digraphs and, as we have noted before, different textbooks employ the same words with slight variations in meaning. The policy in this text is to use the same terms for graphs and digraphs whenever possible, with one exception: In a digraph, we use the term "arc" instead of "edge." An edge is just a set, an unordered pair of vertices $\{u, v\}$; an arc is an ordered pair (u, v) or (v, u) (as ordered pairs, these are different). Just as it has been our custom when naming edges of a graph not to use set notation, we omit parentheses when naming arcs of a digraph, thus referring to "the arc uv" rather than "the arc (u, v)."

As with the term "graph", our definition of "digraph" precludes the existence of loops and multiple arcs. Note the subtle distinction between multiple edges and multiple arcs, however. Since the arcs uv and vu are different, they may both appear in a digraph. All the pictures in Fig 11.4 represent digraphs, in particular

$\mathcal{G}_4$, which, without the arrows, would not be a graph. There are multiple edges but not multiple arcs between u and v.

In the same way that the vertices of a graph have degrees, a vertex of a digraph has an *indegree* and an *outdegree*, these being, respectively, the number of arcs directed into and away from that vertex. Considering the graph $\mathcal{G}_3$ in Fig 11.4, vertices u and x have indegree 2 and outdegree 1, vertex v has indegree 0 and outdegree 2, and vertex w has indegree 1 and outdegree 1. The *indegree sequence* of $\mathcal{G}_3$ is $2, 2, 1, 0$ and the *outdegree sequence* of $\mathcal{G}_3$ is $2, 1, 1, 1$. The following result can be proved with a routine modification of the argument used to prove Proposition 9.2.5.

11.2.2 PROPOSITION ▶ (cf. Proposition 9.2.5) The sum of the indegrees of the vertices of a digraph equals the sum of the outdegrees of the vertices, this common number being the number of arcs.

As might be expected, when considering the possible isomorphism of digraphs, the direction of arcs must be taken into account. In Fig 11.4, $\mathcal{G}_1$ and $\mathcal{G}_2$ are **not** isomorphic because every vertex in $\mathcal{G}_1$ has outdegree 1 whereas in $\mathcal{G}_2$, outdeg $u = 2$. The indegree sequences of isomorphic digraphs must be equal as must be their outdegree sequences. With this small change in property 2, Proposition 9.3.2 is equally valid for digraphs.

By convention, any walk in a digraph respects orientation of arcs; that is, each arc is followed in the direction of its arrow. For example, digraph $\mathcal{G}_1$ in Fig 11.4 is Eulerian because $uvwu$ is an Eulerian circuit. This is also a Hamiltonian cycle. Digraph $\mathcal{G}_2$ has neither an Eulerian circuit nor a Hamiltonian cycle (the arc is uw, not wu), although it does have a Hamiltonian path uvw.

Pause 3 Explain why neither digraph $\mathcal{G}_3$ nor $\mathcal{G}_4$, of Fig 11.4, is Hamiltonian.

In Theorem 10.1.4, we established necessary and sufficient conditions for a graph to be Eulerian. This theorem can be adapted to the situation of digraphs with a suitable strengthening of the concept of connectedness.

11.2.3 DEFINITION ▶ A digraph is called *strongly connected* if and only if there is a walk from any vertex to any other vertex which respects the orientation of each arc.

Imagine a city where every street is one-way. The digraph of streets and intersections of this city is connected if it is possible to drive from any intersection to any other intersection, perhaps by moving the wrong way down a one-way street, and strongly connected if such travel is always possible without breaking the law!

The proof of the following theorem is left to the Exercises.

11.2.4 THEOREM ▶ (cf. Theorem 10.1.4) A digraph is Eulerian if and only if it is strongly connected and, for every vertex, the indegree equals the outdegree.

Pause 4 Is digraph $\mathcal{G}_3$ in Fig 11.4 Eulerian? What about the digraph in Fig 11.5?

The adjacency matrix A of a digraph $\mathcal{G}$ with vertices $v_1, v_2, \ldots, v_n$ is defined by setting $a_{ij} = 1$ if there is an arc from vertex v_i to vertex v_j and 0 otherwise.

Figure 11.5 A digraph associated with $\mathcal{K}_3$.

Unlike the adjacency matrix of a graph (see Section 10.3), the adjacency matrix of a digraph is generally not symmetric, since an arc from v_i to v_j does not imply an arc from v_j to v_i. Most assertions made in Section 10.3 apply, with appropriate changes. For example, the (i, j) entry of A^2 is the number of different walks of length 2 from v_i to v_j (respecting the orientation of arcs). This number need not be the same as the number of walks of length 2 from v_j to v_i, obviously. The ith entry of the diagonal of A^2, being the number of walks of length 2 from v_i to v_i, is the number of vertices v_j such that there exists a circuit of the form $v_i v_j v_i$. (See Fig 11.6.)

Figure 11.6

Theorems 10.3.3 and 10.3.4 hold for digraphs. Digraphs are isomorphic if and only if their vertices can be labeled in such a way that their adjacency matrices are equal. Labeled digraphs $\mathcal{G}_1$ and $\mathcal{G}_2$ with adjacency matrices A_1 and A_2, respectively, are isomorphic if and only if $A_2 = P A_1 P^T$ for some permutation matrix P.

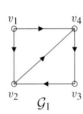

Figure 11.7 Two isomorphic digraphs.

EXAMPLE 2 Consider the digraphs of Fig 11.7. The adjacency matrices of $\mathcal{G}_1$ and $\mathcal{G}_2$ are

$$A_1 = \begin{bmatrix} 0 & 1 & 0 & 1 \\ 0 & 0 & 0 & 1 \\ 0 & 1 & 0 & 0 \\ 0 & 0 & 1 & 0 \end{bmatrix} \quad \text{and} \quad A_2 = \begin{bmatrix} 0 & 0 & 1 & 0 \\ 1 & 0 & 0 & 0 \\ 0 & 1 & 0 & 0 \\ 1 & 0 & 1 & 0 \end{bmatrix}.$$

The map $\varphi: \mathcal{G}_1 \rightarrow \mathcal{G}_2$ defined by

$$\varphi(v_1) = u_4, \quad \varphi(v_2) = u_1, \quad \varphi(v_3) = u_2, \quad \varphi(v_4) = u_3$$

is an isomorphism, and the permutation matrix

$$P = \begin{bmatrix} 0 & 1 & 0 & 0 \\ 0 & 0 & 1 & 0 \\ 0 & 0 & 0 & 1 \\ 1 & 0 & 0 & 0 \end{bmatrix}$$

determined by φ (as in Section 10.3) has the property that $P A_1 P^T = A_2$. ▲

The shortest path algorithms described in Section 10.4 can be applied to digraphs, with obvious modifications which take into account the orientation of arcs. Examples appear in our discussion of scheduling problems in Section 11.5.

Here is another algorithm, due to Richard Bellman (1920–1984) and Lester R. Ford Jr. (1927–) which, for a directed graph with vertices $v_1, v_2, \ldots, v_n$, computes successively the length $d_i(j)$ of a shortest path from v_1 to v_j using at most i arcs. Since no path in such a digraph can have more than $n-1$ arcs, the final values of $d_{n-1}(j)$, $j = 2, \ldots, n$ are the lengths of shortest paths from v_1 to all other vertices.

11.2.5 THE BELLMAN-FORD ALGORITHM ▶

Given a weighted digraph with vertices labeled $v_1, v_2, \ldots, v_n$, to find the shortest distances from v_1 to all other vertices, proceed as follows.

Step 1. for $i, j = 1, 2, \ldots, n$,

- if $i = j$, set $w(i, i) = 0$; if $i \neq j$ and $v_i v_j$ is an arc, set $w(i, j)$ to the weight of $v_i v_j$; otherwise, set $w(i, j) = \infty$.
- set $d_0(1) = 0$ and, for $j = 2, 3, \ldots, n$, set $d_0(j) = \infty$.
- for $i = 2$ to n, set $p(i) = 1$.

Step 2. for $i = 1$ to n
    for $j = 1$ to n
- find the k for which min $= d_{i-1}(k) + w(v_k, v_j)$ is least;
- if min $< d_{i-1}(j)$, set $d_i(j) =$ min and $p(j) = v_k$
  else set $d_i(j) = d_{i-1}(j)$.

Step 3. if $d_n(j) = d_{n-1}(j)$ for $j = 1, 2, \ldots, n$,
    for $j = 1$ to n output $d_n(j)$;
    else output "No shortest paths. There is a negative weight cycle."

If there are no negative weight cycles, the value of $d_n(j) = d_{n-1}(j)$ which is output in Step 3 is the length of a shortest path from v_1 to v_j, this path consisting of vertices $p(2), p(3), \ldots, p(j)$.

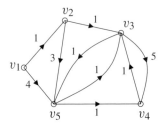

Figure 11.8

We apply this algorithm to the graph shown in Fig 11.8. After initialization in Step 1, the process begins with Step 2. Setting $i = 1$, the algorithm computes the weight of a shortest one arc path to vertices v_2, v_3, v_4, v_5. The shortest path to v_2 has weight 1, and the last vertex on this path is v_1. The shortest path to v_5 has weight 4, and the last vertex on this path is v_1. Shortest paths to v_3 and v_4, using at most one arc, have infinite length. The first column of the table in Table 11.3 records this data.

Now $i = 2$ and the algorithm computes the length of a shortest path to v_2, v_3, v_4, v_5 using at most two arcs. For each $j = 2, 3, 4, 5$, the algorithm first

Table 11.3 Results of applying the Bellman-Ford algorithm to the digraph shown in Fig 11.8. The (i, j) entry is ∞ or, when $d_i(j)$ is finite, the pair $d_i(j), p(j)$.

		Max. no. of arcs				
		1	**2**	**3**	**4**	**5**
Vertices	v_2	$1, v_1$	$1, v_1$	$1, v_1$	$1, v_1$	$1, v_1$
	v_3	∞	$2, v_2$	$2, v_2$	$2, v_2$	$2, v_2$
	v_4	∞	$5, v_5$	$5, v_5$	$4, v_5$	$4, v_5$
	v_5	$4, v_1$	$4, v_1$	$3, v_3$	$3, v_3$	$3, v_3$

computes the minimum $d_1(k) + w(v_k, v_j)$; that is, the length of a shortest path to v_k using one arc plus the weight of arc $v_k v_j$. This number (min) is compared with $d_1(k)$, the length of the shortest one arc path to v_k. The smaller of these two numbers becomes $d_2(j)$. If the shortest path had length min, the algorithm also sets the predecessor vertex $p(2) = v_k$, thus remembering that v_k is the vertex preceding v_j on a shortest at most two arc path to v_j. The shortest at most two arc path to v_2 is still a one arc path, and the same is true for v_5. The shortest at most two arc path to v_3 has length 2, and the last vertex on such a path is v_2. The shortest at most two arc path to v_4 has length 5, and the last vertex on such a path is v_5. We leave it to you to check the remaining columns of the table.

Pause 5 In this example, all vertices eventually have finite labels. Will this always be the case?

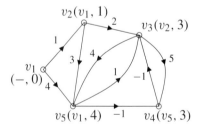

Figure 11.9 An application of Dijkstra's algorithm (first form) to a weighted digraph with some negative weights. The final labeling at v_3 is wrong.

In contrast to the algorithms of Dijkstra, the Bellman-Ford algorithm works perfectly well on weighted digraphs where some arcs have negative weight. In Fig 11.9, we show the result of applying the original form of Dijkstra's algorithm to the digraph of Fig 11.8, with different (and some negative) arc weights. Since neither version of Dijkstra's algorithm examines a vertex again after it has acquired its permanent label, vertex v_3 is incorrectly determined to be at distance 3 from v_1. The correct length of a shortest path from v_1 to v_3 is, of course, 2, along $v_1 v_5 v_4 v_3$, and this is found by the Bellman-Ford algorithm, as shown in Table 11.4.

Any path in a digraph with n vertices cannot contain more than $n - 1$ arcs. So in Step 2 of the Bellman-Ford algorithm, why does the outer for loop run

Table 11.4 Results of applying the Bellman-Ford algorithm to the digraph shown in Fig 11.9.

		1	**2**	**3**	**4**	**5**
	v_2	$1, v_1$	$1, v_1$	$1, v_1$	$1, v_1$	$1, v_1$
Vertices	v_3	∞	$3, v_2$	$2, v_4$	$2, v_4$	$2, v_4$
	v_4	∞	$3, v_5$	$3, v_5$	$3, v_5$	$3, v_5$
	v_5	$4, v_1$	$4, v_1$	$4, v_1$	$4, v_1$	$4, v_1$

Column header group: *Max. no. of arcs*

from $i = 1$ to n and not to $i = n - 1$? (Remember that i is the number of arcs in a path.) The answer is in order to detect negative weight cycles (which can be reached from v_1). This is also the reason for the inclusion of Step 3. If a weighted digraph has a cycle of negative total weight, there is no shortest path to any vertex on this cycle which is reachable from v_1, since by going around the cycle, lengths can be decreased as much as desired. Suppose the value of $d_n(j)$ is different (and hence less) than the value of $d_{n-1}(j)$ for some vertex v_j. Thus there is a walk of n edges from v_1 to v_n of total weight less than any walk which uses fewer than n edges. Such a walk must contain a repeated vertex, and hence a cycle $\mathcal{C}: u_1u_2\cdots u_1$, so the walk is $\mathcal{W}: v_1\cdots\mathcal{C}\cdots v_n$. Notice that the cycle must have negative total weight since otherwise $v_1\cdots u_1\cdots v_n$ (omitting $\mathcal{C}$) would have weight less than or equal to $\mathcal{W}$. It can be shown that any negative weight cycle will be detected by Bellman-Ford since in this case, for some vertex v_j, the value of $d_n(j)$ must be less than $d_{n-1}(j)$. [See Exercise 23(b).]

Pause 6 In Fig 11.9, change the weight of v_3v_5 from 4 to 1. Make a table like that in Table 11.3. Examine the values of $d_4(j)$ and $d_5(j)$ and comment.

Weighted digraphs are helpful in settings where graphs are not appropriate; specifically, in any situation where travel between two vertices can proceed in only one direction. We provide examples in the next three sections.

Answers to Pauses

3. In $\mathcal{G}_3$, vertex v has indegree 0. Since there is no way of reaching v on a walk respecting orientation of edges, no Hamiltonian cycle can exist. In $\mathcal{G}_4$, vertex x has outdegree 0 so no walk respecting orientations can leave x.

4. The digraph $\mathcal{G}_3$ is not Eulerian. It is not strongly connected (there is no way to reach v). Also, the indegrees and outdegrees of three vertices (u, v, and x) are not the same. The digraph in Fig 11.5 is Eulerian, however. It is strongly connected (there is a circuit $uvwu$ which permits travel in the direction of arrows between any two vertices), and the indegree and outdegree of every vertex is 2. (An Eulerian circuit is $uwvuvwu$.)

5. If there is no directed path from v_1 to some vertex v_i, the label for v_i will always be ∞.

6.

Max. no. of arcs

	1	2	3	4	5
v_2	$1, v_1$	$1, v_1$	$1, v_1$	$1, v_1$	$1, v_1$
Vertices v_3	∞	$3, v_2$	$2, v_4$	$2, v_4$	$2, v_4$
v_4	∞	$3, v_5$	$3, v_5$	$3, v_5$	$2, v_5$
v_5	$4, v_1$	$4, v_1$	$4, v_1$	$3, v_3$	$3, v_3$

Since $d_4(v_4) = 3$ and $d_5(v_4) = 2$, the algorithm has detected the negative cycle $v_3 v_5 v_4 v_3$.

EXERCISES

The symbol [BB] means that an answer can be found in the Back of the Book.

1. Which of the following pairs of digraphs are isomorphic? Explain your answers.

(a) [BB]

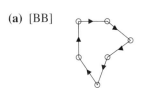

(b)

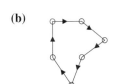

(c)

(d)

2. Must the indegree and the outdegree sequences of a digraph be the same? Explain.

3. [BB] Prove Proposition 11.2.2.

4. Exhibit a digraph which is strongly connected but not Eulerian.

5. [BB] Let $\mathcal{G}$ be a connected graph with all vertices of even degree. Can the edges of $\mathcal{G}$ be oriented so that the resulting digraph is Eulerian? Explain.

6. Let A be the adjacency matrix of a digraph $\mathcal{G}$. What does the sum of the entries in row i of A represent? What about the sum of the entries in column i?

7. [BB] Let A be the adjacency matrix of the digraph pictured at the left.

(a) Find A.

(b) Determine the $(3, 3)$ and $(1, 4)$ entries of A^2 without calculating the matrix A^2 and explain your reasoning.

(c) Determine the $(4, 2)$ entry of A^3 and the $(1, 3)$ entry of A^4 without calculating A^3 or A^4 and explain your reasoning.

(d) Is the digraph strongly connected? Explain.

(e) Is the digraph Eulerian? Explain.

8. Repeat Exercise 7 for the digraph at the right in that Exercise.

9. Let A be the adjacency matrix of a digraph $\mathcal{G}$. Explain why A^T is also the adjacency matrix of a digraph $\mathcal{G}^T$. Is it possible for $\mathcal{G}$ and $\mathcal{G}^T$ to be isomorphic? Explain.

10. [BB; (a)] Label the vertices of each pair of digraphs in Exercise 1 and then give the adjacency matrices corresponding to your labeling. (For any pair of graphs which are isomorphic, label in such a way that the two graphs have the same adjacency matrix.)

11. [BB] Consider the digraphs $\mathcal{G}_1, \mathcal{G}_2$ shown.

(a) Find the adjacency matrix A_1 of $\mathcal{G}_1$ and the adjacency matrix A_2 of $\mathcal{G}_2$.

(b) Explain why the map $\varphi \colon \mathcal{G}_1 \to \mathcal{G}_2$ defined by

$$\varphi(v_1) = u_2, \ \varphi(v_2) = u_1, \ \varphi(v_3) = u_4, \ \varphi(v_4) = u_3$$

is an isomorphism.

(c) Find the permutation matrix P which corresponds to φ and satisfies $P A_1 P^T = A_2$.

(d) Are these digraphs strongly connected?

(e) Are these digraphs Eulerian?

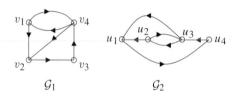

$\mathcal{G}_1$ $\mathcal{G}_2$

12. Repeat Exercise 11 for the digraphs shown and the map $\varphi : \mathcal{G}_1 \to \mathcal{G}_2$ defined by

$$\varphi(v_1) = u_3, \; \varphi(v_2) = u_4, \; \varphi(v_3) = u_2, \; \varphi(v_4) = u_5,$$
$$\varphi(v_5) = u_1.$$

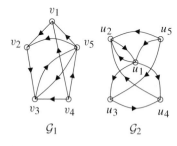

$\mathcal{G}_1$ $\mathcal{G}_2$

13. In each of the following cases, find a permutation matrix P such that $A_2 = PA_1P^T$.

(a) $A_1 = \begin{bmatrix} 0 & 1 & 0 & 1 \\ 0 & 0 & 1 & 0 \\ 0 & 0 & 0 & 1 \\ 0 & 0 & 0 & 0 \end{bmatrix}, \quad A_2 = \begin{bmatrix} 0 & 0 & 1 & 1 \\ 0 & 0 & 1 & 0 \\ 0 & 0 & 0 & 0 \\ 0 & 1 & 0 & 0 \end{bmatrix}$

(b) $A_1 = \begin{bmatrix} 0 & 0 & 1 & 0 \\ 0 & 0 & 0 & 1 \\ 1 & 0 & 0 & 0 \\ 0 & 1 & 0 & 0 \end{bmatrix}, \quad A_2 = \begin{bmatrix} 0 & 0 & 0 & 1 \\ 0 & 0 & 1 & 0 \\ 0 & 1 & 0 & 0 \\ 1 & 0 & 0 & 0 \end{bmatrix}$

(c) $A_1 = \begin{bmatrix} 0 & 1 & 0 & 1 & 1 \\ 0 & 0 & 1 & 0 & 0 \\ 0 & 0 & 0 & 1 & 1 \\ 0 & 0 & 0 & 0 & 1 \\ 0 & 1 & 0 & 0 & 0 \end{bmatrix}, \quad A_2 = \begin{bmatrix} 0 & 0 & 0 & 0 & 1 \\ 0 & 0 & 0 & 1 & 0 \\ 1 & 1 & 0 & 1 & 0 \\ 1 & 0 & 0 & 0 & 0 \\ 0 & 1 & 0 & 1 & 0 \end{bmatrix}$

14. [BB; (a)] Which of the digraphs in Exercise 1 are strongly connected? Explain your answers.

15. An orientation of a graph can be achieved by replacing each edge $\{x, y\}$, an unordered pair of vertices, by one of the ordered pairs (x, y) or (y, x); that is, you put an arrow on each edge of the graph.

(a) [BB] Find all nonisomorphic orientations of $\mathcal{K}_3$.
(b) Find all nonisomorphic orientations of $\mathcal{K}_4$.
(c) Find all nonisomorphic orientations of $\mathcal{K}_{2,3}$.

16. [BB] If a graph $\mathcal{G}$ is connected and some orientation is put on the edges of $\mathcal{G}$, must the resulting digraph be strongly connected? Explain.

17. Prove Theorem 11.2.4.

18. [BB] Answer true or false and explain: If a digraph $\mathcal{G}$ with at least two vertices is strongly connected, then every vertex of $\mathcal{G}$ is part of a circuit (consisting of more than a single vertex).

19. Apply the Bellman-Ford algorithm to the digraphs shown, answering by means of a table like Table 11.3.

(a) [BB]

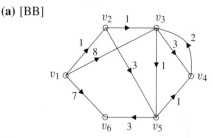

(b)

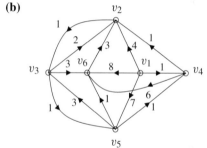

20. [BB] Would it be sensible to use the Bellman-Ford algorithm on undirected graphs? Comment.

21. (a) [BB] Apply the original form of Dijkstra's algorithm. Does it give correct lengths of shortest paths? Explain.

(b) [BB] Would Bellman-Ford work? Explain.

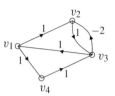

22. When the Bellman-Ford algorithm detects the presence of a negative weight cycle, can it be adapted to exhibit such a cycle? Explain.

23. (a) [BB] The Bellman-Ford algorithm can be terminated as soon as two successive columns are the same; that is, if for two successive values of i, the numbers $d_i(2), \ldots, d_i(n)$ are identical. Explain.

(b) Show that the Bellman-Ford algorithm will detect the presence of any negative weight cycle in a weighted digraph (assuming there is at least one directed path from v_1 to every other vertex).

24. Find a reasonable estimate of the complexity of the Bellman-Ford algorithm, in terms of comparisons.

25. In an attempt to get the algorithms of Dijkstra to work correctly on digraphs with negative weight edges, the following idea is proposed. Determine the highest negative weight k of an arc, add $|k|$ to the weight of every arc (now all arcs have nonnegative weights), use Dijkstra to find shortest paths, and finally subtract from each final weight, $|k|$ times the number of arcs required on a shortest path. Comment.

26. Discover what you can about Bellman and Ford and write a short note about these men.

11.3 RNA CHAINS

In October 1989, it was announced that Sidney Altman, a Canadian-born professor at Yale University, and Thomas Cech of the University of Colorado had won jointly the Nobel Prize in chemistry for research on ribonucleic acid that "could help explain the origins of life" and "probably provide a new tool for gene technology."

Ribonucleic (pronounced "rī-bō-nōo-clā-ĭc") acid (RNA for short) and deoxyribonucleic acid (DNA) are genetic materials carried in the cells of the human body. In fact, RNA is a chain each link of which is one of four chemicals: uracil, cytosine ("sī-tō-zēn"), adenine, and guanine. Throughout this section, we will use the letters U, C, A, and G, respectively, to represent these chemicals. Here, for example, is one RNA chain:

(1)   UCGAGCUAGCGAAG.

There are two kinds of enzymes which break the links in an RNA chain. The first, called a G-enzyme, breaks a chain after each G-link. For example, this enzyme would break the preceding chain into the following *fragments*:

G-fragments: UCG, AG, CUAG, CG, AAG.

The second enzyme, called a U,C-enzyme, breaks the chain after each U- and after each C-link. Thus, it would break the chain in (1) into the fragments

U,C-fragments: U, C, GAGC, U, AGC, GAAG.

It is apparently not possible to discover directly the sequence of U's, C's, A's, and G's which comprise a particular chain, but using the enzymes just described, the collection of fragments of the chain, the so-called *complete enzyme digest*, can be determined. The problem is then how to recover an RNA chain from its complete enzyme digest. In this section, we show how to solve this problem with a method which makes use of graph theory.

There are 5! possible RNA chains with the G-fragments just listed. (In real situations, the number is much larger.[4]) Our aim is to find which of these 120

---

[4] The first chain successfully recovered contained 77 chemicals. See Robert W. Holley, Jean Apgar, George A. Everett, James T. Madison, Mark Marquisse, Susan H. Merrill, John Robert Penswick, and Ada Zamir, "Structure of a Ribonuclei Acid," *Science* **147** (1965), 1462–1465.

chains also gives rise to the given set of U,C-fragments. Our method involves associating with any complete listing of G- and U,C-fragments a directed pseudograph (it is seldom a graph) and hunting for Eulerian circuits, each of which corresponds to an RNA chain with the specified fragments. If the pseudograph is not Eulerian, then no such chain exists. (Perhaps an error was made in listing the fragments.)

In any recovery problem, we may assume that there are at least two G-fragments and two U,C-fragments; otherwise there is no difficulty in obtaining a solution. Also, it is straightforward to determine the fragment with which the chain ends. In the preceding listing of fragments, for example, the U,C-fragment GAAG is **abnormal**: U,C-fragments should end in U or C but GAAG does not. The only way this could happen is for GAAG to be at the end of the chain.

Pause 7   Using the fact that GAAG must come last in the sought after RNA chain, lower the previous estimate of 120 possible chains with the given fragment lists by determining how many RNA chains have the set of U,C-fragments listed previously.

The initial lists of fragments will always contain at least one abnormal fragment (at the end). If this is a G-fragment ending in U or C, or a U,C-fragment ending in G, then there will be only one abnormal fragment and, in that case, the final fragment is uniquely determined. The only way in which there can be two abnormal fragments is if they both end in A. In this case, however, one will be identifiable as contained within the other; it is the longer of the two with which the chain ends. Again, the final fragment is uniquely determined.

Pause 8   There can never be more than two abnormal fragments. Why?

To illustrate the general procedure, we consider the example described previously. We assume that we know only the fragments and, to make things amusing, change the order of the fragments and start with these lists:

G-fragments: CUAG, AAG, AG, UCG, CG
U,C-fragments: U, U, AGC, C, GAAG, GAGC.

Imagine splitting the G-fragments further using the U,C-enzyme and splitting the U,C-fragments further with the G-enzyme. For example, UCG in the first list would become U, C, G, while AGC in the second list would become AG, C. A fragment like AAG wouldn't split further. The resulting bits are called *extended bases*; those extended bases which are neither first nor last in a fragment are called *interior*. Thus, a fragment which splits into $n \geq 3$ extended bases will yield $n - 2$ interior extended bases. In the splitting of the fragment UCG with the U,C-enzyme, U, C and G are extended bases; C is interior. Similarly, the extended base AG from the U,C-fragment GAGC is interior. A fragment must split into at least three extended bases before it gives rise to an interior extended base; there are no interior extended bases from the fragment AGC, for instance.

We now list all the interior extended bases obtained by the above splitting. In our example, we obtain

interior extended bases: U, C, AG,

which arise from the G-fragments CUAG, UCG and the U,C-fragment GAGC, respectively. Next, we list all fragments (G- or U,C-) which are *unsplittable*, that is, fragments which are already extended bases. In our example, we obtain

unsplittable fragments: AAG, AG, U, U, C,

these being the G-fragments which the U,C-enzyme does not split and the U,C-fragments which the G-enzyme does not split.

There are exactly two unsplittable fragments which are not interior extended bases, namely, the fragments AAG and U. These are the fragments with which the original chain begins and ends. Since we have already determined that the chain ends with GAAG, we know that U must be the first fragment in the chain.

Finally, we will see how graph theory comes into play. Consider again the complete enzyme digest:

G-fragments: CUAG, AAG, AG, UCG, CG

U,C-fragments: U, U, AGC, C, GAAG, GAGC.

Having determined the fragments with which the chain begins and ends, we proceed to construct a directed pseudograph. Consider any fragment in one of the preceding lists which is not an extended base (in other words, which splits under the action of the other enzyme), UCG for example. Draw two vertices labeled U and G for the first and last extended bases in this G-fragment and join them by an arc labeled UCG going from the first extended base to the last, as shown in Fig 11.10.

UCG

Figure 11.10   U ○———▸○ G

Similarly, the G-fragment CG determines an arc labeled CG from a vertex labeled C to one labeled G. Repeating for all **normal** fragments which split (**not** just those giving rise to an interior extended base), we obtain all parts of the directed pseudograph shown in Fig 11.11 except for the arc labeled GAAG*U. By what reasoning have we included this additional arc?

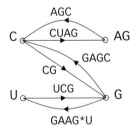

Figure 11.11   A directed pseudograph determining an RNA chain.

The longest abnormal fragment is GAAG and, as we have seen, this is the last fragment in the chain. Draw an arc from G, its first extended base, to U, the first extended base in the chain (determined earlier) and label this arc GAAG*U. To find the RNA chain, we look for Eulerian circuits in the pseudograph which begin with U and end with the arc GAAG*U. We then obtain the original chain by writing down the labels of the arcs on the Eulerian circuit in order, noting that each vertex encountered is listed twice on the arc labels, so one of these occurrences should be discarded. If there is more than one Eulerian circuit, then there is more than one chain with the given fragments. If, by some chance, no Eulerian circuit exists, then the original fragments contradicted each other; there is no RNA chain with these fragments.

In the directed pseudograph of Fig 11.11, there is just one Eulerian circuit: UCG, GAGC, CUAG, AGC, CG, GAAG*U. Eliminating overlap at each vertex, we obtain the RNA chain UCGAGCUAGCGAAG.

Here is a summary of the method we have outlined.

11.3.1 RECOVERY OF AN RNA CHAIN FROM ITS COMPLETE ENZYME DIGEST ▶

1. Determine the fragment with which the chain ends. It's the abnormal fragment if there is only one and otherwise, the longer of the two abnormal fragments.

2. List the interior extended bases and the unsplittable fragments. The chain begins and ends with the two unsplittable fragments which are not interior extended bases. Thus, the first and last fragments of the chain are known.

3. Form a directed pseudograph as follows:

   (a) For each normal fragment which splits under the action of the other enzyme, draw an arc from a vertex labeled with its first extended base to a vertex labeled with its last extended base. Label the arc with the name of the fragment.

   (b) Draw a final arc from the first extended base of the largest abnormal fragment X to the first extended base Y of the chain. Label this arc X*Y.

4. Determine all Eulerian circuits in the pseudograph which end with X*Y and write down in order the labels of the arcs without repeating the label of each vertex on the circuit.

PROBLEM 3. Determine the RNA chain or chains (if any) which give rise to the following complete enzyme digest.

G-fragments: AACUG, UAG, A, AG, AG, AG, G
U,C-fragments: U, AGAAC, AGAGA, GGAGU.

Solution. There are two abnormal fragments, A and AGAGA. The longer is AGAGA, so this is the fragment with which the chain ends.

The interior extended bases and unsplittable fragments are these:

interior extended bases: U, AG, G, AG
unsplittable fragments: A, AG, AG, AG, G, U.

(Notice that the U,C-fragment GGAGU yields **two** interior extended bases.) The two unsplittable fragments which are not extended bases are AG and A. Since the chain ends AGAGA, it begins AG.

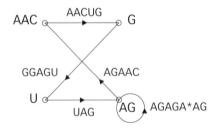

Figure 11.12

In Fig 11.12, we show the directed pseudograph which depicts this situation. The extended base with which the longer of the abnormal fragments begins is AG, which is also the initial fragment. Thus, the final arc, labeled AGAGA*AG, is a loop from AG to itself. The unique answer is AGAACUGGAGUAGAGA. ∎

The connection between graph theory and the problem of recovering RNA chains from their enzyme digests does not seem to be well known. It came first to our attention in a book on combinatorics by Fred Roberts,[5] although the ideas seem to have originated in work of James E. Mosimann.[6] We are indebted to George Hutchinson[7] for making available some additional material to complement his and Mosimann's journal articles.

We would also like to thank Daniel Kobler for making us aware of the fact that graph theory techniques similar to those described in this section have been used recently to study problems of recovering DNA chains from other types of information about fragments.[8]

Answers to Pauses

7. Since GAAG comes last, we seek the number of arrangements of the fragments U,C,GAGC,U,AGC. Since U appears twice, the answer is $\frac{1}{2}5! = 60$. (See Section 7.3.)

8. There can be at most one abnormal G-fragment and at most one abnormal U,C-fragment.

[5] Fred S. Roberts, *Applied Combinatorics*, Prentice Hall, New Jersey, 1984.

[6] James E. Mosimann, *Elementary Probability for the Biological Sciences*, Appleton-Century-Crofts, New York, 1968.

[7] George Hutchinson, "Evaluation of Polymer Sequence Fragment Data Using Graph Theory," *Bulletin of Mathematical Biophysics* **31** (1969), 541–562.

—————————, *Evaluation of Polymer Sequence Data from Two Complete Digests*, Internal Report, Laboratory of Applied Studies, Division of Computer Research and Technology, National Institute of Health, Maryland (1968).

[8] For example, see J. Blazewicz, A. Hertz, D. Kobler, D. de Werra, "On Some Properties of DNA Graphs," Discrete Applied Mathematics **98** (1999), 1–19.

EXERCISES

The symbol [BB] means that an answer can be found in the Back of the Book.

1. Answer the following questions for each pair of fragment lists given.

   i. From how many RNA chains could the given list of G-fragments arise? From how many chains could the given list of U,C-fragments arise? Which of these numbers provides a "better" estimate of the number of RNA chains whose G-fragments and U,C-fragments are as described?

   ii. Find all RNA chains with the given complete enzyme digests.

   (a) [BB] G-fragments: CUG, CAAG, G, UC
   U,C-fragments: C, C, U, AAGC, GGU

   (b) G-fragments: G, UCG, G, G, UU
   U,C-fragments: GGGU, U, GU, C

   (c) G-fragments: UCACG, AA, CCG, AAAG, G
   U,C-fragments: C, GGU, GAAAGAA, C, AC, C

   (d) G-fragments: UUCG, G, ACG, CUAG
   U,C-fragments: G, C, GGAC, U, AGU, C, U

   (e) [BB] G-fragments: UG, CC, ACCAG, G, AUG
   U,C-fragments: C, GAU, C, AGC, GAC, GU

   (f) G-fragments: CACG, AUG, UCAG, AG, UG
   U,C-fragments: AC, C, GU, G, AGAU, GAGC, U

   (g) G-fragments: UCG, AUAG, AG, UCAAUAG, G
   U,C-fragments: C, AGU, AG, GGAU, C, AGU, AAU

   (h) G-fragments: UAG, CUAAG, UA, G, UCAG, UAG
   U,C-fragments: C, U, U, AGGC, AAGU, AGU, AGU, A

2. (a) [BB] Why must there always exist an abnormal fragment?

   (b) [BB] Give a necessary and sufficient condition for there to exist two abnormal fragments.

3. [BB] Find the length of the smallest RNA chain with the property that its G- and U,C-fragments are the same as those of another RNA chain. (Assume that these chains contain no A's and are of length greater than 1.)

4. Consider the two lists:
   - the interior extended bases,
   - the unsplittable fragments.

   Show that

   (a) [BB] every entry on the first list is on the second list;

   (b) [BB] the first and last extended bases in the entire RNA chain are on the second list but not on the first;

   (c) any item on the second list which is not on the first list must be a fragment with which the chain begins or ends.

5. Explain why every RNA chain determines an Eulerian circuit in the corresponding directed pseudograph.

11.4 TOURNAMENTS

In this section, we consider another kind of digraph called a *tournament*, which is just a complete graph with an orientation.

11.4.1 DEFINITION ▶ A *tournament* is a digraph with the property that for every two distinct vertices u and v, exactly one of uv, vu is an arc. The *score* of a vertex in a tournament, denoted $s(v)$, is the outdegree of that vertex. The *score sequence* of a tournament is the list of outdegrees, in nonincreasing order.

Pause 9 Suppose that a tournament T has n vertices, each of which has a different score. What is the score sequence of T?

Tournaments arise in many different situations, the most obvious giving rise to their name. A round-robin competition in sport is one in which every competitor plays every other competitor once. Such a competition can be represented by a tournament in which the players are vertices and an arc directed from vertex v to vertex u indicates that player v defeats player u. In this context, the score of a vertex is the number of wins achieved by that player in the tournament and the score sequence is the sequence of wins by all the players, arranged in decreasing order. An example of a tournament involving five players and with score sequence 3, 3, 2, 2, 0 appears in Fig 11.13.

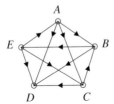

Figure 11.13   A tournament of size five.

There are other less obvious applications of tournaments. In psychology, subjects are sometimes presented with alternatives two at a time and asked to express a preference for one of them. Representing the various alternatives as the vertices of a digraph and assuming all possible pairs of alternatives are presented to the subject, then that person's preferences can be represented as a tournament in a natural way.

Tournaments arise in biology as well. It is sometimes the case that for every pair of animals of a certain species, one is dominant over the other. This relation defines a pecking order in the species, and this can be analyzed by means of tournaments.

It is usually desired to rank the members of a tournament. In the round-robin competition pictured in Fig 11.13, the numbers of wins recorded by A, B, C, D, and E are 3, 2, 2, 0, and 3 respectively. Since A and E have the best records, these players should finish ahead of the rest. Since E defeated A, perhaps E should be ranked first, although A might resent this since A defeated B, who in turn defeated E. A Hamiltonian path provides one way of settling rankings, namely, rank v ahead of u if vertex v occurs before vertex u in the Hamiltonian path. Such a ranking may seem reasonable since a player will always have beaten the person ranked immediately below him or her. In Fig 11.13, for example, $EACBD$ is a Hamiltonian path and E, A, C, B, and D is the corresponding ranking of players. The following theorem, due to the Hungarian mathematician L. Rédei, tells us that this procedure is available in any tournament.

11.4.2 THEOREM ▶   (*Rédei*[9]) Every tournament has a Hamiltonian path.

Proof   Suppose T is a tournament with n vertices. We proceed by induction on n. The result is trivial for $n = 1$ and for $n = 2$ the only tournament is $\circ\!\!\longrightarrow\!\!\circ$, which clearly has a Hamiltonian path. So assume that $k \geq 2$ and that the result is true

[9]L. Rédei, "Ein kombinatorischer Satz," *Acta Litterarum ac Scientiarum Szeged* **7** (1934), 39–43.

for $n = k$. Let $\mathcal{T}$ be a tournament with $k + 1$ vertices, let u be any vertex of $\mathcal{T}$, and let $\mathcal{T} \setminus \{u\}$ be that subgraph of $\mathcal{T}$ obtained by deleting u (and all arcs incident with u). This subgraph has k vertices and it is a tournament since each pair of vertices is still joined by an arc. By the induction hypothesis, we conclude that $\mathcal{T} \setminus \{u\}$ has a Hamiltonian path $v_1 v_2 \cdots v_k$. Now there is an arc between u and v_1 (see Fig 11.14). If this arc is $u v_1$ (that is, directed from u to v_1), then $u v_1 v_2 \cdots v_k$ is a Hamiltonian path in $\mathcal{T}$.

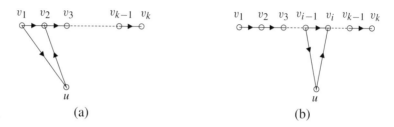

Figure 11.14   (a)   (b)

Suppose, on the other hand, that the arc is $v_1 u$ and consider the arc between u and v_2. If this arc is $u v_2$, then $v_1 u v_2 v_3 \cdots v_k$ is a Hamiltonian path in $\mathcal{T}$ and again we have the desired result. In general, if there is a minimal integer $i > 1$ such that the arc between u and v_i is $u v_i$, then $v_{i-1} u$ is an arc (by minimality of i), and $v_1 v_2 \cdots v_{i-1} u v_i \cdots v_k$ is a Hamiltonian path (see Fig 11.14(b)). If there is no such minimal i, all arcs between u and v_i are of the type $v_i u$ and $v_1 v_2 \cdots v_k u$ is a Hamiltonian path. In any case, there is a Hamiltonian path. ∎

Unfortunately, there is no reason why a Hamiltonian path in a tournament should be unique, so disputes over ranking can still exist. In our earlier example concerning the tournament in Fig 11.13, for instance, $ABECD$ is another Hamiltonian path.

Pause 10   Find a third Hamiltonian path.

The point is that with more than one Hamiltonian path, there is no unique ranking of the players. Certain tournaments, however, do have unique Hamiltonian paths.

11.4.3 DEFINITION ▶   A tournament is *transitive* if and only if whenever uv and vw are arcs, then uw is also an arc.

Figure 11.15   No 3-cycle can appear in a transitive tournament.

Transitive tournaments are precisely those which have no subtournaments of the form illustrated in Fig 11.15. For instance, the tournament in Fig 11.13 is not transitive because $ABEA$ is a 3-cycle like that in Fig 11.15.

It turns out that we can characterize those tournaments with unique Hamiltonian paths in terms of the concepts of transitivity and scores.

11.4.4 THEOREM ▶ The following properties of a tournament $\mathcal{T}$ are equivalent.

(1) $\mathcal{T}$ has a unique Hamiltonian path.
(2) $\mathcal{T}$ is transitive.
(3) Every player in $\mathcal{T}$ has a different score.

Proof We show (1) → (2), that a tournament with a unique Hamiltonian path must be transitive, and leave the rest of the proof to the Exercises (Exercise 9). So assume that $\mathcal{T}$ has a unique Hamiltonian path and order the vertices so that this path is $v_{n-1}v_{n-2}\cdots v_0$. It is sufficient to show that v_jv_i is an arc whenever $j > i$. Suppose, by way of contradiction, that this is not the case. Then there must be some arc v_iv_j with $j > i$. (See Fig 11.16.)

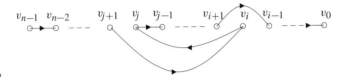

Figure 11.16

Choose i as small as possible such that an arc of this type exists (for some j) and, with that i fixed, choose j as large as possible such that the arc v_iv_j exists. If $j \neq n - 1$ and $i \neq 0$ (as depicted in the figure), then

$$v_{n-1}\cdots v_{j+1}v_iv_jv_{j-1}\cdots v_{i+1}v_{i-1}\cdots v_0$$

is a second Hamiltonian path. If $i = 0$ and $j \neq n - 1$, then

$$v_{n-1}\cdots v_{j+1}v_0v_jv_{j-1}\cdots v_1$$

is a second Hamiltonian path. If $j = n - 1$ and $i \neq 0$,

$$v_iv_{n-1}v_{n-2}\cdots v_{i+1}v_{i-1}\cdots v_0$$

is a second Hamiltonian path and, if $i = 0$ and $j = n - 1$, then

$$v_0v_{n-1}v_{n-2}\cdots v_1$$

is a second Hamiltonian path. In all cases, we contradict the uniqueness of the Hamiltonian path. ∎

Since we don't need graph theory to rank n players all of whose scores are different, the import of Theorem 11.4.4, sadly, is that Hamiltonian paths are not very helpful in ranking tournaments.

Pause 11 In a tournament of n players $v_0, v_1, \ldots, v_{n-1}$ suppose the score of player v_i is i. Find a Hamiltonian path.

Answers to Pauses

9. Since each player in a tournament of n players can achieve at most $n - 1$ victories, the score of each vertex is at most $n - 1$. The score of each vertex is nonnegative, so any score s satisfies $0 \le s \le n - 1$. The only set of n distinct scores in this range is $\{0, 1, 2, \ldots, n - 1\}$, so the score sequence is $n - 1, n - 2, \ldots, 2, 1, 0$.
10. $ACBED$ is a third Hamiltonian path in Fig 11.13.

11. Player v_{n-1} wins all $n-1$ games; so there is an arc from v_{n-1} to each other v_i, in particular to v_{n-2}. Player v_{n-2} wins $n-2$ games (losing only to v_{n-1}) so there is an arc from v_{n-2} to each of $v_0, v_1, \ldots, v_{n-3}$. Continuing in this way, we see that there is an arc from v_i to v_{i-1} for each i, $1 \le i \le n-1$. Thus, $v_{n-1}v_{n-2} \cdots v_0$ is a path and, since it contains all vertices, Hamiltonian.

EXERCISES

The symbol [BB] means that an answer can be found in the Back of the Book.

1. (a) [BB] Draw all nonisomorphic tournaments with three vertices and give the score sequence of each. Which of these are transitive?

   (b) Repeat part (a) for tournaments with four vertices.

2. [BB] Let A be the adjacency matrix of a tournament. Describe $A + A^T$ and explain.

3. (For students of linear algebra)

   (a) Prove that A is the adjacency matrix of a tournament with n players if and only $A + A^T = J - I$, where I is the $n \times n$ identity matrix and J is the $n \times n$ matrix consisting entirely of 1's.

   (b) If A is the adjacency matrix of a tournament, so is its transpose. Why?

   (c) Let P be an $n \times n$ permutation matrix, that is, a matrix obtained from the $n \times n$ identity matrix by rearranging its rows. If A is the adjacency matrix of a tournament, so is PAP^T. Why?

4. (a) Find all Hamiltonian paths in the tournament of Fig 11.13.

   (b) Suppose the arc in Fig 11.13 is DB, not BD. Find all Hamiltonian paths in this tournament.

5. Suppose $s_1, s_2, \ldots, s_n$ are the scores in a tournament.

   (a) [BB] Prove that $\sum_{i=1}^{n} s_i = \sum_{i=1}^{n} (n-1-s_i)$ and interpret this result in the context of a round-robin tournament.

   (b) Prove that $\sum_{i=1}^{n} s_i^2 = \sum_{i=1}^{n} (n-1-s_i)^2$ and interpret this result in the context of a round-robin tournament.

6. [BB] Let $\mathcal{T}$ be a tournament and let v be any vertex of $\mathcal{T}$ having maximum score. Show that the length of the shortest path from v to any other vertex of $\mathcal{T}$ is 1 or 2.

7. (a) [BB] Could $3, 2, 1, 1$ be the score sequence of a tournament?

   (b) Could $6, 6, 6, 6, 1, 1, 1, 1$ be the score sequence of a tournament?

8. Let $\mathcal{S}$ be the set of scores in a tournament $\mathcal{T}$ with n vertices.

   (a) Prove that $\sum_{s \in \mathcal{S}} s = \frac{1}{2}n(n-1)$.

   (b) Let t be a natural number less than n and let $s_1, s_2, \ldots, s_t \in \mathcal{S}$.
   Prove that $\sum_{i=1}^{t} s_i \ge \frac{1}{2}t(t-1)$.

   [*Remark*: H. G. Landau has proved that the two conditions in (a) and (b) are sufficient as well as necessary for the existence of a tournament with a prescribed set of scores.[10]]

9. Complete the proof of Theorem 11.4.4 as follows.

   (a) Prove that the scores in a transitive tournament of n vertices are all different. Thus, $(2) \to (3)$.

   (b) Suppose every player in a tournament $\mathcal{T}$ has a different score. Show that $\mathcal{T}$ has a unique Hamiltonian path. Thus, $(3) \to (1)$. [*Hint*: Pauses 9 and 11]

10. Find all Hamiltonian paths.

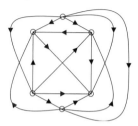

11. [BB; $\to$] Prove that a tournament is transitive if and only if it contains no cycles.

[10]H. G. Landau, "On Dominance Relations and the Structure of Animal Societies, III; The Condition for a Score Structure," *Bulletin of Mathematical Biophysics* **15** (1955), 143–148.

11.5 SCHEDULING PROBLEMS

The shortest path algorithms for weighted graphs discussed in Section 10.4 have many obvious applications. Often, the weight on an edge represents distance (or cost) and the shortest path is just the shortest (or cheapest) route between the distinguished points. There is also a large area of application to weighted digraphs, where the arcs represent tasks which must be performed in a certain order and the weight of an arc represents the time required to carry out the task. The problem in which we are interested is to determine the order in which the tasks should be undertaken in order that they all be completed in the shortest possible time. We present two types of such "scheduling problems" in this section.

We assume throughout that the weights of arcs in digraphs are integers. Such digraphs are also called *directed networks*.

11.5.1 DEFINITION ▶ A *directed network* is a digraph with an integer weight attached to each arc.

11.5.2 A TYPE I SCHEDULING PROBLEM ▶ The construction of a fence involves four tasks: setting posts (Po), cutting wood (C), painting (Pa), and nailing (N). Setting posts must precede painting and nailing, and cutting must precede nailing. Suppose that setting posts takes three units of time, cutting wood takes two units of time, painting takes five units of time for uncut wood and four units of time otherwise, and nailing takes two units of time for unpainted wood and three units of time otherwise. In what order should these tasks be carried out in order to complete the project in the shortest possible time?

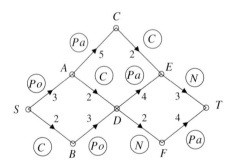

Figure 11.17   A directed network for a scheduling problem.

This is a problem which can be easily analyzed. There are five ways in which the fence can be completed:

$$Po—C—N—Pa$$
$$Po—Pa—C—N$$
$$Po—C—Pa—N$$
$$C—Po—N—Pa$$
$$C—Po—Pa—N$$

and it is a simple matter to compute the time for each. We can also illustrate the situation with the directed network shown in Fig 11.17. The arcs in the digraph represent tasks and the vertices represent stages in the process. The direction of an arc indicates passage from one stage to another. For instance, the arc with weight 2 from D to F represents nailing, while the two arcs leaving vertex A represent cutting and painting. Vertex E represents the stage where all tasks except nailing have been completed.

A path corresponds to a certain order of tasks; thus, the three different paths from S to E represent three different ways to cut, paint, and set posts. Note that the weight given to an arc may depend on which tasks precede it; for example, arcs DF and ET each represent nailing, but they have different weights.

The shortest path from S to T shows us how to complete the project in the least amount of time. Applying Dijkstra's algorithm (first version) to the digraph, we obtain the labels shown in Fig 11.18.

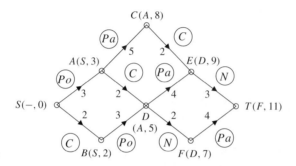

Figure 11.18

One shortest path is $SADFT$, completing the job in 11 units of time. This path corresponds to first setting posts, then cutting, then nailing, and finally painting. Notice that there was a choice in the labeling of vertex D; it could equally well have been labeled $(B, 5)$. There is, therefore, a second shortest path which completes the job in 11 units of time, namely, $SBDFT$. So we could equally well cut, then set posts, then nail, and finally paint.

The procedure just described is important in the management sciences and has been thoroughly studied. Methods for solving such networks are often called *critical path methods* (CPMs) or *project evaluation and review techniques* (PERTs), and a great deal of specialized terminology has been developed. For example, a shortest path such as $SADFT$ (or $SBDFT$) is called a *critical path*.

In the scheduling problem discussed previously, one task had to be completed before another could start. Such would be the case in a one-person project, for example. Also, the time required for a task depended upon what tasks had already been completed. There are other situations, however, where several different tasks can be pursued simultaneously and where the time of a given task is independent of the other tasks. Think of the construction of a house or office building. Such a project often requires the services of a large number of people. While the order of tasks is sometimes important (plumbing before carpets!), sometimes it isn't (plumbing and electrical work). Also, and unlike building a fence, in a large construction project the times of the tasks are often determined only by the nature of the task.

As before, our aim is to plan the project so as to minimize the time required. While strictly speaking, the procedure we outline is not a shortest path method, it does fit naturally into this section.

11.5.3 A TYPE II SCHEDULING PROBLEM ▶

In order to finish a basement, a contractor must arrange for certain tasks to be completed. These are shown in Table 11.5 together with brief codes for referring to them and the times required for each.

Table 11.5

Task	Code	Time
Floor installation	F	4
Plumbing	Pl	3
Electrical work	E	5
Wallboard	W	3
Varnish doors and moldings	V	1
Paint	Pa	2
Install doors and moldings	D	1
Lay carpet	C	2

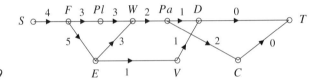

Figure 11.19

We represent this problem by the digraph in Fig 11.19. This time the vertices represent tasks and an arc from u to v means that task u precedes task v. It is necessary to install the floor (F) before any other tasks can commence. The electrical work (E) and plumbing (Pl) must both be completed before the wallboard can go in, but neither the plumbing nor the electrical work is a prerequisite for the other; these two tasks can occur simultaneously. The order in which other tasks are carried out is shown in the digraph. Note that while the electrical work must precede the painting, we do not show an arc from E to Pa since this precedence can be inferred from the arcs EW and WPa.

The weight assigned to an arc is the time it takes to complete the task represented by the terminal vertex of the arc. Both EW and PlW are assigned a weight of 3 since both E and Pl must be completed before W can commence and W requires three units of time. The last two arcs, leading to the finish of the project, are assigned a weight of 0 because the job is completed once D and C are finished.

When the job is completed, all paths from start to finish will have been followed. Thus, in this situation, it is the time required for the longest path which is the shortest time for the entire project. To find this path, we modify in two ways the first version of Dijkstra's algorithm. No vertex is labeled until all prerequisite vertices have been labeled and each vertex is always labeled with the **largest** value of $d + w(e)$. The final labels in our example are shown in Fig 11.20.

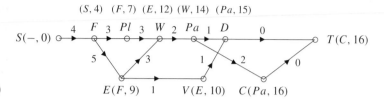

Figure 11.20

The job can be completed in 16 units of time, and $FEWPaC$ is the critical path. Note that the tasks not on the critical path (Pl, V, and D) can be delayed somewhat without delaying the project. For example, plumbing could take five time units instead of three and the label $(E, 12)$ for W would still be correct: The critical path would not be affected. Varnishing could take as much as six units of time without delaying the project because, if this were to occur, only the label for D would change, to $(V, 16)$: The project could still be completed in 16 units of time. On the other hand, if more than six units of time were spent on varnishing, the project would be delayed. This extra $6 - 1 = 5$ units of time which could be spent on V is called the *slack* of the task V. As we have seen, the slack of Pl is 2 and the slack of D is 1. Knowledge of these values allows the contractor to complete the project more quickly by transferring workers to other tasks on the critical path.

EXERCISES

The symbol [BB] means that an answer can be found in the Back of the Book.

1. [BB] Solve the fence problem described in paragraph 11.5.2, assuming that the setting of posts takes three units of time if the wood has not yet been cut and four units of time otherwise. (All other aspects of the project are unchanged.)

2. The construction of a certain part in an automobile engine involves four activities: pouring the mold, calibration, polishing, and inspection. The mold is poured first; calibration must occur before inspection. Pouring the mold takes eight units of time, calibration takes three units, polishing takes six units of time for an uncalibrated product and eight units of time for a calibrated one, and inspection takes two units of time for a polished product and three units of time for an unpolished one.

   (a) What type of scheduling problem is this (type I or II)? Represent the paths to completion of this job by drawing the appropriate directed network.

   (b) What is the shortest time required for this job? Describe the critical path.

3. In Exercise 2, suppose polishing takes nine units of time for an uncalibrated product. (All other conditions remain the same.) Find the new critical path and the length of the revised project.

4. Repeat Exercise 2, assuming that, in addition to the tasks stated, the workers will take a coffee break. The coffee break must take place after calibration and before inspection. If it occurs before polishing, it requires two time units; after polishing, it requires six time units.

5. [BB] Change the basement project described in paragraph 11.5.3 by supposing that

   • plumbing requires four units of time;
   • the doors and moldings will be installed before they are varnished;
   • the carpets cannot be installed until the doors and moldings have been varnished; and

- there are kitchen cabinets to be installed in one unit of time after painting.

Draw the appropriate directed network which depicts the paths to completion of this revised project. Describe all critical paths and the slack in each task.

6. [BB] Suppose we want to build a Klein bottle. The tasks involved in this project and the time required for each are recorded in the following table.

Code	Tasks	Time (in months)
G	blow glass	7
Po	polish	5
Sh	shape	9
A	add hole	4
H	affix handles	5
Tw	twist	2
R	remove hole	4
Pa	paint	6

The glass must be blown before anything else. Polishing precedes all but G and Sh; shaping precedes all but G and Po. The hole must be added before it is removed (!) and before the bottle is twisted. The handles go on before Tw and Pa. Once Tw, R, and Pa are completed, the project is finished.

(a) What type of scheduling problem is this (type I or II)? Draw the appropriate directed network.
(b) Find all critical paths for this project.
(c) Find the slack of Po, Sh, A, Tw, and R for each of the critical paths found in (b).
(d) Suppose that tasks Po and A are each delayed three months. Will the completion of the project be delayed? Explain.

7. Repeat Exercise 6 if, in addition to all the conditions stated there, A must also precede H.

8. Repeat Exercise 6 if A takes six months to complete instead of four and R takes seven months to complete instead of four.

9. What's a Klein bottle? Find out what you can about this thing and write a short note.

10. [BB] To complete her Master's thesis, a geology student must perform field work (F) and laboratory analysis (L), conduct a library search (S) of the literature, create a database of relevant articles (D), and write the thesis (W). The write-up cannot begin until all tasks except for the laboratory analysis are complete. It takes three units of time if it is the last task and otherwise five units of time. The field work, which takes four units of time, must precede both the laboratory analysis and the database creation. The library search takes two units of time. The laboratory analysis takes two units of time if it is undertaken after the library search and otherwise three units of time. Creation of the database takes one unit of time if it is delayed until the library search is complete and otherwise four units of time.

(a) What type of scheduling problem is this (I or II)? Draw the appropriate directed network.
(b) What is the shortest possible time in which this student can complete her thesis?
(c) In what order should she perform her thesis-related tasks in order to achieve this minimum time?

11. Fred and Wilma Noseworthy are planning a dinner party. The things they must do before their guests arrive and the time required for each are shown in the following table. The fish must be caught and the wine purchased before the table is set. The Noseworthys are not fond of raw fish or raw vegetables and do not use wine in their cooking. The table must be dusted before it can be set. Vacuuming is never done until the table is set. Fortunately, Fred and Wilma have the full support of their student daughter and son, who are always very willing to help their parents with whatever jobs have to be done (so up to four jobs can be done simultaneously). The entire family greets their guests together after all tasks have been completed.

Dust house	D	3
Vacuum house	V	2
Set table	T	1
Buy wine	W	4
Catch fish	F	6
Pick vegetables	P	2
Cook food	C	4
Greet guests	G	2

(a) What type of scheduling problem is this (I or II)? Draw the appropriate directed network.
(b) What is the shortest possible time in which dinner preparations can be accomplished? Describe the critical path and illustrate with a directed network, showing all labels.
(c) Find the slack in W, C, and D.

12. Repeat the previous problem under the additional assumption that wine is used in the cooking.

13. Before going to school, John must take a shower, get dressed, eat breakfast, and finish his math assignment. John will take his shower before getting dressed or eating

breakfast and he'll get dressed before finishing his assignment, but there are no other restrictions. Taking a shower requires 9 minutes; getting dressed takes 9 minutes if John has first eaten breakfast but otherwise 12 minutes. Finishing his assignment takes 12 minutes if John has eaten breakfast, but 18 minutes otherwise. Eating breakfast takes 9 minutes if John has finished his assignment and 18 minutes otherwise.

    (a) What type of scheduling problem is this (I or II)? Draw the appropriate directed network.

    (b) Find all possible ways in which John can carry out his tasks before school so that he can leave in the shortest possible time.

14. Repeat Exercise 13, removing the restriction that John must dress before finishing his assignment.

REVIEW EXERCISES FOR CHAPTER 11

1. Solve the Chinese Postman Problem for the weighted graph given in Exercise 17 of the Chapter 10 Review Problems.

2. Solve the Chinese Postman Problem for the unweighted graph $K_{3,6}$.

3. Explain how to solve the Chinese Postman Problem for a graph G which has an Eulerian trail.

4. Let G be a strongly connected digraph with the property that no vertex has indegree equal to its outdegree. Prove that at least two vertices have indegrees greater than their outdegrees.

5. (a) Find two different orientations on the edges of K_4 which lead to nonisomorphic digraphs.

    (b) Find two different orientations on the edges of K_4 which lead to isomorphic digraphs.

6. If A is the adjacency matrix of a digraph and $A^2 = [b_{ij}]$, find $\frac{1}{2} \sum b_{ii}$ and explain.

7. (a) Show that Dijkstra's algorithm fails and explain why. Start at vertex A.

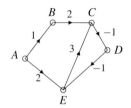

    (b) Apply the Bellman-Ford algorithm to the digraph. Your answer should make clear that you understand this algorithm by showing how the labels on vertices change as the algorithm proceeds.

8. Find all RNA chains with the given complete enzyme digests.

  (a) G-fragments: G, AUG, UCAG, CCUG
          U,C-fragments: G, GGAU, C, C, U, AGC, U

  (b) G-fragments: G, G, AUU, UCG, G, UG, G
          U,C-fragments: GGGGU, C, U, GAU, GU

  (c) G-fragments: UG, CUAAG, AG, CG, UCG
          U,C-fragments: GU, AAG, GAGC, GC, U, U, C

9. Let $n \geq 3$ and assume that the complete graph K_n has vertices labeled $w_1, w_2, \ldots, w_n$ and arc $w_i w_j$ directed w_i to w_j if and only if $i < j$. Show that the tournament on this digraph is transitive.

10. (a) Could $5, 5, 4, 3, 2, 2, 2$ be the score sequence of a tournament?

    (b) Could $8, 8, 8, 6, 6, 3, 3, 1, 1, 1$ be the score sequence of a tournament?

11. Let T be a tournament and let v be any vertex of T having minimum score. Show that the length of the shortest path from any other vertex to v is 1 or 2.

12. In a round-robin tennis tournament, we know that Alice is the clear winner, George is the clear loser, and all other players are tied for second. Show that the number of players competing must have been odd.

13. George has to do four things before his mathematics exam tomorrow morning: study, sleep, eat pizza, and check his e-mail. He intends to spend two different periods of time studying, three hours on each occasion. While the sessions could be back to back, at least one session must occur before he sleeps. Eating pizza requires one hour, but it also takes an hour for the order to arrive (and something else could be done during this interval). Eating pizza must precede sleeping and also at least one of the study sessions. George will get six hours of sleep if he sleeps after all study has been completed, but he's willing to settle for five hours otherwise. If George checks his e-mail between study sessions, he will do so quickly and take only one hour; otherwise, he will spend two hours at this activity.

(a) Find the shortest possible time in which George can complete his activities, and a way in which he can order them and complete them in this time.

(b) Assume that George decides to read his e-mail before eating his pizza. Does this change the answer to (a)? If so, how?

(c) Assume that George decides to read his e-mail before ordering his pizza. Does this change the answer to (a)? If so, how?

14. The following chart lists a number of tasks that must be completed in order for a crew of workers to construct a glynskz.

Task	A B C D E F G H I J
Time (in days)	2 2 3 1 1 2 3 4 3 3

Task A must be carried out before any other tasks can commence. Task B must precede tasks E and F, and both E and F must be completed before H can begin. Tasks C and D must precede task G, which in turn must precede I. Task J must be carried out last. It is assumed that there are enough workers to carry out any number of tasks simultaneously.

(a) What is the fewest number of days needed to construct this glynskz? Find all critical paths.

(b) Find the slack in C, D, and E.

12

Trees

12.1 WHAT IS A TREE?

One of the most commonly occurring kinds of graph is called a *tree*, perhaps because it can be drawn so that it looks a bit like an ordinary tree. In this section, we introduce this concept with a variety of examples.

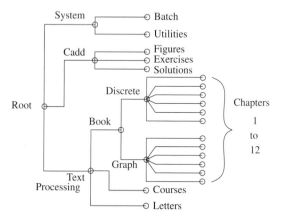

Figure 12.1   A partial computer directory structure.

The operating system on a computer organizes files into directories and subdirectories in the same way that people gather together pieces of paper into folders and put them into the drawers of a filing cabinet. The computer on which this text was first prepared had a directory called "Book" partitioned into two subdirectories called "Discrete" and "Graph," each in turn containing a number of subdirectories, one for each chapter. Each chapter subdirectory had one file for each section in that chapter. Part of the organization of these computer files is depicted in Fig 12.1.

Suppose you wanted to write down all the increasing subsequences which can be formed from the sequence 4, 8, 5, 0, 6, 2:   4, 5, 6 is one such subsequence, as is 0, 2, but not 2, 4 because it is not a subsequence of the given sequence. The order of the elements in a subsequence must be the same as in the sequence itself. To ensure that no subsequences are inadvertently omitted, some systematic method of enumeration is required. The graph drawn in Fig 12.2 indicates such a method. In the figure, the desired subsequences correspond to paths from "start" to each vertex.

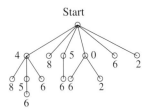

Figure 12.2

Pause 1   List all the monotonically increasing subsequences of the sequence 4, 8, 5, 0, 6, 2.

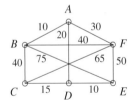

Figure 12.3

Suppose the vertices of the graph shown in Fig 12.3 represent towns, the edges roads, and the label on a road gives the time it takes to travel the road. Is it possible for a salesperson to drive to all the towns and return to his or her starting point having visited each town exactly once? If it is, what is the shortest distance required for such a trip? In Section 10.4, we introduced this "Traveling Salesman's Problem" and mentioned that there is no efficient algorithm known for solving it. In reasonably small graphs, of course, we can exhaustively enumerate all the possibilities. Figure 12.4 indicates how we can use a graph to search systematically for Hamiltonian cycles, if they exist. We assume that the salesperson starts at A. The graph makes it apparent that there are in fact 12 Hamiltonian cycles in the graph of Fig 12.3.

Pause 2   Answer the Traveling Salesman's Problem for the graph of Fig 12.3. Find a shortest Hamiltonian cycle and its length.

John and David are going to play a few games of chess. They agree that the first person to win two games in a row or to win a total of three games will be

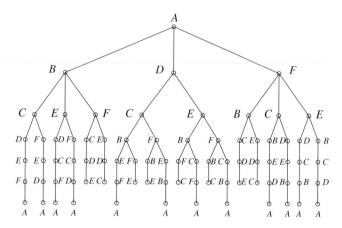

Figure 12.4   A hunt for Hamiltonian cycles in the graph of Fig 12.3.

declared the winner. The possible outcomes can be conveniently pictured with the graph shown in Fig 12.5.

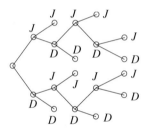

Figure 12.5

How many outcomes are possible? What is the maximum number of games John and David will play?

With the exception of that in Fig 12.3, all the graphs which have appeared in this section are *trees*.

12.1.1 DEFINITION ▶   A *tree* is a connected graph which contains no circuits.

There are several alternative definitions.

12.1.2 PROPOSITION ▶   Let $\mathcal{G}$ be a graph. Then the following statements are equivalent.

(1) $\mathcal{G}$ is a tree.
(2) $\mathcal{G}$ is connected and *acyclic*; that is, without cycles.
(3) Between any two vertices of $\mathcal{G}$ there is precisely one path.

Proof   (1) → (2): Since a cycle is a circuit, statement (1) immediately implies statement (2).

(2) → (3): If there are two different paths $\mathcal{P}_1$, $\mathcal{P}_2$ from some vertex v to another, w, then the closed walk from v to v obtained by following the vertices of $\mathcal{P}_1$ and then those of $\mathcal{P}_2$ in reverse order would contain a cycle, contradicting (2). (See Exercise 12.)

(3) → (1): Since there is a path between any two vertices of $\mathcal{G}$, certainly $\mathcal{G}$ is connected. Moreover, it can contain no circuits. Otherwise, it would contain a cycle (Exercise 14, Section 10.1) and a cycle determines **two** paths between any two vertices of it.

The sequence of implications (1) → (2) → (3) → (1) just established proves the equivalence of (1), (2), and (3). ∎

12.1.3 DEFINITION ▶ A tree is *rooted* if it comes with a specified vertex, called the *root*.

This may not seem like much of a definition, but the concept is useful. Also, there is a convention for picturing rooted trees. Suppose v is a root in a tree $\mathcal{T}$. By Proposition 12.1.2, there is a unique path from v to any other vertex. Thus, the concept of *level*, meaning distance (number of edges on a path) from v, is well defined. When drawing a rooted tree then, we agree to put the root v at the top (in contrast to the way we draw trees that grow out of doors), we put all level one vertices (those of distance 1 from v) on a horizontal line just below v, those of level two (distance 2 from v) on a horizontal line below the previous, and so on. The tree in Fig 12.2 is drawn so as to show the vertex "start" as root. Vertex A in Fig 12.4 is a root. Vertices B, D, and F have level one; the vertices labeled C, E, F, C, E, B, C, E in the next row have level two; and so on.

The way in which a rooted tree is drawn clearly depends upon which vertex is the root. In Fig 12.6, we show two pictures of the same tree, rooted at A on the left and at B on the right.

Figure 12.6   The same tree rooted at A on the left and B on the right.

Answers to Pauses

1. There are 13 in all.

4; $4, 8$; $4, 5$; $4, 5, 6$; $4, 6$; 8; 5; $5, 6$; 0; $0, 6$; $0, 2$; 6; 2

2. A shortest Hamiltonian cycle is $ABCDEFA$, of length 155.

3. There are ten possible outcomes, corresponding to the ten vertices of degree 1 in Fig 12.5. As many as five games could be required to decide a winner.

EXERCISES

The symbol [BB] means that an answer can be found in the Back of the Book.

1. [BB] Suppose eight teams qualify for the playoffs in a local softball league. In the first round of playoffs, Series *A* pits the best team against the team that finished eighth, Series *B* has the second best team playing the team that finished seventh, Series *C* has the third placed team playing the sixth. In Series *D*, the teams that finished fourth and fifth play. In the second round, the winners of Series *A* and *C* compete, as do the winners of Series *B* and *D*, in Series *E* and *F*. Finally, the winners of these series play to determine the league champion. Draw a tree which summarizes the playoff structure in this league.

2. Edward VII, eldest son among Queen Victoria's nine children, finally ascended to the throne of England in 1901 (after 60 years as Prince of Wales). Upon his death in 1910, he was succeeded by King George V. George V had six children, perhaps the most infamous of whom was Edward VIII, who abdicated to marry a twice divorced American. Upon this abdication, George VI, father of the present Queen of England, became king. Queen Elizabeth has four children, Charles, Anne, Andrew, and Edward. Her sister, Princess Margaret, has two, Sarah and David. Prince Charles has two boys, William and Harry, marking the first time since the death of Edward VII that there have been two males in direct succession to the English throne. Draw that portion of Prince William's family tree which shows the people discussed in this short narrative.

3. Make a tree (like that in Fig 12.2) which displays all monotonically increasing subsequences of 3, 2, 8, 0, 9, 1, 5. How many are there altogether? Taking "start" to be the root, how many vertices are at each possible level?

4. John does not play chess very well, so the next time he plays David he says he will quit after winning a game, or after five games in all have been played. Make a graph showing all possible outcomes. How many possible outcomes are there? In how many of these does John win a game?

5. [BB] It's the weekend and Harry has several ways to spend his Friday and Saturday evenings. He can study, read, watch television, or go out with the boys. Draw a tree which shows all possible ways Harry can spend his Friday and Saturday evenings.

6. Mary can travel from St. John's to Corner Brook by car, bus, or plane and from Corner Brook to Goose Bay by plane or boat. Draw a tree showing all the possible ways Mary can go from St. John's to Goose Bay via Corner Brook. In how many of these ways does she avoid the bus?

7. [BB] In how many ways can a committee of three people be chosen from the following group of people: Bruce, Philomena, Irene, Tom, and Dave? Draw a tree which shows all possible committees and the way each was chosen.

8. Draw the tree with root *A* and with root *B*.

9. The vertices in the graph represent towns, the edges roads, and the labels on the roads costs of paving the roads.

(a) Make a tree which shows all paths beginning at vertex *A*. List those paths which terminate at *C*. Indicate which ones, if any, are Hamiltonian.

(b) Is the graph Hamiltonian? Explain.

(c) Which roads should be paved in order that one may drive from *A* along paved roads to as many towns as possible at minimal cost? Justify your answer. What is this minimal cost?

10. Solve the Traveling Salesman's Problem for each of the following graphs by making trees which display all Hamiltonian cycles. (Start at *A* in each case.)

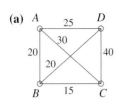

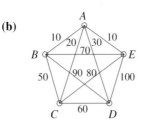

11. A *forest* is a graph with no cycles. Explain why a forest is the disjoint union of trees.

12. [BB] Suppose $\mathcal{P}_1$ and $\mathcal{P}_2$ are two paths from a vertex v to another vertex w in a graph. Prove that the closed walk obtained by following $\mathcal{P}_1$ from v to w and then $\mathcal{P}_2$ in reverse from w to v contains a cycle.

12.2 PROPERTIES OF TREES

In this section, we derive a number of properties of trees and discuss a connection with the isomers of a certain kind of hydrocarbon.

12.2.1 PROPOSITION ▶

Suppose $\mathcal{G}$ is a graph each of whose vertices has degree at least 2. Then $\mathcal{G}$ contains a circuit.

Proof

Let v_1 be any vertex of $\mathcal{G}$. Since $\deg v_1 \neq 0$, we may proceed to an adjacent vertex v_2. Since $\deg v_2 \geq 2$, we may proceed to an adjacent vertex v_3 along a new edge. Note that $v_1 v_2 v_3$ is a path. In general, suppose $k \geq 3$ and that we have found a path through distinct vertices $v_1, v_2, \ldots, v_k$. Since $\deg v_k \geq 2$, v_k is adjacent to a vertex $v_{k+1} \neq v_{k-1}$. If $v_{k+1} \in \{v_1, v_2, \ldots, v_{k-2}\}$, we have a circuit; otherwise, $v_1 v_2 \cdots v_{k+1}$ is a path. Since there are only finitely many vertices in $\mathcal{G}$, we cannot continue to find paths of increasing length, so eventually we find a circuit. ∎

In a trivial way, a graph with a single vertex (of degree 0) is a tree. A tree with more than one vertex can have no vertices of degree 0 (because it is connected) so all its vertices have degree at least 1. If they all had degree at least 2, there would be a circuit. Thus, we obtain a basic property of trees.

12.2.2 COROLLARY ▶

A tree with more than one vertex must contain a vertex of degree 1. (Such a vertex is called a *leaf*.)

Pause 4

What is a leaf?

Pause 5

Modify the argument used to prove Proposition 12.2.1 in order to prove that a tree with more than one vertex must contain at least two leaves.

In Exercise 21(b) of Section 10.1, you were asked to prove that a connected graph with n vertices has at least $n - 1$ edges. It turns out that a connected graph with n vertices and exactly $n - 1$ edges is a tree; moreover, this property characterizes trees.

12.2.3 THEOREM ▶

A connected graph with n vertices is a tree if and only if it has $n - 1$ edges.

Proof

($\rightarrow$) First we prove, by mathematical induction, that a tree with n vertices has $n - 1$ edges. This is clear for $n = 1$, so assume that $k \geq 1$ and that any tree with k vertices has $k - 1$ edges. Let $\mathcal{T}$ be a tree with $k + 1$ vertices. We must prove that $\mathcal{T}$ has k edges. For this, let v be a vertex in $\mathcal{T}$ of degree 1. Remove v and the single edge e with which it is incident. The point to appreciate is that the *pruned tree* $\mathcal{T}_0 = \mathcal{T} \setminus \{v\}$ which remains is itself a tree. It has no circuits since the original tree had no circuits. Why is it connected? If w and u are vertices in $\mathcal{T}_0$, they are joined by a path $\mathcal{P}$ in $\mathcal{T}$. Since you enter and leave each vertex of $\mathcal{P}$ along different edges, each vertex of $\mathcal{P}$ between w and u has degree at least 2. Thus, v and the lone edge e with which it is adjacent are not part of the path $\mathcal{P}$. So $\mathcal{P}$ must lie entirely in $\mathcal{T}_0$. Therefore, $\mathcal{T}_0$ is connected and hence a tree. Since $\mathcal{T}_0$ has k vertices, it has $k-1$ edges, by the induction hypothesis. Since $\mathcal{T}$ has one more edge than $\mathcal{T}_0$, $\mathcal{T}$ has k edges, which is what we wanted to show.

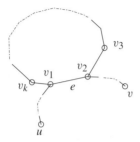

Figure 12.7

($\leftarrow$) For the converse, we suppose that $\mathcal{G}$ is a connected graph with n vertices and $n-1$ edges and prove that $\mathcal{G}$ is a tree. We have only to establish the absence of circuits. Suppose that $\mathcal{G}$ contains a circuit $v_1 v_2 \cdots v_k v_1$. (Since graphs have no loops and no multiple edges, $k \geq 3$.) Remove the edge $e = v_1 v_2$. The remaining graph $\mathcal{G} \setminus \{e\}$ has all n vertices of $\mathcal{G}$ but $n-2$ edges; moreover, $\mathcal{G} \setminus \{e\}$ is connected because any two vertices u, v of $\mathcal{G}$ which were connected via a path which included e are still connected via a walk in which $v_1 v_k v_{k-1} \cdots v_3 v_2$ is substituted for e. (See Fig 12.7.)

If $\mathcal{G} \setminus \{e\}$ contains a circuit we delete an edge from this and obtain a connected graph which still contains n vertices but has just $n-3$ edges. Since $\mathcal{G}$ is finite, eventually all circuits will be deleted and we shall be left with a connected graph with n vertices, fewer than $n-1$ edges, and no circuits. Such a graph is a tree. We showed earlier that a tree with n vertices must have $n-1$ edges. Thus, there can be no circuit in $\mathcal{G}$; $\mathcal{G}$ is indeed a tree. ∎

Pause 6   The tree in Fig 12.1 has 26 vertices and $25 = 26 - 1$ edges, as predicted by Theorem 12.2.3. Verify this theorem for the trees in Figs 12.2 and 12.5.

There are two other properties of trees which follow quickly from Theorem 12.2.3. The first has already been noted.

12.2.4 COROLLARY ▶   A tree with more than one vertex has at least two leaves, that is, at least two vertices of degree 1.

Proof   Suppose $\mathcal{T}$ is a tree with n vertices $v_1, v_2, \ldots, v_n$ and $n-1$ edges. By Proposition 9.2.5, $\sum_{i=1}^{n} \deg v_i = 2(n-1) = 2n - 2$. We already know that some vertex, say v_1, has degree 1. If the remaining $n-1$ vertices each have degree 2 or more, then the sum of all the degrees would be at least $1 + 2(n-1) = 2n - 1$, a contradiction. So there has to be another vertex, in addition to v_1, which has degree 1 too. ∎

In fact, there is a precise formula for the number of leaves, a formula which shows that this number is at least 2. See Exercise 9.

12.2.5 COROLLARY ▶ Any edge added to a tree must produce a circuit.

Proof The new graph remains connected and it is not a tree because it has an equal number of edges and vertices. So it must contain a circuit. ∎

Pause 7 Is it possible for two circuits to be formed when a single edge is added to a tree?

The identification of a tree as an important type of graph is credited to the English mathematician Arthur Cayley (1821–1895), a lawyer, a contemporary of Sir William Rowan Hamilton, and arguably the leading English mathematician of the nineteenth century. Cayley did fundamental work in the theory of matrices and is considered by many to be the founder of abstract group theory. In chemistry, chemical compounds with the formula $C_k H_{2k+2}$ are known as *paraffins*. The molecule $C_k H_{2k+2}$ contains k carbon atoms and $2k + 2$ hydrogen atoms. Each hydrogen atom is bonded to exactly one atom, a carbon atom; each carbon atom is bonded to four atoms. It is not hard to see that there is just one possible structure for each of the first three paraffins; methane, ethane, and propane. (See Fig 12.8.)

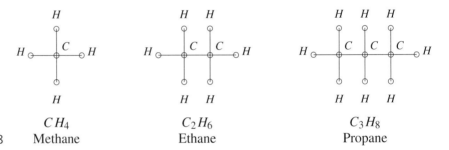

Figure 12.8

| CH_4 | $C_2 H_6$ | $C_3 H_8$ |
| Methane | Ethane | Propane |

On the other hand, for $k \geq 4$, various arrangements of $C_k H_{2k+2}$ are possible. Each arrangement is called an *isomer*. Arthur Cayley wanted to know the number of isomers of each paraffin. Each isomer is clearly associated with a connected graph, a graph which is, in fact, a tree. Here is why.

Each of the k carbon atoms in $C_k H_{2k+2}$ corresponds to a vertex of degree 4 in the associated graph and each of the $2k + 2$ hydrogen atoms to a vertex of degree 1. The total number of vertices in the graph is $k + (2k + 2) = 3k + 2$ and the total number of edges is one half the sum of the degrees; that is, $\frac{1}{2}(4k + (2k + 2)) = 3k + 1$. Since the graph is connected and the number of edges is one less than the number of vertices, it is a tree (by Theorem 12.2.3).

Cayley also noted that an isomer of $C_k H_{2k+2}$ is completely determined by the arrangement of the carbon atoms and that these form a tree with each vertex of degree at most 4. (See Exercise 10.) Note the tree of carbons in each of the molecules in Fig 12.8. To draw an isomer of $C_k H_{2k+2}$, we draw a tree with k vertices labeled C, each of degree at most 4, and then add new edges with end vertices labeled H, of degree 1, at each C vertex where necessary so that each C vertex acquires degree 4. (It can be shown that $2k + 2$ edges must be added. See Exercise 11.)

To enumerate all trees with n vertices, as Cayley was attempting to do, is very difficult unless n is small. The situation for $n \le 4$ is shown in Fig 12.9.

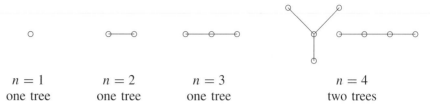

$n = 1$	$n = 2$	$n = 3$	$n = 4$
one tree	one tree	one tree	two trees

Figure 12.9   The unlabeled trees with $n \le 4$ vertices.

Pause 8   Draw graphs of all possible isomers of butane (whose chemical formula is C_4H_{10}). How many are there?

There is no closed formula for the number of trees with n vertices, although these numbers (for various n) appear as the coefficients of a certain generating function which begins

$$x + x^2 + x^3 + 2x^4 + 3x^5 + 6x^6 + 11x^7 + 23x^8,$$

but whose derivation would take us too far afield.[1] On the other hand, as we shall see in the next section, there are precisely n^{n-2} *labeled* trees on n vertices (Theorem 12.3.3). As the name implies, a labeled tree is one in which each vertex has a distinct label attached to it. Figure 12.10 shows graphs of the $3^{3-2} = 3$ different labeled trees with vertices A, B, and C.

$B \quad A \quad C$	$A \quad B \quad C$	$A \quad C \quad B$

Figure 12.10   The three labeled trees on three vertices.

12.2.6   DEFINITION ▶   Trees $\mathcal{T}_1$ and $\mathcal{T}_2$ whose vertices are labeled with the same set of labels are *isomorphic* if and only if, for each pair of labels v and w, vertices v and w are adjacent in $\mathcal{T}_1$ if and only if they are adjacent in $\mathcal{T}_2$.

In Fig 12.10, no two of the labeled trees are isomorphic. The situation is different, however, in Fig 12.11. Here the labeled trees $\mathcal{T}_1$ and $\mathcal{T}_2$ are isomorphic, although neither is isomorphic to $\mathcal{T}_3$. The adjacencies between the labeled vertices of $\mathcal{T}_1$ and $\mathcal{T}_2$ are identical. On the other hand, whereas vertices B and C are adjacent in $\mathcal{T}_1$ (and $\mathcal{T}_2$), such is not the case in $\mathcal{T}_3$.

Answers to Pauses

4. A leaf is a vertex of degree 1.
5. Let v_1 be a leaf in a tree and start a path, as described in Proposition 12.2.1, with v_1. Since there are no circuits in a tree, the path $v_1, v_2, v_3, \ldots$ obtained in the proposition can be extended to a longer path as long as each vertex we meet has degree at least 2. By finiteness, it must eventually terminate with a second vertex of degree 1, that is, another leaf.

---

[1] The interested reader might consult Frank Harary's book, *Graph Theory*, Addison-Wesley (1969).

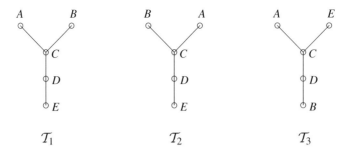

$\mathcal{T}_1$ $\mathcal{T}_2$ $\mathcal{T}_3$

Figure 12.11 $\mathcal{T}_1$ and $\mathcal{T}_2$ are isomorphic labeled trees which are not isomorphic to $\mathcal{T}_3$.

6. The tree in Fig 12.2 has 14 vertices and $13 = 14 - 1$ edges; the tree in Fig 12.5 has 19 vertices and $18 = 19 - 1$ edges.

7. No, it isn't. First note that when an edge is added to a tree, any circuit that is produced must be a cycle. To see why, suppose the addition of edge v_1v_2 produces a circuit $v_1v_2v_3 \cdots v_nv_1$. If this circuit is not a cycle, there is a repeated vertex. This vertex must be v_1 or v_2; otherwise, the original tree contained a circuit, contrary to fact. If the repeated vertex is v_1 (the argument in the case of v_2 is similar), then our circuit has the form $v_1v_2 \cdots v_1w \cdots v_nv_1$. Since the edge v_1v_2 is not repeated in the circuit, $v_1w \cdots v_nv_1$ is a circuit in the original tree, a contradiction. Thus, if an edge v_1v_2 is added to a tree, it is in fact only a cycle which is produced. If two cycles were produced, there would be two paths from v_1 to v_2 in the original graph, contradicting Proposition 12.1.2.

8. Since there are two nonisomorphic trees with four vertices and in each of these, all vertices have degree at most 4, there are two isomers of butane. These are pictured in Fig 12.12. (Note the two trees of four carbon atoms and compare with the two trees on the right of Fig 12.9.)

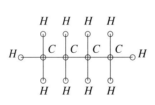

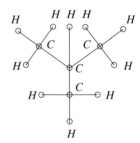

Figure 12.12

EXERCISES

The symbol [BB] means that an answer can be found in the Back of the Book.

1. (a) [BB] Draw the graphs of all nonisomorphic unlabeled trees with five vertices.

(b) [BB] How many isomers does pentane (C_5H_{12}) have? Why?

2. (a) Draw the graphs of all nonisomorphic unlabeled trees with six vertices.

   (b) How many isomers does hexane (C_6H_{14}) have? Why?

3. [BB] Suppose $\mathcal{G}$ is a graph whose vertices represent cities in a country. An edge in $\mathcal{G}$ represents a direct (nonstop) air flight between the corresponding cities. What does a beta index less than 1 say about air travel in the country? There are two possibilities to consider. (See Exercise 5 of Section 9.1.)

4. Prove that a connected graph with n vertices is a tree if and only if the sum of the degrees of the vertices is $2(n-1)$.

5. [BB] Recall that a graph is *acyclic* if it has no cycles. Prove that a graph with n vertices is a tree if and only if it is acyclic with $n-1$ edges.

6. Suppose $\mathcal{G}$ is an acyclic graph with $n \geq 2$ vertices and we remove one edge. Explain why the new graph $\mathcal{G}'$ cannot be connected.

7. Suppose a graph $\mathcal{G}$ has two connected components, $\mathcal{T}_1$, $\mathcal{T}_2$, each of which is a tree. Suppose we add a new edge to $\mathcal{G}$ by joining a vertex of $\mathcal{T}_1$ to a vertex in $\mathcal{T}_2$. Prove that the new graph is a tree.

8. (a) Let e be an edge in a tree $\mathcal{T}$. Prove that the graph consisting of all the vertices of $\mathcal{T}$ but with the single edge e deleted is not connected.

   (b) Without assuming the existence of vertices of degree 1, use the result of (a) and the strong form of mathematical induction to prove that a tree with n vertices has $n-1$ edges.

9. Let $\mathcal{T}$ be a tree with n vertices $v_1, v_2, \ldots, v_n$. Prove that the number of leaves in $\mathcal{T}$ is $2 + \sum_{\deg v_i \geq 3} [\deg v_i - 2]$.

10. [BB] Prove that the subgraph of a C_kH_{2k+2} tree $\mathcal{T}$ consisting of the k carbon vertices and all edges from $\mathcal{T}$ among them is itself a tree.

11. (a) [BB] Suppose $\mathcal{T}$ is a tree with k vertices labeled C, each of degree at most 4. Enlarge $\mathcal{T}$ by adjoining sufficient vertices labeled H so that each vertex C has degree 4 and each vertex H has degree 1. Prove that the number of H vertices adjoined to the graph must be $2k + 2$.

    (b) Can you prove (a) without assuming that $\mathcal{T}$ is a tree?

12. We have noted in this section that there are n^{n-2} labeled trees on n vertices.

    (a) Verify this formula for $n = 4$ by drawing the nonisomorphic unlabeled trees on four vertices and discovering how many different labeled trees each determines. (See Fig 12.9.)

    (b) Repeat (a) for $n = 5$.

13. Prove that a tree with $n \geq 2$ vertices is a bipartite graph.

14. [BB] Find necessary and sufficient conditions for a tree to be a complete bipartite graph. Prove your answer.

15. A *forest* is a graph every component of which is a tree, equivalently, a graph without cycles. (See Exercise 11 of Section 12.1.)

    (a) [BB] Show that a forest with c components, each containing at least two vertices, has at least $2c$ vertices of degree 1.

    (b) Is the result of (a) true without the stipulation that each component contain at least two vertices? Explain.

    (c) Find a formula for the number of edges in a forest with n vertices and c components and prove your answer.

16. (a) [BB] Show that a tree with two vertices of degree 3 must have at least four vertices of degree 1.

    (b) Show that the result of (a) is best possible: A tree with two vertices of degree 3 need not have five vertices of degree 1.

12.3 SPANNING TREES

Figure 12.13 represents a map: The vertices correspond to towns and the labels on the edges represent distances along existing gravel roads between adjacent towns. The province plans to pave certain roads in such a way that one can get between any two towns on pavement. What roads should be paved so as to minimize the total length of pavement required? This problem, known as the *Minimum Connector Problem*, asks, in the language of graph theory, for a *minimum spanning tree* for the graph.

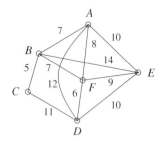

Figure 12.13

12.3.1 DEFINITION ▶ A *spanning tree* of a connected graph G is a subgraph which is a tree and which includes every vertex of G. A *minimum spanning tree* of a weighted graph is a spanning tree of least weight, that is, a spanning tree for which the sum of the weights of all its edges is least among all spanning trees.

The concept of spanning tree exists only for a connected graph G and, if G has n vertices, any spanning tree must necessarily contain $n - 1$ edges.

Finding a spanning tree in a connected graph G is not hard. If G has no circuits, then it is already a tree, so G is a spanning tree for G. If G contains a circuit, then, just as in the proof of the second half of Theorem 12.2.3, we can delete an edge (without deleting any vertex) so as to remove the circuit but leave the graph connected. By repeating this procedure, we eventually find a connected subgraph without circuits containing all the vertices of G, that is, a spanning tree. Figure 12.14 shows a graph and three of its spanning trees. Spanning trees are considered to be "different" if they make use of different edges of the graph. Trees T_1 and T_2 in Fig 12.14 are isomorphic, but they are different spanning trees of G.

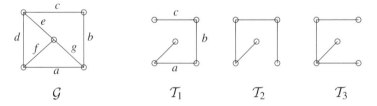

Figure 12.14   A graph G and three of its spanning trees.

There is a procedure for obtaining additional spanning trees from a given spanning tree. In Fig 12.14, T_2 and T_3 were obtained from T_1 as follows: First edge d was added to T_1, giving the circuit $abcd$; removing a gives T_2 and removing b gives T_3. Removing c would give a fourth spanning tree, and, if we wanted more, we could return to T_1 and add edge e, say, thereby completing another circuit $abcef$. Successively removing each of the edges a, b, c, and f of this circuit would give four new spanning trees. Remember that adding an edge to a tree always produces a circuit (Corollary 12.2.5). In Exercise 14, we ask you to show how the procedure described in this paragraph can be adapted to show that all the spanning trees in a graph can be obtained from a given one.

How many spanning trees does a connected graph possess? For this, there is a pretty result established in 1847 by the German physicist Gustav Kirchhoff (1824–1887), who was studying electrical networks.

Readers who have had a first course in linear algebra should know that a *cofactor* of a square matrix A is plus or minus the determinant of a submatrix obtained from A by deleting a row and a column. There is one cofactor for each position in the matrix. The choice of plus or minus for the (i, j) cofactor is determined, respectively, by whether or not there is a plus in position (i, j) of the configuration shown at the right.

$$
\begin{array}{ccccc}
+ & - & + & - & \cdots \\
- & + & - & & \cdots \\
+ & - & + & & \cdots \\
\vdots & \vdots & & &
\end{array}
$$

The "+" in position $(3, 1)$, for example, indicates that the $(3, 1)$ cofactor is the determinant of the matrix obtained by deleting row 3 and column 1, while the "$-$" in position $(2, 3)$ indicates that the $(2, 3)$ cofactor is **minus** the determinant of the matrix obtained by deleting row 2 and column 3.

12.3.2 THEOREM ▶   *(Kirchhoff)* Let M be the matrix obtained from the adjacency matrix of a connected graph $\mathcal{G}$ by changing all 1's to -1's and each diagonal 0 to the degree of the corresponding vertex. Then the number of spanning trees of $\mathcal{G}$ is equal to the value of any cofactor of M.

It is surprising that there is such a simple formula because it is often a difficult problem to count graphs of a specified type. In particular, counting the number of **nonisomorphic** spanning trees in a graph is a complicated affair.

The proof of Kirchhoff's Theorem (which is also known as the *Matrix-Tree Theorem*) will not be included here.[2] Instead, we content ourselves with an illustration.

EXAMPLE 1   Consider the graph whose adjacency matrix is

$$
A = \begin{bmatrix}
0 & 1 & 1 & 1 \\
1 & 0 & 1 & 1 \\
1 & 1 & 0 & 0 \\
1 & 1 & 0 & 0
\end{bmatrix}.
$$

The matrix specified in Kirchhoff's Theorem is

$$
M = \begin{bmatrix}
3 & -1 & -1 & -1 \\
-1 & 3 & -1 & -1 \\
-1 & -1 & 2 & 0 \\
-1 & -1 & 0 & 2
\end{bmatrix}.
$$

---

[2]See Chapter 12 of J. A. Bondy and U. S. R. Murty, *Graph Theory with Applications*, North-Holland, New York, 1981.

The $(1, 1)$ cofactor of M is

$$+ \det \begin{bmatrix} 3 & -1 & -1 \\ -1 & 2 & 0 \\ -1 & 0 & 2 \end{bmatrix} = 3 \begin{vmatrix} 2 & 0 \\ 0 & 2 \end{vmatrix} - (-1) \begin{vmatrix} -1 & 0 \\ -1 & 2 \end{vmatrix} + (-1) \begin{vmatrix} -1 & 2 \\ -1 & 0 \end{vmatrix}$$

$$= 3(4) + 1(-2) + (-1)2 = 8$$

(expanding by cofactors of the first row). Kirchhoff's Theorem guarantees that all the cofactors of M equal 8. For example, the $(2, 3)$ cofactor of M is

$$- \det \begin{bmatrix} 3 & -1 & -1 \\ -1 & -1 & 0 \\ -1 & -1 & 2 \end{bmatrix} = -[(-1)(0) + 2(-3-1)] = -(-8) = 8$$

(expanding by cofactors of the third column). Thus, there are eight spanning trees in the graph. ▲

Pause 9 Draw the graph discussed in the previous example and show each of its eight spanning trees.

Pause 10 How many spanning trees has the graph in Fig 12.14 and why?

Here is a remarkable application of Kirchhoff's Theorem.

12.3.3 THEOREM ▶ The number of labeled trees with n vertices is n^{n-2}.

Proof After the vertices of the complete graph K_n are labeled, any spanning tree of K_n is a labeled tree on n vertices. Conversely, any tree with these n vertices is a spanning tree for K_n because K_n contains all possible edges.[3] Thus, the number of labeled trees on n vertices is the number of spanning trees for K_n, and the latter number can be determined by Kirchhoff's Theorem.

The adjacency matrix of K_n contains 0's on the diagonal and 1's everywhere off the diagonal. So the matrix whose equal cofactors count the number of trees is the $n \times n$ matrix

$$M = \begin{bmatrix} n-1 & -1 & \cdots & -1 \\ -1 & n-1 & \cdots & -1 \\ -1 & -1 & \cdots & -1 \\ \vdots & \vdots & \ddots & \vdots \\ -1 & -1 & \cdots & -1 \\ -1 & -1 & \cdots & n-1 \end{bmatrix}.$$

The $(1, 1)$ cofactor of M is the determinant of an $(n-1) \times (n-1)$ matrix which looks exactly like M. Recall that the determinant of a matrix is not changed if a row is added to another row. Adding to the first row in turn each of the rows

---

[3]Here Fig 12.15 may help. The tree on the left is a spanning tree of K_4 but not of $\mathcal{G}$ since BD is not an edge in $\mathcal{G}$.

below it gives the matrix

$$\begin{bmatrix} 1 & 1 & 1 & \cdots & 1 \\ -1 & n-1 & -1 & \cdots & -1 \\ -1 & -1 & n-1 & & -1 \\ \vdots & \vdots & & \ddots & \\ -1 & -1 & -1 & \cdots & n-1 \end{bmatrix}$$

and then adding the first row of this matrix to each of the rows below it gives the matrix

$$\begin{bmatrix} 1 & 1 & 1 & \cdots & 1 \\ 0 & n & 0 & \cdots & 0 \\ 0 & 0 & n & & 0 \\ \vdots & \vdots & & \ddots & \vdots \\ 0 & 0 & 0 & \cdots & n \end{bmatrix},$$

which is an upper triangular $(n-1) \times (n-1)$ matrix with $n-2$ n's on the diagonal. Its determinant is n^{n-2}. ∎

K_4 $\mathcal{G}$

Figure 12.15

Answers to Pauses

9. The graph and its eight spanning trees are shown.

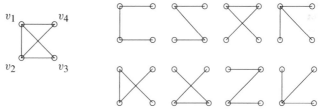

As spanning trees, the eight are all different, although they are partitioned into just two isomorphism classes, the left six in one class, the rightmost two in the other.

10. Labeling the graph as shown, the matrix specified in Kirchoff's Theorem is

$$M = \begin{bmatrix} 3 & -1 & 0 & -1 & -1 \\ -1 & 3 & -1 & 0 & -1 \\ 0 & -1 & 3 & -1 & -1 \\ -1 & 0 & -1 & 2 & 0 \\ -1 & -1 & -1 & 0 & 3 \end{bmatrix}.$$

The value of any cofactor is 24, so the graph in Fig 12.14 has 24 spanning trees.

EXERCISES

The symbol [BB] means that an answer can be found in the Back of the Book.

1. [BB] This exercise concerns the graph G shown in Fig 12.14. Delete edges a, c, and e in order to obtain a new spanning tree T for G. Reinsert edge a to obtain the circuit afg. Draw two other spanning trees for G by removing first f and then g from this circuit.

2. For each of the graphs shown,
   - find three spanning trees representing two isomorphism classes of graphs;
   - find the total number of spanning trees.

   (a)   (b)

   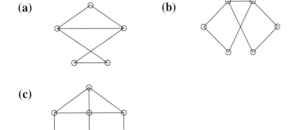

   (c)

3. Let e be an edge of the complete graph K_n. Prove that the number of spanning trees of K_n which contain e is $2n^{n-3}$.

4. [BB] How many labeled trees are there on n vertices, for $1 \leq n \leq 6$?

5. [BB] Draw all the labeled trees on four vertices.

6. How many spanning trees does K_7 have? Why?

7. (a) [BB] Draw all the spanning trees of $K_{2,2}$ and indicate the isomorphism classes of these. How many isomorphism classes are there?
   (b) Repeat (a) for $K_{2,3}$.

8. Determine the number of spanning trees of the complete bipartite graph $K_{2,n}$.

9. [BB] Suppose some edge of a connected graph G belongs to every spanning tree of G. What can you conclude and why?

10. Let G be a tree. Describe all the spanning trees of G and explain.

11. If G is a graph and e is an edge which is not part of a circuit, then e must belong to every spanning tree of G. Why?

12. (a) [BB] Prove that every edge in a connected graph is part of some spanning tree.
    (b) Prove that any two edges of a connected graph are part of some spanning tree.
    (c) [BB] Given three edges in a connected graph, is there always a spanning tree containing these edges? Explain your answer.

13. Let G be a connected graph with at least two vertices. Show that G has a vertex v such that $G \setminus \{v\}$ is still connected. Show that G has at least two vertices with this property.

14. Suppose that T_0 and T are two different spanning trees for a graph G.
    (a) [BB] Let e be an edge in T which is not in T_0. Prove that there is an edge f in T_0 which is not in T such that $(T_0 \cup \{e\}) \setminus \{f\}$ is a spanning tree.
    (b) Describe a procedure for obtaining T from T_0.

15. Let C_n be the cycle with n vertices labeled $1, 2, \ldots, n$ in the order encountered on the cycle.
    (a) Find the number of spanning trees for C_n (in two ways).
    (b) Find a general formula for the number of spanning trees for $C_n \cup \{e\}$, where e joins 1 to a ($3 \leq a \leq n - 1$).

16. [BB] How many graphs have n vertices labeled $v_1, v_2, \ldots, v_n$ and $n - 1$ edges? Compare this number with the number of trees with vertices $v_1, \ldots, v_n$, for $2 \leq n \leq 6$.

17. Show that any shortest path algorithm can be used to construct a spanning tree for an unweighted connected graph. (Caution! A shortest path algorithm requires a weighted graph. How are you going to get started?)

12.4 Minimum Spanning Tree Algorithms

In this section, we answer the Minimum Connector Problem suggested by the town and road map in Fig 12.13. The problem requires us to find a minimum spanning tree, and we'd like to find one efficiently. In theory, we could enumerate all the spanning trees of a weighted graph and simply choose the tree of least weight, but if the graph is at all complicated, this is not an easy chore. In this section, we discuss two better ways discovered in the 1950s by J. B. Kruskal[4] and R. C. Prim.[5] Both algorithms are in some sense "obvious." They produce a minimum weight spanning tree an edge at a time, at each stage making the best choice of next edge without thought to what has happened already or what will ensue. For this reason, they are called *greedy* algorithms.

12.4.1 KRUSKAL'S ALGORITHM ▶

To find a minimum spanning tree in a connected weighted graph with $n > 1$ vertices, carry out the following procedure.

Step 1. Find an edge of least weight and call this e_1. Set $k = 1$.

Step 2. While $k < n$
    if there exists an edge e such that $\{e\} \cup \{e_1, e_2, \dots, e_k\}$ does not contain a circuit, let e_{k+1} be such an edge of least weight and replace k by $k + 1$;
    else output $e_1, e_2, \dots, e_k$ and stop.
end while

EXAMPLE 2   We illustrate with reference to the graph of Fig 12.16. Kruskal's algorithm initially picks an edge of lowest weight. There are several edges of weight 2 and none with less weight. Any of these may be selected. Suppose we begin with edge $e_1 = AJ$. There still remain edges of weight 2, so we choose any other, say $e_2 = GM$. Among those edges which do not form a circuit with e_1 and e_2 (at this point there are no edges which do), we choose one of lowest weight, say DE, and call this e_3. Continuing, we might select edges $AJ, GM, DE, MF, FE, LK, HJ, KM, CK, BC$. At this stage, there remains an edge (GL) of weight 4, but this edge would complete a circuit with LK, KM,

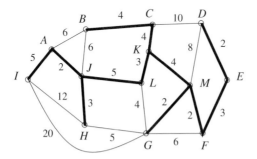

Figure 12.16   An application of Kruskal's algorithm might select, in this order, edges $AJ, GM, DE, MF,$ $FE, LK, HJ, KM, CK, BC, JL, IA.$

[4]J. B. Kruskal, Jr., "On the Shortest Spanning Subtree of a Graph and the Traveling Salesman Problem," *Proceedings of American Mathematical Society* **7** (1956), 48–50.

[5]R. C. Prim, "Shortest Connection Networks and Some Generalizations," *Bell System Technical Journal* **36** (1957), 1389–1401.

GM already selected, and so we choose an edge of weight 5, say JL, and another, IA. Now, any of the remaining edges will complete a circuit with those selected and so we must stop; the algorithm assures us we have the desired minimum spanning tree. It has weight 39. ▲

Why does the algorithm work? Suppose G is the given connected graph and $\mathcal{T}_K$ is the subgraph produced by Kruskal's algorithm. The algorithm terminates only when the inclusion of any edge of G not in $\mathcal{T}_K$ produces a circuit in $\mathcal{T}_K$. We must prove that $\mathcal{T}_K$ is a spanning tree and that it is minimum.

First, note that $\mathcal{T}_K$ contains every vertex of G. Why? Suppose to the contrary that some vertex v is not in $\mathcal{T}_K$. Since G is connected, there exists some edge incident with v. Adding this edge to $\mathcal{T}_K$ cannot produce a circuit in $\mathcal{T}_K$ since none of the edges incident with v is in $\mathcal{T}_K$. Thus, the algorithm could continue, a contradiction. By its construction $\mathcal{T}_K$ contains no circuits. To see that it is connected, let v and w be any two vertices of $\mathcal{T}_K$. There is a path in G from v to w and any edge of this path not in $\mathcal{T}_K$ can be replaced by a path in $\mathcal{T}_K$ since the inclusion of any edge of G not in $\mathcal{T}_K$ produces a circuit. Thus, $\mathcal{T}_K$ is a spanning tree. It is more difficult to establish that $\mathcal{T}_K$ is minimum.

Let $\mathcal{T}_m$ be a minimum spanning tree in G. We wish to show that the weight of $\mathcal{T}_K$ equals the weight of $\mathcal{T}_m$. Suppose that G (hence, also $\mathcal{T}_m$ and $\mathcal{T}_K$) has n vertices. Then $\mathcal{T}_K$ contains $n - 1$ edges. Suppose these were determined by the algorithm in the order $e_1, e_2, \ldots, e_{n-1}$. If $\mathcal{T}_K = \mathcal{T}_m$, there is nothing to prove; otherwise, we may suppose that there are edges of $\mathcal{T}_K$ not in $\mathcal{T}_m$. Let e_k be the first such edge. Since $\mathcal{T}_m$ is also a spanning tree, with an argument similar to that used to solve Exercise 14(a) of Section 12.3, there is an edge e in $\mathcal{T}_m$, different from e_1 and different from any of $e_1, e_2, \ldots, e_{k-1}$ in the case $k > 1$, such that $\mathcal{T} = (\mathcal{T}_m \cup \{e_k\}) \setminus \{e\}$ is a spanning tree. The weight of $\mathcal{T}$ is

(1) $$w(\mathcal{T}) = w(\mathcal{T}_m) + w(e_k) - w(e) \geq w(\mathcal{T}_m)$$

since $\mathcal{T}_m$ has minimum weight. Therefore, $w(e_k) \geq w(e)$.

If $k = 1$, $w(e_k) \leq w(e)$ since e_1 was the first edge chosen and so has minimum weight. So $w(e) = w(e_k)$. If $k > 1$, the edges $e, e_1, e_2, \ldots, e_{k-1}$ do not contain a circuit because they are all part of the tree $\mathcal{T}_m$. Again, $w(e) = w(e_k)$; otherwise, $w(e) < w(e_k)$ and the algorithm would have chosen e instead of e_k. In all cases, this makes $\mathcal{T}$ a minimum spanning tree since, by (1), its weight is the same as the weight of the minimum spanning tree $\mathcal{T}_m$. Also, $\mathcal{T}$ is "closer" to $\mathcal{T}_K$ than was $\mathcal{T}_m$ in the sense that $\mathcal{T}$ has edges $e_1, \ldots, e_k$ in common with $\mathcal{T}_K$. Now if $\mathcal{T} = \mathcal{T}_K$, then $\mathcal{T}_K$ is minimum (which is what we want to show); otherwise, we apply to $\mathcal{T}$ and $\mathcal{T}_K$ the procedure just applied to $\mathcal{T}_m$ and $\mathcal{T}_K$ and obtain another minimum spanning tree with yet another edge in common with $\mathcal{T}_K$. Continuing in this way, we eventually arrive at a minimum spanning tree with edges $e_1, \ldots, e_{n-1}$ in common with $\mathcal{T}_K$ (and so equal to $\mathcal{T}_K$), proving that $\mathcal{T}_K$ is indeed a minimum spanning tree.

What is the complexity of Kruskal's algorithm, in terms of comparisons? To estimate the number of comparisons, we must first decide how we are going to determine, at a particular stage, whether or not a given edge forms a circuit with some of the edges already selected.

At each stage, Kruskal's algorithm has selected certain edges. Think of the subgraph determined by these edges and all n vertices of $\mathcal{G}$. This subgraph has a certain number of connected components, each of which is a tree. The first such subgraph has n components; the final minimum spanning tree has just one component. To begin, assign the labels $1, 2, \ldots, n$ to the vertices of $\mathcal{G}$. If the first edge selected is uv, with u labeled i, v labeled j, and $i < j$, change the label of v to i. The subgraph consisting of this edge and all n vertices has $n - 1$ components and, as at the outset, all the vertices in each component have the same label. If we could preserve this feature of our labeling, then the edges eligible for selection at a certain stage (that is, those edges not forming a circuit with edges already chosen) would be simply those whose end vertices have different labels.

Suppose, at a certain stage, vertices have the same label if and only if they lie in the same component of the subgraph whose vertices are all those of $\mathcal{G}$ and whose edges are those the algorithm has chosen so far, and that each such component is a tree. We seek an edge which does not complete a circuit with previously chosen edges, hence an edge with end vertices lying in different components. Select an edge $e = xy$ whose end vertices x and y have different labels and which has least weight among such edges. If x is labeled k and y is labeled ℓ and $k < \ell$, then x belongs to a component all of whose vertices are labeled k and y belongs to a component all of whose vertices are labeled ℓ. By changing to k the labels on all the vertices labeled ℓ, the next subgraph once again has the property that the vertices have the same label if and only if they lie in the same connected component. Moreover, each new component is again a tree since a circuit cannot be produced by joining two vertices in different components.

Pause 11   Describe the labels on the final minimum spanning tree.

This is our suggested procedure for ensuring no circuits. How many comparisons does it require? First sort the edges $f_1, f_2, \ldots, f_N$ of $\mathcal{G}$ so that $w(f_1) \leq w(f_2) \leq \cdots \leq w(f_N)$. Using an efficient algorithm such as the merge sort discussed in Section 8.3, this can be accomplished with $\mathcal{O}(N \log N)$ comparisons. The algorithm selects $e_1 = f_1$, compares the labels of the ends of e_1, and changes one of these labels. Then $e_2 = f_2$ and, after one more comparison, changes some labels. The algorithm then looks at f_3 and compares the labels of its end vertices to see if it is eligible for selection. It continues successively to examine $f_4, f_5, \ldots$ until $n - 1$ edges have been selected. Each time an edge is examined, one comparison of end labels is required. In all, this part of the algorithm could require as many as N comparisons. Also, for each of the $n - 1$ edges which are eventually selected, it is necessary to check the labels on $n - 2$ vertices to see which must be changed. This contributes a further $(n - 1)(n - 2)$ comparisons. By the approach outlined here, Kruskal's algorithm is $\mathcal{O}(N \log N + N + n^2) = \mathcal{O}(N \log N + n^2)$. In fact, the rule by which the labels of vertices are changed can be modified so that Kruskal's algorithm can be completed with just $\mathcal{O}(N \log N)$ comparisons. (See Exercise 12.)

We turn now our attention to another algorithm, a popular one, for finding a minimum spanning tree. Prim's algorithm is like Kruskal's insofar as it selects an edge of least weight at each stage. This time, however, instead of merely guarding against circuits, the algorithm requires that the subgraph determined by selected edges is always a tree.

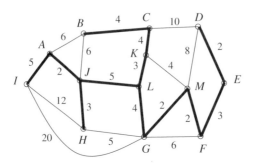

Figure 12.17   An application of Prim's algorithm might begin with vertex K and then select, in this order, edges $KL, LG, GM, MF, FE, DE, KC, BC, JL, AJ, JH, AI$.

EXAMPLE 3   We illustrate Prim's approach as it would apply to the graph of Fig 12.16, which we have shown again in Fig 12.17. Start at any vertex, say K, and select the edge incident with K of least weight. Thus, we select edge KL of weight 3. The next edge chosen is the edge of least weight adjacent to KL. Any of LG, KM, or KC is eligible. Suppose we select LG. Then we select the edge of least weight among those edges forming a tree with the two edges already selected (GM, for instance). We continue to select, in succession, the edge of least weight among those edges, as yet unselected, which form a tree with those already chosen and continue until we have a spanning tree. You should verify that the spanning tree shown in Fig 12.17 is one which Prim's algorithm might identify.   ▲

12.4.2 PRIM'S ALGORITHM ▶

To find a minimum spanning tree in a connected weighted graph with $n > 1$ vertices, proceed as follows.

Step 1. Choose any vertex v and let e_1 be an edge of least weight incident with v. Set $k = 1$.

Step 2. While $k < n$
    if there exists a vertex which is not in the subgraph $\mathcal{T}$ whose edges are
        $e_1, e_2, \ldots, e_k$,
            • let e_{k+1} be an edge of least weight among all edges of the form ux, where u is a vertex of $\mathcal{T}$ and x is a vertex not in $\mathcal{T}$;
            • replace k by $k + 1$;
        else output $e_1, e_2, \ldots, e_k$ and stop.
    end while

Perhaps the best way to see that Prim's algorithm succeeds in finding a minimum spanning tree in a connected graph $\mathcal{G}$ is to prove that at each stage the algorithm has found a tree which is a subgraph of some minimum spanning tree. This is obvious at the beginning since any vertex is a tree and a subgraph of any (minimum) spanning tree of $\mathcal{G}$. Suppose that after k steps, the algorithm has produced a tree T that is a subgraph of a minimum spanning tree T_m for the graph. At the next stage, the algorithm produces another subgraph T_1 by adding a vertex x and an edge ux to T, where the weight of ux is least among all edges joining vertices $x \notin T$ to vertices $u \in T$. Since T contains no circuits and the additional vertex x is incident with only one edge in T_1, the subgraph T_1 also contains no circuits. It is also connected and hence a tree. Why is it contained in a minimum spanning tree for G? This is surely the case if it is contained in T_m. If it is not contained in T_m, then the subgraph T_m with edge ux adjoined contains a circuit, $uxx_1x_2 \cdots x_ru$. Such a circuit must contain an edge of the form yw, with $y \notin T$ and $w \in T$.[6] It follows that $w(ux) \leq w(yw)$, since yw was also eligible for selection by the algorithm when ux was chosen. Thus, the spanning tree $(T_m \cup \{ux\}) \setminus \{yw\}$, having weight not exceeding that of T_m, is also a minimum spanning tree, and it contains T_1. By the Principle of Mathematical Induction, it follows that each of the subgraphs produced by Prim's algorithm is a tree contained in a minimum spanning tree. In particular, the last tree selected is itself a minimum spanning tree.

We now show how to implement Prim's algorithm with complexity $\mathcal{O}(n^3)$ and leave to the exercises an implementation which is $\mathcal{O}(n^2)$. (See Exercise 13.) Recalling that Kruskal's algorithm is $\mathcal{O}(N \log N + n^2)$, where N is the number of edges, we see that the choice between Kruskal's and Prim's algorithms depends primarily on the relative sizes of N and n.

Suppose G is a connected graph with n vertices. In Prim's algorithm, we select a vertex v and find the minimum weight of the potentially $n - 1$ edges incident with v. For this, at most $n - 2$ comparisons are required. Suppose vu is the edge selected. Next, we find the edge of least weight among those which are incident with u or with v (excluding the edge vu). In the worst case, in which every vertex is adjacent to each of these vertices, we would have to find the minimum of $2(n - 2)$ numbers, a process requiring $2(n - 2) - 1$ comparisons. After k edges (and $k + 1$ vertices) have been selected, we seek the edge of least weight among those incident with one of the vertices already selected. In the worst case, the remaining $n - (k + 1)$ vertices are adjacent to each of the first $k + 1$ and we must find the minimum of $(k + 1)(n - k - 1)$ numbers. This would require $(k + 1)(n - k - 1) - 1$ comparisons. In all then, we require

$$\sum_{k=0}^{n-2}[(k + 1)(n - k - 1) - 1] = \sum_{k=0}^{n-2}(nk + n - k^2 - 2k - 2)$$

comparisons. Recalling that

$$\sum_{k=0}^{n}k = \frac{n(n + 1)}{2} \quad \text{and} \quad \sum_{k=0}^{n}k^2 = \frac{n(n + 1)(2n + 1)}{6},$$

---

[6]The edge yw is xx_1 if $x_1 \in T$; otherwise, x_1x_2 if $x_2 \in T$; and so on. This process eventually leads to edge yw, as described, because the last vertex u is in T.

we find that the number of comparisons is

$$n\frac{(n-2)(n-1)}{2} + n(n-1) - \frac{(n-2)(n-1)(2n-3)}{6}$$

$$-2\frac{(n-2)(n-1)}{2} - 2(n-1) = \frac{1}{6}n^3 - \frac{7}{6}n + 1.$$

The procedure is $\mathcal{O}(n^3)$.

A Bound for a Minimum Hamiltonian Cycle

There is an interesting connection between the Traveling Salesman's Problem (Section 10.4) and minimum spanning trees in a weighted graph. Recall that the Traveling Salesman's Problem is to find a Hamiltonian cycle of minimum weight in a weighted Hamiltonian graph. Our ability to find minimum spanning trees gives us a way to determine lower bounds for the weight of a minimum Hamiltonian cycle.

Suppose $\mathcal{G}$ is a weighted Hamiltonian graph with Hamiltonian cycle $\mathcal{H}$. Removing any vertex v of $\mathcal{G}$ (and the edges with which it is incident) gives us a subgraph $\mathcal{G}'$ of $\mathcal{G}$ and a subgraph $\mathcal{H}'$ of $\mathcal{H}$ which is a spanning tree for $\mathcal{G}'$; in fact, $\mathcal{H}'$ is a path through all vertices of $\mathcal{G}$ except v. It follows that the weight of $\mathcal{H}$ is the weight of $\mathcal{H}'$ plus the sum of the weights of the two edges in $\mathcal{H}$ incident with v. Hence,

$$w(\mathcal{H}) \geq w(\text{minimum spanning tree for } \mathcal{G}')$$
$$+ \text{ smallest sum of weights of two edges incident with } v.$$

By choosing several different vertices v, we can obtain various lower bounds for the weight of the Hamiltonian cycle $\mathcal{H}$ and so, in particular, for the weight of a minimum Hamiltonian cycle.

Considering again the graph in Fig 12.16 (which is Hamiltonian), if we remove vertex H (and the three edges with which it is incident), it is easily checked that all the edges of the spanning tree indicated in Fig 12.16, except for HJ, comprise a minimum spanning tree for the subgraph, of weight $39 - 3 = 36$. The smallest two weights of edges incident with H are 3 and 5. Therefore, we obtain $36 + 3 + 5 = 44$ as a lower bound for the weight of any Hamiltonian cycle.

Suppose we remove K instead of H. This time, the heavy edges in Fig 12.18 form a minimum spanning tree in the subgraph, of weight 38. The two smallest weights of edges incident with K are 3 and 4, so we obtain $38 + 3 + 4 = 45$ as a lower bound for the weight of any Hamiltonian cycle, a better estimate than before.

Pause 12 Find a Hamiltonian cycle in the graph of Fig 12.16. What is its weight?

Answers to Pauses 11. Each time an edge is added, we changed the labels on its end vertices to the lower of the two previous labels (and all the vertices in a certain component to the lower as well). All vertices are eventually labeled 1.

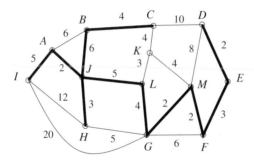

Figure 12.18   A graph $\mathcal{G}$ and a spanning tree for $\mathcal{G} \setminus \{K\}$ with weight 38.

12. $IHJABCDEFMKLGI$ is a Hamiltonian cycle of weight 75. Note how the path $JA \cdots GI$ which results when H is removed is a spanning tree for the subgraph without H, and how the path $LG \cdots FM$ which results when K is removed is a spanning tree for the subgraph without K.

EXERCISES

The symbol [BB] means that an answer can be found in the Back of the Book.

1. Use Kruskal's algorithm to find a spanning tree of minimum total weight in each of the graphs in Fig 12.19. Give the weight of your minimum tree and show your steps.

2. [BB; (a)] For each of the graphs of Exercise 1, apply Prim's algorithm to find a minimum spanning tree. Try to find a tree different from the one found before. Start at vertex E in each graph, explain your reasoning, and draw the tree.

3. [BB] Find a minimum spanning tree for the graph in Fig 12.13. What is the smallest length of pavement required to connect the towns in this graph?

4. (a) Explain how Kruskal's algorithm could be modified so that it finds a spanning tree of **maximum** weight in a connected graph.

   (b) Apply the modified algorithm described in (a) to find a maximum spanning tree for each of the following graphs and give the maximum weight of each.

   i. [BB] the graph in Exercise 1(a)
   ii. the graph in Exercise 1(b)
   iii. the graph in Exercise 1(c)
   iv. the graph in Exercise 1(d)
   v. the graph in Fig 12.16

5. Repeat Exercise 4 for Prim's algorithm. For each part of (b), start at vertex A.

6. [BB] Explain how a minimum spanning tree algorithm can be used to find a spanning tree in a connected graph where the edges are **not** weighted.

7. How could a minimum spanning tree algorithm be used to find a spanning tree in an unweighted graph which **excludes** a given edge, assuming such spanning trees exist?

8. (a) [BB] Suppose we have a connected graph $\mathcal{G}$ and we want to find a spanning tree for $\mathcal{G}$ which contains a given edge e. How could Kruskal's algorithm be used to do this? Discuss both the weighted and unweighted cases.

   (b) Use Kruskal's algorithm to show that if $\mathcal{G}$ is a connected graph, then any (not necessarily connected) subgraph which contains no circuits is part of some spanning tree for $\mathcal{G}$. Consider both the weighted and unweighted cases.

9. Answer Exercise 8 using Prim's algorithm instead of Kruskal's.

10. [BB] Prove that at each vertex v of a weighted connected graph, Kruskal's algorithm always includes an edge of lowest weight incident with v.

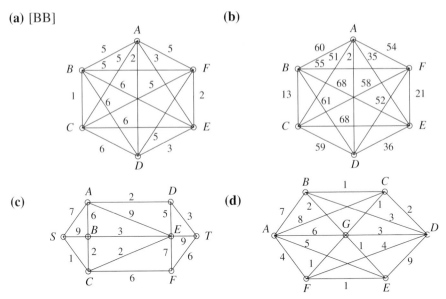

Figure 12.19   Graphs for Exercise 1.

11. (a) Prove that any minimum weight spanning tree in a connected weighted graph can be obtained by Kruskal's algorithm. [*Hint*: Look at the argument on p. 384 showing that Kruskal's algorithm selects a minimum spanning tree.]

(b) [BB] If all the weights in a connected weighted graph are distinct, show that the graph has a unique spanning tree of minimum weight.

12. In our discussion of the complexity of Kruskal's algorithm, we suggested a certain system of labeling vertices (p. 385). Consider the following modification.

Suppose that vertex x is labeled k, vertex y is labeled ℓ, and that we have just selected edge xy. Compare the relative sizes of the components to which x and y belong and change the labels of all vertices in the **smaller** of these components to the label of the larger. (If the components have equal size, label with the smaller number.)

With reference to Example 2, p. 383, we begin with 13 components each containing a single vertex. After the selection of AJ, component 1 contains two vertices— A and J—and component 10 is empty; the remaining components are unchanged. Figure 12.20 shows how the components change as the edges GM, DE, MF, and FE are successively selected.

(a) Continue the scheme shown in this figure to show the relabeling of vertices until the spanning tree of Fig 12.16 is obtained.

(b) [BB] Show that in the general case the number of times the label on any vertex is changed is at most $\log_2 n$, n the number of vertices of $\mathcal{G}$.

(c) [BB] Show that Kruskal's algorithm can be implemented with a number of comparisons which is $\mathcal{O}(N \log N)$, N the number of edges of $\mathcal{G}$.

13. In the implementation of Prim's algorithm discussed on p. 387 it is not really necessary at each stage to find the minimum of so many numbers. Consider the following alternative. Select any vertex v_1 and label every other vertex v with the weight of edge vv_1 if vv_1 is an edge and ∞ otherwise. Then select edge v_1v_2 such that the label on v_2 is a minimum. For each vertex v adjacent to v_2, change the label of v to the minimum of its old label and the weight of edge vv_2. Select v_3 with minimum label and add the corresponding edge (v_3v_1 or v_3v_2). The first graph shows the labels on the vertices of the graph of Fig 12.16 just after $v_1 = K$ has been selected; the second graph depicts the situation just after v_2 has been selected.

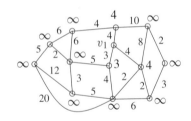

Component	Size	Vertices
1	1	A
2	1	B
3	1	C
4	1	D
5	1	E
6	1	F
7	1	G
8	1	H
9	1	I
10	1	J
11	1	K
12	1	L
13	1	M

add AJ →

Component	Size	Vertices
1	2	A, J
2	1	B
3	1	C
4	1	D
5	1	E
6	1	F
7	1	G
8	1	H
9	1	I
11	1	K
12	1	L
13	1	M

add GM →

Component	Size	Vertices
1	2	A, J
2	1	B
3	1	C
4	1	D
5	1	E
6	1	F
7	2	G, M
8	1	H
9	1	I
11	1	K
12	1	L

add DE →

Component	Size	Vertices
1	2	A, J
2	1	B
3	1	C
4	2	D, E
6	1	F
7	2	G, M
8	1	H
9	1	I
11	1	K
12	1	L

add MF →

Component	Size	Vertices
1	2	A, J
2	1	B
3	1	C
4	2	D, E
7	3	G, M, F
8	1	H
9	1	I
11	1	K
12	1	L

add FE →

Component	Size	Vertices
1	2	A, J
2	1	B
3	1	C
7	5	G, M, F, D, E
8	1	H
9	1	I
11	1	K
12	1	L

Figure 12.20   Components not listed are empty.

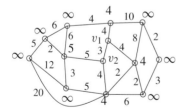

(a) Show the labels just after $v_3 = G$ has been selected.

(b) Explain carefully how the process started here should be continued to produce the next edge. Illustrate by showing the labels on the graph just after v_4 has been chosen.

(c) Show that the implementation of Prim's algorithm described here is $\mathcal{O}(n^2)$ (in terms of comparisons).

14. By considering the subgraphs determined by deleting each of the vertices, find the best (greatest) lower bound for the weight of a minimum Hamiltonian cycle. Find a minimum Hamiltonian cycle.

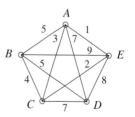

15. This exercise concerns the graph in Exercise 1(a).

   (a) [BB] Try to find a Hamiltonian cycle of lowest weight. What is the weight?

(b) [BB] By removing vertex A, use the method discussed in this section to find a lower bound for the weight of any Hamiltonian cycle.

(c) [BB] Repeat part (b), removing vertex B instead of A.

(d) Try to improve (increase) the lower bounds found in parts (b) and (c) by removing some other vertex.

16. Repeat Exercise 15 for the graph in Exercise 1(b). Use vertices D and F in parts (b) and (c).

17. Repeat Exercise 15 for the graph in Exercise 1(c). Use vertices B and T in parts (b) and (c).

18. Repeat Exercise 15 for the graph in Exercise 1(d). Use vertices E and G in parts (b) and (c).

12.5   ACYCLIC DIGRAPHS AND BELLMAN'S ALGORITHM

A (directed) graph is *acyclic* if it contains no (directed) cycles. For example, the digraph on the left of Fig 12.21 is acyclic, but the one on the right is not—there is a cycle around the outer vertices. Is there any easy way to see that there are no cycles in $\mathcal{G}$, on the left? This is where the vertex labels help. The digraph has been labeled so that $v_i v_j$ is an arc only when $i < j$. For example, there are no arcs of the form $x v_0$, the only arc $x v_1$ is $v_0 v_1$, the only arcs $x v_2$ are with $x = v_0$ and $x = v_1$, and so on. So it is clear why there is no cycle: In any path, the subscripts on the vertices v_i must increase, so there is never an arc on which to return to the first vertex. A labeling of the vertices such as that shown for $\mathcal{G}$ has a special name.

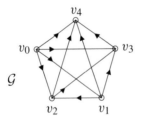

Figure 12.21   The digraph on the left is acyclic; the digraph on the right is not.

12.5.1   DEFINITION ▶   A labeling $v_0, v_1, \ldots, v_{n-1}$ of the vertices of a digraph is called *canonical* if every arc has the form $v_i v_j$ with $i < j$. We also refer to the list $v_0, v_1, \ldots, v_{n-1}$ as a *canonical ordering* of the vertices.

Here's why a digraph with a canonical labeling of vertices must be acyclic: If $v_{i_1} v_{i_2} \cdots v_{i_k}$ is a walk, then $i_1 < i_2 < \cdots < i_k$, so such a walk cannot be closed. It is interesting that the converse of this statement is also true. Any acyclic digraph has a canonical labeling of vertices. To prove this, we require a preliminary result.

12.5.2   LEMMA ▶   Let S be a nonempty set of vertices of an acyclic digraph $\mathcal{G}$. Then there exists $v \in S$ such that $S_v = \{x \in S \mid xv \text{ is an arc}\} = \emptyset$.

Proof   We establish this result by contradiction. Assume the result is false. Thus, for each $v \in S$, there is an arc of the form xv with $x \in S$. Let v_0 be any vertex in S and let $v_1 v_0$ be an arc, $v_1 \in S$. There exists an arc $v_2 v_1$, $v_2 \in S$, and, in general, having found vertices $v_1, v_2, \ldots, v_k$ in $\mathcal{S}$ with $v_{i+1}v_i$ an arc, $i = 1, 2, \ldots, k-1$, there exists an arc $v_{k+1}v_k$ with $v_{k+1} \in \mathcal{S}$. Since $\mathcal{G}$ has only a finite number of vertices, eventually, $v_{k+1} \in \{v_1, v_2, \ldots, v_k\}$, say $v_{k+1} = v_i$, $1 \le i \le k$. If k is minimal with this occurring, $v_{k+1}v_kv_{k-1} \cdots v_i$ is a cycle, contradicting the fact that $\mathcal{G}$ is acyclic.   ∎

This is key to the proof of a theorem which characterizes acyclic digraphs.

12.5.3 THEOREM ▶   A digraph is acyclic if and only if it has a canonical labeling of vertices.

Proof   We gave the proof of sufficiency prior to the statement of Lemma 12.5.2. For necessity, suppose $\mathcal{G}$ is an acyclic digraph with $n > 1$ vertices and vertex set $\mathcal{V}$. Applying the lemma to $\mathcal{S} = \mathcal{V}$, we obtain a vertex v_0 with the property that there are no arcs of the form xv_0. Applying Lemma 12.5.2 to $\mathcal{S} = \mathcal{V} \setminus \{v_0\}$, we obtain a vertex v_1 with the property that there are no arcs of the form xv_1 with $x \in \mathcal{S}$; that is, the only possible arc xv_1 is with $x = v_0$. Now apply the lemma to $\mathcal{S} = \mathcal{V} \setminus \{v_0, v_1\}$. We obtain a vertex v_2 with the property that the only possible arcs of the form xv_2 are v_0v_2 and v_1v_2. Continuing in this manner, we obtain vertices $v_0, v_1, \ldots, v_{n-1}$ with the property that for any j, the only arcs xv_j are with $x = v_i$ and $i < j$. Thus $v_0, v_1, \ldots, v_{n-1}$ is a canonical labeling.   ∎

We want to discuss a second algorithm to which the name of Richard Bellman is attached.

Pause 13   Name the first algorithm described in this book to which Bellman's name is attached.

The algorithm we have in mind takes as input an acyclic digraph whose vertices have been labeled canonically, together with a specified vertex v, and outputs the lengths of shortest paths from v to every other vertex. The algorithm also describes the shortest path from v to any other vertex by means of a *rooted* spanning tree. Remember that a root in a tree is simply a designated vertex. In a digraph, the concept is stronger. There is only one possible root.

12.5.4 DEFINITION ▶   A digraph $\mathcal{T}$ is a *rooted tree* with vertex v as the *root* if the unoriented graph (ignoring orientations of arcs) is a tree, if v has indegree 0, and v is the only such vertex of $\mathcal{T}$.

For example, if the edges of the tree shown in Fig 12.2 are directed down, then we have a rooted tree with "start" as root. On the other hand, the digraph to the right is not a rooted tree because there is more than one vertex of indegree 0.

In Proposition 12.1.2, we showed that any two vertices of a tree are joined by a unique path. This is clearly not the case if the edges of the tree are oriented (and we require paths to respect orientation). There is, however, a unique path from the root to every other vertex.

12.5.5 PROPOSITION ▶ In a rooted (directed) tree with root v, there is a unique path from v to every other vertex.

Proof Uniqueness is straightforward: If there were two different directed paths $\mathcal{P}_1$, $\mathcal{P}_2$ from v to some vertex u, then, ignoring orientations, we would have two different paths from v to u, contradicting the fact that the undirected graph $\mathcal{T}$ is a tree. We now show existence of a directed path from v to any other vertex. So, let u be a vertex of $\mathcal{T}$ different from v. Since $\operatorname{indeg} u \neq 0$, there is a directed edge $u_1 u$. If $u_1 = v$, we have a path from v to u; otherwise, since $\operatorname{indeg} u_1 \neq 0$, we have a directed edge $u_2 u_1$. Note that $u_2 \neq u$ since the arc between u and u_1 is $u_1 u$. Thus $u_2 u_1 u$ is a path and if $u_2 = v$ we are done. In general, suppose $k \geq 2$ and we have found a path $u_k u_{k-1} \cdots u_1 u$ with no $u_i = v$. Since $\operatorname{indeg} u_k \neq 0$, there is a directed arc $u_{k+1} u_k$. Also, $u_{k+1} \notin \{u, u_1, u_2, \ldots, u_k\}$ because the undirected graph $\mathcal{T}$ has no cycles. Thus $u_{k+1} u_k \cdots u_1 u$ is a path and, if $u_{k+1} = v$, we have a directed path from v to u. Since our tree is finite, this process cannot continue indefinitely, so eventually some $u_k = v$. ∎

12.5.6 COROLLARY ▶ In a rooted (directed) tree, every vertex other than the root has indegree 1.

Proof Let $\mathcal{T}$ be a rooted tree with root v and suppose $u \neq v$ is a vertex with $\operatorname{indeg} u \neq 1$. Since $\operatorname{indeg} u \neq 0$, we have $\operatorname{indeg} u \geq 2$, so there are distinct arcs $u_1 u$ and $u_1' u$. The proposition says there is a path $\mathcal{P}$ from v to u_1 and also a path $\mathcal{P}'$ from v to u_1'. Thus we obtain two different paths from v to u (with final arcs $u_1 u$ and $u_1' u$). This contradicts uniqueness of a directed path from v to u. ∎

12.5.7 BELLMAN'S ALGORITHM ▶ Given a weighted acyclic digraph $\mathcal{G}$ with nonnegative weights and a canonical labeling $v_0, v_1, \ldots, v_{n-1}$, to find the shortest paths from v_0 to every other vertex,

Step 1. Set $d_0 = 0$ and $d_i = \infty$ for $i = 1, 2, \ldots, n-1$.
Set $p_i = -1$ for $i = 0, 1, \ldots, n-1$.

Step 2. For t from 1 to $n - 1$,
let $d_t = \min\{d_j + w(v_j, v_t) \mid j = 0, \ldots, t-1\}$ and let p_t be a j which gives this minimum.

As with previous shortest path algorithms, a final value $d_t = \infty$ simply indicates that there is no directed path from v_0 to v_t.

Consider the acyclic canonically labeled graph $\mathcal{G}$ with arcs weighted as shown on the left in Fig 12.22. We show the values of the d_t and p_t (p for predecessor) at the beginning, and after each iteration of the loop in Step 2. The final values of d_1 and p_1 are 2 and 0, respectively. Thus the shortest distance to v_1 has length 2, and the last arc on a shortest path is $v_0 v_1$. The final values of d_2 and p_2 are 4 and 1, respectively, showing that the shortest path to v_2 has length 4, and the last arc on the corresponding path is $v_1 v_2$. The final values of d_3 and p_3 are 2 and 0, respectively, asserting the facts that a shortest path to v_3 has length 2 and $v_0 v_3$ is the last arc on such a path. On the right in Fig 12.22, we show just those arcs which are used last on shortest paths from v_0 to each vertex. It is not a

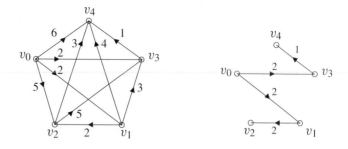

	d_0, p_0	d_1, p_1	d_2, p_2	d_3, p_3	d_4, p_4
Initially	$0, -1$	$\infty, -1$	$\infty, -1$	$\infty, -1$	$\infty, -1$
$t = 1$	$0, -1$	$2, 0$	$\infty, -1$	$\infty, -1$	$\infty, -1$
$t = 2$	$0, -1$	$2, 0$	$4, 1$	$\infty, -1$	$\infty, -1$
$t = 3$	$0, -1$	$2, 0$	$4, 1$	$2, 0$	$\infty, -1$
$t = 4$	$0, -1$	$2, 0$	$4, 1$	$2, 0$	$3, 3$

Figure 12.22

coincidence that these form a spanning tree rooted at v_0. In the Exercises, we ask you to explain why.

Answers to Pauses

13. In Section 11.2, we met the Bellman-Ford shortest path algorithm.

EXERCISES

The symbol [BB] means that an answer can be found in the Back of the Book.

1. For each of the digraphs in Fig 12.23, either show that the digraph is acyclic by finding a canonical labeling of vertices or exhibit a cycle.

2. Suppose $v_0, v_1, \ldots, v_{n-1}$ is a canonical labeling of a digraph $\mathcal{G}$. What can be said about the indegree and outdegree of v_0, if anything? Explain.

3. [BB] Find an $\mathcal{O}(n^2)$ algorithm which computes the indegrees of a digraph with n vertices. (The addition of two numbers is the basic operation.)

4. (a) Describe an algorithm which finds a canonical labeling for the vertices of an acyclic digraph.
   (b) Find a Big Oh estimate for your algorithm in terms of some reasonable basic operation (or operations).
   (c) Write a computer program which implements your algorithm.

5. The algorithm described in the proof of Theorem 12.5.3 can in fact be adapted to find a cycle in a digraph which is not acyclic.

   (a) [BB] Explain how this can be accomplished.
   (b) Describe an algorithm which implements your idea.
   (c) Write computer code which implements your algorithm.

6. How many shortest path algorithms can you name? How many of these can you describe?

7. [BB] Suppose $\mathcal{T}$ is a rooted (directed) tree with n vertices. How many arcs does $\mathcal{T}$ have? Why?

8. Let $\mathcal{G}$ be a connected graph and suppose we orient the edges in such a way that the digraph we obtain has a unique vertex of indegree 0. Must this digraph be a rooted tree?

9. The following digraphs are acyclic, and canonical labelings are shown. Apply Bellman's algorithm to each digraph in order to find the lengths of shortest paths from v_0 to each other vertex. Find a shortest path to v_t and the predecessor vertex v_{p_t}.

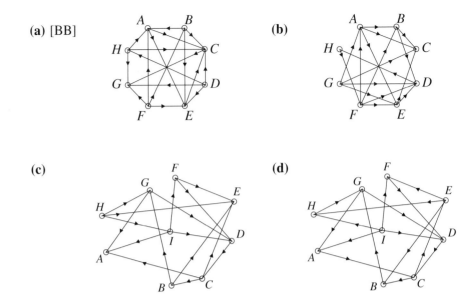

Figure 12.23   Digraphs for Exercise 1.

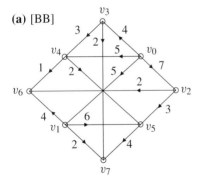

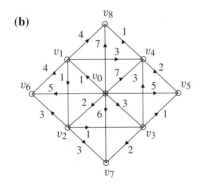

10. Show that Bellman's algorithm works; that is, if $\mathcal{G}$ is an acyclic digraph with vertices $v_0, v_1, \dots, v_{n-1}$ canonically labeled, then the final value of d_t is the length of a shortest path from v_0 to v_t and there is a path of this length whose final arc is $v_{p_t} v_t$.

11. [BB] Suppose we apply Bellman's algorithm to an acyclic digraph with a canonical ordering of vertices. Explain why the arcs $v_{p_t} v_t$ produce a spanning tree rooted at v_0.

12. Find the complexity of Bellman's algorithm (in terms of additions and comparisons).

13. [BB] Explain how Bellman's algorithm can be modified to find the length of a shortest path to each v_t from a given vertex v_k which is not necessarily v_0.

14. [BB; (a)] For each of the digraphs which accompany Exercise 9, find the shortest distance to each v_t and the corresponding value of p_t,

   i. when paths start at v_1 and,
   ii. when paths start at v_2.

REVIEW EXERCISES FOR CHAPTER 12

1. Make a tree which displays all monotonically increasing subsequences of $4, 7, 3, 9, 5, 1, 8$. How many are there?

2. Let $\mathcal{G}$ be a weighted graph with vertices $w_1, w_2, \ldots, w_6$ such that $w_j w_k$ is an edge if and only if $j + k$ is odd, in which case the weight of the edge $w_j w_k$ is $(\max\{j, k\})^2$.

    (a) Identify the unweighted graph described.

    (b) Using a tree, solve the Traveling Salesman's Problem for $\mathcal{G}$.

3. **(a)** Is it possible for a tree to have nine vertices and degree sequence $4, 2, 2, 2, 2, 2, 2, 1, 1$?

    (b) Is it possible for a tree to have nine vertices, two of which have degree 5?

4. Prove that the following conditions are equivalent for a tree $\mathcal{T}$ with $n \geq 2$ vertices.

    i. $\mathcal{T}$ has an Eulerian trail.
    ii. $\mathcal{T}$ has exactly two vertices of degree 1.
    iii. $\mathcal{T}$ has exactly $n - 2$ vertices of degree 2.

5. **(a)** Prove that up to isomorphism there is only one unlabeled tree on seven vertices with a vertex of degree 5 and another of degree 2. Draw a picture of this tree.

    (b) Label the tree obtained in (a) in two different ways so as to produce nonisomorphic labeled trees.

6. How many spanning trees has $\mathcal{K}_8$? Why?

7. Use Kirchoff's Theorem to find the number of spanning trees. Draw pictures of them all.

8. Is it possible for a graph to have exactly two spanning trees? Explain.

9. In Exercise 8 of Section 12.3, you proved that the number of spanning trees of the complete bipartite graph $\mathcal{K}_{2,n}$ is $n2^{n-1}$. If e is an edge of $\mathcal{K}_{2,n}$, how many of these spanning trees contain e? Explain.

10. Use Kruskal's algorithm to find a minimum spanning tree of the weighted graph shown. What is the weight of a minimum spanning tree?

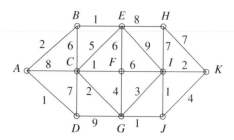

11. Repeat Exercise 10 using Prim's algorithm.

12. Let $v_1, v_2, \ldots, v_n$ denote the vertices of the complete graph $\mathcal{K}_n$, $n \geq 3$, and give edge $v_i v_j$ weight $i + j$. Find a minimum spanning tree for this weighted graph and the weight of such a tree.

13. Determine whether or not each of the digraphs shown is acyclic by finding a canonical labeling of vertices or exhibiting a cycle.

    (a)

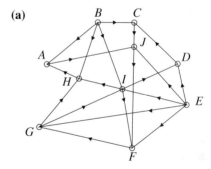

    (b)

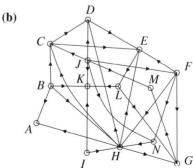

14. Change the direction of the arrows on JF and FG in each digraph of Exercise 13 and answer the questions again.

15. Can an acyclic digraph have two different canonical orderings of vertices? Explain.

16. Given that the digraph is acyclic, find a canonical ordering $v_0, v_1, \ldots$ of the vertices. Then find the lengths of shortest paths from the root to all other vertices, specifying the predecessor vertex (as defined in Bellman's algorithm) in each case when

(a) the root is v_0,
(b) the root is v_2.

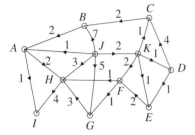

17. Repeat Exercise 16 using the digraph shown and root vertices v_0 and v_1.

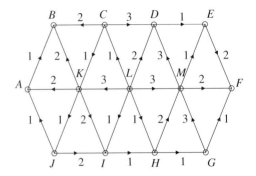

Depth-First Search and Applications

13.1 DEPTH-FIRST SEARCH

In this short chapter, we describe a procedure called *depth-first search*, which is simple, efficient, and the basis for a number of important computer algorithms in graphs. It quickly tests whether or not a graph is connected, it produces a spanning tree in the connected case, and it is the key to solving a problem about turning graphs into digraphs.

13.1.1 THE DEPTH-FIRST SEARCH ALGORITHM ▶

Let G be a graph with n vertices.

Step 1. Choose any vertex, label it 1. Set $k = 1$.

Step 2. While there are unlabeled vertices

    if there exists an unlabeled vertex adjacent to k, assign to it the
       smallest unused label ℓ from the set $\{1, 2, \ldots, n\}$ and set $k = \ell$.
    else if $k = 1$, stop;
        else backtrack to the vertex ℓ from which k was labeled and set
          $k = \ell$.
    end while

Pause 1 Suggest a way that a machine could keep track of the labels $1, 2, \ldots, n$ that have been used as this algorithm proceeds.

When the algorithm terminates, there is a path from vertex 1 to each labeled vertex which uses only edges required by the algorithm (Exercise 3). Thus, if the algorithm successfully labels all n vertices of the graph G, the graph must have been connected. The converse is also true (Exercise 4). Thus, depth-first search is an algorithm which tests connectedness in a graph.

EXAMPLE 1 We consider how the depth-first search algorithm might proceed with the graph in Fig 13.1(a). Starting at vertex 1, one possible forward procedure in the

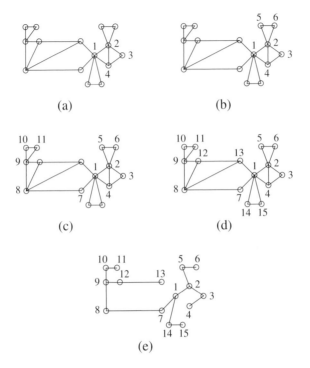

Figure 13.1   A depth-first search and the corresponding spanning tree.

depth-first search yields the labels shown in Fig 13.1(a). All the vertices adjacent to vertex 4 have been labeled and $k = 4 \neq 1$, so the algorithm backtracks first to 3, the vertex from which 4 acquired its label. All vertices adjacent to 3 have also been labeled, so it backtracks to 2. Vertex 2 has an unlabeled adjacent vertex. The smallest unused label to date is 5, so the algorithm would label 5 some vertex adjacent to 2, say the one shown. Vertex 5 has an adjacent unlabeled vertex, so this gets labeled 6. The current assignment of labels is depicted in Fig 13.1(b). The algorithm now backtracks until it finds a vertex with an adjacent unlabeled vertex. It finds vertex 1 in this way and, since this vertex has adjacent unlabeled vertices, the algorithm could conceivably assign labels as far as 11, as shown in Fig 13.1(c). Since all vertices adjacent to 11 have been labeled, the algorithm backtracks through vertex 10 to vertex 9 and then assigns labels 12 and 13. From vertex 13, the algorithm backtracks to 12, then to 9 (from which 12 received its label), then to 8, to 7 and to 1, after which vertices 14 and 15 are labeled. Finally, the algorithm backtracks to 14, then to 1, and stops. The final labels are shown in Fig 13.1(d). Since 15 labels were used and the graph had 15 vertices, the graph was connected, a fact admittedly obvious by sight, but not to a computer.   ▲

In addition to testing for connectedness, depth-first search has another use. If the graph is connected, those edges which were used in the search form a spanning tree for the graph (see Exercise 5). Applied to the graph in the preceding example, the algorithm produces the spanning tree shown in Fig 13.1(e).

Pause 2 For the graph given in Fig 13.1(a), carry out a depth-first search different from the one in the text and draw the corresponding spanning tree. Start at vertex 1 as before.

Pause 3 Show that the complexity function for depth-first search applied to a graph with n vertices is $\mathcal{O}(n^2)$. [*Hint*: Think of edges instead of vertices.]

One critical advantage that a person has over a computer in solving elementary graph problems is sight. The reader with little experience writing computer programs may well wonder from time to time at the rather detailed way we describe some "obvious" procedures. When a person, for instance, applies the depth-first search algorithm to a simple graph which he or she can see, some ways of labeling the vertices are much more sensible than others. Furthermore, a person can **see** when all the vertices are labeled and hence stop without backtracking to vertex 1, as required by the algorithm.

This section concludes by considering a problem which some readers may have seen previously in a different context. Our solution involves moving without thought around the vertices of a graph that is never fully seen, just the way a computer does. On the other hand, when we stop, we can be confident that we have indeed explored **every** vertex.

PROBLEM 2.   An innkeeper has a full eight-pint flagon of wine and empty flagons that hold five and three pints, respectively. The three flagons are unmarked and no other measuring devices are available. Explain how the innkeeper can divide the wine into two equal amounts in the fewest number of steps.

Solution.   We imagine the graph whose vertices and edges are as follows: There is a vertex labeled (a, b, c) if and only if it is possible for the innkeeper to hold, with certain knowledge, a pints of wine in the eight-pint flagon, b pints in the five-pint flagon, and c pints in the three-pint flagon. Thus, vertex $(8, 0, 0)$ corresponds to the initial state. There is an edge joining (a, b, c) and (a', b', c') if it is possible to reach the state (a', b', c') from (a, b, c), or vice versa, by pouring all the wine from one of the three flagons into another. For example, from the initial state, the innkeeper can reach $(3, 5, 0)$ by emptying the eight-pint flagon into the five-pint flagon; thus, there is an edge between $(8, 0, 0)$ and $(3, 5, 0)$. We explore the vertices of this graph with a procedure much like depth-first search. Continue from $(3, 5, 0)$. Emptying the remainder of the eight-pint flagon into the three-pint flagon takes us to $(0, 5, 3)$, emptying the five-pint flagon into the eight-pint flagon takes us to $(5, 0, 3)$, then emptying the three-pint flagon into the five-pint flagon takes us to $(5, 3, 0)$. (See Fig 13.2.)

We continue to visit, in order, vertices $(2, 3, 3)$, $(2, 5, 1)$, $(7, 0, 1)$, $(7, 1, 0)$, $(4, 1, 3)$, and finally $(4, 4, 0)$. This state is what the innkeeper sought; it has been reached in ten steps.

Now we backtrack, first to $(4, 1, 3)$. Emptying the contents of the eight-pint flagon into the five-pint flagon takes us to a previously labeled vertex on the path to $(4, 1, 3)$, namely, $(0, 5, 3)$. If we could get from $(0, 5, 3)$ to $(4, 1, 3)$ in fewer than seven steps, then the path to $(4, 4, 0)$ could be shortened, but that issue will be decided when we have backtracked to $(0, 5, 3)$ and can be ignored for now. Note, however, that we must be careful because an edge in our graph only

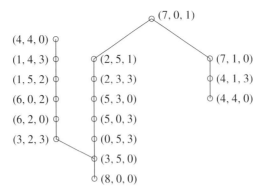

Figure 13.2 The spanning tree for a problem of flagons and wine.

signifies that you can go from one of the end vertices to the other, not necessarily in both directions (see Exercise 9). We continue to investigate vertices adjacent to $(4, 1, 3)$. Emptying the one pint remaining in the five-pint flagon into the eight-pint flagon leads to $(5, 0, 3)$, again a vertex already labeled. There are no other vertices adjacent to $(4, 1, 3)$, so we backtrack to $(7, 1, 0)$. We leave it to you to confirm that all vertices adjacent to those on the route from $(7, 1, 0)$ back to $(3, 5, 0)$ have already been labeled lower on the path. The vertex $(3, 5, 0)$, however, has an adjacent vertex not yet labeled, namely, $(3, 2, 3)$. From this, we move in order to $(6, 2, 0)$, $(6, 0, 2)$, $(1, 5, 2)$, $(1, 4, 3)$, and then $(4, 4, 0)$. We have found a route to the desired state which requires just seven steps. Backtracking again from $(4, 4, 0)$, we come first to $(1, 4, 3)$ and find two adjacent vertices on the first path—$(5, 0, 3)$ and $(0, 5, 3)$—and hence two more routes to $(4, 4, 0)$; for example,

$$(8, 0, 0) \rightarrow (3, 5, 0) \rightarrow (3, 2, 3) \rightarrow \cdots$$
$$\rightarrow (1, 4, 3) \rightarrow (5, 0, 3) \rightarrow (5, 3, 0) \rightarrow \cdots \rightarrow (4, 4, 0).$$

Since each of these routes is (much) longer than the current shortest route, we do not display them on the graph. Continuing to backtrack from $(1, 4, 3)$, we find that $(1, 5, 2)$ is adjacent to $(0, 5, 3)$ and $(6, 0, 2)$ is adjacent to $(5, 0, 3)$, but, as before, we do not display these adjacencies since the routes they give to $(4, 4, 0)$ are longer than the current shortest route. Eventually we reach the initial vertex $(8, 0, 0)$ and discover that all adjacent vertices have been labeled, so we know that all vertices of the graph have been visited. We also discover that we can pass from $(8, 0, 0)$ to $(5, 0, 3)$ in one step, giving yet another route to the desired vertex $(4, 4, 0)$, a route requiring eight steps. The shortest route, however, remains at seven steps. ∎

Answers to Pauses

1. Set up a one-dimensional array of length n (that is, a $1 \times n$ matrix) with all entries initially 0. When label k is selected, change component k of this vector to 1.

2. One possible answer is given, along with the associated spanning tree.

3. Depth-first search proceeds by deciding at each stage either to proceed along a particular edge to the next unlabeled vertex or not to proceed (because the edge joins two labeled vertices). If all edges incident with a given vertex are rejected, then the algorithm backtracks. Hence, focusing our attention on the edges, each edge need be considered at most twice. (Note that once an edge has been rejected or used in backtracking, it is never considered again.) Hence, the total number of steps is at most twice the number of edges, this number being at most $\binom{n}{2} = \frac{1}{2}n(n-1)$. So the complexity of depth-first search is $\mathcal{O}(n^2)$.

EXERCISES

The symbol [BB] means that an answer can be found in the Back of the Book.

1. For each of the following graphs, carry out a depth-first search starting at vertex a. Show all labels and list those used, in order, on the final backtracking. Highlight any resulting spanning tree.

(a) [BB]

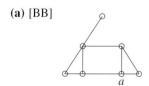

(b)

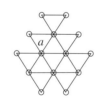

(c)

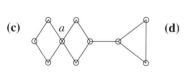

(d)

(a) [BB]

(b)

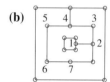

(c)

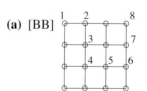

(d)

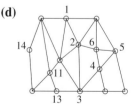

Figure 13.3

2. In each of the graphs shown in Fig 13.3, a depth-first search has labeled several vertices. Show all labels, list in order those used on the final backtracking, and show any resulting spanning tree.

3. Use mathematical induction to show that when the depth-first search algorithm terminates, there is a path from vertex 1 to each labeled vertex which uses only edges required by the algorithm.

4. (a) [BB] Let v be a vertex in a graph $\mathcal{G}$ which is labeled by the depth-first search algorithm. Prove that the algorithm labels all the vertices adjacent to v.

   (b) Prove that if $\mathcal{G}$ is connected, the depth-first algorithm labels every vertex of $\mathcal{G}$.

5. Suppose the depth-first algorithm is applied to a connected graph $\mathcal{G}$. Prove that the vertices of $\mathcal{G}$ together with the edges used by the algorithm form a spanning tree. [*Hint*: Exercises 3 and 4.]

6. [BB] Give an example of a connected graph $\mathcal{G}$ which has a spanning tree not obtainable as a depth-first spanning tree for $\mathcal{G}$.

7. (a) [BB] Explain how to change the depth-first search algorithm so that it can count the number of connected components of a graph.

   (b) Table 13.1 shows the adjacencies in a graph with vertices $v_1, v_2, \ldots, v_{13}$. **Without drawing the graph**, apply a depth-first search and use the method discovered in part (a) to determine the number of components. Give the labels which the search assigns to the vertices.

8. Given a full 12-ounce flagon of wine and two empty flagons holding eight and five ounces respectively, explain how to divide the wine into two equal six-ounce portions in the fewest number of steps. The flagons are unmarked. No other measuring devices are at hand.

9. [BB] In our wine-pouring Problem 2, if it is possible to move from (a, b, c) to (a', b', c') in a single step, is it always possible to move in the other direction, (a', b', c') to (a, b, c), in a single step?

10. [BB] What might the phrase *breadth-first search* mean? Design such a procedure.

11. [BB; (a)] Apply a breadth-first search to each of the graphs shown in Exercise 1 (starting at vertex a in each case).

12. Suppose the breadth-first search algorithm is applied to an arbitrary graph $\mathcal{G}$.

    (a) Prove that when the algorithm terminates, there is a path from vertex 1 to every other labeled vertex using only edges required by the algorithm.

    (b) Prove that if $\mathcal{G}$ is connected, every vertex is labeled.

    (c) Prove that if $\mathcal{G}$ is connected, then the vertices of $\mathcal{G}$ together with the edges used by the algorithm form a spanning tree.

Table 13.1

	1	2	3	4	5	6	7	8	9	10	11	12	13
1	0	1	0	0	0	1	0	0	0	0	1	0	0
2	1	0	0	0	0	0	0	0	0	0	1	0	0
3	0	0	0	0	0	0	0	1	0	0	0	0	1
4	0	0	0	0	1	0	1	0	0	0	0	1	0
5	0	0	0	1	0	0	0	0	0	1	0	0	0
6	1	0	0	0	0	0	0	0	0	0	1	0	0
7	0	0	0	1	0	0	0	0	1	1	0	0	0
8	0	0	1	0	0	0	0	0	0	0	0	0	1
9	0	0	0	0	0	0	1	0	0	0	0	1	0
10	0	0	0	0	1	0	1	0	0	0	0	0	0
11	1	1	0	0	0	1	0	0	0	0	0	0	0
12	0	0	0	1	0	0	0	0	1	0	0	0	0
13	0	0	1	0	0	0	0	1	0	0	0	0	0

13. Let G be a connected graph with $n \geq 2$ vertices, one of which is u. Show that G has a spanning tree with the property that, for any vertex v, the edges of the tree which connect u and v define a shortest path between u and v. [*Hint*: Think of the spanning tree produced by the breadth-first search algorithm. See Exercise 10.]

14. Prove that the complete bipartite graph $\mathcal{K}_{2,n}$ has the property that every spanning tree is obtainable as a depth-first search spanning tree if and only if $n = 2, 3, 4$.

13.2 THE ONE-WAY STREET PROBLEM

Over the years, the amount of traffic on the streets of a small town has increased enormously and the town planners are trying to decide what should be done. One suggestion, looked upon favorably by certain members of the town council, is that all streets be made one-way in order to simplify traffic flow. Certain minimum requirements must be met, however. For example, it must remain possible to travel from any point in town to any other point along one-way streets. Is there a way of putting a direction on each of the streets so that this requirement is met? In the language of graph theory, the one-way street problem asks, "Does a given connected graph have a strongly connected orientation?"

Remember that a digraph is strongly connected if it is possible to move from any vertex to any other vertex along arcs, in the proper direction. (See Section 11.2.) For example, the digraph shown on the right in Fig 13.4 is strongly connected.

Figure 13.4   A graph and a solution to the One-Way Street Problem.

13.2.1 DEFINITIONS ▶   To *orient* or to *assign an orientation* to an edge in a graph is to assign a direction to that edge. To orient or assign an orientation to a graph is to orient every edge in the graph. A graph has a *strongly connected orientation* if it is possible to orient it in such a way that the resulting digraph is strongly connected.

Figure 13.5   A graph for which the One-Way Street Problem cannot be solved.

The graph on the left in Fig 13.4 has a strongly connected orientation (shown beside it). On the other hand, Fig 13.5 displays a graph which does not have a strongly connected orientation. The trouble is the middle edge; once this is given a direction, flow will only be permitted from one side of that edge to the other, but not in reverse. Such an edge is called a *bridge* or *cut edge*.

13.2.2 DEFINITION ▶

An edge e of a connected graph $\mathcal{G}$ is called a *bridge* or a *cut edge* if the subgraph $\mathcal{G} \setminus \{e\}$ is not connected.

Pause 4

Let e be an edge which is not a bridge in a connected graph $\mathcal{G}$. Show that e is part of a circuit.

We shall show that the presence of a bridge in a connected graph $\mathcal{G}$ is exactly what prevents $\mathcal{G}$ from having a strongly connected orientation. First, however, we illustrate a technique that will be useful in our proof.

PROBLEM 3.   Find a strongly connected orientation for the graph on the left of Fig 13.6.

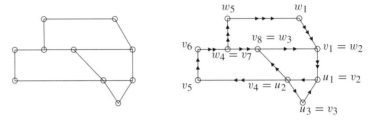

Figure 13.6   A graph and a strongly connected orientation for it.

Solution.   We begin by finding a circuit—for example, $u_1 u_2 u_3 u_1$—and orienting the edges $u_1 \to u_2 \to u_3 \to u_1$ as shown (by single arrows) on the right of Fig 13.6.

Next, we look for a vertex which is not in but which is adjacent to some vertex of this circuit; v_1 is adjacent to u_1, for instance. Now we look for a circuit containing $v_1 u_1$, say, $v_1 v_2 (= u_1) v_3 (= u_3) v_4 (= u_2) v_5 v_6 v_7 v_8 v_1$. We would like to orient the edges of this circuit $v_1 \to v_2 \to \cdots \to v_8 \to v_1$ as before, but we cannot because edges $v_2 v_3$ and $v_3 v_4$ have already been oriented (and in the opposite direction). So we do the best we can, orienting those edges of this circuit not already assigned directions; we orient $v_1 \to v_2$, $v_4 \to v_5$, $v_5 \to v_6$, $v_6 \to v_7$, $v_7 \to v_8$, $v_8 \to v_1$ (with double arrows). So far, our procedure has identified eight vertices and oriented all but one of the edges between pairs of these. We orient the omitted edge $v_8 u_2$ arbitrarily, say $u_2 \to v_8$. It is critical to observe (and we leave it for you to check) that, at this stage, the subgraph whose vertices are the u_i and v_j and whose edges are all edges of the given graph among these vertices is strongly connected; you can get from any u_i or v_j to any other of these vertices following edges in the direction of arrows.

Again, we next look for a vertex which is not in but which is adjacent to some vertex in this subgraph; w_1 is adjacent to v_1, for instance. We find a circuit which includes the edge $w_1 v_1$, for example, $w_1 w_2 (= v_1) w_3 (= v_8) w_4 (= v_7) w_5 w_1$, and orient the edges of this circuit which have not so far been oriented in the direction this sequence of vertices indicates. We orient $w_1 \to w_2$, $w_4 \to w_5$ and $w_5 \to w_1$ as shown (with triple arrows). You should check that a strongly connected orientation has now been assigned to the given graph.   ■

13.2.3 THEOREM ▶ A graph has a strongly connected orientation if and only if it is connected and has no bridges.

Proof ($\rightarrow$) If a graph $\mathcal{G}$ has a strongly connected orientation, then $\mathcal{G}$ is surely connected. Let e be an edge, say with end vertices u and v. If e is oriented in the direction $u \rightarrow v$ when $\mathcal{G}$ is assigned its strongly connected orientation, then in this strongly connected digraph, there is a directed path from v to u. In particular, there is a path $\mathcal{P}$ in $\mathcal{G}$ from v to u which does not use e (definition of path). So any walk in $\mathcal{G}$ which involves e can be replaced by a walk which avoids e, simply by replacing e with $\mathcal{P}$. It follows that $\mathcal{G} \setminus \{e\}$ is connected, so e is not a bridge.

($\leftarrow$) Conversely, suppose $\mathcal{G}$ is connected and has no bridges. We first remark that every edge of $\mathcal{G}$ must be part of a circuit. (See Pause 4.) Now we show that $\mathcal{G}$ can be assigned a strongly connected orientation by mimicking the approach we used to solve Problem 3. Let $\mathcal{C}: u_1 u_2 \cdots u_n u_1$ be a circuit in $\mathcal{G}$. (As noted, every edge is part of a circuit, so some such circuit certainly exists.) For $1 \le i \le n - 1$, assign edge $u_i u_{i+1}$ the orientation $u_i \rightarrow u_{i+1}$. Assign edge $u_n u_1$ the orientation $u_n \rightarrow u_1$. Then orient any other edges between vertices of $\mathcal{C}$ arbitrarily.

If $\mathcal{C}$ contains all the vertices of $\mathcal{G}$, then clearly we have an orientation of $\mathcal{G}$ which is strongly connected, since $\mathcal{C}$ was strongly connected. Suppose, on the other hand, that $\mathcal{C}$ does not contain all the vertices of $\mathcal{G}$. Since $\mathcal{G}$ is connected, there must exist a vertex v_1 not in $\mathcal{C}$ such that $v_1 u_j$ is an edge for some u_j. Since every edge in $\mathcal{G}$ is part of a circuit, $v_1 u_j$ is part of a circuit $v_1 u_j (= v_2) v_3 \cdots v_m v_1$. Assign $v_1 v_2$ the direction $v_1 \rightarrow v_2$ and $v_m v_1$ the direction $v_m \rightarrow v_1$. Leave unchanged the orientation of edges on the circuit which have already been oriented (that is, those which are part of $\mathcal{C}$), but orient any $v_i v_{i+1}$ not yet oriented in the direction $v_i \rightarrow v_{i+1}$. Finally, assign an arbitrary orientation to any remaining edges among vertices considered to this point. We have drawn two possibilities for the oriented subgraph whose vertices are $u_1, \ldots, u_n, v_1, \ldots, v_m$ in Fig 13.7. The subgraph on the right in this figure is intended to serve as a caution that, in addition to v_2, some other of the vertices $v_3, \ldots, v_m$ may coincide with vertices u_j. For example, in the rightmost subgraph, if you wanted to get from v_{k+1} to v_k, you would have to go first to v_1, then to u_j and around $\mathcal{C}$ to v_k.

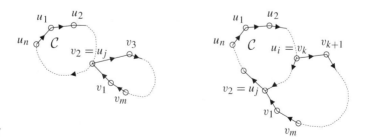

Figure 13.7

In Pause 5, we ask you to verify the following crucial **fact**: The subgraph of vertices and edges identified so far is strongly connected. Thus, if this subgraph is the entire graph, we are finished. Otherwise we find a vertex w_1 which is not part of but which is adjacent to some vertex in this subgraph, find a circuit containing this vertex, and orient edges as before. This procedure must terminate, and with a strongly connected orientation. ▮

Pause 5 Establish the **fact** mentioned toward the end of the preceding proof.

13.2.4 AN ALGORITHM FOR A STRONGLY CONNECTED ORIENTATION ▶

Since the identification of circuits in a graph is difficult without the advantage of sight, the following algorithm, based upon depth-first search, is important.

Suppose G is a connected graph without bridges. To produce a strongly connected orientation, proceed as follows.

Step 1. Carry out a depth-first search in G and let T be the spanning tree which this produces.

Step 2. For each edge e of G, consider the labels i and j of its ends. Suppose $i < j$.

If e is in T, orient it $i \rightarrow j$;
else orient it $j \rightarrow i$.

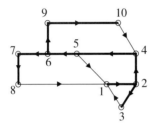

Figure 13.8 A depth-first search, a spanning tree, and a strongly connected orientation.

We apply this algorithm to the graph in Fig 13.6, show in Fig 13.8 the labels which a depth-first search might yield, and, with heavy lines, show the corresponding spanning tree as well. For each edge ij with $i < j$, the algorithm orients $i \rightarrow j$ if the edge is part of the tree and otherwise $j \rightarrow i$. Edge 45, for example, is in the tree, so it is oriented $4 \rightarrow 5$; on the other hand, edge 18 is not in the tree, so it is oriented $8 \rightarrow 1$.

Pause 6 Use Algorithm 13.2.4 to exhibit a strongly connected orientation for the graph in Fig 13.1 based upon the depth-first search shown in that figure.

Suppose G is a connected graph without bridges. To show that our algorithm actually works, it suffices to show that it orients the edges of G in such a way that there is a directed path between 1 and k (in each direction) for all k. For this, we use the strong form of mathematical induction on k.

The result is immediate if $k = 1$, so assume that $k > 1$ and that for all ℓ, $1 \leq \ell < k$, there is a path from 1 to ℓ and from ℓ to 1 which respects the orientation of edges. We must show that such paths exist between 1 and k.

Vertex k acquired its label because it was adjacent to one which was already labeled, say ℓ. Since the depth-first search algorithm always chooses the smallest available label, we have $\ell < k$. By the induction hypothesis, there is a directed path from 1 to ℓ. Since edge ℓk is part of the depth-first search spanning tree and since $\ell < k$, the edge incident with ℓ and k is oriented $\ell \rightarrow k$; thus, there is a directed path from 1 to k. It is more difficult to show that there is also a directed path from k to 1.

First assume the depth-first search algorithm starts backtracking immediately after labeling k (because all vertices adjacent to k have already been labeled). Note that there must be some vertex adjacent to k in addition to ℓ; otherwise, ℓk would be a bridge. Suppose this other vertex has been labeled t where, necessarily, $t < k$. Since edge tk is not part of the spanning tree, Algorithm 13.2.4 orients it $k \rightarrow t$. By the induction hypothesis, there is a directed path from t to 1. Together with kt, we obtain the desired directed path from k to 1.

On the other hand, suppose the algorithm does not backtrack after labeling k. Then it will proceed from k to an adjacent vertex, which will be labeled $k + 1$, and, if there is an unlabeled vertex adjacent to $k + 1$, to $k + 2$. Eventually, the algorithm backtracks from a vertex labeled $k+s$, $s \geq 1$, to k, all vertices adjacent to vertices on this path already labeled. It is important to note that there is a directed path from k to each vertex $k + i$, $0 \leq i \leq s$. Thus, if any of these vertices is adjacent to a vertex labeled t with $t < k$, the induction hypothesis makes it easy to establish the existence of a directed path from k to 1. On the other hand, if the only vertices adjacent to vertices of the form $k + i$ are also of this form, then there is no path in $\mathcal{G}$ from $k + s$ to ℓ which avoids the edge $k\ell$,[1] so $k\ell$ would be a bridge, a contradiction.

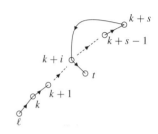

As a concluding remark, we mention in passing that Algorithm 13.2.4 can be applied to give an orientation to any connected graph $\mathcal{G}$ although, if $\mathcal{G}$ has bridges, the result will not be strongly connected.

Answers to Pauses

4. Let u and v denote the end vertices of e. Since $G \setminus \{e\}$ is connected, there is a path in $G \setminus \{e\}$ from v to u. This path, followed by uv, is a circuit in $\mathcal{G}$ containing e.

5. We must show that it is possible to get from any u_i or v_j to any other u_i or v_j following edges in the assigned directions. We can get from any u_i to any other u_j along the circuit $\mathcal{C}$. Next, we show that there is a directed path from each v_i to v_{i+1} and so, since $v_m v_1$ is oriented $v_m \rightarrow v_1$, there is a directed path from any v_i to any v_j. The orientation $v_1 \rightarrow v_2$ provides a directed path from v_1 to v_2 and, for any $i = 2, 3, \ldots, m - 1$, if the orientation is not $v_i \rightarrow v_{i+1}$, then both the vertices v_i and v_{i+1} are part of the circuit $\mathcal{C}$, which we can use to find a directed path from v_i to v_{i+1}. In any case, there is always some directed path from each v_i to v_{i+1} as asserted. This also shows that we can pass from $\mathcal{C}$ to any v_i via the directed edge $u_j(= v_2)v_3$. Since we can move in the other direction via $v_1 v_2(= u_j)$, it follows that the subgraph is strongly connected.

---

[1] The second last vertex on a path to ℓ avoiding $k\ell$ is $k + j$ and, by assumption, ℓ is not adjacent to $k + j$ since $\ell < k + j$.

6. For each edge ij with $i < j$, we orient $i \to j$ if the edge is in the spanning tree and otherwise $j \to i$.

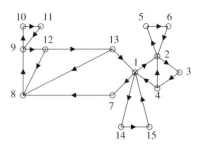

EXERCISES

The symbol [BB] means that an answer can be found in the Back of the Book.

1. [BB; (a)] For each of the graphs in Exercise 1 of Section 13.1, assign an orientation based on your depth-first search and state whether or not this orientation is strongly connected.

2. Find strongly connected orientations of the graphs corresponding to the five so-called *Platonic solids*: the cube, tetrahedron [BB], octahedron, icosahedron, and dodecahedron. (See pp. 314, 317, and 415.)

3. (a) [BB] Does there exist a strongly connected orientation for the Petersen graph? (See p. 312.) If so, find one.

   (b) Must a Hamiltonian graph have a strongly connected orientation? Explain.

   (c) What about the converse of (b)? Explain.

4. How many bridges does a tree with n vertices have? Explain.

5. [BB] Can an Eulerian graph have bridges? Explain.

6. [BB; ($\to$)] Prove that an edge e in a connected graph G is a bridge if and only if there are vertices u and v in G such that every path from u to v requires the edge e.

7. (a) [BB] Show that any connected graph can be oriented so that the resulting digraph is **not** strongly connected.

   (b) Let G be a connected graph with all vertices of even degree. Can G be oriented so that the resulting digraph is strongly connected? Explain.

8. [BB] Let G be a graph with a strongly connected orientation assigned to it. Suppose that the direction of the arrow on each edge of G is reversed. Is the new digraph still strongly connected? Explain.

9. Answer true or false and explain your answers.

   (a) [BB] A graph has a strongly connected orientation if and only if every edge is part of a circuit.

   (b) A graph with a strongly connected orientation has no vertices of degree 1.

   (c) A connected graph with no odd vertices has a strongly connected orientation.

10. Let $e = uv$ be a bridge in a connected graph G. Let G_1 be the subgraph of $G \setminus \{e\}$ whose vertices are those from which there is a path to u and whose edges are all the edges of G among these vertices. Let G_2 be defined analogously interchanging the roles of u and v.

   (a) [BB] Prove that there is no path between u and v in $G \setminus \{e\}$.

   (b) Prove that G_1 and G_2 have no vertices in common.

   (c) Prove that G_1 is a (connected) component of $G \setminus \{e\}$, that is, a maximal connected subgraph of $G \setminus \{e\}$.

   (In Fig 13.5, G_1 and G_2 are the two triangles. This exercise shows that Fig 13.5 shows what happens in general when there is a bridge in a graph.)

11. [BB] Faced with the undeniable fact that his town's street system does indeed have some bridges, Mayor Murphy decides that all bridges should remain two-way streets, but all other streets should be made one-way. Can this be done so as to allow (legal) travel from any intersection to any other?

12. Explain why the depth-first search algorithm can be used to find all bridges in a connected graph. Design a procedure to find all such bridges which is $\mathcal{O}(n^3)$ (where n is the number of vertices).

13. Let G be a graph with n vertices and m edges.

(a) Assuming $m > \frac{(n-1)(n-2)}{2} + 1$, show that G can be given a strongly connected orientation. (You may assume that the condition implies that G is connected. See Exercise 22 of Section 10.1.)

(b) Need G have a strongly connected orientation if $m > \frac{(n-1)(n-2)}{2}$?

REVIEW EXERCISES FOR CHAPTER 13

1. In each of the following graphs, a depth-first search has started and several vertices have been labeled. Finish each search, show all labels, list in order those used on the final backtracking, and show the resulting spanning tree.

(a)   **(b)**

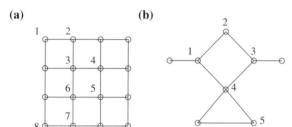

2. Let v be an arbitrary vertex and e an arbitrary edge of a connected graph G. Show that it is always possible to carry out a depth-first search beginning at v which includes e as part of the resulting spanning tree.

3. Assign an orientation to each of the graphs in Exercise 1 based upon your depth-first search and state whether or not this orientation is strongly connected.

4. Which graphs have the property that it is possible to start at some vertex v and carry out a depth-first search which labels all vertices without ever backtracking? Prove your answer.

5. If a connected graph G has a bridge e, show that every spanning tree of G must contain e.

6. Is it possible for a Hamiltonian graph to contain a bridge? Explain.

7. (a) Is the orientation which appears in a tournament always strongly connected? Explain.

(b) Is it possible to change the directions of some of the arrows in a tournament and achieve a strongly connected orientation? Explain. Assume the graph has at least three vertices.

14

Planar Graphs and Colorings

14.1 PLANAR GRAPHS

The Three Houses—Three Utilities Problem posed in Chapter 9 asks whether or not it is possible to draw the complete bipartite graph $\mathcal{K}_{3,3}$ without any crossovers of edges. (See Fig 14.1.)

Figure 14.1   The graph $\mathcal{K}_{3,3}$, representing three houses and three utilities.

Thinking of the edges of this graph as pieces of string and the vertices as knots, the question is whether or not the "net" can be arranged without string crossovers. In the language of graph theory, the problem asks if $\mathcal{K}_{3,3}$ is *planar*.

14.1.1   DEFINITION ▶   A graph is *planar* if it can be drawn in the plane in such a way that no two edges cross.

In this chapter (and abusing terminology) we find it convenient to refer to a picture of a graph in which there are no crossovers of edges as a *plane graph*. Thus, a planar graph is one which can be drawn as a plane graph. There are five places where edges in the graph $\mathcal{G}$ of Fig 14.2(a) cross; nevertheless, $\mathcal{G}$ is planar because it can be drawn as a plane graph, as shown in Fig 14.2(b).

Figure 14.2(c) shows $\mathcal{G}$ with all edges drawn as straight line segments. In fact, this is possible for any planar graph[1] and so we lose no generality if planar graphs are drawn as plane graphs with straight edges.

---

[1]This result is due to I. Fary, "On Straight Line Representation of Planar Graphs," *Acta Scientiarum*

413

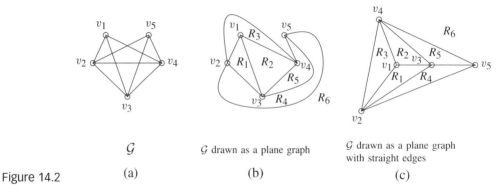

Figure 14.2

$\mathcal{G}$

(a)

$\mathcal{G}$ drawn as a plane graph

(b)

$\mathcal{G}$ drawn as a plane graph with straight edges

(c)

A plane graph divides the plane into various connected regions, one of which is called the exterior region. Every region, including the exterior, is bounded by edges. The graph in Fig 14.2(b) divides the plane into six regions, R_6 being the exterior region. The boundary of R_2, for example, consists of the edges v_1v_3, v_3v_4, and v_4v_1. The boundary of R_6 is v_2v_4, v_4v_5, and v_5v_2. We require that every edge must be the boundary of some region. Thus, in Fig 14.3(a), the edge labeled e is part of the boundary of region R, while in Fig 14.3(b) it is part of the boundary of the exterior region.

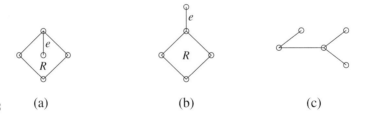

Figure 14.3   (a)   (b)   (c)

A tree determines just one region in the plane, the exterior region, and every edge of the tree is part of the boundary of this region. [See Fig 14.3(c).]

Pause 1   Consider the plane graph shown on the left of Fig 14.4.

(a) How many regions are there?

(b) List the edges which form the boundary of each region.

(c) Which region is exterior?

Answer these same questions for the graph on the right as well.

Planar graphs were first studied by Euler because of their connections with polyhedra. A *convex regular polyhedron* is a geometric solid all of whose faces are congruent. There are in all just five of these—the cube, the tetrahedron, the octahedron, the icosahedron, and the dodecahedron—and they are popularly known as the *Platonic solids* because they were regarded by Plato as symbolizing earth, fire, air, water, and the universe, respectively.

Mathematicarum (Szeged) **11** (1948), 229–233. A proof also appears in *Combinatorial Problems and Exercises* by László Lovász, North-Holland (1979) (Problem 5.38).

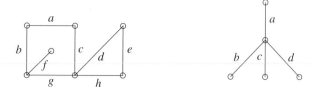

Figure 14.4

When we discussed Sir William Hamilton's "World Tour" in Section 10.2, we showed how to associate with the regular dodecahedron a planar graph whose edges and vertices correspond to the edges and vertices of the solid and whose regions correspond to the faces of the polyhedron. In the exercises of Section 10.2, we introduced planar graphs corresponding to the cube and the regular icosahedron. As we said at the time, we can obtain the planar graph associated with a particular polyhedron by imagining that the bottom face is stretched until the object collapses flat. Figure 14.5 shows the plane graphs associated with the tetrahedron, the cube, and the octahedron.

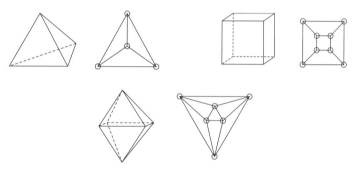

Figure 14.5   The tetrahedron, cube, and octahedron, and their corresponding planar graphs.

In 1752, Euler published the remarkable formula

$$V - E + F = 2,$$

which holds for any convex polyhedron with V vertices, E edges, and F faces.[2] (A polygon is *convex* if the line joining any pair of nonadjacent vertices lies entirely within the polygon.) For the cube, for instance, $V = 8$, $E = 12$, and $F = 6$; for the tetrahedron, $V = F = 4$ and $E = 6$. Euler's formula also works for any plane graph, with regions replacing faces. For example, the plane graph in Fig 14.2 has $V = 5$, $E = 9$, $R = 6$, so, $V - E + R = 5 - 9 + 6 = 2$.

14.1.2 THEOREM ▶   Let $\mathcal{G}$ be a connected plane graph with V vertices, E edges, and R regions. Then $V - E + R = 2$.

Proof   We use induction on E. If $E = 0$, then $V = R = 1$ (because $\mathcal{G}$ is connected) and the formula is true. Now assume the formula holds for connected plane graphs

---

[2]L. Euler, "Demonstratio nonnullarum insignium proprietatum quibus solida hedris planis inclusa sunt praedita," *Novi Commentarii Academiae Scientiarum Petropolitanae* **4** (1752/53), 140–160.

with $E - 1$ edges, where $E \geq 1$, and that $\mathcal{G}$ is a connected plane graph with E edges, V vertices, and R regions. We must show that $V - E + R = 2$. If $\mathcal{G}$ contains no circuits, then $\mathcal{G}$ is a tree and so $E = V - 1$ by Theorem 12.2.3. Since $R = 1$, we have $V - E + R = V - (V - 1) + 1 = 2$ as desired. Suppose, on the other hand, that $\mathcal{G}$ does contain a circuit $\mathcal{C}$. Let e be any edge of $\mathcal{C}$ and consider the subgraph $\mathcal{G} \setminus \{e\}$, which is a plane graph and still connected. The circuit $\mathcal{C}$ determines a region of $\mathcal{G}$ which disappears in $\mathcal{G} \setminus \{e\}$. So $\mathcal{G} \setminus \{e\}$ contains $R - 1$ regions, all V vertices of $\mathcal{G}$, and $E - 1$ edges. By the induction hypothesis, $V - (E - 1) + (R - 1) = 2$; that is, $V - E + R = 2$. ∎

Pause 2   Show that this theorem is not necessarily true if "connected" is omitted from its statement.

We are now ready to solve the Three Houses—Three Utilities problem. It turns out not to be possible to connect three houses to three utilities without crossovers of connection lines.

14.1.3 COROLLARY ▶ $\mathcal{K}_{3,3}$ is not planar.

Proof   We provide a proof by contradiction. The graph $\mathcal{K}_{3,3}$ has six vertices and nine edges. If it is planar, it can be drawn as a plane graph with R regions. Since $V - E + R = 2$, we have $R = 5$. Now count the number of edges on the boundary of each region and sum over all regions. Suppose the sum is N. The point to observe about $\mathcal{K}_{3,3}$ is that it contains no triangles (every edge joins a house to a utility; there are no edges from house to house or utility to utility). Therefore, the boundary of each region contains at least four edges, from which it follows that $N \geq 4R = 20$. On the other hand, in the calculation of N, each edge was counted at most twice, so $N \leq 2E = 18$. This contradiction establishes the corollary. ∎

Pause 3   Find the number N defined in this proof for the graph on the left of Fig 14.4. Verify that $N \leq 2E$. Give an example of an edge which is counted just once.

Intuitively, a graph with a lot of edges cannot be planar since it will be impossible to avoid crossovers when drawing it. Thus, we expect planar graphs to have relatively few edges. The next theorem makes this statement more precise.

14.1.4 THEOREM ▶   Let $\mathcal{G}$ be a planar graph with $V \geq 3$ vertices and E edges. Then $E \leq 3V - 6$.

Proof   We give an argument for the case that $\mathcal{G}$ is connected and ask you in Exercise 19 to prove the general case.

If $V = 3$, then $E \leq 3$ and the theorem holds. So we assume that $V > 3$. Thus, we may also assume that $E \geq 3$ (otherwise, the result is clearly true). Draw $\mathcal{G}$ as a plane graph with R regions. Then, as in the proof of Corollary 14.1.3, we count the number of edges on the boundary of each region, sum these numbers, and denote the sum N. As before, $N \leq 2E$. Also, since each boundary contains at least three edges, $N \geq 3R$. Hence, $3R \leq 2E$. Theorem 14.1.2 says that $V - E + R = 2$. Therefore, $6 = 3V - 3E + 3R \leq 3V - 3E + 2E = 3V - E$ and the result follows. ∎

14.1.5 COROLLARY ▶ K_5 is not planar.

Proof In K_5, $V = 5$ and $E = \binom{5}{2} = 10$ (since every pair of vertices is joined by an edge). Since it is not true that $E \le 3V - 6$, K_5 cannot be planar. ∎

In a graph where the number of edges is not too large, it is reasonable to expect some restrictions on vertex degrees. In Exercise 20, we ask you to show that a planar graph with at least four vertices has at least four vertices of degree $d \le 5$. For the present, we content ourselves with a weaker statement.

14.1.6 COROLLARY ▶ Every planar graph contains at least one vertex of degree $d \le 5$.

Proof Suppose $\deg v_i \ge 6$ for every vertex v_i. Since $\sum \deg v_i = 2E$, we would have $2E \ge 6V$ and hence $E \ge 3V > 3V - 6$. This contradicts $E \le 3V - 6$. ∎

It was the Polish mathematician Kazimierz Kuratowski (1896–1980) who discovered the crucial role played by $K_{3,3}$ and K_5 in determining whether or not a graph is planar. (The K in K_n and $K_{m,n}$ is in Kuratowski's honor.) First, since these graphs are not planar, no graph which contains either of them as a subgraph can be planar. Second, any graph obtained from either $K_{3,3}$ or K_5 simply by adding more vertices to edges cannot be planar either.

14.1.7 DEFINITION ▶ Two graphs are *homeomorphic* if and only if each can be obtained from the same graph by adding vertices (necessarily of degree 2) to edges.

EXAMPLE 1 The graphs $\mathcal{G}_1$ and $\mathcal{G}_2$ in Fig 14.6 are homeomorphic since both are obtainable from the graph $\mathcal{G}$ in that figure by adding a vertex to one of its edges. ▲

$\mathcal{G}_1$ $\qquad$ $\mathcal{G}_2$ $\qquad$ $\mathcal{G}$

Figure 14.6 Two homeomorphic graphs obtained from $\mathcal{G}$ by adding vertices to edges.

EXAMPLE 2 In Fig 14.7, we show two homeomorphic graphs, each obtained from K_5 by adding vertices to edges of K_5. (In each case, the vertices of K_5 are shown with solid dots.) ▲

Pause 4 Any two cycles are homeomorphic. Why?

EXAMPLE 3 Two graphs are homeomorphic if one is simply obtained from the other by adding vertices to edges; that is, the third graph mentioned in Definition 14.1.7 may be one of the two given graphs. For instance, graphs $\mathcal{G}_1$ and $\mathcal{G}_2$ in Fig 14.8 are

Figure 14.7 Two homeomorphic graphs obtained from K_5.

homeomorphic because G_2 is obtained by adding two vertices (the solid dots) to edges of G_1. ▲

When we say "adding vertices to edges," we do not include the possibility of adding a vertex where edges cross over each other, for this is achieved by adding a vertex to **each** edge and then joining these two vertices together. Joining vertices is not allowed. For example, graphs G_1 and G_3 of Fig 14.8 are not homeomorphic.

$$G_1 \qquad\qquad G_2 \qquad\qquad G_3$$

Figure 14.8 G_1 is homeomorphic to G_2 but not to G_3.

The following theorem characterizes planar graphs in a remarkably simple way. The proof in one direction is straightforward and was given earlier: A graph which contains K_5 or $K_{3,3}$ cannot be made planar by adding vertices to edges. The proof of the converse is more complicated and will be omitted.

14.1.8 THEOREM ▶ *(Kuratowski[3])* A graph is planar if and only if it has no subgraph homeomorphic to K_5 or $K_{3,3}$.

EXAMPLE 4 The graph G in Fig 14.9 is not planar. By deleting two edges, we obtain a subgraph S which is $K_{3,3}$ (the bipartition sets are the sets of hollow vertices and solid vertices) except for one vertex (the tiny one) added to an edge. Since S is homeomorphic to $K_{3,3}$, G is not planar, by Kuratowski's Theorem. ▲

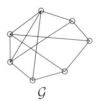
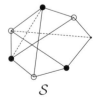

$$G \qquad\qquad\qquad S$$

Figure 14.9 A nonplanar graph G and a subgraph S homeomorphic to $K_{3,3}$.

[3]K. Kuratowski, "Sur le Problème des Courbes Gauches en Topologie," *Fundamenta Mathematicae* **15** (1930), 271–283.

EXAMPLE 5   The graph on the left in Fig 14.10 is not planar because the subgraph shown on the right is homeomorphic to $\mathcal{K}_5$. Notice, for example, that A is adjacent to C and D, and, except for intermediate vertices of degree 2, also to B and E.   ▲

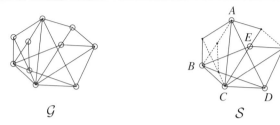

Figure 14.10   A nonplanar graph $\mathcal{G}$ and a subgraph $\mathcal{S}$ homeomorphic to $\mathcal{K}_5$.

Answers to Pauses

1. The graph on the left of Fig 14.4 has three regions whose boundaries are $\{d, e, h\}$, $\{a, b, f, g, c\}$, and $\{a, b, g, c, d, e, h\}$; the last region is exterior. The graph on the right is a tree; it determines only one region, the exterior one, with boundary $\{a, b, c, d\}$.

2. In the graph shown, $V - E + R = 6 - 6 + 3 = 3$.

3. The boundaries of the regions are given in the answer to Pause 1: $N = 3 + 5 + 7 = 15 \le 16 = 2E$. Edge f is counted only once.

4. Any cycle can be obtained from a 3-cycle by adding vertices to edges.

EXERCISES

The symbol [BB] means that an answer can be found in the Back of the Book.

1. (a) [BB] Show that the graph is planar by drawing an isomorphic plane graph with straight edges.

(b) [BB] Label the regions defined by your plane graph and list the edges which form the boundary of each region.

(c) [BB] Verify that $V - E + R = 2$, $N \le 2E$, and $E \le 3V - 6$ (where, as defined in the proof of Corollary 14.1.3, N is the sum of the numbers of edges on the boundaries of all regions).

2. Repeat Exercise 1 for the complete bipartite graph $\mathcal{K}_{2,5}$.

3. [BB] Verify Euler's formula $V - E + F = 2$ for each of the five Platonic solids.

4. One of the two graphs is planar; the other is not. Which is which? Explain. (Note that the graph on the right is the *Petersen Graph*, which was introduced in Section 10.2.)

5. Determine which of the graphs in Fig 14.11 are planar. In each case, either draw a plane graph and a plane graph with straight edges isomorphic to the one presented or exhibit a subgraph homeomorphic to $\mathcal{K}_{3,3}$ or $\mathcal{K}_5$.

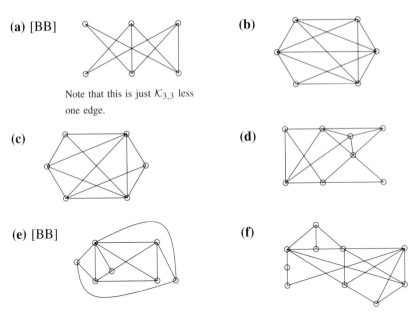

(a) [BB]

Note that this is just $K_{3,3}$ less one edge.

(b)

(c)

(d)

(e) [BB]

(f)

Figure 14.11 Graphs for Exercise 5.

6. Let G be a connected plane graph such that every region of G has at least five edges on its boundary. Prove that $3E \leq 5V - 10$.

7. [BB] If G is a connected plane graph with $V \geq 3$ vertices and R regions, show that $R \leq 2V - 4$.

8. **(a)** [BB] Give an example of a connected planar graph for which $E = 3V - 6$.
   (b) Let G be a connected plane graph for which $E = 3V - 6$. Show that every region of G is a triangle.

9. **(a)** If G is a connected plane graph with at least three vertices such that no boundary of a region is a triangle, prove that $E \leq 2V - 4$.
   (b) Let G be a connected planar bipartite graph with E edges and $V \geq 3$ vertices. Prove that $E \leq 2V - 4$.

10. **(a)** [BB] For which n is K_n planar?
    (b) For which m and n is $K_{m,n}$ planar?

11. [BB] Show that $K_{2,2}$ is homeomorphic to K_3.

12. Show that any graph homeomorphic to K_5 or to $K_{3,3}$ is obtainable from K_5 or $K_{3,3}$, respectively, by addition of vertices to edges.

13. Suppose a graph G_1 with V_1 vertices and E_1 edges is homeomorphic to a graph G_2 with V_2 vertices and E_2 edges. Prove that $E_2 - V_2 = E_1 - V_1$.

14. **(a)** [BB] Let G be a connected graph with V_1 vertices and E_1 edges and let H be a subgraph with V_2 vertices and E_2 edges. Show that $E_2 - V_2 \leq E_1 - V_1$.
    (b) Let G be a connected graph with V vertices, E edges, and $E \leq V + 2$. Show that G is planar.
    (c) Is (b) true if $E = V + 3$? Explain.

15. [BB] Let G be a graph and let H be obtained from G by adjoining a new vertex of degree 1 to some vertex of G. Is it possible for G and H to be homeomorphic? Explain.

16. Discover what you can about Kazimierz Kuratowski and write a short biographical note about this famous Polish mathematician (in good, clear English, of course).

17. Answer true or false and explain.
    (a) If G is Eulerian and H is homeomorphic to G, then H is Eulerian.
    (b) [BB] If G is Hamiltonian and H is homeomorphic to G, then H is Hamiltonian.

18. **(a)** Show that any planar graph all of whose vertices have degree at least 5 must have at least 12 vertices. [*Hint*: It suffices to prove this result for **connected** planar graphs. Why?]
    (b) Find a planar graph each of whose vertices has degree at least 5.

19. (a) [BB] Prove that if $\mathcal{G}$ is a planar graph with n connected components, each component having at least three vertices, then $E \leq 3V - 6n$.

(b) Prove that if $\mathcal{G}$ is a planar graph with n connected components, then it is always true that $E \leq 3V - 3n$. Deduce that Theorem 14.1.4 holds for arbitrary planar graphs.

20. (a) [BB] Prove that every planar graph with $V \geq 2$ vertices has at least two vertices of degree $d \leq 5$.

(b) Prove that every planar graph with $V \geq 3$ vertices has at least three vertices of degree $d \leq 5$.

(c) Prove that every planar graph with $V \geq 4$ vertices has at least four vertices of degree $d \leq 5$.

21. (a) [BB] A connected planar graph $\mathcal{G}$ has 20 vertices. Prove that $\mathcal{G}$ has at most 54 edges.

(b) A connected planar graph $\mathcal{G}$ has 20 vertices, seven of which have degree 1. Prove that $\mathcal{G}$ has at most 40 edges.

22. (a) [BB] Suppose $\mathcal{G}$ is a connected planar graph in which every vertex has degree at least 3. Prove that at least two regions of $\mathcal{G}$ have at most five edges on their boundaries.

(b) Establish (a) for planar graphs which are not connected.

23. Suppose that a convex polygon with n vertices is *triangulated*, that is, partitioned into triangles, possibly by the introduction of new vertices. The graphs illustrate two triangulations of a convex polygon with seven vertices. In the graph at the left, four edges were required and, in the graph at the right, 13. Show that the number of edges added in order to effect a triangulation is at least $n - 3$.

24. [BB] Find a formula for $V - E + R$ which applies to planar graphs which are not necessarily connected.

25. Prove that the Platonic solids are the only regular polyhedra. [*Hint*: In a regular polyhedron, every vertex has the same degree d and every face has the same number a of edges on its boundary. Try to solve $V - E + F = 2$ under such conditions.]

14.2 COLORING GRAPHS

One of the most exciting mathematical developments of the twentieth century was the proof, in 1976, of The Four-Color Theorem, which is easy to state and understand, but whose proof had remained unsolved since 1852 when it was first posed to his brother by Francis Guthrie. Guthrie had discovered that he could color a map of the counties of England with only four colors in such a way that each county had exactly one color and bordering counties had different colors. He guessed that it was possible to color the countries of **any** map with just four colors in such a way that bordering countries have different colors.[4] Both brothers had been students of Augustus De Morgan and so, unable to answer the question themselves, they asked their former teacher, one of the greatest mathematicians of the age. De Morgan was able to show that it is impossible to have five countries each adjacent to all of the others, but this result, while lending support to Francis Guthrie's guess, did not settle it.

It has long been known that five colors are enough, but for well over 100 years, whether one could make do with just four colors was not known. In 1879, the prestigious *American Journal of Mathematics* published a "proof" of the conjecture

---

[4]Our use of the term "map" is not intended to limit us to those maps found between the covers of an atlas. In the context of map colorings, a map just means a plane (possibly pseudo-) graph in which the edges represent borders and vertices are points where borders meet.

by one Alfred B. Kempe[5] (the final "e" is not pronounced), a London barrister, but 11 years later a fatal flaw was discovered. In fact, as pointed out by Percy Heawood,[6] Kempe's argument was a valid one for five colors, but not for four. It turns out that four colors are indeed enough but the proof, by Kenneth Appel and Wolfgang Haken,[7] occupies almost 140 pages of the *Illinois Journal of Mathematics* and will not be presented here. The proof involved showing that any planar graph must contain a subgraph of a certain type. If this subgraph were deleted, one could 4-color the resulting reduced graph. With the assistance of J. Koch and 1200 hours of computer time, it was shown that any 4-coloring of the boundary of any of the subgraphs identified by Appel and Haken could be extended to a 4-coloring of the subgraph; thus, any planar graph could be 4-colored. The final settling of what had been the *Four-Color Conjecture* was considered such an achievement that for a period of time, the postage meters at the University of Illinois bore the inscription, "Four Colors Suffice." Some readers might well be interested in an article by Robin Thomas, which, after a brief history of the Four-Color Problem, describes subsequent efforts to verify independently the proof, this being no small task.[8]

Our modest goal here is to show how graph theory can be used to attack a coloring problem and also to prove that any map can be 5-colored.

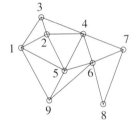

Figure 14.12   A map and an associated planar graph.

Our starting point is the observation that with any map we may associate a planar graph whose vertices correspond to countries and where an edge joins two vertices if the corresponding countries share a common border. There are nine countries in the map depicted in the left of Fig 14.12, so the associated planar graph has nine vertices. Since countries 5 and 6 share a border, vertices 5 and 6 are joined by an edge, and so on. The point is that coloring the countries of the map so that countries with a common border receive different colors is equivalent to coloring the vertices of the associated graph so that adjacent vertices have different colors. For example, the following assignment of four colors to countries

[5]A. B. Kempe, "On the Geographical Problem of the Four Colors," *American Journal of Mathematics* **2** (1879), 193–200.

[6]P. J. Heawood, "Map-Color Theorem," *Quarterly Journal of Mathematics* **24** (1890), 332–339.

[7]K. Appel, W. Haken, J. Koch, "Every Planar Map is Four-Colorable," *Illinois Journal of Mathematics* **21** (1977), 429–567.

[8]Robin Thomas, "An update on the Four-Color Theorem," *Notices of the American Mathematical Society* **45** (1998), no. 7, 848–859.

(or vertices) in Fig 14.12 colors the map (and the graph) in the proper sense:

$$1 - \text{red}; 2 - \text{blue}; 3 - \text{green}; 4 - \text{red}; 5 - \text{green}; 6 - \text{blue};$$

$$7 - \text{green}; 8 - \text{red}; 9 - \text{yellow}.$$

Figure 14.13 shows a map of North America in which most provinces and states have been colored. Starting the coloring with the western states, it is soon apparent that three colors are not enough.

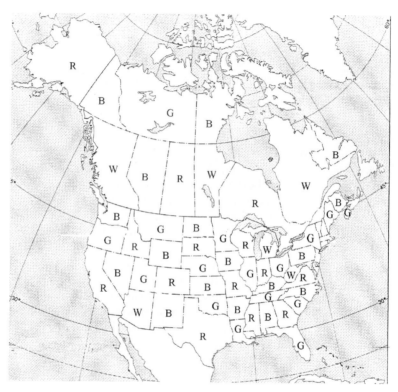

Figure 14.13   A partial 4-coloring of a map of North America.

We formalize the notions we have been discussing.

14.2.1   DEFINITIONS ▶   A *coloring* of a graph is an assignment of colors to the vertices so that adjacent vertices have different colors. An *n-coloring* is a coloring with n colors. The *chromatic number* of a graph G, denoted $\chi(G)$, is the minimum value of n for which an n-coloring of G exists.

Pause 5   Suppose $\chi(G) = 1$ for some graph G. What do you know about G?

The graph on the left of Fig 14.14 has been colored with three colors. Thus, its chromatic number is at most 3. Since it contains several triangles (the vertices of which must be colored differently), at least three colors are required: Its chromatic number is 3. The graph on the right has been colored with four colors, so for this graph G, $\chi(G) \leq 4$. Since G contains $\mathcal{K}_4$ as a subgraph (consider the subgraph determined by the vertices v_3, v_4, v_5, v_6) and since $\mathcal{K}_4$ requires four colors, $\mathcal{G}$ cannot be colored with less than four colors. For this graph then, $\chi(\mathcal{G}) = 4$.

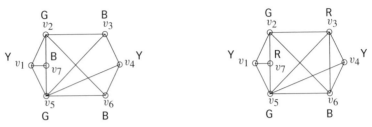

Figure 14.14   The graph on the left has chromatic number 3; the graph on the right has chromatic number 4.

Pause 6   What is the chromatic number of the graph in Fig 14.12?

Pause 7 $\chi(\mathcal{K}_n) = n$, $\chi(\mathcal{K}_{m,n}) = 2$. Why?

14.2.2 THE FOUR-COLOR THEOREM ▶

For any planar graph $\mathcal{G}$, $\chi(\mathcal{G}) \leq 4$.

Because of the Four-Color Theorem, a cartographer with at most four colors at her disposal is certain to be able to color the countries of any map with which she is confronted (so that countries with a common border have different colors) since an n-coloring of the associated (planar) graph with $n \leq 4$ translates into a coloring of the countries of the map.

How can $\chi(\mathcal{G})$ be determined in a specific situation? For a small graph, trial and error is likely the best method, though this approach would surely not be feasible for graphs with a large number of vertices and edges. Actually it is unknown whether or not a "good" (polynomial time) algorithm exists for determining $\chi(\mathcal{G})$; like the Traveling Salesman's Problem, this problem is NP-complete.

One general result that applies to any graph links the chromatic number to the degrees of the vertices.

14.2.3 THEOREM ▶

Let $\Delta(\mathcal{G})$ be the maximum of the degrees of the vertices of a graph $\mathcal{G}$. Then $\chi(\mathcal{G}) \leq 1 + \Delta(\mathcal{G})$.

Proof   The proof is by induction on V, the number of vertices of the graph. When $V = 1$, $\Delta(\mathcal{G}) = 0$ and $\chi(\mathcal{G}) = 1$, so the result clearly holds. Now let k be an integer, $k \geq 1$, and assume that the result holds for all graphs with $V = k$ vertices. Suppose $\mathcal{G}$ is a graph with $k + 1$ vertices. Let v be any vertex of $\mathcal{G}$ and let $\mathcal{G}_0 = \mathcal{G} \setminus \{v\}$ be the subgraph with v (and all edges incident with it) deleted. Note that $\Delta(\mathcal{G}_0) \leq \Delta(\mathcal{G})$. Now $\mathcal{G}_0$ can be colored with $\chi(\mathcal{G}_0)$ colors. Since $\mathcal{G}_0$ has k vertices, we can use the induction hypothesis to conclude that $\chi(\mathcal{G}_0) \leq 1 + \Delta(\mathcal{G}_0)$. Thus, $\chi(\mathcal{G}_0) \leq 1 + \Delta(\mathcal{G})$, so $\mathcal{G}_0$ can be colored with at most $1 + \Delta(\mathcal{G})$ colors. Since there are at most $\Delta(\mathcal{G})$ vertices adjacent to v, one of the available $1 + \Delta(\mathcal{G})$ colors remains for v. Thus, $\mathcal{G}$ can be colored with at most $1 + \Delta(\mathcal{G})$ colors. ∎

The result of Theorem 14.2.3 is best possible as we see by considering the complete graph on n vertices. Since $\chi(\mathcal{K}_n) = n$ (Pause 7) and $\Delta(\mathcal{K}_n) = n - 1$, we have $\chi(\mathcal{K}_n) = 1 + \Delta(\mathcal{K}_n)$.

If $\mathcal{G}$ is an n-cycle, then $\Delta(\mathcal{G}) = 2$ and it is not hard to see that $\chi(\mathcal{G}) = 2$ or 3 according as n is even or odd. Thus, if $\mathcal{G}$ is a cycle with an odd number of vertices, we again have $\chi(\mathcal{G}) = 1 + \Delta(\mathcal{G})$. A theorem of R. L. Brooks[9] asserts that the only connected graphs $\mathcal{G}$ for which $\chi(\mathcal{G}) = 1 + \Delta(\mathcal{G})$ are $\mathcal{K}_n$ and cycles with an odd number of vertices. So for all other connected graphs, we have $\chi(\mathcal{G}) \le \Delta(\mathcal{G})$.

For either graph $\mathcal{G}$ in Fig 14.14, $\Delta(\mathcal{G}) = 5$, and so we can conclude from Theorem 14.2.3 that $\chi(\mathcal{G}) \le 6$ or, from the result of Brooks, that $\chi(\mathcal{G}) \le 5$. Neither observation tells us what is evident from the figure, that the actual chromatic numbers are 3 and 4. Nevertheless, Theorem 14.2.3 and its improvement due to Brooks do give us some information about the chromatic number of a graph and are, therefore, of interest since there is, in general, no easy way to find a chromatic number. Given a graph $\mathcal{G}$, we must simply apply ad hoc methods and hope to find $\chi(\mathcal{G})$ by trial and error.

While we cannot prove the Four-Color Theorem, we can prove the Five-Color Theorem without too much difficulty.

14.2.4 THEOREM ▶ *(Kempe, Heawood)* If $\mathcal{G}$ is a planar graph, then $\chi(\mathcal{G}) \le 5$.

Proof We must prove that any planar graph with V vertices has a 5-coloring. Again we use induction on V and note that if $V = 1$ the result is clear. Let $k \ge 1$ be an integer and suppose that any planar graph with k vertices has a 5-coloring. Let $\mathcal{G}$ be a planar graph with $k + 1$ vertices and assume that $\mathcal{G}$ has been drawn as a plane graph with straight edges. We describe how to obtain a 5-coloring of $\mathcal{G}$.

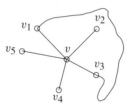

Figure 14.15

First, by Corollary 14.1.6, $\mathcal{G}$ contains a vertex v of degree at most 5. Let $\mathcal{G}_0 = \mathcal{G} \setminus \{v\}$ be the subgraph obtained by deleting v (and all edges with which it is incident). By the induction hypothesis, $\mathcal{G}_0$ has a 5-coloring. For convenience, label the five colors 1, 2, 3, 4, and 5. If one of these colors was not used to color the vertices adjacent to v, then it can be used for v and $\mathcal{G}$ has been 5-colored. Thus, we assume that v has degree 5 and that each of the colors 1 through 5 appears on the vertices adjacent to v. In clockwise order, label these vertices $v_1, v_2, \dots, v_5$ and assume that v_i is colored with color i. (See Fig 14.15.) We show how to recolor certain vertices of $\mathcal{G}_0$ so that a color becomes available for v. There are two possibilities.

Case 1: There is no path in $\mathcal{G}_0$ from v_1 to v_3 through vertices all of which are colored 1 or 3.

[9]R. L. Brooks, "On Coloring the Nodes of a Network," *Proceedings of the Cambridge Philosophical Society* **37** (1941), 194–197.

In this situation, let $\mathcal{H}$ be the subgraph of $\mathcal{G}$ consisting of the vertices and edges of all paths through vertices colored 1 or 3 which start at v_1. By assumption, v_3 is not in $\mathcal{H}$. Also, any vertex which is not in $\mathcal{H}$ but which is adjacent to a vertex of $\mathcal{H}$ is colored neither 1 nor 3. Therefore, interchanging colors 1 and 3 throughout $\mathcal{H}$ produces another 5-coloring of $\mathcal{G}_0$. In this new 5-coloring both v_1 and v_3 acquire color 3, so we are now free to give color 1 to v, thus obtaining a 5-coloring of $\mathcal{G}$.

Case 2: There is a path $\mathcal{P}$ in $\mathcal{G}_0$ from v_1 to v_3 through vertices all of which are colored 1 or 3.

In this case, the path $\mathcal{P}$, followed by v and v_1, gives a circuit in $\mathcal{G}$ which does not enclose both v_2 and v_4. Thus, any path from v_2 to v_4 must cross $\mathcal{P}$ and, since $\mathcal{G}$ is a plane graph, such a crossing can occur only at a vertex of $\mathcal{P}$. It follows that there is no path in $\mathcal{G}_0$ from v_2 to v_4 which uses just colors 2 and 4. Now we are in the situation described in Case 1, where we have already shown that a 5-coloring for $\mathcal{G}$ exists. ∎

Readers may notice that vertex v_5 appears to play no role in this proof. Can our proof be adapted to prove the Four-Color Theorem? Recall that Alfred Kempe thought he had such a proof. See Exercise 16.

We conclude this section with an application of the idea of chromatic number to a problem familiar to most university registrars, the scheduling of exams. At a university of even moderate size, it is not unusual for 500 or 600 exams to be scheduled within a reasonably short period of time. The principal problem is always to minimize "conflicts," that is, to try to avoid situations where a student finds the exams in two of his or her courses scheduled for the same time period.

PROBLEM 6. (Examination Scheduling) Suppose that in one particular semester, there are students taking each of the following combinations of courses.

- Mathematics, English, Biology, Chemistry
- Mathematics, English, Computer Science, Geography
- Biology, Psychology, Geography, Spanish
- Biology, Computer Science, History, French
- English, Psychology, History, Computer Science
- Psychology, Chemistry, Computer Science, French
- Psychology, Geography, History, Spanish

What is the minimum number of examination periods required for exams in the ten courses specified so that students taking any of the given combinations of courses have no conflicts? Find a possible schedule which uses this minimum number of periods.

Solution. In order to picture the situation, we draw a graph (shown on the left of Fig 14.16) with ten vertices labeled $M, E, B, \ldots$ corresponding to Mathematics, English, Biology, and so on, and join two vertices with an edge if exams in the corresponding subjects must not be scheduled together.

The minimal number of examination periods is evidently the chromatic number of this graph. What is this? Since the graph contains $\mathcal{K}_5$ (with vertices

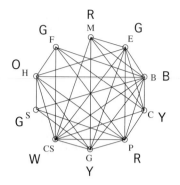

Figure 14.16

Period 1	Mathematics, Psychology
Period 2	English, Spanish, French
Period 3	Biology
Period 4	Chemistry, Geography
Period 5	Computer Science
Period 6	History

M, E, B, G, CS), at least five different colors are needed. (The exams in the subjects which these vertices represent must be scheduled at different times.) Five colors are not enough, however, since P and H are adjacent to each other and to each of E, B, G, and CS. The chromatic number of the graph is, in fact, 6. In Fig 14.16, we show a 6-coloring and the corresponding exam schedule. ∎

Answers to Pauses

5. If $\mathcal{G}$ has an edge, its end vertices must be colored differently, so $\chi(\mathcal{G}) \geq 2$. Thus, $\chi(\mathcal{G}) = 1$ if and only if $\mathcal{G}$ has no edges.

6. A way to 4-color the associated graph was given in the text. From this, we deduce that $\chi(\mathcal{G}) \leq 4$. To see that $\chi(\mathcal{G}) = 4$, we investigate the consequences of using fewer than four colors. Vertices $1, 2, 3$ form a triangle, so three different colors are needed for these. Suppose we use red, blue, and green, respectively, as before. To avoid a fourth color, vertex 4 has to be colored red and vertex 5 green. Thus, vertex 6 has to be blue. Since vertex 9 is adjacent to vertices 1, 5, and 6 of colors red, green, and blue, respectively, vertex 9 requires a fourth color.

7. It takes n colors to color $\mathcal{K}_n$ because any two vertices of $\mathcal{K}_n$ are adjacent: $\chi(\mathcal{K}_n) = n$. On the other hand, $\chi(\mathcal{K}_{m,n}) = 2$; coloring the vertices of each bipartition set the same color produces a 2-coloring of $\mathcal{K}_{m,n}$.

EXERCISES

The symbol [BB] means that an answer can be found in the Back of the Book.

   Assume that all graphs in these Exercises are connected.

1. Name and describe two major mathematical accomplishments of the twentieth century.

2. (a) [BB] Draw a graph corresponding to the map shown at the right and find a coloring which requires the least number of colors. What is the chromatic number of the graph?

   (b) [BB] Answer true or false and explain: The Four-Color Theorem says that the chromatic number of a planar graph is 4.

3. (a) [BB] When we discussed the coloring of maps at the beginning of this section, we assumed implicitly that "bordering" countries were countries which had some positive length of border in common. Suppose we deem countries to border if they merely have a point in common. Will four colors still suffice to color a map?

(b) [BB] We also assumed that a country should consist of a single region. If we drop this restriction, will four colors still suffice to color a map?

4. Compute $\chi(\mathcal{G})$ for each of the graphs shown in Fig 14.17. In each case, explain your answer and exhibit a $\chi(\mathcal{G})$-coloring.

5. Answer Exercise 4 for each of the graphs in Exercise 5, Section 14.1.

6. (a) [BB] Find the chromatic number of each of the graphs in Exercise 4 of Section 14.1.

(b) [BB] State the converse of the Four-Color Theorem. Is it true?

7. (a) [BB] Is a tree planar? Explain.

(b) Use (a) to prove that a tree with n vertices has $n - 1$ edges.

(c) Suppose $\mathcal{T}$ is a tree with n vertices. What is $\chi(\mathcal{T})$ and why?

8. Consider the graph shown on the left.

(a) Is it planar? Either draw an isomorphic plane graph or explain why the graph is not planar.

(b) Find its chromatic number and explain why this piece of information is consistent with the Four-Color Theorem.

9. [BB] Consider the graph shown on the right.

(a) Find a subgraph homeomorphic to $\mathcal{K}_5$.

(b) Find a subgraph homeomorphic to $\mathcal{K}_{3,3}$.

(c) Is the graph planar? Explain.

(d) What's the chromatic number of the graph?

(e) Use this graph to comment on the converse of the Four-Color Theorem.

10. (a) [BB] What is $\chi(\mathcal{K}_{14})$? What is $\chi(\mathcal{K}_{5,14})$? Why?

(b) Let $\mathcal{G}_1$ and $\mathcal{G}_2$ be cycles with 38 and 107 edges, respectively. What is $\chi(\mathcal{G}_1)$? What is $\chi(\mathcal{G}_2)$? Explain.

11. Answer true or false and explain:

(a) [BB] If $\chi(\mathcal{G}) = 3$, then $\mathcal{G}$ contains a triangle.

(b) If $\chi(\mathcal{G}) = 4$, then $\mathcal{G}$ contains $\mathcal{K}_4$.

12. Let $n \geq 4$ be a natural number. Let $\mathcal{G}$ be the graph which consists of the union of $\mathcal{K}_{n-3}$ and a 5-cycle $\mathcal{C}$ together with all possible edges between the vertices of these graphs. Show that $\chi(\mathcal{G}) = n$, yet $\mathcal{G}$ does not have $\mathcal{K}_n$ as a subgraph.

13. Answer true or false and explain. If $\mathcal{G}$ is homeomorphic to $\mathcal{H}$, then $\chi(\mathcal{G}) = \chi(\mathcal{H})$.

(a)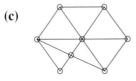

(b) [BB]

(c)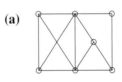

(d)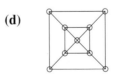

(e) [BB]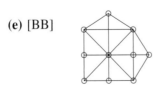

(f)

Figure 14.17   Graphs for Exercise 4.

14. Answer true or false and explain:

(a) [BB] A Hamiltonian graph with chromatic number two must be planar.

(b) A planar graph with chromatic number two must be Hamiltonian.

(c) A graph which is both Hamiltonian and Eulerian and has chromatic number two must be planar.

(d) [BB] A graph which is both Hamiltonian and Eulerian must have chromatic number less than six.

15. [BB] The graph arises in a certain problem of exam scheduling such as that described in Problem 6. Find the minimum number of exam periods required so that conflicts are avoided.

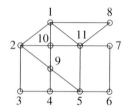

16. Can our proof of the Five Color Theorem be adapted to prove the Four-Color Theorem? Go through the proof presented in this section assuming that only four colors are available and find any possible flaw.

17. In addition to the combinations of courses described in Problem 6, suppose that there are also students taking all of Geography, Computer Science, Spanish, and French. Does this force a change in the exam schedule? If it does, find a new examination schedule which avoids conflicts and uses the fewest number of periods.

18. [BB] There are ten students who, in the coming semester, will be taking the courses shown in the following table. How many time periods must be allowed in order for these students to take the courses they want without conflicts?

Arnold	Physics, Mathematics, English
Bill	Physics, Earth Science, Economics
Carol	Earth Science, Business
Calvin	Statistics, Economics
Eleanor	Mathematics, Business
Frederick	Physics, Earth Science
George	Business, Statistics
Huber	Mathematics, Earth Science
Ingrid	Physics, Water Skiing, Statistics
Jacquie	Physics, Economics, Water Skiing

19. [BB] The following semester, all the students in the previous problem, except Calvin, plan to take second courses in the same subjects. Calvin decides not to take further courses in statistics. How many time periods will then be required?

20. Hubert Noseworthy loves snakes and keeps a dozen different varieties in his apartment (contrary to regulations). Since some varieties of snake attack other varieties, as shown in the following tables, Hubert needs several boxes in which to keep his snakes in order to separate antagonists. What is the minimum number of boxes he needs?

Variety	Attacks Variety	Variety	Attacks Variety
1	3, 4, 5, 8, 10, 12	2	1, 3, 6, 7, 10, 11
3	4, 9, 12	4	5, 8, 9
5	6, 7, 10	6	9, 12
7	10, 12	8	7
9	8, 11	10	11
11	12		

21. The local day care center has a problem because certain children do not get along with certain others. The table shows which of 15 children don't get along with whom. Finally, it is decided that children who do not get along with each other will have to be put into separate rooms. Find the minimum number of rooms required.

Child	doesn't get along with	Child	doesn't get along with
1	2, 6, 9, 10, 11, 13	2	1, 8, 10, 12, 13, 14, 15
3	4, 5, 7, 8, 12	4	3, 5, 6, 8, 11, 14
5	3, 4, 7, 8, 12, 13	6	1, 4, 7, 12, 15
7	3, 5, 6, 8, 11, 14	8	2, 3, 4, 5, 7, 9, 14
9	1, 8, 10, 12, 15	10	1, 2, 9, 11, 15
11	1, 4, 7, 10, 12	12	2, 3, 5, 6, 9, 11, 15
13	1, 2, 5, 14	14	2, 4, 7, 8, 13
15	2, 6, 9, 10, 12		

22. Television channels are to be assigned to stations based in nine cities $A, B, \ldots, I$. Broadcasting regulations require that cities within 150 km of each other be assigned different channels. What is the least number of channels required if the distances between the cities are as given in the table.

	A	B	C	D	E	F	G	H	I
B	85								
C	137	165							
D	123	39	205						
E	164	132	117	171					
F	105	75	235	92	201				
G	134	191	252	223	298	177			
H	114	77	113	117	54	147	247		
I	132	174	22	213	138	237	245	120	

23. Continuing the previous problem, city J is to acquire a television station too. Can it be assigned one of the existing channels or must it have a new one? Its distances from cities $A, \ldots, I$ are as follows.

	A	B	C	D	E	F	G	H	I
J	78	149	101	189	171	183	160	143	94

24. (a) [BB] Let $\mathcal{G}$ be a connected graph with n vertices and n edges. Prove that $\chi(\mathcal{G}) \leq 3$.

(b) Show that (a) remains true if $\mathcal{G}$ has $n + 1$ edges.

(c) Does (a) remain true if $\mathcal{G}$ has $n + 2$ edges? Explain.

connecting vertices if corresponding countries shared a border. By introducing one additional vertex corresponding to the exterior region and additional edges as before, we obtain a graph known as the *dual graph* of the map.

25. (a) [BB] Draw the dual graph of the cube (considered as a map) (Fig 14.5) as a plane graph with straight edges. Identify this dual graph. [*Hint*: It's another of the Platonic solids.]

(b) Repeat (a) for the dodecahedron (Fig 10.13).

26. [BB] Is it possible for a plane graph, considered as a map, to be its own dual?

14.2.5 THE DUAL OF A PLANAR GRAPH ▼

In this section, we showed how to associate a planar graph with a map, by introducing a vertex for each country and

14.3 CIRCUIT TESTING AND FACILITIES DESIGN

In this section, we present two interesting applications of the material in Sections 14.1 and 14.2. The first shows how knowledge of graph colorings can be used to assist in testing printed circuit boards for the existence of possible short circuits. In the second, we apply what we have learned about graph colorings, planarity, and Hamiltonian cycles to the design of floor plans for facilities such as hospitals or shopping malls in order to meet, so far as possible, various requirements of juxtaposition.

Circuit Testing

A printed circuit board can be represented by a finite rectangular grid composed of evenly spaced rows of evenly spaced grid points called *nodes*, which are connected by horizontal and vertical line segments called *grid segments*. On this board, certain vertices are connected via disjoint paths (called *nets*) along grid segments. For example, Fig 14.18 shows a grid with 49 nodes and seven nets labeled $N_1, N_2, \ldots, N_7$.

In general, the grid segments in a particular net pattern correspond to conductor paths in the circuit board. The problem of interest is to determine whether or not any extra conductor paths (short circuits) between nets have been introduced into the board during the manufacturing process. A short circuit need not begin or end at a node.

One definite way to determine whether or not there are short circuits is to take each pair of distinct nets, in turn, and to apply an electrical signal to one member of the pair. If that signal appears in the other net, then there must be a short. This procedure can be extremely time-consuming; for instance, in Fig 14.18, $\binom{7}{2} = 21$

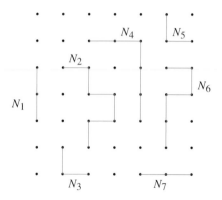

Figure 14.18   A grid of 49 nodes and 7 nets, comprising altogether 24 grid segments.

comparisons would be required. Many of these comparisons may be unnecessary, however. As an example, the left-hand grid in Fig 14.19 shows two nets, A and C, that cannot possibly have a short between them unless there is also a short between each of them and B. Hence, it suffices to test B against each of A and C.

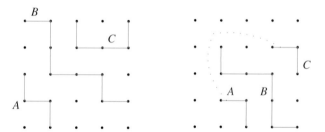

Figure 14.19

In the right-hand grid of Fig 14.19, it is theoretically possible to have a short between A and C which does not touch B, as shown by the dotted line, but this seems unlikely. We shall assume henceforth that any short circuit between nets must be a single straight horizontal or vertical line (of arbitrary length) not touching any other net (but not necessarily through nodes).

An efficient way of testing either grid in Fig 14.19 would be to combine nets A and C into a single set which could be tested against B. In general, a good procedure is to partition the set of nets into subsets such that if two nets N_i, N_j are in the same subset, then N_i and N_j cannot have a short between them. Then we need only test these subsets against each other to determine whether a short circuit exists. For example, in Fig 14.18, the subsets could be $\{N_1, N_3, N_6\}$, $\{N_2, N_5, N_7\}$, $\{N_4\}$, and three tests would suffice (instead of 21). Note that N_1 and N_3 cannot have a short between them, nor can N_3 and N_6, nor N_1 and N_6. The other subsets are obtained by similar reasoning.

Pause 8   Find another partition of the net pattern in Fig 14.18 into three appropriate subsets.

Pause 8 illustrates an important point: The sorts of partitions we have been discussing are not necessarily unique. From an efficiency point of view, however, we don't care about this. What matters is that we find a partition into as few subsets as possible. Here is where graph theory helps.

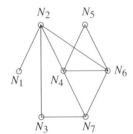

Figure 14.20

Given a grid with a net pattern, we construct the graph which has a vertex corresponding to each net and where two vertices are joined by an edge if and only if there could be a short circuit between the corresponding nets. Figure 14.20 shows the graph corresponding to Fig 14.18. Note that N_1 is joined to N_2 but to no other vertex because the only possible short involving N_1 must be between N_1 and N_2. Other edges are obtained by similar reasoning. Any graph which can be obtained by the procedure described will be called a *line-of-sight graph*.

Pause 9   How would the line-of-sight graph change if net N_2 were not present in the net shown in Fig 14.18? Draw the revised graph.

Since two vertices in a line-of-sight graph are joined by an edge if and only if the corresponding nets could have a short between them, it is easy to see that any coloring of such a graph yields subsets of the type desired for the corresponding nets. For example, a 3-coloring of the graph in Fig 14.20 could be obtained by coloring N_1, N_3, and N_6 blue; N_2, N_5, and N_7 red; and N_4 white. Placing like-colored nets in the same subset gives the partition described earlier.

Another possibility would be to color N_1 and N_6 blue; N_3 and N_4 red; and N_2, N_5, and N_7 white. This yields another suitable partition. Since the graph in Fig 14.20 contains a triangle, we know that at least three colors are required and, hence, we cannot reduce further the number of partitions. In general, the minimum number of subsets obtainable in a suitable partition of nets is the chromatic number of the associated line-of-sight graph.

Pause 10   Find the chromatic number of the graph you obtained in Pause 9, and determine a corresponding partition of the nets.

Is our approach to finding partitions of nets really useful? After all, chromatic numbers are notoriously hard to determine. Fortunately, only certain graphs are obtainable as line-of-sight graphs. Any planar graph is a line-of-sight graph but $\mathcal{K}_9$, for example, is not. It is known that no matter how large the grid is or how many nets are involved, the corresponding graph $\mathcal{G}$ has $\chi(\mathcal{G}) \leq 12$. It follows that the nets can always be partitioned into at most 12 subsets and, hence, we never need to carry out more than $\binom{12}{2} = 66$ tests. Is 12 best possible? Perhaps nets can always be partitioned into at most 11 subsets, or at most 10. The answer is not known, but it is not less than 8 since there exists a line-of-sight graph $\mathcal{G}$ for which $\chi(\mathcal{G}) = 8$. (See Exercise 9.)

The preceding results, as well as many others, appear in a fundamental and very interesting paper by Garey, Johnson, and So.[10] There are still a number of intriguing open problems here—for example, is it possible to characterize precisely those graphs which are line-of-sight graphs? You are encouraged to investigate this area further, first by examining the aforementioned paper.

Facilities Design

As mentioned earlier, we are interested here in the design and layout of physical facilities such as hospital floors. In such problems, there are a number of design areas (rooms perhaps) and it is often desirable (or critical) that certain of these areas be adjacent to each other. The planner of such a project will be presented with a list of relationships from which it must be determined whether or not it is possible to construct such a floor plan and if so how. If it is not possible, the question becomes how close can one come to meeting the requirements.

If it is possible to produce a layout satisfying all the given relationships of one area to another, then we say that the given relationships are *feasible*. If a set of relationships is not feasible, we would like to determine the smallest number of relationships whose omission produces a feasible set.

The first thing to do is to draw a graph (called the *relationship graph*) whose vertices represent design areas and in which an edge between vertices indicates that corresponding areas are to be adjacent. Thinking of a given layout of the design areas as a map whose associated graph (see Fig 14.12) is the relationship graph, we see that a set of relationships will be feasible if and only if the relationship graph is planar.

Theoretically then, Kuratowski's Theorem (14.1.8) could be used to decide whether or not a given set of relationships is feasible. There are, however, two difficulties with this approach. First, Kuratowski's Theorem is difficult to apply, and most architects would prefer an easier algorithm. Second, and much more significant, should the relationship graph not be planar, Kuratowski's Theorem provides no indication of which relationships should be deleted in order to achieve planarity.

What we are looking for, therefore, is a simple test for planarity which will readily identify those edges which are causing a particular graph not to be planar. While this is too much to expect, progress has been made in the case of Hamiltonian graphs.[11] The procedure we outline determines, in a rather nice way, whether or not a Hamiltonian graph is planar. While it does not rigorously identify those edges which should be deleted in the case of nonplanarity, it does model the problem in such a way that it is generally not difficult to locate such edges. We demonstrate this procedure by working through a specific example.

EXAMPLE 7 Consider the relationship graph $\mathcal{G}$ drawn in Fig 14.21. This graph has a Hamiltonian cycle, 1243561. We start by drawing a graph $\mathcal{G}_1$ isomorphic to $\mathcal{G}$ with the

---

[10]Michael R. Garey, David Stifler Johnson, and Hing C. So, "An Application of Graph Coloring to Printed Circuit Testing," *IEEE Transactions on Circuits and Systems*, Vol. CAS-23, (1976), no. 10, 591–599.

[11]G. Demourcron, V. Malgrance, and R. Pertuiset, "Graphes Planaires: Reconnaissance et Construction de Representations Planaires Topologiques," *Recherche Operationelle* **30** (1964), 33.

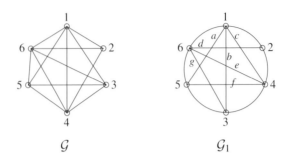

Figure 14.21 $\mathcal{G}$ $\mathcal{G}_1$

vertices of the Hamiltonian cycle in $\mathcal{G}$ appearing in order around the circumference of a circle. We then label arbitrarily those edges of $\mathcal{G}_1$ which are not on the cycle. Next, we draw a new graph $\mathcal{H}$ whose vertices correspond to the edges of $\mathcal{G}_1$ just labeled. Two vertices of $\mathcal{H}$ are joined by an edge if the corresponding edges cross inside the circle. The graph we obtain in this way is shown on the left in Fig 14.22.

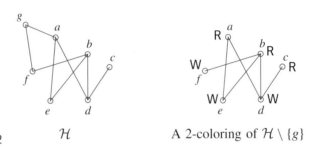

Figure 14.22 $\mathcal{H}$ A 2-coloring of $\mathcal{H} \setminus \{g\}$

Now examine this new graph $\mathcal{H}$. We claim that the original graph $\mathcal{G}$ is planar if and only if $\chi(\mathcal{H}) \leq 2$, and here's why. If $\chi(\mathcal{H}) \leq 2$, then there is a coloring of the vertices of $\mathcal{H}$ requiring at most two colors. Therefore, the interior edges of $\mathcal{G}_1$ could be colored with at most two colors in such a way that edges crossing inside the circle (if any) have different colors. By drawing all the edges of one color outside the circle and leaving alone all edges of the other color, we would then obtain a representation of $\mathcal{G}_1$ with no crossovers of edges; in other words, $\mathcal{G}$ would be drawn as a plane graph. Conversely, if $\mathcal{G}$ is planar and drawn as a plane graph, then coloring all edges inside the cycle with one color and all edges outside with another produces a coloring of the associated graph $\mathcal{H}$ which uses at most two colors.

We return to our example and note from Fig 14.22 that $\chi(\mathcal{H}) \neq 2$ since $\mathcal{H}$ contains a 5-cycle, $adbfga$; hence, the original graph $\mathcal{G}$ is not planar. So we must now look for the smallest number of vertices which, when removed from $\mathcal{H}$, leave a graph whose chromatic number is 2. In our example, if we remove any of a, b, f, or g, we will achieve the desired result.

Suppose we remove g. We picture the resulting graph and a 2-coloring on the right in Fig 14.22. This, in turn, provides a coloring of the interior edges of $\mathcal{G}_1 \setminus \{g\}$. In Fig 14.23(a), we draw $\mathcal{G}_1 \setminus \{g\}$ as a plane graph with all edges of one of the two colors drawn outside the circle. In order to draw a layout with $\mathcal{G}_1 \setminus \{g\}$ as its relationship graph, it is helpful first to draw this plane graph with straight edges, as in Fig 14.23(b). In Fig 14.23(c), we show the corresponding layout. If this were, for example, the floor plan of a hospital, Fig 14.23(c) shows a way that the rooms can be laid out so as to satisfy the aforementioned relationships. Strange-looking rooms, perhaps, but feasible! ▲

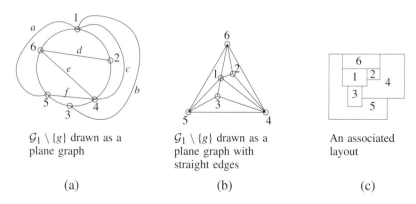

$\mathcal{G}_1 \setminus \{g\}$ drawn as a plane graph

$\mathcal{G}_1 \setminus \{g\}$ drawn as a plane graph with straight edges

An associated layout

(a)   (b)   (c)

Figure 14.23   A plane graph, then drawn with straight edges, and an associated layout of design areas.

Pause 11   Carry out the steps in the previous paragraph, assuming f is removed instead of g.

14.3.1 CLOSING REMARKS ▶

1. We proved in Section 14.1 that connected planar graphs with at least three vertices must satisfy $E \leq 3V - 6$ (Theorem 14.1.4). Thus, $3V - 6$ is an upper bound for the number of relationships which are theoretically possible in a facilities design having V rooms. Despite what the planner might want, it is impossible to do better!

2. In many practical problems, some relationships are more important than others. Certain ones, in fact, might be crucial (in a hospital, for example). Relative importance of relationships can be incorporated into the model by assigning weights to the edges of the relationship graph and then attempting to achieve planarity by removing the least important edges. Kruskal's algorithm, adapted to choose a maximal weight spanning tree (instead of a minimal one), has been applied to this case with some success.

3. The authors first read about this fascinating area in Roberts's book *Applied Combinatorics*.[12] Certainly there is great potential here for further investigation. We refer the interested reader also to Chapter 8 of Chachra, Ghare, and Moore[13] for additional information.

Answers to Pauses

8. $\{N_1, N_6\}$, $\{N_2, N_5, N_7\}$, $\{N_3, N_4\}$.

---

[12]Fred S. Roberts, *Applied Combinatorics*, Prentice Hall, New Jersey, 1984.

[13]V. Chachra, P. M. Ghare, and J. M. Moore, *Applications of Graph Theory Algorithms*, Elsevier, North-Holland, New York, 1979.

9.

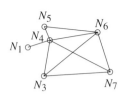

10. The chromatic number of the preceding graph is 4 since it contains $\mathcal{K}_4$ as a subgraph. One partition would be $\{N_1, N_3\}, \{N_4\}, \{N_6\}, \{N_5, N_7\}$.

11. A 2-coloring of $\mathcal{H} \setminus \{f\}$ is shown at the right. On the left in the graphs which follow is $\mathcal{G}_1 \setminus \{f\}$, drawn as a plane graph with all white edges drawn outside the circle. In the middle, it's redrawn with straight edges, and on the right there is a layout whose relationship graph is $\mathcal{G}_1 \setminus \{f\}$.

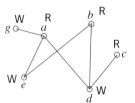

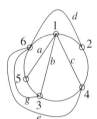

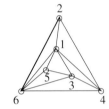

EXERCISES

The symbol [BB] means that an answer can be found in the Back of the Book.

1. [BB] The following questions refer to the net pattern drawn below.

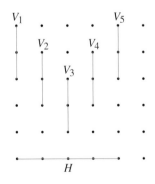

(a) Find the number of nodes, grid segments, and nets.

(b) Give an example of two nets which cannot have a vertical or horizontal short circuit between them unless there is a similar short circuit between one of them and another net.

(c) Draw the associated line-of-sight graph $\mathcal{G}$.

(d) Determine $\chi(\mathcal{G})$ and find a corresponding partition of the nets.

2. Repeat Exercise 1 if the nodes of the far right-hand column are all connected by a vertical line of grid segments, thus forming another net, V_6.

3. [BB] True or false? A line-of-sight graph is always connected.

4. Let $\mathcal{N}$ be the set of nets in a given net pattern. Define a relation $\sim$ on $\mathcal{N}$ by $A \sim B$ if there could be a short circuit between A and B. Determine whether or not $\sim$ is

(a) [BB] reflexive,

(b) symmetric,

(c) antisymmetric,

(d) transitive.

5. Refer to the net pattern shown below.

(a) [BB] Draw the associated line-of-sight graph $\mathcal{G}$.

(b) Determine $\chi(\mathcal{G})$ and find a corresponding partition of the nets.

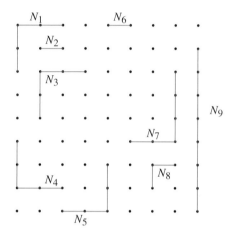

6. Repeat Exercise 5 if net N_3 is removed.

7. [BB] True or false? A line-of-sight graph is always planar.

8. (a) [BB] Assume that the only short circuits in a printed circuit board are **horizontal** straight lines; that is, vertical shorts are not possible. Prove that the associated line-of-sight graph $\mathcal{G}$ is then planar and conclude that $\chi(\mathcal{G}) \leq 4$.

(b) [BB] Let $\mathcal{V}$ and $\mathcal{E}$ be the vertices and edges, respectively, of a general line-of-sight graph $\mathcal{G}$. Show that $\mathcal{E}$ can be partitioned into subsets $\mathcal{E}_1$, $\mathcal{E}_2$ such that the graphs $\mathcal{G}(\mathcal{V}, \mathcal{E}_1)$, $\mathcal{G}(\mathcal{V}, \mathcal{E}_2)$ are both planar.

(c) Using (a) and (b), show that $\chi(\mathcal{G}) \leq 16$ for any line-of-sight graph $\mathcal{G}$.

(d) Using (b), show that $\mathcal{K}_{11}$ is not a line-of-sight graph.

9. [BB] The net pattern shown has a line-of-sight graph $\mathcal{G}$ with $\chi(\mathcal{G}) = 8$. Why? (This example is due to M. R. Garey, D. S. Johnson, and H. C. So, cited in footnote **10**.)

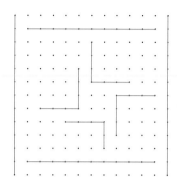

10. Find a best possible feasible relationship graph and draw the corresponding floor plan for $\mathcal{K}_5$ [BB], $\mathcal{K}_{3,3}$, and $\mathcal{K}_6$, each of these considered as a relationship graph.

11. For each of the following relationship graphs, find a best possible feasible relationship graph and draw the corresponding floor plan.

(a)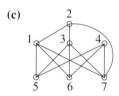

(b)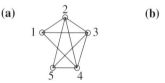

(c)

12. A contractor is building a single-story house for a newly married couple. The house is to consist of a living room, dining room, family room, kitchen, two bedrooms, bathroom, and hall. The couple insists that all rooms border on the hall. The kitchen is to share a wall with the dining room, family room, and bathroom. Both bedrooms should share a wall with the bathroom and the family room. The living room is to border the family room and the dining room. Is it possible for the contractor to meet the couple's demands? If not, how close could the contractor come to meeting all requirements? Justify your answer.

13. Jack is commissioned to build a nine-hole golf course which meets the following specifications.

- The first hole must return to the clubhouse and hence have a common border with the ninth hole.

- To satisfy thirsty patrons, the fifth hole also returns to the clubhouse; it must have a common border with the first hole, but not necessarily with the ninth.

- All even-numbered holes share a common water hazard and have some common border with each other.
- Each hole (after the first) has a common border with the one preceding it.

(a) [BB] Show that it is impossible for Jack to build a golf course meeting all these requirements.

(b) State three different ways in which Jack could satisfy all but one of the requirements.

14. [BB] Apply Kuratowski's Theorem, Theorem 14.1.8, to the graph G in Fig 14.21 to show that G is not planar.

15. [BB] Apply Brooks's Theorem (p. 425) to find the chromatic number of the graph $\mathcal{H}$ in Fig 14.22.

REVIEW EXERCISES FOR CHAPTER 14

1. (a) Show that the graph is planar by drawing an isomorphic plane graph with straight edges.
   (b) Label the regions defined by your plane graph and list the edges which form the boundary of each region.

   (c) Verify that $V - E + R = 2$, $N \le 2E$ and $E \le 3V - 6$. (Recall that N is the sum of the numbers of edges on the boundaries of all regions.)

2. Determine whether or not each of the graphs is planar. In each case, either draw a plane graph with straight edges isomorphic to the one presented, or exhibit a subgraph homeomorphic to $\mathcal{K}_{3,3}$ or $\mathcal{K}_5$.

(a)

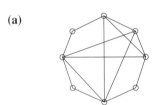

(b)

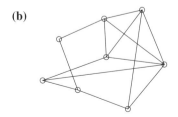

3. If G is a connected plane graph with $V \ge 3$ vertices and R regions, each with at least five edges on its boundary, prove that $3R \le 2V - 4$. (You may use the result of Exercise 6 of Section 14.1, but first be sure you can prove it!)

4. (a) A connected planar graph G has 30 vertices. Prove that G has at most 84 edges.
   (b) A connected planar graph G has 30 vertices, 15 of which have degree 1. Prove that G has at most 54 edges.

5. Show that $\mathcal{K}_{2,3}$ is homeomorphic to a subgraph of $\mathcal{K}_4$.

6. True or false? If G is a tree and $\mathcal{H}$ is homeomorphic to G, then $\mathcal{H}$ is a tree. Explain your answer.

7. Compute $\chi(G)$, where G is the graph in Exercise 1. Explain your answer and exhibit a $\chi(G)$-coloring.

8. Compute $\chi(G)$ for the graphs of Exercise 2. Do your results say anything about the converse of the Four-Color Theorem?

9. Answer true or false and explain your answers.
   (a) A planar graph with chromatic number 3 must be Hamiltonian.
   (b) If G is a planar graph which contains $\mathcal{K}_4$ as a subgraph, then $\chi(G) = 4$.

10. Let G be a connected graph with at least two vertices. Prove that $\chi(G) = 2$ if and only if G is bipartite with at least one edge.

11. (a) Give an example showing that the following theorem, which resembles Theorem 14.2.3, is false:

    Let $\Sigma(G)$ be the minimum of the degrees of the vertices of a graph G. Then $\chi(G) \ge 1 + \Sigma(G)$.

    (b) Is the "theorem" stated in 11(a) true if we restrict our attention to trees with at least two vertices?

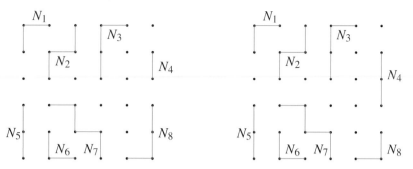

Figure 14.24   Net patterns for Exercises 13 and 14.

12. A town jail contains four holding cells. On a particularly busy night, twelve people are arrested. Certain prisoners do not get along with certain others and must be put into separate cells, as shown in the following tables. If possible, find a way of putting the prisoners into the four cells in such a way as to avoid possible conflicts during the night.

Prisoner	doesn't get along with
1	3, 5, 8, 9, 10, 11
2	3, 4, 6, 7, 9, 11
3	1, 2, 6, 8, 11, 12
4	2, 5, 6, 8, 10, 12
5	1, 4, 7, 9, 10
6	2, 3, 4, 7, 9, 11, 12
7	2, 5, 6, 8, 10
8	1, 3, 4, 7, 12
9	1, 2, 5, 6, 11
10	1, 4, 5, 7, 12
11	1, 2, 3, 6, 9
12	3, 4, 6, 8, 10

13. Draw the line-of-sight graph associated with the net pattern shown on the left in Fig 14.24. Determine $\chi(\mathcal{G})$ and find a corresponding partition of the nets.

14. Repeat Exercise 13 for the net pattern shown on the right of Fig 14.24.

15. Give an example of a net pattern whose line-of-sight graph is Eulerian but not Hamiltonian.

16. Show that the complete bipartite graph $\mathcal{K}_{7,7}$ is **not** a line-of-sight graph. [*Hint*: Use Exercise 8(b) of Section 14.3 and Exercise 9(a) of Section 14.1.]

17. Find a best possible feasible relationship graph and draw the corresponding floor plan for each of the graphs in Exercise 2.

15

The Max Flow—Min Cut Theorem

15.1 FLOWS AND CUTS

In this chapter, we return to the concept of a directed network, introduced in Section 11.5, and prove the very important *Max Flow—Min Cut Theorem*. Recall that a directed network is a digraph in which each arc is assigned an integer weight.

Whereas in Section 11.5, the weights on arcs were viewed as times required to complete certain activities, the point of view we adopt now is rather different. Here, we think of the arcs of a digraph as pipes and the weight on an arc as the *capacity* of the pipe, the maximum amount of some commodity that can flow through it in a given unit of time—liters of oil or gas, kilowatts of electricity, numbers of people, messages, trucks, or letters.

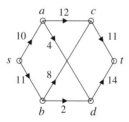

Figure 15.1   A directed network with a source s and a sink t.

Figure 15.1 shows a network typical of the sort we consider in this chapter. It has two distinguished vertices s and t called, respectively, the *source* and the *sink*. Intuitively, we think of the source as the start of the flow and expect that all arcs with source s as an end point should be directed away from s. While this is generally the situation, it is not important for the theory which follows and so we do not impose such a special condition on s. Similarly, while we think of the sink as the place toward which all the flow is directed, we do not make this assumption.

Remember that when we say that uv is an arc in a digraph, we mean that the arrow on the edge joining u and v is in the direction $u \to v$. We denote by c_{uv} the capacity of arc uv in a directed network and, in order that c_{uv} has meaning for any pair of vertices u, v, we define $c_{uv} = 0$ if uv is not an arc. We require that all c_{uv} be nonnegative integers.

In contrast to the capacity of an arc, the actual amount of flow passing through uv is denoted f_{uv}. Again, we define $f_{uv} = 0$ if uv is not an arc so that f_{uv} has a meaning for every u and v. In the physical situations we are trying to model, the flow through an arc cannot exceed its capacity; thus, we require that $0 \le f_{uv} \le c_{uv}$ for every u and v. We also assume a *conservation* law at each vertex (except the source and the sink). This says that for any vertex u (other than s or t) the flow out of u equals the flow into u; symbolically, $\sum_v f_{uv} = \sum_v f_{vu}$. We hope you agree that such a law makes good sense in any physical situation.

15.1.1   DEFINITIONS ▶ Given a directed network with vertex set $\mathcal{V}$, the capacity of arc uv denoted c_{uv}, and given two distinguished vertices s and t, called the *source* and *sink*, respectively, an (s, t)-*flow* is a set $\mathcal{F}$ of numbers $\{f_{uv}\}$ satisfying

1. $0 \le f_{uv} \le c_{uv}$ for all $u, v \in \mathcal{V}$, and
2. $\sum_{v \in \mathcal{V}} f_{uv} = \sum_{v \in \mathcal{V}} f_{vu}$ for all $u \in \mathcal{V} \setminus \{s, t\}$ (*Conservation of Flow*).

Figure 15.2 shows an (s, t)-flow for the directed network of Fig 15.1. The label on each arc shows its capacity and the flow through it, in that order. Thus, the label 12, 10 on arc ac means that this arc has capacity 12 and currently a flow of 10. Note that the flow f_{uv} in each arc never exceeds its capacity c_{uv}. Also observe that the law of conservation of flow is satisfied at each vertex. Consider vertex c, for example. The flow away from c is $\sum_v f_{cv} = f_{ct} = 11$ while the flow into c is $\sum_v f_{vc} = f_{ac} + f_{bc} = 10 + 1 = 11$.

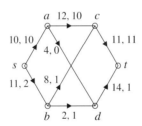

Figure 15.2   An (s, t)-flow.

Pause 1   Verify the law of conservation at vertices a, b, and d.

The flow out of the source s equals $10 + 2 = 12$ and the flow into the sink t is $11 + 1$, which equals 12 as well. Physically this equality makes sense: Since conservation of flow means that no flow is lost at any vertex, whatever flow leaves the source must eventually reach the sink. A general proof follows. We start with the observation that

$$\sum_{u,v \in \mathcal{V}} f_{uv} = \sum_{u,v \in \mathcal{V}} f_{vu}$$

and hence,

$$\sum_{u,v \in \mathcal{V}} f_{uv} - \sum_{u,v \in \mathcal{V}} f_{vu} = 0.$$

Summing first over v and then over u, we obtain

$$\sum_{u \in \mathcal{V}} \left(\sum_{v \in \mathcal{V}} f_{uv} - \sum_{v \in \mathcal{V}} f_{vu} \right) = 0.$$

Conservation of flow says that the term in the parentheses here is 0, except when $u = s$ and $u = t$. Therefore,

$$\sum_{v} f_{sv} - \sum_{v} f_{vs} + \sum_{v} f_{tv} - \sum_{v} f_{vt} = 0.$$

Hence,

(1) $$\sum_{v} f_{sv} - \sum_{v} f_{vs} = \sum_{v} f_{vt} - \sum_{v} f_{tv}.$$

15.1.2 DEFINITION ▶ The *value of a flow* $\mathcal{F} = \{f_{uv}\}$ in a directed network with source s, sink t, and vertex set $\mathcal{V}$ is the integer

$$\text{val}(\mathcal{F}) = \sum_{v \in \mathcal{V}} f_{sv} - \sum_{v \in \mathcal{V}} f_{vs}$$

or, equivalently, by equation (1),

$$\text{val}(\mathcal{F}) = \sum_{v \in \mathcal{V}} f_{vt} - \sum_{v \in \mathcal{V}} f_{tv}.$$

The value of a flow is the **net** amount of flow per unit time leaving the source, or, equivalently, the **net** amount of flow per unit time entering the sink. In the most likely situation where there is no flow into the source and no flow out of the sink, then $f_{vs} = 0 = f_{tv}$ for all vertices v, so the value of the flow is

$$\text{val}(\mathcal{F}) = \sum_{v} f_{sv} = \sum_{v} f_{vt}.$$

This was the case with the network in Fig 15.2; the flow there has value 12.

Our goal in this chapter is to find the (s,t)-flow with the largest possible value through a given directed network. Such a flow is called a *maximum flow*.

Pause 2 With reference to the directed network of Fig 15.2, find a flow whose value exceeds 12.

In addition to (s,t)-flow, the other important concept in directed networks is that of an (s,t)-*cut*.

15.1.3 DEFINITIONS ▶ Let $\mathcal{V}$ be the vertex set of a directed network with source s and sink t. An (s, t)-*cut* is a partition $\{\mathcal{S}, \mathcal{T}\}$ of $\mathcal{V}$ such that $s \in \mathcal{S}$ and $t \in \mathcal{T}$. The *capacity* of an (s, t)-cut $\{\mathcal{S}, \mathcal{T}\}$ is the sum $\mathrm{cap}(\mathcal{S}, \mathcal{T})$ of the capacities of all arcs from $\mathcal{S}$ to $\mathcal{T}$:

$$\mathrm{cap}(\mathcal{S}, \mathcal{T}) = \sum_{u \in \mathcal{S}, v \in \mathcal{T}} c_{uv}.$$

Pause 3 What does it mean to say that $\{\mathcal{S}, \mathcal{T}\}$ is a *partition* of $\mathcal{V}$?

Every path from s to t contains an arc joining a vertex in $\mathcal{S}$ to a vertex in $\mathcal{T}$. Thus, if every such arc is cut, there is no path left from s to t. The flow of goods or information from s to t has been "cut."

With reference to the network in Fig 15.1, the sets $\mathcal{S} = \{s\}$ and $\mathcal{T} = \{a, b, c, d, t\}$ form an $\{\mathcal{S}, \mathcal{T}\}$-cut. The arcs from $\mathcal{S}$ to $\mathcal{T}$ are sa and sb: Severing both of these would eliminate any flow from s to t. Another example is provided by the sets $\mathcal{S} = \{s, a, c\}$ and $\mathcal{T} = \{b, d, t\}$. This time, the arcs from $\mathcal{S}$ to $\mathcal{T}$ are sb, ad, and ct. Arc bc is not included: It goes from $\mathcal{T}$ to $\mathcal{S}$ and so can never affect the flow from s to t.

The capacity of the first cut, $\{\{s\}, \{a, b, c, d, t\}\}$, is $c_{sa} + c_{sb} = 10 + 11 = 21$. The capacity of the cut $\{\{s, a, c\}, \{b, d, t\}\}$ is $c_{sb} + c_{ad} + c_{ct} = 11 + 4 + 11 = 26$. The value of the flow (12) in Fig 15.2 is less than either of these capacities. This is no accident.

15.1.4 THEOREM ▶ In any directed network, the value of an (s, t)-flow never exceeds the capacity of any (s, t)-cut.

Proof Let $\mathcal{F} = \{f_{uv}\}$ be any (s, t)-flow and $\{\mathcal{S}, \mathcal{T}\}$ any (s, t)-cut. Conservation of flow tells us that $\sum_v f_{uv} - \sum_v f_{vu} = 0$ for any $u \in \mathcal{S}$, $u \neq s$. (The possibility $u = t$ is excluded because $t \notin \mathcal{S}$.) Hence,

$$\mathrm{val}(\mathcal{F}) = \sum_{v \in \mathcal{V}} f_{sv} - \sum_{v \in \mathcal{V}} f_{vs}$$

$$= \sum_{u \in \mathcal{S}} \left(\sum_{v \in \mathcal{V}} f_{uv} - \sum_{v \in \mathcal{V}} f_{vu} \right) \quad \text{(since the term in parentheses is 0 except for } u = s)$$

$$= \sum_{u \in \mathcal{S}, v \in \mathcal{V}} f_{uv} - \sum_{u \in \mathcal{S}, v \in \mathcal{V}} f_{vu}.$$

Since $\{\mathcal{S}, \mathcal{T}\}$ is a partition, this last sum can be written

$$\sum_{u \in \mathcal{S}, v \in \mathcal{S}} f_{uv} + \sum_{u \in \mathcal{S}, v \in \mathcal{T}} f_{uv} - \sum_{u \in \mathcal{S}, v \in \mathcal{S}} f_{vu} - \sum_{u \in \mathcal{S}, v \in \mathcal{T}} f_{vu}$$

$$= \sum_{u \in \mathcal{S}, v \in \mathcal{S}} f_{uv} - \sum_{u \in \mathcal{S}, v \in \mathcal{S}} f_{vu} + \sum_{u \in \mathcal{S}, v \in \mathcal{T}} (f_{uv} - f_{vu}).$$

The first two terms in the last line are the same, so we obtain

$$(2) \qquad \mathrm{val}(\mathcal{F}) = \sum_{u \in \mathcal{S}, v \in \mathcal{T}} (f_{uv} - f_{vu}).$$

But $f_{uv} \leq c_{uv}$ and $f_{vu} \geq 0$, so $f_{uv} - f_{vu} \leq c_{uv}$ for all u and v. Therefore,

$$\text{val}(\mathcal{F}) \leq \sum_{u \in \mathcal{S}, v \in \mathcal{T}} c_{uv} = \text{cap}(\mathcal{S}, \mathcal{T})$$

as desired. ∎

Equation (2) relates the value of an (s, t)-flow to **any** (s, t)-cut. We highlight this rather surprising fact.

15.1.5 COROLLARY ▶ If $\mathcal{F}$ is any (s, t)-flow and $\{\mathcal{S}, \mathcal{T}\}$ is any (s, t)-cut, then

$$\text{val}(\mathcal{F}) = \sum_{u \in \mathcal{S}, v \in \mathcal{T}} (f_{uv} - f_{vu}).$$

With reference to the network in Fig 15.2 and the cut $\mathcal{S} = \{s, a, c\}$, $\mathcal{T} = \{b, d, t\}$, the sum specified in the corollary is

$$\sum_{u \in \mathcal{S}, v \in \mathcal{T}} (f_{uv} - f_{vu}) = f_{sb} + f_{ad} - f_{bc} + f_{ct} = 2 + 0 - 1 + 11 = 12,$$

which is the value of the flow in this network. We have also the following fundamental result about directed networks.

15.1.6 COROLLARY ▶ Suppose there exists some (s, t)-flow $\mathcal{F}$ and some (s, t)-cut $\{\mathcal{S}, \mathcal{T}\}$ such that the value of $\mathcal{F}$ equals the capacity of $\{\mathcal{S}, \mathcal{T}\}$. Then $\text{val}(\mathcal{F})$ is the maximum value of any flow and $\text{cap}(\mathcal{S}, \mathcal{T})$ is the minimum capacity of any cut.

Proof Let $\mathcal{F}_1$ be any flow. To see that $\text{val}(\mathcal{F}_1) \leq \text{val}(\mathcal{F})$, note that the theorem says that $\text{val}(\mathcal{F}_1) \leq \text{cap}(\mathcal{S}, \mathcal{T})$ and, by hypothesis, $\text{cap}(\mathcal{S}, \mathcal{T}) = \text{val}(\mathcal{F})$. So $\text{val}(\mathcal{F})$ is maximum, as asserted. The proof that $\text{cap}(\mathcal{S}, \mathcal{T})$ is minimum is similar and left to the Exercises. ∎

It turns out that the hypotheses of Corollary 15.1.6 can always be satisfied; that is, in any directed network, there is always a flow and a cut such that the value of the flow is the capacity of the cut. By the corollary, such a flow has maximum value.

The Construction of Flows

We conclude this section with an indication as to how we can construct flows in a directed network and continue to center our discussion around the network of Fig 15.1. First locate a path $\mathcal{P}$ from s to t (which follows the direction specified by the arrows on arcs) and define a flow by setting

$$f_{uv} = \begin{cases} 1 & \text{if } uv \in \mathcal{P} \\ 0 & \text{if } uv \notin \mathcal{P}. \end{cases}$$

Such a flow is called a *unit flow* because it has value 1. In Fig 15.1, *sact* is a path $\mathcal{P}$ from s to t for which the unit flow just defined is shown in Fig 15.3.

It is not difficult to see that conservation of flow is guaranteed: If v is a vertex not on $\mathcal{P}$, then all arcs entering and leaving v have flow 0; if v is on $\mathcal{P}$, there is

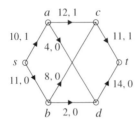

Figure 15.3 A unit flow in a network.

precisely one arc entering v which has nonzero flow and precisely one arc leaving with nonzero flow, and each of these arcs has a flow of 1.

Having obtained one flow, we can continue to increment, by 1 at a time, the flow on the arcs of $\mathcal{P}$ until we reach the smallest capacity of an arc on the path; that is, until the path has a *saturated arc*, meaning one carrying a flow equal to its capacity.

In our example, we can achieve a flow of 10 on each arc of $\mathcal{P}$, but no larger number, since at 10, arc sa is saturated. This flow is not by any means maximum, however: We showed a flow of value 12 in Fig 15.2. By what method can we obtain such a flow, or any flow of value greater than 10? We search for another path, $\mathcal{P}'$, from s to t, and a flow in $\mathcal{P}'$ such that all f_{uv} are equal and some arc is saturated.

In Fig 15.4, we show, in addition to a flow through $\mathcal{P}$ with a saturated arc, a flow through the path $\mathcal{P}'$: $sbdt$ with the capacity of 2 in each arc and arc bd saturated. We continue to check every path from s to t. For any such path, we increase the flow in each arc one unit at a time until the path contains a saturated arc. We observe that the flow in each arc of the path $\mathcal{P}''$: $sbct$ from s to t can be incremented by 1, as indicated in Fig 15.5.

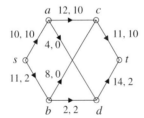

Figure 15.4

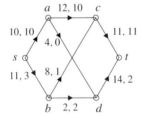

Figure 15.5

At this stage, every path between s and t contains a saturated arc, and we have a flow of value 13. Unfortunately, this is still not a maximum. In fact, there exists a flow of value 17 (see Fig 15.7) which, as we shall soon see, is maximum for this network. By what process might this maximum flow be obtained? We require, and describe in the next section, a refinement of the current technique.

Answers to Pauses

1. The law of conservation holds at a because

$$\sum_v f_{va} = f_{sa} = 10 \text{ and } \sum_v f_{av} = f_{ac} + f_{ad} = 10 + 0 = 10.$$

It holds at b because $\sum_v f_{vb} = f_{sb} = 2$ and $\sum_v f_{bv} = f_{bc} + f_{bd} = 1 + 1 = 2$.

It holds at d because $\sum_v f_{vd} = f_{ad} + f_{bd} = 0 + 1 = 1$ and $\sum_v f_{dv} = f_{dt} = 1$.

2. A flow with value 13 appears in Fig 15.5 and one with value 17 is shown in Fig 15.7.
3. To say that sets $\mathcal{S}$ and $\mathcal{T}$ comprise a partition of $\mathcal{V}$ is to say that $\mathcal{S}$ and $\mathcal{T}$ are disjoint subsets of $\mathcal{V}$ whose union is $\mathcal{V}$.

EXERCISES

The symbol [BB] means that an answer can be found in the Back of the Book.

1. **(a)** [BB] Verify the law of conservation of flow at a, e, and d.
   (b) [BB] Find the value of the indicated flow.
   (c) [BB] Find the capacity of the (s, t)-cut defined by $\mathcal{S} = \{s, a, b\}$ and $\mathcal{T} = \{c, d, e, t\}$.

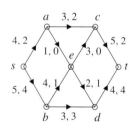

   (d) [BB] Can the flow be increased along the path $sbedt$? If so, by how much?
   (e) [BB] Is the given flow maximum? Explain.
   (f) [BB] Illustrate Corollary 15.1.5 for $\mathcal{S}$ and $\mathcal{T}$ as in (c).

2. **(a)** Verify the law of conservation of flow at a, e, and d.
   (b) [BB] Find the value of the indicated flow.
   (c) Find the capacity of the (s, t)-cut defined by $\mathcal{S} = \{s, a, b, c, d, e\}$ and $\mathcal{T} = \{t\}$.
   (d) Name all saturated arcs.
   (e) Is the given flow maximum? Explain.

   (f) Illustrate Corollary 15.1.5 for $\mathcal{S}$ and $\mathcal{T}$ as in (c).

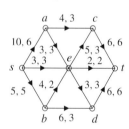

3. **(a)** Verify the law of conservation of flow at a, c, and d.
   (b) Find the value of the indicated flow.
   (c) [BB] Find the capacity of the (s, t)-cut defined by $\mathcal{S} = \{s, a, b, d\}$ and $\mathcal{T} = \{c, e, f, t\}$.

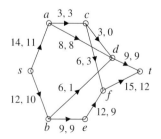

   (d) Is the given flow maximum? Explain.
   (e) Illustrate Corollary 15.1.5 for $\mathcal{S}$ and $\mathcal{T}$ as in (c).

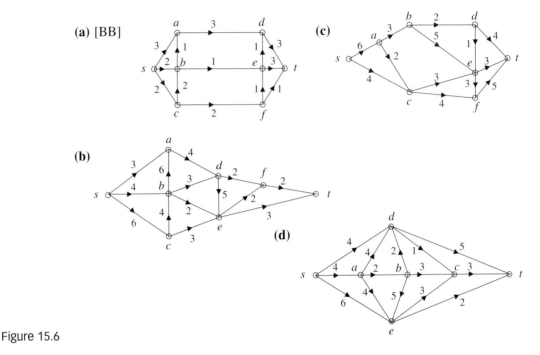

Figure 15.6

4. (a) Verify the law of conservation of flow at a, b, and e.
(b) Find the value of the indicated flow.
(c) Find the capacity of the (s, t)-cut defined by $\mathcal{S} = \{s, a, c, d\}$ and $\mathcal{T} = \{b, e, t\}$.

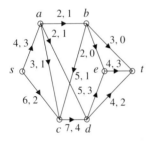

(d) Can the flow be increased along the path $sadt$? If so, by how much? Can it be increased along $scdt$? If so, by how much?

(e) Is the given flow maximum? Explain.
(f) Illustrate Corollary 15.1.5 for $\mathcal{S}$ and $\mathcal{T}$ as in (c).

5. Answer the following questions for each of the networks shown in Fig 15.6.

   i. Exhibit a unit flow.
   ii. Exhibit a flow with a saturated arc.
   iii. Find a "good" and, if possible, a maximum flow in the network. State the value of your flow.

6. [BB] Give an example of a flow $\{f_{uv}\}$ in a directed network which has the following properties:

- there is no flow into the source or out of the sink;
- not all $f_{uv} = 0$;
- the flow has value 0.

7. [BB] Complete the proof of Corollary 15.1.6 by showing that if a network has an (s, t)-flow $\mathcal{F}$ and an (s, t)-cut $\{\mathcal{S}, \mathcal{T}\}$ such that $\mathrm{val}(\mathcal{F}) = \mathrm{cap}(\mathcal{S}, \mathcal{T})$, then $\mathrm{cap}(\mathcal{S}, \mathcal{T})$ is the minimum capacity of any cut.

15.2   CONSTRUCTING MAXIMAL FLOWS

Our goal now is to modify the procedure for constructing flows given at the end of the previous section in order to construct a maximum flow for a directed network.

In the theory of directed networks, *chain* is a more common term than *trail* to describe a walk with distinct arcs which can be followed in either direction.

An arc of a chain is a *forward arc* if it is followed in the direction of the arrow and a *backward arc* if it is followed in a direction opposite that of the arrow. In Fig 15.5, the chain *sbcadt* has forward arcs *sb*, *bc*, *ad*, and *dt*, and backward arc *ca*.

15.2.1 DEFINITION ▶ Given an (s, t)-flow $\{f_{uv}\}$ in a network, a chain is called *flow-augmenting* if and only if $f_{uv} < c_{uv}$ for each forward arc uv of the chain and $f_{uv} > 0$ for each backward arc.

The chain $\mathcal{C}$: *sbcadt* from s to t in Fig 15.5 is flow-augmenting because, checking forward arcs, $3 = f_{sb} < c_{sb} = 11$; $1 = f_{bc} < c_{bc} = 8$; $0 = f_{ad} < c_{ad} = 4$; $2 = f_{dt} < c_{dt} = 14$; and, checking the backward arc, $10 = f_{ac} > 0$.

The term "flow-augmenting" chain derives from the fact that such a chain can be used to increase flow: increase the flow on each forward arc by some fixed amount and decrease the flow on each backward arc by the same amount. The largest "amount" that can be used here is a number called the *slack* of the chain. The slack is determined as follows.

> Calculate the unused capacity $c_{uv} - f_{uv}$ of each forward arc and the flow f_{uv} in each backward arc. The *slack* of the chain is the minimum of all these numbers.

In our example, the slack is 4, the value of $c_{ad} - f_{ad}$. Increasing by 4 the flow in the forward arcs of $\mathcal{C}$ and decreasing by 4 the flow in the backward arcs, we obtain the network and flow shown in Fig 15.7.

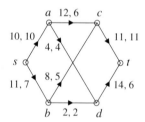

Figure 15.7   A maximum flow.

Pause 4   Why does the procedure just described of adding an amount q to the forward arcs of a chain and subtracting the same amount from the backward arcs preserve conservation of flow at each vertex?

By checking all possibilities, it can be seen that there are no further flow-augmenting chains in this network, and, marvelously, this means that a maximum flow has been achieved! We shall see why shortly, but for now, we content ourselves with showing that the flow in Fig 15.7 is maximum. For this, we consider the (s, t)-cut given by $\mathcal{S} = \{s, a, b, c\}$ and $\mathcal{T} = \{d, t\}$, which has capacity $c_{ad} + c_{bd} + c_{ct} = 4 + 2 + 11 = 17$. Since the displayed flow has value 17, Corollary 15.1.6 guarantees that it is maximum.

It is important to observe that the vertices of $\mathcal{S}$ are exactly those which can be reached from s by flow-augmenting chains. In addition to s, $\mathcal{S}$ includes b (since $f_{sb} = 7 < 11 = c_{sb}$) and, therefore, c (since $f_{bc} = 5 < 8 = c_{bc}$). From c, we are able to continue on a backward arc to a (since $f_{ac} = 6 > 0$). From a, a backward arc leads to s, which is already in $\mathcal{S}$, and the forward arc is saturated. The vertices which can be reached from s by flow-augmenting chains are $\{s, a, b, c\}$, precisely the vertices of $\mathcal{S}$. This observation is key to the proof of the famous Max Flow—Min Cut Theorem of Lester R. Ford Jr. and D. R. Fulkerson.[1]

Pause 5   Where have you heard Ford's name before?

15.2.2 MAX FLOW — MIN CUT THEOREM ▶

In a directed network, the maximum value of an (s, t)-flow equals the minimum capacity of an (s, t)-cut.

Proof   Using the procedure just described, we construct an (s, t)-flow $\mathcal{F}$ that does not contain any flow-augmenting chains from s to t. Corollary 15.1.6 tells us that the proof will be complete if we can construct an (s, t)-cut $\{\mathcal{S}, \mathcal{T}\}$ such that $\mathrm{val}(\mathcal{F}) = \mathrm{cap}(\mathcal{S}, \mathcal{T})$.

With the hindsight of our earlier example, we let $\mathcal{S}$ be the set of all vertices which can be reached by flow-augmenting chains starting at s (including s itself) and we let $\mathcal{T}$ be the set of all remaining vertices. Since there are no flow-augmenting chains from s to t, it cannot be that $t \in \mathcal{S}$, hence $t \in \mathcal{T}$. Thus, $\{\mathcal{S}, \mathcal{T}\}$ is an (s, t)-cut.

Let u be any vertex of $\mathcal{S}$ and let v be any vertex of $\mathcal{T}$. Because $u \in \mathcal{S}$, there is a flow-augmenting chain from s to u. Since $v \notin \mathcal{S}$, we conclude that $f_{uv} = c_{uv}$ and $f_{vu} = 0$, for otherwise the flow-augmenting chain from s to u (which exists by definition of $\mathcal{S}$) could be extended to a flow-augmenting chain from s to v. Now Corollary 15.1.5 tells us that

$$\mathrm{val}(\mathcal{F}) = \sum_{u \in \mathcal{S}, v \in \mathcal{T}} (f_{uv} - f_{vu})$$

$$= \sum_{u \in \mathcal{S}, v \in \mathcal{T}} (c_{uv} - 0) = \sum_{u \in \mathcal{S}, v \in \mathcal{T}} c_{uv} = \mathrm{cap}(\mathcal{S}, \mathcal{T}).$$

Hence, we have found a cut whose capacity equals $\mathrm{val}(\mathcal{F})$, proving that $\mathcal{F}$ is maximum. ∎

The proof of this theorem shows that if an (s, t)-flow has the property that there are no flow-augmenting chains from s to t, then the flow is maximum (a fact to which we earlier alluded).

Also observe that we have, in this section, provided both a formal algorithm for constructing a maximum flow and a method for checking that it is maximum (via the explicit construction of a cut with capacity the value of the flow). We illustrate with a final example.

PROBLEM 1.   Find a maximum flow in the directed network shown in Fig 15.8 and prove that it is a maximum.

---

[1] L. R. Ford, Jr. and D. R. Fulkerson, "Maximal flow through a network," *Canadian Journal of Mathematics* **8** (1956), 399–404.

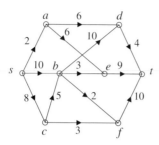

Figure 15.8   A directed network.

Solution.   We start by sending a flow of 2 units through the path *sadt*, a flow of 3 units through *sbet*, and a flow of 3 units through *scft*, obtaining the flow shown on the left in Fig 15.9. Continue by sending flows of 2 units through *sbdt* and 2 units through *sbft*, obtaining the flow shown on the right.

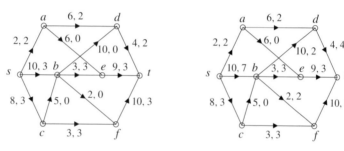

Figure 15.9   Two flows.

At this point, there are no further flow-augmenting chains from *s* to *t* involving only forward arcs. However, we can use the backward arc *da* to obtain a flow-augmenting chain *scbdaet*. Since the slack of this chain is 2, we add a flow of 2 to *sc*, *cb*, *bd*, *ae*, and *et*, and subtract 2 from *ad*. The result is shown in Fig 15.10.

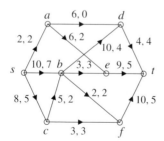

Figure 15.10

A search for further flow-augmenting chains takes us from *s* to *c* or *b* and on to *d*, where we are stuck. This tells us that the current flow (of value 14) is maximum. It also presents us with a cut verifying maximality, namely, $\mathcal{S} = \{s, b, c, d\}$ (those vertices reachable from *s* by flow-augmenting chains) and $\mathcal{T} = \{a, e, f, t\}$ (the complement of $\mathcal{S}$). The capacity of this cut is

$$c_{sa} + c_{be} + c_{bf} + c_{cf} + c_{dt} = 2 + 3 + 2 + 3 + 4 = 14.$$

Since this is the same as the value of the flow, we have verified that our flow is maximum. ∎

Rational Weights

Throughout this chapter there has been the underlying assumption that the weights attached to arcs are nonnegative integers. In fact, there is no additional complication if capacities are arbitrary rational numbers since any problem related to such a network is equivalent to a problem in the related network (where capacities are integral) obtained by multiplying all capacities by the least common multiple of the denominators.

Even if some capacities are not rational, the Max Flow—Min Cut Theorem is still true, but the proof we have given breaks down. In fact, Ford and Fulkerson[2] exhibit a small network with irrational capacities and the following properties:

- there is a flow, not maximum, for which the process we have described of obtaining flows of higher value via flow-augmenting chains never terminates;
- the infinite sequence of flows (of higher and higher value) does, however, converge, to a flow of some value Q; but
- the maximum value of a flow is $4Q$.

4. The flow on the arcs incident with a vertex not on the chain are not changed, so conservation of flow continues to hold at such a vertex. What is the situation at a vertex on the chain? Remember that a chain in a directed network is just a trail whose edges can be followed in either direction; thus, each vertex on a chain is incident with exactly two arcs. Suppose a chain contains the arcs uv, vw (in that order) and that the flows on these arcs before changes are f_{uv} and f_{vw}. There are essentially two cases to consider.

   Case 1: Suppose the situation at vertex v in the network is $u \to v \to w$. In this case, both uv and vw are forward arcs, so each has the flow increased by q. The total flow into v increases by q, but so does the total flow out of v, so there is still conservation of flow at v. (The analysis is similar if the situation at v is $u \leftarrow v \leftarrow w$.)

   Case 2: The situation at v is $u \to v \leftarrow w$.
   Here the flow on the forward arc uv is increased by q and the flow on the backward arc wv is decreased by q. There is no change in the flow out of v. Neither is there any change in the flow into v since the only terms in the sum $\sum_r f_{rv}$ which change occur with $r = u$ and $r = w$, and these become, respectively, $f_{uv} + q$ and $f_{vw} - q$. (The analysis is similar if the situation at v is $u \leftarrow v \to w$.)

5. Ford was one of the co-discovers of the "Bellman-Ford" shortest path algorithm. See Section 11.2.

---

[2]L. R. Ford, Jr. and D. R. Fulkerson, *Flows in Networks*, Princeton (1962), p. 21.

EXERCISES

The symbol [BB] means that an answer can be found in the Back of the Book.

1. Answer the following two questions for each of the directed networks shown.

   i. Show that the given flow is not maximum by finding a flow-augmenting chain from s to t. What is the slack in your chain?

   ii. Find a maximum flow, give its value, and prove that it is maximum by appealing to Theorem 15.2.2.

(a) [BB]

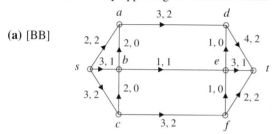

(b)

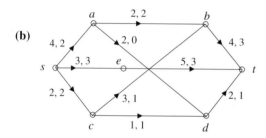

(c)

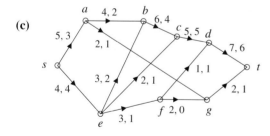

(d) [BB]

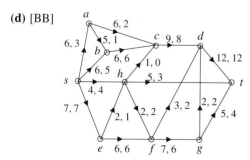

(e)

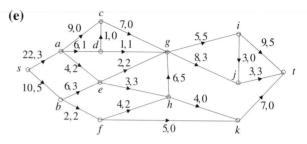

2. [BB; (a)] Find a maximum flow for each of the networks in Exercise 5 of Section 15.1. In each case, verify that your flow is maximum by finding a cut whose capacity equals the value of the flow.

3. Find a maximum flow for each of the networks shown in Fig 15.11. In each case, verify your answer by finding a cut whose capacity equals the value of the flow.

4. Shown are two networks whose arc capacities are rational but not always integral. Use the idea suggested at the end of this section to find a maximum flow in each of these networks. In each case, state the maximum flow value.

(a) [BB]

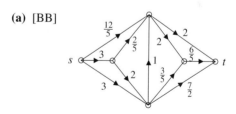

(b)

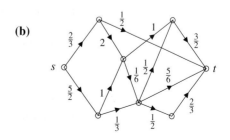

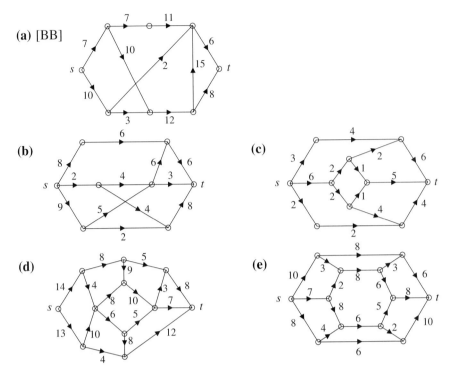

Figure 15.11 Graphs for Exercise 3.

15.3 APPLICATIONS

Obvious examples of directed networks include the intake and waste systems of plumbing in a house, oil and natural gas pipelines, telephone and electrical networks, mail courier services, and the flow of people through some system of tunnels. There are, however, less obvious applications, some of which we shall discuss here. One, a problem of supply and demand, will be discussed in detail; the others are merely sketched.

Multiple Sources and/or Sinks

Some directed networks have several sources and/or several sinks (an electronic mail network, for instance). Such networks can be handled easily by adding two new vertices, s and t, drawing an arc from s to each source and an arc from each sink to t, and putting a sufficiently large capacity on each of these new arcs so that their presence does not change the various possibilities for flow through the original network. A maximum (s, t)-flow in this new network will give a maximum flow in the other network, simply by forgetting about the added arcs. We shall see a worked example of this sort in conjunction with our next application.

Supply and Demand Problems

A small Alberta department store chain maintains three warehouses A, B, and C in the south of the province and three stores D, E, and F in rural communities in the far north. The warehouses have, respectively, 500, 500, and 900 snow blowers in stock on October 1. The first blizzard of winter is forecast to start near midnight, October 2, and there is an immediate demand from the stores for 700, 600, and 600 snow blowers, respectively. Various routes are available for shipping merchandise to the stores. These are shown in Fig 15.12, where the capacity of an arc uv is the largest number of snow blowers that can be shipped from u to v in the course of a single day.

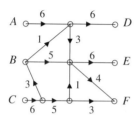

Figure 15.12   A supply and demand network (capacities are in hundreds).

The vertices in the middle should be thought of as "middlemen," baggage handlers, for example, who can only cope with so many snow blowers at once. We will assume that freight can cover as many arcs as necessary in a single day. Can all the required snow blowers reach the stores before the blizzard arrives? If not, how close can the company come to satisfying demand?

The first step in solving this problem is to view it as one with three sources and three sinks. Following the plan proposed earlier, we add a new source s joined to each of A, B, C and a new sink t to which D, E, and F are joined. We assign to each of the arcs incident with s a capacity equal to the number of snow blowers present at the given warehouse and to each arc incident with t a capacity equal to the number of snow blowers demanded at the given store. The new network is shown on the left in Fig 15.13.

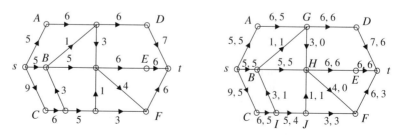

Figure 15.13   A supply and demand network and a maximum flow (capacities and flows in hundreds).

The network we have constructed has a nice physical interpretation. Imagine a new giant warehouse s containing all 1900 snow blowers. The labels on the arcs from s indicate the number of snow blowers that can be transferred to each warehouse if need be. Similarly, we have created a single sink which could take up to 1900 snow blowers. The routing of these machines will indicate which demands are actually met.

Applying the Max Flow—Min Cut procedure of the previous section, we obtain the maximum flow for our network shown on the right in Fig 15.13. We know that the flow is maximum because the cut

$$S = \{s, B, C, I, J\}, \quad T = \{A, G, H, D, E, F, t\}$$

has capacity $\text{cap}(S, T) = 500 + 100 + 500 + 100 + 300 = 1500$, which is the value of the flow.

We conclude that the snow blowers cannot all be delivered before the blizzard arrives; in fact, the best one can do is to deliver 1500. On the right of Fig 15.13, we see one routing by which this occurs, one in which stores D and E receive 600 snow blowers each while store F receives 300. This solution is not unique, however. For example, some of the snow blowers destined for E could be diverted to F via arc HF. Such a modification in shipping arrangements could be justified on the grounds that F is currently receiving only half of its requirements while the demands of E have been completely met.

Undirected Edges

Suppose a network contains some edges which are undirected. (Some of the pipes allow flow in two directions.) This turns out to be an easy situation to handle. Replace each undirected edge xy by two directed arcs, xy and yx, each having the same capacity as the original undirected edge. Two examples are presented in the exercises (Exercise 5). For these, it is easiest just to "imagine" two arcs, one going each way, rather than to draw them physically on the graph.

Edge Disjoint Paths

In order to deliver an important message, messengers are to be sent from s to t via a network of roads. Because terrorists are active in the area, it is prudent for the messengers to use different roads. What is the largest number of messengers that can be sent, subject to this constraint? Equivalently, how many edge disjoint paths are there from s to t? A beautiful theorem of Karl Menger, which is essentially a corollary of the Max Flow—Min Cut result (Theorem 15.2.2) (though it was first established much earlier), answers this question.

15.3.1 THEOREM ▶ **(Menger[3])** The maximum number of edge disjoint paths between two vertices s and t in a graph is the minimum number of edges whose removal leaves no path between s and t.

---

[3]K. Menger, "Zur allgemeinen Kurventheorie," *Fundamenta Mathematicae* **10** (1927), 95–115.

EXAMPLE 2   Consider the graph shown to the right. The removal of edges *ab* and *cf* severs all paths from *s* to *t* and, by inspection, we see that the removal of no single edge achieves this. Thus, the minimum number of edges required to cut all paths from *s* to *t* is two. Note that *sabt* and *scfbgt* are two edge disjoint paths from *s* to *t*. According to Menger's Theorem, there do not exist more than two edge disjoint paths, a fact easily checked.   ▲

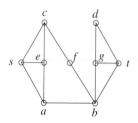

To obtain this result from the Max Flow—Min Cut Theorem, we convert the given graph to a network by replacing each edge between vertices *v* and *w* by arcs *vw* and *wv*, each of capacity 1, and then argue that a maximum flow in this network is precisely the number of edge disjoint paths from *s* to *t*. We refer you to Dantzig and Fulkerson[4] for details.

In Section 15.2, we developed a procedure for finding a maximum flow in a network. Therefore, we have an algorithm for finding all the edge disjoint paths in a graph. Be forewarned that the application of this algorithm without computer assistance may not be easy. While it is simple for a computer to take account of arcs in each direction between two vertices (via the adjacency matrix, for instance), as we mentioned in connection with undirected edges, it is better when proceeding by hand just to imagine the arcs rather than trying to draw them.

Job Assignments

Suppose there are *m* people $P_1, P_2, \ldots, P_m$, and *n* jobs $J_1, J_2, \ldots, J_n$, and that only certain people are qualified for certain jobs. Is it possible to assign to each person a job for which he or she is qualified? If not, how close can we come? We represent this problem as a network in Fig 15.14.

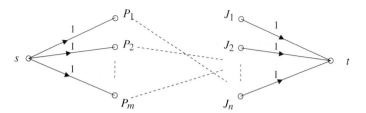

Figure 15.14

A source *s* is joined to each person by an arc having capacity 1 and each job is joined to *t* by an arc having capacity 1 also. If person P_i is qualified for job J_k, then these vertices are joined and given any nonzero integer capacity. (A capacity of 1 is fine, but in the next section, we will see that other capacities are

[4]G. B. Dantzig and D. R. Fulkerson, "On the Max Flow—Min Cut Theorem of Networks," *Linear Inequalities and Related Systems*, 215–221. *Annals of Mathematics Studies*, no. 38, Princeton University Press, Princeton, N. J., 1956.

sometimes useful as well.) A maximum flow through this network will provide the best possible set of job assignments.

EXERCISES

The symbol [BB] means that an answer can be found in the Back of the Book.

1. [BB] In the figure, warehouses a, b, and c have supplies of 30, 20, and 10 Klein bottles, respectively; retail outlets d and e require 30 and 25 Klein bottles, respectively. The capacities on arcs represent the maximum number of Klein bottles which can be shipped along that path. Can the demand be met? If not, find the maximum number of Klein bottles which can be sent.

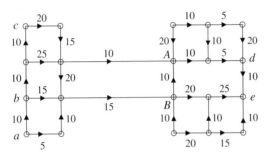

2. Repeat the previous question after reversing the arrow on edge AB.

3. [BB] Four warehouses A, B, C, and D with monthly shipping capacities of 4, 10, 3, and 2 units, respectively, are linked via the network shown to three retail outlets E, F, G whose monthly requirements are 4, 5, and 6 units, respectively. The monthly capacity of each link in the network is as indicated.

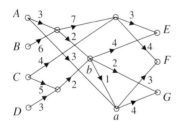

Can these warehouses meet the needs of the retail outlets? If not, find the maximum number of units which the three outlets (combined) can receive in a month. Explain your answer.

4. Answer the previous question again, this time assuming that the arc is ab, not ba.

5. Find a maximum flow for each of the following networks. (Note that some edges are undirected.)

   (a) [BB]

   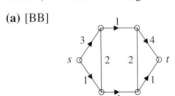

   (b)

   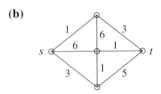

6. Verify Menger's Theorem, Theorem 15.3.1, for the planar graph of the cube with s and t as shown.

7. [BB] The diagram represents a system of roads from A to B. The mayor of town A wishes to send a message to the mayor of town B, and since all other means of communication are tapped, he sends out a certain number of trusted individuals on foot. If all paths taken by the individuals are to be disjoint, how many messengers can he employ?

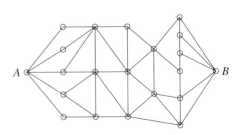

8. Repeat Exercise 7 for the following network of roads.

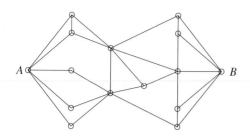

A   B

9. [BB] Verify Menger's Theorem for the complete graph $\mathcal{K}_n$.

10. Verify Menger's Theorem for the complete bipartite graph $\mathcal{K}_{n,m}$ in each of the following cases.

 (a) vertices s and t lie in the bipartition set of order n

 (b) vertices s and t lie in different bipartition sets

15.4 MATCHINGS

In Section 15.3, we saw how the procedure we have described for finding a maximum flow could be applied to the problem of assigning jobs to job applicants. Actually, this is just one example of a whole host of problems concerning *matchings* in graphs. We take a brief look at this area in this section.

15.4.1 DEFINITIONS ▶

A *matching* in a graph is a set of edges with the property that no vertex is incident with more than one edge in the set. A vertex which is incident with an edge in the set is said to be *saturated*. A matching is *perfect* if and only if every vertex is saturated; that is, if and only if every vertex is incident with precisely one edge of the matching.

EXAMPLES 3

In the graph shown on the left in Fig 15.15,

- the single edge bc is a matching which saturates b and c, but neither a nor d;
- the set $\{bc, bd\}$ is not a matching because vertex b belongs to two edges;
- the set $\{ab, cd\}$ is a perfect matching. ▲

Figure 15.15   Edge set $\{ab, cd\}$ is a perfect matching in the graph on the left. In the graph on the right, edge set $\{u_1v_2, u_2v_4, u_3v_1\}$ is a matching which is not perfect.

It is worth noticing that if a matching is perfect, the vertices of the graph can be partitioned into two sets of equal size and the edges of the matching provide a one-to-one correspondence between these sets. In the graph on the left in Fig 15.15, for instance, the edges of the perfect matching $\{ab, cd\}$ establish a one-to-one correspondence between $\{a, c\}$ and $\{b, d\}$: $a \mapsto b$, $c \mapsto d$.

EXAMPLES 4

In the graph on the right of Fig 15.15,

- the set of edges $\{u_1v_2, u_2v_4, u_3v_1\}$ is a matching which is not perfect but which saturates $\mathcal{V}_1 = \{u_1, u_2, u_3\}$;
- no matching can saturate $\mathcal{V}_2 = \{v_1, v_2, v_3, v_4\}$ since such a matching would require four edges, but then at least one u_i would be incident with more than one edge. ▲

Any assignment of workers to jobs in the job assignment application of Section 15.3 is a matching since no worker is given two jobs and each job is filled by at most one worker. In this case, we are thinking of the graph whose vertices correspond to workers and jobs and where edges join workers to jobs for which they are qualified. This graph is bipartite, and our discussion here will initially be restricted just to this type of graph. Later we will comment briefly on matchings for arbitrary graphs.

Let $\mathcal{G}$ be a bipartite graph with bipartition sets $\mathcal{V}_1$ and $\mathcal{V}_2$ and assume that $|\mathcal{V}_1| \leq |\mathcal{V}_2|$. Thinking of $\mathcal{V}_1$ as "workers" and $\mathcal{V}_2$ as "jobs," we can follow the procedure outlined at the end of Section 15.3 to obtain a maximal matching for $\mathcal{G}$, that is, a matching which saturates as many vertices as possible. One question remains unanswered, however—is every vertex in $\mathcal{V}_1$ saturated by this maximal matching? In the language of job assignments, was every worker successful in finding a job?

Clearly, if there are more workers than jobs—$|\mathcal{V}_1| > |\mathcal{V}_2|$—it would be impossible to employ everybody. Thus, the condition $|\mathcal{V}_1| \leq |\mathcal{V}_2|$ is necessary for $\mathcal{V}_1$ to be saturated. More generally, if for any subset of k workers, the number of jobs for which they are collectively qualified is less than k, then again it would be impossible to employ everybody. In other words, if X is any subset of $\mathcal{V}_1$ and we define $A(X)$ to be the set of all vertices in $\mathcal{V}_2$ which are adjacent to a vertex in X,

$$A(X) = \{v \in \mathcal{V}_2 \mid vx \text{ is an edge for some } x \in X\},$$

then we require $|X| \leq |A(X)|$.

EXAMPLE 5   In the figure to the right, if $X = \{u_1, u_2, u_4\}$, then $A(X) = \{v_3, v_4\}$. Since $|X| \not\leq |A(X)|$, the workers in X cannot all find jobs for which they are qualified. There is no matching in this graph which saturates $\mathcal{V}_1$. ▲

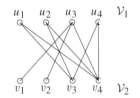

It is surprising that the so obviously necessary condition $|X| \leq |A(X)|$ for all $X \subseteq \mathcal{V}_1$ is also sufficient for the existence of a matching which saturates $\mathcal{V}_1$. This was discovered in 1935 by the English mathematician Philip Hall.

15.4.2   THEOREM ▶   **(Hall[5])** If $\mathcal{G}$ is a bipartite graph with bipartition sets $\mathcal{V}_1$ and $\mathcal{V}_2$, then there exists a matching which saturates $\mathcal{V}_1$ if and only if, for every subset X of $\mathcal{V}_1$, $|X| \leq |A(X)|$.

Proof   It remains to prove that the given condition is sufficient, so we assume that $|X| \leq |A(X)|$ for all subsets X of $\mathcal{V}_1$. In particular, this means that every vertex in $\mathcal{V}_1$ is joined to at least one vertex in $\mathcal{V}_2$ and also that $|\mathcal{V}_1| \leq |\mathcal{V}_2|$. Assume that there is no matching which saturates all vertices of $\mathcal{V}_1$. We derive a contradiction.

We turn $\mathcal{G}$ into a directed network in exactly the same manner as with the job assignment application of Section 15.3. Specifically, we adjoin two vertices s and t to $\mathcal{G}$ and draw directed arcs from s to each vertex in $\mathcal{V}_1$ and from each vertex in $\mathcal{V}_2$ to t. Assign a weight of 1 to each of these new arcs. Orient each edge of

G from its vertex in V_1 to its vertex in V_2, and assign a large integer $I > |V_1|$ to each of these edges. As noted before, there is a one-to-one correspondence between matchings of G and (s, t)-flows in this network, and the value of the flow equals the number of edges in the matching.

Since we are assuming that there is no matching which saturates V_1, it follows that every flow has value less than $|V_1|$ and, hence, by the Max Flow–Min Cut Theorem, there exists an (s, t)-cut $\{S, T\}$ ($s \in S$, $t \in T$) whose capacity is less than $|V_1|$.

Now every original edge of G has been given a weight larger than $|V_1|$. Since the capacity of our cut is less than $|V_1|$, no edge of G can join a vertex of S to a vertex of T. Letting $X = V_1 \cap S$, we have $A(X) \subseteq S$. Since each vertex in $A(X)$ is joined to $t \in T$, each such vertex contributes 1 to the capacity of the cut. Similarly, since s is joined to each vertex in $V_1 \setminus X$, each such vertex contributes 1. Since $|X| \leq |A(X)|$, we have a contradiction to the fact that the capacity is less than $|V_1|$. ∎

Theorem 15.4.2 is often called *Hall's Marriage Theorem* and stated in the following way:

> Given a set of boys and a set of girls some of whom know some of the boys, then the girls can get married, each to a boy she knows, if and only if any k of the girls collectively know at least k boys.

PROBLEM 6. Let G be a bipartite graph with bipartition sets V_1, V_2 in which every vertex has the same degree k. Show that G has a matching which saturates V_1.

Solution. Let X be any subset of V_1 and let $A(X)$ be as defined earlier. We count the number of edges joining vertices of X to vertices of $A(X)$. On the one hand (thinking of X), this number is $k|X|$. On the other hand (thinking of $A(X)$), this number is at most $k|A(X)|$ since $k|A(X)|$ is the total degree of all vertices in $A(X)$. Hence, $k|X| \leq k|A(X)|$, so $|X| \leq |A(X)|$ and the result follows from Hall's Theorem. ∎

Pause 6   Can you conclude from this problem that G also has a matching which saturates V_2? More generally, does G have a matching which saturates both V_1 and V_2 at the same time (a perfect matching)?

Imagine a police captain standing in front of all the police constables in her precinct. She wishes to divide her troops into pairs for patrol duty, but there are restrictions on how this can be carried out because certain constables do not work well with certain other constables. Is it possible for the captain to pair her constables in such a way that all restrictions are met and, if not, how close can she come?

It is easy to see how graph theory can be used here. We make a graph in which vertices represent police constables and an edge indicates that the corresponding constables can be paired. A perfect matching of this graph would give a pairing of the type desired. Failing this, we seek a matching containing as many pairings as possible. Notice that in this situation, there is no reason for the graph to be bipartite.

We are most interested in determining whether or not an arbitrary graph G has a perfect matching, that is, one which saturates every vertex. A theorem giving necessary and sufficient conditions, similar in spirit to Theorem 15.4.2, was established by Tutte in 1947,[6] but we content ourselves here with one example of a sufficient condition. For this, we require the following proposition, the proof of which is left to the exercises.

15.4.3 PROPOSITION ▶   Let $\mathcal{G}$ be a graph with vertex set $\mathcal{V}$.

1. If $\mathcal{G}$ has a perfect matching, then $|\mathcal{V}|$ is even.
2. If $\mathcal{G}$ has a Hamiltonian path or cycle, then $\mathcal{G}$ has a perfect matching if and only if $|\mathcal{V}|$ is even.

It is now easy to prove our sample result.

15.4.4 THEOREM ▶   If $\mathcal{G}$ is a graph with vertex set $\mathcal{V}$, $|\mathcal{V}|$ is even, and each vertex has degree $d \geq \frac{1}{2}|\mathcal{V}|$, then $\mathcal{G}$ has a perfect matching.

Proof   If $|\mathcal{V}| = 2$, then $\mathcal{G}$ is ○—○, which clearly has a perfect matching. If $|\mathcal{V}| \geq 4$, then Dirac's Theorem (10.2.4) tells us that $\mathcal{G}$ is Hamiltonian, and part 2 of Proposition 15.4.3 gives us the answer. ∎

Answers to Pauses

6. Yes, the same argument works. But more easily, note that since $\mathcal{G}$ is bipartite, the sum of the degrees of vertices in $\mathcal{V}_1$ must equal the sum of the degrees of vertices in $\mathcal{V}_2$. Since all vertices have the same degree, we conclude that $|\mathcal{V}_1| = |\mathcal{V}_2|$, so a matching which saturates $\mathcal{V}_1$ must automatically saturate $\mathcal{V}_2$ as well, and vice versa.

EXERCISES

The symbol [BB] means that an answer can be found in the Back of the Book.

1. [BB] Five ladies have men friends as shown in the following table.

Lady	Men Friends
Karen	David, Stuart, Paul, Roger
Mary	Stuart, Philip
Aurie	David, Paul
Pamela	Paul, Roger
Lynne	Stuart, Paul

Draw a bipartite graph depicting this situation. Find a way in which each lady can marry a man she knows, or use Hall's Marriage Theorem to explain why no such matching exists.

2. Repeat the previous question with reference to the following table.

Lady	Men Friends
Karen	Stuart, Paul
Mary	Philip, Stuart
Aurie	David, Paul, Roger, Stuart
Pamela	Philip, Paul
Lynne	Stuart, Philip, Paul

3. [BB] Determine whether or not the graph has a perfect matching.

---

[6]See J. A. Bondy and U. S. R. Murty, *Graph Theory with Applications*, North-Holland, New York (1976), 76ff.

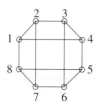

4. Angela, Brenda, Christine, Helen, Margaret, Phyllis, and Renée are seeking to fill eight vacant managerial positions A, B, C, D, E, F, G, and H. Angela can fill A, B, and F; Brenda can fill B, C, and F; Christine can fill C, E, and A; Helen can fill D, G, and H; Margaret can fill E, A, and F; Phyllis can fill F, A, and C; Renée can fill C, E, and F.

(a) [BB] Show that it is not possible for all seven women to fill managerial positions by showing that the condition in Theorem 15.4.2 is not satisfied.

(b) Find a matching which allows as many women as possible to fill positions.

5. (Based upon a problem of Tucker[7]) New Horizons Dating Service has 13 male and 13 female clients. Each of the male clients selects five females that he would like to date, and each female selects five males. By coincidence, it is discovered that every choice is mutual; that is, a male selected a certain female if and only if that female selected that male.

(a) Show that it is possible to match the couples into 13 different pairs, following the aforementioned selections.

(b) Show that it is possible to repeat the matching procedures given in (a) on five consecutive nights such that no individual dates the same person twice.

6. Bruce, Edgar, Eric, Herb, Maurice, Michael, Richard, and Roland decide to go on a week-long canoe trip in Labrador. They must divide themselves into pairs, one pair for each of four canoes. The following table indicates those people who are willing to paddle with each other.

(a) [BB] Find a way of dividing the campers into four canoes.

(b) Explain why it is impossible to find a pairing in which Roland and Bruce share a canoe.

(c) Halfway through the trip, Roland decides that he cannot share a canoe with Eric and Edgar refuses to share a canoe with Herb. Is it now possible to find a suitable matching?

	Bruce	Edgar	Eric	Herb	Maurice	Michael	Richard	Roland
Bruce				✓	✓			✓
Edgar			✓	✓	✓			
Eric				✓			✓	✓
Herb		✓	✓				✓	
Maurice	✓	✓						✓
Michael	✓	✓						✓
Richard			✓	✓				
Roland	✓		✓		✓	✓		

7. (a) [BB] Given a set S and n subsets $A_1, A_2, \ldots, A_n$ of S, it is possible to select distinct elements s_1, $s_2, \ldots, s_n$ of S such that $s_1 \in A_1$, $s_2 \in A_2, \ldots$, $s_n \in A_n$ if and only if, for each subset X of $\{1, 2, \ldots, n\}$ the number of elements in $\bigcup_{x \in X} A_x$ is at least $|X|$. Why?

(b) In a certain game of solitaire, the 52 cards of a standard deck are dealt face up into 13 columns of 4 cards each. Show that it is always possible to select 13 cards, one of each of the 13 denominations $2, 3, \ldots, 10$, Jack, Queen, King, Ace, including at least one card from each column.

[This problem comes from an article entitled "Marriage, Magic and Solitaire" by David B. Leep and Gerry Myerson, which appeared in the *American Mathematical Monthly*, Vol. 106 (1999), no. 5, 419–429.]

8. Suppose $\mathcal{V}_1$, $\mathcal{V}_2$ are the bipartition sets of a bipartite graph $\mathcal{G}$. Let m be the smallest of the degrees of the vertices in $\mathcal{V}_1$ and M the largest of the degrees of the vertices in $\mathcal{V}_2$. Prove that if $m \geq M$, then $\mathcal{G}$ has a matching which saturates $\mathcal{V}_1$.

9. (a) [BB] Suppose $\mathcal{V}_1$ and $\mathcal{V}_2$ are the bipartition sets in a bipartite graph $\mathcal{G}$. If $|\mathcal{V}_1| > |\mathcal{V}_2|$, then it is clearly impossible to find a matching which saturates $\mathcal{V}_1$. State a result which is applicable to this case and give a necessary and sufficient condition for your result to hold.

(b) How should your result in (a) be interpreted when applied to the job assignment problem?

10. (a) [BB] Determine necessary and sufficient conditions for the complete bipartite graph $\mathcal{K}_{m,n}$ to have a perfect matching.

(b) In those cases where $\mathcal{K}_{m,n}$ does not have a perfect matching, find a matching which saturates as many vertices as possible.

---

[7]See Chapter 10 of Alan Tucker, *Applied Combinatorics*, Wiley (New York), 1980.

11. Let G be a bipartite graph with bipartition sets V_1, V_2 and assume that G has a perfect matching. Add two vertices x and y to G such that x is adjacent to all vertices in V_1, y is adjacent to all vertices in V_2, and x is not adjacent to y.

(a) Show that $G \cup \{x, y\}$ is bipartite.

(b) Show that $G \cup \{x, y\}$ has a perfect matching.

12. (a) [BB] Show that the complete graph K_n has a perfect matching if and only if n is even.

(b) Let e be a fixed edge in K_n, where $n > 2$. Does the result in (a) still hold for $K_n \setminus \{e\}$? Explain. What if $n = 2$?

13. Prove Proposition 15.4.3.

14. [BB] Is the converse of Proposition 15.4.3(1) true? Explain.

15. [BB] Is the converse of Theorem 15.4.4 true? Explain.

16. Prove that if a tree has a perfect matching, then that matching is unique.

REVIEW EXERCISES FOR CHAPTER 15

1. (a) Verify the law of conservation of flow at a and at e.

(b) Find the value of the indicated flow.

(c) Find the capacity of the (s, t)-cut defined by $S = \{s, a, b\}$, $T = \{c, d, e, f, t\}$. Illustrate Corollary 15.1.5 for this cut.

(d) Can the flow be increased along the path $sbft$?

(e) Is the given flow maximum?

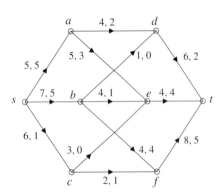

2. In the directed network of Exercise 1, assume that the flow along each of the paths $sbdt$ and $scft$ has been increased by one unit.

(a) Show that every path from s to t now contains a saturated arc.

(b) Despite (a), show that $sbeadt$ is a flow-augmenting chain from s to t. What is the slack in this chain?

(c) Find a maximum flow for this directed network, give its value, and prove that it is maximum by appealing to Theorem 15.2.2.

3. Find a maximal (s, t)-flow. Verify your answer by finding an (s, t)-cut whose capacity equals the value of the flow.

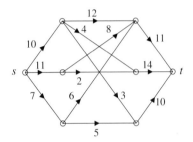

4. One example of an (s, t)-cut in a directed network with vertex set V is the case $S = \{s\}$, $T = V \setminus \{s\}$. Assume you have a flow in a directed network and you discover that the capacity of the cut just mentioned equals the value of the flow. Prove that every arc leaving s must be saturated. (You may assume no flow is going into s.)

5. Four warehouses A, B, C, D with shipping capacities 6, 8, 7, 9, respectively, are linked via the network shown to four retail outlets E, F, G, H with requirements 5, 4, 9, 7, respectively. Can these requirements be met? If not, find the maximum number of units which the four outlets (combined) can receive.

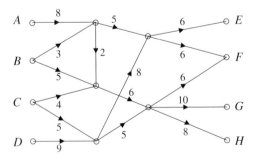

6. Find a maximum flow.

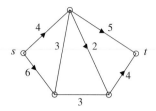

7. (a) Which graphs have the property that for any pair of vertices s and t, the number highlighted by Menger's Theorem is 1?

(b) Which Hamiltonian graphs have the property that for any pair of vertices s and t, the number highlighted by Menger's Theorem is 2?

8. The six teams entering the final round of the World Hockey Championships are Canada, Finland, Russia, Slovakia, Romania, and the United States. Albert, Bruce, Craig, Camelia, Oana and Yuri wish to bet with each other on who the winner will be and, ideally, they would like to all select different teams. The teams that each bettor is willing to support are shown in the following table.

Bettor	Teams
Albert	Canada, Finland, Slovakia, USA
Bruce	Slovakia, Romania
Craig	Russia, Slovakia
Camelia	Russia, Romania
Oana	Canada, Finland, Russia, Romania, USA
Yuri	Russia, Slovakia, Romania

Can these people indeed bet on different teams? Give such a selection or use Hall's Marriage Theorem to explain why such a selection is impossible.

9. Do Exercise 8 again with assumptions as before but for three exceptions:

- Albert is now happy not to bet on Finland,
- Oana will no longer bet on Russia or Romania, and
- Bruce is now willing to bet on Canada.

10. Let G be a connected graph, every vertex of which has degree 2. Prove that G has a perfect matching if and only if G has an even number of vertices.

11. Let G be a connected graph every vertex of which has degree 3.

(a) Show that G must have an even number of vertices.

(b) Must G have a perfect matching?

Solutions to Selected Exercises

Exercises 1.1

1. (a) True, because both $4 = 2 + 2$ and $7 < \sqrt{50}$ are true statements.

(c) True, because both hypothesis and conclusion are true.

(e) True, because this is an implication with false hypothesis.

(g) False, because the hypothesis is true but the conclusion is false.

2. (a) This is true: The hypothesis is true only when $a \geq b$ and $b \geq a$, that is, when $a = b$, and then the conclusion is also true.

3. (a) $a^2 \leq 0$ and a is a real number (more simply, $a = 0$).

(c) $x \neq 1$ and $x \neq -1$.

(e) There exists a real number x such that $n \leq x$ for every integer n.

(g) Every planar graph can be colored with at most four colors.

(i) There exist integers a and b such that for all integers q and r, $b \neq qa + r$.

4. (a) Converse: If $\frac{a}{c}$ is an integer, then $\frac{a}{b}$ and $\frac{b}{c}$ are also integers.

Contrapositive: If $\frac{a}{c}$ is not an integer, then $\frac{a}{b}$ is not an integer or $\frac{b}{c}$ is not an integer.

(c) Converse: A connected graph is Eulerian.

Contrapositive: If a graph is not connected, then it is not Eulerian.

(e) Converse: A four-sided figure is a square.

Contrapositive: If a figure does not have four sides, then it is not a square.

5. (a) There exists a continuous function which is not differentiable.

(c) For every real number x, there exists a real number y such that $y > x$.

(e) For every positive integer n, there exist primes $p_1, p_2, \ldots, p_t$ such that $n = p_1 p_2 \cdots p_t$.

Exercises 1.2

1. (a) Hypothesis: a and b are positive numbers; Conclusion: $a + b$ is positive.

2. This statement is true. Suppose the hypothesis, x is an even integer, is true. Then $x = 2k$ for some other integer k. Then $x + 2 = 2k + 2 = 2(k + 1)$ is also twice an integer. So $x + 2$ is even. The conclusion is also true.

6. The converse is the statement, "A continuous function is differentiable." This is false.

8. $\mathcal{A}$ is true. It expresses the fact that every real number lies between two consecutive integers. Statement $\mathcal{B}$ is most definitely false. It asserts that there is a remarkable integer n with the property that every real number lies in the unit interval between n and $n + 1$.

11. (a) Case 1: a is even. In this case, we have one of the desired conclusions.

Case 2: a is odd. In this case, $a = 2m + 1$ for some integer m, so $a + 1 = 2m + 2 = 2(m + 1)$ is even, another desired result.

(b) $n^2 + n = n(n + 1)$ is the product of consecutive integers one of which must be even; so $n^2 + n$ is even.

13. $2x^2 - 4x + 3 = 2(x^2 - 2x) + 3 = 2[(x - 1)^2 - 1] + 3 = 2(x - 1)^2 + 1$ is the sum of 1 and a nonnegative number. So it is at least 1 and hence positive.

15. ($\rightarrow$) To prove this direction, we establish the contrapositive, that is, we prove that n odd implies n^2 odd. For this, if n is odd, then $n = 2m + 1$ for some integer m. Thus $n^2 = 4m^2 + 4m + 1 = 2(2m^2 + 2m) + 1$ is odd.

($\leftarrow$) Here we assume that n is even. Therefore, $n = 2m$ for some integer m. So $n^2 = (2m)^2 = 4m^2 = 2(2m^2)$ which is even, as required.

17. Since n is odd, $n = 2k + 1$ for some integer k.

Case 1: k is even.

In this case $k = 2m$ for some integer m, so $n = 2(2m) + 1 = 4m + 1$.

Case 2: k is odd.

In this case, $k = 2m + 1$ for some integer m, so $n = 2(2m + 1) + 1 = 4m + 3$.

Since each case leads to one of the desired conclusions, the result follows.

20. Since 0 is an eigenvalue of A, there is a nonzero vector x such that Ax $= 0$. Now suppose that A is invertible. Then $A^{-1}(A$x$) = A^{-1}0 = 0$, so x $= 0$, a contradiction.

22. Observe that $(1 + a)(1 + b) = 1 + a + b + ab = 1$. Thus $1 + a$ and $1 + b$ are integers whose product is 1. There are two possibilities: $1 + a = 1 + b = 1$, in which case $a = b = 0$, or $1 + a = 1 + b = -1$, in which case $a = b = -2$.

24. We begin by assuming the negation of the desired conclusion; in other words, we assume that there exist real numbers x, y, z which simultaneously satisfy each of these three equations. Subtracting the second equation from the first we see that $x + 5y - 4z = -2$. Since the third equation we were given says $x + 5y - 4z = 0$, we have $x + 5y - 4z$ equal to both 0 and to -2. Thus, the original assumption has led us to a contradiction.

25. (a) False: $x = y = 0$ is a counterexample.

(c) False: $x = 0$ is a counterexample.

(e) False: $a = b = \sqrt{2}$ is a counterexample.

27. (a) Since $n^2 + 1$ is even, n^2 is odd, so n must also be odd. Writing $n = 2k + 1$, then $n^2 + 1 = 2m$ says $4k^2 + 4k + 2 = 2m$, so $m = 2k^2 + 2k + 1 = (k + 1)^2 + k^2$ is the sum of two squares as required.

(b) We are given that $n^2 + 1 = 2m$ for $n = 4373$ and $m = 9561565$. Since $n = 2(2186) + 1$, our solution to (a) shows that $m = k^2 + (k + 1)^2$ where $k = 2186$. Thus, $9561565 = 2186^2 + 2187^2$.

Exercises 1.3

1. (a)

p	q	$\neg q$	$(\neg q) \vee p$	$p \wedge ((\neg q) \vee p)$
T	T	F	T	T
T	F	T	T	T
F	T	F	F	F
F	F	T	T	F

(d)

p	q	r	$\neg q$	$p \vee (\neg q)$	$\neg (p \vee (\neg q))$	$\neg p$	$(\neg p) \vee r$	$(\neg (p \vee (\neg q))) \wedge ((\neg p) \vee r)$
T	T	T	F	T	F	F	T	F
T	T	F	F	T	F	F	F	F
T	F	T	T	T	F	F	T	F
T	F	F	T	T	F	F	F	F
F	T	T	F	F	T	T	T	T
F	T	F	F	F	T	T	T	T
F	F	T	T	T	F	T	T	F
F	F	F	T	T	F	T	T	F

3.

p	q	r	s	¬r	q ∧ (¬r)	p → [q ∧ (¬r)]	¬s	(¬s) ∨ q
T	T	T	T	F	F	F	F	T

r ↔ [(¬s) ∨ q]	[p → (q ∧ (¬r))] ∨ [r ↔ ((¬s) ∨ q)]
T	T

6. (a)

p	q	r	p → q	q → r	(p → q) ∧ (q → r)	p → r	[(p → q) ∧ (q → r)] → [p → r]
T	T	T	T	T	T	T	T
T	T	F	T	F	F	F	T
T	F	T	F	T	F	T	T
T	F	F	F	T	F	F	T
F	T	T	T	T	T	T	T
F	T	F	T	F	F	T	T
F	F	T	T	T	T	T	T
F	F	F	T	T	T	T	T

Since $[(p \to q) \land (q \to r)] \to [p \to r]$ is true for all values of p, q and r, this statement is a tautology.

(b) If p implies q which, in turn, implies r, then certainly p implies r.

8. (a) We are given that $\mathcal{A}$ is false for any values of its variables. An implication $p \to q$ is false only if p is true and q is false. Since $\mathcal{A}$ is always false, $\mathcal{A} \to \mathcal{B}$ is always true. So it is a tautology.

10. (a)

p	q	p ⊻ q
T	T	F
T	F	T
F	T	T
F	F	F

(c)

p	q	p ⊻ q	p ∨ q	(p ⊻ q) → (p ∨ q)
T	T	F	T	T
T	F	T	T	T
F	T	T	T	T
F	F	F	F	T

The truth table shows that $(p \veebar q) \to (p \lor q)$ is true for all values of p and q, so it is a tautology.

Exercises 1.4

1. 1. (Idempotence) The truth tables at the right show that $p \lor p \Longleftrightarrow p$ and $p \land p \Longleftrightarrow p$.

p	p ∨ p
T	T
F	F

p	p ∧ p
T	T
F	F

3. (Associativity) The equality of the fifth and seventh columns in the truth table shows that $((p \lor q) \lor r) \Longleftrightarrow (p \lor (q \lor r))$.

p	q	r	p ∨ q	(p ∨ q) ∨ r	q ∨ r	p ∨ (q ∨ r)
T	T	T	T	T	T	T
T	T	F	T	T	T	T
T	F	T	T	T	T	T
F	T	T	T	T	T	T
T	F	F	T	T	F	T
F	T	F	T	T	T	T
F	F	T	F	T	T	T
F	F	F	F	F	F	F

The equality of the fifth and seventh columns in the truth table shows that $((p \wedge q) \wedge r) \iff (p \wedge (q \wedge r))$.

p	q	r	p ∧ q	(p ∧ q) ∧ r	q ∧ r	p ∧ (q ∧ r)
T	T	T	T	T	T	T
T	T	F	T	F	F	F
T	F	T	F	F	F	F
F	T	T	F	F	T	F
T	F	F	F	F	F	F
F	T	F	F	F	F	F
F	F	T	F	F	F	F
F	F	F	F	F	F	F

5. (Double negation) The equality of the first and third columns in the truth table shows that $p \iff \neg(\neg p)$.

p	¬p	¬ (¬p)
T	F	T
F	T	F

7. The two truth tables show, respectively, that $p \vee 1 \iff 1$ and $p \wedge 1 \iff p$.

p	1	p ∨ 1
T	T	T
F	T	T

p	1	p ∧ 1
T	T	T
F	T	F

9. The two truth tables show, respectively, that $(p \vee (\neg p)) \iff 1$ and $(p \wedge (\neg p)) \iff 0$.

p	¬p	p ∨ (¬p)	1
T	F	T	T
F	T	T	T

p	¬p	p ∧ (¬p)	0
T	F	F	F
F	T	F	F

11. The third and sixth columns of the truth table show that $(p \rightarrow q) \iff ((\neg q) \rightarrow (\neg p))$.

p	q	p → q	¬q	¬p	(¬q) → (¬p)
T	T	T	F	F	T
T	F	F	T	F	F
F	T	T	F	T	T
F	F	T	T	T	T

13. The third and fifth columns of the truth table show that $(p \rightarrow q) \iff ((\neg p) \vee q)$.

p	q	p → q	¬p	(¬p) ∨ q
T	T	T	F	T
T	F	F	F	F
F	T	T	T	T
F	F	T	T	T

2. (a) Using one of the laws of De Morgan and one distributive property, we obtain

$$[(p \wedge q) \vee (\neg((\neg p) \vee q))] \iff [(p \wedge q) \vee (p \wedge (\neg q))] \iff [p \wedge (q \vee (\neg q))] \iff (p \wedge \mathbf{1}) \iff p.$$

3. (a)

p	q	$p \wedge q$	$p \vee (p \wedge q)$
T	T	T	T
T	F	F	T
F	T	F	F
F	F	F	F

4. (a) Distributivity gives $[(p \vee q) \wedge (\neg p)] \iff (p \wedge (\neg p)) \vee (q \wedge (\neg p)) \iff [\mathbf{0} \vee ((\neg p) \wedge q)] \iff ((\neg p) \wedge q).$

(d) $\neg[(p \leftrightarrow q) \vee (p \wedge (\neg q))] \iff [\neg(p \leftrightarrow q) \wedge \neg(p \wedge (\neg q))] \iff [(p \leftrightarrow (\neg q)) \wedge ((\neg p) \vee q)]$, using Exercise 4(c).

6. 1. The truth table shows that idempotence fails.

p	$p \veebar p$
T	F
F	F

3. Associativity holds.

p	q	r	$p \veebar q$	$(p \veebar q) \veebar r$	$q \veebar r$	$p \veebar (q \veebar r)$
T	T	T	F	T	F	T
T	T	F	F	F	T	F
T	F	T	T	F	T	F
F	T	T	T	F	F	F
T	F	F	T	T	F	T
F	T	F	T	T	T	T
F	F	T	F	T	T	T
F	F	F	F	F	F	F

7. The truth table shows that $p \veebar \mathbf{1}$ is no longer $\mathbf{1}$:

p	$\mathbf{1}$	$p \veebar \mathbf{1}$
T	T	F

9. The truth table shows that $(p \veebar (\neg p)) \iff \mathbf{1}$:

p	$\neg p$	$p \veebar (\neg p)$	$\mathbf{1}$
T	F	T	T
F	T	T	T

13. This is no longer true:

p	q	$p \to q$	$\neg p$	$(\neg p) \veebar q$
F	T	T	T	F

7. (b) $(p \wedge q) \vee ((\neg p) \wedge (\neg q))$ is in disjunctive normal form.

(c) $p \vee ((\neg p) \wedge q)$ is not in disjunctive normal form: not all variables are included in the first term.

8. (a) This is already in disjunctive normal form!   (b) $(p \wedge q) \vee (p \wedge (\neg q))$

Exercises 1.5

1. (a) Since $[p \to (q \to r)] \iff [p \to ((\neg q) \vee r)] \iff [(\neg p) \vee (\neg q) \vee r]$,
the given argument can be rewritten as shown and this is valid by
disjunctive syllogism.

$$\frac{[(\neg p) \vee r] \vee (\neg q)}{q}$$
$$\overline{(\neg p) \vee r}$$

(b) We analyze with a truth table. In row one, the premises are true but the conclusion is not. The argument is not valid.

p	q	r	$p \to q$	$q \vee r$	$\neg q$	$r \to (\neg q)$	
T	T	T	T	T	F	F	★
T	T	F	T	T	F	T	★
T	F	T	F	T	T	T	
F	T	T	T	T	F	F	★
T	F	F	F	F	T	T	
F	T	F	T	T	F	T	★
F	F	T	T	T	T	T	★
F	F	F	T	F	T	T	

3. (a) We analyze with a truth table. There are five rows when the premises are all true and in each case the conclusion is also true. The argument is valid.

p	q	r	$p \vee q$	$p \to r$	$q \to r$	$(p \vee q) \to r$	
T	T	T	T	T	T	T	★
T	T	F	T	F	F	F	
T	F	T	T	T	T	T	★
F	T	T	T	T	T	T	★
T	F	F	T	F	T	F	
F	T	F	T	T	F	F	
F	F	T	F	T	T	T	★
F	F	F	F	T	T	T	★

4. (a) Since $[(\neg r) \vee (\neg q)] \iff [q \to (\neg r)]$, the first two premises give $p \to (\neg r)$ by the chain rule. Now $\neg p$ follows by modus tollens.

(b) This argument is not valid. If q and r are false, p and s are true, and t takes on any truth value, then all premises are true, yet the conclusion is false.

5. (a) Let p and q be the statements
p :   I stay up late at night
q :   I am tired in the morning.

The given argument is
$$\frac{\begin{array}{c} p \to q \\ p \end{array}}{q}.$$
This is valid by modus ponens.

(b) Let p and q be the statements
p :   I stay up late at night
q :   I am tired in the morning.

The given argument is
$$\frac{\begin{array}{c} p \to q \\ q \end{array}}{p}.$$

This is not valid, as the truth table shows. In row three, the two premises are true but the conclusion is false.

p	q	$p \to q$	
T	T	T	★
T	F	F	
F	T	T	★
F	F	T	

(e) Let p and q be the statements
p :   I wear a red tie
q :   I wear blue socks.

The given argument is
$$\frac{\begin{array}{c} p \vee q \\ \neg q \end{array}}{p}.$$
This is valid by disjunctive syllogism.

(g) Let p, q, and r be the statements

p: I work hard
q: I earn lots of money
r: I pay high taxes.

The given argument is
$$\frac{\begin{array}{c} p \to q \\ q \to r \end{array}}{r \to p}.$$

This is not valid, as the truth table shows. In row five, the two premises are true but the conclusion is false.

p	q	r	$p \to q$	$q \to r$	$r \to p$	
T	T	T	T	T	T	★
T	T	F	T	F	T	
T	F	T	F	T	T	
T	F	F	F	T	T	
F	T	T	T	T	F	★
F	T	F	T	F	T	
F	F	T	T	T	F	★
F	F	F	T	T	T	★

(l) Let p, q, and r be the statements

p: I like mathematics
q: I study
r: I pass mathematics
s: I graduate.

The given argument is
$$\frac{\begin{array}{c} p \to q \\ (\neg q) \vee r \\ (\neg s) \to (\neg r) \end{array}}{s \to q}.$$
This is the same as
$$\frac{\begin{array}{c} p \to q \\ q \to r \\ r \to s \end{array}}{s \to q}$$

which is certainly not valid, as the following partial truth table shows.

p	q	r	s	$p \to q$	$q \to r$	$r \to s$	$s \to q$
F	F	F	T	T	T	T	F

6. $r \vee q$ is logically equivalent to $[\neg(\neg r) \vee q] \iff [(\neg r) \to q]$ so, with $p \to \neg r$, we get $p \to q$ by the chain rule.

7. We will prove by contradiction that no such conclusion is possible. Say to the contrary that there is such a conclusion $\mathcal{C}$. Since $\mathcal{C}$ is not a tautology, some set of truth values for p and q must make $\mathcal{C}$ false. But if r is true, then both the premises $(\neg p) \to r$ and $r \vee q$ are true regardless of the values of p and q. This contradicts $\mathcal{C}$ being a valid conclusion for this argument.

8. (a) $p \wedge q$ is true precisely when p and q are both true.

10. In Latin, *modus ponens* means "method of affirming" and *modus tollens* means "method of denying". This is a reflection of the fact that *modus tollens* has a negative $\neg p$ as its conclusion, while *modus ponens* affirms the truth of a statement q.

Exercises 2.1

1. (a) $\{-\sqrt{5}, \sqrt{5}\};$ (c) $\{0, -\frac{3}{2}\}$

2. (a) For example, $1 + i$, $1 + 2i$, $1 + 3i$, $-8 - 5i$ and $17 - 43i$.

3. (a) $\{1, 2\}, \{1, 2, 3\}, \{1, 2, 4\}, \{1, 2, 3, 4\}$

4. Only (c) is true. The set A contains one element, $\{a, b\}$.

5. (a) True;   **(d)** False.

6. (a) $\{\emptyset\}$

7. (a) True;   **(e)** False;   **(h)** False.

8. Yes it is; for example, let $x = \{1\}$ and $A = \{1, \{1\}\}$.

9. (a) ii. $\{a, b, c\}, \{a, b, d\}, \{a, c, d\}, \{b, c, d\}$

11. (a) 4;   **(b)** 8;

   (c) There are 2^n subsets of a set of n elements. (See Exercise 16 in Section 5.1 for a proof.)

12. (a) False;   **(d)** True.

14. (a) True. ($\rightarrow$) If $C \in \mathcal{P}(A)$, then by definition of power set, C is a subset of A; that is, $C \subseteq A$.
   ($\leftarrow$) If $C \subseteq A$, then C is a subset of A and so, again by definition of power set, $C \in \mathcal{P}(A)$.

Exercises 2.2

1. (a) $A = \{1, 2, 3, 4, 5, 6\}$, $B = \{-1, 0, 1, 2, 3, 4, 5\}$, $C = \{0, 2, -2\}$.

2. (a) $S \cap T = \{\sqrt{2}, 25\}$, $S \cup T = \{2, 5, \sqrt{2}, 25, \pi, \frac{5}{2}, 4, 6, \frac{3}{2}\}$,

   $T \times (S \cap T) = \{(4, \sqrt{2}), (4, 25), (25, \sqrt{2}), (25, 25), (\sqrt{2}, \sqrt{2}), (\sqrt{2}, 25), (6, \sqrt{2}), (6, 25), (\frac{3}{2}, \sqrt{2}), (\frac{3}{2}, 25)\}$.

  (b) $\mathbb{Z} \cup S = \{\sqrt{2}, \pi, \frac{5}{2}, 0, 1, -1, 2, -2, \ldots\}$; $\mathbb{Z} \cap S = \{2, 5, 25\}$;

   $\mathbb{Z} \cup T = \{\sqrt{2}, \frac{3}{2}, 0, 1, -1, 2, -2, \ldots\}$; $\mathbb{Z} \cap T = \{4, 25, 6\}$.

3. (a) $\{1, 9, 0, 6, 7\}$

5. (a) $\{c, \{a, b\}\}$;   **(e)** $\emptyset$

6. (a) $A^c = (-2, 1]$

7. (a) 2^{n-2}

8. $(a, b)^c = (-\infty, a] \cup [b, \infty)$, $[a, b)^c = (-\infty, a) \cup [b, \infty)$, $(a, \infty)^c = (-\infty, a]$ and $(-\infty, b]^c = (b, \infty)$.

9. (a) $T \subseteq CS$;   **(b)** $M \cap P = \emptyset$.

11. (a) $E \cap P \neq \emptyset$

12. (a) $A_3 \cup A_{-3} = A_3$

13. Region 2 represents $(A \cap C) \setminus B$. Region 3 represents $A \cap B \cap C$; region 4 represents $(A \cap B) \setminus C$.

14. (a) $A \subseteq B$, by Problem 7.

15. Think of listing the elements of the given set. There are n pairs of the form $(1, b)$, $n - 1$ pairs of the form $(2, b)$, $n - 2$ pairs of the form $(3, b)$, and so on until finally we list the only pair of the form (n, b). The answer is $1 + 2 + 3 + \cdots + n = \frac{1}{2}n(n + 1)$.

17. (b) Yes. Given $A \cap B = A \cap C$ and $A^c \cap B = A^c \cap C$, certainly we have $A \cap B \subseteq C$ and $A^c \cap B \subseteq C$ so, from (a), we have that $B \subseteq C$. Reversing the roles of B and C in (a), we can also conclude that $C \subseteq B$; hence, $B = C$.

20. Using the fact that $X \setminus Y = X \cap Y^c$, we have

$$(A \setminus B) \setminus C = (A \cap B^c) \cap C^c = A \cap (B^c \cap C^c) = A \cap (B \cup C)^c = A \setminus (B \cup C).$$

23. (a) $(A \cup B \cup C)^c = [A \cup (B \cup C)]^c = A^c \cap (B \cup C)^c = A^c \cap (B^c \cap C^c) = A^c \cap B^c \cap C^c$.

   $(A \cap B \cap C)^c = [A \cap (B \cap C)]^c = A^c \cup (B \cap C)^c = A^c \cup (B^c \cup C^c) = A^c \cup B^c \cup C^c$.

24. (a) Looking at the Venn diagram at the right, $A \oplus B$ consists of the points in regions 1 and 3. To have $A \oplus B = A$, we must have both regions 2 and 3 empty; that is, $B = \emptyset$. On the other hand, since $A \oplus \emptyset = A$, this condition is necessary and sufficient.

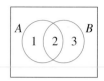

25. (a) This does not imply $B = C$. For example, let $A = \{1, 2\}$, $B = \{1\}$, $C = \{2\}$. Then $A \cup B = A \cup C$, but $B \neq C$.

26. (d) True. Let $x \in A$. Then $x \in A \cup B$, so $x \in A \cap B$ and, in particular, $x \in B$. Thus, $A \subseteq B$. Similarly, we have $B \subseteq A$, so $A = B$.

28. (c) True. Let $(x, y) \in (A \oplus B) \times C$. This means that $x \in A \oplus B$ and $y \in C$; that is, $x \in A \cup B$, $x \notin A \cap B$, $y \in C$. If $x \in A$, then $x \notin B$, so $(x, y) \in (A \times C) \setminus (B \times C)$. If $x \in B$, then $x \notin A$, so $(x, y) \in (B \times C) \setminus (A \times C)$. In either case, $(x, y) \in (A \times C) \oplus (B \times C)$. So $(A \oplus B) \times C \subseteq (A \times C) \oplus (B \times C)$.

Now, let $(x, y) \in (A \times C) \oplus (B \times C)$. This means that $(x, y) \in (A \times C) \cup (B \times C)$, but $(x, y) \notin (A \times C) \cap (B \times C)$. If $(x, y) \in A \times C$, then $(x, y) \notin B \times C$, so $x \in A$, $y \in C$ and, therefore, $x \notin B$. If $(x, y) \in B \times C$, then $(x, y) \notin A \times C$, so $x \in B$, $y \in C$ and, therefore, $x \notin A$. In either case, $x \in A \oplus B$ and $y \in C$, so $(x, y) \in (A \oplus B) \times C$. Therefore, $(A \times C) \oplus (B \times C) \subseteq (A \oplus B) \times C$ and we have equality, as claimed.

Exercises 2.3

1. $S \times B$ is the set of ordered pairs (s, b), where s is a student and b is a book; thus, $S \times B$ represents all possible pairs of students and books. One sensible example of a binary relation is $\{(s, b) \mid s$ has used book $b\}$.

3. (a) not reflexive, not symmetric, not transitive.

(c) not reflexive, not symmetric, but it is transitive.

4. (a)

	a	b	c	d
a	×	×		
b	×	×		
c			×	
d				×

5. (a) $\{(1, 1), (1, 2), (2, 3)\}$;   (g) $\{(1, 2), (2, 1), (1, 1), (2, 2)\}$.

7. The argument assumes that for $a \in \mathcal{R}$ there exists a b such that $(a, b) \in \mathcal{R}$. This need not be the case: See Exercise 5(g).

8. (a) Reflexive since every word has at least one letter in common with itself.

Symmetric since if a and b have at least one letter in common, then so do b and a.

Not antisymmetric. (cat, dot) and (dot, cat) are both in the relation but dot $\neq$ cat!!

Not transitive. (cat, dot) and (dot, mouse) are both in the relation but (cat, mouse) is not.

9. (b) Not reflexive: $(2, 2) \notin \mathcal{R}$.

Not symmetric: $(3, 4) \in \mathcal{R}$ but $(4, 3) \notin \mathcal{R}$.

Not antisymmetric: $(1, 2)$ and $(2, 1)$ are both in $\mathcal{R}$ but $1 \neq 2$.

Not transitive: $(2, 1)$ and $(1, 2)$ are in $\mathcal{R}$ but $(2, 2)$ is not.

(c) Reflexive: For any $a \in \mathbb{Z}$, it is true that $a^2 \geq 0$. Thus, $(a, a) \in \mathcal{R}$.

Symmetric: If $(a, b) \in \mathcal{R}$, then $ab \geq 0$, so $ba \geq 0$ and hence, $(b, a) \in \mathcal{R}$.

Not antisymmetric: $(5, 2) \in \mathcal{R}$ because $5(2) = 10 \geq 0$ and similarly $(2, 5) \in \mathcal{R}$, but $5 \neq 2$.

Not transitive: $(5, 0) \in \mathcal{R}$ because $5(0) = 0 \geq 0$ and similarly, $(0, -6) \in \mathcal{R}$; however, $(5, -6) \notin \mathcal{R}$ because $5(-6) \ngeq 0$.

(i) Reflexive: If $(x, y) \in \mathbb{R}^2$, then $x + y \leq x + y$, so $((x, y), (x, y)) \in \mathcal{R}$.

Not symmetric: $((1, 2), (3, 4)) \in \mathcal{R}$ since $1 + 2 \leq 3 + 4$, but $((3, 4), (1, 2)) \notin \mathcal{R}$ since $3 + 4 \nleq 1 + 2$.

Not antisymmetric: $((1, 2), (0, 3)) \in \mathcal{R}$ since $1 + 2 \leq 0 + 3$ and $((0, 3), (1, 2)) \in \mathcal{R}$ since $0 + 3 \leq 1 + 2$, but $(1, 2) \neq (0, 3)$.

Transitive: If $((a, b), (c, d))$ and $((c, d), (e, f))$ are both in $\mathcal{R}$, then $a + b \leq c + d$ and $c + d \leq e + f$, so $a + b \leq e + f$ (by transitivity of $\leq$) which says $((a, b), (e, f)) \in \mathcal{R}$.

10. (a) Reflexive: For any set X, we have $X \subseteq X$.

Not symmetric: Let $a, b \in S$. Then $\{a\} \subseteq \{a, b\}$ but $\{a, b\} \nsubseteq \{a\}$.

Antisymmetric: If $X \subseteq Y$ and $Y \subseteq X$, then $X = Y$.

Transitive: If $X \subseteq Y$ and $Y \subseteq Z$, then $X \subseteq Z$.

11. (a) Reflexive: Any book has price $\geq$ its own price and length $\geq$ its own length, so $(a, a) \in \mathcal{R}$ for any book a.

Not symmetric: $(Y, Z) \in \mathcal{R}$ because the price of Y is greater than the price of Z and the length of Y is greater than the length of Z, but for these same reasons, $(Z, Y) \notin \mathcal{R}$.

Antisymmetric: If (a, b) and (b, a) are both in $\mathcal{R}$, then a and b must have the same price and length. This is not the case here unless $a = b$.

Transitive: If (a, b) and (b, c) are in $\mathcal{R}$, then the price of a is $\geq$ the price of b and the price of b is $\geq$ the price of c, so the price of a is $\geq$ the price of c. Also the length of a is $\geq$ the length of b and the length of b is $\geq$ the length of c, so the length of a is $\geq$ the length of c. Hence, $(a, c) \in \mathcal{R}$.

Exercises 2.4

2. (a) This is not reflexive: $(2, 2) \notin \mathcal{R}$.

3. Equality! The equivalence classes specify that $x \sim y$ if and only if $x = y$.

4. (a) Reflexive: If $a \in \mathbb{R} \setminus \{0\}$, then $a \sim a$ because $\frac{a}{a} = 1 \in \mathbb{Q}$.

Symmetric: If $a \sim b$, then $\frac{a}{b} \in \mathbb{Q}$ and this fraction is not zero (because $0 \notin A$). So it can be inverted and we see that $\frac{b}{a} = 1/\frac{a}{b} \in \mathbb{Q}$ too. Therefore, $b \sim a$.

Transitive: If $a \sim b$ and $b \sim c$, then $\frac{a}{b} \in \mathbb{Q}$ and $\frac{b}{c} \in \mathbb{Q}$. Since the product of rational numbers is rational, $\frac{a}{c} = \frac{a}{b}\frac{b}{c}$ is in $\mathbb{Q}$, so $a \sim c$.

(b) $\overline{1} = \{a \mid a \sim 1\} = \{a \mid \frac{a}{1} \in \mathbb{Q}\} = \{a \mid a \in \mathbb{Q}\} = \mathbb{Q} \setminus \{0\}$.

(c) $\frac{\sqrt{12}}{\sqrt{3}} = \frac{2\sqrt{3}}{\sqrt{3}} = 2 \in \mathbb{Q}$, so $\sqrt{3} \sim \sqrt{12}$ and hence $\overline{\sqrt{3}} = \overline{\sqrt{12}}$.

6. (a) Reflexive: For any $a \in \mathbb{R}$, $a \sim a$ because $a - a = 0 \in \mathbb{Z}$.

Symmetric: If $a \sim b$, then $a - b \in \mathbb{Z}$, so $b - a \in \mathbb{Z}$ (because $b - a = -(a - b)$) and, hence, $b \sim a$.

Transitive: If $a \sim b$ and $b \sim c$, then both $a - b$ and $b - c$ are integers; hence, so is their sum, $(a - b) + (b - c) = a - c$. Thus, $a \sim c$.

(b) $\overline{5} = \{x \in \mathbb{R} \mid x \sim 5\} = \{x \mid x - 5 \in \mathbb{Z}\} = \mathbb{Z}$;

$\overline{5\frac{1}{2}} = \{x \in \mathbb{R} \mid x \sim 5\frac{1}{2}\} = \{x \mid x = n + \frac{1}{2}, \text{ for some } n \in \mathbb{Z}\}$

(c) For each $a \in \mathbb{R}$, $0 \leq a < 1$, there is one equivalence class $\overline{a} = \{x \in \mathbb{R} \mid x = a + n \text{ for some integer } n\}$. The quotient set is $\{\overline{a} \mid 0 \leq a < 1\}$.

7. Reflexive: For any $a \in \mathbb{Z}$, $a \sim a$ because $2a + 3a = 5a$.

Symmetric: If $a \sim b$, then $2a + 3b = 5n$ for some integer n. So $2b + 3a = (5a + 5b) - (2a + 3b) = 5(a + b) - 5n = 5(a + b - n)$. Since $a + b - n$ is an integer, $b \sim a$.

Transitive: If $a \sim b$ and $b \sim c$, then $2a + 3b = 5n$ and $2b + 3c = 5m$ for integers n and m. Therefore, $(2a + 3b) + (2b + 3c) = 5(n + m)$ and $2a + 3c = 5(n + m) - 5b = 5(n + m - b)$. Since $n + m - b$ is an integer, $a \sim c$.

10. (a) Reflexive: If $a \in Z \setminus \{0\}$, then $aa = a^2 > 0$, so $a \sim a$.

Symmetric: If $a \sim b$, then $ab > 0$. So $ba > 0$ and $b \sim a$.

Transitive: If $a \sim b$ and $b \sim c$, then $ab > 0$ and $bc > 0$. Also $b^2 > 0$ since $b \neq 0$. Hence, $ac = \frac{(ac)b^2}{b^2} = \frac{(ab)(bc)}{b^2} > 0$ since $ab > 0$ and $bc > 0$. Hence, $a \sim c$.

(b) $\begin{aligned} \overline{5} &= \{x \in Z \setminus \{0\} \mid x \sim 5\} = \{x \mid 5x > 0\} = \{x \mid x > 0\} \\ \overline{-5} &= \{x \in Z \setminus \{0\} \mid x \sim -5\} = \{x \mid -5x > 0\} = \{x \mid x < 0\} \end{aligned}$

(c) This equivalence relation partitions $Z \setminus \{0\}$ into the positive and the negative integers.

11. (a) Reflexive: For any $a \in Z$, $a^2 - a^2 = 0$ is divisible by 3, so $a \sim a$.

Symmetric: If $a \sim b$, then $a^2 - b^2$ is divisible by 3, so $b^2 - a^2$ is divisible by 3, so $b \sim a$.

Transitive: If $a \sim b$ and $b \sim c$, then $a^2 - b^2$ is divisible by 3 and $b^2 - c^2$ is divisible by 3, so $a^2 - c^2 = (a^2 - b^2) + (b^2 - c^2)$ is divisible by 3.

12. (a) Yes, this is an equivalence relation.

Reflexive: Note that if a is any triangle, $a \sim a$ because a is congruent to itself.

Symmetric: Assume $a \sim b$. Then a and b are congruent. Therefore, b and a are congruent, so $b \sim a$.

Transitive: If $a \sim b$ and $b \sim c$, then a and b are congruent and b and c are congruent, so a and c are congruent. Thus, $a \sim c$.

13. (a) $\{(1, 1), (1, 2), (2, 1), (2, 2), (3, 3), (3, 4), (3, 5), (4, 4), (4, 5), (5, 5), (4, 3), (5, 3), (5, 4)\}$

14. (c) As suggested in the text, a good way to list the equivalence relations on $\{a, b, c\}$ is to list the partitions of this set. Here they are: $\{ \{a\}, \{b\}, \{c\} \}$; $\{ \{a, b, c\} \}$; $\{ \{a, b\}, \{c\} \}$; $\{ \{a, c\}, \{b\} \}$; $\{ \{b, c\}, \{a\} \}$. There are five in all.

15. (a) The given statement is an implication which concludes "$x - y = x - y$," whereas what is required is a logical argument which concludes "so $\sim$ is reflexive."

A correct argument is this: For any $(x, y) \in R^2$, $x - y = x - y$; thus, $(x, y) \sim (x, y)$. Therefore, $\sim$ is reflexive.

16.   Reflexive: If $(x, y) \in R^2$, then $x^2 - y^2 = x^2 - y^2$, so $(x, y) \sim (x, y)$.

Symmetric: If $(x, y) \sim (u, v)$, then $x^2 - y^2 = u^2 - v^2$, so $u^2 - v^2 = x^2 - y^2$ and $(u, v) \sim (x, y)$.

Transitive: If $(x, y) \sim (u, v)$ and $(u, v) \sim (w, z)$, then $x^2 - y^2 = u^2 - v^2$ and $u^2 - v^2 = w^2 - z^2$, so $x^2 - y^2 = u^2 - v^2 = w^2 - z^2$; $x^2 - y^2 = w^2 - z^2$ and $(x, y) \sim (w, z)$.

$$\overline{(0, 0)} = \{(x, y) \mid (x, y) \sim (0, 0)\} = \{(x, y) \mid x^2 - y^2 = 0^2 - 0^2 = 0\} = \{(x, y) \mid y = \pm x\}$$

Thus, the equivalence class of $(0, 0)$ is the pair of lines with equations $y = x$, $y = -x$.

$$\overline{(1, 0)} = \{(x, y) \mid (x, y) \sim (1, 0)\} = \{(x, y) \mid x^2 - y^2 = 1^2 - 0^2 = 1\}$$

Thus, the equivalence class of $(1, 0)$ is the hyperbola whose equation is $x^2 - y^2 = 1$.

19. (a) The ordered pairs defined by $\sim$ are $(1, 1), (1, 4), (1, 9), (2, 2), (2, 8), (3, 3), (4, 1), (4, 4), (4, 9), (5, 5), (6, 6), (7, 7), (8, 2), (8, 8), (9, 1), (9, 4), (9, 9)$.

(b) $\overline{1} = \{1, 4, 9\} = \overline{4} = \overline{9}$; $\overline{2} = \{2, 8\} = \overline{8}$; $\overline{3} = \{3\}$; $\overline{5} = \{5\}$; $\overline{6} = \{6\}$; $\overline{7} = \{7\}$.

(c) Since the sets $\{1, 4, 9\}$, $\{2, 8\}$, $\{3\}$, $\{5\}$, $\{6\}$ and $\{7\}$ partition A, they determine an equivalence relation, namely, that equivalence relation in which $a \sim b$ if and only if a and b belong to the same one of these sets. This is the given relation.

20.   Reflexive: If $a \in A$, then a^2 is a perfect square, so $a \sim a$.

Symmetric: If $a \sim b$, then ab is a perfect square. Since $ba = ab$, ba is also a perfect square, so $b \sim a$.

Transitive: If $a \sim b$ and $b \sim c$, then ab and bc are each perfect squares. Thus $ab = x^2$ and $bc = y^2$ for integers x and y. Now $ab^2c = x^2y^2$, so $ac = \dfrac{x^2y^2}{b^2} = \left(\dfrac{xy}{b}\right)^2$. Because ac is an integer, so also $\dfrac{xy}{b}$ is an integer. Therefore, $a \sim c$.

Exercises 2.5

1. (a) This defines a partial order.

Reflexive: For any $a \in \mathsf{R}$, $a \geq a$.

Antisymmetric: If $a, b \in \mathsf{R}$, $a \geq b$ and $b \geq a$, then $a = b$.

Transitive: If $a, b, c \in \mathsf{R}$, $a \geq b$ and $b \geq c$, then $a \geq c$.

This partial order is a total order because for any $a, b \in \mathsf{R}$, either $a \geq b$ or $b \geq a$.

(b) This is not a partial order because the relation is not reflexive; for example, $1 < 1$ is not true.

2. (a) 1, 10, 100, 1000, 1001, 101, 1010, 11, 110, 111

3. (a) $(a, b), (a, c), (a, d), (b, c), (b, d), (c, d)$ (b) $(a, b), (c, d)$

4. (a) a is minimal and minimum; d is maximal and maximum.

(b) a and c are minimal; b and d are maximal; there are no minimum nor maximum elements.

7. $A \subsetneq B$ and the set B contains exactly one more element than A.

9. (a) Let $(A, \preceq)$ be a finite poset and let $a \in A$. If a is not maximal, there is an element a_1 such that $a_1 \succ a$. If a_1 is not maximal, there is an element a_2 such that $a_2 \succ a_1$. Continue. Since A is finite, eventually this process must stop, and it stops at a maximal element. A similar argument shows that $(A, \preceq)$ must also contain minimal elements.

11. (a) Suppose that a and b are two maximum elements in a poset $(A, \preceq)$. Then $a \preceq b$ because b is maximum and $b \preceq a$ because a is maximum, so $a = b$ by antisymmetry.

12. (a) Assuming it exists, the greatest lower bound G of A and B has two properties:

(1) $G \subseteq A$, $G \subseteq B$;

(2)   if $C \subseteq A$ and $C \subseteq B$, then $C \subseteq G$.

We must prove that $A \cap B$ has these properties. Note first that $A \cap B \subseteq A$ and $A \cap B \subseteq B$, so $A \cap B$ satisfies (1). Also, if $C \subseteq A$ and $C \subseteq B$, then $C \subseteq A \cap B$, so $A \cap B$ satisfies (2) and $A \cap B = A \wedge B$.

13. (a) $a \vee b = b$ and here is why. We are given $a \preceq b$ and have $b \preceq b$ by reflexivity. Thus b is an upper bound for a and b. It is least because if c is any other upper bound, then $a \preceq c$, $b \preceq c$; in particular, $b \preceq c$.

14. (a) Suppose x and y are each glbs of two elements a and b. Then $x \preceq a$, $x \preceq b$ implies $x \preceq y$ because y is a **greatest** lower bound, and $y \preceq a$, $y \preceq b$ implies $y \preceq x$ because x is greatest. So, by antisymmetry, $x = y$.

16. (a) $(\mathcal{P}(S), \subseteq)$ is not totally ordered provided $|S| \geq 2$ (since $\{a\}$ and $\{b\}$ are not comparable if $a \neq b$). But $\emptyset$ is a minimum because $\emptyset$ is a subset of any set and the set S itself is a maximum because any of its subsets is contained in it.

18. (a) We have to prove that if $b \preceq a$, then $b = a$. So suppose $b \preceq a$. Since a is minimum, we have also $a \preceq b$. By antisymmetry, $b = a$.

Exercises 3.1

1. (a) Not a function; f contains two different pairs of the form $(3, -)$.   (c) This is a function.

2. (a) This is not a function unless each student at the University of Calgary has just one professor, for if student a is taking courses from professors b_1 and b_2, the given set contains (a, b_1) and (a, b_2).

3. $A \times B$ is a function $A \to B$ if and only if B contains exactly one element.

To see why, first note that if $B = \{b\}$, then $A \times B = \{(a, b) \mid a \in A\}$ is certainly a function.

Conversely, if $A \times B$ is a function but B contains two elements b_1, b_2, then for any $a \in A$, (a, b_1) and (a, b_2) are both in $A \times B$, so $A \times B$ is not a function.

4. (a) the function defined by $f(n) = 2n$, for example.

5. (a) If $x \in X$, then x is a country in the British Commonwealth with a uniquely determined Prime Minister y who lives in that country; that is, $y \in Y$.

   If $y \in Y$, then y is a person living in one of the countries in the British Commonwealth. Thus, the domicile of y is a uniquely determined element $x \in X$.

6. Parking rates, bus fares, admission prices are several common examples.

8. (a) add is not one-to-one since, for example, add$(1, 1)$=add$(0, 2)$ while $(1, 1) \neq (0, 2)$. It is onto, however, because, for any $y \in \mathsf{R}$, the equation $y = \text{add}(x)$ has the solution $x = (y, 0)$.

9. (a) g is not one-to-one since, for example, $g(1) = g(-1) = 2$. Neither is g onto: For any $x \in \mathsf{Z}$, $|x| \geq 0$, so $|x| + 1 \geq 1$. Thus, for example, $0 \notin \text{rng } g$.

   (b) g is not one-to-one as in (a). This time it is onto, however, because for $n \in \mathsf{N}$, the equation $n = g(x)$ has the solution $x = n - 1$. (Note that $n > 0$ for $n \in \mathsf{N}$, so $|n - 1| = n - 1$.)

10. (a) If $f(x_1) = f(x_2)$, then $3x_1 + 5 = 3x_2 + 5$, so $x_1 = x_2$ which proves that f is one-to-one. Also f is onto, since given $y \in \mathsf{Q}$, $y = f(\frac{1}{3}(y - 5))$ with $\frac{1}{3}(y - 5) \in \mathsf{Q}$.

13. (a) Note that $f(x) = (x + 7)^2 - 100$. If $f(x_1) = f(x_2)$ it follows that $(x_1 + 7)^2 - 100 = (x_2 + 7)^2 - 100$, so $(x_1 + 7)^2 = (x_2 + 7)^2$ and, taking square roots, $|x_1 + 7| = |x_2 + 7|$. Since $x_1, x_2 \in \mathsf{N}$, we know $x_1 + 7 > 0$ and $x_2 + 7 > 0$. Thus, $x_1 + 7 = x_2 + 7$ and $x_1 = x_2$. Thus, f is one-to-one, but it is not onto: For example, $1 \in B$, but there is no $x \in \mathsf{N}$ with $f(x) = 1$ since $(x + 7)^2 - 100 = 1$ implies $(x + 7)^2 = 101$ and this equation has no solution in the natural numbers.

14. (a) The domain of f is R. Its range is also R because every $y \in \mathsf{R}$ can be written $y = f(x)$ for some x; namely, $x = y^{1/3}$. The function is, therefore, onto. It is also one-to-one: If $f(x_1) = f(x_2)$, then $x_1^3 = x_2^3$ and this implies $x_1 = x_2$.

15. (a) The graph of f shown at the right makes it clear that f is one-to-one and onto.

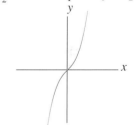

   (b) Since $f: \mathsf{R} \to \mathsf{R}$ is one-to-one by part (a), it is also one-to-one as a function with domain Z. Here, however, f is not onto for we note that $f(0) = 0$, $f(1) = 4$ and f is increasing, so 1 is not in the range of f.

17. (a) This is not one-to-one since, for example, $f(1, 3) = f(4, 1) = 11$ while $(1, 3) \neq (4, 1)$. The function is not onto since, for example, the equation $f(x) = 1$ has no solution $x = (n, m)$ (because $2n + 3m \geq 5$ for every $n, m \in \mathsf{N}$).

   (b) This is not one-to-one since, for example, $f(1, 3) = f(4, 1) = 11$ while $(1, 3) \neq (4, 1)$. The function is onto, however, since for $k \in \mathsf{Z}$, the equation $k = f(x)$ has the solution $x = (-k, k)$.

18. (a) $A = \{x \in \mathsf{R} \mid x \neq 3\}$; rng $f = \{y \mid y \neq 0\}$.

19. (b) Note that $f(x) = -(2x - 3)^2$. This function is not one-to-one since, for example, $f(1) = f(2) (= -1)$. Restrict the domain to $\{x \mid x \geq 3/2\}$ (or to $\{x \mid x \leq \frac{3}{2}\}$).

22. (a) Here are the functions $X \to Y$:

$$\begin{array}{lll} \{(a, 1), (b, 1)\} & \{(a, 1), (b, 2)\} & \{(a, 1), (b, 3)\} \\ \{(a, 2), (b, 1)\} & \{(a, 2), (b, 2)\} & \{(a, 2), (b, 3)\} \\ \{(a, 3), (b, 1)\} & \{(a, 3), (b, 2)\} & \{(a, 3), (b, 3)\}. \end{array}$$

Here are the functions $Y \rightarrow X$:

$$\{(1, a), (2, a), (3, a)\} \quad \{(1, a), (2, a), (3, b)\}$$
$$\{(1, a), (2, b), (3, a)\} \quad \{(1, a), (2, b), (3, b)\}$$
$$\{(1, b), (2, a), (3, a)\} \quad \{(1, b), (2, a), (3, b)\}$$
$$\{(1, b), (2, b), (3, a)\} \quad \{(1, b), (2, b), (3, b)\}.$$

(b) There are no one-to-one functions $Y \rightarrow X$. The one-to-one functions from $X \rightarrow Y$ are

$$\{(a, 1), (b, 2)\} \quad \{(a, 1), (b, 3)\} \quad \{(a, 2), (b, 1)\}$$
$$\{(a, 2), (b, 3)\} \quad \{(a, 3), (b, 1)\} \quad \{(a, 3), (b, 2)\}.$$

(c) There are no onto functions $X \rightarrow Y$. The onto functions $Y \rightarrow X$ are

$$\{(1, a), (2, a), (3, b)\} \quad \{(1, a), (2, b), (3, a)\}$$
$$\{(1, a), (2, b), (3, b)\} \quad \{(1, b), (2, a), (3, a)\}$$
$$\{(1, b), (2, a), (3, b)\} \quad \{(1, b), (2, b), (3, a)\}.$$

24.

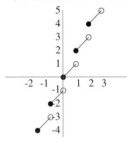

		1	2	3	4
	1	1	2	3	4
m	2	1	4	9	16
	3	1	8	27	64
	4	1	16	81	256

(top label: n)

We would guess that for $|X| = m$ and $|Y| = n$, the number of functions $X \rightarrow Y$ is n^m.

26. (a) ($\rightarrow$) Suppose A and B each contain n elements. Assume that $f: A \rightarrow B$ is one-to-one and let $C = \{f(a) \mid a \in A\}$. Since $f(a_1) \neq f(a_2)$ if $a_1 \neq a_2$, C is a subset of B containing n elements; so $C = B$. Therefore, f is onto.

27. (a)

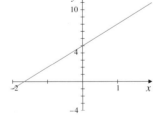

(b) The domain of f is R. The range is the set of all real numbers of the form $2n + a$, where $n \in$ Z and $0 \le a < 1$.

Exercises 3.2

1. (a) $f^{-1} = \{(1, 5), (2, 1), (3, 2), (4, 3), (5, 4)\}$ (b) $f^{-1} = \{(1, 4), (2, 1), (3, 3), (4, 2), (5, 5)\}$

2. (a) $f^{-1}: \text{R} \rightarrow \text{R}$ is defined by $f^{-1}(x) = \frac{1}{3}(x - 5)$.

3. (a) If $f(x_1) = f(x_2)$, then $1 + \dfrac{1}{x_1 - 4} = 1 + \dfrac{1}{x_2 - 4}$, $\dfrac{1}{x_1 - 4} = \dfrac{1}{x_2 - 4}$ and $x_1 - 4 = x_2 - 4$. Thus $x_1 = x_2$ and f is one-to-one.

rng $f = B = \{y \in \text{R} \mid y \neq 1\}$ and $f^{-1}(x) = 4 + \dfrac{1}{x - 1}$.

5. (a) maternal grandmother

6. (a) $f \circ g = \{(1, 1), (3, 8), (2, 1), (4, 9), (5, 1)\}$

$g \circ f$ is not defined because rng $f = \{1, 2, 3, 8, 9\} \nsubseteq \text{dom } g = \{1, 2, 3, 4, 5\}$.

$f \circ f$ is not defined because rng $f = \{1, 2, 3, 8, 9\} \nsubseteq \text{dom } f = \{1, 2, 3, 4, 5\}$.

$g \circ g = \{(1, 2), (2, 2), (3, 2), (4, 1), (5, 2)\}$.

7. (a) $g^{-1} \circ f \circ g = \{(1, 3), (2, 1), (3, 2), (4, 4)\}$

8. $g \circ f(x) = \dfrac{1}{(x + 2)^2 + 1} = \dfrac{1}{x^2 + 4x + 5}; \quad f \circ g(x) = \dfrac{1}{x^2 + 1} + 2 = \dfrac{2x^2 + 3}{x^2 + 1};$

$h \circ g \circ f(x) = 3; \ g \circ h \circ f(x) = g(3) = \frac{1}{10}.$

Since $f^{-1} \circ f(x) = x$, we have $g \circ f^{-1} \circ f(x) = g(x) = \dfrac{1}{x^2 + 1}.$

Since $f^{-1}(x) = x - 2$, we have

$$f^{-1} \circ g \circ f(x) = f^{-1}\left(\frac{1}{(x + 2)^2 + 1}\right) = \frac{1}{(x + 2)^2 + 1} - 2 = \frac{-2(x + 2)^2 - 1}{(x + 2)^2 + 1}.$$

10. $(g \circ f)(x) = g(f(x)) = f(x) - c$. Thus the graph of $g \circ f$ is the graph of f translated vertically c units down if $c > 0$ and $-c$ units up if $c < 0$. The graphs are identical if $c = 0$.

12. (a) Since $-|x| = \begin{cases} -x & \text{if } x \geq 0 \\ x & \text{if } x < 0, \end{cases}$ we have $f \circ g(x) = f(-|x|) = \begin{cases} f(-x) & \text{if } x \geq 0 \\ f(x) & \text{if } x < 0. \end{cases}$

So the graph of $f \circ g$ is the same as the graph of f to the left of the y-axis (where $x < 0$) while to the right of the y-axis, the graph of $f \circ g$ is the reflection (mirror image) of the left half of the graph of f in the y-axis. We call $f \circ g$ an *even* function since it is symmetric with respect to the y-axis: $f \circ g(-x) = f \circ g(x)$.

15. (a) $f \circ g = \{(1, 4), (2, 3), (3, 2), (4, 1), (5, 5)\}; \ g \circ f = \{(1, 5), (2, 3), (3, 2), (4, 4), (5, 1)\}$ Clearly, $f \circ g \neq g \circ f$.

16. (a) For $x \in B$, $(f \circ g)(x) = f\left(\frac{2x}{x-1}\right) = \dfrac{\frac{2x}{x-1}}{\frac{2x}{x-1} - 2} = x.$

(b) For $x \in A$, $(g \circ f)(x) = g\left(\frac{x}{x-2}\right) = \dfrac{2\left(\frac{x}{x-2}\right)}{\frac{x}{x-2} - 1} = x$ and so, by Proposition 3.2.7, f and g are inverses.

18. (a) Suppose $g(b_1) = g(b_2)$ for $b_1, b_2 \in B$. Since f is onto, $b_1 = f(a_1)$ and $b_2 = f(a_2)$ for some $a_1, a_2 \in A$. Thus, $g(f(a_1)) = g(f(a_2))$; that is, $g \circ f(a_1) = g \circ f(a_2)$. Since $g \circ f$ is one-to-one, $a_1 = a_2$. Therefore, $b_1 = f(a_1) = f(a_2) = b_2$ proving that g is one-to-one.

19. (a) Suppose $f: A \to B$ and $g: B \to C$ are one-to-one. We prove that $g \circ f: A \to C$ is one-to-one. For this, suppose $(g \circ f)(a_1) = (g \circ f)(a_2)$ for some $a_1, a_2 \in A$. Then $g(f(a_1)) = g(f(a_2))$ [an equation of the form $g(b_1) = g(b_2)$]. Since g is one-to-one, we conclude that $f(a_1) = f(a_2)$, and then, since f is one-to-one, that $a_1 = a_2$.

20. (b) With $A = \{1, 2\}$, $B = \{-1, 1, 2\}$, $C = \{5, 10\}$, $f = \{(1, 1), (2, 2)\}$ and $g = \{(-1, 5), (1, 5), (2, 10)\}$ we have $g \circ f = \{(1, 5), (2, 10)\}$. Then $g \circ f: A \to C$ is onto but f is not.

21. Since a bijective function is, by definition, a one-to-one onto function, we conclude, by the results of part (a) of the previous two exercises, that indeed the composition of bijective functions is bijective.

24. (b) Writing $x = \lfloor x \rfloor + a$, $0 \leq a < 1$, we have $t(x) = \lfloor x \rfloor - a$. Now it is straightforward to see that t is one-to-one: Suppose $t(x_1) = t(x_2)$, where $x_1 = \lfloor x_1 \rfloor + a_1$, $x_2 = \lfloor x_2 \rfloor + a_2$ and $0 \leq a_1, a_2 < 1$. Then $\lfloor x_1 \rfloor - a_1 = \lfloor x_2 \rfloor - a_2$, so $\lfloor x_1 \rfloor - \lfloor x_2 \rfloor = a_1 - a_2$. The left side is an integer; hence, so is the right. Because of the restrictions on a_1 and a_2, the only possibility is $a_1 = a_2$. Hence, also $\lfloor x_1 \rfloor = \lfloor x_2 \rfloor$, so $x_1 = x_2$.

Exercises 3.3

1. Ask everyone to find a seat.

2. The two lists $1^2, 2^2, 3^2, 4^2, \ldots$ and $1, 2, 3, 4, \ldots$ obviously have the same length; $a^2 \mapsto a$ is a one-to-one correspondence between the set of perfect squares and N.

3. (c) The function $f: \mathsf{N} \times \mathsf{N} \to \mathsf{C}$ defined by $f(m, n) = m + ni$ for all $m, n \in \mathsf{N}$.

6. This is false. For example, $|\mathsf{N}| = |\mathsf{N} \cup \{0\}|$, as shown in the text.

9. (a) False. Let $X = \{1\}$, $Y = \{2\}$, $Z = \{3\}$. Then $((1, 2), 3) \in (X \times Y) \times Z$ but $((1, 2), 3) \notin X \times (Y \times Z)$ (a set whose **second** coordinates are ordered pairs).

(b) Define $f: (X \times Y) \times Z \to X \times (Y \times Z)$ by $f((x, y), z) = (x, (y, z))$.

11. (a) $f(x) = x + 2$ is a one-to-one correspondence between $(1, \infty)$ and $(3, \infty)$. We conclude that these sets have the same cardinality.

12. (a) The function defined by $f(x) = x + 1$ is a one-to-one correspondence between $(0, 1)$ and $(1, 2)$.

13. (a) The function defined by $g(x) = \dfrac{1}{x} + 9$ is a one-to-one correspondence between $(0, 1)$ and $(10, \infty)$. This is just $f(x) + 10$, where f is the function given in Problem 26.

15. f is certainly a function from R to R^+ since $3^x > 0$ for all $x \in \mathsf{R}$. If $3^x = 3^y$, then $x \log 3 = y \log 3$ (any base), so $x = y$. Thus, f is one-to-one. If $r \in \mathsf{R}^+$, then $3^{\log_3 r} = r$, so f is onto. We conclude that R and R^+ have the same cardinality.

16. Since (a, b) has the same cardinality as R^+ by Exercise 14 and since R^+ and R have the same cardinality by Exercise 15, the result follows by transitivity of the notion of "same cardinality"—see Exercise 10.

17. (d) Follow the procedure given in the text for all rational numbers and omit those with even denominators. The listing starts $1, 2, \frac{1}{3}, 3, 4, \frac{2}{3}, \frac{1}{5}, 5, 6, \frac{4}{3}, \frac{2}{5}, \frac{1}{7}, \frac{3}{5}, \frac{5}{3}, 7, \ldots$.

18. (a) This set is uncountable. The function defined by $f(x) = x - 1$ gives a one-to-one correspondence between it and $(0, 1)$, which we showed in the text to be uncountable.

(e) This set is countably infinite. In Exercise 3(c) we showed it is in one-to-one correspondence with $\mathsf{N} \times \mathsf{N}$.

20. (a) Impossible. To the contrary, suppose that the union were a finite set S. Since S has only finitely many subsets (the precise number is $2^{|S|}$), there could not have been infinitely many sets at the outset.

(c) Impossible. If even one infinite set is contained in a **union**, then the union must be infinite.

21. Imagine S_1 sitting inside S_2, both spheres with the same center. Rays emanating from this common point establish a one-to-one correspondence between the points on S_1 and the points on S_2.

22. Let $s_1, s_2, s_3, \ldots$ be a countably infinite subset of S. Define $f: S \to S \cup \{x\}$ by

$$f(s_1) = x$$
$$f(s_{k+1}) = s_k \quad \text{for } k \geq 1$$
$$f(s) = s \quad \text{if } s \notin \{s_1, s_2, s_3, \ldots\}.$$

Then f is a one-to-one correspondence.

24. We are given that $A = \{a_1, a_2, \ldots, a_n\}$ for some $n \in \mathsf{N}$ and that $B = \{b_1, b_2, b_3, \ldots\}$. Then $A \cup B$ is countably infinite because it is infinite and its elements can be listed $a_1, a_2, \ldots, a_n, b_1, b_2, b_3, \ldots$. The function $f: \mathsf{N} \to A \cup B$ corresponding to this listing is defined by $f(i) = \begin{cases} a_i & \text{if } i \leq n \\ b_{i-n} & \text{if } i > n. \end{cases}$

Exercises 4.1

1. $(a+b)c = c(a+b)$ (by commutativity) $= ca + cb$ (by the first distributive law) $= ac + bc$ (by commutativity again).

2. (a) True.

4. (a) $q = 29$; $r = 7$;   (b) $q = -30$; $r = 10$ (c) $q = -29$; $r = 7$ (d) $q = 30$; $r = 10$

7. (a) The domain of f is $\mathbb{Z}$; its range is $\mathbb{Z}$ as well, because given $q \in \mathbb{Z}$, we have $q = f(qn)$.

(b) f is not one-to-one since $f(0) = f(1)$. (Note that $n > 1$ guarantees that the quotient is 0 when either 0 or 1 is divided by n.)

(c) f is onto, as shown in (a).

9. By definition, $x \leq \lceil x \rceil < x + 1$ for any x. Letting $k = \lceil a/b \rceil$, we have $a/b \leq k < (a/b) + 1$. Multiplying by the negative number b, we obtain $a \geq kb > a + b$. Thus, $0 \leq a - kb < -b$. Letting $r = a - kb$, we have $a = kb + r$ with $0 \leq r < |b|$, so, by uniqueness, $k = q$ as asserted.

10. (a) $4034 = (111,111,000,010)_2 = (7702)_8 = (FC2)_{16}$.

Exercises 4.2

1. (a) Not totally ordered; for example, 6 and 21 are not comparable.

(b) 1 is a minimum element since $1 \mid n$ for all $n \in \mathbb{N}$, but there is no maximum element since given any $n \in \mathbb{N}$, $n \mid 2n$, so n can't be maximum.

3. (a)

5. As in Problem 6, write $a = nq + r$, where $0 \leq r < n$. Thus, $r = 0, 1, 2, \ldots$, or $n - 1$. If $r = 0$, then $a = qn$ is divisible by n. If $r = 1$, then $a = nq + 1$, so $a + n - 1$ is divisible by n. If $r = 2$, then $a = nq + 2$ and $a + n - 2$ is divisible by n. In general, if $r = k$, then $a + n - k$ is divisible by n.

6. Assume, to the contrary, that for some integer n, we have $4 \mid (n^2 - 2)$; that is, $n^2 - 2 = 4x$ for some integer x. We derive a contradiction. First consider the case that n is even; that is, $n = 2y$ for some integer y. Then $(2y)^2 - 2 = 4x$ says $4(y^2 - x) = 2$. Since $y^2 - x$ is an integer, this says $4 \mid 2$, which is impossible.

Next suppose n is odd; that is, $n = 2y + 1$ for some integer y. Then $(2y + 1)^2 - 2 = 4x$ says $4(y^2 + y - x) = 1$. Since $y^2 + y - x$ is an integer, this says $4 \mid 1$, another impossibility. In either case we reach a contradiction, so $4 \mid (n^2 - 2)$ can never be true.

8. We have $a = q_1 n + r$ and $b = q_2 n + r$ for some integers q_1, q_2 and r. Subtracting, $a - b = (q_1 n + r) - (q_2 n + r) = (q_1 - q_2)n$. Thus, $n \mid (a - b)$ as required.

9. (a) True. Since $a \mid b$, $b = ax$ for some integer x. Since $b \mid (-c)$, $-c = by$ for some integer y. Thus, $c = -by = -axy = a(-xy)$. Since $-xy$ is an integer, $a \mid c$.

11. (a) $\gcd(93, 119) = 1 = (-25)(119) + 32(93)$.

(b) $\gcd(-93, 119) = 1 = (-25)(119) + (-32)(-93)$.

(c) $\gcd(-93, -119) = 1 = (25)(-119) + (-32)(-93)$.

13. Since a and b are relatively prime, we have $ma + nb = 1$ for some integers m and n, so $2 = 2ma + 2nb$. Now suppose $x \mid (a + b)$ and $x \mid (a - b)$. Then, by Proposition 4.2.2, $x \mid [(a + b) + (a - b)]$; that is, $x \mid 2a$. Also $x \mid [(a + b) - (a - b)]$; that is, $x \mid 2b$. Again, by Proposition 4.2.2, $x \mid (2ma + 2nb)$, so $x \mid 2$ and we conclude that $x = \pm 1$ or $x = \pm 2$. The result follows.

14. If g_1 and g_2 are each greatest common divisors of a and b, then $g_1 \leq g_2$ (because g_2 is greatest) and $g_2 \leq g_1$ (because g_1 is greatest), so $g_1 = g_2$.

16. (a) $17369(-10588) + (5472)(33608) = 4$.

    (c) Consider the possibility that $154x + 260y = 3$ for certain integers x and y. The left side of this equation is even while the right side is odd, a contradiction.

17. (a) We know that $\gcd(x, y)\,|\,x$ and $\gcd(x, y)\,|\,y$ and so $\gcd(x, y)\,|\,(mx + ny)$ (Proposition 4.2.2).

19. Let $g_1 = \gcd(a, b)$ and $g_2 = \gcd(ac, bc)$. We have to show that $g_2 = cg_1$. First note that $cg_1\,|\,ac$ since $g_1\,|\,a$, and, similarly, $cg_1\,|\,bc$. Thus, cg_1 is a common divisor of ac and bc, hence, $cg_1 \leq g_2$, the largest of the common divisors of ac, bc. On the other hand, since $g_1 = am + bn$ for some integers m, n, $cg_1 = acm + bcn$. Since $g_2\,|\,acm$ and $g_2\,|\,bcn$, it must be that $g_2\,|\,cg_1$, and so, since c, g_1, g_2 are all positive, $g_2 \leq cg_1$. We conclude that $g_2 = cg_1$, as required.

21. Say $x\,|\,n$ and $x\,|\,(n + 1)$. Then $x\,|\,[(n + 1) - n]$, so $x\,|\,1$ and $\gcd(n, n + 1) = 1$. We have $(-1)n + (1)(n + 1) = 1$.

24. As suggested by the hint, consider the set S of all positive linear combinations of a and b. Since S contains $a = 1a + 0b$ if $a > 0$ and $-a$ otherwise, S is not empty so, by the Well-Ordering Principle, S contains a smallest element g. Since $g \in S$, we know that $g = ma + nb$ for integers a and b; hence, we have only to prove that g is the greatest common divisor of a and b. First we prove that $g\,|\,a$. Write $a = qg + r$ with $0 \leq r < g$. Note that r is a linear combination of a and b since $r = a - qg = a - q(ma + nb) = (1 - qa)m + (-qn)b$. Since g is the smallest positive linear combination of a and b, we have $r = 0$, so $g\,|\,a$ as desired. Similarly, $g\,|\,b$. Finally, if $c\,|\,a$ and $c\,|\,b$, then $c\,|\,(ma + nb)$, so $c\,|\,g$, so $c \leq g$.

25. $\mathrm{lcm}(63, 273) = \frac{63(273)}{21} = 819$; $\mathrm{lcm}(56, 200) = \frac{56(200)}{8} = 1400$.

26. (a) $\mathrm{lcm}(93, 119) = (93 \cdot 119)/1 = 11067$

28. $\gcd(a, b)\,|\,a$ and $a\,|\,\mathrm{lcm}(a, b)$ so $\gcd(a, b)\,|\,\mathrm{lcm}(a, b)$ by transitivity of $|$. (See Proposition 4.2.3.)

30. (a) Let $x = (a_n a_{n-1} \ldots a_0)_{10}$ be the integer. Thus, $x = a_0 + 10a_1 + \cdots + 10^n a_n$.

    Now $x = (a_0 + a_1 + \cdots + a_n) + (9a_1 + 99a_2 + \cdots + \underbrace{99 \cdots 9}_{n\text{ times}} a_n)$

$$= (a_0 + a_1 + \cdots + a_n) + 3(3a_1 + 33a_2 + \cdots + \underbrace{33 \cdots 3}_{n\text{ times}} a_n).$$

    Since $a_0 + a_1 + \cdots + a_n$ is divisible by 3 and 3 clearly divides the second term on the right, $3\,|\,x$ as required.

    (b) Reverse the argument in (a); that is, note that if $3\,|\,x$, then we can conclude from the preceding that $3\,|\,(a_0 + a_1 + \cdots + a_n)$.

33. (b)

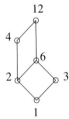

34. (a) By Theorem 4.2.9, we know there is at least one pair of integers a, b such that $g = am + bn$. Now observe that for any integer k, it is also true that $g = (a + kn)m + (b - km)n$.

Exercises 4.3

1. (a) 157 is prime;   **(b)** 9831 is not prime; $3\,|\,9831$.

3. (a) Note that $n = p(n/p)$. Now if n/p is not prime, then it has a prime factor $q < \sqrt{n/p}$ by Lemma 4.3.4. Since a prime factor of n/p is a prime factor of n, q would then be a prime factor of n not exceeding $\sqrt{n/p}$, contradicting the fact that the smallest prime factor of n is bigger than this.

(b) Note that $16{,}773{,}121 = 433(38{,}737)$ and that $\sqrt{38{,}737} \approx 197$. Since 433, the smallest prime dividing 38,737, is larger than 197, 38,737 is prime, by part (a). Thus, $16{,}773{,}121 = 433(38{,}737)$ is the representation of 16,773,121 as the product of primes.

4. (a) $856 = 2^3 \cdot 107;$ **(e)** $(2^8 - 1)^{20} = 3^{20} \cdot 5^{20} \cdot 17^{20}$.

5. If 14^n terminates in 0, then it is divisible by 10 and, hence, by the prime 5; that is, there is a 5 in the factorization of 14^n. This contradicts unique factorization because $14^n = 2^n 7^n$ says that the only primes dividing 14^n are 2 and 7.

6. (b) We have to prove that $a, b \in A$ implies $a + b \in A$. Let $a, b \in A$. Then $a = \frac{m}{n}$ and $b = \frac{k}{\ell}$, where $3 \nmid n$, $3 \nmid \ell$. Then $a + b = \dfrac{m\ell + nk}{n\ell}$. Since 3 is prime, if $3 \mid n\ell$, then $3 \mid n$ or $3 \mid \ell$, neither of which is true. Thus, $3 \nmid n\ell$, so $a + b \in A$.

8. (b) False. The prime described here has the miraculous property that it divides all natural numbers greater than 1. Certainly this prime cannot be 2 since $2 \nmid 3$, neither can it be odd since if p is an odd prime, $p \nmid 2$.

10. (a) This function is one-to-one. If $f(n_1, m_1) = f(n_2, m_2)$, then $2^{m_1} 6^{n_1} = 2^{m_2} 6^{n_2}$, and so $2^{m_1} 2^{n_1} 3^{n_1} = 2^{m_2} 2^{n_2} 3^{n_2}$. By the Fundamental Theorem of Arithmetic, the powers of 3 and 2 on each side of this equation are the same. So $n_1 = n_2$, hence $m_1 = m_2$ too. Therefore, $(n_1, m_1) = (n_2, m_2)$.

11. (a) $\pi(10) = 4$, $10/\ln 10 \approx 4.343$ and $\frac{\pi(10)}{10/\ln 10} \approx \frac{4}{4.343} \approx 0.921$.

14. This is a special case of Exercise 10, Section 4.2 since if p and q are distinct primes, then p and q are relatively prime. The result also follows quickly from the Fundamental Theorem of Arithmetic since, writing $n = p_1 p_2 \cdots p_r$, the hypotheses say that one of the p_i is p and some other p_j is q.

15. Without loss of generality, we may assume that $a < b$. Since there are infinitely many primes, there exists an integer $n \geq 0$ such that $b + n$ is prime. Since $a + n < b + n$, $b + n$ cannot divide $a + n$, so $a + n$ and $b + n$ are relatively prime.

16. (b) First note that, if p is prime, $d(p^2) = 3$ since the only divisors of p^2 are 1, p and p^2. We claim that only integers of the form p^2 satisfy the condition. To see this, notice that if n is divisible by distinct primes p and q, then n is divisible by each of 1, p, q and pq (see Exercise 14) and so $d(n) \geq 4$. Thus, n has just one prime factor. Also, if n is divisible by the cube of a prime, p^3, then n is divisible by each of 1, p, p^2, p^3 so again, $d(n) \geq 4$. We conclude that $n = p^2$ for some prime p.

17. (a) No. $2^{15} - 1 = (2^5 - 1)(2^{10} + 2^5 + 1)$.

18. (a) $2^6 + 1 = (2^2 + 1)(2^4 - 2^2 + 1)$.

20. Suppose, to the contrary, that p and q are consecutive primes (in that order) such that $p + q = 2r$, where r is a prime. Then $p = \frac{p+p}{2} < \frac{p+q}{2} = r < \frac{q+q}{2} = q$; that is, $p < r < q$. This contradicts the fact that p and q are **consecutive** primes.

22. This is true, and the proof has given Andrew Wiles a place in history. This is Fermat's Last Theorem (4.3.14)!

23. The condition tells us that in the prime decompositions of a, b and c, no prime appears in two of the decompositions. Hence, $a = p_1^{\alpha_1} p_2^{\alpha_2} \cdots p_r^{\alpha_r}$, $b = q_1^{\beta_1} q_2^{\beta_2} \cdots q_s^{\beta_s}$, $c = r_1^{\gamma_1} r_2^{\gamma_2} \cdots r_t^{\gamma_t}$ with the primes p_i, q_i, r_i all different. Thus,

$$ab = p_1^{\alpha_1} \cdots p_r^{\alpha_r} q_1^{\beta_1} \cdots q_s^{\beta_s}$$
$$ac = p_1^{\alpha_1} \cdots p_r^{\alpha_r} r_1^{\gamma_1} \cdots r_t^{\gamma_t}$$
$$bc = q_1^{\beta_1} \cdots q_s^{\beta_s} r_1^{\gamma_1} \cdots r_t^{\gamma_t}.$$

Suppose $\gcd(ab, bc, ac) \neq 1$. Then there must exist a prime s such that $s \mid ab$, $s \mid bc$, $s \mid ac$. But there is no prime common to the decompositions of ab, ac, bc. Hence, no such s exists and $\gcd(ab, bc, ac) = 1$.

25. (a) $\gcd(ab, p^4) = p^3$.

27. (a) Writing $a = p^\alpha r$ and $b = p^\beta s$ with $\gcd(p, r) = 1$ and $\gcd(p, s) = 1$, there are two possibilities: If $\beta = 1$, then $\gcd(a^2, b) = p$, while if $\beta > 1$, then $\gcd(a^2, b) = p^2$.

28. Let the prime decompositions of a and b be

$$a = p_1^{\alpha_1} p_2^{\alpha_2} \cdots p_r^{\alpha_r} \quad \text{and} \quad b = q_1^{\beta_1} q_2^{\beta_2} \cdots q_s^{\beta_s}.$$

Since $\gcd(a, b) = 1$, we know that $p_i \neq q_j$ for any i and j. Thus, $ab = p_1^{\alpha_1} \cdots p_r^{\alpha_r} q_1^{\beta_1} \cdots q_s^{\beta_s}$ with no simplification possible. Since $ab = x^2$, it follows that each α_i and each β_i must be even. But this means that a and b are perfect squares.

30. Let $a = p_1^{\alpha_1} p_2^{\alpha_2} \cdots p_r^{\alpha_r}$. Since a is a square, all the α_i are even. Since a is a cube, all the α_i are divisible by 3. It follows from Exercise 14 that $6 \,|\, \alpha_i$ for all i. This means that a is a 6th power.

31. (a) True. Write $a = p_1^{\alpha_1} p_2^{\alpha_2} \cdots p_r^{\alpha_r}$ as the product of powers of primes. Then $a^{11} = p_1^{11\alpha_1} \cdots p_r^{11\alpha_r}$. If $p \,|\, a^{11}$, then p must be one of the p_i, so $p \,|\, a$.

32. Whenever a, b, c satisfy $x^2 + y^2 = z^2$, so also do ka, kb, kc satisfy the equation, for any integer k. Thus, for example, all triples of the form $3k, 4k, 5k$ satisfy $a^2 + b^2 = c^2$.

33. (a) Write $3n + 2 = p_1 p_2 \cdots p_m$ as the product of (necessarily odd) primes. Each of the p_i is either 3 or of the form $3k + 1$ or $3k + 2$. Now we notice that the product of integers of the form $3k + 1$ is of the same form and the product of 3 with such an integer is a multiple of 3. So if none of the p_i is of the form $3k + 2$, the product cannot be either. If we drop the word odd, the result is false. For example, $3(2) + 2 = 8$ is divisible by only the prime 2, which is not of the form $3n + 2$ for $n \in \mathbb{N}$.

35. (a) $f(8) = 64$;   **(b)** $x = 1000$.

Exercises 4.4

1. (a) $5, 12, 19, -2, -9, -16$ are in $\overline{5}$. $4, 11, 18, -3, -10, -17$ are in $\overline{-3}$.

(b) The general element of $\overline{5}$ is an integer of the form $7k + 5$ for some integer k. The general form of an element in $\overline{-3}$ is an integer of the form $7k - 3$ for some integer k.

3. (a) $1286 \pmod{39} = 38$;   **(c)** $-545{,}608 \pmod{51} = 41$.

4. (a) False: $18 - 2 = 16$ is not divisible by 10.   **(c)** True: $44 - (-8) = 52$ is divisible by 13.

5. (a) $\overline{0} = \overline{3}, \overline{1} = \overline{4}, \overline{2} = \overline{5}$.

6. (a) $4 \pmod 6$;   **(e)** $2, 4, 8, 5, 10, 9, 7, 3, 6, 1$

8. Observe that $5^0 \equiv 1 \pmod 7$, $5^1 \equiv 5 \pmod 7$, $5^2 \equiv 4 \pmod 7$, $5^3 \equiv 6 \pmod 7$, $5^4 \equiv 2 \pmod 7$ and $5^5 \equiv 3 \pmod 7$. Thus, each of the integers $1, 2, 3, 4, 5, 6$ is congruent mod 7 to some power of 5. By Proposition 4.4.5, any integer a is congruent mod 7 to one of $0, 1, 2, 3, 4, 5, 6$. So the result follows.

9. (a) No x exists. The values of $3x \bmod 6$ are 0 and 3.   **(e)** $x = 9, 34$;   **(f)** $x = 9$;   **(k)** $x = 56$.

10. (b) i. $x = 2, 11$.   iv. No x exists. The values of $x^2 \pmod 6$ are 0, 1, 3 and 4.

11. (a) There is just one solution mod 6; namely, $x = 0$, $y = 1$.

(b) Subtracting the first congruence from the second gives $3x \equiv -2 \equiv 7 \pmod 9$. But the values of $3x \bmod 9$ are just 0, 3 and 6. There is no solution.

12. Since $\gcd(c, n) = 1$, we have $cx + ny = 1$ for some integers x and y. We also know that $ac - bc = nk$ for some integer k. So $(a - b)cx = nkx$; that is, $(a - b)(1 - ny) = nkx$, or, $a - b = n(kx + (a - b)y)$. Thus, $n \,|\, (a - b)$ and $a \equiv b \pmod n$, as desired.

13. Since $n \,|\, (a - b)$, we have $a - b = qn$ for some integer q, hence, $a - qn = b$. Let $g = \gcd(a, n)$. Then $g \,|\, a$ and $g \,|\, n$. Therefore, $g \,|\, b$ (and $g \,|\, n$), so $g \,|\, \gcd(b, n)$. Similarly $\gcd(b, n) \,|\, g$ and so these natural numbers are equal.

14. (a) This is false. Consider $r = 5$, $s = 1$, $a = 2$, $n = 4$. Then $5 \equiv 1 \pmod 4$, but $2^5 \not\equiv 2^1 \pmod 4$ because $2^5 - 2^1 = 30$ and $4 \nmid 30$.

15. (a) False! For example, with $a = b = 1$ and $n = 4$, we have $a \equiv b \pmod 4$, but $3a = 3$, $b^2 = 1$ so $3 \not\equiv 1 \pmod 4$.

16. (a) Suppose $\sqrt{2} = a/b$ for some integers a and b. If a and b have any factors in common, these can be canceled leaving us with an equation of the form $\sqrt{2} = a/b$ where a and b have no common factors (except ± 1). So we now make this assumption. Then $a^2 = 2b^2$, so $a^2 + b^2 = 3b^2 \equiv 0 \pmod 3$. By Problem 23, a and b have 3 as a common factor, which is not true.

17. (b) $x = 1$ and $x = p - 1$.

18. (a) i. $x = 1$ and $x = 24$.

20. (a) By Fermat's Little Theorem, $18^{8970} \equiv 1 \pmod{8971}$. So $18^{8971} \equiv 18$ and $18^{8972} \equiv 18^2 = 324 \pmod{8971}$.

21. The key to this problem is the observation that 97 is a prime. Thus, by Fermat's Little Theorem, if $x \not\equiv 0 \pmod{97}$, then $x^{97} \equiv x \pmod{97}$, so $x^{97} - x + 1 \equiv 1 \not\equiv 0 \pmod{97}$. Since $x \equiv 0 \pmod{97}$ is also not a solution, there are no solutions at all.

Exercises 4.5

1. (a) This number is valid.   (b) This number is not valid.

2. (a) $A = 9$;   (d) $A = 7$.

4. $a_1 + 2a_2 + \cdots + 9a_9 + 10a_{10} \equiv 0 \pmod{11} \leftrightarrow a_1 + 2a_2 + \cdots + 9a_9 \equiv -10a_{10} \leftrightarrow a_1 + 2a_2 + \cdots + 9a_9 \equiv a_{10}$ since $-10a_{10} \equiv a_{10} \pmod{11}$.

6. (a) For example, 2-41971‌3-29-0 is not valid because $a_1 + 2a_2 + \cdots + 10a_{10} = 208 \not\equiv 0 \pmod{11}$.

7. (a) The given rule is $\mathsf{w} \cdot \mathsf{a} \equiv 0 \pmod{11}$ for $\mathsf{a} = (a_1, a_2, \ldots, a_{10})$ and $\mathsf{w} = (1, 1, 1, 1, 1, 1, 1, 1, 1, 1)$.

8. (a) 1-13579-02468-8;   (d) 0-63042-00635-5.

9. (a) Not valid: $3(0 + 2 + 4 + 6 + 8 + 0) + (1 + 3 + 5 + 7 + 9 + 1) = 60 + 26 = 86 \not\equiv 0 \pmod{10}$.

   (c) Valid: $3(2 + 2 + 9 + 1 + 3 + 4) + (5 + 9 + 8 + 7 + 9 + 9) = 63 + 47 = 110 \equiv 0 \pmod{10}$.

10. (a) $x = 1$.

11. (a) Suppose the sum of the digits in the odd positions is a and the sum of the digits in the even positions is b. If the check digit in our bar code is correct, $3a + b \equiv 0 \pmod{10}$. An error in an even position digit changes b to $c \not\equiv b \pmod{10}$. For the new number, we compute $3a + c \not\equiv 3a + b$ and thus, $3a + c \not\equiv 0 \pmod{10}$. So we know an error has been made.

   (b) Here is a valid number: 3-12498-66132-9. Changing the 1 and 4 in positions two and four to 8 and 7 respectively produces the number 3-82798-66132-9 which is also valid.

13. (a) 1-23586-98732-6;   (b) 1-23006-98732-6.

15. (b) Single-digit errors are not detected. Since $10a_3 \equiv 0 \pmod{10}$, the third digit can in fact be changed arbitrarily and the new number will still pass the test.

16. (a) If $n \leq 9$, then any a_i in the range $0 \leq a_i \leq 9 - n$ could be replaced by $n + a_i$ and the test would miss the error.

17. (a) $x = 18$;   (d) $x = 142$.

18. (a) $x = 53$;   (d) $x = 260$.

19. The number was 340.

20. (a) $x \equiv 297 \pmod{900}$. Since there are many integers less than 50,000 which are congruent to 297 (mod 900), it is impossible to determine ab exactly.

   (e) $ab \equiv 27{,}887 \pmod{34{,}300}$; $ab = 27{,}887$.

21. First we address the hint. Suppose you want to write $7s + 22t = 1$. The "obvious" numbers are $s = -3$, $t = 1$: $(-3)7 + (1)22 = 1$. But suppose you want $s > 0$. This can be achieved by adding and subtracting $22(7)$ in the equation $(-3)7 + 1(22)$: $(-3 + 22)7 + (1 - 7)22 = 1$; that is, $19(7) + (-6)22 = 1$. In general, we know that $1 = sm + tn$ for some integers s and t by Theorem 4.2.9. Now if $s \not> 0$, choose k such that $s + kn > 0$ and note that $(s + kn)m + (t - km)n = 1$.

22. (a) 300;   **(c)** 259.

23. (a) Z;   **(c)** G.

24. (a) HI.

Exercises 5.1

1. (a) $\sum_{i=1}^{5} i^2 = 1^2 + 2^2 + 3^2 + 4^2 + 5^2 = 55$ **(b)** $\sum_{i=1}^{4} 2^i = 2^1 + 2^2 + 2^3 + 2^4 = 30$

   (c) $\sum_{t=1}^{1} \sin \pi t = \sin \pi(1) = \sin \pi = 0$

4. (a) If $n = 1$, $n^2 + n = 1^2 + 1 = 2$, which is divisible by 2. Suppose $k \geq 1$ and the result is true for $n = k$; that is, $2 \,|\, (k^2 + k)$. We wish to prove that $2 \,|\, [(k+1)^2 + (k+1)]$. But $(k+1)^2 + (k+1) = k^2 + 2k + 1 + k + 1 = k^2 + k + 2(k+1)$. Since $2 \,|\, (k^2 + k)$ (by the induction hypothesis) and since $2 \,|\, 2(k + 1)$, 2 divides the sum of $k^2 + k$ and $2(k + 1)$; that is, $2 \,|\, [(k + 1)^2 + (k + 1)]$ as required. By the Principle of Mathematical Induction, the result holds for all $n \geq 1$.

   (c) If $n = 1$, $n^3 + (n + 1)^3 + (n + 2)^3 = 1^3 + 2^3 + 3^3 = 36$ and $9 \,|\, 36$. Suppose the result is true for $n = k \geq 1$; that is, $9 \,|\, [k^3 + (k + 1)^3 + (k + 2)^3]$. Then $(k + 1)^3 + (k + 2)^3 + (k + 3)^3 = (k + 1)^3 + (k + 2)^3 + k^3 + 9k^2 + 27k + 27 = [k^3 + (k + 1)^3 + (k + 2)^3] + [9k^2 + 27k + 27]$. Since 9 divides the first term by the induction hypothesis and since 9 clearly divides the second term, 9 divides $(k + 1)^3 + (k + 2)^3 + (k + 3)^3$, as desired. By the Principle of Mathematical Induction, the result holds for all $n \geq 1$.

5. (a) For $n = 1$, $1 + 2 + 3 \cdots + n$ is 1, by definition, and $\frac{n(n+1)}{2} = \frac{2}{2} = 1$ also. Thus, the result is true for $n = 1$. Now suppose that $k \geq 1$ and the result is true for k; that is, assume that $1 + 2 + 3 + \cdots + k = \frac{k(k+1)}{2}$.
Then $1 + 2 + \cdots + (k + 1) = (1 + 2 + \cdots + k) + (k + 1)$
$$= \frac{k(k + 1)}{2} + (k + 1) \quad \text{(by the induction hypothesis)}$$
$$= \frac{k(k + 1) + 2(k + 1)}{2} = \frac{(k + 1)(k + 2)}{2},$$
which is the given statement with $n = k + 1$. So, by the Principle of Mathematical Induction, we conclude that $1 + 2 + \cdots + n = \frac{1}{2}n(n + 1)$ for all $n \geq 1$.

6. (a) If $n = 1$, $1 + 2^1 + 2^2 + 2^3 + \cdots + 2^n = 1 + 2 = 3$, while $2^{1+1} - 1 = 4 - 1 = 3$, so the result is true if $n = 1$. Suppose the result is true for $n = k \geq 1$; that is, suppose $1 + 2 + 2^2 + 2^3 + \cdots + 2^k = 2^{k+1} - 1$. We wish to prove the result for $n = k + 1$; that is, we wish to prove that $1 + 2 + 2^2 + 2^3 + \cdots + 2^{k+1} = 2^{(k+1)+1} - 1$. Now $1 + 2 + 2^2 + 2^3 + \cdots + 2^{k+1} = (1 + 2 + 2^2 + 2^3 + \cdots + 2^k) + 2^{k+1} = (2^{k+1} - 1) + 2^{k+1}$ (by the induction hypothesis) $= 2 \cdot 2^{k+1} - 1 = 2^{k+2} - 1 = 2^{(k+1)+1} - 1$ as required. By the Principle of Mathematical Induction, we conclude that the given assertion is true for all $n \geq 1$.

   (b) If $n = 1$, $1^2 - 2^2 + 3^2 - 4^2 + \cdots + (-1)^{n-1}n^2 = 1^2 = 1$, by definition, while $(-1)^{n-1}\frac{n(n+1)}{2} = (-1)^{1-1}\frac{1(1+1)}{2} = 1$, and so the result is true for $n = 1$. Now suppose that $k \geq 1$ and the result is true for $n = k$; that is, suppose that
$$1^2 - 2^2 + 3^2 - 4^2 + \cdots + (-1)^{k-1}k^2 = (-1)^{k-1}\frac{k(k + 1)}{2}.$$

We wish to prove that $1^2 - 2^2 + 3^2 - 4^2 + \cdots + (-1)^{(k+1)-1}(k+1)^2 = (-1)^{(k+1)-1}\dfrac{(k+1)[(k+1)+1]}{2}$.

Now $1^2 - 2^2 + 3^2 - 4^2 + \cdots + (-1)^{(k+1)-1}(k+1)^2$

$$= [1^2 - 2^2 + 3^2 - 4^2 + \cdots + (-1)^{k-1}k^2] + (-1)^{(k+1)-1}(k+1)^2$$

$$= (-1)^{k-1}\frac{k(k+1)}{2} + (-1)^k(k+1)^2 \qquad \text{(using the induction hypothesis)}$$

$$= (-1)^{k-1}\frac{k^2+k}{2} + (-1)^{k-1}(-1)(k^2+2k+1)$$

$$= (-1)^{k-1}\left(\frac{k^2+k}{2} - k^2 - 2k - 1\right)$$

$$= (-1)^{k-1}\frac{k^2+k-2k^2-4k-2}{2}$$

$$= (-1)^{k-1}\left(\frac{-k^2-3k-2}{2}\right) = (-1)^k\frac{(k+1)(k+2)}{2}$$

as desired. By the Principle of Mathematical Induction, the result is true for all $n \geq 1$.

7. (a) $\displaystyle\sum_{k=0}^{n} 2^k$

8. (a) For $n = 1$, $\displaystyle\sum_{i=1}^{1}(i+1)2^i = 2(2^1) = 4$, while $n2^{n+1} = 1(2^2) = 4$. Thus, the formula is correct for $n = 1$. Now suppose

that $k \geq 1$ and the formula is correct for $n = k$; thus, we suppose that $\displaystyle\sum_{i=1}^{k}(i+1)2^i = k2^{k+1}$.

We must prove that the formula is correct for $n = k + 1$; that is, we must prove that $\displaystyle\sum_{i=1}^{k+1}(i+1)2^i = (k+1)2^{k+2}$.

Now $\displaystyle\sum_{i=1}^{k+1}(i+1)2^i = \sum_{i=1}^{k}(i+1)2^i + (k+2)2^{k+1} = k2^{k+1} + (k+2)2^{k+1}$ (using the induction hypothesis)

$$= (k+k+2)2^{k+1} = (2k+2)2^{k+1} = 2(k+1)2^{k+1} = (k+1)2^{k+2}$$

as required. By the Principle of Mathematical Induction, the given assertion is true for all $n \geq 1$.

9. (a) If $n = 5$, $2^5 = 32$, $5^2 = 25$. Since $32 > 25$, the result is true for $n = 5$. Now suppose that $k \geq 5$ and the result is true for $n = k$; that is, suppose that $2^k > k^2$. We must prove that the result is true for $n = k + 1$; that is, we must prove that $2^{k+1} > (k+1)^2$. Now $2^{k+1} = 2 \cdot 2^k > 2k^2$ by the induction hypothesis, and $2k^2 = k^2 + k^2 \geq k^2 + 5k$ (since $k \geq 5$), and $k^2 + 5k = k^2 + 4k + k > k^2 + 2k + 1$ (since $k \geq 1$), and $k^2 + 2k + 1 = (k+1)^2$. By the Principle of Mathematical Induction, the result is true for all $n \geq 5$.

10. (a) For $n = 1$, $\displaystyle\sum_{i=1}^{1}(x_i + y_i)$ is, by convention, just the single term $x_1 + y_1$. Since this equals $\displaystyle\sum_{i=1}^{1}x_1 + \sum_{i=1}^{1}y_1$ the formula

holds for $n = 1$. Now suppose $k \geq 1$ and the formula holds for $n = k$. Then

$$\sum_{i=1}^{k+1}(x_i + y_i) = \left[\sum_{i=1}^{k}(x_i + y_i)\right] + (x_{k+1} + y_{k+1})$$

$$= \sum_{i=1}^{k}x_i + \sum_{i=1}^{k}y_i + (x_{k+1} + y_{k+1}) \quad \text{(by the induction hypothesis)}$$

$$= \left[\left(\sum_{i=1}^{k}x_i\right) + x_{k+1}\right] + \left[\left(\sum_{i=1}^{k}y_i\right) + y_{k+1}\right] = \sum_{i=1}^{k+1}x_i + \sum_{i=1}^{k+1}y_i,$$

which is the formula with $n = k+1$. Thus, the result holds for all $n \geq 1$ by the Principle of Mathematical Induction.

11.   The k to $k+1$ step does not apply to the case $k = 1$. When $k = 1$, $G = \{a_1, a_2\}$. Observe that the groups $\{a_1\}$, $\{a_2\}$ have no member in common.

12.   The induction was not started properly. When $n = 1$, the left side is 1, while the right side is 9/8. The statement is not true when $n = 1$.

15. (a) We leave the primality checking of $f(1), \ldots, f(39)$ to the reader, but note that $f(40) = 41^2$.

   (b) $f(k^2 + 40) = (k^2 + 41 + k)(k^2 + 41 - k)$.

16.   If a set A contains $n = 0$ elements, then $A = \emptyset$ and A has exactly 1 subset, namely, $\emptyset$. Since $1 = 2^0$, the statement is true for $n = 0$. Now assume that a set of k elements has 2^k subsets. Let $A = \{a_1, a_2, \ldots, a_{k+1}\}$ be a set of $k+1$ elements. We show that A must have 2^{k+1} subsets. This follows because every subset of A either contains a_{k+1} or it doesn't. By the induction hypothesis there are 2^k sets which do not contain a_{k+1}; but, by the induction hypothesis, there are also 2^k subsets which **do** contain a_{k+1} since these are precisely those sets obtained by forming the union of a subset of $\{a_1, a_2, \ldots, a_k\}$ and $\{a_{k+1}\}$. Altogether A has $2^k + 2^k = 2(2^k) = 2^{k+1}$ subsets. So, by the Principle of Mathematical Induction, we conclude that a set of n elements contains 2^n subsets.

17.   Let S be a nonempty set of natural numbers. Let a be any element of S. Then the smallest element of S is the smallest element of the nonempty set $\{x \in S \mid 1 \leq x \leq a\}$ (which is finite).

19.   For $n = 1$, $\bigcap_{i=1}^{n} B_i = B_1$, and so $A \cup \left(\bigcap_{i=1}^{n} B_i\right) = A \cup B_1 = \bigcap_{i=1}^{n}(A \cup B_i)$ and the result holds for $n = 1$. Now assume that $A \cup \left(\bigcap_{i=1}^{k} B_i\right) = \bigcap_{i=1}^{k}(A \cup B_i)$ for $k \geq 1$. Then, given a set A and $k+1$ sets $B_1, B_2, \ldots, B_{k+1}$, we have

$$A \cup \left(\bigcap_{i=1}^{k+1} B_i\right) = A \cup \left(\left(\bigcap_{i=1}^{k} B_i\right) \cap B_{k+1}\right)$$

$$= \left(A \cup \left(\bigcap_{i=1}^{k} B_i\right)\right) \cap \left(A \cup B_{k+1}\right) \quad \text{since } A \cup (B \cap C) = (A \cup B) \cap (A \cup C)$$

$$= \left(\bigcap_{i=1}^{k}(A \cup B_i)\right) \cap (A \cup B_{k+1}) \quad \text{(by the induction hypothesis)}$$

$$= \bigcap_{i=1}^{k+1}(A \cup B_i),$$

giving the result for $k+1$. By the Principle of Mathematical Induction, we conclude that $A \cup \left(\bigcap_{i=1}^{n} B_i\right) = \bigcap_{i=1}^{n}(A \cup B_i)$, for all $n \geq 1$.

23.   When $n = 2$, $\gcd(a_1, a_2) = s_1 a_1 + s_2 a_2$ by Theorem 4.2.9. Now suppose that $k \geq 2$ and the statement is true for $n = k$; that is, suppose that the gcd of any $k \geq 2$ integers is an integral linear combination of them. We have to show the statement is true for $n = k+1$, that is, that the gcd of $k+1$ integers is an integral linear combination of them. Now $\gcd(a_1, a_2, \ldots, a_{k+1}) = \gcd(a_1, \gcd(a_2, \ldots, a_{k+1}))$. By the induction hypothesis, there are k integers $r_2, r_3, \ldots, r_{k+1}$ such that $\gcd(a_2, \ldots, a_{k+1}) = \sum_{i=2}^{k+1} r_i a_i$. Then Theorem 4.2.9 tells us that

$$\gcd\left(a_1, \sum_{2}^{k+1} r_i a_i\right) = x a_1 + y\left(\sum r_i a_i\right) = x a_1 + (y r_2)a_2 + \cdots (y r_{k+1})a_{k+1}$$

as desired ($s_1 = x$, $s_2 = y r_2, \ldots, s_{k+1} = y r_{k+1}$). By the Principle of Mathematical Induction, the statement is true for all $n \geq 1$.

28. When $n = 2$, we have two distinct points in the plane joined by exactly one line. Since $\frac{2(2-1)}{2} = 1$, the result holds in this case. Now suppose the result is true for $n = k \geq 2$; that is, suppose that the number of lines obtained by joining k distinct points in the plane, no three of which are collinear, is $\frac{1}{2}k(k-1)$. We must prove the result for $n = k+1$; that is, we must prove that the number of lines obtained by joining $k+1$ distinct points, no three of which are collinear, is $\frac{1}{2}(k+1)[(k+1)-1] = \frac{1}{2}k(k+1)$. Let P be one of the $k+1$ points. By the induction hypothesis, the remaining k points are joined by $\frac{1}{2}k(k-1)$ lines. On the other hand, P can be joined to each of those remaining k points, giving k additional lines. (Since no three of the $k+1$ points are collinear, these additional lines are different from the lines determined by the other k points.) The total number of lines is

$$\frac{k(k-1)}{2} + k = \frac{k(k-1) + 2k}{2} = \frac{k^2 - k + 2k}{2} = \frac{k^2 + k}{2} = \frac{k(k+1)}{2}$$

as desired. We conclude that the given assertion is true for all $n \geq 2$, by the Principle of Mathematical Induction.

31. (a) First set aside two coins; then separate the remaining six into two piles of three and compare weights. If the weights are equal, the only possible light coin is one of the two set aside and this can be found in a second weighing. If one set of three is lighter than the other, then there is indeed a light coin and it is one of a known three. Remove one of these three and compare the weights of the remaining two coins (with a second weighing). Either one of these coins will be found to be lighter or the one set aside is the light one.

32. (a) False. If $n = 3$, $5^n + n + 1 = 129$ and $7 \nmid 129$.

36.
$$\lim_{x \to \pm 1} \frac{1 - x^{2^{r+1}}}{1 - x^2} = \lim_{x \to \pm 1} \prod_{r=1}^{n}(1 + x^{2^r}) \quad \text{(using the result of Exercise 35)}$$
$$= \prod_{r=1}^{n}(1 + (\pm 1)^{2^r}) = 2^n.$$

Exercises 5.2

1. (a) $a_1 = 1$; $a_{k+1} = 5a_k$ for $k \geq 1$.

2. (a) $16, 8, 4, 2, 1, 1, 1$

4. When $n = 1$, the formula becomes $\frac{(1-1)(1+2)(1^2+1+2)}{4} = 0$, which is a_1. Now assume that the formula is correct when $n = k$. When $n = k + 1$,

$$a_{k+1} = (k+1)^3 + a_k = (k+1)^3 + \frac{(k-1)(k+2)(k^2+k+2)}{4}$$

$$= \frac{4k^3 + 12k^2 + 12k + 4 + k^4 + 2k^3 + k^2 - 4}{4}$$

$$= \frac{k^4 + 6k^3 + 13k^2 + 12k}{4}$$

$$= \frac{k(k+3)(k^2+3k+4)}{4} = \frac{k(k+3)[(k+1)^2 + (k+1) + 2]}{4}$$

as desired. By the Principle of Mathematical Induction, we conclude that the formula is correct for all $n \geq 1$.

6. The first six terms are $1, 3, 7, 15, 31, 63$. Our guess is that $a_n = 2^n - 1$. When $n = 1$, $2^1 - 1 = 1$, agreeing with a_1. Now assume that $k \geq 1$ and the result is true for $n = k$; that is, assume that $a_k = 2^k - 1$. We must prove the result is true for $n = k + 1$; that is, we must prove that $a_{k+1} = 2^{k+1} - 1$. Now $a_{k+1} = 2a_k + 1 = 2(2^k - 1) + 1$ (by the induction hypothesis) $= 2^{k+1} - 2 + 1 = 2^{k+1} - 1$, as required. By the Principle of Mathematical Induction, we conclude that the result is true for all $n \geq 1$.

9. $a_1 = 1, a_2 = 2^2 - a_1 = 4 - 1 = 3, a_3 = 3^2 - a_2 = 9 - 3 = 6. a_4 = 4^2 - a_3 = 16 - 6 = 10, a_5 = 5^2 - a_4 = 25 - 10 = 15$, $a_6 = 6^2 - a_5 = 36 - 15 = 21$. Thus, the first six terms are $1, 3, 6, 10, 15, 21$. We recognize these numbers as the six sums $1, 1+2, 1+2+3, 1+2+3+4, 1+2+3+4+5$ and $1+2+3+4+5+6$ and so guess that $a_n = n(n+1)/2$. (See Exercise 5(a) of Section 5.1.)

For $n = 1, n(n+1)/2 = 1(2)/2 = 1$, which agrees with a_1. Now assume the formula holds for $n = k$; that is, assume $a_k = \frac{k(k+1)}{2}$. We must prove the formula holds for $n = k + 1$; that is, we must prove that $a_{k+1} = \frac{(k+1)(k+2)}{2}$. Now

$$a_{k+1} = (k + 1)^2 - a_k = (k + 1)^2 - \frac{k(k + 1)}{2} \quad \text{using the induction hypothesis}$$

$$= \frac{2(k + 1)^2 - k(k + 1)}{2} = \frac{(k + 1)[2(k + 1) - k]}{2} = \frac{(k + 1)(k + 2)}{2}$$

as required. By the Principle of Mathematical Induction, we conclude that the result is true for all $n \geq 1$.

11. The first few terms are $0, 1, 0, 4, 0, 16, \dots$. Our guess is that

$$a_n = \begin{cases} 0 & \text{if } n \text{ is odd} \\ 4^{\frac{n}{2}-1} & \text{if } n \text{ is even.} \end{cases}$$

We will prove this using the strong form of mathematical induction. Note first that $a_1 = 0$ and $4^{\frac{2}{2}-1} = 4^0 = 1 = a_2$ so that our guess is correct for a_1 and a_2. Now let $k > 2$ and assume the result is true for all n, $1 \leq n < k$. We must show that the result is true if $n = k$. If k is odd, then $a_k = 4a_{k-2} = 4(0) = 0$ since $a_{k-2} = 0$ by the induction hypothesis, $k - 2$ being odd. If k is even, then

$$a_k = 4a_{k-2} = 4(4^{\frac{k-2}{2}-1}) \text{ (using the induction hypothesis) } = 4^{\frac{k-2}{2}} = 4^{\frac{k}{2}-1}$$

as desired. By the Principle of Mathematical Induction, the result is true for all $n \geq 1$.

15. (a) The first ten terms are $2, 5, 8, 11, 14, 17, 20, 23, 26, 29$. The 123rd term is $2 + 122(3) = 368$.

(b) We attempt to solve $752 = 2 + (n - 1)3$ and discover $n - 1 = 250$. So 752 does belong to the sequence; it is the 251st term.

(d) The sum of 75 terms is $\frac{75}{2}[2(2) + (74)3] = 8475$.

16. (a) $a_{17} = -1; a_{92} = -38\frac{1}{2};$ **(b)** $S = -\frac{171}{2}$.

19. For $n = 1, a + (1 - 1)d = a$, which agrees with $a_1 = a$. Now assume that the formula is correct for the integer k, that is, that $a_k = a + (k - 1)d$. Then $a_{k+1} = a_k + d = a + (k - 1)d + d = a + kd$ shows that the formula is true also for $k + 1$. So, by the Principle of Mathematical Induction, the formula is correct for all $n \geq 1$.

20. (a) The first ten terms are $59049, -19683, 6561, -2187, 729, -243, 81, -27, 9, -3$. The 33rd term is $59049(-\frac{1}{3})^{32} = \frac{3^{10}}{3^{32}} = (\frac{1}{3})^{22}$.

(b) The sum of the first 12 terms is $59049\left(\frac{1-(-\frac{1}{3})^{12}}{1-(-\frac{1}{3})}\right) = (3^{10})\frac{3}{4}(1 - (\frac{1}{3})^{12}) = \frac{1}{4}(3^{11}) - \frac{1}{12} = \frac{3^{12} - 1}{12}$.

23. (a) $a_{129} = (-.00001240)(-1.1)^{128} \approx -2.4643$. **(b)** $S = \frac{-.00001240(1-(-1.1)^{129})}{2.1} \approx -1.2908$.

24. This is straightforward to prove by mathematical induction but we give an alternative proof. Letting $S = a + ar + ar^2 + \cdots + ar^{n-1}$, we have $rS = ar + ar^2 + \cdots + ar^n$. Subtraction gives $(1 - r)S = S - rS = a - ar^n = a(1 - r^n)$ from which the formula follows.

26. (b) If $|r| < 1$, then $\lim_{n \to \infty} r^n = 0$. Hence, $\lim_{n \to \infty} a \frac{1 - r^n}{1 - r} = \frac{a}{1 - r}$. **(c)** 6.

28. (a) This is the sum of an arithmetic sequence with $a = 75, d = -4$. Solving $-61 = a + (n - 1)d = 75 + (n - 1)(-4)$, we obtain $n = 35$. So the sum is $\frac{35}{2}[2(75) + 34(-4)] = 245$.

29. Suppose the arithmetic sequence $a_0, a_0 + d, a_0 + 2d, \ldots$ is also the geometric sequence $a_1, a_1r, a_1r^2, \ldots$. Then $a_0 = a_1$. Let's call this number a. Note that one and only one possibility occurs when $a = 0$ (namely, $0, 0, 0, \ldots$) so assume henceforth that $a \neq 0$. Since the sequence is both arithmetic and geometric, it follows that all terms of this sequence are nonzero. We have $a + d = ar$ and $a + 2d = ar^2$, so

$$\frac{a+d}{a} = \frac{ar}{a} = r = \frac{a+2d}{a+d}.$$

So $(a + d)^2 = a(a + 2d)$, giving $a^2 + 2ad + d^2 = a^2 + 2ad$ and hence $d = 0$. Since $ar = a + d = a$, $r = 1$. The only arithmetic sequences which are also geometric are constant, of the form $a, a, a, \ldots$ for some number a.

31. (a) $\$1000(1.15)^2 = \$1322.50$.

(b) We seek t such that $1000(1.15)^t = 2000$. Thus, $1.15^t = 2$, $t = \log 2 / \log 1.15 \approx 4.9595$ years.

36. Rewriting $f_{k+1} = f_k + f_{k-1}$ as $f_{k-1} = f_{k+1} - f_k$, we see that $f_0 = f_2 - f_1 = 1 - 1 = 0$. Similarly, $f_{-1} = 1$, $f_{-2} = -1$, $f_{-3} = 2$, $f_{-4} = -3$, $f_{-5} = 5$, $f_{-6} = -8$. In general, $f_{-n} = (-1)^{n+1} f_n$.

37. The Fibonacci sequence is the sequence $1, 1, 2, 3, 5, 8 \ldots$, each term after the first two being the sum of the two previous terms. For $n = 1$, $f_2 f_1 = 1(1) = 1^1 = f_1^2$. So the assertion is correct for $n = 1$. Let $k \geq 1$ and assume the assertion is correct for $n = k$; that is, assume that $f_{k+1} f_k = \sum_{i=1}^{k} f_i^2$. We wish to prove the assertion is correct for $n = k + 1$; that is, we wish to prove that $f_{k+2} f_{k+1} = \sum_{i=1}^{k+1} f_i^2$. But $f_{k+2} = f_{k+1} + f_k$, so, $f_{k+2} f_{k+1} = (f_{k+1} + f_k) f_{k+1} = f_{k+1}^2 + f_{k+1} f_k = f_{k+1}^2 + \sum_{i=1}^{k} f_i^2$ (using the induction hypothesis) $= \sum_{i=1}^{k+1} f_i^2$ as desired. By the Principle of Mathematical Induction, we conclude that the assertion is correct for all $n \geq 1$.

40. The induction hypothesis is valid only for n in the interval $3 \leq n < k$ because all integers in this problem are at least as large as $n_0 = 3$. But the induction hypothesis is applied to the integer $k - 2$; this is not valid if $k = 4$.

42. (a) $a_1 = 1$, $a_2 = 2$ (2 and $1 + 1$), $a_3 = 4$, as given. We can write $4 = 4 = 3 + 1 = 1 + 3 = 2 + 1 + 1 = 1 + 2 + 1 = 1 + 1 + 2 = 2 + 2 = 1 + 1 + 1 + 1$ so $a_4 = 8$. Careful counting shows that $a_5 = 16$.

44. (a) $a_n = 2a_{n-1}$. The sequence is the sequence $2, 4, 8, \ldots$ of powers of 2.

47. (b) $f^1(1) = f(1) = [4(1) - 1]/3 = 1$. $f^2(1) = f \circ f(1) = f(f(1)) = f(1) = 1$. This continues: $f^n(1) = 1$ for all n.

Exercises 5.3

1. $a_n = 3^n$

3. $a_n = -5(3^n) + 6n(3^n)$

5. $a_n = \frac{1}{2\sqrt{15}}(-4 + \sqrt{15})^n - \frac{1}{2\sqrt{15}}(-4 - \sqrt{15})^n$

7. $a_n = -2(-1)^n + 2(3^n) = 2[(-1)^{n+1} + 3^n]$

10. (a) $a_n = \frac{1}{2}(3^n) + \frac{1}{2}(-5)^n$;   **(b)** $a_n = 2(3^n) + (-5)^n - 2$.

13. (a) $a_n = 4^n$;   **(b)** $a_n = 2(8^n) - 4^n$.

(c) For $n = 0$, $2(8^0) - 4^0 = 2 - 1 = 1$, as required.

Let $k \geq 0$ and assume that $a_k = 2(8^k) - 4^k$. We wish to prove that $a_{k+1} = 2(8^{k+1}) - 4^{k+1}$. Now

$$a_{k+1} = 4a_k + 8^{k+1} = 4[2(8^k) - 4^k] + 8^{k+1} \quad \text{(by the induction hypothesis)}$$

$$= 8^{k+1} - 4^{k+1} + 8^{k+1} = 2(8^{k+1}) - 4^{k+1}$$

as desired. By the Principle of Mathematical Induction, the result is true for all $n \geq 0$.

16. (a) $a_n = 4^n + 3$;   **(b)** $a_n = 10(4^n) - (3n + 6)2^n$.

18. The nth term of the Fibonacci sequence is $a_{n-1} = \frac{1}{\sqrt{5}}\left(\frac{1+\sqrt{5}}{2}\right)^n - \frac{1}{\sqrt{5}}\left(\frac{1-\sqrt{5}}{2}\right)^n$. Thus, $\left|a_{n-1} - \frac{1}{\sqrt{5}}\left(\frac{1+\sqrt{5}}{2}\right)^n\right| = \left|\frac{1}{\sqrt{5}}\left(\frac{1-\sqrt{5}}{2}\right)^n\right|$. Now $\left|\frac{1}{\sqrt{5}}\left(\frac{1-\sqrt{5}}{2}\right)^n\right| \leq \frac{1}{\sqrt{5}} < \frac{1}{2}$ since $\left|\frac{1-\sqrt{5}}{2}\right| < 1$, so $\left|a_{n-1} - \frac{1}{\sqrt{5}}\left(\frac{1+\sqrt{5}}{2}\right)^n\right| < \frac{1}{2}$. Remembering that

a_{n-1} is an integer, the result follows from the fact that there is precisely one integer within $\frac{1}{2}$ of any given real number.

22. (a) Since x is a root of the characteristic polynomial, $x^2 = rx + s$. Hence, $ra_{n-1} + sa_{n-2} = r(cx^{n-1}) + s(cx^{n-2}) = cx^{n-2}(rx + s) = cx^{n-2}(x^2) = cx^n = a_n$.

(b) We know that $p_n = rp_{n-1} + sp_{n-2}$ and $q_n = rq_{n-1} + sq_{n-2}$. Thus, $p_n + q_n = rp_{n-1} + sp_{n-2} + rq_{n-1} + sq_{n-2} = r(p_{n-1} + q_{n-1}) + s(p_{n-2} + q_{n-2})$.

(c) If x_1 and x_2 are the characteristic roots, part (a) tells us that $c_1 x_1^n$ and $c_2 x_2^n$ both satisfy the recurrence relation while part (b) says that $c_1 x_1^n + c_2 x_2^n$ is also a solution. If either x_1 or x_2 is 0, one initial condition determines the single unknown constant. Otherwise, two initial conditions will determine c_1 and c_2 for they determine two linear equations in the unknowns c_1, c_2 which, because $x_1 \neq x_2$, must have a unique solution.

23. (a) The characteristic polynomial is $x^2 - rx - s$ with characteristic roots $x = \frac{r \pm \sqrt{r^2 + 4s}}{2}$. Since there is only one root, we must have $r^2 = -4s$ in which case $x = r/2$. Hence, $r = 2x$ and $s = -r^2/4 = -4x^2/4 = -x^2$.

Exercises 5.4

1. (a) $4, -12, 9, 0, 0, \ldots$;   (c) $1, -6, 27, -108, \ldots, (n+1)(-3)^n, \ldots$.

2. (a) $1 + 2x + 5x^2$;   (f) $\dfrac{1}{1 - x^2}$.

3. $a_n = 2^n$

5. $a_n = -2^n + 2 \cdot 3^n$

7. $a_n = 2^{n+1} + 3$

9. (a) $a_n = 2(-5)^n$

11. $a_n = -\frac{1}{2} + (-\frac{1}{6})(-1)^n + \frac{5}{3}(2^n)$

12. $a_n = \frac{1}{4}(2n + 1) + \frac{7}{4}(-1)^n$

Exercises 6.1

1.   Let B be the set of connoisseurs of Canadian bacon, and A the set of those who like anchovies.

(a) $|B \cup A| = |B| + |A| - |B \cap A| = 10 + 7 - 6 = 11$;   (b) $|B \setminus A| = |B| - |B \cap A| = 4$;

(c) $|B \oplus A| = |B \cup A| - |B \cap A| = 5$;   (d) $|U| - |B \cup A| = 4$.

4.   Let A be the set of people with undergraduate degrees in arts, S those with undergraduate degrees in science, and G those with graduate degrees.

(a) $|A \cup S \cup G| = 300$;   (b) $|S \setminus (A \cup G)| = 35$.

7.   Let D be the set of delegates who voted to decrease the deficit, E be the set of delegates who voted in favor of the motion concerning environmental issues, and T be the set of delegates who voted in favor of not increasing taxes.

(a) $|(D \cup E \cup T)^c| = 200$;   (b) $|T \setminus (D \cup E)| = 316$.

9.   Let O, A, G, and C denote the sets of people who bought orange juice, apple juice, grapefruit juice, and citrus punch, respectively.

(a) $|O \cap A \cap G \cap C| = 1$

10. (a) $75 - (4(28) - 6(12) + 4(5) - 1) = 75 - 59 = 16$

11. (a) Let A and B be the set of integers between 1 and 500 which are divisible by 3 and 5, respectively. The question asks for $|A \cup B|$. This number is $|A| + |B| - |A \cap B| = \lfloor \frac{500}{3} \rfloor = 233$.

(b) The question asks for $|A \setminus (B \cup C)| = 66$.

13. Let A, B, and C be the sets of integers between 1 and 250 which are divisible by 4, by 6, and by 15, respectively. We want $|A \cup B \cup C| = 91$.

14. (a) Let A, B, C, D be the sets of integers between 1 and 1000 (inclusive) which are divisible by 2, by 3, by 5, and by 7, respectively. We want $|A \cup B \cup C \cup D|^c = 1000 - 772 = 228$.

17. Let A, B, C, D, E, F be the sets of natural numbers between 1 and 200 (exclusive) which are not prime and divisible by 2, 3, 5, 7, 11, and 13, respectively. Let A, B, C, D, E, F be the sets of natural numbers between 1 and 200 (exclusive) which are not prime and divisible by 2, 3, 5, 7, 11, and 13, respectively. $|A \cup B \cup C \cup D \cup E \cup F| = 152$, so the number of primes less than 200 is $198 - 152 = 46$.

18. Write $a = qb + r$ with $0 \le r < b$. Then $\lfloor \frac{a}{b} \rfloor = q$. Clearly the positive integers $b, 2b, \ldots, qb$ are all less than or equal to a and are all divisible by b. On the other hand, if $sb \le a$, then s must belong to the set $\{1, 2, \ldots, q\}$. Hence, there are exactly q such natural numbers, as required.

22. (a) $|(A \oplus B) \cap C| = \left|((A \setminus B) \cup (B \setminus A)) \cap C\right| = \left|((A \setminus B) \cap C) \cup ((B \setminus A) \cap C)\right|$
$$= |(A \cap C) \setminus B| + |(B \cap C) \setminus A| \quad \text{since } (A \setminus B) \cap (B \setminus A) = \emptyset$$
$$= |A \cap C| - |A \cap B \cap C| + |B \cap C| - |A \cap B \cap C|$$
$$= |A \cap C| + |B \cap C| - 2|A \cap B \cap C|$$

Exercises 6.2

1. (a) 12; (b) 7.

3. $10 \times 9 \times 8 = 720$

5. (a) $9 \times 26 \times 26 \times 26 \times 10 \times 10 \times 10 = 158,184,000$ (b) $9 \times 26 \times 10^5 = 23,400,000$

(c) $158,184,000 + 23,400,000 = 181,584,000$

7. (a) $13 \times 6 \times 2 \times 4 = 624$; (b) $13 + 4 = 17$; (c) 25.

9. (a) $4 \times 4 = 16$; (b) $4 \times 4 \times 4 = 64$; (c) $16 + 64 = 80$.

11. $60 \times 60 \times 60 = 216,000$

13. $500 + (500)(499) + (500)(499)(498) = 1.24501 \times 10^8$

16. (a) $52 \times 3 \times 2 \times 1 = 312$

18.

Total	2	3	4	5	6	7	8	9	10	11	12
No. of ways	1	2	3	4	5	6	5	4	3	2	1

20. (a) HHHH, HHHT, HHTH, HTHH, THHH, HHTT, HTHT, HTTH, THHT, THTH, TTHH, HTTT, THTT, TTHT, TTTH, TTTT. There are 16 possibilities in all.

22. (a) $26 \times 26 \times 26 \times 10 \times 9 \times 8 = 12,654,720$

25. (a) The easiest way to see that there are 2^n functions from A to B is to note that $\{a_i \mid (a_i, 0) \in f\} \leftrightarrow f$ is a one-to-one correspondence between the (2^n) subsets of A and the set of functions $f : A \to B$.

(b) Of the 2^n functions $A \to B$, precisely two are not onto,
$$\{(a_1, 0), (a_2, 0), \ldots, (a_n, 0)\} \quad \text{and} \quad \{(a_1, 1), (a_2, 1), \ldots, (a_n, 1)\}.$$
Thus, the number of onto functions is $2^n - 2$ as claimed.

Exercises 6.3

1.   There are seven days of the week. Hence, we wish to put eight objects (people) into seven boxes (days). By the Pigeon-Hole Principle, some day must have at least two people corresponding to it.

4. $\lceil \frac{100}{12} \rceil = 9$

6. (a) Any given processor is connected to at least one of 19 other processors. There are 20 processors, so the Pigeon-Hole Principle assures us that at least two are connected to the same number.

   (b) The result is still true, though for a somewhat more subtle reason. The number of processors to which a given processor is connected is in the range 0–19 (inclusive). On the other hand, if 0 occurs (that is, some processor is not connected to any other), then 19 cannot. So, as in part (a), there are at most 19 possibilities for the number of processors connected to a given processor. By the Pigeon-Hole Principle, at least two processors are connected to the same number.

9.   Mimicking the solution to Problem 14, we let a_i be the number of sets Martina plays on day i. Then we have $1 \le a_1 < a_1 + a_2 < \cdots < a_1 + a_2 + \cdots + a_{77} \le 132$. Now it is not true that the only integer in the range 1–77 which is divisible by 21 is 21 itself. Thus, while two of these sums must leave the same remainder upon division by 21, we can conclude this time only that the difference of these sums is divisible by 21, **not** that the difference **is** 21.

10.   Suppose Brad and his mother drive a_1 quarter hours on day one, a_2 quarter hours on day two, and so on, a_{35} quarter hours on day 35. The list

$$a_1, \ a_1 + a_2, \ a_1 + a_2 + a_3, \ \ldots, \ a_1 + a_2 + \cdots + a_{35}$$

consists of 35 natural numbers between 1 and $15 \times 4 = 60$ (15 hours = 60 quarter hours) and so, if any of them is divisible by 35 ($8\frac{3}{4}$ hrs. = 35 quarter hrs.), it is 35 and we are done. Otherwise, each of these numbers leaves a remainder between 1 and 34 upon division by 35, and so two leave the same remainder. Suppose

$$a_1 + a_2 + \cdots + a_s = 35q + r$$
$$a_1 + a_2 + \cdots + a_t = 35q' + r.$$

Then (assuming $s > t$) $a_{t+1} + a_{t+2} + \cdots + a_s$ is divisible by and, hence, equal to 35; that is, Brad drives $8\frac{3}{4}$ hours on days $t+1, t+2, \ldots, s$.

13.   Let box 1 correspond to days 1, 2, and 3, box 2 to days 4, 5, and 6, box 3 to days 7, 8, and 9, and box 4 to days 10, 11, and 12. Putting each bill into the box corresponding to the day it was mailed, we see that one box must contain at least $\lceil \frac{195}{4} \rceil = 49$ bills, by the general form of the Pigeon-Hole Principle. This gives the desired result.

16.   Divide the rectangle into 25 rectangles, each 3×4. By the Pigeon-Hole Principle, at least two points are within, or on the boundary, of one of these smaller rectangles. This gives the result since the maximum distance between two points of such a rectangle is $\sqrt{3^2 + 4^2} = 5$.

20.   As suggested, we consider the sequence of natural numbers $M_1 = 3, M_2 = 33, \ldots, M_n = 33 \cdots 3$ (n 3's). If one of these is divisible by n, we have the desired result. Otherwise, at least two of these natural numbers leave the same (nonzero) remainder upon division by n; that is, for some $i \ne j$, $M_i = qn + r$, $M_j = q'n + r$ for the same r. Assuming, without loss of generality, that $i > j$, $M_i - M_j = (q - q')n$ is divisible by n and has only 3's and 0's in its base 10 representation.

22.   After finitely many steps in the long division process, we will be adjoining a "0" at each stage to the remainder from the division at the previous stage. Since only finitely many remainders are possible (each remainder is less than the divisor), two must be the same and from the first repetition, all steps (and corresponding decimal places) will repeat.

24. (a) Suppose no two people have the same age so that there are at least 51 different ages in the room. Let box 1 correspond to integers 1 and 2, box 2 correspond to integers 3 and 4, ..., box 50 to 99 and 100. Assigning ages to boxes, by the Pigeon-Hole Principle, some box contains two ages. This says that ages of two people are consecutive integers, as required.

26. (a) Focus on one of the ten people, say Hilda. Suppose first that at least four of the remaining nine people are strangers to Hilda. If these four are mutual friends we are done; otherwise two of these are strangers and, hence, together with

Hilda, we have three mutual strangers and again we are done. Thus, we suppose that less than four people are strangers to Hilda; so Hilda has at least six friends in the group. By Problem 17, this group of six contains either three mutual strangers (in which case we are done) or three mutual friends who, together with Hilda, give four mutual friends.

Exercises 7.1

1. $13 \cdot 12 \cdot 11 \cdots 6 = P(13, 8) = 51,891,840$

3. $10 \cdot 9 \cdot 8 \cdot 7 = P(10, 4) = 5040$

5. $3 \times 5! = 360$

7. (b) $10!4!2!$

9. (a) $3! \times 4! = 144$;   (b) $P(4, 2) \times 4! = 288$.

10. (a) $m!$

(b) If $m > n$, there are 0 injective functions $X \to Y$.
If $m \le n$, the number is $\underbrace{n(n - 1)(n - 2) \cdots (n - m + 1)}_{m \text{ factors}} = P(n, m)$.

12. Let N be the set of lines in which the Noseworthys are beside each other and A the set of lines in which the Abbotts are beside each other.

(a) $|N| = 2 \times 3! = 12$;   (b) $|N^c| = 24 - 12 = 12$;   (c) $|N \cap A| = 2 \times 2 \times 2 = 8$;

(d) 4;   (e) $|N \cup A| = |N| + |A| - |N \cap A| = 12 + 12 - 8 = 16$;

(f) $|N \oplus A| = |N| + |A| - 2|N \cap A| = 12 + 12 - 2(8) = 8$.

14. (a) $2 \times 5! = 240$

16. (a) $9 \times 9 \times 8 \times 7 \times 6 \times 5 \times 4 = 9 \times P(9, 6) = 9 \times 60,480 = 544,320$.

17. (b) $7 \times 6 \times P(7, 5) = 42 \times 2520 = 105,840$;   (d) $P(7, 7) = 5040$.

Exercises 7.2

1. (a) $\binom{6}{2}\binom{7}{2}\binom{8}{2} = 8820$;   (b) $\binom{6}{3}\binom{7}{2}\binom{8}{2} + \binom{6}{2}\binom{7}{3}\binom{8}{2} + \binom{6}{2}\binom{7}{2}\binom{8}{3} = 44,100$.

3. $\binom{12}{5} = \frac{12!}{7!5!} = 792$

5. (a) $\binom{15}{5}\binom{10}{5}$;   (b) $\binom{15}{7}\binom{8}{4}$;   (c) $3\binom{15}{7}\binom{8}{4}$.

6. $\binom{10}{0} + \binom{10}{1} + \binom{10}{2} + \binom{10}{3} = 176$

8. (a) $4\binom{13}{5} = 5148$

10. (a) 2^{10};   (b) $\binom{10}{4} = 210$.

11. (a) $\binom{25}{5} = 53,130$;   (b) $\binom{15}{5} = 3003$;   (c) $\binom{15}{3}\binom{10}{2} = 20,475$.

13. (a) $\binom{100}{20}$;   (b) $\binom{100}{20}$;   (c) $\binom{100}{20} \times \binom{100}{20}$.

(d) The answer is $\binom{100}{5} \times \binom{95}{15} \times \binom{80}{15}$. This is also $\binom{100}{20}\binom{20}{5}\binom{80}{15}$. (Why?)

17. (a) $\binom{10}{8} = 45$;   (c) $\binom{5}{3} + \binom{5}{4}\binom{5}{4} = 10 + 25 = 35$.

19. (a) $\binom{10}{2} = 45$

20. (a) $\binom{12}{3} = 220$

21. (a) $\binom{8}{2} - 8 = 28 - 8 = 20$

23. The product of n consecutive natural numbers starting at m is

$$m(m+1)(m+2)\cdots(m+n-1) = P(m+n-1, n) = n!\binom{m+n-1}{n}.$$

So $\frac{m(m+1)(m+1)\cdots(m+n-1)}{n!}$ is the integer $\binom{m+n-1}{n}$.

Exercises 7.3

1. $\binom{30+7-1}{30} = \binom{36}{30}$

3. $\binom{6+5-1}{5} = \binom{10}{5} = 252$

4. $\binom{10+4-1}{10} = \binom{13}{10} = 286$

6. (a) $\binom{27}{8} = 2,220,075$

7. (c) $\frac{12!}{3!} = 79,833,600$

8. (a) $\binom{12}{3}\binom{9}{3}\binom{6}{3}\binom{3}{3} = 369,600$;   (b) $\frac{369,600}{4!} = 15,400.$

12. (a) $\binom{60}{10}\binom{50}{10}\binom{40}{10} = \frac{60!}{10!10!10!30!}$

14. $\binom{13}{1}\binom{12}{2}\binom{10}{5} = \frac{13!}{1!2!5!5!} = 216,216$

15. (a) $\frac{10!}{2!3!} = 302,400$

16. $\frac{30!}{10!5!7!8!}$

19. (a) Given any solution (x_1, x_2, x_3, x_4) of (*) with $x_1 \geq 8$, then $(x_1 - 8, x_2, x_3, x_4)$ is a solution of (**) with $x_1 - 8, x_2, x_3, x_4$ nonnegative integers. Conversely, given a solution (x_1, x_2, x_3, x_4) of (**) in nonnegative integers, then $(x_1 + 8, x_2, x_3, x_4)$ is a solution to (*) with $x_1 + 8 \geq 8$. The number of solutions to (**) in nonnegative integers corresponds to the number of ways to put 13 identical marbles into four boxes labeled x_1, x_2, x_3, x_4. This number is $\binom{16}{13} = 560.$

Exercises 7.4

1. $D_6 = 265$; $D_7 = 1854$; $D_8 = 14,833.$

3. $D_{11} = 11!\left(1 - \frac{1}{1!} + \frac{1}{2!} - \frac{1}{3!} + \cdots - \frac{1}{11!}\right)$

5. (b) $7! - D_7$

6. (c) $20D_{19}$

7. (a) $9! - [5(8!) - 10(7!) + 10(6!) - 5(5!) + 4!] = 205,056$

10. (a) Using Proposition 7.4.2,

$$D_n = n! - n! + \left[n(n-1)(n-2)\cdots 3\right] - \left[n(n-1)(n-2)\cdots 4\right] + \cdots + (-1)^{n-1}n + (-1)^n$$

$$= n[(n-1)(n-2)\cdots 3 - (n-1)(n-2)\cdots 4 + \cdots + (-1)^{n-1}(n-1)] + (-1)^n \equiv (-1)^n \pmod{n}.$$

(b) This follows immediately from part (a).

11. (a) $(n-1)(D_{n-1} + D_{n-2})$

$$= (n-1)\left\{ (n-1)!\left[1 - \frac{1}{1!} + \frac{1}{2!} - \cdots + (-1)^{n-1}\frac{1}{(n-1)!}\right]\right.$$

$$\left. + (n-2)!\left[1 - \frac{1}{1!} + \frac{1}{2!} - \cdots + (-1)^{n-2}\frac{1}{(n-2)!}\right]\right\}$$

$$= n(n-1)!\left[1 - \frac{1}{1!} + \frac{1}{2!} - \cdots + (-1)^{n-1}\frac{1}{(n-1)!}\right] - (n-1)!\left[1 - \frac{1}{1!} + \frac{1}{2!} - \cdots + (-1)^{n-1}\frac{1}{(n-1)!}\right]$$

$$+ (n-1)!\left[1 - \frac{1}{1!} + \frac{1}{2!} - \cdots + (-1)^{n-2}\frac{1}{(n-2)!}\right]$$

$$= n!\left(1 - \frac{1}{1!} + \frac{1}{2!} - \cdots + (-1)^n\frac{1}{n!}\right) - n!(-1)^n\frac{1}{n!} - (n-1)!(-1)^{n-1}\frac{1}{(n-1)!}$$

$$= D_n - (-1)^n - (-1)^{n-1} = D_n$$

Exercises 7.5

1. (a) $(x+y)^6 = x^6 + 6x^5y + 15x^4y^2 + 20x^3y^3 + 15x^2y^4 + 6xy^5 + y^6$

(b) $(2x+3y)^6 = 64x^6 + 576x^5y + 2160x^4y^2 + 4320x^3y^3 + 4860x^2y^4 + 2916xy^5 + 729y^6$

3. (b) $(2x^3 - x^2)^8 = 256x^{24} - 1024x^{23} + 1792x^{22} - 1792x^{21} + 1120x^{20} - 448x^{19} + 112x^{18} - 16x^{17} + x^{16}$

4. $\binom{12}{3}(x^3)^9(-2y^2)^3 = -1760x^{27}y^6$

5. (c) The seventh term is $\binom{20}{6}x^{14}y^6 = 38,760x^{14}y^6$. The fifteenth term is $\binom{20}{14}x^6y^{14} = 38,760x^6y^{14}$.

6. (a) 17 terms

(b) There is a middle term (since 16 is even). This term is $\binom{16}{8}(2x)^8(-y)^8 = 3,294,720x^8y^8$.

8. The required term is $\binom{10}{7}(4x)^3(5y)^7 = 120(64x^3)(78125y^7)$. The coefficient is 600,000,000.

9. The term in question is $\binom{17}{4}x^{13}(3y^2)^4 = 2380x^{13}(81y^8)$. The coefficient is $2380(81) = 192,780$.

11. The general term is $\binom{18}{k}\left(\frac{3}{x}\right)^{18-k}(x^2)^k = \binom{18}{k}3^{18-k}x^{3k-18}$. We want $3k - 18 = 27$, so $k = 15$ and the coefficient is $\binom{18}{15}3^3 = (816)(27) = 22,032$.

14. (a) If $n = 0$, $(1+\sqrt{2})^0 = 1 = 1 + 0\sqrt{2}$, so $x_0 = 1$, $y_0 = 0$.

If $n = 1$, $(1+\sqrt{2})^1 = 1 + \sqrt{2}$, so $x_1 = 1$, $y_1 = 1$.

If $n = 2$, $(1+\sqrt{2})^2 = 1 + 2\sqrt{2} + 2 = 3 + 2\sqrt{2}$ so $x_2 = 3$, $y_2 = 2$.

If $n = 5$, $(1+\sqrt{2})^5 = 1 + 5\sqrt{2} + 10(\sqrt{2})^2 + 10(\sqrt{2})^3 + 5(\sqrt{2})^4 + (\sqrt{2})^5$

$$= 1 + 5\sqrt{2} + 20 + 20\sqrt{2} + 20 + 4\sqrt{2} = 41 + 29\sqrt{2}$$

and so $x_5 = 41$, $y_5 = 29$.

16. With the given interpretation, row n of Pascal's triangle becomes the single number

$$10^n\binom{n}{0} + 10^{n-1}\binom{n}{1} + \cdots + 10\binom{n}{n-1} + 1 = \sum_{r=0}^{n} 10^{n-r}\binom{n}{r}.$$

So we are trying to prove that $\sum_{r=0}^{n} 10^{n-r}\binom{n}{r} = 11^n$. This follows immediately from the Binomial Theorem since

$$11^n = (10 + 1)^n = \sum_{r=0}^{n}\binom{n}{r}(10)^{n-r}(1)^r.$$

20. Consider $(x - y)^n = \sum_{k=0}^{n} \binom{n}{k} x^{n-k}(-y)^k = \sum_{k=0}^{n} \binom{n}{k} x^{n-k}(-1)^k y^k$.

Setting $x = y = 1$, we obtain $0 = (1 - 1)^n = \sum_{k=0}^{n} \binom{n}{k} 1^{n-k}(-1)^k 1^k = \sum_{k=0}^{n} \binom{n}{k}(-1)^k$

$$= \binom{n}{0} - \binom{n}{1} + \binom{n}{2} - \cdots + (-1)^n.$$

22. (a) $\displaystyle\sum_{k=1}^{n} k \binom{n}{k} = \binom{n}{1} + 2\binom{n}{2} + 3\binom{n}{3} + 4\binom{n}{4} + \cdots + (n-1)\binom{n}{n-1} + n$

$$= n + n(n-1) + \frac{n(n-1)(n-2)}{2} + \frac{n(n-1)(n-2)(n-3)}{3!} + \cdots + n(n-1) + n$$

$$= n\left[1 + (n-1) + \frac{(n-1)(n-2)}{2} + \frac{(n-1)(n-2)(n-3)}{3!} + \cdots + n - 1 + 1\right]$$

$$= n\left[1 + \binom{n-1}{1} + \binom{n-1}{2} + \binom{n-1}{3} + \cdots + 1\right] = n2^{n-1}$$

using the result of Problem 18 in the last step.

Exercises 8.1

1. To find the mid-point of a line segment AB, choose a radius r (for example, $r = |AB|$) such that the arcs with centers A and B and radius r meet at two distinct points, P and Q. The point of intersection, M, of AB and PQ is the required mid-point.

To see why this works, consider the diagram to the right, in which the labels are given as before and line segments AP, BP, AQ, BQ are joined. Note that $\angle PAB = \angle PBA$ since $\triangle APB$ is isosceles. Since $\triangle BPQ$ is congruent to $\triangle APQ$ (three pairs of sides of equal length), $\angle BPQ = \angle APQ$. Thus, triangles PAM and PBM are congruent (two equal pairs of sides and equal contained angles), so $|AM| = |BM|$ as required.

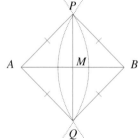

3. (a) Let A be any point on ℓ and let $r = |AP|$. Draw the circle with center P and radius r. If this meets ℓ in only the single point A, then PA is the desired perpendicular to ℓ. Otherwise, the circle meets ℓ in two points, A and B. Draw arcs with centers A and B and radius r meeting in P and Q. Then PQ is the desired perpendicular to ℓ.

To see this, consider the diagram at the right, with M the point of intersection of PQ and ℓ. Since triangles PAQ and PBQ are congruent, $\angle APM = \angle BPM$. Then it follows that $\triangle APM \equiv \triangle BPM$ and thus, $\angle AMP = \angle BMP$. Since these angles have sum $180°$, each is a right angle.

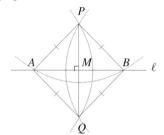

5. The basic idea behind this is the simple fact that $xy = (\frac{x}{2})(2y)$. When x is even, the product xy can be equally well determined as the product of $x/2$ and $2y$. This explains why one crosses out a line where the first term is even: The product of the two numbers in a crossed out line equals the product of the numbers in the line below.

When x is odd, however, the entry below x is $\frac{x-1}{2}$ while the entry below y is $2y$. Their product is $(\frac{x-1}{2})(2y) = xy - y$, so the product xy is the product of the numbers in the line below, **plus y**. Note that y is the number in the right column corresponding to the odd x on the left. Thus, we do not cross out lines where x is odd; the y terms must be added at the end to give the correct product.

7. We find the sum of the numbers and divide by n. Thus, to find the average of n numbers $a_1, a_2, \ldots, a_n$,

   Step 1. set $S = 0$;

   Step 2. for $i = 1$ to n, replace S by $S + a_i$;

   Step 3. output $\frac{S}{n}$.

8. To find the maximum of n numbers $a_1, a_2, \ldots, a_n$,

   Step 1. let $M = a_1$;

   Step 2. for $i = 2$ to n, if $a_i > M$, replace M by a_i;

   Step 3. output M.

   The value of M output in Step 3 is the maximum of the a_i.

11. If $x \leq a_1$, we wish to output $x, a_1, a_2, \ldots, a_n$. If $a_1 < x \leq a_2$, we wish to output $a_1, x, a_2, \ldots, a_n$, and so on. If $x \geq a_n$, we wish to output $a_1, a_2, \ldots, a_n, x$. Here is one suitable algorithm.

    To insert x into its correct position in the ordered list $a_1 \leq a_2 \leq \cdots \leq a_n$,

    Step 1. set $F = 0$;

    Step 2. for $i = 1$ to n,
          if $x \leq a_i$ and $F = 0$, output x and set $F = 1$;
          else output a_i.

    Step 3. if $F = 0$, output x.

    The variable F introduced here is a *flag*, whose purpose is to tell us whether or not x has been output. At the end of the loop in Step 2, if x has not been output (that is, x is larger than all the a_i's), then we will know this, because F will not have changed from its initial value of 0; hence, we must output x as the final element. Note that the "and" in Step 2 is the logical "and" introduced in Section 1.1. We output x and set $F = 1$ only if both $x \leq a_i$ and $F = 0$ are true.

13. Here's the idea. Divide a by 2, thereby obtaining $a = 2q_1 + a_1$, $0 \leq a_1 < 2$. Then divide q_1 by 3, obtaining $q_1 = 3q_2 + a_2$, $0 \leq a_2 < 3$. Note that, at this point, we have $a = 2(3q_2 + a_2) + a_1 = 3!q_2 + 2!a_2 + a_1$. Next, divide q_2 by 4, obtaining $q_2 = 4q_3 + a_3$ and $a = 4!q_3 + 3!a_3 + 2!a_2 + a_1$. Eventually we have

    $$a = n!q_{n-1} + a_{n-1}(n-1)! + \cdots + a_2 2! + a_1,$$

    which, with $a_n = q_{n-1}$, is the required expression. (Note that if at some stage $q_{k-1} = 0$, we have $a_n = a_{n-1} = \cdots = a_k = 0$, but the procedure is still valid.) Here is the algorithm for writing $a = a_n n! + a_{n-1}(n-1)! + \cdots + a_2 2! + a_1$.

    Step 1. Let $q_0 = a$.

    Step 2. For $i = 1$ to $n - 1$, write $q_{i-1} = (i+1)q_i + a_i$ with $0 \leq a_i < i + 1$.

    Step 3. Let $a_n = q_{n-1}$.

15. A number between 0 and $2^n - 1$, when expressed in base 2, is a string of at most n 0's and 1's. By "padding" with initial 0's, if necessary, we can assume that all such strings have length exactly n. Then such a string, $\epsilon_1 \epsilon_2 \cdots \epsilon_n$, each $\epsilon_i = 0$ or 1, determines the subset A of $\{a_1, a_2, \ldots, a_n\}$ as follows: $a_i \in A \leftrightarrow \epsilon_i = 1$. Thus, the string $00\ldots0$ of n 0's corresponds to the empty set, the string $11\ldots1$ of n 1's to the entire set $\{a_1, a_2, \ldots, a_n\}$, the string $11010\ldots0$ to

the subset $\{a_1, a_2, a_4\}$, and so on. Enumerating the subsets of $\{a_1, \ldots, a_n\}$ now simply amounts to listing the integers between 0 and $2^n - 1$ in base 2. Here is an algorithm.

Step 1. Let $M = 0$. Output the words "empty set."

Step 2. for $i = 1$ to $2^n - 1$

- replace M by $M + 1$;
- write $M = \epsilon_1\epsilon_2\cdots\epsilon_n$ as an n-digit number in base 2;
- for $k = 1$ to n, if $\epsilon_k = 1$, output a_k.

Each value of i in Step 2 yields one subset of $\{a_1, \ldots, a_n\}$ and, when Step 2 is complete, all subsets have been output.

16. (a) i. $S = 1$; $S = 1 + (-3)(2) = -5$; $S = -5 + 2(2^2) = -5 + 8 = 3$.

ii. $S = 2$; $S = -3 + 2(2) = 1$; $S = 1 + 1(2) = 3$.

17. The first value of S is $S = a_n$.

With $i = 1$ in Step 2, the value of S is $a_{n-1} + Sx = a_{n-1} + a_n x$.

With $i = 2$ in Step 2, the value of S is $a_{n-2} + Sx = a_{n-2} + (a_{n-1} + a_n x)x = a_{n-2} + a_{n-1}x + a_n x^2$.

With $i = 3$ in Step 2, the value of S is $a_{n-3} + Sx = a_{n-3} + a_{n-2}x + a_{n-1}x^2 + a_n x^3$.

With $i = n$, the value of S is $a_{n-n} + a_{n-(n-1)}x + a_{n-(n-2)}x^2 + \cdots + a_n x^n = a_0 + a_1 x + a_2 x^2 + \cdots + a_n x^n$ as desired.

Exercises 8.2

1. $f(n) = n + 2n + n - 1 + 20 = 4n + 19$

3. Consider the division of a number a, which has at most two digits, by a single-digit number b, where $a < 10b$.

$$b \,\overline{)\,\begin{matrix} q \\ a \end{matrix}}$$
$$\frac{qb}{\times}$$

Since the quotient q has just one digit, the division counts as one operation, the product qb counts as another, and since $a - qb < b$, the difference $a - qb$ has just a single digit, so the subtraction requires a third operation. In the division of an n-digit number by a single-digit number, this process is repeated at most $n - 1$ or n times, depending on whether or not b is greater than the first digit of a. So at most $3n$ operations are required.

6. (a) With reference to Horner's algorithm as described in (8.1.1), each iteration of Step 2 requires two operations. Since Step 2 is repeated n times, this method requires $2n$ operations. This is fewer operations than are needed by the obvious method of polynomial evaluation, as we now show.

Assuming i multiplications to compute $a_i x^i$, $1 + 2 + \cdots + n = \frac{1}{2}n(n+1)$ multiplications are involved in the evaluation of $a_0 + a_1 x + \cdots + a_n x^n$. As well, n additions are required, for a total of $\dfrac{n(n+1)}{2} + n = \dfrac{n^2 + 3n}{2}$ operations. This estimate can be improved. Problem 18 shows that x^i can be computed with approximately $\log i$ multiplications. So $1 + \log i$ multiplications are required for $a_i x^i$ and

$$1 + (1 + \log 2) + (1 + \log 3) + \cdots + (1 + \log n) = n + \log n!$$

for the polynomial. Including additions, approximately $2n + \log n!$ operations are required.

7. (a) Since $5n \le n^3$ for all $n \ge 3$, we can take $c = 1$, $n_0 = 3$, or $c = 5$, $n_0 = 1$.

(c) For $n \ge 1$, we have $8n^3 + 4n^2 + 5n + 1 \le 9n^4 + 18n^2 + 24n + 6 = 3(3n^4 + 6n^2 + 8n + 2)$, so we can take $c = 3$, $n_0 = 1$.

8. (a) By Proposition 8.2.7, for instance, $5n \asymp n$, and by Proposition 8.2.8, $n \prec n^3$. Thus, $5n \prec n^3$.

(c) By Proposition 8.2.7, $f(n) \asymp n^3$ and $g(n) \asymp n^4$. By Proposition 8.2.8, $n^3 \prec n^4$, so $f \prec g$.

10. (b) Suppose $f \prec g$ and $g \prec h$. We must prove $f \prec h$. By part (a), $f = \mathcal{O}(h)$, so it remains only to prove that $h \neq \mathcal{O}(f)$. Assume to the contrary that $h = \mathcal{O}(f)$. Then, since $f = \mathcal{O}(g)$, using part (a) again, we would have that $h = \mathcal{O}(g)$, contradicting $g \prec h$.

11.   If $f \asymp g$, then $f = \mathcal{O}(g)$. Since $g \prec h$, $g = \mathcal{O}(h)$ and so, by Exercise 10(a), $f = \mathcal{O}(h)$. To obtain $f \prec h$, we have now to prove that h is not $\mathcal{O}(f)$. But if $h = \mathcal{O}(f)$, then, because $f = \mathcal{O}(g)$, we would have that $h = \mathcal{O}(g)$, contradicting $g \prec h$.

12.   We have $a^n < b^n$ for $n \geq 1$ since $a < b$, so $a^n = \mathcal{O}(b^n)$. We must prove $b^n \neq \mathcal{O}(a^n)$. If $b^n = \mathcal{O}(a^n)$, then there would exist a constant c such that $b^n \leq ca^n$ for all suitably large n. Dividing by a, we obtain $(\frac{b}{a})^n \leq c$, for all suitably large n, but this is not true because $\frac{b}{a} > 1$ implies that $(\frac{b}{a})^n$ grows without bound as n increases.

13. (a) We established $2^n < n!$ for $n \geq 4$ in Problem 6 of Chapter 5. Thus, $2^n = \mathcal{O}(n!)$ ($c = 1$, $n_0 = 4$). On the other hand, $n! \neq \mathcal{O}(2^n)$, for consider

$$\frac{n!}{2^n} = \frac{1}{2}\frac{2}{2}\frac{3}{2}\frac{4}{2}\cdots\frac{n}{2}.$$

The product of the first three factors on the right is $\frac{3}{4}$ and each of the remaining $n - 3$ terms is bigger than 2. So $\frac{n!}{2^n} > \frac{3}{4}(2^{n-3})$ for $n > 3$. If $n! = \mathcal{O}(2^n)$, then $n! < c2^n$ for some constant c and all suitably large n and so $\frac{n!}{2^n} < c$. We have shown this is not possible. Since $2^n = \mathcal{O}(n!)$ but $n! \neq \mathcal{O}(2^n)$, we have $2^n \prec n!$ as required.

15.   For any $n \geq 1$, $|kf(n)| = |k||f(n)|$. Taking $n_0 = 1$ and $c = |k|$ in Definition 8.2.1, we see that kf is $\mathcal{O}(f)$. Also, $|f(n)| = \frac{1}{|k|}|k(f(n))|$ says that $f = \mathcal{O}(kf)$ and so $f \asymp kf$.

18.   Since $\log_2 n = \mathcal{O}(n)$ by Proposition 8.2.9 and $n = \mathcal{O}(n)$, we have $n\log_2 n = \mathcal{O}(n^2)$ by Proposition 8.2.3. Thus, it remains only to show that $n^2 \neq \mathcal{O}(n\log_2 n)$. Assume to the contrary that for some positive constant c, we have $n^2 \leq cn\log_2 n$ for sufficiently large n. Then, for large n, we would have $n \leq c\log_2 n$. This says $n = \mathcal{O}(\log_2 n)$, contradicting Proposition 8.2.9.

19. (b) $f \asymp n^2$

20. (a) n^5

21.   Since $\log_a n = (\log_a b)(\log_b n)$, we have $|\log_a n| = |\log_a b||\log_b n|$ for all $n \geq 1$. With $c = \log_a b$ and $n_0 = 1$ in Definition 8.2.1 we see that $\log_a n$ is $\mathcal{O}(\log_b n)$. By symmetry, $\log_b n = \mathcal{O}(\log_a n)$; hence, $\log_a n \asymp \log_b n$.

23.   Since $n! = n(n - 1)(n - 2)\cdots 3 \cdot 2 \cdot 1 < n^n$, $\log n! \leq n\log n$. With $c = 1$, $n_0 = 1$ in Definition 8.2.1, we see that $\log n! = \mathcal{O}(n\log n)$ as required.

Exercises 8.3

1. (a) Since $n = 9 \neq 1$, we set $m = \lfloor\frac{9}{2}\rfloor = 4$. Since $2 = x = a_4 = 4$ is false and $x < a_4$ is true, we set $n = m = 4$ and change the list to 1, 2, 3, 4.

Since $4 = n \neq 1$, we set $m = \lfloor\frac{4}{2}\rfloor = 2$. Since $2 = x = a_2$ we output "true" and stop.

This search required three comparisons of 2 with an element in the list—$x = a_4$, $x < a_4$ and $x = a_2$. A linear search would have used two comparisons.

3. (a) To find the complement of $A = \{a_1, a_2, \ldots, a_k\}$ with respect to $U = \{1, 2, \ldots, 100\}$,

Step 1. For $i = 1$ to 100, search A for i and if i is not found, output i.

The output numbers are the elements of A^c.

4.
2, 3, 4, 5	2, 4, 6, 8, 10	1
3, 4, 5	2, 4, 6, 8, 10	1, 2
3, 4, 5	4, 6, 8, 10	1, 2, 2
4, 5	4, 6, 8, 10	1, 2, 2, 3
5	4, 6, 8, 10	1, 2, 2, 3, 4
5	6, 8, 10	1, 2, 2, 3, 4, 4
	6, 8, 10	1, 2, 2, 3, 4, 4, 5
		1, 2, 2, 3, 4, 4, 5, 6, 8, 10

The algorithm required seven comparisons.

6. (a) Here's the bubble sort:

$k = 5$: $3, 1, 7, 2, 5, 4 \to 1, \underline{3, 7}, 2, 5, 4 \to 1, 3, \underline{7, 2}, 5, 4 \to 1, 3, 2, \underline{7, 5}, 4 \to 1, 3, 2, 5, \underline{7, 4}$
$k = 4$: $1, 3, 2, 5, 4, 7 \to 1, \underline{3, 2}, 5, 4, 7 \to 1, 2, \underline{3, 5}, 4, 7 \to 1, 2, 3, \underline{5, 4}, 7$
$k = 3$: $1, 2, 3, 4, 5, 7 \to 1, \underline{2, 3}, 4, 5, 7 \to 1, 2, \underline{3, 4}, 5, 7$
$k = 2$: $1, \underline{2}, 3, 4, 5, 7 \to 1, \underline{2, 3}, 4, 5, 7$
$k = 1$: $1, \underline{2}, 3, 4, 5, 7 \to 1, 2, 3, 4, 5, 7.$

This required a total of $5 + 4 + 3 + 2 + 1 = 15$ comparisons.

(b) Here's the merge sort.

Step 2.   1;   3;   2;   7;   4;   5
Step 3.   1, 3;   2, 7;   4, 5
Step 3.   1, 2, 3, 7;   4, 5
Step 3.   1, 2, 3, 4, 5, 7

Merging two lists of length 1 to one of length 2 requires $1 + 1 - 1 = 1$ comparison. Thus, the initial merging of six lists of length 1 to three lists of length 2 requires $1 + 1 + 1 = 3$ comparisons. Merging three lists of length 2 to one of length 2 and one of length 2 requires $2 + 2 - 1 = 3$ comparisons. The final merging of lists of lengths 4 and 2 requires $4 + 2 - 1 = 5$ comparisons, for a total of $3 + 3 + 5 = 11$ comparisons.

8.   Sort the list into increasing order $a_1 \le a_2 \le \cdots \le a_n$. If $n = 2m + 1$ is odd, the median is a_m; if $n = 2m$ is even, the median is $\frac{1}{2}(a_m + a_{m+1})$. Since determining the parity of n adds only another operation, the complexity of this algorithm is the complexity of the sort, at best $\mathcal{O}(n \log n)$.

9. (a) Here's the bubble sort.

$k = 7$: $a, b, c, d, u, v, w, x \to a, \underline{b, c}, d, u, v, w, x \to a, c, \underline{b, d}, u, v, w, x \to a, c, d, \underline{b, u}, v, w, x$
$\quad\quad \to a, c, d, u, \underline{b, v}, w, x \to a, c, d, u, b, \underline{v, w}, x \to a, c, d, u, b, v, \underline{w, x}$
$k = 6$: $a, c, d, u, b, v, x, w \to a, \underline{c, d}, u, b, v, x, w \to a, d, \underline{c, u}, b, v, x, w \to a, d, u, \underline{c, b}, v, x, w$
$\quad\quad \to a, d, u, c, \underline{b, v}, x, w \to a, d, u, c, b, \underline{v, x}, w$
$k = 5$: $a, d, u, c, b, x, v, w \to d, \underline{a, u}, c, b, x, v, w \to d, a, \underline{u, c}, b, x, v, w \to d, a, u, \underline{c, b}, x, v, w$
$\quad\quad \to d, a, u, c, \underline{b, x}, v, w$
$k = 4$: $d, a, u, c, x, b, v, w \to d, \underline{a, u}, c, x, b, v, w \to d, a, \underline{u, c}, x, b, v, w \to d, a, u, \underline{c, x}, b, v, w$
$k = 3$: $d, a, u, c, x, b, v, w \to d, \underline{a, u}, c, x, b, v, w \to d, a, \underline{u, c}, x, b, v, w$
$k = 2$: $d, a, u, c, x, b, v, w \to d, \underline{a, u}, c, x, b, v, w$
$k = 1$: $d, a, u, c, x, b, v, w \to d, a, u, c, x, b, v, w$

A total of $7 + 6 + 5 + 4 + 3 + 2 + 1 = 28$ comparisons is needed.

10. (a) The lists 1 and 2, 3, 4 can be merged with one comparison.

(b) The lists 5 and 2, 3, 4 can be merged with three comparisons.

(c) The lists 1, 3, 5 and 2, 4, 6 can be merged with $3 + 3 - 1 = 5$ comparisons.

11. (a) $c, a, e, b, d \rightarrow a, \underline{c, e}, b, d \rightarrow a, c, \underline{e, b}, d \rightarrow a, c, b, \underline{e, d}$
$\underline{a, c}, b, d, e \rightarrow a, \underline{c, b}, d, e \rightarrow a, b, \underline{c, d}, e$
$\underline{a, b}, c, d, e \rightarrow a, \underline{b, c}, d, e$
$\underline{a, b}, c, d, e \rightarrow a, b, c, d, e$

14. The answer is $\min\{s, t\}$. It is impossible to have fewer than this number of comparisons since until $\min\{s, t\}$ of comparisons have been made, elements remain in each list. To see that $\min\{s, t\}$ can be achieved, consider ordered lists $a_1, a_2, \ldots, a_s$ and $b_1, b_2, \ldots, b_t$ where $s < t$ and $a_s < b_1$. After s comparisons, the first list is empty.

16. Using the efficient binary search, $\mathcal{O}(\log k)$ comparisons are needed to search the predecessors of a_k. In all, the number of comparisons is $\mathcal{O}(\log 1 + \log 2 + \cdots + \log n)$. Since $\log a + \log b = \log ab$, we have $\log 1 + \log 2 + \cdots + \log n = \log n!$. The result follows.

18. This is less efficient. A merge sort is $\mathcal{O}(n \log n)$ and a binary search is $\mathcal{O}(\log n)$. Since $n + n \log n \sim n \log n$, the suggested procedure is $\mathcal{O}(n \log n)$, while a linear search is $\mathcal{O}(n)$, which is better.

Exercises 8.4

1. (a)

t	Perm(t)	j	m	S
1	$\begin{array}{cccc} 1 & 2 & 3 & 4 \\ \pi_1 & \pi_2 & \pi_3 & \pi_4 \end{array}$	3	4	$\{1, 2, 4\}^c = \{3\}$
2	$\begin{array}{cccc} 1 & 2 & 4 & 3 \\ \pi_1 & \pi_2 & \pi_3 & \pi_4 \end{array}$	2	3	$\{1, 3\}^c = \{2, 4\}$
3	$\begin{array}{cccc} 1 & 3 & 2 & 4 \\ \pi_1 & \pi_2 & \pi_3 & \pi_4 \end{array}$	3	4	$\{1, 3, 4\}^c = \{2\}$
4	$\begin{array}{cccc} 1 & 3 & 4 & 2 \\ \pi_1 & \pi_2 & \pi_3 & \pi_4 \end{array}$	2	4	$\{1, 4\}^c = \{2, 3\}$
5	$\begin{array}{cccc} 1 & 4 & 2 & 3 \\ \pi_1 & \pi_2 & \pi_3 & \pi_4 \end{array}$	3	3	$\{1, 4, 3\}^c = \{2\}$
6	$\begin{array}{cccc} 1 & 4 & 3 & 2 \\ \pi_1 & \pi_2 & \pi_3 & \pi_4 \end{array}$	1	2	$\{2\}^c = \{1, 3, 4\}$
7	$\begin{array}{cccc} 2 & 1 & 3 & 4 \\ \pi_1 & \pi_2 & \pi_3 & \pi_4 \end{array}$	3	4	$\{2, 1, 4\}^c = \{3\}$
8	$2 \quad 1 \quad 4 \quad 3$			

2. (b) $42531, 43125, 43152, 43215, 43251$

3. (a) We consider each part of Step 2. Finding the largest j such that $\pi_j < \pi_{j+1}$ requires at most n comparisons. At most another n comparisons are needed to find the minimum of $\{\pi_i \mid i < j, \pi_i > \pi_j\}$. To find the complement of a subset A of $\{1, 2, \ldots, n\}$ requires searching A for each of the elements $1, 2, \ldots, n$ and noting those which are not in A. Using the efficient binary search, each search is $\mathcal{O}(\log_2 n)$, adding another $n \log_2 n$ comparisons. An efficient merge sort adds another $n \log_2 n$ comparisons, so each pass through Step 2 requires at most $n + n + n \log_2 n + n \log_2 n = 2(n + n \log_2 n)$ comparisons. Since this step is executed $n! - 1$ times and $n + n \log_2 n \sim n \log_2 n$, the algorithm is $\mathcal{O}(n!n \log_2 n)$.

5. (a)

1234	1235	1236	1245	1246
1256	1345	1346	1356	1456
2345	2346	2356	2456	3456

6. (a) $23459, 23467, 23468$ precede; $23478, 23479, 23489$ follow.

7. (b) First list the combinations of $1, 2, \ldots, n$ taken r at a time by the method of Proposition 8.4.3. Then, for each of these combinations, enumerate all permutations of its elements using Algorithm 8.4.2.

8. (a)
123456	123457	123458	123467	123468	123478	123567
123568	123578	123678	124567	124568	124578	124678
125678	134567	134568	134578	134678	135678	145678
234567	234568	234578	234678	235678	245678	345678

9. If $a_1 a_2 \ldots a_r$ is one of the combinations of $1, 2, \ldots, n$ taken r at a time, a_1 cannot be less than 1, a_2 cannot be less than 2 and, in general, a_i cannot be less than i. Thus, $123 \ldots r$ is the smallest combination.

Now let π be a combination and π' the combination determined by π as in the proposition. We show that π' is the immediate successor of π with respect to lexicographic order. First, since all numbers to the left of $k - 1$ are the same in both combinations, while $k - 1$ is increased to k, we conclude that $\pi \prec \pi'$. Now suppose that $\pi \prec \sigma \preceq \pi'$ for some combination σ. All integers to the left of $k - 1$ (in π) are the same in π' as well, and hence also in σ. If $k - 1$ were also the same in σ, then we would have $\sigma = \pi$ since no number to the right of $k - 1$ can be increased in π. Thus, in σ, $k - 1$ must be increased to k (and no more, since $\sigma \prec \pi'$). It then follows that $\sigma = \pi'$ since π' is the smallest sequence whose initial segment (up to k) consists of the integers of π'.

10. (a) n; $n - r + 1$; $n - r + j$.

(b) It must be $123 \ldots r$.

(c) It must be $(n - r + 1)(n - r + 2) \ldots n$.

11. Given $n \geq r > 0$, to enumerate the $\binom{n}{r}$ combinations of $1, 2, \ldots, n$ taken r at a time, proceed as follows.

Step 1. Set $t = 1$. Output Comb$(1) = 123 \ldots r$. If $r = n$, stop.

Step 2. For $t = 1$ to $\binom{n}{r} - 1$, given combination Comb$(t) = a_1 a_2 \ldots a_r$, determine the next combination Comb$(t + 1)$ as follows.

(i) Find the largest j such that $a_j < n - r + j$.

(ii) Output Comb$(t + 1) = a_1 a_2 \ldots a_{j-1} a_j + 1 a_j + 2 \ldots a_j + r - j + 1$.

Exercises 9.1

1.

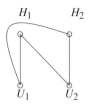

2.

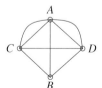

4. (a) We present the graph which corresponds to the cubes, and two edge disjoint subgraphs.

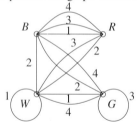

Here is the solution indicated.

	F	B	R	L
Cube 4	B	G	W	G
Cube 3	R	B	R	W
Cube 2	W	R	G	B
Cube 1	G	W	B	R

(e) Here are the graph corresponding to the cubes and two edge disjoint subgraphs.

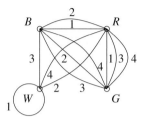

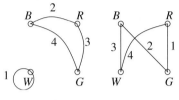

Here is the solution indicated.

	F	B	R	L
Cube 4	G	B	W	R
Cube 3	R	G	B	W
Cube 2	B	R	G	B
Cube 1	W	W	R	G

5. (a) Beta index $= \frac{7}{6}$

6.

```
        | 6 |
   | 3 |
 1 | 4 | 5 |
   | 2 |
```

9. If there are four or more vertices of one color, say red, then there are at least $\binom{4}{3} = 4$ red triangles. Otherwise, there are three red and three white vertices, hence, one triangle of each color.

Exercises 9.2

1. Here is one possibility.

3.

5. 10 edges. This is $\mathcal{K}_5$, the complete graph on five vertices.

9. (a) i.

	v_1	v_2	v_3	v_4	v_5	v_6
v_1	0	1	2	2	2	1
v_2	1	0	1	2	3	2
v_3	2	1	0	1	2	1
v_4	2	2	1	0	2	1
v_5	2	3	2	2	0	1
v_6	1	2	1	1	1	0

Vertex	v_1	v_2	v_3	v_4	v_5	v_6
column total	8	9	7	8	10	6
vertex degree	2	2	3	2	1	4
accessibility index	4	4.5	2.$\dot{3}$	4	10	1.5

ii.

City v_6 has the lowest accessibility index; it is the most accessible. City v_5 is least accessible.

iii. Joining v_1 and v_3 gives now eight edges, with still six vertices.

New beta index $= \frac{8}{6} = 1.\dot{3}$.

New accessibility indices are 2.$\dot{3}$, 4.5, 1.5, 4, 10, 1.5.

Cities v_6 and v_3 are now tied for most accessible; v_5 is least accessible.

iv. Joining v_2 and v_6 gives eight edges and still six vertices.

New beta index $= \frac{8}{6} = 1.\dot{3}$.

New accessibility indices are 4, 2.$\dot{3}$, 2.$\dot{3}$, 4, 9, 1.

City v_6 is most accessible; v_5 is least accessible.

10. $\mathcal{K}_n$ has $\binom{n}{2} = \frac{n(n-1)}{2}$ edges. Each of the n vertices has degree $n - 1$, so the sum of the degrees is $n(n - 1)$. This is twice the number of edges, as asserted by Proposition 9.2.5. If n is even there are n odd vertices; if n is odd, there are 0 odd vertices. In either case the number of odd vertices is even, in accordance with Corollary 9.2.7. The beta index is $\frac{n(n-1)/2}{n} = \frac{n-1}{2}$.

12. Consider the graph in which the vertices correspond to the people at the party and an edge between vertices indicates that the corresponding people shook hands. The degree of a vertex in this graph is the number of hands that person shook. Thus, the result is an immediate consequence of Corollary 9.2.7.

14. (a) Yes, as indicated.

15. (a)

$\mathcal{G} \setminus \{e\}$

$\mathcal{G} \setminus \{v\}$

$\mathcal{G} \setminus \{u\}$

16. (a) Degree of v_1 is 1; degree of v_2 is 3; degree of v_3 is 4; degree of v_4 is 2.

(b) No. The maximum degree of a vertex in a graph with four vertices is 3. (Loops are not allowed in graphs.)

17. (a) No such graph exists. The sum of the degrees of the vertices is an odd number, 17, which is impossible.

(c) Impossible. A vertex of degree 5 in a graph with six vertices must be adjacent to all other vertices. Two vertices of degree 5 means all other vertices have degree at least 2, but the given degree sequence contains a 1.

19. (a) 12

20. (a) This is not bipartite because it contains a triangle.

(c) This is bipartite with bipartition sets indicated R and W.

21. At least two of the three vertices must lie in one of the bipartition sets. Since these two are joined by an edge, the graph cannot be bipartite.

23.   Let x be the number of vertices in one of the bipartition sets. Then $n - x$ is the number of vertices in the other. The largest number of edges occurs when all x vertices in one set are joined to all $n - x$ vertices in the other; so the number of edges is at most $x(n - x)$. The function $f(x) = x(n - x)$ (whose graph is a parabola) has a unique maximum at $(\frac{n}{2}, \frac{n^2}{4})$, so $x(n - x) \leq \frac{n^2}{4}$ for all x, and the result follows.

25.   (a) $2^{\binom{n}{2}}$ (c) $\binom{n}{3} 2^{\binom{n}{2} - 3}$

27.   23

29. $\sum \deg v_i = k|\mathcal{V}|$. But also, $\sum \deg v_i = 2|\mathcal{E}|$. Therefore, $2|\mathcal{E}| = k|\mathcal{V}|$ and so k divides $2|\mathcal{E}|$. But k is odd, so $k \,\big|\, |\mathcal{E}|$.

30.   There are n vertices and n possible degrees for the vertices; namely, $0, 1, 2, \ldots, n - 1$. If, however, we have a vertex of degree 0, then it is not possible to have another vertex of degree $n - 1$. Hence, there are really only $n - 1$ possible "holes" into which the n vertices can fit. Hence, some vertex degree is repeated.

Exercises 9.3

1.   (i) and (ii) are not isomorphic because (i) has five edges and (ii) has four.

(i) and (iii) are isomorphic, as shown by the labeling.

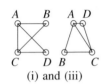
(i) and (iii)

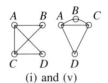
(i) and (v)

(i) and (iv) are not isomorphic because (iv) has a vertex of degree 1 and (i) does not. (Also, (iv) has only four edges.)

(i) and (v) are isomorphic, as shown by the labeling.

(ii) and (iii) are not isomorphic because (ii) has four edges and (iii) has five edges.
(ii) and (iv) are not isomorphic because (iv) has a vertex of degree 1 and (ii) does not.
(ii) and (v) are not isomorphic because (v) has five edges and (ii) has four edges.
(iii) and (iv) are not isomorphic because (iv) has a vertex of degree 1 and (iii) does not.
(iii) and (v) are isomorphic, as shown by the labeling to the right.
(iv) and (v) are not isomorphic because (iv) has a vertex of degree 1 [or because (v) has five edges].

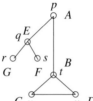

3. (a)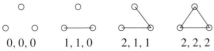

$0, 0, 0 \qquad 1, 1, 0 \qquad 2, 1, 1 \qquad 2, 2, 2$

4. (b)   These graphs are isomorphic. One possible isomorphism is given by

$$\varphi(A) = p, \quad \varphi(B) = t, \quad \varphi(C) = u, \quad \varphi(D) = v, \quad \varphi(E) = Q, \quad \varphi(F) = s, \quad \varphi(G) = r$$

as illustrated.

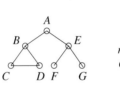

5. (a)   No; it has a vertex of degree 5.

6.   Any graph $\mathcal{G}$ with n vertices is a subgraph of $\mathcal{K}_n$, as is easily seen by joining any pair of vertices of $\mathcal{G}$ where there is not already an edge.

7. (a) Suppose that $\mathcal{G}$ and $\mathcal{H}$ are isomorphic graphs and $\varphi: V(\mathcal{G}) \to V(\mathcal{H})$ is the isomorphism of vertex sets given by Definition 9.3.1. Label the edges of $\mathcal{G}$ arbitrarily, $g_1, g_2, \ldots, g_m$, and then use φ to label the edges of $\mathcal{H}$ as $h_1, h_2, \ldots, h_m$; that is, if edge g_1 has end vertices v and w in $\mathcal{G}$, let h_1 be the edge joining $\varphi(v)$ and $\varphi(w)$ in $\mathcal{H}$. Repeat to obtain $h_2, \ldots, h_m$. Now a triangle in $\mathcal{G}$ is a set of three edges $\{g_i, g_j, g_k\}$ each two of which are adjacent. It follows from the way we labeled the edges of $\mathcal{H}$ that $\{g_i, g_j, g_k\}$ is a triangle in $\mathcal{G}$ if and only if $\{h_i, h_j, h_k\}$ is a triangle in $\mathcal{H}$. Thus, the number of triangles in each graph is the same.

10. (a) No.

$\mathcal{G}$ $\mathcal{H}$

Exercises 10.1

1. (a) $\circ\!\!-\!\!\!-\!\!\circ$

2. Since the graph describing the Königsberg Bridge Problem (Fig 9.2) has several odd vertices, it is not Eulerian. It is not possible to walk over the bridges of Königsberg exactly once and return to the starting position.

3. (a) The graph is Eulerian because it is connected and each vertex has even degree 4. Piecing the circuits $ABCDEFGA$ and $ACFBDGEA$ together at A gives the Eulerian circuit $ABCDEFGACFBDGEA$.

4. (a)

6. (a) Yes, there is because A and B are the only vertices of odd degree.

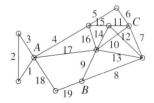

7. (a) No. In the graph representing the modified Königsberg Bridge Problem there are two vertices of odd degree.

(b) This question asks about the possibility of an Eulerian trail and there is one, between the two vertices A and B of odd degree. One possibility is $ACBDADCAB$.

9. Yes. In this case, both $\mathcal{G}$ and $\mathcal{H}$ must be cycles.

11. (a) $\mathcal{K}_n$ is Eulerian $\leftrightarrow$ n is odd.

(b) $\mathcal{K}_n$ has an Eulerian trail if and only if $n = 2$. For $n = 2$, certainly $\circ\!\!-\!\!\circ$ has an Eulerian trail. For $n > 2$, if two vertices have odd degree, then there are other vertices of odd degree so no Eulerian trail can exist.

14. Suppose the vertices of the circuit are $v_0, v_1, \ldots, v_n, v_0$. Consider all subcircuits of the form $v_i, v_{i+1}, v_{i+2}, \ldots, v_i$. (There is at least one such, taking $i = 0$.) That subcircuit $v_i, v_{i+1}, v_{i+2}, \ldots, v_i$ which uses the fewest number of vertices is a cycle. If the original circuit was not a cycle, then those vertices not on the first chosen subcircuit

$v_i, v_{i+1}, v_{i+2}, \ldots, v_i$, together with v_i, form another subcircuit, and the same argument as before shows that this contains a second cycle.

15. False. In the graph shown, $v_1 v_2 v_3 v_2 v_1$ is a closed walk that does not contain a cycle.

17. By definition, $u \sim u$, so $\sim$ is reflexive. If $u \sim v$, then there is a walk $u = u_0, u_1, \ldots, u_k = v$ from u to v. But then $v = u_k, u_{k-1}, \ldots, u_1, u_0 = u$ is a walk from v to u, so $v \sim u$ and $\sim$ is symmetric. Finally, if $u \sim v$ and $v \sim w$, then there is a walk $u = u_0, u_1, \ldots, u_k = v$ from u to v and a walk $v = v_0, v_1, \ldots, v_\ell = w$ from v to w. But then $u = u_0, u_1, \ldots, u_k = v = v_0, v_1, \ldots, v_\ell = w$ is a walk from u to w, proving $u \sim w$ and establishing transitivity.

20. We must show that there is a walk between any two vertices x and y of $\mathcal{G}$. We show that there is, in fact, a walk of length at most 2 between x and y. If xy is an edge, then obviously there is a walk from x to y, so suppose that xy is not an edge. Let S be the set of vertices adjacent to x and T be the set of vertices adjacent to y. Thus $x \notin S \cup T$ and $y \notin S \cup T$ so the number of vertices in $\mathcal{G}$ is $20 \geq |S \cup T| + 2$. Using the Principle of Inclusion-Exclusion, $20 \geq |S| + |T| - |S \cap T| + 2 = \deg x + \deg y - |S \cap T| + 2 \geq 21 - |S \cap T|$. It follows that $S \cap T \neq \emptyset$, so there is a vertex u in both S and T and, thus, a walk xuy from x to y.

21. (a) Since $\mathcal{G}$ is connected and $n > 1$, no vertices have degree zero. Therefore, if there are no vertices of degree one, every vertex of $\mathcal{G}$ has degree at least 2. Using $\mathcal{E}$ to denote the set of edges of $\mathcal{G}$, it follows that $2|\mathcal{E}| = \sum \deg v_i \geq 2n$ (by Proposition 9.2.5) and the number of edges $|\mathcal{E}| \geq n$ as required.

24. Suppose that $\mathcal{G}_1$ has n_1 components and that $\mathcal{G}_2$ has $n_2 < n_1$ components. Let $v_1, \ldots, v_{n_1}$ be vertices of $\mathcal{G}_1$ each in a different component. Then there is no walk between any pair of these vertices. On the other hand, if $w_1, \ldots, w_{n_1}$ are any n_1 vertices of $\mathcal{G}_2$, at least two of these must lie in the same component (by the Pigeon-Hole Principle) and hence there is a walk between these two. Thus, the vertices v_i in $\mathcal{G}_1$ do not correspond to any n_1 vertices of $\mathcal{G}_2$, so these graphs cannot be isomorphic.

Exercises 10.2

1. This graph is not Hamiltonian. To see this, suppose $\mathcal{H}$ were a Hamiltonian cycle. Since vertices A and B have degree 2, the two edges incident with each of these vertices would be in $\mathcal{H}$. Thus, $\mathcal{H}$ would contain the cycle $ACBDA$, which cannot be the case since this does not contain all vertices of the graph. The graph is not Eulerian because it contains vertices of odd degree.

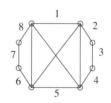

2. (b) This is not Hamiltonian, since it isn't connected.　　　　**(d)**

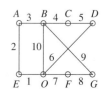

4. (a) Yes, it is Hamiltonian: $ABDCEA$ is a Hamiltonian cycle.

　(c) The graph is not Eulerian. Vertices A and E have odd degree.

7. In the graph to the right, vertices correspond to rooms and edges to doorways.

(a) No, because the graph is not Hamiltonian. Since vertices A, C, D, and E have degree 2, the edges labeled 3, 4, 5, 6, 1, and 2 would have to be part of any Hamiltonian cycle. These edges, however, define a proper cycle, which is not allowed.

8. The result is obvious if $n = 1$ since in this case there are just two people who are friends. So we assume that $n > 1$. Consider the graph whose vertices correspond to people and where an edge between vertices v and u signifies that v and u are friends. The question asks us to prove that this graph is Hamiltonian. This is an immediate consequence of Dirac's Theorem since the graph in question has $2n \geq 3$ vertices each of degree $d \geq n = \frac{2n}{2}$.

9. (a) n edges.

(c) $\mathcal{K}_n$ has $\frac{n(n-1)}{2}$ edges, so the maximum is $\frac{n-1}{2}$ edge disjoint cycles.

10. The cube is indeed Hamiltonian; the labels $1, \ldots, 8$ on the vertices exhibit a Hamiltonian cycle.

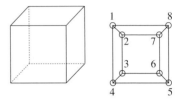

13. (a) As suggested, add an extra vertex v to $\mathcal{G}$ and join it to all other vertices. Then $\deg v = n \geq \frac{n+1}{2}$, and $\deg w \geq \frac{n-1}{2} + 1 = \frac{n+1}{2}$ for all other vertices. By Theorem 10.2.4, this new graph with $n+1$ vertices has a Hamiltonian cycle. Deleting v and all the new edges incident with v leads to a Hamiltonian path in our original graph.

(e) $\circ\!\!-\!\!\circ$ is an example!

14. As in the proof of Dirac's Theorem all vertices adjacent to v_1 are in $\mathcal{P}$.

(a) If v_1 and v_t are adjacent, $\mathcal{C}_1 : v_1 v_2 \cdots v_t v_1$ is a cycle. If $t < n$, then there exists a vertex w not in $\mathcal{C}_1$. As noted, w is not adjacent to v_1, hence the hypothesis tells us that $\deg v_1 + \deg w \geq n$. It follows that there exists some vertex u which is adjacent to both v_1 and w; otherwise, the graph has at least $n+2$ vertices, counting v_1 and w and the disjoint sets of vertices adjacent to v_1 and to w. Since u is adjacent to v_1, $u = v_i$ for some i. Now $w v_i v_{i+1} \cdots v_t v_1 \cdots v_{i-1}$ is a path longer than $\mathcal{P}$, a contradiction. Thus $t = n$, so $\mathcal{C}_1$ is a Hamiltonian cycle and we're done.

15. Add a new vertex adjacent to all existing vertices and apply the result of Exercise 14.

16. (c) False. ⊠ is Eulerian but not Hamiltonian.

17. (a) Yes! Since there is a path between any two vertices, the graph is connected. Thus, there exists an edge e in the graph, joining, say, vertices v and w. Now let $\mathcal{P}$ be a Hamiltonian path from v to w. Then starting at v, following $\mathcal{P}$ to w and then e to v produces a Hamiltonian cycle from v to v.

Exercises 10.3

1. $A = \begin{bmatrix} 0 & 1 & 0 & 1 & 0 & 0 \\ 1 & 0 & 1 & 1 & 0 & 0 \\ 0 & 1 & 0 & 1 & 1 & 1 \\ 1 & 1 & 1 & 0 & 0 & 0 \\ 0 & 0 & 1 & 0 & 0 & 1 \\ 0 & 0 & 1 & 0 & 1 & 0 \end{bmatrix}$

3. (a) The $(3, 5)$ entry of A^3 is 5. The $(2, 2)$ entry of A^3 is 2.

4.   Each 1 represents an edge. Each edge $v_i v_j$ contributes two 1's to the matrix, in positions (i, j) and (j, i). The number of 1's is twice the number of edges.

6. (a) The (i, j) entry in A^2 is the number of walks of length 2 from i to j. Hence, the sum of all such entries is the total number of walks of length 2.

8. (a) $A_1 = \begin{bmatrix} 0 & 1 & 0 & 1 & 1 \\ 1 & 0 & 1 & 0 & 0 \\ 0 & 1 & 0 & 0 & 1 \\ 1 & 0 & 0 & 0 & 0 \\ 1 & 0 & 1 & 0 & 0 \end{bmatrix}$, $A_2 = \begin{bmatrix} 0 & 1 & 0 & 0 & 1 \\ 1 & 0 & 0 & 1 & 0 \\ 0 & 0 & 0 & 1 & 0 \\ 0 & 1 & 1 & 0 & 1 \\ 1 & 0 & 0 & 1 & 0 \end{bmatrix}$.

(b) The function φ is an isomorphism because, if the vertices of $\mathcal{G}_1$ are relabeled, v_i being replaced by $\varphi(v_i) = u_i$, then the adjacency matrix of $\mathcal{G}_1$ relative to the u_i's is A_2. (See Theorem 10.3.3.)

(c) $P = \begin{bmatrix} 0 & 0 & 1 & 0 & 0 \\ 0 & 0 & 0 & 0 & 1 \\ 0 & 0 & 0 & 1 & 0 \\ 1 & 0 & 0 & 0 & 0 \\ 0 & 1 & 0 & 0 & 0 \end{bmatrix}$

11. (a) $PA = \begin{bmatrix} p & q & r \\ x & y & z \\ a & b & c \end{bmatrix}$ is A, but with rows written in the order $2, 3, 1$, the order in which the rows of I were rearranged to give P.

13. (a) The matrices are the adjacency matrices of graphs $\mathcal{G}_1$, $\mathcal{G}_2$, respectively. Since $\mathcal{G}_2$ has three vertices of degree 1 while $\mathcal{G}_1$ has only one, the graphs are not isomorphic, so no such P exists, by Theorem 10.3.4.

14.   The ith entry on the diagonal of A^{37} is the number of walks of length 37 from v_i to itself. But in a bipartite graph, you can only get from v_i back to itself in an even number of steps. Hence, the entry is 0.

15. (a) A^2 is an adjacency matrix $\leftrightarrow$ A is the 0 matrix.

Proof. For A^2 to be an adjacency matrix, it must have all diagonal entries equal to 0. But the ith diagonal entry of A^2 is the number of walks of length 2 from v_i to itself. Now, if $v_i v_j$ is an edge of $\mathcal{G}$, then $v_i v_j v_i$ is a walk of length 2 from v_i to itself, and the ith diagonal entry would not be 0. We conclude that $\mathcal{G}$ cannot have any edges; that is, A is the zero matrix. On the other hand, if A is the zero matrix, certainly $A^2 = A$ is an adjacency matrix.

18.   We could store the *incidence matrix*, whose columns correspond to edges and rows to vertices. The (i, j) entry is 1 if vertex i is incident with edge j, and otherwise 0. We could also simply store the edges as a linear list, perhaps of numbers: Assuming less than 100 edges, we could store the edge ij as $100i + j$ and recover i as $\lfloor \frac{100i+j}{100} \rfloor$ and j as $(100i + j) - 100i$.

Exercises 10.4

2. (a) [BB] The final labeling starting at A is shown to the right. The shortest path from A to E is $ABCJIGFKE$ and has length 13.

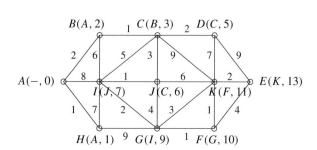

3. If we start at A, permanent labels will be assigned in the order A, H, B, C, D, J, I, G, F, K, E.

5. A shortest path has length 11. One way in which permanent labels might be assigned is in the order A, B, D, C, S, J, I, H, E, L, G, F, K, M, O, P, N, Q, R.

7. The shortest path has length 13, as shown.

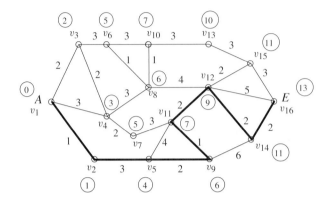

9. In each case the answer is yes, if A and E were in different components of a graph which was not connected.

11. (a) Assign each edge a weight of 1.

12. As explained in the text, the complexity function for determining the shortest distance from a given vertex to each of the others is $\mathcal{O}(n^2)$; that is, for sufficiently large n and some constant c, the algorithm requires at most cn^2 comparisons. Applying the algorithm to each of the n vertices (after which all shortest distances are known) requires at most $n(cn^2) = cn^3$ comparisons. Thus, this process is $\mathcal{O}(n^3)$.

13. (b) $d(1, 5) = 27$; $d(1, 6)$ is still ∞; $d(3, 4)$ is 8; $d(8, 5) = 24$.

16. It is indeed necessary to continue. An identical set of $d(i, j)$ for $k = r$ and $k = r + 1$ simply indicates that the shortest path from each v_i to each v_j passing through $v_1, \ldots, v_r$ has the same length as the shortest path through $v_1, \ldots, v_{r+1}$. In the graph shown to the right, for instance, the values of $d(i, j)$ do not change until $k = 3$ since the shortest path between pairs of vertices cannot be reduced until vertex v_3 is used.

Exercises 11.1

1. (a)

2. K_5 is Eulerian, so no additional edges are needed.

3. Here is the unique solution.

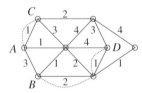

4. We show one of several solutions in each case.

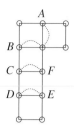

6. The shortest route from X to X is a circuit (perhaps in a pseudograph) passing through Y. Hence, it can also be viewed as a circuit from Y to Y. Any other route from Y to Y could also be viewed as a route from X to X. So there cannot be any shorter route from Y to Y.

9. Each odd vertex in $\mathcal{G}$ is the end vertex of precisely one new path constructed by the algorithm, so it is even in $\mathcal{G}'$. Each even vertex in $\mathcal{G}$ either has unchanged degree in $\mathcal{G}'$ or, as an intermediate point on one or more paths between odd vertices, has its degree increased by a multiple of 2, and so remains even in $\mathcal{G}'$.

Exercises 11.2

1. (a) These digraphs are not isomorphic because one has six vertices while the other has five.

3. Every arc comes out of one vertex and goes into another, hence, adds one to the sum of all indegrees and one to the sum of all outdegrees.

5. The answer is yes. Since $\mathcal{G}$ is an Eulerian graph, there exists an Eulerian circuit. Now just orient the edges of this circuit in the direction of a walk along it.

7. (a) $A = \begin{bmatrix} 0 & 1 & 1 & 0 \\ 1 & 0 & 1 & 0 \\ 0 & 1 & 0 & 1 \\ 1 & 1 & 1 & 0 \end{bmatrix}$

(b) The $(3, 3)$ entry of A^2 is 2 because there are two directed walks of length 2 from vertex 3 to vertex 3; namely, 323 and 343. The $(1, 4)$ entry of A^2 is 1 because the only directed walk of length 2 from vertex 1 to vertex 4 is 134.

(c) The $(4, 2)$ entry of A^3 is 4 because there are four directed walks of length 3 from vertex 4 to vertex 2; namely, 4212, 4232, 4132, and 4342. The $(1, 3)$ entry of A^4 is 6. There are six walks of length 4 from vertex 1 to vertex 3; namely, 13423, 13413, 13213, 12123, 12323, and 12343.

(d) The digraph is strongly connected; 12341 is a directed circuit which permits travel in the right direction between any two vertices.

(e) The digraph is not Eulerian. Vertices 2, 3, and 4 have different indegree and outdegree; vertex 3, for instance, has indegree 3 and outdegree 2.

10. (a) With the graphs labeled as shown, the adjacency matrices are

$$A_1 = \begin{bmatrix} 0 & 1 & 0 & 0 & 0 & 0 \\ 0 & 0 & 1 & 0 & 0 & 0 \\ 0 & 0 & 0 & 1 & 0 & 0 \\ 0 & 0 & 0 & 0 & 1 & 0 \\ 0 & 0 & 0 & 0 & 0 & 1 \\ 1 & 0 & 0 & 0 & 0 & 0 \end{bmatrix} \text{ and } A_2 = \begin{bmatrix} 0 & 0 & 0 & 0 & 1 \\ 1 & 0 & 0 & 0 & 0 \\ 0 & 1 & 0 & 0 & 0 \\ 0 & 0 & 1 & 0 & 0 \\ 0 & 0 & 0 & 1 & 0 \end{bmatrix}.$$

11. (a) $A_1 = \begin{bmatrix} 0 & 1 & 0 & 1 \\ 0 & 0 & 1 & 0 \\ 0 & 0 & 0 & 1 \\ 1 & 1 & 0 & 0 \end{bmatrix}$ and $A_2 = \begin{bmatrix} 0 & 0 & 0 & 1 \\ 1 & 0 & 1 & 0 \\ 1 & 1 & 0 & 0 \\ 0 & 0 & 1 & 0 \end{bmatrix}$.

(b) With the vertices of $\mathcal{G}_1$ relabeled according to φ, its adjacency matrix becomes that of $\mathcal{G}_2$.

(c) $P = \begin{bmatrix} 0 & 1 & 0 & 0 \\ 1 & 0 & 0 & 0 \\ 0 & 0 & 0 & 1 \\ 0 & 0 & 1 & 0 \end{bmatrix}$

(d) The digraphs are strongly connected: In $\mathcal{G}_1$, for instance, $v_1v_2v_3v_4v_1$ is a circuit which respects arrows and $\mathcal{G}_2$ is isomorphic to $\mathcal{G}_1$ hence, also strongly connected.

(e) The digraphs are not Eulerian. In $\mathcal{G}_1$, for instance, vertex v_2 has indegree 2 but outdegree 1.

14. (a) Each of these graphs is strongly connected; each is a cycle.

15. (a) There are just two possibilities for the outdegree sequence; $1, 1, 1$ and $2, 1, 0$. The corresponding graphs are shown at the right.

16. No. $\mathcal{K}_3$ is connected, but 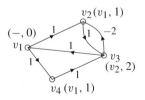 is not strongly connected.

18. True. Let v be a vertex. Since $\mathcal{G}$ has at least two vertices and $\mathcal{G}$ is strongly connected there is some arc of the form vw. Since $\mathcal{G}$ is strongly connected, there is a path from w to v. This path does not use arc vw since all vertices of a path are distinct. For the same reason, all arcs on this (or any) path are distinct. Thus, arc vw followed by the path from w to v gives the desired circuit.

19. (a)

	Max. no. of arcs					
	1	2	3	4	5	6
v_2	$1, v_1$	$1, v_1$	$1, v_1$	$1, v_1$	$1, v_1$	$1, v_1$
v_3	$8, v_1$	$2, v_2$	$2, v_2$	$2, v_2$	$2, v_2$	$2, v_2$
v_4	∞	$11, v_3$	$5, v_3$	$4, v_5$	$4, v_5$	$4, v_5$
v_5	∞	$4, v_2$	$3, v_3$	$3, v_3$	$3, v_3$	$3, v_3$
v_6	$7, v_1$	$7, v_1$	$7, v_1$	$6, v_5$	$6, v_5$	$6, v_5$

20. Bellman-Ford works fine on undirected graphs without negative edges. It wouldn't make sense to apply Bellman-Ford to an undirected graph with a negative edge weight, since any walk could be shortened by passing up and down that edge as often as desired.

21. (a) Dijkstra incorrectly determines that the length of a shortest path to v_2 is 1. Dijkstra does not always work when applied to digraphs which have arcs of negative weight.

(b) No shortest path algorithm will work. There is a negative weight cycle, hence no shortest distance to v_2, for example.

23. (a) As Step 2 of the algorithm shows, for each j, the values $d_i(j)$ depend only on the values $d_{i-1}(k)$ and the arc weights. It follows that if the values $d_{i-1}(j) = d_i(j)$ are identical, then $d_{i-1}(j) = d_i(j) = d_{i+1}(j) = \cdots = d_{n-1}(j)$.

Exercises 11.3

1. (a) Since the chain ends UC, the given G-fragments arise from any of $3! = 6$ chains. The given U,C-fragments arise from any of $5!/2! = 60$ chains. The better estimate is 6. The abnormal fragment is UC, and this ends the chain. The interior extended bases are U, G. The unsplittable fragments are G, C, C, U. The chain starts and ends with C. The only answer is CAAGCUGGUC.

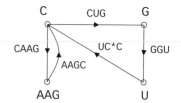

(e) Since the chain ends CC, the G-fragments arise from any of $4! = 24$ chains. The U,C-fragments arise from any of $6!/2! = 360$ chains. The better estimate is 24. The only interior extended base is C. The unsplittable fragments are G, C, C. This chain starts with G and ends with C. There are two answers: GUGAUGACCAGCC and GAUGUGACCAGCC.

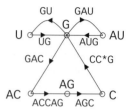

2. (a) The last letter in the chain is U, C, G, or A. If it is U or C, the last G-fragment will be abnormal. If it is G, the last U,C-fragment will be abnormal. If it is A, the last G and the last U,C-fragments will each be abnormal.

(b) There are two abnormal fragments if and only if the chain ends with A.

3. The chains GGUGU and GUGGU have the same G- and U,C-fragments. We leave it to you to check that no shorter chains exist.

4. (a) Let B be an interior extended base. Then B came from some fragment, say a U,C-fragment. (The argument which follows easily adapts to the case of a G-fragment.) Thus our fragment has the form XBY, and the only U or C here is possibly at the end of Y. Since the G-enzyme splits off B, both X and B end with a G and B contains no other G. Thus B is itself a G-fragment. Since B contains no U, C or G, it's unsplittable.

(b) The first extended base is certainly an unsplittable fragment: It ends at the first U, C or G. Clearly, it isn't interior. (It's first!) The last extended base contains at most one U, C or G (and this at its end) so it is also unsplittable. Again, it is clearly not interior. (It is last!)

Exercises 11.4

1. (a) See Exercise 15(a) of Section 11.2. The score sequences (left to right) are 1, 1, 1 and 2, 1, 0. The tournament on the left is not transitive; the one on the right is.

2. Since for each pair of (distinct) vertices v_i, v_j precisely one of v_iv_j, v_jv_i is an arc, in the adjacency matrix A, for each $i \neq j$, precisely one of a_{ij}, a_{ji} is 1. Thus, $A + A^T$ has 0's on the diagonal and 1's in every off-diagonal position.

5. (a) The sum of the scores is the number of arcs, by Proposition 11.2.2; thus, $\sum_{i=1}^{n} s_i = \binom{n}{2} = \frac{1}{2}n(n-1)$. Since $\sum_{i=1}^{n}(n-1-s_i) = \sum_{i=1}^{n}(n-1) - \sum_{i=1}^{n} s_i = n(n-1) - \frac{1}{2}n(n-1) = \frac{1}{2}n(n-1)$, we have the desired result. In a tournament with n players, each player plays $n-1$ games, so if a player wins s_i games, he loses $n-1-s_i$ games. The result says that the sum of the numbers of wins equals the sum of the numbers of losses.

6. Let w be any other vertex. If v beats w, there is a path of length 1 from v to w and we're fine. Hence, assume that w beats v, that is, $w \longrightarrow v$. Among those vertices that v beats there must be one, say x, which beats w, since otherwise $s(w) \geq s(v) + 1$ (recall that w beats v), contradicting the maximum score of v. Hence, for some x, we have

, but then there is a path of length 2 from v to w.

7. (a) No. The sum of the scores, $\sum s(v)$, is the number of arcs. Here, $1 + 1 + 2 + 3 = 7$, but the number of arcs is $\binom{4}{2} = 6$.

11. Suppose T is a transitive tournament that contains the cycle $v_1 v_2 \ldots v_n v_1$. Since T contains no 3-cycle, upon considering the vertices v_1, v_2 and v_3, we see that $v_1 v_3$ must be an arc. Then, considering v_1, v_3, v_4, we see that $v_1 v_4$ is an arc. Continuing, we eventually have an arc $v_1 v_n$, contradicting the fact that $v_n v_1$ is an arc. (In a tournament, for any pair u, v of vertices, precisely one of uv, vu is an arc.)

Exercises 11.5

1. With reference to Fig 11.18, the arc BD now has weight 4. This does not change any of the labels shown, but it does eliminate the possibility of a second shortest path. Now the only shortest path is $SADFT$, requiring 11 units of time as before.

5. The new digraph is shown. The project now requires 18 units of time. The critical path is $SFEWPaDVCT$. The slack in Pl is now 1; the slack in the installation of kitchen cabinets is 3. All other tasks have slack 0 since they lie on the critical path.

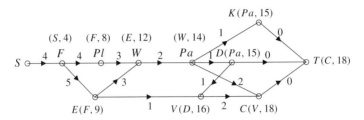

6. Since the time required for a task does not depend on which other tasks have been completed and since some tasks can occur simultaneously, this a Type II scheduling problem.

(a)

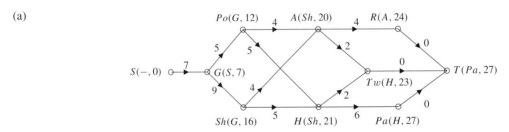

(b) The only critical path is $SGShHPaT$, taking a total of 27 months.

(c) The slack of Po is 4. This job could take as much as 9 units of time without affecting the time (21) by which the task H, subsequent to Po and on the critical path, would be accomplished. There is no slack in Sh. It's on the critical path. The slack in A is 3. If this task took 7 units of time, its label would be $(Sh, 23)$ and the label on R would become $(A, 27)$. There would be another critical path, $SGShART$, but the time for the job would be unaffected. The slack in Tw is 4. Twisting could take as long as 6 units of time and the project would still be completed in 27 units. The slack in R is 3.

(d) No, the project will not be delayed. Delaying Po by three months changes Po's label to $(G, 15)$, but this does not affect any other labels on the digraph. Delaying A by three months changes A's label to $(Sh, 23)$, but does not affect the critical path, or its length.

10. (a) This type I scheduling problem is described by the following digraph.

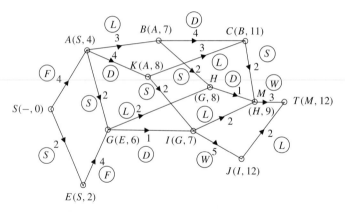

(b) The shortest time is 12 units.

(c) One of several ways in which this time can be achieved is to first conduct the library search, then do the field work, then the laboratory analysis, then create the database and finally do the write-up.

Exercises 12.1

1.

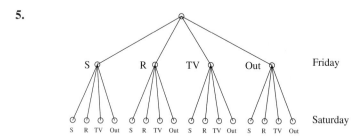

5.

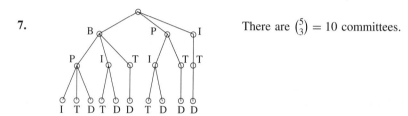

7.

There are $\binom{5}{3} = 10$ committees.

12. We think of a path as a sequence of vertices. There are at least two vertices of $\mathcal{P}_2$ which are also in $\mathcal{P}_1$ (v and w, for example). Let $u \neq w$ be the first vertex of $\mathcal{P}_1$ which is encountered on the reversed path $\mathcal{P}_2$ from w back to v.

(It is possible that u might equal v.) It follows that the path from u to w along $\mathcal{P}_1$ followed by the path back to u along $\mathcal{P}_2$ in reverse from w is a cycle.

Exercises 12.2

1. (a)

(b) Since each tree with five vertices has all vertices of degree at most four, there is one isomer for each such tree, the C atoms corresponding to the vertices. There are three isomers of C_5H_{12}.

3. A beta index less than 1 says that there are fewer edges than vertices. One possibility is that the graph is not connected; in other words, there exist two cities such that it is impossible to fly from one city to the other. If the graph is connected, then it must be a tree by Theorem 12.2.3. This means there is a unique way of flying from any city to any other city.

5. ($\rightarrow$) A tree with n vertices has $n - 1$ edges by Theorem 12.2.3 and no cycles by Proposition 12.1.2.

($\leftarrow$) Suppose $\mathcal{G}$ is an acyclic graph with n vertices and $n - 1$ edges. Since $\mathcal{G}$ has no cycles, Proposition 12.1.2 shows that we have only to prove that $\mathcal{G}$ is connected. Let then $C_1, C_2, \ldots, C_k$ be the connected components of $\mathcal{G}$ and suppose that C_i has n_i vertices. (Thus, $\sum n_i = n$.) Since $\mathcal{G}$ has no cycles, there are no cycles within each C_i. It follows that each C_i is a tree with $n_i - 1$ edges. The number of edges in $\mathcal{G}$ is, therefore, $\sum_1^k (n_i - 1) = (\sum_1^k n_i) - k = n - k$. So $n - k = n - 1$, $k = 1$, $\mathcal{G}$ has only one component; that is, $\mathcal{G}$ is connected.

10. There are no circuits in the subgraph since there are no circuits in $C_k H_{2k+2}$. Also, given any two C vertices, there is a path between them in $C_k H_{2k+2}$ (because $C_k H_{2k+2}$ is connected). Any H vertex on this path would have degree two. Thus, there is none; the path consists entirely of C vertices and hence lies within the subgraph. Thus, the subgraph is connected, hence a tree.

11. (a) Let x be the number of H vertices adjoined. Since T had $k - 1$ edges, and one new edge is added for each H, $\mathcal{G}$ has $(k - 1) + x$ edges. Therefore, $\sum \deg v_i = 2(k - 1 + x)$. But $\sum \deg v_i = 4k + x$ since each C has degree 4 and each H has degree 1. Therefore, $4k + x = 2k - 2 + 2x$ and $x = 2k + 2$.

14. A tree is a complete bipartite graph if and only if it is $\mathcal{K}_{1,n}$ for some n.

Proof. Certainly $\mathcal{K}_{1,n}$ is a tree. Conversely, if a tree is complete bipartite, then it is $\mathcal{K}_{m,n}$ for some m and n. But such a graph has no vertices of degree one unless m or n is 1. The result follows.

15. (a) By Corollary 12.2.4, each of the c components has at least two vertices of degree 1. So there are at least $2c$ vertices of degree 1 altogether.

16. (a) Using Corollary 12.2.4, we have $\sum \deg(v_i) \geq 8$, so the tree has at least four edges and hence at least five vertices. If the result is not true, then there are at most three vertices of degree one while the rest have degree at least two. Then

$$\sum \deg v_i \geq 2(3) + 3(1) + (n - 5)2 = 2n - 1,$$

contradicting the fact that $\sum \deg v_i = 2(n - 1)$.

Exercises 12.3

1. We show T and two other spanning trees found by adding a and then successively deleting f and g.

4. By Theorem 12.3.3, the numbers are $1^{-1} = 1$, $2^0 = 1$, $3^1 = 3$, $4^2 = 16$, $5^3 = 125$, and $6^4 = 1296$.

5.

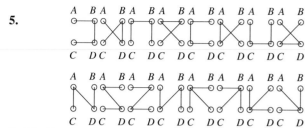

7. (a) $\mathcal{K}_{2,2}$ has four spanning trees (obtained by deleting each edge in succession). They are all isomorphic to ○—○—○—○.

9. The edge in question is a *bridge* (see Definition 13.2.2); that is, its removal disconnects the graph. To see why, call the edge e. If $\mathcal{G} \setminus \{e\}$ were connected, it would have a spanning tree. However, since $\mathcal{G} \setminus \{e\}$ contains all the vertices of $\mathcal{G}$, any spanning tree for it is also a spanning tree for $\mathcal{G}$. We have a contradiction.

12. (a) Say the edge is e and $\mathcal{T}$ is any spanning tree. If e is not in $\mathcal{T}$, then $\mathcal{T} \cup \{e\}$ must contain a circuit. Deleting any edge of this circuit other than e gives another spanning tree which includes e.

(c) No. If the three edges form a circuit, no spanning tree can contain them.

14. (a) The subgraph $\mathcal{T}_0 \cup \{e\}$ is connected because $\mathcal{T}_0$ is. Since it has the same number of vertices and edges it cannot be a tree; therefore, it contains a circuit. Let f be an edge on this circuit such that $f \notin \mathcal{T}$. (Since $\mathcal{T}$ does not contain a circuit, this is possible.) Then $(\mathcal{T}_0 \cup \{e\}) \setminus \{f\}$ is still connected (you can go either way around a circuit), and it has n vertices and $n - 1$ edges, so it's a spanning tree.

16. There are $\binom{n}{2}$ possible edges from which we choose $n - 1$. The number of graphs is, therefore, $\binom{\binom{n}{2}}{n-1}$. The number of trees on n labeled vertices is n^{n-2}. The number of trees on n labeled vertices is n^{n-2}. For $n \leq 6$, the table shows the numbers of trees vs. graphs.

n	No. of trees	No. of graphs
2	1	1
3	3	3
4	16	20
5	125	210
6	1296	3003

Exercises 12.4

1. (a) We want five edges (since there are six vertices). Choose BC, then AD, FE, and DE. We would like next to choose AE, but this would complete a circuit with AD and DE, so we choose AC and obtain the spanning tree shown, of weight 13.

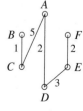

2. (a) The edge of least weight incident with E is EF. The least weight of those edges adjacent to EF is 3; we choose one of them, say AE (in an effort to obtain a different tree from before). There is just one edge of least weight among whose which, together with EF and AE, form a tree; namely, AD. Now those edges which together with EF, AE, and AD form a tree have weights 5 and 6. We choose one of least weight, say CF. Finally, we choose BC, obtaining the tree of weight 13 shown to the right.

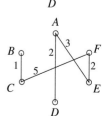

3. We need five edges because there are six vertices in the graph. We use Kruskal's algorithm. First select BC of weight 5, then DF of weight 6. Next we select BF and AB, both of weight 7. We would like to select AF next, but cannot since it completes the circuit $ABFA$, so we select EF of weight 9 next. The five edges we have selected comprise a minimum spanning tree. The smallest length of pavement required is $5 + 6 + 7 + 7 + 9 = 34$.

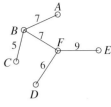

4. (b) i. Choose BD, BE, and CE each of weight 6. The remaining edge, CD, of maximum weight (6) cannot be chosen because it would complete the circuit $BDCEB$ with the previously chosen edges. So we choose DF and then AB obtaining the spanning tree shown at the right of maximum weight 28.

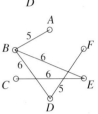

6. Assign all edges of the graph a weight of 1. Then carry out the algorithm.

8. (a) If the graph is unweighted, put a weight of 1 on edge e and 2 on every other edge. If the graph is weighted, ensure (by temporarily changing weights if necessary) that the weights of the edges different from e are all larger than the weight of e. In either case, Kruskal's algorithm will select e first.

10. Let d be the lowest weight among the edges incident with vertex v. The first time that an edge incident with v is considered for selection (because it has least weight among remaining edges), no edge incident with v will complete a circuit with edges previously selected. Always seeking edges of lowest weight, the algorithm must select an edge of weight d.

11. (b) If the weights of a graph with n vertices are distinct, then Kruskal's algorithm selects at each stage a unique edge of lowest weight. Since Kruskal's algorithm yields only one tree, the graph has only one minimum spanning tree, by (a).

12. (b) Each time a vertex is relabeled, the component to which it belongs contains at least twice as many vertices as before. Thus, if the label on a vertex changes t times, $2^t \leq n$; so $t \leq \log_2 n$ as required.

(c) Since the initial sorting of edges requires $\mathcal{O}(N \log N)$ comparisons, we have only to show that the relabeling process described in (a) and (b) can also be accomplished within this bound. By (b), the total number of vertex relabelings is $\mathcal{O}(n \log n)$. If $n \leq N$, we are done. If $n = N + 1$, then $n \log n \leq (2N) \log N^2 = 4N \log N$ and again we are done, except when $n = 1$ or 2 when no steps are needed. Finally, if $N \leq n - 2$, no spanning tree is possible and a check for this could be included at the beginning of the algorithm.

15. (a) The graph is complete, so an obvious approach is to try to choose the lowest weight available edge at each vertex. Such a cycle is $ADEFBCA$, which has weight $2 + 3 + 2 + 5 + 1 + 5 = 18$.

(b) As shown on the left, the minimum weight of a spanning tree after A is removed is 11. The two edges of least weight at A have weights 2 and 3 so we obtain an estimate of $11 + 2 + 3 = 16$ as a lower bound for the weight of any Hamiltonian cycle.

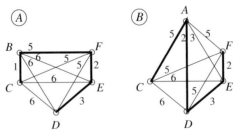

(c) As shown on the right, the minimum weight of a spanning tree after B is removed is 12. The two least edges at B have weights 1 and 5, so we obtain $12 + 1 + 5 = 18$ as a lower bound.

Exercises 12.5

1. (a) This is acyclic: F, B, D, H, A, E, C, G is a canonical labeling.

3. Let $\mathcal{G}$ be a digraph with n vertices and let A be the adjacency matrix of $\mathcal{G}$. The indegree of vertex i is the sum of the entries in column i. This requires $n - 1$ two-number additions. Repeating for n vertices involves $n(n - 1) = \mathcal{O}(n^2)$ additions.

5. (a) We use the notation and ideas of the proof of Theorem 12.5.3. If $\mathcal{G}$ has a cycle, then we can construct a (possibly empty) set of vertices $\{v_0, v_1, \dots, v_k\}$ such that some $\mathcal{S} = \mathcal{V} \setminus \{v_0, v_1, v_2, \dots, v_k\}$ is nonempty and has the property that $\mathcal{S}_v \neq \emptyset$ for all $v \in \mathcal{S}$; that is, for every $v \in \mathcal{S}$, there exists an arc of the form xv with $x \in \mathcal{S}$. Let $u_0 \in \mathcal{S}$ be arbitrary and choose $u_1 \in \mathcal{S}$ such that $u_1 u_0$ is an arc. Choose $u_2 \in \mathcal{S}$ such that $u_2 u_1$ is an arc. If $u_2 = u_0$, then $u_2 u_1 u_0$ is a cycle. Otherwise, choose $u_3 \in \mathcal{S}$ so that $u_3 u_2$ is an arc. In general, having chosen distinct vertices $u_0, u_1, \dots, u_k \in \mathcal{S}$ such that $u_{i+1} u_i$ is an arc for $i = 0, 1, \dots k - 1$, choose $u_{k+1} \in \mathcal{S}$ such that $u_{k+1} u_k$ is an arc. Since $\mathcal{S}$ is finite, there exists k such that $u_{k+1} = u_\ell$ with $\ell < k$. Then $u_{k+1} u_k u_{k-1} \cdots u_\ell$ is a cycle.

7. Since the undirected graph is a tree with n vertices, $\mathcal{T}$ has $n - 1$ arcs, by Theorem 12.2.3.

9. (a) Here are the distances d_t and the corresponding values of p_t.

t	0	1	2	3	4	5	6	7
d_t	0	5	7	4	5	7	6	6
p_t	-1	0	0	0	0	4	4	3

11. The arc $v_{p_t} v_t$ is the last on a shortest path which the algorithm has found from v_0 to v_t. Let $\mathcal{G}$ be the digraph which consists of precisely those arcs (and their end vertices). Vertex v_0 has indegree 0 since there is no arc $v_i v_0$ with $i > 0$. Every other vertex in $\mathcal{G}$ has indegree 1 since it is the last on a shortest directed path. Suppose $\mathcal{G}$ has r vertices. Then it has $r - 1$ edges, one for each of its vertices other than v_0. By Theorem 12.2.3, it suffices to show that $\mathcal{G}$ is connected. This will certainly be the case if we show that every vertex of $\mathcal{G}$ is connected to v_0 by some path in $\mathcal{G}$. Suppose this is not the case and let t be minimal with the property that there is no directed path from v_0 to v_t. Clearly $t > 0$. Then $\mathcal{G}$ contains an arc $v_{p_t} v_t$, the last on a directed path to v_t. Since $p_t < t$, there is a directed path from v_0 to v_{p_t}, and the arc $v_{p_t} v_t$ is directed $v_{p_t} \to v_t$. Thus there is a path from v_0 to v_t.

13. It is sufficient simply to modify Step 1 by setting $d_k = 0$ and $d_i = \infty$ for $i < k$. (We could also modify Step 2 by starting the for loop at $t = k + 1$, but there is little change in efficiency unless k is roughly the size of n.)

14. (a) i. Here are the distances d_t from v_1 and the corresponding values of p_t.

t	0	1	2	3	4	5	6	7
d_t	∞	0	∞	∞	∞	6	4	2
p_t	-1	-1	-1	-1	-1	1	1	1

ii. Here are the distances d_t from v_2 and the corresponding values of p_t.

t	0	1	2	3	4	5	6	7
d_t	∞	∞	0	∞	∞	3	2	7
p_t	-1	-1	-1	-1	-1	2	2	5

Exercises 13.1

1. (a) The final backtracking is 7, 2, 1.

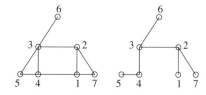

2. (a)

The final backtracking 16, 15, 14, 13, 12, 11, 6, 5, 4, 3, 2, 1.

4. (a) Consider the status of the algorithm at the time that v is labeled. If there are no unlabeled vertices adjacent to v at this stage, we have nothing to prove. So we assume that there are some. Let $\mathcal{G}'$ denote the subgraph of $\mathcal{G}$ consisting of those vertices as yet unlabeled which are connected to v via a path of unlabeled vertices, and all the edges among these vertices. After v is labeled, the algorithm moves to an unlabeled vertex adjacent to v and hence into the subgraph $\mathcal{G}'$. It then either returns to v or moves to another vertex in $\mathcal{G}'$ and gives it a label.

Now the depth-first search algorithm covers any edge at most twice, once when it labels a vertex and once when it backtracks. Since there are only finitely many edges in $\mathcal{G}'$, the algorithm must eventually backtrack to v. If there remain any adjacent unlabeled vertices, the algorithm chooses one to label and moves again into $\mathcal{G}'$. As before, it must eventually backtrack to v. The process continues until all vertices adjacent to v are labeled.

6. If $\mathcal{G}$ is [figure], one spanning tree, namely [figure], is not obtainable by depth-first search.

7. (a) Each time the procedure terminates, start it again with an unlabeled vertex. The number of connected components is equal to the number of times the procedure terminates.

9. No. For instance, you can get from $(7, 1, 0)$ to $(3, 5, 0)$ in one step, but not back.

10. Breadth-first search is another way of systematically moving through all the vertices of a graph. Suppose a graph has n vertices. The procedure assigns to these vertices labels from the set $\{1, 2, \dots , n\}$. Assign label 1 to any vertex. Then label **all** vertices adjacent to vertex 1 with consecutive labels starting at 2. Then pass to vertex 2 and label **all** vertices adjacent to 2 which have not yet been labeled with consecutive labels from the unused members of $\{1, 2, \dots , n\}$, always using the smallest available integer first. Then pass to vertex 3 and so on, at each stage moving to the next highest labeled vertex and labeling all adjacent vertices not yet labeled with the unused members of $\{1, 2, \dots , n\}$, smallest unused integer first. Continue until some vertex acquires label n or all vertices adjacent to the vertex with the highest label are already labeled.

11. (a)

Exercises 13.2

1. (a) Not strongly connected; edge 36 is a bridge.

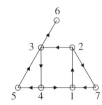

2.

3. (a) A strongly connected orientation for the Petersen graph is shown to the right.

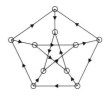

5. No, it cannot. Every edge of an Eulerian graph is part of a circuit; the removal of an edge of a circuit certainly does not disconnect a graph.

6. If e is a bridge, we can let u and v be its ends. If there were a path from u to v which did not require e, together with e, we would have a circuit containing e. Since the deletion of an edge which is part of a circuit cannot disconnect a graph, we have a contradiction.

7. (a) Choose a vertex v. For each vertex u adjacent to v, orient the edge uv in the direction $u \to v$.

8. Yes. To get from a to b in the new orientation, just find the path from b to a in the old orientation and follow it in reverse.

9. (a) False. The graph shown at the right cannot be given a strongly connected orientation, yet every edge is part of a circuit.

10. (a) If there were a path $\mathcal{P}$ between u and v in $\mathcal{G} \setminus \{e\}$, then $\mathcal{P} \cup \{e\}$ would be a circuit, contradicting the fact that e is a bridge.

11. Yes it can. Cutting all the bridges divides the city into connected components without bridges. Each component, therefore, has a strongly connected orientation (by Theorem 13.2.3), so it is possible to assign directions to the streets of each component which allows the possibility of (legal) travel between any two points. Now note that making each bridge a two-way street allows arbitrary travel between components.

Exercises 14.1

1. (a) We draw the graph quickly as a planar and then, after some thinking, as a plane graph with straight edges.

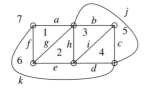   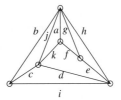

(b) There are seven regions, numbered $1, 2, \ldots, 7$, with boundaries afg, ghe, hbi, icd, bjc, $fedk$, and ajk, respectively.

(c) $E = 11$, $V = 6$, $R = 7$, $N = 22$; so $V - E + R = 6 - 11 + 7 = 2$; $N = 22 \le 22 = 2E$ and $E = 11 \le 12 = 3V - 6$.

3.

Solid	V	E	F	V − E + F
tetrahedron	4	6	4	2
cube	8	12	6	2
octahedron	6	12	8	2
dodecahedron	20	30	12	2
icosahedron	12	30	20	2

5. (a) This is planar. Here it is drawn as a plane graph (with straight edges).

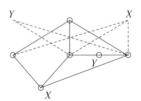

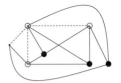

(e) This is not planar. There is a subgraph homeomorphic to $\mathcal{K}_{3,3}$ as shown.

7.   We know from Theorem 14.1.4 that $E \le 3V − 6$. Substituting in $E = V + R − 2$, we obtain $V + R − 2 \le 3V − 6$, or $R \le 2V − 4$, as required.

8. (a) $E = 3 = 3(3) − 6 = 3V − 6.$

10. (a) $\mathcal{K}_n$ is planar if and only if $n \le 4$.

11.

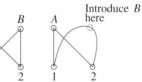

14. (a) Assume the result is not true. Then there is some counterexample $\mathcal{G}$ and subgraph $\mathcal{H}$; that is, for these graphs, $E_1 − V_1 < E_2 − V_2$. (In particular, $V_1 \ne V_2$.) Choose $\mathcal{H}$ such that $V_1 − V_2$ is as small as possible. Since $\mathcal{G}$ is connected, we can find a vertex v which is in $\mathcal{G}$, but not $\mathcal{H}$, which is joined to some vertex in $\mathcal{H}$. Let $\mathcal{K}$ be that subgraph of $\mathcal{G}$ consisting of $\mathcal{H}$, v and all edges joining v to vertices in $\mathcal{H}$. Letting V_3 and E_3 denote the numbers of vertices and edges, respectively, in $\mathcal{K}$, we have $V_3 = V_2 + 1$, while $E_3 \ge E_2 + 1$. Hence, $E_2 − V_2 \le E_3 − V_3$ and so $E_1 − V_1 < E_3 − V_3$. Thus, $\mathcal{G}$ and its subgraph $\mathcal{K}$ provide another counterexample, but this contradicts the minimality of $V_1 − V_2$ since $V_1 − V_3 < V_1 − V_2$.

15.   Yes. An example is shown to the right. The graphs are homeomorphic since the one on the right is obtainable from the other by adding a vertex of degree 2.

17. (b) False, as shown by the homeomorphic graphs at the right. The graph on the left is Hamiltonian, but the one on the right is not.

19. (a) Let $\mathcal{G}_1, \mathcal{G}_2, \ldots, \mathcal{G}_n$ be the connected components of $\mathcal{G}$. Since $\mathcal{G}_i$ has at least three vertices, we have $E_{\mathcal{G}_i} \le 3V_{\mathcal{G}_i} − 6$. Hence, $\sum E_{\mathcal{G}_i} \le 3 \sum V_{\mathcal{G}_i} − 6n$, so $E \le 3V − 6n$ as required.

20. (a) We may assume that $\mathcal{G}$ is connected. Say there is only one vertex of degree at most 5. Then $\sum \deg v_i \ge 6(V − 1) = 6V − 6$, contradicting $\sum \deg v_i = 2E \le 6V − 12$.

21. (a) By Theorem 14.1.4, $E \le 3V − 6$, so $E \le 3(20) − 6 = 54$.

22. (a) Say at most one region has at most five edges on its boundary. Then, with N as in the proof of Corollary 14.1.3, $N \geq 6(R-1)$. But $N \leq 2E$, so $2E \geq 6R-6$, $3R \leq E+3$. Since $V-E+R=2$, $6=3V-3E+3R \leq 3V-2E+3$; that is, $2E \leq 3V-3$. But $2E = \sum \deg v_i \geq 3V$ by assumption, and this is a contradiction.

24. Letting x denote the number of connected components of $\mathcal{G}$, we have $V-E+R=1+x$.

Proof. For each component $\mathcal{C}$, $V_{\mathcal{C}} - E_{\mathcal{C}} + R_{\mathcal{C}} = 2$. Adding, we get $\sum V_{\mathcal{C}} - \sum E_{\mathcal{C}} + \sum R_{\mathcal{C}} = 2x$. We have $\sum V_{\mathcal{C}} = V$ and $\sum E_{\mathcal{C}} = E$, but $\sum R_{\mathcal{C}} = R + (x-1)$ since the exterior region is common to all components. Thus, $V - E + R + x - 1 = 2x$, $V - E + R = x + 1$.

Exercises 14.2

2. (a) We show the graph superimposed over the given map. Since this graph contains triangles, at least three colors are necessary. A 3-coloring is shown, so the chromatic number is 3.

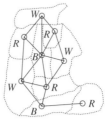

(b) False! The Four-Color Theorem says that the chromatic number of a planar graph is **at most** 4. The planar graph in (a) has chromatic number 3.

3. In neither case, will four colors necessarily suffice as the pictures to the right demonstrate.

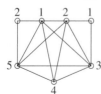

4. (b) A 5-coloring is shown; hence, $\chi(\mathcal{G}) \leq 5$. Since $\mathcal{K}_5$ is a subgraph, $\chi(\mathcal{G}) = 5$.

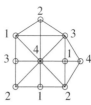

(e) A 4-coloring is shown; hence, $\chi(\mathcal{G}) \leq 4$. Since $\mathcal{G}$ contains $\mathcal{K}_4$ as a subgraph, $\chi(\mathcal{G}) = 4$.

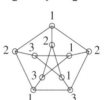

6. (a) The graph $\mathcal{G}_1$ on the left has a 3-coloring, as shown. Since it contains a triangle, $\chi(\mathcal{G}_1) = 3$. The graph $\mathcal{G}_2$ on the right has a 3-coloring, as shown. By trying to label alternately the vertices of the outer pentagon, we see that two colors will not suffice, so $\chi(\mathcal{G}_2) = 3$ too.

(b) The converse of the Four-Color Theorem states that if the chromatic number of a graph is at most four, then the graph is planar. This result is false. The Petersen graph is not planar (Exercise 4 of Section 14.1), but, as we saw in part (a), $\chi = 3$.

7. (a) Yes, a tree is planar and we can prove this by induction on n, the number of vertices. Certainly a tree with one vertex is planar. Then, given a tree with n vertices, removing a vertex of degree 1 (and the edge with which it is incident) leaves a tree with $n - 1$ vertices which is planar by the induction hypothesis. The deleted vertex and edge can now be reinserted without destroying planarity.

9. (a) 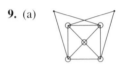   (b)

(c) The graph isn't planar, by Kuratowski's Theorem, as the results of (a) and (b) each illustrate.

(d) A 3-coloring is shown. Since the graph contains triangles, fewer than three colors will not suffice. The chromatic number is three.

(e) The converse of the Four-Color Theorem says that a graph with $\chi \leq 4$ is planar. This is not true, as this graph illustrates: The chromatic number is 3, but the graph is not planar.

10. (a) By Pause 7 of this section, for any n, $\chi(\mathcal{K}_n) = n$ and for any m, n, $\chi(\mathcal{K}_{m,n}) = 2$. Thus, $\chi(\mathcal{K}_{14}) = 14$ and $\chi(\mathcal{K}_{5,14}) = 2$.

11. (a) False. If $\mathcal{G}$ is a 5-cycle, then $\chi(\mathcal{G}) = 3$ since 5 is odd but $\mathcal{G}$ contains no triangle.

14. (a) False. $\mathcal{K}_{3,3}$ is Hamiltonian and has chromatic number two (by Pause 7) but is not planar.

(d) False. $\mathcal{K}_7$ is Eulerian (since the degree of every vertex is six) and Hamiltonian (since every complete graph is Hamiltonian), yet $\chi(\mathcal{K}_7) = 7$.

15.   Three exam periods are required since this is the chromatic number of the graph.

18.   We draw a graph $\mathcal{G}$. The vertices are courses, and two vertices are joined if somebody is taking both courses. Since $\chi(\mathcal{G}) = 4$, four time periods are required.

19.   This means we can delete the edge joining "Econ" to "Stat," so "Stat" can be colored 3, $\chi(\mathcal{G}) = 3$ and three time periods now suffice.

24. (a) We assume $\mathcal{G}$ has more than three vertices for otherwise there is nothing to prove. Let $\mathcal{T}$ be a spanning tree for $\mathcal{G}$. By Exercise 7(c) we know that $\chi(\mathcal{T}) = 2$. Since $\mathcal{G}$ has n vertices, $\mathcal{G} = \mathcal{T} \cup \{e\}$ for some edge e. If the ends of e have different colors in $\mathcal{T}$, then $\chi(\mathcal{G}) = 2$; otherwise, one of these ends must be given a third color and $\chi(\mathcal{G}) = 3$. In either case, $\chi(\mathcal{G}) \leq 3$ as required.

25. (a) The dual of the cube is the octahedron whose graph appears in Fig 14.5.

26.   Yes, it is. Both the tetrahedron and its dual are $\mathcal{K}_4$.

Exercises 14.3

1. (a) 36 nodes, 14 grid segments, 6 nets.   (c)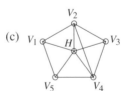

(b) V_1 and V_3, for example.

(d) We know $\chi(\mathcal{G}) \geq 4$ since $\mathcal{G}$ contains $\mathcal{K}_4$ (vertices H, V_2, V_3, V_4). On the other hand, a 4-coloring is given by $\{V_1, V_3\}, \{V_2, V_5\}, \{V_4\}, \{H\}$. Hence, $\chi(\mathcal{G}) = 4$. The above 4-coloring is also a partition of the nets.

3. False. An example is the net pattern for which the line-of-sight graph is ○ ○.

4. (a) ∼ need not be reflexive. Consider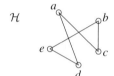

5. (a)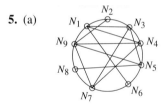

7. False. The graph in Exercise 5 isn't planar. The subgraph with vertices N_1, N_3, N_4, N_5, N_6, N_7 and with edge $N_3 N_4$ removed is isomorphic to $K_{3,3}$.

8. (a) Imagine the possible shorts between nets as the edges in a graph: They won't cross since they're all parallel. Now compress each net to a point, and we have a plane graph. By the Four Color Theorem, $\chi(\mathcal{G}) \leq 4$.

   (b) Draw separate graphs $\mathcal{G}(\mathcal{V}, \mathcal{E}_1)$, $\mathcal{G}(\mathcal{V}, \mathcal{E}_2)$ for horizontal lines of sight [as in (a)] and vertical lines of sight. These graphs have the same vertex set, so combine them. If an edge is repeated (that is, both horizontal and vertical lines exist between nets), then omit one of the original occurrences of it.

9. It is possible for a short to exist between any pair of nets here. Hence, $\mathcal{G} \cong K_8$, and $\chi(\mathcal{G}) = 8$, as required.

10. K_5: $\mathcal{G}_1$ $\mathcal{H}$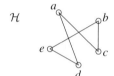

$\mathcal{H}$ cannot be 2-colored since it is a cycle of odd length. If we delete any vertex, however, the resulting graph can be 2-colored. If we delete e, for example, we obtain the graph at the right. Shown are $\mathcal{G}'$ and a corresponding floor plan.

 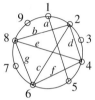

13. (a) We can immediately draw $\mathcal{G}_1$ for this problem (using the obvious Hamiltonian cycle that a golfer must follow). Both $\mathcal{G}_1$ and $\mathcal{H}$ are shown.

$\mathcal{G}_1$ 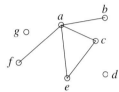 $\mathcal{H}$

Since $\mathcal{H}$ contains a triangle, $\chi(\mathcal{H}) = 3$, so such a course is impossible.

14. Deleting vertex 2 (and the three edges incident with 2) leaves K_5.

15. Since the graph is not K_n for any n, nor an odd cycle, Brooks's Theorem says $\chi(\mathcal{G}) \leq \Delta(\mathcal{G})$. Here $\Delta(\mathcal{G}) = 3$, so we have $\chi(\mathcal{G}) \leq 3$. But this graph contains cycles of odd length, so $\chi(\mathcal{G}) \geq 3$. We conclude that $\chi(\mathcal{G}) = 3$.

Exercises 15.1

1. (a) At a, $\sum_v f_{va} = f_{sa} = 2$ and $\sum_v f_{av} = f_{ac} + f_{ae} = 2 + 0 = 2$. At e, $\sum_v f_{ve} = f_{ae} + f_{be} = 0 + 1 = 1$ and $\sum_v f_{ev} = f_{ec} + f_{ed} = 0 + 1 = 1$. At d, $\sum_v f_{vd} = f_{bd} + f_{ed} = 3 + 1 = 4$ and $\sum_v f_{dv} = f_{dt} = 4$.

(b) The value of the flow is 6.

(c) The capacity of the cut is $c_{ac} + c_{ae} + c_{be} + c_{bd} = 3 + 1 + 4 + 3 = 11$.

(d) No. Arc dt is saturated.

(e) The flow is not maximum. For instance, it can be increased by adding 1 to the flow in the arcs along $sact$.

(f) $(f_{ac} - f_{ca}) + (f_{ae} - f_{ea}) + (f_{be} - f_{eb}) + (f_{bd} - f_{db}) = (2 - 0) + (0 - 0) + (1 - 0) + (3 - 0) = 6$.

2. (b) The value of the flow is 14.

3. (c) $c_{ac} + c_{be} + c_{dt} = 3 + 9 + 9 = 21$.

5. (a)   i. Send one unit through the path $sbet$.

   ii. The flow in (a) has a saturated arc, be.

   iii. Here is a maximum flow, of value 6. (Did you get it?) To see that the flow is maximum, consider the cut $S = \{s, a, b\}$, $T = \{c, d, e, f, t\}$.

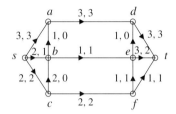

6.

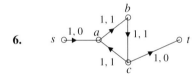

7.   Let $\{S_1, T_1\}$ be any cut. By Theorem 15.1.4, $\mathrm{val}(\mathcal{F}) \le \mathrm{cap}(S_1, T_1)$. But $\mathrm{val}(\mathcal{F}) = \mathrm{cap}(S, T)$. So $\mathrm{cap}(S, T) \le \mathrm{cap}(S_1, T_1)$. Since $\{S_1, T_1\}$ was an arbitrary cut, the result is proven.

Exercises 15.2

1. (a)   i. One flow-augmenting chain is $sbadt$ in which the slack is 1.

   ii. Here is a maximum flow, of value 7. We can see this is maximum by examining the cut $S = \{s, a, b\}$, $T = \{c, d, e, f, t\}$, of capacity $c_{sc} + c_{ad} + c_{be} = 3 + 3 + 1 = 7$, the value of the flow.

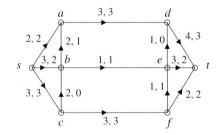

(d)   i. One flow-augmenting chain is $sacdfgt$, which has slack 1.

   ii. Here is a maximum flow, of value 20. We can see this is maximum by examining the cut $S = \{s, a, b, c\}$, $T = \{d, e, f, g, h, t\}$. This has capacity $c_{se} + c_{sh} + c_{cd} = 7 + 4 + 9 = 20$, the value of the flow.

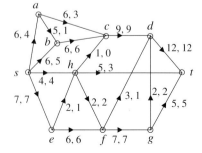

2. (a) A flow of value 6 is given in Exercise 5(a) of Section 15.1. To see this is maximum, consider the cut $S = \{s, a, b\}$, $T = \{c, d, e, f, t\}$.

3. (a) To see that the pictured flow is maximum, consider the cut $S = \{s, b\}$, $T = \{a, c, d, e, f, t\}$.

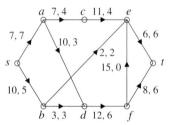

4. (a) We multiply the capacities by 10, the greatest common divisor of the denominators, and find a maximum flow in the new network, as shown on the left. Then, dividing by 10, we find a maximum flow in the given network, of value 67/10, as shown on the right. (Only the flow in each arc is shown.)

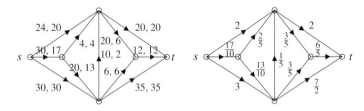

Exercises 15.3

1.

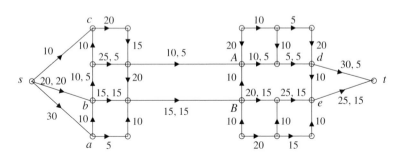

The demand cannot be met; the best that can be done is as shown. Warehouse b sends out 20 Klein bottles and the others send out none. Five go to retail outlet d and 15 go to e.

3. The needs of the retail outlets cannot be met. They can receive at most 14 units a month, as shown. To see that the flow there is a maximum, consider the cut in which $T = \{a, F, G, t\}$ and S is the set of all other vertices.

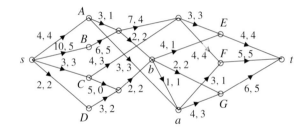

5. (a) A maximum flow is shown, with direction added to the undirected edges.

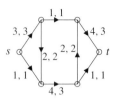

7. Four messengers can be sent.

9. Let u and v be two vertices of $\mathcal{K}_n$. If x is any other vertex, u, x, v is a path from u to v, while u, v is also a path. These $n - 1$ paths are edge disjoint. Also, there cannot be more than $n - 1$ paths from u to v since $\deg u = n - 1$ and edge disjoint paths must start with different edges. Thus, the maximum number of edge disjoint paths between u and v is $n - 1$. But this is also the minimum number of edges which must be removed in order to sever all paths from u to v (in accordance with Menger's Theorem). To see this, note that deleting all edges incident with u certainly severs all paths, so at most $n - 1$ edges need be deleted. On the other hand, if we remove fewer than $n - 1$ edges, then some path from u to v remains because we have already exhibited $n - 1$ edge disjoint paths.

Exercises 15.4

1. One possible matching is shown.

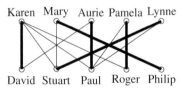

3. $\{12, 34, 56, 78\}$ is a perfect matching.

4. (a) Let $X = \{$Angela, Brenda, Christine, Margaret, Phyllis, Renée$\}$. Then $A(X) = \{A, B, C, E, F\}$, so $|X| > |A(X)|$ and the condition fails.

6. (a) Bruce $\leftrightarrow$ Maurice, Edgar $\leftrightarrow$ Michael, Eric $\leftrightarrow$ Roland, Herb $\leftrightarrow$ Richard.

7. (a) Construct a bipartite graph with vertex sets $\mathcal{V}_1$ and $\mathcal{V}_2$, where $\mathcal{V}_1$ has n vertices corresponding to $A_1, \ldots, A_n$, $\mathcal{V}_2$ has one vertex for each element of S and there is an edge joining A_i to s if and only if $s \in A_i$. Given a subset X of $\mathcal{V}_1$, the set $A(X)$ (notation as in the text) is precisely the set of elements in $\bigcup_{x \in X} A_x$. Thus this question is just a restatement of Hall's Marriage Theorem.

9. (a) Exchange the roles of $\mathcal{V}_1$ and $\mathcal{V}_2$ in Theorem 15.4.2. It may be possible to find a matching which saturates $\mathcal{V}_2$. Here's a necessary and sufficient condition for this to happen: If X is any subset of $\mathcal{V}_2$ and $A(X)$ is the set of all vertices of $\mathcal{V}_1$ which are adjacent to some vertex of X, then $|A(X)| \geq |X|$.

10. (a) $\mathcal{K}_{m,n}$ has a perfect matching if and only if $m = n$. To see this, first assume that $m = n$ and let the vertex sets be $\mathcal{V}_1 = \{u_1, u_2, \ldots, u_m\}$ and $\mathcal{V}_2 = \{v_1, v_2, \ldots, v_m\}$. Then $\{u_1v_1, u_2v_2, \ldots, u_mv_m\}$ is a perfect matching. Conversely, say we have a perfect matching and $m \leq n$. Since each edge in a matching must join a vertex of $\mathcal{V}_1$ to a vertex of $\mathcal{V}_2$, there can be at most m edges. If $m < n$, some vertex in $\mathcal{V}_2$ would not be part of any edge in the matching, a contradiction. Thus, $m = n$.

12. (a) By Proposition 15.4.3, if $\mathcal{K}_n$ has a perfect matching, then n must be even. Conversely, assume that $n = 2m$ is even and label the vertices $u_1, u_2, \ldots, u_{2m}$. Since every pair of vertices is joined by an edge, $\{u_1u_2, u_3u_4, \ldots, u_{2m-1}u_{2m}\}$ will give us a perfect matching.

14. No. The graph ⬡ has $|\mathcal{V}|$ even, but it has no perfect matching.

15. No. The graph ⬡ has a perfect matching, but each vertex has degree 2, which is less than $\frac{1}{2}|\mathcal{V}| = 3$.

Glossary

Without a firm grasp of vocabulary, it is impossible to speak, let alone understand, a language. The same applies to any science. If you aren't comfortable with its terminology, you can't expect to understand the subject.

What follows is a vocabulary list of all the technical terms discussed in this book together with the page where each was first introduced. In most cases, each definition is followed by an example.

A

absolute value The *absolute value* of a number x, denoted $|x|$, is defined to be x if $x \geq 0$ and $-x$ if $x < 0$. For example, $|0| = 0$, $|5| = 5$ and $|-5| = 5$. *p. 76*

acyclic A graph is *acyclic* if it contains no cycles. *p. 369*

adjacency matrix The *adjacency matrix* of a graph with n vertices $v_1, v_2, \ldots, v_n$ is the $n \times n$ matrix $A = [a_{ij}]$ whose (i, j) entry a_{ij} is 1 if vertices v_i and v_j are adjacent and otherwise, $a_{ij} = 0$. *p. 318*

adjacent Two vertices v, w of a pseudograph are *adjacent* if $\{v, w\}$ is an edge. *p. 286*

algorithm An *algorithm* is a procedure for carrying out some process. Its ingredients are an input, an output, and a sequence of precise steps for converting the input to the output. For example, the method by which a perpendicular to a line can be constructed at a specified point is a familiar algorithm of Euclidean geometry. *p. 240*

antisymmetric A binary relation $\sim$ on a set A is *antisymmetric* if $a \sim b$ and $b \sim a$ imply $a = b$. For example, $\leq$ is antisymmetric on R but $\sim$ defined by $a \sim b$ if $ab > 0$ is not. *p. 53*

arc An *arc* is a directed edge in a digraph. *p. 342*

argument An *argument* is a finite collection of statements $\mathcal{A}_1, \mathcal{A}_2, \ldots, \mathcal{A}_n$ called *premises* or *hypotheses* followed by a statement $\mathcal{B}$ called the *conclusion*. Such an argument is *valid* if, whenever $\mathcal{A}_1, \mathcal{A}_2, \ldots, \mathcal{A}_n$ are all true, then $\mathcal{B}$ is also true. *p. 28*

arithmetic sequence The *arithmetic sequence* with first term a and *common difference* d is the sequence a, $a+d, a+2d, \ldots$. For example, $11, 8, 5, 2, \ldots$ is an arithmetic sequence with first term 11 and common difference -3. *p. 166*

B

base b representation The *base b representation* of a natural number N is the expression $(a_{n-1}a_{n-2} \ldots a_0)_b$, where the integers $a_0, a_1, \ldots, a_{n-1}$ are all in the range $0 \leq a_i < b$ and $N = a_{n-1}b^{n-1} + a_{n-2}b^{n-2} + \cdots + a_1b + a_0$. For example, $100 = (10201)_3$ because $100 = 1(3^4) + 0(3^3) + 2(3^2) + 0(3^1) + 1(3^0)$. *p. 102*

Big Oh For functions $f, g \colon \mathsf{N} \to \mathsf{R}$, we say that f is *Big Oh* of g, written $f = \mathcal{O}(g)$, if there is an integer n_0 and a positive real number c such that $|f(n)| \leq c|g(n)|$ for all $n \geq n_0$. For example, $\ln n = \mathcal{O}(n)$. *p. 247*

bijection A *bijection* is a one-to-one onto function. For example, $n \mapsto 2n$ defines a bijection from Z to 2Z. *p. 73*

binary relation A *binary relation* from a set A to a set B is any subset of $A \times B$. A *binary relation on A* is a binary relation from A to A. For example, $\{(1, a), (2, a)\}$ is a binary relation from N to the letters of the English alphabet, and $\{(x, y) \in \mathsf{R}^2 \mid x^2 + y^2 = 1\}$ is a binary relation on R. *p. 51*

bipartite graph A *bipartite graph* is a graph whose vertices can be partitioned into two disjoint sets in such a way that every edge joins a vertex in one set to a vertex in the other. *p. 288*

bipartition set The *bipartition sets* of a bipartite graph are the two disjoint sets into which the vertices are partitioned so that any edge joins a vertex in one set to a vertex in the other. *p. 288*

bridge An edge e of a connected graph $\mathcal{G}$ is a *bridge* if and only if the subgraph $\mathcal{G} \setminus \{e\}$ is not connected. *p. 406*

C

canonical labeling A labeling $v_0, v_1, \ldots, v_{n-1}$ of the vertices of a digraph is *canonical* if every arc is of the form $v_i v_j$ with $i < j$. *p. 392*

capacity The *capacity* of an (s, t)-cut $\{\mathcal{S}, \mathcal{T}\}$ ($s \in \mathcal{S}, t \in \mathcal{T}$) is the sum of the capacities of all arcs from $\mathcal{S}$ to $\mathcal{T}$. *p. 444*

cardinality The *cardinality* of a finite set is the number of elements in the set. While this book does not define the term *cardinality* for infinite sets, it does discuss the concept of *same cardinality* (see definition). *pp. 88 and 89*

Cartesian product The *Cartesian product* of sets A and B—denoted $A \times B$ or A^2 if $B = A$—is the set of ordered pairs (a, b), where $a \in A$ and $b \in B$. For example, $\{1, 2\} \times \{x\} = \{(1, x), (2, x)\}$.   *p. 48*

ceiling   The *ceiling* of a real number x, denoted $\lceil x \rceil$, is the least integer greater than or equal to x. For example, $\lceil 3\frac{1}{4} \rceil = 4$, $\lceil 0 \rceil = 0$ and $\lceil -3\frac{1}{4} \rceil = -3$.   *p. 76*

chain   A *chain* is a *trail* in a directed network, that is, a walk in which the arcs are distinct and can be followed in either direction.   *p. 448*

chromatic number   The *chromatic number* of a graph is the smallest natural number n for which an n-coloring exists.   *p. 423*

circuit   A *circuit* in a pseudograph is a closed trail.   *p. 303*

coloring   A *coloring* of a graph is an assignment of colors to vertices so that adjacent vertices have different colors. An n-coloring is a coloring which uses n colors.   *p. 423*

combination   A *combination* of objects is a subset of them. An *r-combination* is a subset of r objects. For example, {Charles, Andrew} is a 2-combination of the members of the British royal family.   *p. 218*.

common difference   See the definition of *arithmetic sequence*.

common ratio   See the definition of *geometric sequence*.

comparable   Two elements a and b are *comparable* with respect to a partial order $\preceq$ if and only if either $a \preceq b$ or $b \preceq a$. For example, with respect to "divides" on N, 2 and 6 are comparable, but 2 and 7 are not.   *p. 64*

complement   The *complement* of a set A, written A^c, is the set of elements which belong to some universal set, defined by the context, but which do not belong to A. For example, if the universal set is R, then $\{x \mid x < 1\}^c = \{x \mid x \geq 1\}$.   *p. 45*

complete bipartite   The *complete bipartite* graph on two sets $\mathcal{V}_1$ and $\mathcal{V}_2$ is that bipartite graph whose vertices are the union of $\mathcal{V}_1$ and $\mathcal{V}_2$ and whose edges consist of all possible edges between these sets.   *p. 288*

complete graph   The *complete graph on n vertices* is that graph which has n vertices, each pair of which are adjacent.   *p. 288*

complexity   The *complexity* of an algorithm is a function which gives an upper bound for the number of operations required to carry out the algorithm. It is usually specified in general terms, by giving another function of which it is Big Oh. The binary search algorithm, for example, has complexity $\mathcal{O}(\log_2 n)$.   *p. 246*

component   A *component* of a graph is a connected subgraph which is properly contained in no connected subgraph with a larger vertex set or larger edge set.   *p. 310*

composite   A *composite* number is a natural number larger than 1 which is not prime; for example, 6 and 10.   *p. 114*

composition   The *composition* of functions $f : A \to B$ and $g : B \to C$ is the function $g \circ f : A \to C$ defined by $g \circ f(a) = g(f(a))$ for $a \in A$. For example, if f and g are the functions R $\to$ R defined by $f(x) = x + 2$ and $g(x) = 2x - 3$, then $g \circ f(x) = 2x + 1$.   *p. 82*

conclusion   See *argument*.

congruence class   The *congruence class* of an integer a (mod n) is the set $\{b \in Z \mid a \equiv b \pmod{n}\}$ of all integers to which a is congruent *mod n*; for example, the congruence class of 5 (mod 7) is the set $7Z + 5$ of integers of the form $7k + 5$.   *p. 126*

congruence mod n Integers a and b are *congruent* modulo n, where $n > 1$ is a natural number—and we write $a \equiv b$ (mod n)—if $a - b$ is divisible by n. For example, $-2 \equiv 10$ (mod 4), but $7 \not\equiv -19$ (mod 12).   *p. 126*

connected   A pseudograph is *connected* if there is a walk between any two vertices.   *p. 305*

contradiction   A *contradiction* is a compound statement that is always false. For example, $((\neg p) \wedge q) \wedge (p \vee (\neg q))$ is a contradiction. (Examine the truth table.)   *p. 20*

contrapositive   The *contrapositive* of the implication "$p \to q$" is the implication "$(\neg q) \to (\neg p)$." For example, the contrapositive of "If a graph is planar, then it can be colored with at most four colors" is "If a graph cannot be colored with at most four colors, then it is not planar."   *p. 5*

converse   The *converse* of the implication $p \to q$ is the implication $q \to p$. For example, the converse of "If a graph is planar, then it can be colored with at most four colors" is "If a graph can be colored with at most four colors, then it is planar."   *p. 4*

countable   A set is *countable* if and only if it is finite or countably infinite. For example, the rational numbers and the set of natural numbers not exceeding 1000 are both countable sets.   *p. 90*

countably infinite   A *countably infinite* set is an infinite set which has the cardinality of the natural numbers. For example, the rational numbers form a countably infinite set.   *p. 90*

cut   An (s, t)-*cut* in a directed network with vertex set $\mathcal{V}$, two of whose members are s and t, is a pair of disjoint sets $\mathcal{S}$ and $\mathcal{T}$, $s \in \mathcal{S}, t \in \mathcal{T}$, whose union is $\mathcal{V}$.   *p. 444*

cycle   A *cycle* in a pseudograph is a circuit in which the first vertex appears exactly twice (at the beginning and the end) and in which no other vertex appears more than once. An n-*cycle* is a cycle with n vertices.   *p. 303*

D

degree   The *degree* of a vertex in a pseudograph is the number of edges incident with that vertex.   *p. 286*

degree sequence   The *degree sequence* of a pseudograph is a list of the degrees of the vertices in nonincreasing order. *p. 290*

derangement   A *derangement* of n distinct symbols which have some natural order is a permutation of them in which no symbol is in its correct position. For example, 312 is a derangement of $1, 2, 3$.   *p. 228*

digraph   A *digraph* is a pair $(\mathcal{V}, \mathcal{E})$ of sets, $\mathcal{V}$ nonempty and each element of $\mathcal{E}$ an ordered pair of distinct elements of $\mathcal{V}$. The elements of $\mathcal{V}$ are called vertices, and the elements of $\mathcal{E}$ are called arcs.   *p. 342*

direct product   This is another name for the *Cartesian product*, whose definition appeared previously.   *p. 48*

directed network   A *directed network* is a digraph with an integer weight assigned to each arc.   *p. 360*

disjunctive normal form   A compound statement based on variables $x_1, x_2, \ldots, x_n$, $n \geq 1$, is in *disjunctive normal form* if and only if it has the form $(a_{11} \wedge a_{12} \wedge \cdots \wedge a_{1n}) \vee (a_{21} \wedge a_{22} \wedge \cdots \wedge a_{2n}) \vee \cdots \vee (a_{m1} \wedge a_{m2} \wedge \cdots \wedge a_{mn})$, where, for each i and j, $1 \leq i \leq m$, $1 \leq j \leq n$, either $a_{ij} = x_j$ or $a_{ij} = \neg x_j$ and all *minterms* $a_{i1} \wedge a_{i2} \wedge \cdots \wedge a_{in}$ are distinct. For example, the statement $(x_1 \wedge x_2 \wedge x_3) \vee (x_1 \wedge (\neg x_2) \wedge (\neg x_3))$ is in disjunctive normal form on the variables x_1, x_2, x_3.   *p. 26*

divides   Given integers a and b, $b \neq 0$, we say that b *divides* a—written $b \mid a$—and say that b is a *factor* or a *divisor* of a if and only if $a = qb$ for some integer q. For example, $3 \mid 18$ but $5 \nmid 18$.   *p. 104*

divisor   See the definition of *divides*.

domain   The *domain* of a function $f: A \to B$ is the set A.   *p. 73*

E

edge   An *edge* in a graph is a set of two distinct vertices. *p. 286*

equivalence class   If $\sim$ denotes an *equivalence relation* (see definition) on a set A and $a \in A$, then the *equivalence class* of a is the set of all elements $x \in A$ with $x \sim a$. For example, for the equivalence relation "congruence mod 2" on Z, the equivalence class of 17 is the set of all odd integers.   *p. 57*

equivalence relation   An *equivalence relation* on a set is a binary relation which is reflexive, symmetric, and transitive. For example, congruence modulo a natural number $n > 1$ is an equivalence relation on Z but divides is not. *p. 56*

Eulerian circuit   An *Eulerian circuit* in a pseudograph is a circuit which contains every vertex and every edge of the pseudograph.   *p. 304*

Eulerian pseudograph   An *Eulerian pseudograph* is a pseudograph which possesses an Eulerian circuit. *p. 304*

even cycle   An *even cycle* in a pseudograph is a cycle which contains an even number of edges.   *p. 303*

even vertex   An *even* vertex of a pseudograph is one of even degree.   *p. 286*

F

factor   See the definition of *divides*.

factorial   For a natural number $n \geq 1$, n *factorial* means the product of the first n natural numbers. It's denoted $n!$. For example, $3! = 6$ and $5! = 120$. By convention, $0! = 1$.   *p. 155*

Fibonacci sequence   The *Fibonacci sequence* is the sequence which begins $1, 1$, and each of whose subsequent terms is the sum of the preceding two. Thus, the first ten terms are $1, 1, 2, 3, 5, 8, 13, 21, 34, 55$. *p. 168*

floor   The *floor* of a real number x, denoted $\lfloor x \rfloor$, is the greatest integer less than or equal to x. For example, $\lfloor 3\frac{1}{4} \rfloor = 3$, $\lfloor 0 \rfloor = 0$, and $\lfloor -3\frac{1}{4} \rfloor = -4$.   *p. 76*

flow   Given a directed network with vertex set $\mathcal{V}$, the weight (or *capacity*) of arc uv denoted c_{uv}, and two distinguished vertices s and t called the source and the sink, respectively, an (s, t)-*flow* is a set of numbers $\{f_{uv}\}$ satisfying

1. $0 \leq f_{uv} \leq c_{uv}$ for all $u, v \in \mathcal{V}$
2. $\sum_{v \in \mathcal{V}} f_{uv} = \sum_{v \in \mathcal{V}} f_{vu}$ for all $u \in \mathcal{V} \setminus \{s, t\}$.   *p. 442*

flow-augmenting chain   A *flow-augmenting chain* in a network with (s, t)-flow $\{f_{uv}\}$ is a chain in which $f_{uv} < c_{uv}$ for each forward arc uv of the chain and $f_{uv} > 0$ for each backward arc.   *p. 449*

function   A *function* from a set A to a set B is a binary relation f from A to B with the property that for every $a \in A$, there is exactly one $b \in B$ such that $(a, b) \in f$. We write $f: A \to B$ to indicate that f is a function from A to B. For example, a sequence of integers is just a function $\mathsf{N} \to \mathsf{Z}$.   *p. 71*

G

generating function   The *generating function* of a sequence $a_0, a_1, a_2, \ldots$ is the expression $f(x) = a_0 + a_1 x + a_2 x^2 + \cdots$. The generating function of the natural numbers, for instance, is the expression $1 + 2x + 3x^2 + \cdots$.   *p. 178*

geometric sequence   The *geometric sequence* with first term a and *common ratio* r is the sequence $a, ar, ar^2, ar^3, \ldots$. For example, $1, \frac{1}{2}, \frac{1}{4}, \frac{1}{8}, \ldots$ is the geometric sequence with first term 1 and common ratio $\frac{1}{2}$. *p. 167*

graph   A *graph* is a pair $(\mathcal{V}, \mathcal{E})$ of sets, $\mathcal{V}$ nonempty and each element of $\mathcal{E}$ an unordered pair of distinct elements of $\mathcal{V}$. The elements of $\mathcal{V}$ are called *vertices*; the elements of $\mathcal{E}$ are called *edges*.   *p. 286*

greatest common divisor   The *greatest common divisor* (abbreviated gcd) of integers a and b not both of which are zero is the largest common divisor of a and b; that is, a number g such that

1. $g \mid a$, $g \mid b$ (it's a common divisor) and,
2. if $c \mid a$ and $c \mid b$, then $c \leq g$ (it's the largest common divisor).

In general, for $n > 2$, the gcd of n nonzero integers $a_1, a_2, \ldots, a_n$ can be defined either as the largest of the common divisors of these integers or, inductively, by

$$\gcd(a_1, a_2, \ldots, a_n) = \gcd(a_1, \gcd(a_2, \ldots, a_n)).$$

For example, $\gcd(4, 18) = 2$, $\gcd(18, -30) = 6$ and $\gcd(-6, 10, -15) = 1$.   *p. 105*

greatest lower bound   A *greatest lower bound* (abbreviated glb) for two elements a, b in a partially ordered set $(A, \preceq)$ is an element $g \in A$ satisfying

1. $g \preceq a$, $g \preceq b$; and
2. if $c \preceq a$ and $c \preceq b$ for some $c \in A$ then $c \preceq g$.

If a and b have a glb, then this element is unique and denoted $a \wedge b$. For example, the greatest lower bound of two natural numbers with respect to "divides" is just their greatest common divisor.   *p. 66*

H

Hamiltonian cycle   A *Hamiltonian cycle* in a graph is a cycle in which every vertex of the graph appears.   *p. 310*

Hamiltonian graph   A *Hamiltonian graph* is one which has a Hamiltonian cycle.   *p. 310*

homeomorphic   Two graphs are *homeomorphic* if and only if each can be obtained from the same graph by adding vertices to edges.   *p. 417*

hypothesis   See *argument*.

I

ideal   An *ideal* of Z is a subset A of Z with three properties: (i) $0 \in A$; (ii) $a \in A$ implies $-a \in A$; (iii) $a, b \in A$ implies $a + b \in A$. For example, the set of even integers is an ideal of Z.   *p. 162*

identity function   The *identity function* on a set A is the function $\iota \colon A \to A$ defined by $\iota(a) = a$ for any $a \in A$. Sometimes we write ι_A to emphasize that the domain of this function is the set A.   *p. 77*

incident   A vertex v of a pseudograph is *incident* with an edge e if and only if $e = \{v, w\}$ for some vertex w.   *p. 286*

indegree   The *indegree* of a vertex in a directed graph is the number of arcs directed towards that vertex.   *p. 343*

injective   This is another word for *one-to-one* (see definition).   *p. 73*

intersection   The *intersection* of sets—denoted $A \cap B$ for two sets, $A \cap B \cap C$ for three, and $\cap_{i=1}^{n} A_i$ for n sets— is the set of elements which belong to each of the sets. For example, $\{0, 1, 2\} \cap \{0, 1, 3\} = \{0, 1\}$; $\{0\} \cap \mathsf{N} = \varnothing$.   *p. 43*

inverse function   The *inverse* of a one-to-one onto function $f \colon A \to B$ is the function $B \to A$ obtained by reversing the pairs of f. (Only a one-to-one onto function has an inverse.) For example, if $f = \{(1, a), (2, b)\}$, then $f^{-1} = \{(a, 1), (b, 2)\}$. The inverse of the exponential function $\mathsf{R} \to (0, \infty)$ defined by $f(x) = e^x$ is the natural logarithm $(0, \infty) \to \mathsf{R}$ defined by $f^{-1}(x) = \ln x$.   *p. 80*

isolated vertex   An *isolated* vertex in a pseudograph is one of degree 0.   *p. 286*

isomorphic   Graphs are *isomorphic* if and only if there is an *isomorphism* (see definition) from one to the other.   *p. 295*

isomorphic labeled trees   Labeled trees $\mathcal{T}_1$ and $\mathcal{T}_2$ are *isomorphic* if and only if they have the same number of vertices, they are labeled with the same set of labels, and vertices labeled v and w in $\mathcal{T}_1$ are adjacent if and only if v and w are adjacent in $\mathcal{T}_2$.   *p. 375*

isomorphism   An *isomorphism* from a graph $\mathcal{G}_1(\mathcal{V}_1, \mathcal{E}_1)$ to a graph $\mathcal{G}_2(\mathcal{V}_2, \mathcal{E}_2)$ is a one-to-one function φ from $\mathcal{V}_1$ onto $\mathcal{V}_2$ such that if vw is an edge in $\mathcal{E}_1$, then $\varphi(v)\varphi(w)$ is an edge is $\mathcal{E}_2$, and such that every edge in $\mathcal{E}_2$ has the form $\varphi(v)\varphi(w)$ for some edge $vw \in \mathcal{E}_1$.   *p. 295*

L

lattice   A *lattice* is a partially ordered set in which every two elements have a greatest lower bound and a least upper bound. For example, the set of subsets of a set is a lattice.   *p. 67*

leaf   A *leaf* in a graph is a vertex of degree 1.   *p.372*

least common multiple   The *least common multiple* (lcm) of nonzero integers a and b is the smallest of the positive common multiples of a and b; that is, it's an integer $\ell > 0$ such that

1. $a \mid \ell$, $b \mid \ell$ (it's a common multiple) and,
2. if $m > 0$, $a \mid m$ and $b \mid m$, then $\ell \leq m$ (it's the smallest positive common multiple).

For example, $\mathrm{lcm}(10, 12) = 60$ and $\mathrm{lcm}(18, 30) = 90$.   *p. 110*

least upper bound   A *least upper bound* (abbreviated lub) for two elements a, b in a partially ordered set $(A, \preceq)$ is an element $\ell \in A$ satisfying

1. $a \preceq \ell$, $b \preceq \ell$ and,
2. if $a \preceq c$ and $b \preceq c$ for some $c \in A$, then $\ell \preceq c$.

If a and b have a lub, then this element is unique and denoted $a \vee b$. For example, the least upper bound of two natural numbers with respect to "divides" is just their least common multiple.   *p. 66*

length The *length* of a walk in a pseudograph is the number of edges it contains.   *p. 303*

logically equivalent Statements $\mathcal{A}$ and $\mathcal{B}$ are *logically equivalent* and we write "$\mathcal{A} \Longleftrightarrow \mathcal{B}$" if $\mathcal{A}$ and $\mathcal{B}$ have identical truth tables. For example, $\mathcal{A} \colon p \to (\neg q)$ is logically equivalent to $\mathcal{B} \colon \neg(p \wedge q)$.   *p. 20*

loop A *loop* is an edge of a pseudograph which is incident with only one vertex.   *p. 286*

M

map This is another word for *function*, which was defined previously.   *p. 71*

matching A *matching* in a graph is a set of edges with the property that no vertex is incident with more than one edge in the set.   *p. 459*

maximal A *maximal* element of a partially ordered set $(A, \preceq)$ is an element a with the property that $a \preceq b$ implies $a = b$. For example, 4 is a maximal element of $\{1, 2, 3, 4, 5, 6\}$ with respect to "divides."   *p. 66*

maximum A *maximum* element of a partially ordered set $(A, \preceq)$ is an element a with the property that $a \succeq b$ for every $b \in A$. For example, N is a maximum element with respect to $\subseteq$ on N; also, 10 is not a maximum element with respect to "divides" on N.   *p. 65*

minimal A *minimal* element of a partially ordered set $(A, \preceq)$ is an element a with the property that $b \preceq a$ implies $a = b$. For example, 3 is a minimal element of $\{2, 3, 4, 5, 6\}$ with respect to "divides."   *p. 66*

minimum A *minimum* element of a partially ordered set $(A, \preceq)$ is an element a with the property that $a \preceq b$ for every $b \in A$. For example, 1 is a minimum element with respect to "divides" on N but 3 is not.   *p. 65*

Minimum Connector Problem Find the spanning tree of least weight in a weighted connected graph.   *p. 377*

minimum spanning tree A *minimum spanning tree* in a weighted graph is a spanning tree for which the sum of the weights of the edges is least among all spanning trees.   *p. 378*

multiple edges Multiple edges are edges (by implication, more than one) in a pseudograph which are incident with the same two vertices.   *p. 286*

multiplicity See the definition of *prime decomposition*.

N

n-**coloring** See *coloring*.

n-**cycle** See *cycle*.

negation The *negation* of the statement p is the statement $\neg p$ which asserts that p is not true. For example, the negation of "x equals 4" is the statement "x does not equal 4."   *p. 5*

O

odd cycle An *odd cycle* is a cycle which contains an odd number of edges.   *p. 303*

odd vertex An *odd* vertex in a pseudograph is a vertex of odd degree.   *p. 286*

one-to-one A function $f \colon A$ implies B is *one-to-one*, or 1–1, if and only if $f(a_1) = f(a_2)$ implies $a_1 = a_2$. For example, the function $f \colon Z$ implies Z defined by $f(n) = 2n$ is one-to-one, but the floor function $\lfloor \ \rfloor \colon R$ implies Z is not.   *p. 73*

one-to-one correspondence This is the same as *bijection* and is the preferred term in the context of *cardinality*. The function $x \mapsto e^x$ defines a one-to-one correspondence between the reals and the positive reals showing that these sets have the same cardinality.   *p. 89*

onto A function $f \colon A \to B$ is *onto* if and only if for each $b \in B$, there exists $a \in A$ such that $b = f(a)$. For example, the function $f \colon (0, \infty) \to R$ defined by $f(x) = \ln x$ is onto.   *p. 73*

order See *same order* or *smaller order* or *total order*.

orient To *orient* an edge in a pseudograph is to assign a direction to it.   *p. 405*

outdegree The *outdegree* of a vertex in a directed graph is the number of arcs directed away from that vertex.   *p. 343*

P

partially ordered set A *partially ordered set* is a pair $(A, \preceq)$, where A is a set on which there is defined the partial order $\preceq$. For example, $(R, \leq)$ is a partially ordered set.   *p. 63*

partial order A *partial order* on a set A is a binary relation on A which is reflexive, antisymmetric, and transitive. For example, $\leq$ defines a partial order on R.   *p. 63*

partition A *partition* of a set A is a collection of disjoint subsets (sometimes called *cells*) of A whose union is A. For example, the positive integers, the negative integers, and $\{0\}$ form a partition of Z.   *p. 59*

path A *path* in a pseudograph is a walk in which all vertices are distinct.   *p. 303*

perfect matching A *perfect matching* in a graph is a set of edges which have the property that each vertex is incident with exactly one edge in the set.   *p. 459*

permutation A *permutation* of a set of distinct symbols is an arrangement of them in a line in some order. For example, *cab* is a permutation of a, b, c.   *p. 213*

planar A *planar* graph is one that can be drawn in the plane in such a way that no two edges cross. *p. 413*

poset This is an abbreviation for partially ordered set. *p. 63*

power set The *power set* $\mathcal{P}(A)$ of a set A is the set of all subsets of A. For example, if $A = \{1, 2\}$, then $\mathcal{P}(A) = \{\emptyset, A, \{1\}, \{2\}\}$. *p. 41*

premise See *argument*.

prime A *prime* or *prime number* is a natural number $p \geq 2$ whose only positive divisors are itself and 1; for example, 2, 3, and 5. *p. 114*

prime decomposition The *prime decomposition* of an integer $n \geq 2$ is its representation in the form $n = q_1^{\alpha_1} q_2^{\alpha_2} \cdots q_s^{\alpha_s}$, where $q_1, \ldots, q_s$ are distinct primes and $\alpha_1, \ldots, \alpha_s$ are natural numbers called the *multiplicities* of $q_1, \ldots, q_s$, respectively. For example, the prime decomposition of 200 is $200 = 2^3 5^2$. The prime factors of 200 are 2 and 5, with multiplicities 3 and 2, respectively. *p. 118*

prime divisor See *prime factor*.

prime factor The *prime factors* (or *prime divisors*) of an integer $n \geq 2$ are the prime numbers which divide it. These are the p_i or q_i when n is expressed in the form $n = p_1 p_2 \cdots p_r$ or $n = q_1^{\alpha_1} q_2^{\alpha_2} \cdots q_s^{\alpha_s}$ by the Fundamental Theorem of Arithmetic. For example, the prime factors of 15 are 3 and 5; the prime factors of 36 are 2 and 3. *p. 118*

Principle of Mathematical Induction Given a statement $\mathcal{P}$ concerning the integer n, suppose

1. $\mathcal{P}$ is true for some particular integer n_0, and

2. if $\mathcal{P}$ is true for some particular integer $k \geq n_0$, then it is true for the next integer $k + 1$.

Then $\mathcal{P}$ is true for all integers $n \geq n_0$. *p. 151*

proper subset: A set A is a *proper subset* of a set B and we write $A \subsetneq B$ if and only if every element of A is also an element of B, but $A \neq B$. For example, $\mathsf{N} \subsetneq \mathsf{Z}$. *p. 40*

pseudograph A *pseudograph* is like a graph, but it may contain multiple edges or loops. *p. 287*

Q

quotient If a and b are integers, $b \neq 0$, and $a = qb + r$ with $0 \leq r < b$, then q is called the *quotient* and r the *remainder* when a is divided by b. For example, when 26 is divided by 7, the quotient is 3 and the remainder is 5, because $26 = 3(7) + 5$. *p. 99*

quotient set The *quotient set* of an equivalence relation is the set of all equivalence classes. For example, if the equivalence relation on Z is defined by $a \sim b$ if and only if $a - b$ is divisible by 2, there are two equivalence classes, the even integers and the odd integers, so the quotient set is the set {evens, odds}. *p. 57*

R

range The *range* of a function $f : A \to B$ is the set of elements $b \in B$ which are of the form $b = f(a)$ for some $a \in A$. For example, the range of the floor function $\lfloor \ \rfloor : \mathsf{R} \to \mathsf{R}$ is the set of integers. *p. 73*

r-**combination** See *combination*.

reflexive A binary relation $\sim$ on a set A is *reflexive* if and only if $a \sim a$ for all $a \in A$. For example, $\leq$ is reflexive on R but $<$ is not. *p. 52*

relatively prime Integers are *relatively prime* if and only if they are nonzero and the only positive integer which divides each of them is 1; equivalently, their greatest common divisor is 1. For example, 2 and 3 are relatively prime, as are the pairs $-16, 21$ and $-15, -28$. *p. 108*

remainder See the definition of *quotient*.

r-**permutation** An *r-permutation* of n symbols (r and n positive integers, $r \leq n$) is an arrangement of r of them in a line in some order. For example, *caz* is a 3-permutation of the letters of the English alphabet. *p. 213*

root, rooted tree A tree is *rooted* if it comes with a specified vertex, called the *root*. A digraph is a *rooted tree* if the underlying graph is a tree and there is a unique vertex called the *root*, of indegree 0. *p. 370 and p.393*

S

same cardinality Two sets A and B have the *same cardinality* if and only if there is a one-to-one correspondence between them. For example, the integers and the even integers have the same cardinality. *p. 89*

same order Functions f and $g : \mathsf{N} \to \mathsf{R}$ have the *same order*, and we write $f \asymp g$, if and only if each is Big Oh of the other. For example, $15n^3 \asymp 4n^3$. *p. 248*

saturated A *saturated* vertex in a graph is a vertex which is incident with one of the edges in a given matching. *p. 459*

score The *score* of a vertex in a tournament is the outdegree of that vertex. *p. 355*

score sequence The *score sequence* of a tournament is the list of outdegrees, in nonincreasing order. *p. 355*

sequence A *sequence* is a function with domain some set of integers (often N). It is usually described by listing the elements of its range. For example, the sequence which is the function $f : \mathsf{N} \to \mathsf{R}$ defined by $f(n) = n^2$ would likely be described by the list $1, 4, 9, 16, \ldots$. *p. 163*

set difference The *set difference* $A \setminus B$ of sets A and B is the set of elements which are in A but not in B. For example, $\{0, 1, 2, 3\} \setminus \{0, 1, 4\} = \{2, 3\}$. *p. 45*

sink This is a distinguished vertex in a directed network. *p. 441*

smaller order A function $f : \mathsf{N} \to \mathsf{R}$ has *smaller order* than another function g, written $f \prec g$, if and only if f is Big Oh of g but g is not Big Oh of f. For example, $n^7 \prec 2^n$. *p. 248*

source This is a distinguished vertex in a directed network. *p. 441*

spanning tree A *spanning tree* for a connected graph $\mathcal{G}$ is a subgraph which is a tree and which includes every vertex of $\mathcal{G}$. *p. 378*

strongly connected A digraph is *strongly connected* if and only if there is a walk from any vertex to any other vertex which respects orientation of arcs. *p. 343*

strongly connected orientation A *strongly connected orientation* of a graph is an assignment of orientations (directions) to all edges in such a way that the resulting digraph is strongly connected. *p. 405*

subgraph A *subgraph* of a pseudograph $\mathcal{G}$ is a pseudograph whose vertices and edges are subsets of the vertices and edges, respectively, of $\mathcal{G}$. *p. 287*

subset A set A is a *subset* of a set B, denoted $A \subseteq B$, if and only if every element of A is also an element of B. For example, $\{0, 1\} \subseteq \{0, 1, 2\}$ but $\mathsf{Z} \not\subseteq \mathsf{N}$. *p. 40*

surjective Another word for *onto*. *p. 73*

symmetric A binary relation $\sim$ on a set A is *symmetric* if and only if $a \sim b$ implies $b \sim a$. For example, $a \sim b$ if and only if $ab = 1$ defines a symmetric relation on Q, but $\le$ does not. *p. 52*

symmetric difference The *symmetric difference* of sets A and B, denoted $A \oplus B$, is the set of elements which belong to precisely one of the sets. More generally, the symmetric difference $A_1 \oplus A_2 \oplus A_3 \oplus \cdots \oplus A_n$ of n sets $A_1, A_2, A_3, \dots, A_n$ is the set of those elements which are members of an odd number of the sets A_i. For example, $\{0, 1, 2, 3\} \oplus \{0, 1, 4, 5\} = \{2, 3, 4, 5\}$. *pages 47 and 161.*

T

tautology A *tautology* is a compound statement that is always true. For example, $(p \wedge q) \to (p \vee q)$ is a tautology. (Examine the truth table.) *p. 20*

term The *terms* of a sequence are the numbers of its range or, thinking of a sequence as a list, the numbers in the list. For example, the first five terms of the Fibonacci sequence are $1, 1, 2, 3, 5$. *p. 164*

total order A *total order* on a set is a partial order in which every two elements are comparable. For example, $\le$ is a total order on R, but $\subseteq$ is not a total order on $\mathcal{P}(\mathsf{R})$. *p. 64*

tournament A *tournament* is a digraph with the property that for every two distinct vertices u, v, exactly one of uv, vu is an arc. *p. 355*

trail A *trail* in a pseudograph is a walk in which all edges are distinct. *p. 303*

transitive A binary relation $\sim$ on a set A is *transitive* if and only if $a \sim b$ and $b \sim c$ implies $a \sim c$. For example, the binary relation $\sim$ defined by $a \sim b$ if $ab \ge 0$ is a transitive relation on $\mathsf{Z} \setminus \{0\}$ but not on Z. *p. 54*

transitive tournament A *transitive tournament* is a tournament with the property that whenever uv and vw are arcs, so is uw an arc. *p. 357*

Traveling Salesman's Problem The Traveling Salesman's Problem is to find a Hamiltonian cycle of least weight in a weighted connected graph. *p. 325*

tree A *tree* is a connected graph which contains no circuits. *p. 369*

triangle A *triangle* in a graph is a 3-cycle. *p. 299*

U

uncountable An *uncountable* set is an infinite set which does not have the cardinality of N. For example, the real numbers form an uncountable set. *p. 90*

union The *union* of sets—denoted $A \cup B$ for two sets, $A \cup B \cup C$ for three, and $\bigcup_{i=1}^{n} A_i$ for n—is the set of elements which belong to at least one of them. For example, $\{0, 1, 2\} \cup \{0, 1, 3\} = \{0, 1, 2, 3\}$. *p. 43*

V

valid argument See *argument*.

value of a flow The *value of a flow* $\{f_{uv}\}$ in a directed network with source s, sink t, and vertex set $\mathcal{V}$ is the integer

$$\text{val}(\mathcal{F}) = \sum_{v \in \mathcal{V}} f_{sv} - \sum_{v \in \mathcal{V}} f_{vs};$$

equivalently,

$$\text{val}(\mathcal{F}) = \sum_{v \in \mathcal{V}} f_{vt} - \sum_{v \in \mathcal{V}} f_{tv}.$$ *p. 443*

W

walk A *walk* in a pseudograph is an alternating sequence of vertices and edges, beginning and ending with a vertex, in which each vertex (except the last) is incident with the edge which follows it and the last vertex is incident with the preceding edge. It is *closed* if and only if the first vertex is the same as the last and otherwise *open*. *p. 303*

weighted graph A *weighted graph* is a graph in which there is a nonnegative real number associated with each edge. *p. 325*

Well-Ordering Principle This principle asserts that every nonempty set of natural numbers has a smallest element. *p. 98*

Index